T0189156

Lecture Notes in Computer Science 10212

Commenced Publication in 1973
Founding and Former Series Editors:
Gerhard Goos, Juris Hartmanis, and Jan van Leeuwen

More information about this series at http://www.springer.com/series/7410

Jean-Sébastien Coron · Jesper Buus Nielsen (Eds.)

Advances in Cryptology – EUROCRYPT 2017

36th Annual International Conference on the Theory
and Applications of Cryptographic Techniques
Paris, France, April 30 – May 4, 2017
Proceedings, Part III

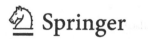 Springer

Editors
Jean-Sébastien Coron
University of Luxembourg
Luxembourg
Luxembourg

Jesper Buus Nielsen
Aarhus University
Aarhus
Denmark

ISSN 0302-9743 ISSN 1611-3349 (electronic)
Lecture Notes in Computer Science
ISBN 978-3-319-56616-0 ISBN 978-3-319-56617-7 (eBook)
DOI 10.1007/978-3-319-56617-7

Library of Congress Control Number: 2017936355

LNCS Sublibrary: SL4 – Security and Cryptology

Printed on acid-free paper

This Springer imprint is published by Springer Nature
The registered company is Springer International Publishing AG
The registered company address is: Gewerbestrasse 11, 6330 Cham, Switzerland

Preface

Eurocrypt 2017, the 36th annual International Conference on the Theory and Applications of Cryptographic Techniques, was held in Paris, France, from April 30 to May 4, 2017. The conference was sponsored by the International Association for Cryptologic Research (IACR). Michel Abdalla (ENS, France) was responsible for the local organization. He was supported by a local organizing team consisting of David Pointcheval (ENS, France), Emmanuel Prouff (Morpho, France), Fabrice Benhamouda (ENS, France), Pierre-Alain Dupoint (ENS, France), and Tancrède Lepoint (SRI International). We are indebted to them for their support and smooth collaboration.

The conference program followed the now established parallel track system where the works of the authors were presented in two concurrently running tracks. Only the invited talks spanned over both tracks.

We received a total of 264 submissions. Each submission was anonymized for the reviewing process and was assigned to at least three of the 56 Program Committee members. Submissions co-authored by committee members were assigned to at least four members. Committee members were allowed to submit at most one paper, or two if both were co-authored. The reviewing process included a first-round notification followed by a rebuttal for papers that made it to the second round. After extensive deliberations the Program Committee accepted 67 papers. The revised versions of these papers are included in these three-volume proceedings, organized topically within their respective track.

The committee decided to give the Best Paper Award to the paper "Scrypt Is Maximally Memory-Hard" by Joël Alwen, Binyi Chen, Krzysztof Pietrzak, Leonid Reyzin, and Stefano Tessaro. The two runners-up to the award, "Computation of a 768-bit Prime Field Discrete Logarithm," by Thorsten Kleinjung, Claus Diem, Arjen K. Lenstra, Christine Priplata, and Colin Stahlke, and "Short Stickelberger Class Relations and Application to Ideal-SVP," by Ronald Cramer, Léo Ducas, and Benjamin Wesolowski, received honorable mentions. All three papers received invitations for the *Journal of Cryptology*.

The program also included invited talks by Gilles Barthe, titled "Automated Proof for Cryptography," and by Nigel Smart, titled "Living Between the Ideal and Real Worlds."

We would like to thank all the authors who submitted papers. We know that the Program Committee's decisions, especially rejections of very good papers that did not find a slot in the sparse number of accepted papers, can be very disappointing. We sincerely hope that your works eventually get the attention they deserve.

We are also indebted to the Program Committee members and all external reviewers for their voluntary work, especially since the newly established and unified page limits and the increasing number of submissions induce quite a workload. It has been an honor to work with everyone. The committee's work was tremendously simplified by Shai Halevi's submission software and his support, including running the service on IACR servers.

Finally, we thank everyone else —speakers, session chairs, and rump session chairs — for their contribution to the program of Eurocrypt 2017. We would also like to thank Thales, NXP, Huawei, Microsoft Research, Rambus, ANSSI, IBM, Orange, Safran, Oberthur Technologies, CryptoExperts, and CEA Tech for their generous support.

May 2017   Jean-Sébastien Coron
                                                              Jesper Buus Nielsen

Eurocrypt 2017

The 36th Annual International Conference on the Theory and Applications of Cryptographic Techniques

Sponsored by *the International Association for Cryptologic Research*

30 April – 4 May 2017
Paris, France

General Chair

Michel Abdalla   ENS, France

Program Co-chairs

Jean-Sébastien Coron   University of Luxembourg
Jesper Buus Nielsen   Aarhus University, Denmark

Program Committee

Gilad Asharov	Cornell Tech, USA
Nuttapong Attrapadung	AIST, Japan
Fabrice Benhamouda	ENS, France and IBM, USA
Nir Bitansky	MIT, USA
Andrey Bogdanov	Technical University of Denmark
Alexandra Boldyreva	Georgia Institute of Technology, USA
Chris Brzuska	Technische Universität Hamburg, Germany
Melissa Chase	Microsoft, USA
Itai Dinur	Ben-Gurion University, Israel
Léo Ducas	CWI, Amsterdam, The Netherlands
Stefan Dziembowski	University of Warsaw, Poland
Nicolas Gama	Inpher, Switzerland and University of Versailles, France
Pierrick Gaudry	CNRS, France
Peter Gaži	IST Austria, Austria
Niv Gilboa	Ben-Gurion University, Israel
Robert Granger	EPFL, Switzerland
Nathan Keller	Bar Ilan University, Israel
Aggelos Kiayias	University of Edinburgh, UK
Eike Kiltz	Ruhr-Universität Bochum, Germany

Vladimir Kolesnikov	Bell Labs, USA
Ranjit Kumaresan	MIT, USA
Eyal Kushilevitz	Technion, Israel
Gregor Leander	Ruhr-University Bochum, Germany
Tancrède Lepoint	SRI International, USA
Benoît Libert	ENS de Lyon, France
San Ling	Nanyang Technological University, Singapore
Anna Lysyanskaya	Brown University, USA
Tal Malkin	Columbia University, USA
Willi Meier	FHNW, Switzerland
Florian Mendel	Graz University of Technology, Austria
Bart Mennink	K.U. Leuven, Belgium
Ilya Mironov	Google, USA
María Naya-Plasencia	Inria, France
Ivica Nikolić	Nanyang Technological University, Singapore
Miyako Ohkubo	NICT, Japan
Rafail Ostrovsky	UCLA, USA
Omkant Pandey	Stony Brook University, USA
Omer Paneth	Boston University, USA
Chris Peikert	University of Michigan, USA
Thomas Peters	UCL, Belgium
Krzysztof Pietrzak	IST Austria, Austria
Emmanuel Prouff	Morpho, France
Leonid Reyzin	Boston University, USA
Louis Salvail	University of Montreal, Canada
Yu Sasaki	NTT Secure Platform Laboratories, Japan
Abhi Shelat	University of Virginia, USA
Elaine Shi	Cornell University, USA
Martijn Stam	University of Bristol, UK
Damien Stehlé	ENS de Lyon, France
John P. Steinberger	Tsinghua University, China
Ingrid Verbauwhede	K.U. Leuven, Belgium
Brent Waters	University of Texas, USA
Daniel Wichs	Northeastern University, USA
Mark Zhandry	Princeton University, USA

Additional Reviewers

Michel Abdalla	Martin Albrecht	Daniel Apon
Masayuki Abe	Ghada Almashaqbeh	Benny Applebaum
Aysajan Abidin	Jacob Alperin-Sheriff	Christian Badertscher
Hamza Abusalah	Joël Alwen	Saikrishna
Divesh Aggarwal	Abdelrahaman Aly	Badrinarayanan
Shashank Agrawal	Elena Andreeva	Shi Bai
Navid Alamati	Yoshinori Aono	Josep Balasch

Foteini Baldimtsi
Marshall Ball
Valentina Banciu
Subhadeep Banik
Razvan Barbulescu
Guy Barwell
Carsten Baum
Anja Becker
Christof Beierle
Amos Beimel
Sonia Belaïd
Shalev Ben-David
Iddo Bentov
Jean-François Biasse
Begul Bilgin
Olivier Blazy
Xavier Bonnetain
Joppe Bos
Christina Boura
Florian Bourse
Luis Brandao
Dan Brownstein
Chris Campbell
Ran Canetti
Anne Canteaut
Angelo De Caro
Ignacio Cascudo
David Cash
Wouter Castryck
Hubert Chan
Nishanth Chandran
Jie Chen
Yilei Chen
Nathan Chenette
Mahdi Cheraghchi
Alessandro Chiesa
Ilaria Chillotti
Sherman S.M. Chow
Kai-Min Chung
Michele Ciampi
Ran Cohen
Craig Costello
Alain Couvreur
Claude Crépeau
Edouard Cuvelier
Guillaume Dabosville

Ivan Damgård
Jean Paul Degabriele
Akshay Degwekar
David Derler
Apoorvaa Deshpande
Julien Devigne
Christoph Dobraunig
Frédéric Dupuis
Nico Döttling
Maria Eichlseder
Keita Emura
Xiong Fan
Pooya Farshim
Sebastian Faust
Omar Fawzi
Dario Fiore
Ben Fisch
Benjamin A. Fisch
Nils Fleischhacker
Georg Fuchsbauer
Eiichiro Fujisaki
Steven Galbraith
Chaya Ganesh
Juan Garay
Sumegha Garg
Romain Gay
Ran Gelles
Mariya Georgieva
Benedikt Gierlichs
Oliver W. Gnilke
Faruk Göloğlu
Sergey Gorbunov
Dov Gordon
Rishab Goyal
Hannes Gross
Vincent Grosso
Jens Groth
Daniel Gruss
Jian Guo
Siyao Guo
Qian Guo
Benoît Gérard
Felix Günther
Britta Hale
Carmit Hazay
Felix Heuer

Shoichi Hirose
Viet Tung Hoang
Justin Holmgren
Fumitaka Hoshino
Pavel Hubáček
Ilia Iliashenko
Laurent Imbert
Takanori Isobe
Tetsu Iwata
Malika Izabachene
Kimmo Jarvinen
Eliane Jaulmes
Dimitar Jetchev
Daniel Jost
Marc Joye
Herve Kalachi
Seny Kamara
Chethan Kamath
Angshuman Karmakar
Pierre Karpman
Nikolaos Karvelas
Marcel Keller
Elena Kirshanova
Fuyuki Kitagawa
Susumu Kiyoshima
Thorsten Kleinjung
Lars Knudsen
Konrad Kohbrok
Markulf Kohlweiss
Ilan Komargodski
Venkata Koppula
Thomas Korak
Lucas Kowalczyk
Thorsten Kranz
Fabien Laguillaumie
Kim Laine
Virginie Lallemand
Adeline Langlois
Hyung Tae Lee
Jooyoung Lee
Kwangsu Lee
Troy Lee
Kevin Lewi
Huijia (Rachel) Lin
Jiao Lin
Wei-Kai Lin

Feng-Hao Liu
Atul Luykx
Vadim Lyubashevsky
Xiongfeng Ma
Houssem Maghrebi
Mohammad Mahmoody
Daniel Malinowski
Alex Malozemoff
Antonio Marcedone
Daniel P. Martin
Daniel Masny
Takahiro Matsuda
Christian Matt
Alexander May
Sogol Mazaheri
Peihan Miao
Kazuhiko Minematsu
Ameer Mohammed
Tal Moran
Fabrice Mouhartem
Pratyay Mukherjee
Elke De Mulder
Pierrick Méaux
Michael Naehrig
Yusuke Naito
Kashif Nawaz
Kartik Nayak
Khoa Nguyen
Ryo Nishimaki
Olya Ohrimenko
Elisabeth Oswald
Ayoub Otmani
Giorgos Panagiotakos
Alain Passelègue
Kenneth G. Paterson
Serdar Pehlivanoglou
Alice Pellet–Mary
Pino Persiano
Cécile Pierrot
Rafaël Del Pino
Bertram Poettering
David Pointcheval
Antigoni Polychroniadou

Romain Poussier
Thomas Prest
Erick Purwanto
Carla Rafols
Ananth Raghunathan
Srinivasan Raghuraman
Sebastian Ramacher
Somindu Ramanna
Francesco Regazzoni
Ling Ren
Oscar Reparaz
Silas Richelson
Thomas Ricosset
Thomas Ristenpart
Florentin Rochet
Mike Rosulek
Yannis Rouselakis
Sujoy Sinha Roy
Michal Rybár
Carla Ràfols
Robert Schilling
Jacob Schuldt
Nicolas Sendrier
Yannick Seurin
Ido Shahaf
Sina Shiehian
Siang Meng Sim
Dave Singelee
Luisa Siniscalchi
Daniel Slamanig
Benjamin Smith
Akshayaram Srinivasan
François-Xavier Standaert
Ron Steinfeld
Noah
    Stephens-Davidowitz
Katerina Stouka
Koutarou Suzuki
Alan Szepieniec
Björn Tackmann
Stefano Tessaro
Adrian Thillard
Emmanuel Thomé

Mehdi Tibouchi
Elmar Tischhauser
Yosuke Todo
Ni Trieu
Roberto Trifiletti
Yiannis Tselekounis
Furkan Turan
Thomas Unterluggauer
Margarita Vald
Prashant Vasudevan
Philip Vejre
Srinivas Vivek Venkatesh
Daniele Venturi
Frederik Vercauteren
Ivan Visconti
Vanessa Vitse
Damian Vizár
Petros Wallden
Michael Walter
Lei Wang
Huaxiong Wang
Mor Weiss
Weiqiang Wen
Mario Werner
Benjamin Wesolowski
Carolyn Whitnall
Friedrich Wiemer
David Wu
Keita Xagawa
Sophia Yakoubov
Shota Yamada
Takashi Yamakawa
Avishay Yanay
Kan Yasuda
Eylon Yogev
Kazuki Yoneyama
Henry Yuen
Thomas Zacharias
Karol Zebrowski
Rina Zeitoun
Bingsheng Zhang
Ryan Zhou
Dionysis Zindros

Contents – Part III

Contents – Part I

Contents – Part II

Blockchain

Memory Hard Functions

Depth-Robust Graphs and Their Cumulative Memory Complexity

Joël Alwen[1]([✉]), Jeremiah Blocki[2], and Krzysztof Pietrzak[1]

[1] IST, Klosterneuburg, Austria
jalwen@ist.ac.at
[2] Purdue University, West Lafayette, USA

Abstract. Data-independent Memory Hard Functions (iMHFS) are finding a growing number of applications in security; especially in the domain of password hashing. An important property of a concrete iMHF is specified by fixing a directed acyclic graph (DAG) G_n on n nodes. The quality of that iMHF is then captured by the following two pebbling complexities of G_n:

- The parallel cumulative pebbling complexity $\Pi_{cc}^{\parallel}(G_n)$ must be as high as possible (to ensure that the amortized cost of computing the function on dedicated hardware is dominated by the cost of memory).
- The sequential space-time pebbling complexity $\Pi_{st}(G_n)$ should be as close as possible to $\Pi_{cc}^{\parallel}(G_n)$ (to ensure that using many cores in parallel and amortizing over many instances does not give much of an advantage).

In this paper we construct a family of DAGs with best possible parameters in an asymptotic sense, i.e., where $\Pi_{cc}^{\parallel}(G_n) = \Omega(n^2/\log(n))$ (which matches a known upper bound) and $\Pi_{st}(G_n)$ is within a constant factor of $\Pi_{cc}^{\parallel}(G_n)$.

Our analysis relies on a new connection between the pebbling complexity of a DAG and its depth-robustness (DR) – a well studied combinatorial property. We show that high DR is *sufficient* for high $\Pi_{cc}^{\parallel}$. Alwen and Blocki (CRYPTO'16) showed that high DR is *necessary* and so, together, these results fully characterize DAGs with high $\Pi_{cc}^{\parallel}$ in terms of DR.

Complementing these results, we provide new upper and lower bounds on the $\Pi_{cc}^{\parallel}$ of several important candidate iMHFs from the literature. We give the first lower bounds on the memory hardness of the Catena and Balloon Hashing functions in a parallel model of computation and we give the first lower bounds of any kind for (a version) of Argon2i.

Finally we describe a new class of pebbling attacks improving on those of Alwen and Blocki (CRYPTO'16). By instantiating these attacks we upperbound the $\Pi_{cc}^{\parallel}$ of the Password Hashing Competition winner Argon2i and one of the Balloon Hashing functions by $O\left(n^{1.71}\right)$. We also show an upper bound of $O(n^{1.625})$ for the Catena functions and the two remaining Balloon Hashing functions.

© International Association for Cryptologic Research 2017
J.-S. Coron and J.B. Nielsen (Eds.): EUROCRYPT 2017, Part III, LNCS 10212, pp. 3–32, 2017.
DOI: 10.1007/978-3-319-56617-7_1

1   Introduction

Moderately Hard Functions. Functions which are "moderately" hard to compute have found a variety of practical applications including password hashing, key-derivation and for proofs of work. In the context of password hashing, the goal is to minimize the damage done by a security breach where an adversary learns the password file; Instead of storing $(login, password)$ tuples in the clear, one picks a random salt and stores a tuple $(login, f(password, salt), salt)$, where $f(.)$ is a moderately hard function $f(.)$. This comes at a price, the server verifying a password must evaluate $f(.)$, which thus cannot be too hard. On the other hand, if a tuple $(login, y, salt)$ is leaked, an adversary who tries to find the password by a dictionary attack must evaluate $f(.)$ for every attempt. A popular moderately hard function is PBKDF2 (Password Based Key Derivation Function 2) [Kal00], which basically just iterates a cryptographic hash function H several times (1024 is a typical value).

Unfortunately a moderately hard function like PBKDF2 offers much less protection against adversaries who can build customized hardware to evaluate the underlying hash function than one would hope for. The reason is that the cost of computing a hash function H like SHA256 or MD5 on an ASIC (Application Specific Integrated Circuit) is orders of magnitude smaller than the cost of computing H on traditional hardware [DGN03, NBF+15].

Memory-Bound and Memory-Hard Functions. [ABW03] recognized that cache-misses are more egalitarian than computation, in the sense that they cost about the same on different architectures. They propose "memory-bound" functions, which are functions that will incur many expensive cache-misses. This idea was further developed by [DGN03].

Along similar lines, Percival [Per09] observes that unlike computation, memory costs tend to be relatively stable across different architectures, and suggests to use memory-hard functions (MHF) for password hashing. [Per09] also introduced the scrypt MHF which has found a variety of applications in practice. Very recently it has been proven to indeed offer optimal time/space trade-offs in the random oracle model [ACP+17, ACK+16].

MHFs come in two flavours, data-dependent MHFs (dMHF) such as scrypt, and data independent MHFs (iMHF). The former are potentially easier to construct and allow for more extreme memory-hardness [ACP+17, AB16], but they leave open the possibility of side-channel attacks [FLW13], thus iMHFs are preferable when the inputs are sensitive, as in the case of password hashing. We shortly discuss the state of the art for dMHFs at the end of this section.

iMHF as Graphs. An iMHF comes with an algorithm that computes the function using a fixed memory access pattern. In particular the pattern is independent of the input. Such functions can thus be described by a directed acyclic graph (DAG) G, where each node v of the graph corresponds to some intermediate value ℓ_v that appears during the computation of the function, and the edges capture the computation: if ℓ_v is a function of previously computed values $\ell_{i_1}, \ldots, \ell_{i_\delta}$, then the nodes $i_1, \ldots, i_\delta$ are parents of v in G. For an iMHF F, we'll denote with $G(F)$ the underlying graph. For example $G(\text{PBKDF2})$ is simply a path.

Graph Labeling Functions. Not only can an iMHF be captured by a graph as just outlined, we will actually construct iMHFs by first specifying a graph, and then defining a "labeling function" on top of it: Given a graph G with node set $V = [n] = \{1, 2, \ldots, n\}$, a hash function $H : \{0,1\}^* \to \{0,1\}^w$ and some input x, define the labeling of the nodes of G as follows: a source (a node v with indegree 0) has label $\ell_v(x) = H(v, x)$, a node v with parents $v_1 < v_2 < \cdots < v_\delta$ has label $\ell_v(x) = H(v, \ell_{v_1}(x), \ldots, \ell_{v_\delta}(x))$. For a DAG G with a unique sink s we define the labeling function of G as $f_G(x) = \ell_s(x)$. Note that using the convention from the previous paragraph, we have $G(f_G) = G$.

The Black Pebbling Game One of the main techniques for analyzing iMHF is to use pebbling games played on graphs. First introduced by Hewitt and Paterson [HP70] and Cook [Coo73] the (sequential) black pebbling game (and its relatives) have been used to great effect in theoretical computer science. Some early applications include space/time trade-offs for various computational tasks such as matrix multiplication [Tom78], the FFT [SS78, Tom78], integer multiplication [SS79b] and solving linear recursions [Cha73, SS79a]. More recently, pebbling games have been used for various cryptographic applications including proofs of space [DFKP15, RD16], proofs of work [DNW05, MMV13], leakage-resilient cryptography [DKW11a], garbled circuits [HJO+16], one-time computable functions [DKW11b], adaptive security proofs [HJO+16, JW16] and memory-hard functions [FLW13, AS15, AB16, AGK+16]. It's also an active research topic in proof complexity (cf. the survey on http://www.csc.kth.se/~jakobn/research/PebblingSurveyTMP.pdf).

The black pebbling game is played over a fixed directed acyclic graph (DAG) $G = (V, E)$ in rounds. The goal of the game is to pebble all sink nodes of G (not necessarily simultaneously). Each round $i \geq 1$ is characterized by its pebbling configuration $P_i \subseteq V$ which denotes the set of currently pebbled nodes. Initially $P_0 = \emptyset$, i.e., all nodes are unpebbled. P_i is derived from the previous configuration P_{i-1} according to two simple rules. (1) A node v may be pebbled (added to P_i) if, in the previous configuration all of its parents were pebbled, i.e., parents$(v) \subseteq P_{i-1}$. (2) A pebble can always be removed from P_i. In the sequential version rule (1) may be applied at most once per round while in the parallel version no such restriction applies. A sequence of configurations $P = (P_0, P_1, \ldots)$ is a (sequential) pebbling of G if it adheres to these rules and each sink node of G is contained in at least one configuration.

From a technical perspective, in this paper we investigate upper and lower bounds on various pebbling complexities of graphs, as they can be related to the cost of evaluating the "labeling function" f_G (to be defined below) in various computational models. In particular, let $\mathcal{P}_G$ and $\mathcal{P}_G^\|$ denote all valid sequential and parallel pebblings of G, respectively. We are interested in the *parallel* cumulative pebbling complexity of G, denoted $\Pi_{cc}^\|(G)$, and the sequential space-time complexity of G, denoted $\Pi_{st}(G)$, which are defined as

$$\Pi_{cc}^\|(G) = \min_{(P_1,\ldots,P_t)\in\mathcal{P}_G^\|} \sum_{i=1}^{t} |P_i| \qquad \Pi_{st}(G) = \min_{(P_1,\ldots,P_t)\in\mathcal{P}_G} t \cdot \max_i(|P_i|).$$

A main technical result of this paper is a family of graphs with high (in fact, as we'll discuss below, maximum possible) $\Pi_{cc}^{\parallel}$ complexity, and where the Π_{st} complexity is not much higher than the $\Pi_{cc}^{\parallel}$ complexity.[1] Throughout, we'll denote with $\mathbb{G}_n$ the set of all DAGs on n nodes and with $\mathbb{G}_{n,d} \subseteq \mathbb{G}_n$ the DAGs where each node has indegree at most d.

Theorem 1. *There exists a family of DAGs $\{G_n \in \mathbb{G}_{n,2}\}_{n\in\mathbb{N}}$ where*

1. the parallel cumulative pebbling complexity is

$$\Pi_{cc}^{\parallel}(G_n) \in \Omega(n^2/\log(n))$$

2. and where the sequential space-time complexity matches the parallel cumulative pebbling complexity up to a constant

$$\Pi_{st}(G_n) \in O(n^2/\log(n)).$$

The lower bound on $\Pi_{cc}^{\parallel}$ in item *1.* above is basically optimal due to the following bound from [AB16].[2]

Theorem 2 ([AB16, Theorem 8]). *For any constant $\epsilon > 0$ and sequence of DAGs $\{G_n \in \mathbb{G}_{n,\delta_n}\}_{n\in\mathbb{N}}$ it holds that*

$$\Pi_{cc}^{\parallel}(G_n) = o\left(\frac{\delta_n n^2}{\log^{1-\epsilon}}\right).$$

In particular if $\delta_n = O(\log^{1-\epsilon})$ then $\Pi_{cc}^{\parallel}(G_n) = o(n^2)$, and

$$\text{if } \delta_n = \Theta(1) \text{ then } \Pi_{cc}^{\parallel}(G_n) = o(n^2/\log^{1-\epsilon}(n)). \tag{1}$$

Pebbling vs. Memory-Hardness. The reason to focus on the graph $G = G(F)$ underlying an iMHF F is that clean combinatorial properties of G – i.e., bounds on the pebbling complexities – imply both upper and lower bounds on the cost of evaluating F in various computational models. For upper bounds (i.e., attacks), no further assumption on F are required to make the transition from pebbling to computation cost. For lower bounds, we have to assume that there's no "short-cut" in computing F, and the only way is to follow the evaluation sequence as given by G. Given the current state of complexity theory, where not even superlinear lower bounds on evaluating any function in $\mathcal{NP}$ are known, we cannot hope to exclude such shortcuts unconditionally. Instead, we assume that the

---

[1] Note that $\Pi_{cc}^{\parallel}(G) \leq \Pi_{st}(G)$ as parallelism can only help, and space-time complexity (i.e., number of rounds times the size of the largest state) is always higher than cumulative complexity (the sum of the sizes of all states).

[2] The statement below is obtained from the result in [AB16] by treating the core-memory ratio as a constant and observing that, trivially, at most n pebbles are on G during a balloon phase and at most n pebbles are placed in one step during a balloon phase.

underlying hash function H is a random oracle and circuits are charged unit cost for queries to the random oracle.

For our lower bounds, we must insist on G having constant indegree. The reason is that in reality H must be instantiated with some cryptographic hash function like SHA1, and the indegree corresponds to the input length on which H is invoked. To evaluate H on long inputs, one would typically use some iterated construction like Merkle-Damgard, and the assumption that H behaves like a black-box that can only be queried once the entire input is known would be simply wrong in this case.

As advocated in [AS15], bounds on $\Pi_{cc}^{\parallel}(G)$ are a reasonable approximation for the cost of evaluating f_G in dedicated hardware, whereas a bound on $\Pi_{st}(G)$ gives an upper bound on the cost of evaluating f_G on a single processor machine. The reason [AS15] consider cumulative complexity for lower and space-time complexity for the upper bound is that when lower bounding the cost of evaluating f_G we do want to allow for amortization of the cost over arbitrary many instances,[3] whereas for our upper bound we don't want to make such an assumption. The reason we consider parallel complexity for the lower and only sequential for the upper bound is due to the fact that an adversary can put many (cheap) cores computing H on dedicated hardware, whereas for the upper bound we only want to consider a single processor machine.

If $\Pi_{cc}^{\parallel}(G)$ is sufficiently larger than $|V(G)|$ (in Theorem 1 it's almost quadratic), then the cost of evaluating f_G in dedicated hardware is dominated by the memory cost. As memory costs about the same on dedicated hardware and general purpose processors, if our G additionally satisfies $\Pi_{cc}^{\parallel}(G) \approx \Pi_{st}(C)$, then we get a function f_G whose evaluation on dedicated hardware is not much cheaper than evaluating it on an off the shelf machine (like a single core x86 architecture). This is exactly what the family from Theorem 1 achieves. We elaborate on these computational models and how they are related to pebbling in the full version.

On the positive side, previous to this work, the construction with the best asymptotic bounds was due to [AS15] and achieved $\Pi_{cc}^{\parallel}(G_n) \in \Omega(n^2/\log^{10}(n))$. However the exponent 10 (and the complexity of the construction) makes this construction uninteresting for practical purposes.

On the negative side [AB16, ACK+16, AB17] have broken many popular iMHFs in a rather strong asymptotic sense. For example, in [AB16], the graph underlying Argon2i-A [BDK16], the winner of the recent Password Hashing Competition[4], was shown to have $\Pi_{cc}^{\parallel}$ complexity $\tilde{O}(n^{1.75})$. For Catena [FLW13] the

---

[3] $\Pi_{cc}^{\parallel}$ satisfies a direct product property: pebbling k copies of G cost k times as much as pebbling G, i.e., $\Pi_{cc}^{\parallel}(G^k) = k \cdot \Pi_{cc}^{\parallel}(G)$, but this is not true for Π_{st} complexity.

[4] The Argon2 specification [BDK16] has undergone several revisions all of which are regularly referred to as "Argon2." To avoid confusion we follow [AB17] and use Argon2i-A [BDK15] to denote the version of Argon2i from the password hashing competition [PHC] and we use Argon2i-B [BDKJ16] to refer the version of Argon2 that is currently being considered for standardization by the Cryptography Form Research Group (CFRG) of the IRTF. We conjecture that the techniques introduced in this paper could also be used to establish tighter bounds for Argon2i-B. However, we leave this as an open challenge for future work.

upper bound $O(n^{5/3})$ is shown in [AB16]. In [AB17] these results were extended to show that Argon2i-B [BDKJ16] has $\Pi_{cc}^{\parallel}(G) = O(n^{1.8})$. Moreover [AB17] show that for random instances of these functions (which is how they are supposed to be used in practice) the attacks actually have far lower $\Pi_{cc}^{\parallel}$ than these asymptotic analyses indicate.

A New Generic Attack and Its Applications. In this work we improve on the attacks of [BK15, AS15, AB16] (Sect. 6). We give a new parallel pebbling strategy for pebbling DAGs which lack a generalization of depth-robustness. Next we investigate this property for the case of Argon2i-A, the three Balloon-Hashing variants and both Catena variants to obtain new upper bounds on their respective $\Pi_{cc}^{\parallel}$. For example, we further improve the upper bound on $\Pi_{cc}^{\parallel}$ for Argon2i-A and the Single Buffer variant of Balloon-Hashing from $\tilde{O}(n^{1.75})$ to $O(n^{1.708})$.

New Security Proofs. Complementing these results, in Sect. 5, we give the first security proofs for a variety of iMHFs. Hitherto the only MHF with a full security proof in a parallel computational model was [AS15] which employed relatively construction specific techniques. When restricted to sequential computation the results of [LT82, AS15] show that Catena has Π_{st} complexity $\Omega(n^2)$. Similar results are also shown for Argon2i-A and Balloon Hashing in [BCGS16].

In this work we introduce two new techniques for proving security of iMHFs. In the case of Argon2i-A and Argon2i-B we analyze its depth-robustness to show that its $\Pi_{cc}^{\parallel}$ is at least $\tilde{\Omega}(n^{5/3})$. The second technique involves a new combinatorial property called dispersion which we show to imply lower bounds on the $\Pi_{cc}^{\parallel}$ of a graph. We investigate the dispersion properties of the Catena and Balloon Hashing variants to show their $\Pi_{cc}^{\parallel}$ to be $\tilde{\Omega}(n^{1.5})$. Previously no (non-trivial) lower bounds on $\Pi_{cc}^{\parallel}$ were known for Argon2i-A, Catena or Balloon Hashing. Interestingly, our results show that Argon2i-A and Argon2i-B have better asymptotic security guarantees than Catena since $\Pi_{cc}^{\parallel} = \Omega(n^{5/3})$ for Argon2i-A and $\Pi_{cc}^{\parallel} = O(n^{13/8})$ for Catena.

While these lower bounds are significantly worse than what we might ideally hope for in a secure iMHF (e.g., $\Pi_{cc}^{\parallel} \geq \Omega(n^2/\log(n))$), we observe that, in light of our new attacks in Sect. 6, they are nearly tight. Unfortunately, together with the bounds on the sequential complexity of these algorithms our results do highlight a large asymptotic gap between the memory needed when computing the functions on parallel vs. sequential computational devices.

A table summarizing the asymptotic cumulative complexity of various iMHFs can be found in Table 1.

Depth-Robust Graphs. The results in this work rely on a new connection between the depth-robustness of a DAG and its $\Pi_{cc}^{\parallel}$ complexity. A DAG G is (e, d)-depth-robust if, after removing any subset of at most e nodes there remains a directed path of length at least d. First investigated by Erdös, Graham and Szemerédi [EGS75], several such graphs enjoying low indegree and increasingly extreme depth-robustness have been constructed in the past [EGS75, PR80, Sch82, Sch83, MMV13] mainly in the context of proving

Table 1. Overview of the asymptotic cumulative complexity of various iMHF.

Algorithm	Lowerbound	Upperbound	Appearing in
Argon2i-A		$\tilde{O}\left(n^{1.75}\right)$	[AB16]
Argon2i-A	$\tilde{\Omega}\left(n^{1.6}\right)$	$\tilde{O}\left(n^{1.708}\right)$	This work
Argon2i-B		$O\left(n^{1.8}\right)$	[AB17]
Argon2i-B	$\tilde{\Omega}\left(n^{1.\bar{6}}\right)$		This work
Balloon-Hashing: Linear and Double Buffer (DB)		$O\left(n^{1.67}\right)$	[AB16]
Balloon-Hashing: Linear and Double Buffer (DB)	$\tilde{\Omega}\left(n^{1.5}\right)$	$\tilde{O}\left(n^{1.625}\right)$	This work
Balloon-Hashing: Single Buffer (SB)		$\tilde{O}\left(n^{1.75}\right)$	[AB16]
Balloon-Hashing: Single Buffer (SB)	$\tilde{\Omega}\left(n^{1.\bar{6}}\right)$	$\tilde{O}\left(n^{1.708}\right)$	This work
Catena: Dragonfly		$O\left(n^{1.67}\right)$	[AB16]
Catena: Dragonfly	$\tilde{\Omega}\left(n^{1.5}\right)$	$\tilde{O}\left(n^{1.625}\right)$	This work
Catena: Butterfly		$O\left(n^{1.67}\right)$	[AB16]
Catena: Butterfly	$\tilde{\Omega}\left(n^{1.5}\right)$	$o\left(n^{1.625}\right)$	This work
[AS15]	$\Omega\left(\frac{n^2}{\log^{10} n}\right)$		[AS15]
Theorem 1	$\Omega\left(\frac{n^2}{\log n}\right)$		This work
Arbitrary iMHF		$O\left(\frac{n^2 \log \log n}{\log n}\right)$	[AB16]

lower-bounds on circuit complexity and Turing machine time. Depth-robustness has been used as a key tool in the construction of cryptographic objects like proofs of sequential work [MMV13]. In fact depth-robust graphs were already used as a building block in the construction of a high $\Pi_{cc}^{\parallel}$ graph in [AS15].

Depth-Robustness and $\Pi_{cc}^{\parallel}$. While the flavour of the results in this work are related to those of [AS15] the techniques are rather different. As mentioned above already, they stem from a new tight connection between depth-robustness and $\Pi_{cc}^{\parallel}$. A special case of this connection shows that if G is (e, d)-depth-robust, then its $\Pi_{cc}^{\parallel}$ can be lower bounded as

$$\Pi_{cc}^{\parallel}(G) \geq e \cdot d.$$

This complements a result from [AB16], which gives a pebbling strategy that is efficient for graphs of low depth-robustness. Thus a DAG has high $\Pi_{cc}^{\parallel}$ if and only if it is very depth-robust.

Moreover, we give a new tool for reducing the indegree of a DAG while not reducing the $\Pi_{cc}^{\parallel}$ of the resulting graph (in terms of its size). Together these results directly have some interesting consequences

- The family of DAGs $\{G_n \in \mathbb{G}_{n,\log(n)}\}_{n \in \mathbb{N}}$ from Erdös et al. [EGS75] have optimally high $\Pi_{cc}^{\parallel}(G_n) \in \Omega(n^2)$.

– Using our indegree reduction we can turn the above family of $\log(n)$ indegree into a family of indegree 2 DAGs $\{G'_n \in \mathbb{G}_{(n,2)}\}_{n \in \mathbb{N}}$ with $\Pi_{cc}^{\parallel}(G'_n) \in \Omega(n^2/\log(n))$, which by Theorem 2 is optimal for constant indegree graphs.

Data-Dependent MHFs. One can naturally extend the $\Pi_{cc}^{\parallel}$ notion also to "dynamic" graphs – where some edges are only revealed as some nodes are pebbled – in order to analyse data-dependent MHFs (dMHF) like scrypt. In this model, [ACK+16] show that $\Pi_{cc}^{\parallel}(\text{scrypt}) = \Omega(n^2/\log^2(n))$. Unfortunately unlike for iMHFs, for dMHFs we do not have a proof that a lower bound on $\Pi_{cc}^{\parallel}$ implies roughly the same lower bound on the cumulative memory complexity in the random oracle model.[5] Recently a "direct" proof (i.e., avoiding pebbling arguments) – showing that scrypt has optimal cumulative memory complexity $\Omega(n^2)$ – has been announced, note that this bound is better than what we can hope to achieve for iMHFs (as stated in Theorem 2). Unfortunately, the techniques that have now been developed to analyse dMHFs seem not to be useful for the iMHF setting.

2   Pebbling Complexities and Depth-Robustness of Graphs

We begin by fixing some common notation. We use the sets $\mathbb{N} = \{0, 1, 2, \ldots\}$, $\mathbb{N}^+ = \{1, 2, \ldots\}$, and $\mathbb{N}_{\geq c} = \{c, c+1, c+2, \ldots\}$ for $c \in \mathbb{N}$. Further, we also use the sets $[c] := \{1, 2, \ldots, c\}$ and $[b, c] = \{b, b+1, \ldots, c\}$ where $b \in \mathbb{N}$ with $b \leq c$. For a set of sets $A = \{B_1, B_2, \ldots, B_z\}$ we use the notation $||A|| := \sum_i |B_i|$.

2.1   Depth-Robust Graphs

We say that a directed acyclic graph (DAG) $G = (V, E)$ has *size* n if $|V| = n$. A node $v \in V$ has indegree $\delta = \mathsf{indeg}(v)$ if there exist δ incoming edges $\delta = |(V \times \{v\}) \cap E|$. More generally, we say that G has indegree $\delta = \mathsf{indeg}(G)$ if the maximum indegree of any node of G is δ. A node with indegree 0 is called a *source* node and one with no outgoing edges is called a *sink*. We use $\mathsf{parents}_G(v) = \{u \in V : (u, v) \in E\}$ to denote the parents of a node $v \in V$. In general, we use $\mathsf{ancestors}_G(v) = \bigcup_{i \geq 1} \mathsf{parents}_G^i(v)$ to denote the set of all ancestors of v — here, $\mathsf{parents}_G^2(v) = \mathsf{parents}_G(\mathsf{parents}_G(v))$ denotes the grandparents of v and $\mathsf{parents}_G^{i+1}(v) = \mathsf{parents}_G(\mathsf{parents}_G^i(v))$. When G is clear from context we will simply write $\mathsf{parents}$ ($\mathsf{ancestors}$). We denote the set of all sinks of G with $\mathsf{sinks}(G) = \{v \in V : \nexists(v, u) \in E\}$ — note that $\mathsf{ancestors}(\mathsf{sinks}(G)) = V$. We often consider the set of all DAGs of equal size $\mathbb{G}_n = \{G = (V, E) : |V| = n\}$ and often will bound the maximum indegree $\mathbb{G}_{n,\delta} = \{G \in \mathbb{G}_n : \mathsf{indeg}(G) \leq \delta\}$.

---

[5] [ACK+16] introduces a combinatorial conjecture, which if true, means that lower bounds on $\Pi_{cc}^{\parallel}$ translate to cumulative memory complexity. At this point a strong variant of the conjecture has already been refuted, the state of the conjecture is updated in the eprint version [ACK+16] of the paper.

For directed path $p = (v_1, v_2, \ldots, v_z)$ in G its length is the number of nodes it traverses $\mathsf{length}(p) := z$. The depth $d = \mathsf{depth}(G)$ of DAG G is the length of the longest directed path in G.

We will often consider graphs obtained from other graphs by removing subsets of nodes. Therefore if $S \subset V$ then we denote by $G - S$ the DAG obtained from G by removing nodes S and incident edges. The following is a central definition to our work.

Definition 1 (Depth-Robustness). *For $n \in \mathbb{N}$ and $e, d \in [n]$ a DAG $G = (V, E)$ is (e, d)-depth-robust if*

$$\forall S \subset V \quad |S| \leq e \Rightarrow \mathsf{depth}(G - S) \geq d.$$

We will make use of the following lemma due to Erdös, Graham and Szemerédi [EGS75], who showed how to construct a family of log indegree DAGs with extreme depth-robustness.

Theorem 3 ([EGS75]). *For some fixed constants $c_1, c_2, c_3 > 0$ there exists an infinite family of DAGs $\{G_n \in \mathbb{G}_{n,c_3 \log(n)}\}_{n=1}^{\infty}$ such that G_n is $(c_1 n, c_2 n)$-depth-robust.*

2.2   Graph Pebbling

We fix our notation for the parallel graph pebbling game following [AS15].

Definition 2 (Parallel/Sequential Graph Pebbling). *Let $G = (V, E)$ be a DAG and let $T \subseteq V$ be a target set of nodes to be pebbled. A pebbling configuration (of G) is a subset $P_i \subseteq V$. A legal parallel pebbling of T is a sequence $P = (P_0, \ldots, P_t)$ of pebbling configurations of G where $P_0 = \emptyset$ and which satisfies conditions 1 & 2 below. A sequential pebbling additionally must satisfy condition 3.*

1. *At some step every target node is pebbled (though not necessarily simultaneously).*

$$\forall x \in T \; \exists z \leq t \quad : \quad x \in P_z.$$

2. *Pebbles are added only when their predecessors already have a pebble at the end of the previous step.*

$$\forall i \in [t] \quad : \quad x \in (P_i \setminus P_{i-1}) \Rightarrow \mathsf{parents}(x) \subseteq P_{i-1}.$$

3. *At most one pebble placed per step.*

$$\forall i \in [t] \quad : \quad |P_i \setminus P_{i-1}| \leq 1.$$

We denote with $\mathcal{P}_{G,T}$ and $\mathcal{P}_{G,T}^{\parallel}$ the set of all legal sequential and parallel pebblings of G with target set T, respectively. Note that $\mathcal{P}_{G,T} \subseteq \mathcal{P}_{G,T}^{\parallel}$. We will be mostly interested in the case where $T = \mathsf{sinks}(G)$ and then will simply write $\mathcal{P}_G$ and $\mathcal{P}_G^{\parallel}$.

Definition 3 (Time/Space/Cumulative Pebbling Complexity). *The time, space, space-time and cumulative complexity of a pebbling* $P = \{P_0, \ldots, P_t\} \in \mathcal{P}_G^\parallel$ *are defined to be:*

$$\Pi_t(P) = t \qquad \Pi_s(P) = \max_{i \in [t]} |P_i| \qquad \Pi_{st}(P) = \Pi_t(P) \cdot \Pi_s(P) \qquad \Pi_{cc}(P) = \sum_{i \in [t]} |P_i|.$$

For $\alpha \in \{s, t, st, cc\}$ and a target set $T \subseteq V$, the sequential and parallel pebbling complexities of G are defined as

$$\Pi_\alpha(G, T) = \min_{P \in \mathcal{P}_{G,T}} \Pi_\alpha(P) \qquad and \qquad \Pi_\alpha^\parallel(G, T) = \min_{P \in \mathcal{P}_{G,T}^\parallel} \Pi_\alpha(P).$$

When $T = \mathsf{sinks}(G)$ we simplify notation and write $\Pi_\alpha(G)$ and $\Pi_\alpha^\parallel(G)$.

It follows from the definition that for $\alpha \in \{s, t, st, cc\}$ and any G the parallel pebbling complexity is always at most as high as the sequential, i.e., $\Pi_\alpha(G) \geq \Pi_\alpha^\parallel(G)$, and cumulative complexity is at most as high as space-time complexity, i.e., $\Pi_{st}(G) \geq \Pi_{cc}(G)$ and $\Pi_{st}^\parallel(G) \geq \Pi_{cc}^\parallel(G)$.

In this work we will consider constant in-degree DAGs $\{G_n \in \mathbb{G}_{n,\Theta(1)}\}_{n \in \mathbb{N}}$, and will be interested in the complexities $\Pi_{st}(G_n)$ and $\Pi_{cc}^\parallel(G_n)$ as these will capture the cost of evaluating the labelling function derived from G_n on a single processor machine (e.g. a x86 processor on password server) and amortized AT complexity (which is a good measure for the cost of evaluating the function on dedicated hardware), respectively.

Before we state our main theorem let us observe some simple facts. Every n-node graph can be pebbled in n steps, and we cannot have more than n pebbles on an n node graph, thus

$$\forall G_n \in \mathbb{G}_n \ : \ \Pi_{cc}^\parallel(G_n) \ \leq \ \Pi_{st}(G_n) \ \leq \ n^2.$$

This upper bound is basically matched for the complete graph $K_n = (V = [n], E = \{(i, j) : 1 \leq i < j \leq n\})$ as

$$n(n-1)/2 \ \leq \ \Pi_{cc}^\parallel(K_n) \ \leq \ \Pi_{st}(K_n) \ \leq \ n^2.$$

Graph K_n has the desirable properties that its Π_{st} is within a constant factor to its $\Pi_{cc}^\parallel$ complexity and that its $\Pi_{cc}^\parallel$ complexity is maximally high. Unfortunately, K_n very high indegree, which makes it useless for our purpose to construct memory-hard functions. The path $Q_n = (V = [n], E = \{(i, i+1) : 1 \leq i \leq n-1\})$ on the other hand has indegree 1 and its Π_{st} is even exactly as large as its $\Pi_{cc}^\parallel$ complexity. Unfortunately it has very low pebbling complexity

$$\Pi_{cc}^\parallel(Q_n) \ = \ \Pi_{st}(Q_n) \ = \ n$$

which means that in the labelling function we get from Q_n (which is basically PBKDF2 discussed in the introduction) the evaluation cost will not be dominated by the memory cost even for large n. As stated in Theorem 1, in this paper we construct a family of graphs $\{G_n \in \mathbb{G}_{n,2}\}_{n \in \mathbb{N}}$ which satisfies all three properties at once: (1) the graphs have indegree 2 (2) the parallel cumulative pebbling complexity is $\Pi_{cc}^{\parallel}(G_n) \in \Omega(n^2/\log(n))$, which by Theorem 2 is optimal for constant indegree graphs, and (3) $\Pi_{st}(G_n)$ is within a constant factor of $\Pi_{cc}^{\parallel}(G_n)$.

3 Depth-Robustness Implies High $\Pi_{cc}^{\parallel}$

In this section we state and prove a theorem which lowerbounds the $\Pi_{cc}^{\parallel}$ of a given DAG G in terms of its depth robustness.

Theorem 4. *Let G be an (e,d)-depth-robust DAG, then $\Pi_{cc}^{\parallel}(G) > ed$.*

Proof. Let $(P_1, \ldots, P_m)$ be a parallel pebbling of minimum complexity, i.e., $\sum_{i=1}^{m} |P_i| = \Pi_{cc}^{\parallel}(G)$. For any d, we'll show that there exists a set B of size $|B| \leq \Pi_{cc}^{\parallel}(G)/d$ such that there's no path of length d in $G - B$, or equivalently, G is not $(\Pi_{cc}^{\parallel}(G)/d, d)$-depth-robust, note that this implies the theorem.

For $i \in [d]$ define $B_i = P_i \cup P_{i+d} \cup P_{i+2d} \ldots$. We observe that by construction $\sum_{i=0}^{d-1} |B_i| \leq \sum_{i=1}^{m} |P_i| = \Pi_{cc}^{\parallel}(G)$, so the size of the B_i's is $\leq \Pi_{cc}^{\parallel}(G)/d$ on average, and the smallest B_i has size at most this. Let B be the smallest B_i, as just outlined $|B| \leq \Pi_{cc}^{\parallel}(G)/d$.

It remains to show that $G - B$ has no path of length d. For this consider any path $v_1, \ldots, v_d$ of length d in G. Let j be minimal such that $v_d \in P_j$ (so v_d is pebbled for the first time in round j of the pebbling). It then must be the case that $v_{d-1} \in P_{j-1}$ (as to pebble v_d in round j there must have been a pebble on v_{d-1} in round $j-1$). In round $j-2$ either the pebble on v_{d-1} was already there, or there was a pebble on v_{d-2}. This argument shows that each of the pebbling configurations $\{P_{j-d+1}, \ldots, P_j\}$ must contain at least one node from $v_1, \ldots, v_d$. As B contains each dth pebbling configuration, B contains at least one of these pebbling configurations $\{P_{j-d+1}, \ldots, P_j\}$. Specifically we can find $j - d + 1 \leq k \leq j$ s.t $P_k \subseteq B$, thus the path $v_1, \ldots, v_d$ is not contained entirely in $G - B$. $\square$

An immediate implication of Theorems 4 and 3 is that there is an infinite family of DAGs with maximal $\Pi_{cc}^{\parallel}(G) = \Omega(n^2)$ whose indegree scales with $\log n$. Note that this means that allowing indegree as small as $O(\log(n))$ is sufficient to get DAGs whose $\Pi_{cc}^{\parallel}$ is within a constant factor of the n^2 upper bound on $\Pi_{cc}^{\parallel}$ for any n node DAG. In the next section we will show how to reduce the indegree to $O(1)$ while only reducing $\Pi_{cc}^{\parallel}(G)$ by a factor of $O(\log(n))$.

Corollary 1 (of Theorems 4 and 3). *For some constants $c_1, c_2 > 0$ there exists an infinite family of DAGs $\{G_{n,\delta} \in \mathbb{G}_{n,\delta}\}_{n=1}^{\infty}$ with $\delta \leq c_1 \log(n)$ and $\Pi_{cc}^{\parallel}(G) \geq c_2 n^2$. This is optimal in the sense that for any family $\{\delta_n \in [n]\}_{n=1}^{\infty}$ and $\{J_n \in \mathbb{G}_{n,\delta_n}\}_{n=1}^{\infty}$ it holds that $\Pi_{cc}^{\parallel}(J_n) \in O(n^2)$. Moreover if $\delta_n = o(\log(n)/\log\log(n))$ then $\Pi_{cc}^{\parallel}(J_n) = o(n^2) = o(\Pi_{cc}^{\parallel}(G_n))$.*

Corollary 2 lower bounds the cost of pebbling a target set T given a starting pebbling configuration S. In particular, if the ancestors of T in $G - S$ induce an (e, d)-depth-robust DAG then the pebbling cost is at least $\Pi_{cc}^{\parallel}(G - S, T) \geq ed$. We will use Corollary 2 to lower bound $\Pi_{cc}^{\parallel}$ for iMHFs like Argon2i and SB.

Corollary 2 (of Theorem 4). *Given a DAG $G = (V, E)$ and subsets $S, T \subset V$ such that $S \cap T = \emptyset$ let $G' = G - (V \setminus \mathsf{ancestors}_{G-S}(T))$. If G' is (e, d)-depth robust then the cost of pebbling $G - S$ with target set T is $\Pi_{cc}^{\parallel}(G - S, T) > ed$.*

Proof. Note that $\Pi_{cc}^{\parallel}(G - S, T) \geq \Pi_{cc}^{\parallel}(G')$ since we will need to pebble every node in the set $\mathsf{ancestors}_{G-S}(T) = V(G')$ to reach the target set T in $G - S$. By Theorem 4, we have $\Pi_{cc}^{\parallel}(G') > ed$. □

Corollary 3 states that it remains expensive to pebble any large enough set of remaining nodes in a depth-robust graph even if we are permitted to first remove an arbitrary node set of limited size. An application of Corollary 3 might involve analysing the cost of pebbling stacks of depth-robust graphs. For example if there are not enough pebbles on the graph at some point in time then there must be some layers with few pebbles. If we can then show that many of the nodes on those layers will eventually need to be (re)pebbled then we can use this lemma to show that the remaining pebbling cost incurred by these layers is large.

Corollary 3 (of Theorem 4). *Let DAG $G = (V, E)$ be (e, d)-depth-robust and let $S, T \subset V$ such that*

$$|S| \leq e \quad and \quad T \cap S = \emptyset.$$

Then the cost of pebbling $G - S$ with target set T is $\Pi_{cc}^{\parallel}(G - S, T) > (e - |S|)(d - |\mathsf{ancestors}_{G-S}(T)|)$.

Proof. Let $G' = G - (V - \mathsf{ancestors}_{G-S}(T))$ and observe that G' is, at minimum, $(e - |S|, d - |\mathsf{ancestors}_{G-S}(T)|)$-depth robust. By Corollary 2 we have $\Pi_{cc}^{\parallel}(G - S, T) \geq \Pi_{cc}^{\parallel}(G') > (e - |S|)(d - |\mathsf{ancestors}_{G-S}(T)|)$. □

We remark that Theorem 4 is a special case of Corollary 3 by setting $S = \emptyset$ letting $T = \mathsf{sinks}(G)$. Recall that $\Pi_{cc}^{\parallel}(G)$ is the parallel pebbling of minimal cumulative cost when pebbling all sinks of G, this requires pebbling all nodes of G at least once.

4 Indegree Reduction: Constant Indegree with Maximal $\Pi_{cc}^{\parallel}$

In this section we use the result from the previous section to show a new, more efficient, degree-reduction lemma. We remark that Lemma 1 is similar to [AS15, Lemma 9] in that both reductions replace high indegree nodes v in G with a path.

However, we stress two key differences between the two results. First, our focus is on reducing the indegree while preserving depth-robustness. By contrast, [AS15, Lemma 9] focuses directly on preserving $\Pi_{cc}^{\parallel}$. Second, we note that the guarantee of [AS15, Lemma 9] is weaker in that it yields a reduced indegree graph G' whose size grows by a factor of indeg ($n' \leq n \times$ indeg) while $\Pi_{cc}^{\parallel}$ can drop by a factor of indeg—[AS15, Lemma 9] shows that $\Pi_{cc}^{\parallel}(G') \geq \frac{\Pi_{cc}^{\parallel}(G)}{\text{indeg}-1}$. By contrast, setting $\gamma =$ indeg in Lemma 1 yields a reduced indegree graph G' whose size grows by a factor of $2 \times$ indeg ($n' \leq 2n \times$ indeg) and better depth-robustness $(e', d') = (e, d \times \text{indeg})$. In particular, when we apply Theorem 4 the lower-bound $\Pi_{cc}^{\parallel}(G') \geq ed \times$ indeg *improves* by a factor of indeg when compared with the original graph G.

Lemma 1. *Let G be a (e, d)-depth-robust DAG. For $\gamma \in \mathbb{Z}_{\geq 0}$ there exists a $(e, d\gamma)$-depth-robust DAG G' with*

$$\text{size}(G') \leq (\text{indeg}(G) + \gamma) \cdot \text{size}(G) , \qquad \text{indeg}(G') = 2 \quad and \quad \Pi_{st}(G') \leq \frac{\text{size}(G')^2}{\gamma} .$$

Proof. Fix a $\gamma \in \mathbb{Z}_{\geq 0}$ and let $\delta = \text{indeg}(G)$. We identify each node in V' with an element of the set $V \times [\delta + \gamma]$ and we write $\langle v, j \rangle \in V'$. For every node $v \in V$ with $\alpha_v := \text{indeg}(v) \in [0, \delta]$ we add the path $p_v = (\langle v, 1 \rangle, \langle v, 2 \rangle, \ldots, \langle v, \alpha_v + \gamma \rangle)$ of length $\alpha_v + \gamma$. We call v the *genesis node* and p_v its *metanode*. In particular $V' = \cup_{v \in V} p_v$. Thus G has size at most $(\delta + \gamma)n$.

Next we add the remaining edges. Intuitively, for the ith incoming edge (u, v) of v we add an edge to G' connecting the end of the metanode of u to the ith node in the metanode of v. More precisely, for every $v \in V$, $i \in [\text{indeg}(v)]$ and edge $(u_i, v) \in E$ we add edge $(\langle u_i, \text{indeg}(u_i) + \gamma \rangle, \langle v, i \rangle)$ to E'. It follows immediately that G' has indegree (at most) 2.

Fix any node set $S \subset V'$ of size $|S| \leq e$. Then at most e metanodes can share a node with S. For each such metanode remove its genesis node in G. As G is (e, d)-depth-robust we are still left with a path p of length (at least) d in G. But that means that after removing S from G' there must remain a path p' in G' running through all the metanodes of p and $|p'| \geq |p|\gamma \geq d\gamma$. In other words G' is $(e, d\gamma)$-depth-robust.

To see that $\Pi_{st}(G') \leq \text{size}(G')^2/\gamma$ we simply pebble G' in topological order. We note that we never need to keep more than one pebble on any metanode $p_v = (\langle v, 1 \rangle, \langle v, 2 \rangle, \ldots, \langle v, \alpha_v + \gamma \rangle)$ with $\alpha_v = \text{indeg}(v)$. Once we pebble the last node $\langle v, \alpha_v + \gamma \rangle$ we can permanently discard any pebbles on the rest of p_v since $\langle v, \alpha_v + \gamma \rangle$ is the only node with outgoing edges. $\square$

Proof of Theorem 1. Theorem 1 follows by applying Lemma 1 to the family from Theorem 3 with $\gamma =$ indeg $= \log n$. We get that for some fixed constants $c_1, c_2 > 0$ there exists an infinite family of indegree 2 DAGs $\{G_n \in \mathbb{G}_{n,2}\}_{n=1}^{\infty}$ where G_n is $(c_1 n/\log n, c_2 n)$-depth robust and $\Pi_{st}(G_n) \leq O(n^2/\log(n))$. By Theorem 4 then $\Pi_{cc}^{\parallel}(G_n) > (c_1 c_2)n^2/\log(n)$, which is basically optimal for constant indegree DAGs by Theorem 2. $\square$

5 Security Proofs of Candidate iMHFs

On the surface, in this and the next section we give both security proofs and nearly optimal attacks for several of the most prominent iMHF proposals. That is we show both lower and (relatively tight) upperbounds on their asymptotic memory-hardness in the PROM. However, more conceptually, we also introduce two new proof techniques for analysing the depth-robustness of DAGs as well as a new very memory-efficient class of algorithms for pebbling a DAG improving on the techniques used in [AB16]. Indeed for all candidates considered the attack in the next section is almost optimal in light of the accompanying security proofs in this section.

More specifically, in the first subsection we prove bounds for a class of random graphs which generalize the Argon2i-A construction [BDK16] and the Single Buffer (SB) variant of Balloon Hashing [BCGS16]. To prove the lowerbound we use a simple and clean new technique for bounding the depth-robustness of a random DAG. In particular, we show that a random DAG is almost certainly $\left(e, \tilde{\Omega}\left(n^2/e^2\right)\right)$-depth robust for any $e > \sqrt{n}$. Combined with Theorem 4 we could immediately obtain a lower bound of $\tilde{\Omega}\left(n^{1.5}\right)$. We can improve the lower bound to $\tilde{\Omega}\left(n^{5/3}\right)$ by introducing a stronger notion of depth-robustness that we call block depth-robustness.

In the second subsection we prove bounds for a family of layered graphs which generalize both of the Catena constructions [FLW13] as well as Linear (Lin) and Double Buffer (DB) variants of Balloon Hashing [BCGS16]. In particular, we introduce a new technique for proving lowerbounds on the cumulative pebbling complexity of a graph without going through the notion of depth-robustness. For example the (single layer) version of the Catena Dragonfly graph has the worst possible depth-robustness of any graph of linear depth. This shows that (in the lower but still non-trivial regimes of) cumulative complexity alternative combinatorial structures exist besides depth-robustness that can also confer some degree of pebbling complexity.

5.1 Lowerbounding the CC of Random DAGs

We begin by defining a (n, δ, w)-random DAG, the underlying DAGs upon which Argon2i-A and SB are based. The memory window parameter w specifies the intended memory usage and throughput of the iMHF—the cost of the naïve pebbling algorithm is $\Pi_{cc}^{\parallel}(\mathcal{N}) = wn$. In particular, a t-pass Argon2i-A iMHF is based on a $(n, 2, n/t)$-random DAG. Similarly, a t-pass Single-Buffer (SB) iMHF [BCGS16] is based on a $(n, 20, n/t)$-random DAG. In this section we focus on the $t = 1$-pass variants of the Argon2i-A and [BCGS16] iMHFs.

Definition 4 ((n, δ, w)-random DAG). *Let $n \in \mathbb{N}$, $1 < \delta < n$, and $1 \leq w \leq n$ such that w divides n. An (n, δ, w)-random DAG is a randomly generated directed acyclic (multi)graph with n nodes $v_1, \ldots, v_n$ (which we identify with the set $[n]$ according to there topological order) and with maximum in-degree δ for each node. The graph has directed edges (v_i, v_{i+1}) for $1 \leq i < n$ and random forward* edges $(v_{r(i,1)}, v_i), \ldots, (v_{r(i,\delta-1)}, v_i)$ *for each node* v_i. *Here,* $r(i,j)$ *is independently chosen uniformly at random from the set* $[\max\{0, i - w\}, i - 1]$.

Theorem 5 states that for a (n, δ, n)-random DAG G such as Argon2i-A or SB we almost certainly have $\Pi_{cc}^{\parallel}(G) = \tilde{\Omega}\left(n^{5/3}\right)$.

Theorem 5. *Let* G *be a* (n, δ, n)-*random DAG then, except with probability* $o(n^{-7})$, *we have*

$$\Pi_{cc}^{\parallel}(G) = \tilde{\Omega}\left(n^{5/3}\right).$$

Security Lower Bound. To prove the lower bound we rely on a slightly stricter notion of depth robustness. Given a node v let $N(v, b) = \{v - b + 1, \ldots, v\}$ denote a segment of b consecutive nodes ending at v and given a set $S \subseteq V(G)$ let $N(S, b) = \bigcup_{v \in S} N(v, b)$. We say that a DAG G is (e, d, b)-block depth-robust if for every set $S \subseteq V(G)$ of size $|S| \leq e$ we have $\mathsf{depth}(G - N(S, b)) \geq d$. Notice that when $b = 1$ (e, d, b)-block-depth robustness is equivalent to (e, d)-depth-robustness. However, when $b > 1$ (e, d, b)-block-depth robustness is a strictly stronger notion since the set $N(S, b)$ may have size as large as $|N(S, b)| = eb$. [6]

The proof of Theorem 5 relies on Lemma 2, which states that for any $e \geq \sqrt{n}$, with high probability, a $(n, 2, n)$-random DAG G will be (e, d, b)-block depth-robust with $d = \frac{n^2}{e^2 \mathsf{polylog}(n)}$ and $b = n/(20e)$. By contrast Lemma 9 states that G will be (e, d)-reducible with $d = \tilde{O}\left(n^2/e^2\right)$.

Lemma 2. *For any* $e \geq \sqrt{n}$ *any* $\delta \geq 2$ *a* (n, δ, n)-*random DAG will be* $\left(e, \Omega\left(\frac{n^2}{e^2 \log(n)}\right), \frac{n}{20e}\right)$-*block depth robust except with negligible probability in* n.

Setting $e = \sqrt{n}$ in Lemma 2 and applying Theorem 4 already implies that $\Pi_{cc}^{\parallel}(G) = \tilde{\Omega}\left(n^{1.5}\right)$. To obtain the stronger bound in Theorem 5 we rely on Corollary 2 combined with a more sophisticated argument exploiting block depth-robustness.

In more detail, let G be an (n, δ, n)-random DAG and let t_j denote the first time we place a pebble on node j. Observe that, since G contains all edges of the form $(j, j + 1)$ it must be that $t_{j+i} - t_j \geq i$ in any legal pebbling of G. We will show that for any $j > n/2$ a legal pebbling must (almost certainly) incur a cost of $\tilde{\Omega}\left(n^{4/3}\right)$ between pebbling steps t_j and t_{j+2k} where $k = \tilde{\Theta}\left(n^{2/3}\right)$. That is $\sum_{t=t_j}^{t_{j+2k}} |P_t| = \tilde{\Omega}\left(n^{4/3}\right)$ for any legal pebbling of G. Thus, $\sum_{t=t_{n/2+1}}^{t_n} |P_t| = \tilde{\Omega}\left(n^{4/3}\frac{n/2}{k}\right) = \tilde{\Omega}\left(n^{5/3}\right)$. In the remaining discussion we set $e = \tilde{\Omega}\left(n^{2/3}\right)$, $d = \tilde{\Omega}\left(n^{2/3}\right)$ and $b = \tilde{\Omega}\left(n^{1/3}\right)$.

To show that $\sum_{t=t_j}^{t_{j+2k}} |P_t| = \tilde{\Omega}\left(n^{4/3}\right)$ we consider two cases: we either have $|P_t| \geq e/2 = \tilde{\Omega}\left(n^{2/3}\right)$ pebbles on the DAG during each round $t_j \leq t \leq t_{j+k}$, or we do not. In the first case we trivially have $\sum_{t=t_j}^{t_{j+2k}} |P_t| \geq ke/2 = \tilde{\Omega}\left(n^{4/3}\right)$.

---

[6] In particular, $(e, d, b \geq 1)$-block depth robustness implies (e, d)-depth robustness. However, (e, d)-depth robustness only implies $(e/b, d, b)$-block depth robustness.

The second case is the trickier one to handle. To address it we essentially show that if, at some moment t', few pebbles are left on G then between t' and t_{j+2k} it must be that (in particular) a depth-robust sub-graph of G was pebbled which we know requires a high pebbling cost. In more detail, suppose at some moment $t' \in [t_j, t_{j+k}]$ only $|P_t| < e/2$ pebbles remain on G. Then we consider the sub-graph H induced by the node set $\mathsf{ancestors}_{G_1 - N(P_{t'}, b)} ([j + k + 1, j + 2k])$. We observe that, on the one hand, H must be fully pebbled during the interval $[t', t_{j+2k}]$. On the other hand, we observe that $G_1 = G - \{n/2 + 1, \ldots, n\}$ is a $(n/2, \delta, n/2)$-random DAG and, hence, by Lemma 2, G_1 is (almost certainly) (e, d, b)-block depth robust with $e = \tilde{\Omega}\left(n^{2/3}\right)$, $d = \Omega\left(\frac{n^{2/3}}{\log(n)}\right)$ and $b = \tilde{\Omega}\left(n^{1/3}\right)$. By exploiting the block depth robustness of G_1 we can show that H must itself be $\left(\tilde{\Omega}\left(n^{2/3}\right), \tilde{\Omega}\left(n^{2/3}\right)\right)$-depth robust. But then by Corollary 2 we get that H has cumulative complexity $\tilde{\Omega}(n^{4/3})$ and so have

$$\sum_{t=t_{j+k+1}}^{t_{j+2k}} |P_t| \geq \Pi_{cc}^{\parallel} (G_1 - P_{t'}, [j + k + 1, j + 2k]) \geq \tilde{\Omega}\left(n^{4/3}\right).$$

The proofs of Lemma 2 and Theorem 5 can be found in the full version. We now make a couple of observations about Lemma 2 and Theorem 5.

1. The lower bounds from Lemma 2 and Theorem 5 also apply to Argon2i-B. An Argon2i-B DAG G is similar to an (n, δ, n)-random DAG except that the randomly chosen forward edge $(r(i), i)$ for each node i is not chosen from the uniform distribution. However, these edges are still chosen independently and for each pair $j < i$ we still have $\Pr[r(i) = j] = \Omega(1/i)$. These are the only properties we used in the proofs of Lemma 2 and Theorem 5. Thus, essentially the same analysis shows that (whp) an Argon2i-B DAG G is $\left(e, \Omega\left(n^2/e^2\right), \frac{n}{20e}\right)$-block depth robust and that $\Pi_{cc}^{\parallel}(G) = \tilde{\Omega}\left(n^{5/3}\right)$.

2. The lower bound from Lemma 2 is tight up to polylogarithmic factors. In particular, a generalization of an argument of Alwen and Blocki [AB16] shows that a (n, δ, n)-random DAG is $\left(e, \tilde{\Omega}\left(\frac{n^2}{e^2}\right)\right)$-reducible—see Lemma 9. However, this particular upper bound does not extend to Argon2i-B.

3. The lower bound from Theorem 5 might be tight. Alwen and Blocki [AB16] gave an attack $\mathcal{A}$ such that $\Pi_{cc}^{\parallel}(\mathcal{A}) = O\left(n^{1.75} \delta \log n\right)$ for a (n, δ, t)-random DAG. In the following section we reduce the gap of $\tilde{O}(n^{1/12})$ further by developing an improved recursive version of the attack of Alwen and Blocki [AB16]. In particular, we show that for any $\epsilon > 0$ we have $\Pi_{cc}^{\parallel}(\mathcal{A}) = o\left(n^{1+\sqrt{1/2}+\epsilon}\right) = o\left(n^{1.708}\right)$. Our modified attack also improves the upper bound for other iMHF candidates like Catena [FLW13].

4. Theorem 4 alone will not yield any meaningful lower bounds on the $\Pi_{cc}^{\parallel}$ of the Catena iMHFs [FLW13]. In particular, the results from Alwen and Blocki [AB16] imply that for any t-pass variant of Catena the corresponding DAG is (e, d)-reducible for $ed \geq nt$ (typically, $t = O(\mathsf{polylog}(n))$). However,

in the remainder of this section, we use an alternative techniques to prove that $\Pi_{cc}^{\parallel}(G) = \Omega(n^{1.5})$ for the both Catena iMHFs and the Linear and DB iMHFs of [BCGS16].

5.2   Lowerbounding Dispersed Graphs

In this section we define dispersed graphs and prove a lowerbound on their CC. Next we show that several of the iMHF constructions from the literature are based on such graphs. Thus we obtain proofs of security for each of these constructions (albeit for limited levels of security). In the subsequent section we give an upperbound on the CC of these constructions showing that the lowerbounds in this section are relatively tight.

Generic Dispersed Graphs. Intuitively a (g,k)-dispersed DAG is a DAG ending with a path ϕ of length k which has widely dispersed dependencies. The following definitions make this concept precise.

Definition 5 (Dependencies). *Let $G = (V,E)$ be a DAG and $L \subseteq V$. We say that L has a (z,g)-dependency if there exist node disjoint paths $p_1, \ldots, p_z$ each ending in L and with length (at least) g.*

We are interested in graphs with long paths with many sets of such dependencies.

Definition 6 (Dispersed Graph). *Let $g \leq k$ be positive integers. A DAG G is called (g,k)-dispersed if there exists a topological ordering of its nodes such that the following holds. Let $[k]$ denote the final k nodes in the ordering of G and let $L_j = [jg, (j+1)g-1]$ be the jth subinterval. Then $\forall j \in [\lfloor k/g \rfloor]$ the interval L_j has a (g,g)-dependency.*

More generally, let $\epsilon \in (0,1]$. If each interval L_j only has an $(\epsilon g, g)$-dependency then G is called (ϵ, g, k)-dispersed.

We show that many graphs in the literature consist of a *stack* of dispersed graphs. Our lowerbound on the CC of a dispersed graph grows in the height of this stack. The next definition precisely captures such stacks.

Definition 7 (Stacked Dispersed Graphs). *A DAG $G = (V, E)$ is called $(\lambda, \epsilon, g, k)$-dispersed if there exist $\lambda \in \mathbb{N}^+$ disjoint subsets of nodes $\{L_i \subseteq V\}$, each of size k with following two properties.*

1. *For each L_i there is a path running through all nodes of L_i.*
2. *Fix any topological ordering of G. For each $i \in [\lambda]$ let G_i be the sub-graph of G containing all nodes of G up to the last node of L_i. Then G_i is an (ϵ, g, k)-dispersed graph.*

We denote the set of $(\lambda, \epsilon, g, k)$-dispersed graphs by $\mathbb{D}_{\epsilon,g}^{\lambda,k}$.

We are now ready to state and prove the lowerbound on the CC of stacks of dispersed graphs.

Theorem 6.

$$G \in \mathbb{D}_{\epsilon,g}^{\lambda,k} \quad \Rightarrow \quad \Pi_{cc}^{\parallel}(G) \geq \epsilon\lambda g \left(\frac{k}{2} - g\right).$$

Intuitively we sum the CC of pebbling the last k nodes L_i of each sub-graph G_i. For this we consider any adjacent intervals A of $2g$ nodes in L_i. Let p be a path in the $(\epsilon g, g)$-dependency of the second half of A. Either at least one pebble is always kept on p while pebbling the first half of A (which takes time at least g since a path runs through L_i) or p must be fully pebbled in order to finish pebbling interval A (which also takes time at least g). Either way pebbling A requires an additional CC of g per path in the $(\epsilon g, g)$-dependency of the second half of A. Since there are $k/2g$ such interval pairs each with ϵg incoming paths in their dependencies we get a total cost for that layer of $kg\epsilon/2$. So the cost for all layer of G is at least $\lambda kg\epsilon/2$. The details (for the more general case when g doesn't divide n) can be found in the full version.

The Graphs of iMHFs. We apply Theorem 6 to some important iMHFs from the literature. For this we first describe the particular DAGs (or at least their salient properties) underlying the iMHF candidates for which we prove lowerbounds in this section. Then we state a theorem summarizing our lowerbounds for these graphs. Finally we prove the theorem via a sequence of lemma; one per iMHF being considered.

Catena Dragonfly. We begin with the Catena Dragonfly graph. We briefly recall the properties of the DFG_λ^n construction, relevant to our proof, summarized in the following lemma which follows easily from the definition of DFG_λ^n in [FLW13, Definitions 8 & 9].

For this we describe the "bit-reversal" function (from which the underlying bit-reversal graph derives its name). Let $k \in \mathbb{N}^+$ such that $c = \log_2 k$ is an integer. On input $x \in [k]$ the *bit-reversal function* $\mathbf{br}(\cdot) : [k] \to [k]$ returns $y + 1$ such that the binary representation of $x - 1$ using c bits is the reverse of the binary representation of y using c bits.

Lemma 3 (Catena Dragonfly). *Let $\lambda, n \in \mathbb{N}^+$ be such that $k = n/(\lambda+1)$ is a power of 2. Let $G = \mathsf{DFG}_\lambda^n$ be the Catena Bit Reversal graph. Then the following holds:*

1. *G has n nodes.*
2. *Number them in topological order with the set $[n]$ and $\forall i \in [0, \lambda]$ let node set $L_i = [1 + ik, (i+1)k]$. A path runs through all nodes in each set L_i.*
3. *Node $ki + x \in L_i$ has an incoming edge from $k(i-1) + \mathbf{br}(x) \in L_{i-1}$.*

Catena Butterfly. Next describe the graph underlying the Catena Butterfly graph. We summarize its key properties relevant to our proof in the following lemma (which follows immediately by inspection of the Catena Butterfly definition [FLW13, Def. 10 & 11]).

Lemma 4 (Catena Butterfly Graph). *Let $\lambda, n \in \mathbb{N}^+$ such that $n = \bar{n}(\lambda(2c - 1) + 1)$ where $\bar{n} = 2^c$ for some $c \in \mathbb{N}^+$. Then the Catena Butterfly Graph BFG_λ^n consists of a stack of λ sub-graphs such that the following holds.*

1. *The graph BFG_λ^n has n nodes in total.*
2. *The graph BFG_λ^n is built as a stack of λ sub-graphs $\{G_i\}_{i \in [\lambda]}$ each of which is a superconcentrator[7]. In the unique topological ordering of BFG_λ^n denote the first and final $\bar{n}$ nodes of each G_i as $L_{i,0}$ and $L_{i,1}$ respectively. Then there is a path running through all nodes in each $L_{i,1}$.*
3. *Moreover, for any $i \in [\lambda]$ and subsets $S \subset L_{i,0}$ and $T \subset L_{i,1}$ with $|S| = |T| = h \le \bar{n}$ there exist h node disjoint paths $p_1, \ldots, p_h$ of length $2c$ from S to T.*

Balloon Hashing Linear. Finally we describe the graph underlying both the Linear and DB construction [BCGS16]. The graph $G = \mathsf{Lin}_\tau^\sigma$ is a pseudo-randomly constructed τ-layered graph with $\mathsf{indeg}(G) = 21$. It is defined as follows:

- $G = (V, E)$ has $n = \sigma\tau$ nodes $V = [n]$, and G contains a path $1, 2, \ldots, n$ running through V.
- For $i \in [0, \tau - 1]$ let $L_i = [i\sigma + 1, (i + 1)\sigma]$ denote the i'th layer. For each node $x \in L_i$, with $i > 0$, we select 20 nodes $y_1, \ldots, y_{20} \in L_{i-1}$ (uniformly at random) and add the directed edges $(y_1, x), \ldots, (y_{20}, x)$ to E.

5.3 The Lowerbounds

Now that we have our lowerbound for stacks of dispersed graphs it remains to analyse for which parameters each of the above three graphs can be viewed as being dispersed graphs. The results of this analysis are summarized in the theorem bellow.

Theorem 7 (iMHF Constructions Based on Dispersed Graphs).

- *If $\lambda, n \in \mathbb{N}^+$ such that $n = \bar{n}(\lambda(2c - 1) + 1)$ where $\bar{n} = 2^c$ for some $c \in \mathbb{N}^+$ then it holds that*

$$\mathsf{BFG}_\lambda^n \in \mathbb{G}_{n,3} \qquad \mathsf{BFG}_\lambda^n \in \mathbb{D}_{1,\lceil\sqrt{\bar{n}}\rceil}^{\lambda,\bar{n}} \qquad \Pi_{cc}^\|(\mathsf{BFG}_\lambda^n) = \Omega\left(\frac{n^{1.5}}{c\sqrt{c\lambda}}\right).$$

- *If $\lambda, n \in \mathbb{N}^+$ such that $k = n/(\lambda + 1)$ is a power of 2 then it holds that*

$$\mathsf{DFG}_\lambda^n \in \mathbb{G}_{n,2} \qquad \mathsf{DFG}_\lambda^n \in \mathbb{D}_{1,\lceil\sqrt{k}\rceil}^{\lambda,k} \qquad \Pi_{cc}^\|(\mathsf{DFG}_\lambda^n) = \Omega\left(\frac{n^{1.5}}{\sqrt{\lambda}}\right).$$

- *If $\sigma, \tau \in \mathbb{N}^+$ such that $n = \sigma * \tau$ then with high probability it holds that*

$$\mathsf{Lin}_\tau^\sigma \in \mathbb{G}_{n,21} \qquad \mathsf{Lin}_\tau^\sigma \in \mathbb{D}_{0.25,\sqrt{\sigma}/2}^{\tau-1,\sigma} \qquad \Pi_{cc}^\|(\mathsf{Lin}_\tau^\sigma) = \Omega\left(\frac{n^{1.5}}{\sqrt{\tau}}\right).$$

---

[7] A superconcentrator is a DAG with m inputs and outputs such that any subset of $s \in [m]$ inputs and outputs are connected by s node disjoint paths.

The theorem is proven in the following three lemma bellow (one lemma per graph). We begin with the graph for Catena Dragonfly.

Lemma 5. *It holds that* $\mathsf{DFG}_\lambda^n \in \mathbb{D}_{1,\sqrt{k}}^{\lambda,k}$ *where* $k = \frac{n}{(\lambda+1)}$ *and* $\Pi_{cc}^{\|}(\mathsf{DFG}_\lambda^n) = \Omega\left(\frac{n^{1.5}}{\sqrt{\lambda}}\right).$

Proof of Lemma 5. Let $G = \mathsf{DFG}_\lambda^n$ and set $k = n/(\lambda+1)$, $c = \log_2 k$ and $g = \sqrt{k}$. By construction c is an integer. For simplicity assume c is even and so $g \in \mathbb{N}^+.$[8] Number the nodes of G according to (the unique) topological order with the set $[0, n-1]$. It suffices to show that for all $i \in [\lambda]$ the sub-graph G_i consisting of nodes $[(i+1)k - 1]$ is (g,k)-dispersed (with probability $\epsilon = 1$). If this holds then Theorem 6 immediately implies that $\Pi_{cc}^{\|}(\mathsf{DFG}_\lambda^n) = \Omega\left(\frac{n^{1.5}}{\sqrt{\lambda}}\right).$

Recall that G consists of layerls L_i of length k. For each $j \in [k/2g]$ let $L_{i,j}$ be the jth interval of $2g$ nodes of L_i. Let $R_{i,j}$ be the second half of $L_{i,j}$. We will show that there are g node-disjoint paths each terminating in $R_{i,j}$ whose remaining nodes are all in layer L_{i-1}. Let node set $S_x = [s + y - (g-2), s + y]$ where $y = \mathrm{br}(x)$ and $s = (i-1)k$. The next three properties follow immediately from Lemma 3 and they imply the lemma.

- $\forall x \in R$ it holds that $S_x \subset L_{i-1}$.
- $\forall x \in R$ there is a path of length g going through the nodes of S_x and ending in x.
- $\forall$ distinct $x, x' \in R$ sets S_x and $S_{x'}$ are disjoint.   □

Next we turn to the Catena Dragonfly graph.

Lemma 6. *Let* $\lambda, n \in \mathbb{N}^+$ *such that* $n = \bar{n}(\lambda(2c-1)+1)$ *with* $\bar{n} = 2^c$ *for some* $c \in \mathbb{N}^+$. *It holds that* $\mathsf{BFG}_\lambda^n \in \mathbb{D}_{1,g}^{\lambda,\bar{n}}$ *for* $g = \lceil\sqrt{\bar{n}}\rceil$ *and* $\Pi_{cc}^{\|}(\mathsf{BFG}_\lambda^n) = O\left(\frac{n^{1.5}}{c\sqrt{c\lambda}}\right).$

Proof of Lemma 6. Let $G = \mathsf{BFG}_\lambda^n$ and let $G_1, G_2, \ldots, G_\lambda$ be the sub-graphs of G described in Lemma 4. We will show that each G_i is $(g, \bar{n})$-dispersed for $g = \lfloor\sqrt{\bar{n}}\rfloor$. Fix arbitrary $i \in [\lambda]$ and L_1 be the last $\bar{n}$ nodes in the (the unique) topological ordering of G_i. We identify the nodes in L_1 with the set $\{1\} \times [\bar{n}]$ such that the second component follows their topological ordering. Let $\bar{g} = \lfloor\bar{n}/g\rfloor$ and for each $j \in [\bar{g}]$ let $L_{1,j} = \{\langle 1, jg + x\rangle : x \in [0, g-1]\}$. We will show that $L_{1,j}$ has a (g,g)-dependency.

Let L_0 be the first $\bar{n}$ nodes of G_i which we identify with the set $\{0\} \times [\bar{n}]$ (again with the second component respecting their topological ordering). Notice that for $n > 1$ and $g = \lfloor\sqrt{\bar{n}}\rfloor$ it holds that $g(g - 2c + 1) \leq n$. Thus the set $S = \{\langle 0, i(g - 2c + 1)\rangle : i \in [g]\}$ is fully contained in L_0. Property (3) of Lemma 4 implies there exist g node disjoint paths from S to $L_{1,j}$ of length $2c$. In particular $L_{1,j}$ has a $(g, 2c)$-dependency.

---

[8] The odd case is identical but with messy but inconsequential rounding terms.

We extend this to a (g,g)-dependency. Let path p, beginning at node $\langle 0, v \rangle \in S$, be a path in the $(g, 2c)$-dependency of $L_{1,j}$. Prepend to p the path traversing

$$(\langle 0, v - (g - 2c - 1) \rangle, \langle 0, v - (g - 2c - 2) \rangle, \ldots, \langle 0, v \rangle)$$

to obtain a new path p^+ of length g. As this is a subinterval of L_0 property (2) of Lemma 4 implies this prefix path always exists. Moreover since any paths $p \neq q$ in a $(g, 2c)$-dependency of $L_{1,i}$ are node disjoint they must, in particular, also begin at distinct nodes $\langle 0, v_p \rangle \neq \langle 0, v_q \rangle$ in S. But by construction of S any such pair of nodes is separated by $g - 2c$ nodes. In particular paths p^+ and q^+ are also node disjoint and so by extending all paths in a $(2c, g)$-dependency we obtain a (g, g)-dependency for $L_{1,i}$. This concludes the first part of the lemma.

It remains to lowerbound $\Pi_{cc}^{\|}(\mathsf{BFG}_\lambda^n)$ using Theorem 6.

$$\Pi_{cc}^{\|}(\mathsf{BFG}_\lambda^n) \geq \lambda g \left(\frac{k}{2} - g \right) \geq \lambda \lfloor \sqrt{\bar{n}} \rfloor \left(\frac{\bar{n}}{2} - \lfloor \sqrt{\bar{n}} \rfloor \right)$$

$$= \lambda \sqrt{\bar{n}} \left(\frac{\bar{n}}{2} - \sqrt{\bar{n}} \right) - O(\bar{n}) = \Omega \left(\lambda \bar{n}^{1.5} \right)$$

$$= \Omega \left(\frac{n^{1.5}}{c\sqrt{c}\lambda} \right).$$

$\square$

Finally we prove a lowerbound for the Linear and DB variants of Balloon Hashing.

Lemma 7. *If $\sigma, \tau \in \mathbb{N}^+$ such that $n = \sigma\tau$ then with high probability it holds that*

$$\mathsf{Lin}_\tau^\sigma \in \mathbb{G}_{n,21} \qquad \mathsf{Lin}_\tau^\sigma \in \mathbb{D}_{0.25, \sqrt{\sigma}/2}^{\tau-1, \sigma} \qquad \Pi_{cc}^{\|}(\mathsf{Lin}_\tau^\sigma) = \Omega \left(\frac{n^{1.5}}{\sqrt{\tau}} \right).$$

Proof. *(sketch)* It suffices to show that $\mathsf{Lin}_\tau^\sigma \in \mathbb{D}_{0.25, \sqrt{\sigma}/2}^{\tau-1, \sigma}$. By Theorem 6 it immediately follows that

$$\Pi_{cc}^{\|}(\mathsf{Lin}_\tau^\sigma) \geq \frac{(\tau - 1)\sqrt{\sigma}/2}{4} \left(\sigma/2 - \sqrt{\sigma}/2 \right) = \Omega \left(\frac{n^{1.5}}{\sqrt{\tau}} \right).$$

Fix any $i \in [0, \tau - 1]$. Consider layer L_i and given set $S_x = [x, x + \sqrt{\sigma}/2 - 1] \subset L_i$ denoting an interval of $\sqrt{\sigma}/2$ nodes in L_i begining at node x. Without loss of generality we suppose that each node in S_x only has one randomly chosen parent in L_{i-1}—adding additional edges can only improve dispersity. We partition L_{i-1} into $\sqrt{\sigma}$ intervals of length $\sqrt{\sigma}$. We say that an interval $[u, u + \sqrt{\sigma} - 1] \subset L_{i-1}$ is *covered* by S_x if there is exists edge (y, v) from the second half of the interval to a node in S_x; that is if $y \in [u + \sqrt{\sigma}/2, u + \sqrt{\sigma} - 1]$ and $v \in S_x$. In this case the path $(u, u + 1, \ldots, y, v)$ has length $\geq \sqrt{\sigma}/2$ and this path will not intersect the corresponding paths from any of the other (disjoint) intervals in L_{i-1} (recall that we are assuming that $v \in S_x$ only has one parent in L_{i-1}). The probability that an interval $[u, u + \sqrt{\sigma} - 1] \subset L_{i-1}$ is covered by S_x is at least

$$1 - \left(1 - \frac{\sqrt{\sigma}/2}{\sigma}\right)^{\sqrt{\sigma}} \approx 1 - \sqrt{1/e} \; .$$

Thus, in expectation we will have at least $\mu = \sqrt{\sigma}\left(1 - \sqrt{1/e}\right) \geq 0.39 \times \sqrt{\sigma}$ node disjoint paths of length $\sqrt{\sigma}/2$ ending in S_x. Standard concentration bounds imply that we will have at least $\sqrt{\sigma}/4$ such paths with high probability.   □

6 New Memory-Efficient Evaluation Algorithm and Applications

In this section we introduce a new generic parametrized pebbling algorithm for DAGs (i.e. an evaluation algorithm for an arbitrary iMHF). We upperbound the pebbling strategy's cumulative pebbling complexity in terms of its parameters. In particular we see that for graphs which are not depth-robust there exist parameter settings for which the algorithm results in low CC pebbling strategies. Next we instantiate the parameters to obtain attacks on the random graphs defined in the previous section. By "attack" we mean that, for Argon2i-A and SB, the algorithm has significantly less asymptotic memory-hardness in the PROM than both that of their naïve algorithms, and even that of the attack in [AB16].

Review of [AB16]. In order to describe the results in this section we first review the generic pebbling algorithm PGenPeb of [AB16] which produces a pebbling $P_1, P_2, \ldots, P_n$ of G as follows. PGenPeb takes as input a node set $S \subset V$ of size $|S| = e$ such that removing S reduces the depth of the DAG depth$(G - S) \leq d$. Intuitively, keeping pebbles on S compresses G in the sense that G can now quickly be entirely (re)pebbled within d (parallel) steps. This is because when S is already pebbled then no remaining unpebbled path has length greater than d. Algorithm PGenPeb never removes pebbles from nodes in S and its goal is to always pebble node i at time i so as to finish in $n = $ size(G) steps.[9] To ensure that parents of node i are all pebbled at time i algorithm PGenPeb sorts nodes in topological order and partitions them into consecutive intervals of g nodes (where $g \in [d, n]$ is another input parameter). Nodes in interval $I_c = [(c-1)g + 1, cg] \cap [n] \subset V$ are pebbled during "light phase" Λ_c which runs for g time steps. To ensure that the result is a legal pebbling, PGenPeb guarantees the following invariant $\mathcal{I}$: just before light phase Λ_c begins (i.e. at time $(c-1)g$) we have $X_{cg} = $ parents$(I_c) \cap [(c-1)g] \subset P_{(c-1)g}$ so that we begin Λ_c with all of the necessary pebbles. Now, in phase Λ_c algorithm PGenPeb simply places a pebble on node $i \in I_c$ at time i.

Notice that for $c = 1$, $X_1 = \emptyset$ and so $\mathcal{I}$ is trivially satisfied. Let $c > 1$. Partition X_{cg} into $X_{cg}^- = X_{cg} \cap [(c-1)g - d + 1]$ and $X_{cg}^+ = X_{cg} \setminus X_{cg}^-$. Since X_{cg}^+ is pebbled in the final d steps of light phase Λ_{c-1}, PGenPeb can simply not

---

[9] Formally, $i \in P_i$ and $S \cap [i] \subseteq P_i$ for each $i \leq n$.

remove those until time step $(c-1)g$. In order to ensure that X_{cg}^- is pebbled at that time PGenPeb also runs a "balloon phase" B_{c-1} in parallel with the final d steps of Λ_{c-1}.[10] Intuitively in phase B_{c-1} all nodes in $[(c-1)g] \subseteq V$ are quickly "decompressed" by greedily re-pebbling everything possible in parallel. Recall that pebbles are never removed from nodes in S. So at time j all of $S \cap [j]$ is already pebbled. Therefore, at time $(c-1)g$, there is no unpebbled path longer than d nodes within the first $[(c-1)g]$ nodes and so B_{c-1} can indeed entirely (and legally) repebble those nodes (and so in particular X_{cg}^-). Thus, together with the nodes in X_{cg}^+ pebbled in the final d steps of Λ_{c-1} it follows that $\mathcal{I}$ also holds for Λ_c.

The runtime of PGenPeb is n. Thus the cost is at most $\Pi_{cc}^{\parallel}(\text{PGenPeb}) \leq en + \delta gn + \lceil n/g \rceil (dn)$ where $\delta = \text{indeg}(G)$. The en term upper bounds the cost of always keeping pebbles on S, δgn bounds the cost of all light phases, and the third term upper bounds the cost of all balloon phases — each balloon phase costs at most dn and at most $\lceil n/g \rceil$ balloon phases are run.

Notice that (for constant δ) we would like to set $g \leq e$ so that the second term doesn't dominate the first. Conversely, to keep the number of (expensive) balloon phases at a minimum we also want g to be large. Therefore, as long as $e \geq d$, the asymptotically minimal complexity is obtained when $g = e$.

Recursive Attack: Intuition. Our new algorithm relies on the following key insight. Algorithm PGenPeb can actually pebble, not just the sink with the above complexity, but instead any target set $T \subseteq V$ simultaneously.[11] This more general view allows us to recast the task of the balloon phase as such a pebbling problem. The graph being pebbled is $G' = G - (S \cup [(c-1)g - d + 1])$ and the target set is X_c^-. So instead of implementing balloon phases with an expensive greedy pebbling strategy as in PGenPeb we can apply the same strategy as (the generalized version of) PGenPeb recursively. This is the approach of the new algorithm RGenPeb. (The complete pseudocode of RGenPeb can be found in the full version. For this approach to work we need that not only is G (e,d)-reducible via some set S but that there is also a set S' of size $e' > e$ such that $\text{depth}(G - S') = d' < d$. Only when these conditions can no longer be met do we have to resort to greedy pebbling for the balloon phases. As we show below, it turns out that RGenPeb leads to improved attacks compared to PGenPeb for the DAGs underlying key iMHFs like Argon2i, Catena and Balloon Hashing.

Outline. The remainder of this section has the following structure. First, in Lemma 8 we generalize the results of [AB16] to upperbound the CC of a graph by the cost of pebbling all light phases plus the CC of the pebblings problems solved by balloon phases. Next we define a generalization of (e,d)-reducible graphs called f-reducible graphs; namely graphs which are $(f(d), d)$-reducible for all $d \in [n]$. This allows us to state the main theorem of this section. It considers a certain class of functions f and upper bounds the complexity of RGenPeb on

---

[10] Recall that $g \geq d$ so Λ_{c-1} lasts long enough to accommodate B_{c-1}.

[11] For example, in the final d steps of the execution one last balloon phase can be run to (re)pebble all of G including T at no added cost to the asymptotic complexity.

such f-reducible graphs using any number k levels of recursion. To apply the theorem to the iMHFs from the literature we prove Lemma 9 which describes the f-reducibility of their underlying DAGs. Thus we obtain the final corollary of the section describing new upperbounds on those iMHFs. At the end of this section we give a more detailed description of RGenPeb and the proof of Lemma 8.

Generalizing [AB16]. In order to derive the new pebbling strategy we first generalize the results of [AB16]. Given a DAG $G = (V, E)$, node set $T \subseteq V$ and integer t we define $\mathcal{P}^{\parallel}_{G,T,t} \subseteq \mathcal{P}^{\parallel}_{G,T}$ to be the set of all parallel pebblings $(P_1, \ldots, P_z)$ of G such that $z \leq t$. Analogously we let $\Pi^{\parallel}_{cc}(G, T, t) = \min_{P \in \mathcal{P}^{\parallel}_{G,T,t}} \Pi^{\parallel}_{cc}(P)$.

We remark that if $\mathsf{depth}(G) = d$ then $\Pi^{\parallel}_{cc}(G, T, d) \leq dn$ since we can greedily pebble G in topological order in time $\mathsf{depth}(G)$. Lemma 8 provides an alternative upper bound on $\Pi^{\parallel}_{cc}(G, T, 2d)$.

Lemma 8. *Let $G = (V, E)$ be a DAG of size n, indegree δ and $\mathsf{depth}(G) \leq d_0$. If G is (e_1, d_1)-reducible with parameters e_1, d_1 such that $2d_1 n \leq e_1 d_0$ and $d_1 \leq d_0$ then for any target set $T \subseteq V$ we have*

$$\Pi^{\parallel}_{cc}(G, T, 2d_0) \leq (4\delta + 4)\, e_1 d_0 + \frac{n}{e_1} \left(\max_{\substack{T' \subseteq V - S_1 \\ |T'| \leq \delta \cdot e_1}} \Pi^{\parallel}_{cc}(G - S_1, T', 2d_1) \right),$$

where $S_1 \subseteq V$ has size $|S_1| \leq e_1$ such that $\mathsf{depth}(G - S_1) \leq d_1$.

To prove the lemma we first define the RGenPeb algorithm and argue the legality of the pebbling it produces at the end of this section. Armed with this, it remains only to upperbound the complexity of a call to RGenPeb in terms of the complexity of the recursive call it makes. This involves a relatively straightforward (but somewhat tedious) counting of the pebbles placed by RGenPeb, the details of which can be found in the full version.

We observe that Lemma 8 generalizes the main result of [AB16] as that work only considered the special case where balloon phases are implemented with a greedy pebbling strategy. The advantage of the above formulation (and the more general RGenPeb) is that now we can be apply the lemma (and algorithm) recursively.

In order to apply this lemma repeatedly we will need graphs which are reducible for a sequence of points parameters (e, d) satisfying the conditions laid out in Lemma 8 relating consecutive parameters. To help characterize such graphs we generalize the notion of reducibility as follows.

Definition 8. *Let $G = (V, E)$ be a DAG with n nodes and let $f : \mathbb{N} \to \mathbb{N}$ be a function. We say that G is f-reducible if for every positive integer $n \geq d > 0$ there exists a set $S \subseteq V$ of $|S| = f(d)$ nodes such that $\mathsf{depth}(G - S) \leq d$.*

Next we state the main theorem of this section which, for a certain class of natural functions f, upperbounds the CC of any f-reducible graph.

Theorem 8. *Let G be a f-reducible DAG on n nodes then if $f(d) = \tilde{O}\left(\frac{n}{d^b}\right)$ for some constant $0 < b \leq 2/3$ and let $a = \frac{1-2b+\sqrt{1+4b^2}}{2}$. Then for any constant $\epsilon > 0$*

$$\Pi_{cc}^{\parallel}(G) \leq O\left(n^{1+a+\epsilon}\right).$$

The proof of Theorem 8 can be found in the full version. We briefly sketch the intuition here. We define a sequence $e_1, e_2, \ldots$ and $d_1, d_2, \ldots$ such that G is (e_i, d_i)-reducible for each i, $e_i = n^{a_i + \epsilon/3}$ and $d_i = n^{\frac{1-a_i}{b}}$ with

$$a_{i+1} = 1 + \frac{(a-1)(1-a_i)}{b}, \quad \text{where} \qquad a_1 = a = \frac{1-2b+\sqrt{1+4b^2}}{2}.$$

If $b \leq a$ we have $e_{i+1}d_i \geq nd_{i+1}$ for every i so we can repeatedly invoke Lemma 9 as many times as we desire. By exploiting several key properties of the sequence $\{a_i\}_{i=1}^{\infty}$ we can show that unrolling the recurrence k times yields a pebbling with cost at most $k(4\delta+2)n^{1+a+\epsilon/3} + n^{1+a+\epsilon/3}d_k$. For any $\epsilon > 0$ we can select the constant k sufficiently large that $d_k \leq n^{\epsilon/3}$. Thus, the pebbling cost is $o\left(n^{1+a+\epsilon}\right)$.

Analysing Existing iMHFs. We can now turn to applying Theorem 8 to iMHFs from the literature. Lemma 9 below states that an (n, δ, n)-random DAGs and λ-layered DAGs are f-reducible. In particular these are the types of DAGs underlying all of the iMHFs considered in the previous section.

Lemma 9. *Let $f_b(d) = \tilde{O}\left(\frac{n}{d^b}\right)$ then*

1. *Let $\delta = O(\mathrm{polylog}(n))$ then a (n, δ, n)-random DAG is $f_{0.5}$-reducible with high probability.*
2. *The Catena DAGs DFG_λ^n and BFG_λ^n are both f_1-reducible for $\lambda = O(\mathrm{polylog}(n))$.*
3. *The Balloon Hashing Linear (and the DB) graph Lin_τ^σ is f_1-reducible for $\tau = O(\mathrm{polylog}(n))$.*

The proof generalizes the arguments used in [AB16] to first establish a particular pair (e, d) for which the graphs are reducible. It can be found in the full version.

Together with Theorem 8 and Lemma 9 we now obtain the main application of RGenPeb which is described in the following corollary upperbounding the memory-hardness of each of the considered iMHFs.

Corollary 4. *Let $\epsilon > 0$ be any constant*

1. *Let $\delta = O(\mathrm{polylog}(n))$ then an (n, δ, n)-random DAG G has $\Pi_{cc}^{\parallel}(G) = O\left(n^{1+\sqrt{1/2}+\epsilon}\right) \approx O\left(n^{1.707+\epsilon}\right)$.*
2. *Both $\Pi_{cc}^{\parallel}(\mathsf{DFG}_\lambda^n)$ and $\Pi_{cc}^{\parallel}(\mathsf{BFG}_\lambda^n)$ are in $\tilde{O}\left(n^{\frac{13}{8}}\right) = \tilde{O}\left(n^{1.625}\right)$.*
3. *$\Pi_{cc}^{\parallel}(\mathsf{Lin}_\tau^\sigma) = \tilde{O}\left(n^{\frac{13}{8}}\right) = \tilde{O}\left(n^{1.625}\right)$, where Lin_τ^σ has $n = \tau\sigma$ nodes.*

We remark that Theorem 8 does not yield tighter bounds for Catena iMHFs DFG_λ^n or BFG_λ^n or for Lin_τ^σ. Each DAG is indeed f_b reducible for any $b \leq 2/3$ (even for $b \leq 1$), but for $b \leq 2/3$ it follows that $a = \frac{1-2b+\sqrt{1+4b^2}}{2} \geq 2/3$. Thus, Theorem 8 yields an attack with cost $O\left(n^{\frac{5}{3}+\epsilon}\right)$, which does not improve on the non-recursive $\mathsf{PGenPeb}$ attack in [AB16] as that has cost $O\left(n^{\frac{5}{3}}\right)$. However, we can set $e_1 = n^{5/8}, e_2 = n^{7/8}$ and exploit the fact that the DAGs are (e_i, d_i)-reducible with $d_i = \tilde{O}(n/e_i)$. Applying Lemma 8 twice we have $\Pi_{cc}^{\parallel}(G) = O\left(e_1 n + \frac{n}{e_1}\left(e_2 d_1 + \frac{n}{e_2} n d_2\right)\right) = O\left(n^{13/8} + n^{3/8+7/8}d_1 + n^{3/8+1/8+1}d_2\right) = \tilde{O}\left(n^{\frac{13}{8}}\right)$. Note that $e_2 d_1 = \tilde{O}(n^{10/8}) > \tilde{O}(n^{9/8}) = n d_2$ so it is legal to invoke Lemma 8 for sufficiently large n.

The $\mathsf{RGenPeb}$ *Algorithm.* In the remainder of this section we sketch $\mathsf{RGenPeb}$ algorithm and justify that it produces a legal pebbling. The analysis of its complexity in terms of the complexity of its recursive call is contained in the proof of Lemma 8. The final complexity of an execution requires unravelling the recursive statement of Lemma 8 which is done in the proof of Theorem 8.

In the following we will ignore rounding errors here as they are inconsequential for the asymptotic behaviour while adding needless complexity to the exposition. For completeness we observe that if $\mathsf{RGenPeb}$ finishes a light phase and there is not enough steps left to complete a full light phase then it can simply runs the next light phase as far as it can (and completely omits any further balloon phases). This affects neither the legality of the resulting pebbling nor its asymptotic complexity.

Algorithm $\mathsf{RGenPeb}$ takes input a DAG $G = (V, E)$, sets $S_1 \subseteq S_2 \subseteq \ldots \subseteq S_k \subseteq V$, integers $d_1, d_2, \ldots, d_k$ and a target set T such that $\forall i \in [k] : e_i d_{i-1} \geq n d_i$ and $d_i \geq \mathsf{depth}(G - S_i)$ where $n = |V|$, $e_i = |S_i|$ and $d_0 = \mathsf{depth}(G)$. For this $\mathsf{RGenPeb}$ makes use of an arbitrary partition of the nodes of G into $2d_0$ into sets $D_1, D_2, \ldots, D_{2d_0}$ such that the following properties hold:[12]

TOPOLOGICALLY ORDERED: $\forall j \in [2d_0 - 1]$ $\mathsf{parents}(D_{j+1}) \subseteq \bigcup_{y \in [j]} D_y$,

MAXIMUM SIZE: $\forall j \leq 2d_0$ $|D_j| \leq \frac{n}{d_0}$.

Intuitively, the set D_j is the set of nodes that will be pebbled by a light phase in the jth step. So for $\mathsf{PGenPeb}$ we would simply have $D_j = \{j\}$.

At the top level of the recursion $\mathsf{RGenPeb}$ looks relatively similar to $\mathsf{PGenPeb}$. The goal is to pebble $G_0 = G$ with the target set $T_0 = \mathsf{sinks}(G_0)$ in at most $2d_0$ steps which is done by executing a sequence of light phases lasting $m = e_2 d_0/n$ steps and balloon phases lasting $2d_1$ steps. The requirement that $e_2 d_0 \geq 2d_1 n$ ensures that $m \geq 2d_1$ so that we can complete each balloon phase in time for the upcoming light phase. For $t \in [2d_0]$ let $U_t = \bigcup_{j \in [t]} D_j$ be all nodes pebbled

---

[12] For example we can sort the nodes in topological order and divided them up into the partition. Whenever a set is larger than n/d_0 we insert a new set into the partition with the overflow.

by light phases up to step t. Then for $c \in [2d_0/m]$ the light phase Λ_c runs during time interval $I_c = [(c-1)m+1, cm]$ during which it will pebble nodes $U_{cm} \setminus U_{(c-1)m}$. It never removes pebbles from S_1 and, at each time step t it keeps pebbles on $\mathsf{parents}(U_{cm} \setminus U_t)$ as it will still need those to finish the light phase.

As for PGenPeb the light phase Λ_1 is trivially a legal pebbling. Let $X_{cm} = \mathsf{parents}(U_{cm+m} \setminus U_{cm+1}) \cap U_{cm}$ and $X_{cm}^- = (X_{cm} \cap U_{cm-2d_1}) \setminus S_1$ and $X_{cm}^+ = X_{cm} \setminus X_{cm}^-$. To ensure that all pebbles placed by during light phase Λ_{c+1} are done so legally it suffices for RGenPeb to ensure that X_{cm} is fully pebbled at time cm. This is done by balloon phase running in parallel to the final $2d_1$ steps of Λ_c; that is during the interval $[cm - 2d_1, cm]$. The pebbling for the balloon phase may be obtained by a recursive call to RGenPeb for the graph $G' = G - S - (V \setminus U_{cm-2d_1})$ (G' is the DAG induced by nodes $U_{cm-2d_1} - S$) with target set X_{cm}^- as well as parameters $S_2 \subseteq S_3 \subseteq \ldots \subseteq S_k$ and $d_2, d_3, \ldots, d_k$ (both lists now have length $k-1$ and clearly still satisfy the conditions on parameters stated above). If RGenPeb is ever called with empty lists $\bar{S} = \emptyset$ and $\bar{d} = \emptyset$ (i.e., $k = 0$) then it simply greedy pebbles G. The result of the recursive call is added to the final $2d_1$ steps of light phase Λ_c. Finally the pebbling is modified to never remove pebbles from X_{cm} during the those final steps of Λ_{c-1}. Notice that each node in X_{cm}^+ is either in S or is pebbled at some point during the final $2d_1$ steps of Λ_{c+1}. Thus we are guaranteed that $X_{cm} \subseteq P_{(c-1)m}$ as desired.

To see why RGenPeb produces a legal pebbling it suffices to observe that pebbles placed during light phases always have their parents already pebbled. So if the recursive call returns a legal pebbling for the balloon phase then the final result is also legal. But at the deepest level of the recursion RGenPeb resorts to a greedy pebbling which is trivially legal. Thus, by induction, so is the pebbling at the highest level of the recursion.

7 Open Questions

We conclude with several open questions for future research.

- We showed that for some constant $c \geq 0$ we can find a DAG G on n nodes with $\Pi_{cc}^{\|}(G) \geq cn^2/\log(n)$ and $\mathsf{indeg}(G) = 2$. While this result is asymptotically optimal the constant terms are relevant for practical applications to iMHFs. How big can this constant c be? Can we find explicit constructions of constant-indegree, $(c_1 n/\log(n), c_2 n)$-depth robust DAGs that match these bounds?
- Provide tighter upper and lower bounds on $\Pi_{cc}^{\|}(G)$ for Argon2i-B [BDKJ16], the most recent version of Argon2i which was submitted to IRTF for standardization.
- Another interesting direction concerns understanding the cumulative pebbling complexity of generic graphs. Given a graph G is it computationally tractable to (approximately) compute $\Pi_{cc}^{\|}(G)$? An efficient approximation algorithm for $\Pi_{cc}^{\|}(G)$ would allow us to quickly analyze candidate iMHF constructions. Conversely, as many existing iMHF constructions are based on fixed random graphs, [BDK16,BCGS16] showing that approximating such

a graphs complexity is hard would provide evidence that an adversary will likely not be able to leverage properties of the concrete instance to improve their evaluation strategy for the iMHF. Indeed, it may turn out that the most effective way to construct depth-robust graphs with good constants is via a randomized construction.

Acknowledgments. The authors would like to thank Pierrick Gaudry for his careful reading and many helpful suggestions. The first and third authors were supported by the European Research Council, ERC consolidator grant (682815 - TOCNeT).

References

[AB16] Alwen, J., Blocki, J.: Efficiently computing data-independent memory-hard functions. In: Robshaw, M., Katz, J. (eds.) CRYPTO 2016. LNCS, vol. 9815, pp. 241–271. Springer, Heidelberg (2016). doi:10.1007/978-3-662-53008-5_9

[AB17] Alwen, J., Blocki, J.: Towards practical attacks on Argon2i and balloon hashing. In: Proceedings of the 2nd IEEE European Symposium on Security and Privacy (EuroS&P 2017). IEEE (2017, to appear). http://eprint.iacr.org/2016/759

[ABW03] Abadi, M., Burrows, M., Wobber, T.: Moderately hard, memory-bound functions. In: Proceedings of the Network and Distributed System Security Symposium, NDSS, San Diego, California, USA (2003)

[ACK+16] Alwen, J., Chen, B., Kamath, C., Kolmogorov, V., Pietrzak, K., Tessaro, S.: On the complexity of RYPT and proofs of space in the parallel random oracle model. In: Advances in Cryptology - EUROCRYPT 2016 - Vienna, Austria, May 8–12, 2016, Part II, pp. 358–387 (2016). http://eprint.iacr.org/2016/100

[ACP+17] Alwen, J., Chen, B., Pietrzak, K., Reyzin, L., Tessaro, S.: RYPT is Maximally Memory-Hard. In: Advances in Cryptology-EUROCRYPT 2017. Springer (2017, to appear). http://eprint.iacr.org/2016/989

[AGK+16] Alwen, J., Gai, P., Kamath, C., Klein, K., Osang, G., Pietrzak, K., Reyzin, L., Rolnek, M., Rybr. M.: On the memory-hardness of data-independent password-hashing functions. Cryptology ePrint Archive, Report 2016/783 (2016). http://eprint.iacr.org/2016/783

[AS15] Alwen, J., Serbinenko, V.: High parallel complexity graphs and memory-hard functions. In: Proceedings of the Eleventh Annual ACM Symposium on Theory of Computing, STOC 2015 (2015). http://eprint.iacr.org/2014/238

[BCGS16] Boneh, D., Corrigan-Gibbs, H., Schechter, S.: Balloon hashing: provably space-hard hash functions with data-independent access patterns. Cryptology ePrint Archive, Report 2016/027, Version: 20160601:225540 (2016). http://eprint.iacr.org/

[BDK15] Biryukov, A., Dinu, D., Khovratovich, D.: Fast and tradeoff-resilient memory-hard functions for cryptocurrencies and password hashing. Cryptology ePrint Archive, Report 2015/430 (2015). http://eprint.iacr.org/2015/430

[BDK16] Biryukov, A., Dinu, D., Khovratovich, D.: Argon2 password hash. Version 1.3 (2016). https://www.cryptolux.org/images/0/0d/Argon2.pdf

[BDKJ16] Biryukov, A., Dinu, D., Khovratovich, D., Josefsson, S.: The memory-hard Argon2 password hash and proof-of-work function. Internet-Draft draft-irtf-cfrg-argon2-00, Internet Engineering Task Force, March 2016

[BK15] Biryukov, A., Khovratovich, D.: Tradeoff cryptanalysis of memory-hard functions. Cryptology ePrint Archive, Report 2015/227 (2015). http://eprint.iacr.org/

[Cha73] Chandra, A.K.: Efficient compilation of linear recursive programs. In: SWAT (FOCS), pp. 16–25. IEEE Computer Society (1973)

[Coo73] Cook, S.A.: An observation on time-storage trade off. In: Proceedings of the Fifth Annual ACM Symposium on Theory of Computing, STOC 1973, pp. 29–33. ACM, New York (1973)

[DFKP15] Dziembowski, S., Faust, S., Kolmogorov, V., Pietrzak, K.: Proofs of space. In: Gennaro, R., Robshaw, M. (eds.) CRYPTO 2015. LNCS, vol. 9216, pp. 585–605. Springer, Heidelberg (2015). doi:10.1007/978-3-662-48000-7_29

[DGN03] Dwork, C., Goldberg, A., Naor, M.: On memory-bound functions for fighting spam. In: Boneh, D. (ed.) CRYPTO 2003. LNCS, vol. 2729, pp. 426–444. Springer, Heidelberg (2003). doi:10.1007/978-3-540-45146-4_25

[DKW11a] Dziembowski, S., Kazana, T., Wichs, D.: Key-evolution schemes resilient to space-bounded leakage. In: Rogaway, P. (ed.) CRYPTO 2011. LNCS, vol. 6841, pp. 335–353. Springer, Heidelberg (2011). doi:10.1007/978-3-642-22792-9_19

[DKW11b] Dziembowski, S., Kazana, T., Wichs, D.: One-time computable self-erasing functions. In: Ishai, Y. (ed.) TCC 2011. LNCS, vol. 6597, pp. 125–143. Springer, Heidelberg (2011). doi:10.1007/978-3-642-19571-6_9

[DNW05] Dwork, C., Naor, M., Wee, H.: Pebbling and proofs of work. In: Shoup, V. (ed.) CRYPTO 2005. LNCS, vol. 3621, pp. 37–54. Springer, Heidelberg (2005). doi:10.1007/11535218_3

[EGS75] Paul, E., Graham, R.L., Szemeredi, E.: On sparse graphs with dense long paths. Technical report, Stanford, CA, USA (1975)

[FLW13] Christian, F., Stefan, L., Jakob, W.: Catena: a memory-consuming password scrambler. IACR Cryptology ePrint Archive 2013, 525 (2013)

[HJO+16] Hemenway, B., Jafargholi, Z., Ostrovsky, R., Scafuro, A., Wichs, D.: Adaptively secure garbled circuits from one-way functions. In: Robshaw, M., Katz, J. (eds.) CRYPTO 2016. LNCS, vol. 9816, pp. 149–178. Springer, Heidelberg (2016). doi:10.1007/978-3-662-53015-3_6

[HP70] Hewitt, C.E. Paterson, M.S.: Comparative Schematology. In: Record of the Project MAC Conference on Concurrent Systems and Parallel Computation, pp. 119–127. ACM, New York (1970)

[JW16] Jafargholi, Z., Wichs, D.: Adaptive security of Yao's garbled circuits. Cryptology ePrint Archive, Report 2016/814 (2016). http://eprint.iacr.org/2016/814

[Kal00] Kaliski, B.: PKCS#5: password-based cryptography specification version 2.0 (2000)

[LT82] Lengauer, T., Tarjan, R.E.: Asymptotically tight bounds on time-space trade-offs in a pebble game. J. ACM 29(4), 1087–1130 (1982)

[MMV13] Mahmoody, M., Moran, T., Vadhan, S.P.: Publicly verifiable proofs of sequential work. In: Kleinberg, R.D. (ed.) ITCS 2013, pp. 373–388. ACM, January 2013

[NBF+15] Narayanan, A., Bonneau, J., Felten, E.W., Miller, A., Goldfeder, S.: Bitcoin and Cryptocurrency Technology (manuscript) (2015). Accessed 8 June 2015

[Per09] Percival, C.: Stronger key derivation via sequential memory-hard functions. In: BSDCan 2009 (2009)

[PHC] Password hashing competition. https://password-hashing.net/

[PR80] Paul, W.J., Reischuk, R.: On alternation II. A graph theoretic approach to determinism versus nondeterminism. Acta Inf. **14**, 391–403 (1980)

[RD16] Ren, L., Devadas, S.: Proof of space from stacked bipartite graphs. Cryptology ePrint Archive, Report 2016/333 (2016). http://eprint.iacr.org/

[Sch82] Schnitger, G.: A family of graphs with expensive depth reduction. Theor. Comput. Sci. **18**, 89–93 (1982)

[Sch83] Schnitger, G.: On depth-reduction and grates. In: 24th Annual Symposium on Foundations of Computer Science, Tucson, Arizona, USA, 7–9 November 1983, pp. 323–328. IEEE Computer Society (1983)

[SS78] Savage, J.E., Swamy, S.: Space-time trade-offs on the FFT algorithm. IEEE Trans. Inf. Theory **24**(5), 563–568 (1978)

[SS79a] Savage, J.E., Swamy, S.: Space-time tradeoffs for oblivious integer multiplication. In: Maurer, H.A. (ed.) ICALP 1979. LNCS, vol. 71, pp. 498–504. Springer, Heidelberg (1979). doi:10.1007/3-540-09510-1_40

[SS79b] Swamy, S., Savage, J.E.: Space-time tradeoffs for linear recursion. In: Aho, A.V., Zilles, S.N., Rosen, B.K. (eds) POPL, pp. 135–142. ACM Press (1979)

[Tom78] Tompa, M.: Time-space tradeoffs for computing functions, using connectivity properties of their circuits. In: Proceedings of the Tenth Annual ACM Symposium on Theory of Computing, STOC 1978, pp. 196–204. ACM, New York (1978)

Scrypt Is Maximally Memory-Hard

Joël Alwen[1], Binyi Chen[2], Krzysztof Pietrzak[1], Leonid Reyzin[3(✉)],
and Stefano Tessaro[2]

[1] IST Austria, Klosterneuburg, Austria
{jalwen,pietrzak}@ist.ac.at
[2] UC Santa Barbara, Santa Barbara, USA
{binyichen,tessaro}@cs.ucsb.edu
[3] Boston University, Boston, USA
reyzin@cs.bu.edu

Abstract. Memory-hard functions (MHFs) are hash algorithms whose evaluation cost is dominated by memory cost. As memory, unlike computation, costs about the same across different platforms, MHFs cannot be evaluated at significantly lower cost on dedicated hardware like ASICs. MHFs have found widespread applications including password hashing, key derivation, and proofs-of-work.

This paper focuses on scrypt, a simple candidate MHF designed by Percival, and described in RFC 7914. It has been used within a number of cryptocurrencies (e.g., Litecoin and Dogecoin) and has been an inspiration for Argon2d, one of the winners of the recent password-hashing competition. Despite its popularity, no rigorous lower bounds on its memory complexity are known.

We prove that scrypt is *optimally memory-hard*, i.e., its cumulative memory complexity (cmc) in the parallel random oracle model is $\Omega(n^2 w)$, where w and n are the output length and number of invocations of the underlying hash function, respectively. High cmc is a strong security target for MHFs introduced by Alwen and Serbinenko (STOC '15) which implies high memory cost even for adversaries who can amortize the cost over many evaluations and evaluate the underlying hash functions many times in parallel. Our proof is the first showing optimal memory-hardness for any MHF.

Our result improves both quantitatively and qualitatively upon the recent work by Alwen *et al.* (EUROCRYPT '16) who proved a *weaker* lower bound of $\Omega(n^2 w / \log^2 n)$ for a *restricted* class of adversaries.

Keywords: Scrypt · Memory-hard functions · Password hashing

1 Introduction

Several applications rely on so-called "moderately-hard tasks" that are not infeasible to solve, but whose cost is non-trivial. The cost can for example be the hardware or electricity cost of computation as in proofs of work [13,14,18] or

© International Association for Cryptologic Research 2017
J.-S. Coron and J.B. Nielsen (Eds.): EUROCRYPT 2017, Part III, LNCS 10212, pp. 33–62, 2017.
DOI: 10.1007/978-3-319-56617-7_2

time-lock puzzles [23], the cost for disk storage space as in proofs of space [15], the cost of "human attention" in captchas [24] or the cost of memory in memory-bound [2,12] or memory-hard functions [8,20], the latter being the topic of this work. Applications of such tasks include the prevention of spam [13], protection against denial-of-service attacks [19], metering client access to web sites [16], consensus protocols underlying decentralized cryptocurrencies [10] (knows as blockchains) or password hashing [22], which we'll discuss in more detail next.

In the setting of *password hashing*, a user's password (plus a salt and perhaps other system-dependent parameters) is the input to a moderately-hard function f, and the resulting output is the password hash to be kept in a password file. The hope is that even if the password file is compromised, a brute-force dictionary attack remains costly as it would require an attacker to evaluate f on every password guess. Traditional approaches for password hashing have focused on iterating a hash function a certain number (typically a few thousands) of times, as for instance in PBKDF2. An advantage for the honest user results from the fact that he or she needs to compute f only once on the known password, while an attacker is forced to compute f on a large number of passwords. However, this advantage can be eroded, because in constrast to honest users, who typically use general-purpose hardware, attackers may invest into special-purpose hardware like ASICs (Application Specific Integrated Circuits) and recoup the investment over multiple evaluations. Moreover, such special-purpose hardware may exploit parallelism, pipelining, and amortization in ways that the honest user's single evaluation of f cannot. Consequently, the adversary's cost per evaluation can be several orders of magnitude lower than that for the honest user.

MEMORY-HARD FUNCTIONS. To reduce the disparity between per-evaluation costs of the honest user and a potential attacker, Percival [20] suggested measuring cost by the amount of space used by the algorithm multiplied by the amount of time. A *memory-hard function* (MHF), in Percival's definition, is one where this evaluation cost measure is high not only for the honest user's sequential algorithm, but also no parallel algorithm can do much better. In particular, if a parallel algorithm can cut the time to evaluate f by some factor, it must come at the cost of increase in space by roughly the same factor. Since memory is inherently general-purpose, measuring cost in terms of space provides a reasonably accurate comparison of resources used by different implementations. We stress that memory hardness is a very different notion than that of memory-*bound* functions proposed by Abadi, Burrows, Manasse, and Wobber [1,2], which maximize the number of memory accesses at unpredictable locations so that the inherent memory-access latency (resulting from frequent cache misses) imposes a lower bound on the *time* needed to evaluate the function which is independent from the actual CPU power. Memory hardness also does not guarantee that a lot of memory will be required, because it allows trading memory for time.

Alwen and Serbinenko [8] observed that Percival's notion of cost is not robust to amortization: it may be that an algorithm uses a large amount of memory at its peak, but a much smaller amount on average; pipelining multiple evaluations (by multiple CPUs using shared memory) in such a way that peaks occur at different times can thus reduce the per-evaluation cost of f. They propose the notion of *cumulative memory complexity* (abbreviated cc_{mem}), which is robust to amortization. It is defined as the sum of memory actually used at each point in time (rather than the product of peak memory and time). We will use this notion in our work. The cc_{mem} of a function is defined as the lowest cc_{mem} of all algorithms that evaluate the function.

BEST-POSSIBLE HARDNESS. Given the state of computational complexity theory, where we cannot even prove superlinear lower bounds for problems in NP, all cc_{mem} lower-bound results so far necessarily make use of idealized models of computation, like the random oracle model.

Many candidate memory-hard functions (including scrypt) can be viewed as a mode of operation for an underlying building block like a cryptographic hash-function h. Such MHFs come with an evaluation algorithm—which we'll call "the naïve algorithm"—which makes only sequential access to h. Note that after t steps (each involving at most one query to h), even a naïve algorithm which stores all t outputs it received from h : $\{0,1\}^* \rightarrow \{0,1\}^w$ will not use more that $O(t \cdot w)$ memory; therefore, if the naïve algorithm for an MHF f^h runs for n steps total, the cc_{mem} of f^h will be in $O(n^2 \cdot w)$. Thus a lower bound on the cc_{mem} of $\Omega(n^2 \cdot w)$ is the best we can hope for in any model which captures at least the naïve algorithm.

If the naïve algorithm has the feature that its memory access pattern (addresses read and written) is independent of the input, the MHF is called *data-independent*. A data-independent f^h can be represented as a directed acyclic graph, with a unique source corresponding to the input, a unique sink corresponding to the final output, and the other nodes indicating intermediary values where the value of a node is computed as a function of the nodes of its parents (using one invocation of h). To derive meaningful bounds, we require that this graph has constant in-degree, so computing an intermediate value takes constant time. The evaluation of f^h can now be cast as a graph pebbling problem [8]. Any constant in-degree graph can be pebbled (in the so called parallel black pebbling game) using "cumulative pebbling complexity" $cc_{peb} = O(n^2/\log n)$.[1] As any such pebbling implies an evaluation algorithm with $cc_{mem} \approx cc_{peb} \cdot w$, we get an $O(w \cdot n^2/\log n)$ upper bound on cc_{mem} for any data-indepedent MHF [3]. This upper bound is matched by [4], who construct a data-independent function with $cc_{mem} = \Omega(w \cdot n^2/\log(n))$ in the parallel random oracle model. This calls for the question of whether the lower bound of $\Omega(w \cdot n^2)$ is achievable at all; the above discussion shows that to achieve this lower bound, it will not be sufficient to only consider data-independent MHFs.

THE Scrypt MHF. Percival [20] proposed a candidate (data-dependent) memory-hard function called scrypt.[2] On input X, the $scrypt^h$ function–where

---

[1] Technically, the bound is marginally worse, $O(n^2/\log^{1-\epsilon}(n))$ for any $\epsilon > 0$.

[2] In fact, what we discuss in the following is Percival's ROMix construction, which constitutes the core of the actual scrypt function. We use the two names interchangeably.

h is a cryptographic hash function modeled as a random oracle for the lower bound proof – computes values $X_0, X_1, \ldots, X_{n-1}, S_0, \ldots, S_n$ as defined below, and finally outputs S_n

- $X_0 = X$ and for $i = 1, \ldots, n - 1$: $X_i = h(X_{i-1})$
- $S_0 = h(X_{n-1})$ and for $i = 1, \ldots, n$: $S_i = h(S_{i-1} \oplus X_{S_{i-1} \bmod n})$

Scrypt has found widespread popularity: it is used in proofs-of-work schemes for cryptocurrencies (most notably Litecoin [10], but also Tenebrix or Dogecoin), is described by an RFC [21], and has inspired the design of one of the Password-hashing Competition's [22] winners, Argon2d [9].

An intuitive explanation for why scrypt was conjectured to be memory-hard is as follows. View the first portion of scrypt as an n-node line graph, with nodes labeled by $X_0, \ldots, X_{n-1}$. To compute S_{i+1}, an algorithm needs $X_{S_i \bmod n}$, whose index ($S_i \bmod n$) is random and unknown until S_i is computed. If the algorithm stores a subset of the X values of size p before S_i is known, then the label of a random node in the line graph will be on average $n/(2p)$ steps from a stored label, and will therefore take $n/(2p)$ sequential evaluations of h to compute, for a total memory $\cdot$ time cost of $p \cdot n/(2p) = n/2$. Since there are n S_i values to compute, this strategy has $\mathsf{cc_{mem}}$ of $w \cdot n \cdot n/2 = \frac{1}{2} w n^2$.

This simple argument, however, does not translate easily into a proof. The two main challenges are as follows. First, in general an algorithm computing scrypt is not restricted to just store labels of nodes, but can compute and store arbitrary information. Surprisingly, f for which storing information other than just labels provably decreases $\mathsf{cc_{mem}}$ have been constructed in [5, Appendix A]. Second, an algorithm is not compelled to keep all p labels in memory after the index $S_i \bmod n$ is known. In fact, [8] show that if one is given the indices $S_i \bmod n$ in advance, an evaluation algorithm exists which evaluates scrypt^h with $\mathsf{cc_{mem}}$ only $O(w \cdot n^{1.5})$, because knowing the future indices enables the algorithm to keep or recompute the labels that will be needed in the near future, and delete those that won't.

PREVIOUS WORK ON scrypt. Percival's original paper [20] proposed an analysis of scrypt, but his analysis is incorrect, as we point out in Appendix A, in addition to not targeting $\mathsf{cc_{mem}}$. Recent progress toward proving that scrypt is memory-hard was made by Alwen et al. [6]. They lower bound the $\mathsf{cc_{mem}}$ of scrypt by $\Omega(w \cdot n^2 / \log^2 n)$, albeit only for a somewhat restricted class of adversaries (informally, adversaries who can store secret shares of labels, but not more general functions). We'll compare their work with ours in more detail below.

OUR RESULTS. We give the first non-trivial unconditional lower bound on $\mathsf{cc_{mem}}$ for scrypt^h in the parallel random oracle model, and our bound already achieves optimal $\mathsf{cc_{mem}}$ of $\Omega(w \cdot n^2)$.

We'll give the exact theorem statement and an overview of the proof in Sect. 3. However, to appreciate the novelty of our results, we note that the only existing proofs to lower bound $\mathsf{cc_{mem}}$ of MHFs go through some kind of lower bounds for pebbling.

For *data independent* MHFs [8] there is an elegant argument (known as "ex post facto") stating that a lower bound on the cumulative complexity for the parallel black pebbling game translates directly into a lower bound for cc_{mem}. Thus, the problem is reduced to a purely combinatorial problem of proving a pebbling lower bound on the graph underlying the MHF.

For *data dependent* MHFs no such result, showing that pebbling lower bounds imply cc_{mem} lower bounds for general adversaries, is known.[3] The lower bound on cc_{mem} for scrypt from [6] was also derived by first proving a lower bound on the pebbling complexity, but for a more powerful pebbling adversary that can use "entangled" pebbles. This lower bound then translated into a lower bound for cc_{mem} for a limited class of adversaries who, apart from labels, can also store "secret shares" of labels. It was conjectured [6] that lower bounds for this entangled pebbling game already imply lower bounds on cc_{mem} for *arbitrary* adversaries, and a combinatorial conjecture was stated which, if true, would imply this. Unfortunately the strongest (and simplest) version of this conjecture has already been refuted. A weaker version of the conjecture has been "weakly" refuted, in the sense that, even if it was true, one would lose a factor of at least $\log(n)$ by going from pebbling to memory lower bounds. (The current state of the conjecture is available on the eprint version of the paper [5].)

In this work, in Sect. 5, we also prove an optimal $\Omega(n^2)$ lower bound on the parallel cumulative pebbling complexity for a game which abstracts the evaluation of scrypt: we consider a path of length n, and an adversary must pebble n randomly chosen nodes on this graph, where the ith challenge node is only revealed once the node of challenge $i-1$ is pebbled. This already gives an optimal $\Omega(n^2 \cdot w)$ lower bound on cc_{mem} for scrypt for adversaries who are only allowed to store entire labels, but not any functions thereof. This improves on the $\Omega(n^2/\log^2(n))$ lower bound from [6], who use a rather coarse potential argument which roughly states that, for any challenge, either we pay a lot for pebbling the next challenge node, or the "quality" of our pebbling configuration decreases. As this quality cannot decrease too many times, at least every $\log(n)$'th challenge will cost $n/\log(n)$ in cumulative complexity, giving the overall $\Omega(n^2/\log^2(n))$ lower bound after n challenges. In this work we introduce a new technique for analyzing the cumulative pebbling cost where—for every challenge—we take into account the cumulative cost of the pebbling configurations *before* this challenge is revealed. Both the potential argument from [6], as well as our new proof, rely on the generalization of the fact that given a configuration with p pebbles, and a random challenge, with good probability (say at least $\frac{1}{2}$), an adversary also needs also at least (roughly) n/p steps to pebble the challenge.

As discussed above, pebbling lower bounds are not known to directly imply cc_{mem} lower bounds for data dependent MHFs, so to prove our main result in Sect. 6, we in some sense emulate our proof for the pebbling game directly in the parallel random oracle model. However, there are two problems we will

---

[3] A lower bound on the parallel cumulative pebbling complexity is only known to imply a lower bound on cc_{mem} for a very restricted class of adversaries who are allowed to store only labels, but not any function thereof.

need to overcome. The first is that the adversary's state is not made of labels (corresponding to pebbles), but could be any function thereof. Still, we will want to show that in order to compute a challenge, an adversary storing any $p \cdot w$ bits of information about the random oracle, will need to take with good probability (say at least $\frac{1}{2}$) at least (roughly) n/p steps. We will show this using a careful compression argument in Sect. 4. The second problem is the fact that in scrypt the challenges are not randomly and externally generated, but come from the random oracle.

2   Preliminaries

We review basic notation and concepts from the literature on memory-hard functions. We will also define the scrypt function as needed further below.

THE PARALLEL-RANDOM ORACLE MODEL. We first define the parallel random-oracle model (pROM), essentially following the treatment from [8], with some highlighted differences.

Concretely, we consider an oracle-aided deterministic[4] algorithm A which runs in rounds, starting with round 1. Let h denote an oracle with w-bit outputs. It does not matter for our model whether oracle inputs are restricted in length, but it will be simpler to assume a general upper bound (even very large) on the length of its inputs to make the set of oracles finite.

In general, a *state* is a pair $(\tau, \mathbf{s})$ where *data* τ is a string and $\mathbf{s}$ is a tuple of strings. In an execution, at the end of round i, algorithm A produces as output an *output state* $\bar{\sigma}_i = (\tau_i, \mathbf{q}_i)$ where $\mathbf{q}_i = [q_i^1, \ldots, q_i^{z_i}]$ is a tuple of *queries* to h. At the begining of next round $i + 1$, algorithm A gets as input the corresponding *input state* $\sigma_i = (\tau_i, \mathsf{h}(\mathbf{q}_i))$ where $\mathsf{h}(\mathbf{q}_i) = [\mathsf{h}(q_i^1), \ldots, \mathsf{h}(q_i^{z_i})]$ is the tuple of *responses* from h to the queries $\mathbf{q}_i$. In particular, since A is deterministic, for a given h the input state σ_{i+1} is a function of the input state σ_i.

The initial input state σ_0 is normally empty with length 0 (though in the proof we will also need to consider a non-empty initial input state); an input X is given together with σ_0 in the first round. We require that A eventually terminates and denote its output by $A^{\mathsf{h}}(X)$.

COMPLEXITY MEASURE. For a given execution the complexity measure we are going to be concerned with is the sum of the bit-lengths of the input states. To that make this precise we introduce the following notation. For a string x we denote its bit-length by $|x|$. For state $\sigma = (\tau, \mathbf{s})$ where $\mathbf{s} = [s_1, \ldots, s_z]$ we denote the bit-length (or size) of σ by $|\sigma| = |\tau| + \sum_{j=1}^{z} |s_j|$. We can now define the *cumulative (memory) complexity* of an execution of algorithm A on input X using oracle h resulting in *input* states $\sigma_0, \sigma_1, \ldots$ as

$$\mathsf{cc_{mem}}(A^{\mathsf{h}}(X)) = \sum_{i \geq 0} |\sigma_i|.$$

---

[4] Considering deterministic algorithms is without loss of generality as we can always fix the randomness of A to some optimal value.

We will assume without loss of generality that at each round, the query tuple **q** contains at least one query, for otherwise A can proceed directly to the next round where it issues a query, without increasing its cumulative complexity. In particular, this implies $|\sigma_i| \geq w$ for $i > 0$.

Note that $\mathsf{cc_{mem}}$ does not charge anything for computation or memory used within each round itself. We are also allowing inputs to **h** to be arbitrary long without extra memory cost—only the output length w is charged to the cumulative complexity. This only makes our *lower bound* stronger. Note however that this also means that $\mathsf{cc_{mem}}$ gives a good *upper bound* only when computation is dominated by the memory cost (as is the case for the naïve evaluation algorithm of $\mathsf{scrypt^h}$, which, aside from querying **h** sequentially, performs only a few trivial computations, such as exlusive-ors and modular reductions).

THE scrypt MHF. We will consider the $\mathsf{scrypt^h}$ function throughout this paper (more specifically, we study its core, ROMix, as defined in [20]). Recall that for a hash function $\mathsf{h} : \{0,1\}^* \rightarrow \{0,1\}^w$, $\mathsf{scrypt^h}$ on input $X \subset \{0,1\}^w$ and parameter $n \in \mathbb{N}$ computes values $X_0, X_1, \ldots, X_{n-1}, S_0, \ldots, S_n$ and outputs S_n, where

- $X_0 = X$ and for $i = 1, \ldots, n-1$: $X_i = \mathsf{h}(X_{i-1})$
- $S_0 = \mathsf{h}(X_{n-1})$ and for $i = 1, \ldots, n$: $S_i = \mathsf{h}(S_{i-1} \oplus X_{S_{i-1} \bmod n})$

We will also define intermediate variables $T_0, \ldots, T_n$ with $T_0 = X_{n-1}$ and $T_i = S_{i-1} \oplus X_{S_{i-1} \bmod n}$ for $1 \leq i \leq n$, so that $S_i = \mathsf{h}(T_i)$.

Note that one may not want to restrict X to w bits. In this case, one can replace X with $\mathsf{h}(X)$ in the above construction. For notational simplicity, we will only analyze the w-bit input case in this paper, but the general analysis is very similar.

GRAPH AND PEBBLING PRELIMINARIES. For some of our partial results below, we will adopt the graph-pebbling view on computing candidate MHFs, following [8]. A parallel black pebbling considers a direct acyclic graph $G = (V, E)$. At each time step t starting with $t = 0$, the adversary maintains a subset P_t of nodes ("pebbles"). A node v is allowed (but not required) to get a pebble at time t if there is a pebble on all of its predecessors (i.e., all v' such that $(v', v) \in E$), or if there was a pebble on v itself at time $t - 1$. Formally, define $\mathsf{pre}(v)$ to be the set of all predecessors of v, and for $U \subseteq V$, define $U^+ = \{v \in V : \mathsf{pre}(v) \subseteq U\}$. Then, at time $t > 0$, the set P_t must be a subset of $P_{t-1} \cup P_{t-1}^+$.

We define $\mathsf{p}_i = |P_i| \geq 1$. The (parallel) *cumulative pebbling complexity* of a sequence of pebbling configuration $P_0, P_1, \ldots, P_t$ is $\sum_{i=0}^{t} \mathsf{p}_i$. We remark that we modify the pebbling rules slightly from [8] by not permitting the adversary to put a pebble on the source for free: v_0 is contained in P_0 and cannot be added to P_t if it is absent in P_{t-1} (this change will simplify calculations, and only increase the size of each set by 1).

PEBBLING WITH CHALLENGES. Normally, the goal of pebbling games is to place a pebble on the sink of the graph. Here, we are going to consider pebbling games with Q *challenges* on a graph $G = (V, E)$, where the adversary proceeds in rounds, and in each round i, it receives a random challenge $c_i \in V$ (usually uniform from a subset $V' \subseteq V$), and the goal is to place a pebble on c_i, which enables the adversary to move to the next round (unless this was the last challenge c_Q, in which case the game terminates.) For instance, the core of the evaluation of scrypt is captured by the *line graph* with vertices $v_0, \dots, v_{n-1}$ and edges (v_i, v_{i+1}) for $i = 0, \dots, n-2$, and we will study this pebbling game in detail below.

3   Main Result and Overview

In this section, we state our main result, and give a brief high-level overview of the next sections.

Theorem 1 (Memory-hardness of Scrypt, main theorem). *For any $X \in \{0,1\}^w$ and $n \geq 2$, if $A^h(X, n)$ outputs $S_n = \text{scrypt}^h(X, n)$ with probability χ, where the probability is taken over the choice of the random oracle h, then with probability (over the choice of h) at least $\chi - .08n^6 \cdot 2^{-w} - 2^{-n/20}$,*

$$\text{cc}_{\text{mem}}(A^h(X)) > \frac{1}{25} \cdot n^2 \cdot (w - 4\log n).$$

We note that if w is large enough in terms of n (say, $4\log n \leq w/2$, which clearly holds for typical values $w = 256, n = 2^{20}$), then $\text{cc}_{\text{mem}}(A^h(X))$ is in $\Omega(n^2 w)$. As discussed in the introduction, this is the best possible bound up to constant factors, as already the (sequential) naïve algorithm for evaluating scrypt^h has $\text{cc}_{\text{mem}} \in O(n^2 w)$. We discuss the constants following Theorem 5.

PROOF OUTLINE. The proof consists of three parts outlined below. The first two parts, in fact, will give rise to statements of independent interest, which will then be combined into the proof of our main theorem.

– Section 4: Single-shot time complexity. To start with, we consider a pROM game where the adversary $A^h(X)$ starts its execution with input X and an M-bit state σ_0 that can depend *arbitrarily* on h and X. Then, $A^h(X)$ is given a random challenge $j \in \{0, \dots, n-1\}$ and must return $X_j = h^j(X)$.

Clearly, σ_0 may contain X_j, and thus in the best case, A may answer very quickly, but this should not be true for all challenges if $M \ll nw$. We will prove a lower bound on the *expected* time complexity (in the pROM) of answering such a challenge. We will show that with good probability (e.g., $\frac{1}{2}$) over the choice of j, $A^h(X)$ needs at least (roughly) nw/M steps.

This validates in particular the intuition that the adversary in this game cannot do much better than an adversary in the corresponding pebbling game on the line graph with vertices $v_0, v_1, \dots, v_{n-1}$, where the adversary gets to choose an initial configuration with $p = M/w$ pebbles, and is then asked to put a pebble on v_j for a random $j \in \{0, 1, \dots, n-1\}$. Here, one can show that at least n/p steps are needed with good probability. In fact, this pebbling game is equivalent to a variant of the above pROM game where the adversary only stores random-oracle output labels, and thus our result shows that an adversary cannot do much better than storing whole labels.

– Section 5: Multi-challenge cumulative pebbling complexity. In the above scenario, we have only considered the *time* needed to answer a challenge. There is no guarantee, a priori, that the cumulative complexity is also high: An optimal adversary, for instance, stores p labels corresponding to equidistant pebbles, and then computes the challenge from the closest label, dropping the remainder of the memory contents.

Here, for the randomized pebbling game with Q challenges on the line graph, we will show a lower bound of $\Omega(nQ)$ on the cumulative pebbling complexity. Our argument will use in particular a (generalization) of the above single-shot trade-off theorem, i.e., the fact that whenever p pebbles are placed on the line, at least n/p steps are needed with good probability to pebble a randomly chosen node. We will use this to lower bound the cumulative complexity *before* each particular challenge is answered. Our proof gives a substantial quantitative improvement over the looser lower bound of [6].

– *Section 6*: $\mathsf{cc_{mem}}$ *of* scrypt. Finally, we lower bound the cumulative memory complexity of scrypt^h as stated in Theorem 1. Unfortunately, this does not follow by a reduction from the pebbling lower bound directly. Indeed, as discussed in the introduction (and as explained in [6]), unlike for data-independent MHFs, for data-dependent MHFs like scrypt it is an open problem whether one can translate lower bounds on the cumulative *pebbling* complexity to lower bounds for cumulative *memory* complexity. Fortunately, however, through a careful analysis, we will be able to employ the same arguments as in the proof of Sect. 5 in the pROM directly.

In particular, we will use our result from Sect. 4 within an argument following the lines to that of Sect. 5 in the pROM. One particularly delicate technical issue we have to address is the fact that in scrypt^h the challenges are not sampled randomly, but will depend on the random oracle h, which the adversary can query. We will provide more intuition below in Sect. 6.

Remark 1. Note that in Theorem 1 above, the random oracle h is sampled uniformly *after* the input X is chosen arbitrarily. This is equivalent to saying that X and h are independent. In practice this assumption is usually (nearly) satisfied. For example, when used in a blockchain, X will be the output of h on some previous inputs, typically a hash of the last block and a public-key. This doesn't make X independent of h, but its distribution will be dense in the uniform distribution even conditioned on h, which is means it is very close to independent. When used for password hashing, $X = \mathsf{h}(pwd, S)$ for a password pwd and a random salt S. For a sufficiently long salt, this will make X as good as uniform [11]. We defer more rigorous treatment of this issue to the full version of this paper.

4 Time Complexity of Answering a Single Challenge in the Parallel Random Oracle Model

We prove the following theorem, and below discuss briefly how this result can be extended beyond the setting of scrypt.

Fix positive integers n, u and w, a string $X \in \{0,1\}^u$, a finite domain $\mathcal{D}$ that contains at least $\{X\} \cup \{0,1\}^w$, and let $\mathcal{R} = \{0,1\}^w$. Given a function $h : \mathcal{D} \to \mathcal{R}$, define $X_i = h^i(X)$. Let A be any oracle machine (in the parallel random oracle model as defined in Sect. 2) that on any input and oracle makes at most $q - 1$ total queries to its oracle. Suppose $A^h(X, j)$ starts on input state σ_0 with the goal of eventually querying X_j to h. Let t_j be the number of the earliest round in which $A^h(X, j)$ queries X_j to h (with $t_j = \infty$ if never). We show that A cannot do much better than if it were doing the following in the corresponding random challenge pebbling game on the line graph: initially placing $p \approx M/w$ equidistant pebbles, and then pebbling the challenge from the closest pebble preceding it.

Theorem 2 (Single-Challenge Time Lower Bound). *There exists a set of random oracles good_h such that $\Pr_{h \in \mathcal{R}^\mathcal{D}}[h \notin \mathrm{good}_h] \leq qn^3 2^{-w}$, and for every $h \in \mathrm{good}_h$, the following holds: for every memory size M, and every input state σ_0 of length at most M bits,*

$$\Pr_{j \leftarrow \{0,\dots,n-1\}}\left[t_j > \frac{n}{2p}\right] \geq \frac{1}{2},$$

where the probability is taken over only the challenge j and $p = \lceil (M+1)/(w - 2\log n - \log q) + 1\rceil$.

We will actually prove a slightly more general result: for any $0 \leq \mathrm{pr}_{\mathrm{hard}} \leq 1$,

$$\Pr_{j \leftarrow \{0,\dots,n-1\}}\left[t_j > \frac{n(1 - \mathrm{pr}_{\mathrm{hard}})}{p}\right] \geq \mathrm{pr}_{\mathrm{hard}}.$$

Proof. Recall that for each j, A performs t_j rounds of the following process. At round k read an input state containing oracle responses $h(\mathbf{q}_{k-1})$ (except for $k = 1$, when A reads σ_0). Then (after arbitrary computation) produce an output state containing oracle queries $\mathbf{q}_k$. We count rounds starting from 1. Consider the sequence of such tuples of queries and responses to and from h. If the first appearance of X_i in this sequence is a query to h in round k ($k > 0$ is minimal such that $X_i \in \mathbf{q}_k$), then we assign X_i position $\pi_{ij} = k$. If instead the first appearance of X_i is a response from h to query X_{i-1} made at round k ($k > 0$ is minimal such that $X_{i-1} \in \mathbf{q}_k$), then we assign X_i position $\pi_{ij} = k + 1/2$. In all other cases (i.e., if X_i does not appear, or appears only because of a hash collision in response to some query that is not X_{i-1}), let $\pi_{ij} = \infty$.

Let "best position" correspond to the earliest time, over all j, that X_i appears during the computation of X_j: $\beta_i := \min_j \pi_{ij}$; let "best challenge" $\mathrm{bestchal}_i$ be $\mathrm{argmin}_j \pi_{ij}$ (if argmin returns a set, pick one element arbitrarily). Let i be "blue" if β_i is an integer (i.e., it was produced "out of the blue" by A as a query to h).

Let $B = \{i$ s.t. $i > 0$ and i is blue$\}$ (that is, all the blue indices except X_0). In the rest of the proof, we will show that the size of B cannot exceed $p - 1$ (for most h), where p, as defined in the theorem statement, is proportional to the memory size M; and that the amount of time to answer the challenge is at least its distance from the preceding blue index. Thus, blue indices effectively act like pebbles, and the bounds on the time to reach a random node in the line graph by moving pebbles apply.

Claim 1. *Given adversary A and input X, there exists a predictor algorithm $\mathcal{P}$ (independent of h, but with oracle access to it) with the following property: for every h, every M, and every length M input state σ_0 of A, there exists a hint of length $|B|(2 \log n + \log q)$ such that given σ_0 and the hint, $\mathcal{P}$ outputs every X_i for $i \in B$ without querying X_{i-1} to h.*

Moreover, if we want fewer elements, we can simply give a shorter hint: there exists a predictor algorithm that similarly outputs p elements of B whenever $p \leq |B|$, given σ_0 and an additional $p(2 \log n + \log q)$-bit hint.

Note that the inputs to $\mathcal{P}$ can vary in size; we assume that the encoding of inputs is such that the size is unambiguous.

Proof. We will focus on the first sentence of the claim and address the second sentence at the end.

$\mathcal{P}$ depends on input label $X = X_0$ and algorithm A (which are independent of h). $\mathcal{P}$ will get the state σ_0 of A (which may depend on h) as input, and, for every $i \in B$, a hint containing the challenge bestchal$_i$ for which X_i appears earliest, and the sequential order (among all the $q - 1$ queries A makes in answering bestchal$_i$) of the first query to X_i (using the value q to indicate that this query never occurs). This hint (which depends on h) will thus consist of a list of $|B|$ entries, each containing $i \in B$, bestchal$_i$, and $\log q$ bits identifying the query number, for a total of $|B|(2 \log n + \log q)$ bits.

$\mathcal{P}$ will build a table containing X_i for $i \geq 0$ (initializing $X_0 = X$). To do so, $\mathcal{P}$ will run A on every challenge in parallel, one round at a time. After each round k, $\mathcal{P}$ will obtain, from the output states of A, all the queries A makes for all the challenges in round k. Then $\mathcal{P}$ will fill in some spots in its table and provide answers to these queries as input states for round $k + 1$ by performing the following three steps:

Step k. put any blue queries into its table (blue queries and their positions in the table can easily be recognized from the hint);

Step k+1/4. answer any query that can be answered using the table (i.e., any query that matches X_{i-1} in the table for some filled positions $i - 1$ and i);

Step k+1/2. send remaining queries to h, return the answers to A, and fill in any new spots in the table that can be filled in (i.e., for every query that matches X_{i-1} in the table for some filled-in position $i - 1$, fill in position i with the answer to that query).

Once every X_i for $i \in B$ is in the table, $\mathcal{P}$ queries h to fill in the missing positions in the table, and outputs the prediction that $h(X_{i-1}) = X_i$ for $i \in B$.

To prove that $\mathcal{P}$ simulates h correctly to A, it suffices to show that the table contains correct labels. This can be easily argued by induction on i. Assume all the labels in the table are correct up to now. A new label X_i enters the table either because it is marked as blue (and thus correct by the hint) or is obtained as an answer from h to the query that $\mathcal{P}$ identified as X_{i-1} using the table (which is correct by inductive hypothesis).

The above also shows that $\mathcal{P}$ will not output an incorrect prediction. It remains to show that $\mathcal{P}$ did not query to h the value X_{i-1} for any $i \in B$. To prove this, we first show that X_i is placed into the table no later than step β_i of $\mathcal{P}$, by induction on β_i. The base case is X_0, which is in the table at step 0. If β_i is an integer, then $i \in B$ and this is true because of step β_i of $\mathcal{P}$. If β_i is not an integer, then $\beta_{i-1} < \beta_i$ (because X_{i-1} appears as a query at round $\lfloor \beta_i \rfloor$), so at the beginning of step β_i of $\mathcal{P}$, by the inductive hypothesis, position $i - 1$ in the table will already contain X_{i-1}, and thus position i will get filed in when X_{i-1} gets queried to h.

Note also that X_i cannot be placed into the table earlier than step β_i, so it is placed in the table exactly at step β_i (as long as $\beta_i \neq \infty$, in which case it is placed into the table at the end, when $\mathcal{P}$ fills in the missing positions).

Now suppose, for purposes of contradiction, that $\mathcal{P}$ queries h for some value X_{i-1} for some $i \in B$. That can happen only if at the end of some round k, X_{i-1} is queried by A as part of the output state, but either X_{i-1} or X_i are not in the table at that time.

- If X_{i-1} is not in the table at the beginning of step $k + 1/2$ of $\mathcal{P}$, then $\beta_{i-1} \geq k + 1/2$; but since X_{i-1} is being queried at the end of round k, $\beta_{i-1} \leq k$, which is a contradiction.
- If X_i is not in the table at the beginning of step $k + 1/2$ of $\mathcal{P}$, then $\beta_i \geq k + 1$ (because β_i is an integer); but since X_{i-1} appears as query in the output state of round k, $\beta_i \leq k + 1/2$, which is also a contradiction.

Thus, $\mathcal{P}$ always achieves its goal.

For the second sentence of the claim, observe that we can simply give $\mathcal{P}$ the hint for the p blue labels with the smallest β values.   □

In the next claim, we show that for every input to $\mathcal{P}$, the algorithm $\mathcal{P}$ cannot be correct for too many oracles h.

Claim 2. *Fix an algorithm $\mathcal{P}$ and fix its input, a positive integer p, some domain $\mathcal{D}$, and range $\mathcal{R}$. For $h : \mathcal{D} \to \mathcal{R}$, call $\mathcal{P}^h$ successful if $\mathcal{P}$ with oracle access to h outputs p distinct values $x_1, \ldots, x_p \in \mathcal{D}$ and $h(x_1), \ldots, h(x_p)$ without querying h on any of $x_1, \ldots, x_p$. Then $\Pr_{h \in \mathcal{R}^{\mathcal{D}}}[\mathcal{P}^h \text{ is successful}] \leq |\mathcal{R}|^{-p}$.*

Proof. Instead of choosing h all at once, consider the equivalent view of choosing answers to fresh queries of $\mathcal{P}$ uniformly at random, and then choosing the remainder of the h uniformly at random after $\mathcal{P}$ produces its output. Since $\mathcal{P}$ does not query $x_1, \ldots, x_p$, the choices of h on those points will agree with $y_1, \ldots, y_p$ with the probability at most $|\mathcal{R}|^{-p}$.   □

Using the previous two claims, we now bound the number of random oracles for which the size of the blue set is too large. Recall B is the blue set minus X_0.

Claim 3. *Given adversary A, there exists a set of random oracles* $\mathsf{good_h}$ *such that* $\Pr[\mathsf{h} \notin \mathsf{good_h}] \leq qn^3 2^{-w}$, *and for every* $\mathsf{h} \in \mathsf{good_h}$, *every* M, *and every initial state* σ_0 *of* A *of size at most* M *bits,* $|B| \leq p - 1$, *where* $p = \lceil (M + 1)/(w - 2\log n - \log q) + 1 \rceil$.

Proof. The intuition is as follows: if for some h and some initial input state of length M, $|B| > p - 1$, then either $\mathcal{P}$ successfully predicts the output of h on p distinct inputs (by Claim 1), or some of the values among $X_0, \ldots, X_{n-1}$ are not distinct. We will define $\mathsf{bad_h}$ as the set of random oracles for which this can happen, and then bound its size.

Let $\mathcal{S}$ be the size of the space of all possible random oracles h. There are at most $\frac{1}{2}\mathcal{S}n^2 2^{-w}$ random oracles for which some of the values among $X_0, \ldots, X_{n-1}$ are not distinct; (suppose the first collision pair is $i, j < n$, thus $X_{i-1} \neq X_{j-1}$, and the probability that $X_i = X_j$ is 2^{-w}; then the bound is given by taking union bound over at most $n^2/2$ pairs of (i, j).) Call this set of random oracles $\mathsf{colliding}$.

In the next paragraph, we will formally define the set $\mathsf{predictable}$ as the set of random oracles for which $\mathcal{P}$ correctly predicts the output on p distinct inputs given the M-bit input state of A and an additional $p(2\log n + \log q)$-bit hint. We will bound the size of $\mathsf{predictable}$ by bounding it for every possible memory state of A and every possible hint, and then taking the union bound over all memory states and hints.

Consider a particular input state of length M for A; recall that $p = \lceil (M + 1)/(w - 2\log n - \log q) + 1 \rceil$. Assume $1 \leq p \leq n - 1$ (otherwise, the statement of Claim 3 is trivially true). Fix a particular value of the hint for $\mathcal{P}$ for predicting p elements of B. (Recall that the hint was previously defined dependent on the random oracle; we are now switching the order of events by fixing the hint first and then seeing for how many random oracles this hint can work.) Since the input to $\mathcal{P}$ is now fixed, there are most $\mathcal{S}2^{-pw}$ random oracles for which it can correctly output p distinct values without querying them, by Claim 2. The set $\mathsf{predictable}$ consists of all such random oracles, for every value of M such that $p \leq n$, every M-bit input state σ_0, and every hint.

To count how many random oracles are in $\mathsf{predictable}$, first fix p. Let M_p be the largest input state length that gives this particular p. Take all input state lengths that give this p, all possible input states of those lengths (there are at most $2^{M_p} + 2^{M_p-1} + \cdots + 1 < 2^{M_p+1}$ of them), and all possible hints for extracting p values (there are at most $2^{p(2\log n + \log q)}$ of them). This gives us at most $\mathcal{S}2^{(M_p+1)+p(2\log n+\log q-w)}$ random oracles in $\mathsf{predictable}$. Since $(M_p + 1) \leq (p - 1)(w - 2\log n - \log q)$ by definition of p, this number does not exceed $\mathcal{S}2^{(2\log n+\log q-w)} = \mathcal{S}n^2 q 2^{-w}$. Now add up over all possible values of p (from 2 to n), to get $|\mathsf{predictable}| \leq \mathcal{S}(n - 1)n^2 q 2^{-w}$.

Set $\mathsf{bad_h} = \mathsf{colliding} \cup \mathsf{predictable}$ and let $\mathsf{good_h}$ be the complement of $\mathsf{bad_h}$. $\qquad\qquad\square$

Claim 4. *For every* i, $0 \leq i < n$, *the value* t_i *is at least* $1 + i - j$, *where* $j = max\{a \leq i \mid a \text{ is blue}\}$.

Proof. If i is blue, we are done, since $t_i \geq 1$ simply because we start counting rounds from 1.

We will first show that $\lceil \beta_i - \beta_j \rceil \geq i - j$. Fix a blue j and proceed by induction on i such that $i > j$ and there are no blue indices greater than j and less than i.

For the base case, suppose $i = j + 1$. Recall that β_i is not an integer because i is not blue. Then $\beta_{i-1} \leq \beta_i - 1/2$, because X_{i-1} is present as the query to h that produces response X_i in the sequence of queries that A makes when responding to the challenge bestchal$_i$, and we are done. For the inductive case, it suffices to show that $\beta_{i-1} \leq \beta_i - 1$, which is true by the same argument as for the base case, except that we add that β_{i-1} is also not an integer (since $i - 1$ is also not blue).

Therefore, $\lceil \beta_i \rceil \geq i - j + 1$, because $\beta_j \geq 1$. We thus have $\pi_{ii} = t_i \geq \lceil \beta_i \rceil \geq i - j + 1$. $\qquad\square$

The number of blue indices (namely, $|B| + 1$, because X_0 is blue but not in B) is at most p if h $\in$ good$_h$. Since at most d indices are within distance $d - 1$ of any given blue index, and there are at most p blue indices, we can plug in $d = n(1 - \text{pr}_{\text{hard}})/p$ to get

$$\Pr_i \left[t_i \leq \frac{n(1 - \text{pr}_{\text{hard}})}{p} \right] \leq 1 - \text{pr}_{\text{hard}}.$$

This concludes the proof of Theorem 2. $\qquad\square$

GENERALIZING TO OTHER GRAPHS. In general, every single-source directed acyclic graph G defines a (data-independent) function whose evaluation on input X corresponds to labeling G as follows: The source is labeled with X, and the label of every node is obtained by hashing the concatenation of the labels of its predecessors. Rather than evaluating this function, one can instead consider a game with challenges, where in each round, the adversary needs to compute the label of a random challenge node from G. Theorem 2 above can be seen as dealing with the special case where G is a line graph.

Theorem 2 can be generalized, roughly as follows. We replace $\log q$ with $\log q + \log d$ (where d is the degree of the graph), because identifying blue nodes now requires both the query number and the position within the query. We modify the proof of Claim 1 to account for the fact that the random oracle query resulting in response X_i is not necessarily X_{i-1}, but a concatenation of labels. The only conceptually significant change due to this generalization is in Claim 4, whose generalized statement is as follows. For every node i, define the "limiting depth of i" to be the length of a longest possible path that starts at a blue node, goes through no other blue nodes, and ends at i. The generalized version of Claim 4 states that the amount of time required to query X_i is at least one plus the limiting depth of node i.

With this more general claim in place, it follows that

$$\Pr_i [t_i > m] \geq \frac{1}{2},$$

where m defined as follows. Let S denote set of blue nodes, and let m_S denote the median limiting depth of the nodes in G. We define m to be the minimum m_S over all S such that the origin is in S and $|S| = p$.

Of course, other statistical properties of the distribution of t_i can also be deduced from this claim if we use another measure instead of the median. Essentially, the generalized theorem would show that the best the adversary can do is place p pebbles on the graph and use parallel pebbling.

5 Cumulative Complexity of Answering Repeated Challenges in the Parallel Pebbling Model

In the previous part, we showed that, in the parallel random oracle model, an adversary with memory (input state) of size M cannot do much better when answering a random challenge than placing $p \approx M/w$ pebbles on the graph and pebbling. In this section, we prove a lower bound on the cumulative complexity of *repeated* random challenges in the pebbling model. While the result in this section does not directly apply to the random oracle model for reasons explained in Sect. 5.1, all of the techniques are used in the proof of our main theorem in Sect. 6.

THE SINGLE CHALLENGE PEBBLING GAME. Consider now the pebbling game for the line graph G consisting of nodes $v_0, \ldots, v_{n-1}$ and edges (v_i, v_{i+1}) for every $0 \le i < n$. Recall that in this game, at each time step t starting with $t = 0$, the adversary maintains a subset P_t of nodes ("pebbles"). If there is a pebble on a node at time $t - 1$, its successor is allowed (but not required) to get a pebble at time t. Formally, at time $t > 0$, the set P_t must be a subset of $P_{t-1} \cup \{v_{i+1} : v_i \in P_{t-1}\}$. Also recall that we modify the game of [8] slightly by not permitting the adversary to put a pebble on the source for free: v_0 is contained in P_0 and cannot be added to P_t if it is absent in P_{t-1} (this change simplifies calculations). Let $p_i = |P_i| \ge 1$.

We will say that the adversary answers a challenge chal (for $0 \le$ chal $< n$) in t steps if $t > 0$ is the earliest time when $v_{\mathsf{chal}} \in P_{t-1}$ (note that a pebble needs to be on v_{chal} at time $t - 1$—think of time t as the step when the output to the challenge is presented; this convention again simplifies calculations, and intuitively corresponds to the scrypt evaluation, in which the "output" step corresponds to querying $X_{\mathsf{chal}} \oplus S_i$ in order to advance to the next challenge).

It is easy to see that t is at least one plus the distance between chal and the nearest predecessor of chal in P_0. Therefore, for the same reason as in the proof of Theorem 2 (because at most $n/(2p_0)$ challenges are within $n/(2p_0) - 1$ distance to a particular node in P_0 and there are p_0 nodes in P_0).

$$\Pr_{\mathsf{chal}} \left[t > \frac{n}{2p_0} \right] \ge \frac{1}{2}.$$

More generally, the following is true for any $0 \leq \mathrm{pr_{hard}} \leq 1$:

Fact 3

$$\Pr_{\mathrm{chal}} \left[t > \frac{c}{\mathrm{p_0}} \right] \geq \mathrm{pr_{hard}} \quad with \quad c = n(1 - \mathrm{pr_{hard}}).$$

REPEATED CHALLENGES PEBBLING GAME. We now consider repeated challenges. At time $s_1 = 0$, the adversary receives a challenge c_1, $0 \leq c_1 < n$. The adversary answers this challenge at the earliest moment $s_2 > s_1$ when P_{s_2-1} contains X_{c_1}; after P_{s_2} is determined, the adversary receives the next challenge c_2, and so on, for Q challenges, until challenge c_Q is answered at time s_{Q+1}. We are interested in the cumulative pebbling complexity $\mathrm{cc_{peb}} = \sum_{t=0}^{s_{Q+1}} \mathrm{p}_t$.

Note that the adversary can adaptively vary the number of pebbles used throughout the game, while Fact 3 above addresses only the number of pebbles used before a challenge is known. Nevertheless, we are able to show the following.

Theorem 4 (Cumulative pebbling complexity of repeated challenges game). *The cumulative pebbling complexity of the repeated challenges pebbling game is with high probability* $\Omega(nQ)$.

More precisely, suppose the adversary never has fewer than $\mathrm{p_0}$ *pebbles. Then for any* $\epsilon > 0$, *with probability at least* $1 - e^{-2\epsilon^2 Q}$ *over the choice of the* Q *challenges,*

$$\mathrm{cc_{peb}} \geq \mathrm{p_0} + \frac{n}{2} \cdot Q \cdot \left(\frac{1}{2} - \epsilon \right) \cdot \ln 2.$$

More generally, we replace the condition that the adversary never has fewer than $\mathrm{p_0}$ *pebbles with the condition* $\mathrm{p}_t \geq \mathrm{p_{min}}$ *for some* $\mathrm{p_{min}}$ *and every* $t \geq 1$, *we need to replace* $\ln 2$ *with*

$$\ln \left(1 + \left(\frac{\mathrm{p_{min}}}{\mathrm{p_0}} \right)^{\frac{1}{Q(\frac{1}{2}-\epsilon)}} \right).$$

This result improves [6, Theorem 1] by eliminating the $\log^2 n$ factor from the cumulative memory complexity of the pebbling game. In the full version [7], we discuss the general case of $\mathrm{p}_t \geq \mathrm{p_{min}}$ and show an attack showing that a bound as the above is necessary (up to constant factors in the exponent).

Our approach is general enough to apply to space-time tradeoffs other than inverse proportionality, to other graphs, and even to some other models of computation that do not deal with pebbling. However, we will explain in Sect. 5.1 why it cannot be used without modification in the parallel random oracle model and other models where space is measured in bits of memory.

Proof. Recall time starts at 0, p_t denotes the number of pebbles at time t, and s_i denotes the moment in time when challenge number i (with $1 \leq i \leq Q$) is issued. Let t_i denote the amount of time needed to answer challenge number i (thus, $s_1 = 0$ and $s_{i+1} = s_i + t_i$; let $s_{Q+1} = s_Q + t_Q$). Let $\mathrm{cc}(t_1, t_2)$ denote $\sum_{t=t_1}^{t_2} \mathrm{p}_t$.

The Main Idea of the Proof. The difficulty in the proof is that we cannot use t_i to infer anything about the number of pebbles used during each step of answering challenge i. All we know is that the number of pebbles has to be inversely proportional to t_i immediately before the challenge was issued—but the adversary can then reduce the number of pebbles used once the challenge is known (for, example by keeping pebbles only on v_0 and on the predecessor of the challenge).

The trick to overcome this difficulty is to consider how many pebbles the adversary has to have in order to answer the next challenge not only immediately before the challenge, but one step, two steps, three steps, etc., earlier.

Warm-Up: Starting with a Stronger Assumption. For a warm-up, consider the case when the pebbles/time tradeoff is guaranteed (rather than probabilistic, as in Fact 3): assume, for now, that in order to answer the next random challenge in time t, it is necessary to have a state of size c/t right before the challenge is issued. Now apply this stronger assumption not only to the moment s in time when the challenge is issued, but also to a moment in time some j steps earlier. The assumption implies that the number of pebbles needed at time $s - j$ is at least $c/(j + t)$ (because the challenge was answered in $j + t$ steps starting from time $s - j$, which would be impossible with a lower number of pebbles even if the challenge had been already known at time $s - j$).

We will use this bound for every challenge number $i \geq 2$, and for every $j = 0$ to t_{i-1}, i.e., during the entire time the previous challenge is being answered. Thus, cumulative pebbling complexity during the time period of answering challenge $i - 1$ is at least

$$\mathsf{cc}(s_{i-1} + 1, s_i) \geq \sum_{j=0}^{t_{i-1}-1} \mathsf{p}_{s_i-j} \geq c\left(\frac{1}{t_i} + \frac{1}{t_i + 1} + \cdots + \frac{1}{t_i + t_{i-1} - 1}\right)$$

$$\geq c\int_{t_i}^{t_{i-1}+t_i} \frac{dx}{x} = c(\ln(t_{i-1} + t_i) - \ln t_i).$$

Then adding these up for each i between 2 and Q, we get the cumulative pebbling complexity of

$$\mathsf{cc}(1, s_{Q+1}) \geq c\sum_{i=2}^{Q}(\ln(t_{i-1} + t_i) - \ln t_i).$$

If all t_i are equal (which is close to the minimum, as we will show below), this becomes $c(Q - 1) \cdot \ln 2$.

Back to the Actual Assumption. The proof is made messier by the fact that the bound in the assumption is not absolute. Moreover, the bound does not give the number of pebbles in terms of running time, but rather running time in terms of the number of pebbles (it makes no sense to talk probabilistically of the number of pebbles, because the number of pebbles is determined by the adversary before the challenge is chosen). To overcome this problem, we look at the number of pebbles at all times before s_i and see which one gives us the best lower bound on t_i.

Consider a point $t \leq s_i$ in time. We can apply Fact 3 to the size p_t of the set of pebbles P_t at time t, because the ith challenge is selected at random after the adversary determines P_t. The ith challenge will be answered $t_i + (s_i - t)$ steps after time t; thus, with probability at least $\mathrm{pr}_{\mathrm{hard}}$ over the choice of the ith challenge, $t_i + (s_i - t) > c/\mathsf{p}_t$, i.e., $t_i > c/\mathsf{p}_t - (s_i - t)$. Let r_i be a moment in time that gives the best bound on t_i:

$$r_i = \operatorname*{argmax}_{0 \leq t \leq s_i} \left(\frac{c}{\mathsf{p}_t} - (s_i - t) \right).$$

Call the ith challenge "hard" if $t_i + (s_i - r_i) > c/\mathsf{p}_{r_i}$. We claim that if challenge i is hard, then the same fact about the number of pebbles j steps before the challenge as we used in the warm-up proof holds.

Claim 5. *If challenge i is hard, then for any j, $0 \leq j \leq s_i$, $\mathsf{p}_{s_i - j} > c/(t_i + j)$.*

Proof. Indeed, let $t = s_i - j$. Then $c/\mathsf{p}_{s_i - j} - j = c/\mathsf{p}_t - (s_i - t) \leq c/\mathsf{p}_{r_i} - (s_i - r_i)$ by the choice of r_i. This value is less than t_i by definition of a hard challenge. Therefore, $c/\mathsf{p}_{s_i - j} - j < t_i$ and the result is obtained by rearranging the terms. $\square$

We now claim that with high probability, the number of hard challenges is sufficiently high.

Claim 6. *For any $\epsilon > 0$, with probability at least $1 - e^{-2\epsilon^2 Q}$, the number of hard challenges is at least $H \geq Q(\mathrm{pr}_{\mathrm{hard}} - \epsilon)$.*

Proof. The intuitive idea is to apply Hoeffding's inequality [17], because challenges are independent. However, the hardness of challenges is not independent, because it may be (for example) that one particular challenge causes the adversary to slow down for every subsequent challenge. Fortunately, it can only be "worse than independent" for the adversary. Specifically, for any fixing of the first $i - 1$ challenges $c_1, \ldots, c_{i-1}$, we can run the adversary up to time s_i; at this point, time r_i is well defined, and we can apply Fact 3 to r_i to obtain that $\Pr[c_i \text{ is hard} \mid c_1, \ldots, c_{i-1}] \geq \mathrm{pr}_{\mathrm{hard}}$. This fact allows us to apply the slightly generalized version of Hoeffding's inequality stated in Claim 7 (setting $V_i = 1$ if c_i is hard and $V_i = 0$ otherwise) to get the desired result. $\square$

Claim 7 (Generalized Hoeffding's inequality). *If $V_1, V_2, \ldots, V_Q$ are binary random variables such that for any i $(0 \leq i < Q)$ and any values of $v_1, v_2, \ldots, v_i$, $\Pr[V_{i+1} = 1 \mid V_1 = v_1, \ldots, V_i = v_i] \geq \rho$, then for any $\epsilon > 0$, with probability at least $1 - e^{-2\epsilon^2 Q}$, $\sum_{i=1}^{Q} V_i \geq Q(\rho - \epsilon)$.*

Proof. For $0 \leq i < Q$, define the binary random variable F_{i+1} as follows: for any fixing of $v_1, \ldots, v_i$ such that $\Pr[V_1 = v_1, \ldots, V_i = v_i] > 0$, let $F_{i+1} = 1$ with probability $\rho / \Pr[V_{i+1} = 1 \mid V_1 = v_1, \ldots, V_i = v_i]$ and 0 otherwise, independently of $V_{i+1}, \ldots, V_Q$. Let $W_{i+1} = V_{i+1} \cdot F_{i+1}$. Note that $\Pr[W_{i+1} = 1] = \rho$

regardless of the values of $V_1, \ldots, V_i$, and thus W_{i+1} is independent of $V_1, \ldots, V_i$. Since $F_1, \ldots, F_i$ are correlated only with $V_1, \ldots, V_i$, we have that W_{i+1} is independent of $(V_1, \ldots, V_i, F_1, \ldots, F_i)$, and thus independent of $W_1, \ldots, W_i$. Therefore, $W_1, \ldots, W_Q$ are mutually independent (this standard fact can be shown by induction on the number of variables), and thus $\sum_{i=1}^{Q} V_i \geq \sum_{i=1}^{Q} W_i \geq Q(\rho - \epsilon)$ with probability at least $1 - e^{-2\epsilon^2 Q}$ by Hoeffding's inequality. $\qquad\square$

Now assume H challenges are hard. What remains to show is a purely algebraic statement about the sum of p_i values when $H \geq Q(\mathrm{pr}_{\mathrm{hard}} - \epsilon)$ of challenges satisfy Claim 5.

Claim 8. *Let c be a real value. Let $t_1, \ldots, t_Q$ be integers, $s_1 = 0$, and $s_i = s_{i-1} + t_{i-1}$ for $i = 2, \ldots, Q+1$. Let $p_0, \ldots, p_Q$ be a sequence of real values with $p_t > p_{\min}$ for every $t \geq 1$. Suppose further that there exist at least H distinct indices i, with $1 \leq i \leq Q$ (called "hard indices") such that for any $0 \leq j \leq s_i$, $p_{s_i - j} \geq c/(t_i + j)$. Then*

$$
\sum_{i=1}^{s_{Q+1}} p_i \geq c \cdot H \cdot \ln\left(1 + \left(\frac{p_{\min}}{p_0}\right)^{\frac{1}{H}}\right).
$$

Proof. Let $i_1 < i_2 < \cdots < i_H$ be the hard indices. Recall the notation $cc(i,j) = \sum_{t=i}^{j} p_t$. Then for $k \geq 2$,

$$
cc(s_{i_{k-1}} + 1, s_{i_k}) \geq cc(s_{i_k} - t_{i_{k-1}} + 1, s_{i_k}) = \sum_{j=0}^{t_{i_{k-1}} - 1} p_{s_{i_k} - j}
$$

$$
\geq \sum_{j=0}^{t_{i_{k-1}} - 1} \frac{c}{t_{i_k} + j} \geq c \cdot (\ln(t_{i_{k-1}} + t_{i_k}) - \ln t_{i_k}),
$$

(the last inequality follows by the same reasoning as in the warm-up). To bound $cc(1, Q+1)$, we will add up the pebbling complexity during these nonoverlapping time periods for each k and find the minimum over all sets of values of t_{i_k}. Unfortunately, the result will decrease as t_{i_1} decreases and as t_{i_H} increases, and we have no bounds on these values. To get a better result, we will need to consider special cases of $k = 2$ (to replace t_{i_1} with c/p_0) and $k = H+1$ (to add another term with $t_{i_H+1} = c/p_{\min}$).

For $k = 2$, we will bound $cc(1, s_{i_2})$ by noticing that $s_{i_2} \geq t_{i_1} + s_{i_1} \geq c/p_0$ (where the second step follows by Claim 5 with $j = s_{i_1}$), and therefore

$$
cc(1, s_{i_2}) \geq \sum_{j=0}^{s_{i_2} - 1} p_{s_{i_2} - j} \geq c \sum_{j=0}^{s_{i_2} - 1} \frac{1}{t_{i_2} + j} \geq c \int_{t_{i_2}}^{s_{i_2} + t_{i_2}} \frac{dx}{x}
$$

$$
= c(\ln(s_{i_2} + t_{i_2}) - \ln t_{i_2}) \geq c \cdot (\ln(c/p_0 + t_{i_2}) - \ln t_{i_2}).
$$

For $k = H + 1$,

$$cc(s_{i_H} + 1, s_{H+1}) \geq \mathsf{p}_{\min} \cdot t_{i_H} \geq c \cdot \left(\frac{1}{c/\mathsf{p}_{\min}} t_{i_H} \right)$$

$$\geq c \cdot \left(\frac{1}{c/\mathsf{p}_{\min}} + \cdots + \frac{1}{c/\mathsf{p}_{\min} + t_{i_H} - 1} \right)$$

$$\geq c \cdot (\ln(t_{i_h} + c/\mathsf{p}_{\min}) - \ln c/\mathsf{p}_{\min}).$$

Adding these up, we get

$$cc(0, s_{i+1}) = \mathsf{p}_0 + cc(1, s_{i_2}) + cc(s_{i_2} + 1, s_{i_3}) + \cdots$$

$$+ cc(s_{i_{H-1}} + 1, s_{i_H}) + cc(s_{i_H} + 1, s_{i_{H+1}})$$

$$\geq \mathsf{p}_0 + c \cdot \sum_{i=1}^{H} (\ln(x_i + x_{i+1}) - \ln x_{i+1}),$$

where $x_1 = c/\mathsf{p}_0$, $x_2 = t_{i_2}$, $x_3 = t_{i_3}$, ..., $x_H = t_{i_H}$, and $x_{H+1} = c/\mathsf{p}_{\min}$.

To find the minimum of this function, observe that the first derivative with respect to x_i is $\frac{1}{x_i + x_{i-1}} + \frac{1}{x_i + x_{i+1}} - \frac{1}{x_i}$, which, assuming all the x_is are positive, is zero at $x_i = \sqrt{x_{i-1} x_{i+1}}$, is negative for $x_i < \sqrt{x_{i-1} x_{i+1}}$, and is positive for $x_i > \sqrt{x_{i-1} x_{i+1}}$. Therefore, the minimum of this function occurs when each x_i, for $2 \leq i \leq H$, is equal to $\sqrt{x_{i-1} x_{i+1}}$, or equivalently, when $x_i = c/(\mathsf{p}_{\min}^{i-1} \mathsf{p}_0^{H-i+1})^{1/H}$. This setting of x_i gives us $\ln(x_i + x_{i+1}) - \ln x_{i+1} = \ln(1 + (\mathsf{p}_{\min}/\mathsf{p}_0)^{1/H})$, which gives the desired result. □

Plugging in $Q \cdot (\mathsf{pr}_{\mathsf{hard}} - \epsilon)$ for H and $\mathsf{pr}_{\mathsf{hard}} = \frac{1}{2}$ concludes the proof of Theorem 4. □

5.1 Why This Proof Needs to be Modified for the Parallel Random Oracle Model

The main idea of the proof above is to apply the space-time tradeoff of Fact 3 to every point in time before the challenge is known, arguing that delaying the receipt of the challenge can only hurt the adversary. While this is true in the pebbling game (via an easy formal reduction—because getting bits does not help get pebbles), it's not clear why this should be true in the parallel random oracle game, where space is measured in bits rather than pebbles. A reduction from the adversary A who does not know the challenge to an adversary B who does would require B to store the challenge in memory, run A until the right moment in time, and then give A the challenge. This reduction consumes memory of B—for storing the challenge and keeping track of time. In other words, A can save on memory by not knowing, and therefore not storing, the challenge. While the amount of memory is small, it has to be accounted for, which makes the formulas even messier. Things get even messier if A is not required to be 100% correct, because, depending on the exact definition of the round game, B may not know when to issue the challenge to A.

Nevertheless, this difficulty can be overcome when challenges come from the random oracle, as they do in scrypt. We do so in the next section.

6   Main Result: Memory Hardness of scrypt in the Parallel Random Oracle Model

We are ready to restate our main theorem, which we now prove extending the techniques from the two previous sections. We state it with a bit more detail than Theorem 1.

Fix positive integers $n \geq 2$ and w, a string $X \in \{0,1\}^w$, a finite domain $\mathcal{D}$ that contains at least $\{0,1\}^w$, and let $\mathcal{R} = \{0,1\}^w$.

Theorem 5. *Let A be any oracle machine (in the parallel random oracle model as defined in Sect. 2) with input X. Assume $A^h(X)$ outputs $S_n^h = \mathsf{scrypt}^h(X)$ correctly with probability χ, where the probability is taken over the choice of $h : \mathcal{D} \to \mathcal{R}$. Then for any $\epsilon > 0$ and $q \geq 2$, with probability (over the choice h) at least $\chi - 2qn^4 \cdot 2^{-w} - e^{-2\epsilon^2 n}$ one of the following two statements holds: either $A^h(X)$ makes more than q queries (and thus $\mathsf{cc_{mem}}(A^{hn}) > qw$ by definition) or*

$$\mathsf{cc_{mem}}(A^h(X)) \geq \frac{\ln 2}{6} \cdot \left(\frac{1}{2} - \epsilon\right) \cdot n^2 \cdot (w - 2\log n - \log q - 1).$$

To get the statement of Theorem 1, we set $\epsilon = 1/7$ and observe that then $e^{-2\epsilon^2 n} = 2^{-\frac{2n}{49\ln 2}} < 2^{-n/20}$ and $\frac{\ln 2}{6}(\frac{1}{2} - \frac{1}{7}) > \frac{1}{25}$. We also plug in $q = \min(2, \frac{n^2}{25})$ (and therefore $\log q \leq 2\log n - 1$), thus removing the "either/or" clause (this setting of q requires us to manually check that the probability statement is correct when $q > \frac{n^2}{25}$, i.e., $2 \leq n \leq 7$—a tedious process that we omit here).

The multiplicative constant of $\ln(2)(1/2 - \epsilon)/6$, which becomes $1/25$ in Theorem 1 (and can be as small as $\approx 1/18$ if we use a smaller ϵ), is about 9–12 times worse than the constant in the naïve scrypt algorithm described on p. 4, which has $\mathsf{cc_{mem}}$ of $n^2 w/2$. We describe approaches that may improve this gap to a factor of only about $1/\ln 2 \approx 1.44$ in the full version [7]. We note that there is also gap between w in the naïve algorithm and $(w - 2\log n - \log q - 1)$ in the lowerbound, which matters for small w (as values of 20–30 for $\log n$ and $\log q$ are reasonable).

The rest of this section is devoted to the proof of this theorem.

6.1   Outline of the Approach

Before proceeding with the proof, we justify our proof strategy by highlighting the challenges of extending Theorems 2 and 4 to this setting. Theorem 2 applies to a fixed random oracle h and a random challenge. In fact, the proof relies crucially on the ability to try every challenge for a given oracle. However, in the present proof, once the random oracle is fixed, so is every challenge. Moreover, Theorem 4 crucially relies on the uniformity and independence of each challenge, which is issued only when the previous challenge is answered. In contrast, here, again, once the oracle is fixed, the challenges are fixed, as well. Even if we think of the oracle as being lazily created in response to queries, the challenges implicitly contained in the answers to these queries are not necessarily independent once we condition (as we need to in Theorem 2) on the oracle not being in $\mathsf{bad_h}$. We resolve these issues by working with multiple carefully chosen random oracles.

Recall our notation: $X_0 = X$, $X_1 = \mathsf{h}(X_0), \ldots, X_{n-1} = \mathsf{h}(X_{n-2})$; $T_0 = X_{n-1}$, $S_0 = \mathsf{h}(T_0)$, and for $i = 1, \ldots, n$, $T_i = S_{i-1} \oplus X_{S_{i-1} \bmod n}$ and $S_i = \mathsf{h}(T_i)$. Because we will need to speak of different random oracles, we will use notation X_i^{h}, T_i^{h}, and S_i^{h} when the label values are being computed with respect to the random oracle h (to avoid clutter, we will omit the superscript when the specific instance of the random oracle is clear from the context). We will denote by A^{h} the adversary running with oracle h. (To simplify notation, we will omit the fixed argument X to the adversary A for the remainder of this section.)

Let $\mathrm{changeModn}(S, i)$ be the function that keeps the quotient $\lfloor S/n \rfloor$ but changes the remainder of S modulo n to i. Consider the following process of choosing a random oracle (this process is described more precisely in the following section). Choose uniformly at random an oracle h_0. Choose uniformly at random challenges $c_1, \ldots, c_n$, each between 0 and $n-1$. Let h_1 be equal to h_0 at every point, except $\mathsf{h}_1(T_0^{\mathsf{h}_0}) = \mathrm{changeModn}(S_0^{\mathsf{h}_0}, c_1)$. Similarly, let h_2 be equal to h_1 at every point, except $\mathsf{h}_2(T_1^{\mathsf{h}_1}) = \mathrm{changeModn}(S_1^{\mathsf{h}_1}, c_2)$, and so on, until h_n, which is our final random oracle.

This method of choosing h_n is close to uniform, and yet explicitly embeds a uniform random challenge. Unless some (rare) bad choices have been made, each challenge has about a $\frac{1}{2}$ probability of taking a long time to answer, by the same reasoning as in Theorem 2. And since the challenges are independent (explicitly through the choices of c_i values), we can use the same reasoning as in Theorem 4 to bound the cumulative complexity.

The main technical difficulty that remains is to define exactly what those bad choices are and bound their probability without affecting the independence of the challenges. In particular, there are n^n possible challenge combinations, and the probability that all of them yield random oracles that are acceptable (cause no collisions and cannot be predicted) is not high enough. We have to proceed more carefully.

The first insight is that if h_{k-1} is not predictable (i.e., a predictor $\mathcal{P}$ with a short input cannot correctly extract many oracle values), and no oracle queries up to T_{k-1} collide, then $\Pr_{c_k}[\text{time between queries } T_{k-1}^{\mathsf{h}_k} \text{ and } T_k^{\mathsf{h}_k} \text{ is high}] \geq \frac{1}{2}$, by the same reasoning as in Theorem 2 (except predictor needs an extra $\log q$ bits of hint to know when query T_{k-1} occurs, so as to substitute the answer to T_{k-1} with $\mathrm{changeModn}(S_{k-1}, c_k)$ for every possible c_k). This allows us to worry about only n random oracles avoiding collisions and the set $\mathsf{predictable}$ (instead of worrying about n^n random oracles) to ensure that the time between consecutive challenges is likely to be high.

However, the reasoning in the previous paragraph bounds the time required to answer the challenge c_k only with respect to oracle h_k. In order to reason about A interacting with oracle h_n, we observe that if for every k, A_k^{h} asks the queries $X_0, \ldots, X_{n-1} = T_0, T_1, \ldots, T_n$ in the correct order, then the computation of A^{h_n} is the same as the computation of A^{h_k} until the kth challenge is answered—i.e., until T_k is queried. Thus, results about each of the oracles h_k apply to h_n.

The rest of the work involves a careful probability analysis to argue that the challenges $c_1, \ldots, c_n$ are almost independent even when conditioned on all the bad events not happenning, and to bound the probability of these events.

6.2   The Detailed Proof

Recall that we assume that the adversary A is deterministic without loss of generality (this fact will be used heavily throughout the proof). In particular, the randomness of the experiment consists solely of the random oracle A is given access to.

Following up on the above high-level overview, we now make precise the definition of h_k. Let h_0 be a uniformly chosen random oracle. Let changeModn(S, i) be a function that keeps the quotient $\lfloor S/n \rfloor$ but changes the remainder of S modulo n to i if possible: it views S as an integer in $[0, 2^w - 1]$, computes $S' = \lfloor S/n \rfloor \cdot n + i$, and outputs S' (viewed as a w-bit string) if $S' < 2^w$, and S otherwise (which can happen only if n is not a power of 2, and even then is very unlikely for a random S).

Definition 1. *Let* roundingProblem$_k$ *be the set of all random oracles* h *for which the value of at least one of* $S_0^{\mathsf{h}}, \ldots, S_k^{\mathsf{h}}$ *is greater than* $\lfloor 2^w/n \rfloor \cdot n - 1$ *(i.e., those for which* changeModn *does not work on some* S *value up to* S_k*).*

Definition 2. *Let* colliding$_k^*$ *be the set of all* h *which there is at least one collision among the values* $\{X_0, X_1^{\mathsf{h}}, X_2^{\mathsf{h}}, \ldots, X_{n-2}^{\mathsf{h}}, T_0^{\mathsf{h}}, T_1^{\mathsf{h}}, \ldots, T_k^{\mathsf{h}}\}$. *Let* colliding$_k$ = roundingProblem$_{\min(k,n-1)}$ $\cup$ colliding$_k^*$.

Definition 3. *For every* k $(0 \leq k < n)$*, let* $h_{k+1} = h_k$ *if* $h_k \in$ colliding$_k$*; else, choose* c_{k+1} *uniformly at random between* 0 *and* $n-1$*, let* $h_{k+1}(T_k^{\mathsf{h}_k}) =$ changeModn$(S_k^{\mathsf{h}_k}, c_{k+1})$*, and let* $h_{k+1}(x) = h_k(x)$ *for every* $x \neq T_k^{\mathsf{h}_k}$*. (Recall that* h_0 *is chosen uniformly.)*

Note that this particular way of choosing h_{k+1} is designed to ensure that it is uniform, as we argue in the full version.

THE SINGLE CHALLENGE ARGUMENT. In the argument in Theorem 2, the predictor issues different challenges to A. Here, the predictor will run A with different oracles. Specifically, given $1 \leq k \leq n$ and a particular oracle $h_{k-1} \notin$ colliding$_{k-1}$, consider the n oracles $h_{k,j}$ for each $0 \leq j < n$, defined to be the same as h_{k-1}, except $h_{k,j}(T_{k-1}^{\mathsf{h}_{k-1}}) =$ changeModn$(S_{k-1}^{\mathsf{h}_{k-1}}, j)$ (instead of $S_{k-1}^{\mathsf{h}_{k-1}}$).

Since $h_{k-1} \notin$ colliding$_{k-1}$, $T_{k-1}^{\mathsf{h}_{k-1}}$ is not equal to $X_i^{\mathsf{h}_{k-1}}$ for any $0 \leq i < n-1$ and $T_i^{\mathsf{h}_{k-1}}$ for any $0 \leq i < k-1$. Therefore (since h_{k-1} and $h_{k,j}$ differ only at the point $T_{k-1}^{\mathsf{h}_{k-1}}$), we have $X_i^{\mathsf{h}_{k-1}} = X_i^{\mathsf{h}_{k,j}}$ for every $0 \leq i \leq n-1$ and $T_i^{\mathsf{h}_{k-1}} = T_i^{\mathsf{h}_{k,j}}$ for any $0 \leq i \leq k-1$. In particular, the execution of A with oracle $h_{k,j}$ will proceed identically for any j (and identically to the execution of $A^{\mathsf{h}_{k-1}}$) up to the point when the query T_{k-1} is first made (if ever). We will therefore omit the superscript on T_{k-1} for the remainder of this argument.

The observation is that the moment T_{k-1} is queried is the moment when the predictor argument of Theorem 2 can work, by having the predictor substitute different answers to this query and run A on these different answers in parallel. However, since Sect. 5 requires a time/memory tradeoff for every point in time before the challenge is given, we will prove a more general result for any point in time before T_{k-1} is queried.

We number all the oracle queries that A makes across all rounds, sequentially. We will only care about the first q oracle queries that A makes, for some q to be set later (because if q is too large, then cc_{mem} of A is automatically high). Note that q here is analogous to $q - 1$ in Theorem 2.

Let $s_k > 0$ be the round in which T_{k-1} is first queried, i.e., contained in q_{s_k}. For an integer $r \le s_k$, consider the output state $\bar{\sigma}_r$ of $A^{h_{k-1}}$ from round r. Given $\bar{\sigma}_r$, consider n different continuations of that execution, one for each oracle $h_{k,j}, 0 \le j < n$. For each of these continuations, we let $t_j > 0$ be the smallest value such that such $r + t_j > s_k$ and the query $T_k^{h_{k,j}}$ is contained in q_{r+t_j} (if ever before query number $q + 1$; else, set $t_j = \infty$). We can thus define π_{ij}, β_i, and $bestchal_i$, blue nodes, and the set B the same way as in Theorem 2, by counting the number of rounds after round r (instead of from 0) and substituting, as appropriate "challenge j" with $h_{k,j}$ and "query X_j" with "query X_j or $T_k^{h_{k,j}}$" (note that because $h_{k-1} \notin roundingProblem_{k-1}$, $S_{k-1}^{h_{k,j}} \mod n = j$, and so $T_k^{h_{k,j}} = X_j \oplus S_{k-1}^{h_{k,j}}$). (We stop the execution of A after q total queries in these definitions.)

We now show that, similarly to Claim 1, we can design a predictor algorithm $\mathcal{P}$ that predicts every $X_i^{h_{k-1}}$ in B by interacting with h_{k-1} but not querying it at the predecessors of points in B. The difference is that instead of running $A^{h_{k-1}}$ on σ_0 and giving A different challenges j, $\mathcal{P}$ will run A with initial input state σ_r, simulating different oracles $h_{k,j}$ (which differ from h_{k-1} on only one point—namely, the output on input T_{k-1}). $\mathcal{P}$ gets, as input, σ_r and the same hint as in Claim 1. $\mathcal{P}$ also needs an additional hint: an integer between 1 and q indicating the sequential number (across all queries made in round r or later) of the first time query T_{k-1} occurs, in order to know when to reply with $S_{k-1}^{h_{k,j}} = changeModn(S_{k-1}^{h_{k-1}}, j)$ instead of $S_{k-1}^{h_{k-1}}$ itself. Note that this substitution will require $\mathcal{P}$ to modify the input state σ_{s_k}. If $s_k > r$, then $\mathcal{P}$ will not only be able to answer with $S_{k-1}^{h_{k,j}}$, but will also see the query T_{k-1} itself as part of the output state $\bar{\sigma}_{s_k}$, and will therefore be able to answer subsequent queries to T_{k-1} consistently. However, if $s_k = r$, then we need to give T_{k-1} to $\mathcal{P}$ to ensure subsequent queries to T_{k-1} are answered consistently. In order to do so without lengthening the input of $\mathcal{P}$, we note that in such a case we do not need $S_{k-1}^{h_{k-1}}$ in σ_r (since $\mathcal{P}$ can obtain it by querying h_{k-1}), and so we can take out $S_{k-1}^{h_{k-1}}$ and replace it with T_{k-1} ($\mathcal{P}$ will recognize that this happened by looking at the additional hint that contains the query number for T_{k-1} and noticing that it is smaller than the number of queries z_r in round r).

There is one more small modification: if $X_j \in B$ and $bestchal_j = j$, then in order to correctly predict X_j itself (assuming $X_j \in B$), $\mathcal{P}$ will need one additional bit of hint, indicating whether X_j is first queried by itself or as part of the "next round," i.e., as part of the query $T_k^{h_{k},j} = S_{k-1}^{h_{k},j} \oplus X_j$ (in which case $\mathcal{P}$ will need to xor the query with $S_{k-1}^{h_{k},j}$, which $\mathcal{P}$ knows, having produced it when answering the query T_{k-1}). Finally, note that $\log q$ bits suffice for the query number of X_i on challenge bestchal_i, because it is not the same query number as T_{k-1}, because $h_{k-1} \notin \text{colliding}_{k-1}$, so there are $q-1$ possibilities plus the possibility of "never".

We thus need to give (in addition to σ_r) $\log q + |B|(1 + 2\log n + \log q)$ bits of hint to $\mathcal{P}$, and $\mathcal{P}$ is guaranteed to be correct as long as $h_{k-1} \notin \text{colliding}_{k-1}$.

Suppose σ_r has m_r bits. Claim 2 does not change. We modify Claim 3 as follows. We replace p with a function p_r of the memory size m_r, defined as

$$p_r = \lceil (m_r + 1 + \log q)/(w - 2\log n - \log q - 1) + 1 \rceil \qquad (1)$$

(note that it is almost the same as the definition of p, but accounts for the longer hint). We now redefine predictable according to our new definition of $\mathcal{P}$, p_r, and hint length.

Definition 4. *The set* predictable *consists of all random oracles* h *for which there exists an input state* σ_r *of size* m_r *(such that* $1 \leq p_r \leq n-1$*) and a hint of length* $\log q + p_r(1 + 2\log n + \log q)$*, given which* $\mathcal{P}$ *can correctly output* p_r *distinct values from among* $X_1^h, \ldots, X_{n-1}^h$ *without querying them.*

Finally, we replace bad_h with $\text{colliding}_{k-1} \cup \text{predictable}$. As long as $h_{k-1} \notin \text{colliding}_{k-1} \cup \text{predictable}$, we are guaranteed that $\Pr_j[t_j > n/(2p_r)] \geq 1/2$, like in Theorem 2.

The discussion above gives us the following lemma (analogous to Theorem 2).

Lemma 1. *Fix any* k *(*$1 \leq k \leq n$*). Assume* $h_{k-1} \notin \text{colliding}_{k-1} \cup \text{predictable}$. *Let* $s_k > 0$ *be the smallest value such that* $T_{k-1}^{h_{k-1}}$ *is among the queries* $\mathbf{q}_{s_k}$ *during the computation of* $A^{h_{k-1}}$. *Let* $r \leq s_k$ *and* m_r *be the bit-length of the input state* σ_r *of* $A^{h_{k-1}}$ *in round* $r + 1$. *Let* $t_{k,j,r} > 0$ *be such that the first time* $T_k^{h_{k},j}$ *is queried by* $A^{h_{k},j}$ *after round* s_k *is in round* $r + t_{k,j,r}$ *(let* $t_{k,j,r} = \infty$ *if such a query does not occur after round* σ_k *or does not occur among the first* q *queries, or if* $T_{k-1}^{h_{k-1}}$ *is never queried). Call* j *"hard" for time* r *if* $t_{k,j,r} > n/(2p_r)$*, where* $p_r = \lceil (m_r + 1 + \log q)/(w - 2\log n - \log q - 1) + 1 \rceil$. *We are guaranteed that*

$$\Pr_j[j \text{ is hard for time } r] \geq \frac{1}{2}.$$

HARDNESS OF CHALLENGE c_k. We continue with the assumptions of Lemma 1. In order to get an analogue of Claim 5, we need to define what it means for a challenge to be hard. Consider running A^{h_k}. Let $t_k > 0$ be such that $T_k^{h_k}$ is queried for the first time in round $s_k + t_k$ (again, letting $t_k = \infty$ if this query does not occur among the first q queries). Find the round $r_k \leq s_k$ such that bit-length m_r of the input state σ_r in round $r_k + 1$ gives us the best bound on t_k using the equation of Lemma 1 (i.e., set $r_k = \arg\max_{0 \leq r \leq s_k}(n/(2p_r) - (s_k - r))$, where m_r denotes the size of the state σ_r at the end of round r, and p_r is the function of m_r defined by Eq. 1), and define c_k to be "hard" if it is hard for time r_k.

Definition 5. *A challenge c_k is hard if for $r_k = \text{argmax}_{0 \leq r \leq s_k}(n/(2p_r)) - (s_k - r))$ we have $t_{k,c_k,r_k} > n/(2p_r)$, where $s_k, t_{k,j,r}$ and p_r are as defined in Lemma 1.*

THE MULTIPLE CHALLENGES ARGUMENT. So far, we considered hardness of c_k during the run of A with the oracle h_k. We now need to address the actual situation, in which A runs with h_n. We need the following claim, which shows that the actual situation is, most of the time, identical. Define wrongOrder$_k$ as the set of all random oracles h for which the values $\{T_0^h, T_1^h, \ldots, T_k^h\}$ are not queried by A^h in the same order as they appear in the correct evaluation of scrypt (when we look at first-time queries only, and only up to the first q queries).

Definition 6. wrongOrder$_k$ *consists of all h for which there exist i_1 and i_2 such that $0 \leq i_1 < i_2 \leq k$ and, in the run of A^h, query $T_{i_2}^h$ occurs, while query $T_{i_1}^h$ does not occur before query $T_{i_2}^h$ occurs.*

Claim 9. *If for every j $(0 \leq j \leq n)$, $h_j \notin \text{colliding}_j \cup \text{wrongOrder}_j$, then for every k and $i \leq k$, $T_i^{h_n} = T_i^{h_k}$, and the execution of A^{h_n} is identical to the execution of A^{h_k} until the query T_k is first made, which (for $1 \leq k \leq n$) happens later than the moment when query $T_{k-1}^{h_n} = T_{k-1}^{h_k}$ is first made.*

Proof. To prove this claim, we will show, by induction, that for every $j \geq k$ and $i \leq k$, $T_i^{h_k} = T_i^{h_j}$, and the execution of A^{h_j} is identical to the execution of A^{h_k} until the query T_k is first made.

The base of induction $(j = k)$ is simply a tautology.

The inductive step is as follows. Suppose the statement is true for some $j \geq k$. We will show it for $j + 1$. We already established that if $h_j \notin \text{colliding}_j$, then $T_i^{h_j} = T_i^{h_{j+1}}$ for every $i \leq j$, and is therefore equal to $T_i^{h_k}$ by the inductive hypothesis. Since h_j and h_{j+1} differ only in their answer to the query $T_j^{h_j} = T_j^{h_{j+1}}$, the execution of $A^{h_{j+1}}$ proceeds identically to the execution of A^{h_j} until this query is first made. Since $h_j \notin \text{wrongOrder}_j$, this moment is no earlier than when the query T_k is made; therefore, until the point the query T_k is first made, the execution of $A^{h_{j+1}}$ proceeds identically to the execution of A^{h_j} and thus (by the inductive hypothesis) identically to the execution of A^{h_k}.

The last part of the claim follows because $h_n \notin \text{wrongOrder}_n$. $\square$

We therefore get the following analogue of Claim 5.

Claim 10. *Given adversary A, assume for every k $(0 \leq k \leq n)$, $h_k \notin \text{colliding}_k \cup \text{wrongOrder}_k$. Let $c = n/2$. If challenge i is hard (i.e., $t_i + (s_i - r_i) > c/p_{r_i}$), then, during the run of A^{h_n}, for any $0 \leq j \leq s_i$, $p_{s_i - j} \geq c/(t_i + j)$.*

Definition 7. *Let E_1 be the event that there are at least $H \geq n(\frac{1}{2} - \epsilon)$ hard challenges (as defined in Definition 5). Let E_2 be the event that $h_k \notin \text{colliding}_k \cup \text{wrongOrder}_k$ (see Definitions 2 and 6) for every k, and A^{h_n} queries $T_n^{h_n}$. Let E_q be the event that A^{h_n} makes no more than q total queries.*

Claim 11. *If $E_1 \cap E_2 \cap E_q$, then*

$$\sum_{r=1}^{s_{n+1}} p_r \geq \ln 2 \cdot \left(\frac{1}{2} - \epsilon\right) \cdot \frac{1}{2} \cdot n^2.$$

Proof. Since E_2 holds, every query $T_0, \ldots, T_n$ gets made, in the correct order. Since E_q holds, all these queries happen no later than query q, thus ensuring that Claim 10 applies and each t_k is finite. Moreover, by definition of p_r in Eq. 1, $p_r \geq 1$ and $p_0 = 1$. Therefore, we can apply Claim 8 to the execution of A^{h_n} to get the desired result. □

CONVERTING FROM $\sum p_r$ TO $\mathsf{cc}_{\mathsf{mem}}$ Now we need to convert from $\sum p_r$ to $\sum m_r$.

Claim 12. *For every $r > 0$,*

$$m_r \geq p_r \cdot (w - 2\log n - \log q - 1)/3.$$

Proof. By definition of p_r, we have that

$$p_r = \left\lceil \frac{m_r + 1 + \log q}{w - 2\log n - \log q - 1} + 1 \right\rceil \leq \frac{m_r + 1 + \log q}{w - 2\log n - \log q - 1} + 2,$$

because the ceiling adds at most 1. Therefore,

$$(p_r - 2) \cdot (w - 2\log n - \log q - 1) \leq m_r + 1 + \log q,$$

(because we can assume $(w - 2\log n - \log q - 1) > 0$ — otherwise, Theorem 5 is trivially true) and thus

$$m_r \geq (p_r - 2) \cdot (w - 2\log n - \log q - 1) - \log q - 1 \tag{2}$$
$$= p_r \cdot (w - 2\log n - \log q - 1) - 2 \cdot (w - 2\log n - \log q - 1) - \log q - 1$$
$$= p_r \cdot (w - 2\log n - \log q - 1) - 2 \cdot (w - 2\log n - 0.5\log q - 0.5). \tag{3}$$

Since $m_r \geq w$ (see our complexity measure definition in Sect. 2), $m_r \geq w - 2\log n - 0.5\log q - 0.5$ and therefore we can increase the left-hand side by $2 \cdot m_r$ and the right-hand side by $2 \cdot (w - 2\log n - 0.5\log q - 0.5)$ and the inequality still holds; and therefore $3m_r \geq p_r \cdot (w - 2\log n - \log q - 1)$. □

Lemma 2. *Assuming $E_1 \cap E_2$ (see Definition 7), for any integer q, either A^{h_n} makes more than q queries (and thus $\mathsf{cc}_{\mathsf{mem}}(A^{h_n}) > qw$ by definition) or*

$$\mathsf{cc}_{\mathsf{mem}}(A^{h_n}(X)) \geq \frac{\ln 2}{6} \cdot \left(\frac{1}{2} - \epsilon\right) \cdot n^2 \cdot (w - 2\log n - \log q - 1) .$$

Proof. We observe that if A^{h_n} makes no more than q queries, then $E_1 \cap E_2 \cap E_q$ hold, and we can combine Claims 11 and 12 to get

$$\mathsf{cc}_{\mathsf{mem}}(A^{h_n}(X)) = \sum_{r=1}^{s_{n+1}} m_r \geq \frac{1}{3} \cdot \sum_{r=1}^{s_{n+1}} p_r \cdot (w - 2\log n - \log q - 1)$$

$$\geq \frac{\ln 2}{3} \cdot \left(\frac{1}{2} - \epsilon\right) \cdot \frac{1}{2} \cdot n^2 \cdot (w - 2\log n - \log q - 1).$$

This concludes the proof of Lemma 2. □

All that remains is to show a lower bound for the probability of $(E_1 \cap E_2 \cap E_q) \cup \bar{E}_q$, and to argue that h_n is uniform, because the statement we are trying to prove is concerned with A^h for uniform h rather than with A^{h_n}. The details are deferred to the full version [7].

Acknowledgments. We thank Chethan Kamath and Jeremiah Blocki for helpful discussions. We are also grateful to anonymous referees and Jeremiah Blocki for their careful reading of our proof and detailed suggestions.

Leonid Reyzin gratefully acknowledges the hospitality and support of IST Austria, where most of this work was performed, and the hospitality of École normale supérieure, Paris.

The work was partially supported by the following U.S. NSF grants: Binyi Chen by CNS-1423566, CNS-1528178, and CNS-1514526; Stefano Tessaro by CNS-1423566, CNS-1528178, CNS-1553758 (CAREER), and IIS-1528041; Leonid Reyzin by 1012910, 1012798, and 1422965. Moreover, Joël Alwen and Krzysztof Pietrzak were supported by the European Research Council consolidator grant 682815-TOCNeT.

A   On Percival's Proof

We note that Percival [20] claims a weaker result than the one in our main theorem, similar in spirit to our single-shot trade-off theorem (Theorem 2 above), in that it considers only a single random challenge, as well as an overall upper bound on the size of the initial state. Also, the proof technique of [20] may, at first, somewhat resemble the one used in Theorem 2, where multiple copies of the adversary are run on all possible challenges. In contrast to both this work and that of [6], however, Percival considers adversaries with only a limited amount a parallelism.

Upon closer inspection, however, we have found serious problems with the proof in [20]. In more detail, the proof considers an adversary running in two stages. In the preprocessing stage the adversary gets input B and access to h and must eventually output an arbitrary state (bit-string) σ. In the second phase n copies of the adversary are run in parallel. For $x \in [0, n-1]$ the xth copy is given challenge x, state σ and access to h. Its goal is to produce output $h^x(B)$. The main issue with the proof stems from the fact that information about h contained within σ is never explicitly handled. Let us be a bit more concrete.

The proof looks in particular at the set $\overline{R}_i$ of all $i \in [n]$ of all values U for which some copy of the adversary queries $h(U)$ within the first i steps. Here, some key aspects remain undefined. For instance, it is unclear whether the initial time step in the second phase is 0 or 1, and consequently, there is also no clear definition of the contents of the set $\overline{R}_0$. We briefly discuss now why, no matter how we interpret $\overline{R}_0$, the technique does not imply the desired statement.

Suppose we assume that $\overline{R}_0$ is the set of queries to h made by the adversary in this first step of the second stage. In particular, for all i, the set $\overline{R}_i$ contains only queries to h made during the second phase of the execution. However this creates a serious problem. At a (crucial) later step in the proof it is claimed that if $h^{x-1}(B) \notin \overline{R}_{i-1}$, then the probability that $h^x(B)$ is queried at the i-th step is the same as simply guessing $h^x(B)$ out of the blue (a highly unlikely event). But this statement is now incorrect as it ignores potential information contained in the state σ. For example σ may even contain $h^x(B)$ explicitly making it trivial to query h at that point at any time i regardless of the contents of $\overline{R}_{i-1}$.

Suppose instead that we assume the time of the second phase begins at 1 leaving $\overline{R}_0$ open to interpretation. Setting $\overline{R}_0 = \emptyset$ leads to the exact same problem as before. So instead, in an attempt to avoid this pitfall, we could let $\overline{R}_0$ be the set of queries made during the pre-computation stage. Indeed, if $h^{x-1}(B) \notin \overline{R}_i$ then that means $h^{x-1}(B)$ was not queried while σ was being prepared and so (whp) σ contains no information about $h^x(B)$ avoiding the previous problem. Yet here too we run into issues. Consider the following adversary $\mathcal{A}$: In the pre-processing stage $\mathcal{A}$ makes all queries $h^x(B)$ for $x \in [0, n-1]$ and then generates some state σ (what this state really is, and how the adversary proceeds in the second stage is somewhat irrelevant). In particular for this adversary, for all i the set $\overline{R}_i$ already contains all relevant queries $\{h^x(B) : x \in [0, n-1]\}$. Most of the remainder of the proof is concerned with upper bounding the expected size of $\overline{R}_i$. But in the case of $\mathcal{A}$ for each i we now have $|\overline{R}_i| \geq n$ which contradicts the bounds shown in the proof. Worse, when plugging in this new upper bound into the remaining calculations in the proof we would get that the expected runtime of each instance of $\mathcal{A}$ in the second phase is at least 0; an uninteresting result. Thus this too can not be the right interpretation. Unfortunately, we were unable to come up with any reasonable interpretation which results in an interesting statement being proven.

In conclusion, we note that the proof can be adapted to the randomized pebbling setting, as considered in [6]. However, we note that for this setting, [6] already contains a much simpler proof of such a single-shot trade-off theorem. We also note that Theorem 2 confirms that Percival's statement is in fact true, although using a very different proof technique.

References

1. Abadi, M., Burrows, M., Manasse, M.S., Wobber, T.: Moderately hard, memory-bound functions. ACM Trans. Internet Technol. 5(2), 299–327 (2005)
2. Abadi, M., Burrows, M., Wobber, T.: Moderately hard and memory-bound functions. In: NDSS 2003. The Internet Society, February 2003
3. Alwen, J., Blocki, J.: Efficiently computing data-independent memory-hard functions. In: Robshaw, M., Katz, J. (eds.) CRYPTO 2016. LNCS, vol. 9815, pp. 241–271. Springer, Heidelberg (2016). doi:10.1007/978-3-662-53008-5_9
4. Alwen, J., Blocki, J., Pietrzak, K.: Depth-robust graphs and their cumulative memory complexity. In: EUROCRYPT (2017)
5. Alwen, J., Chen, B., Kamath, C., Kolmogorov, V., Pietrzak, K., Tessaro, S.: On the complexity of scrypt and proofs of space in the parallel random oracle model. Cryptology ePrint Archive, report 2016/100 (2016). http://eprint.iacr.org/2016/100
6. Alwen, J., Chen, B., Kamath, C., Kolmogorov, V., Pietrzak, K., Tessaro, S.: On the complexity of scrypt and proofs of space in the parallel random oracle model.

In: Fischlin, M., Coron, J.-S. (eds.) EUROCRYPT 2016. LNCS, vol. 9666, pp. 358–387. Springer, Heidelberg (2016). doi:10.1007/978-3-662-49896-5_13

7. Alwen, J., Chen, B., Pietrzak, K., Reyzin, L., Tessaro, S.: Scrypt is maximally memory-hard. Cryptology ePrint Archive, report 2016/989 (2016). http://eprint.iacr.org/2016/989

8. Alwen, J., Serbinenko, V.: High parallel complexity graphs and memory-hard functions. In: Servedio, R.A., Rubinfeld, R. (eds.) 47th ACM STOC, pp. 595–603. ACM Press, New York (2015)

9. Biryukov, A., Dinu, D., Khovratovich, D.: Argon2 password hash. Version 1.3 (2016). https://www.cryptolux.org/images/0/0d/Argon2.pdf

10. Lee, C.: Litecoin (2011)

11. Dodis, Y., Guo, S., Katz, J.: Random oracles with auxiliary input, revisited. In: EUROCRYPT, Fixing cracks in the concrete (2017)

12. Dwork, C., Goldberg, A., Naor, M.: On memory-bound functions for fighting spam. In: Boneh, D. (ed.) CRYPTO 2003. LNCS, vol. 2729, pp. 426–444. Springer, Heidelberg (2003). doi:10.1007/978-3-540-45146-4_25

13. Dwork, C., Naor, M.: Pricing via processing or combatting junk mail. In: Brickell, E.F. (ed.) CRYPTO 1992. LNCS, vol. 740, pp. 139–147. Springer, Heidelberg (1993). doi:10.1007/3-540-48071-4_10

14. Dwork, C., Naor, M., Wee, H.: Pebbling and proofs of work. In: Shoup, V. (ed.) CRYPTO 2005. LNCS, vol. 3621, pp. 37–54. Springer, Heidelberg (2005). doi:10.1007/11535218_3

15. Dziembowski, S., Faust, S., Kolmogorov, V., Pietrzak, K.: Proofs of space. In: Gennaro, R., Robshaw, M. (eds.) CRYPTO 2015. LNCS, vol. 9216, pp. 585–605. Springer, Heidelberg (2015). doi:10.1007/978-3-662-48000-7_29

16. Franklin, M.K., Malkhi, D.: Auditable metering with lightweight security. In: Hirschfeld, R. (ed.) FC 1997. LNCS, vol. 1318, pp. 151–160. Springer, Heidelberg (1997). doi:10.1007/3-540-63594-7_75

17. Hoeffding, W.: Probability inequalities for sums of bounded random variables. J. Am. Stat. Assoc. **58**(301), 13–30 (1963)

18. Jakobsson, M., Juels, A.: Proofs of work, bread pudding protocols. In: Proceedings of the IFIP TC6/TC11 Joint Working Conference on Secure Information Networks: Communications and Multimedia Security, CMS 1999, pp. 258–272. Kluwer, B.V., Deventer (1999)

19. Juels, A., Brainard, J.G.: Client puzzles: a cryptographic countermeasure against connection depletion attacks. In: NDSS 1999. The Internet Society, February 1999

20. Percival, C.: Stronger key derivation via sequential memory-hard functions. In: BSDCan (2009)

21. Percival, C., Josefsson, S.: The scrypt password-based key derivation function. RFC 7914 (Informational), August 2016

22. Password hashing competition. https://password-hashing.net/

23. Rivest, R.L., Shamir, A., Wagner, D.A.: Time-lock puzzles and timed-release crypto. Technical report, Massachusetts Institute of Technology, Cambridge, MA, USA (1996)

24. von Ahn, L., Blum, M., Hopper, N.J., Langford, J.: CAPTCHA: using hard AI problems for security. In: Biham, E. (ed.) EUROCRYPT 2003. LNCS, vol. 2656, pp. 294–311. Springer, Heidelberg (2003). doi:10.1007/3-540-39200-9_18

Symmetric-Key Constructions

Quantum-Secure Symmetric-Key Cryptography Based on Hidden Shifts

Gorjan Alagic[1(✉)] and Alexander Russell[2]

[1] QMATH, Department of Mathematical Sciences,
University of Copenhagen, Copenhagen, Denmark
galagic@gmail.com
[2] Department of Computer Science and Engineering,
University of Connecticut, Mansfield, USA
acr@cse.uconn.edu

Abstract. Recent results of Kaplan et al., building on work by Kuwakado and Morii, have shown that a wide variety of classically-secure symmetric-key cryptosystems can be completely broken by *quantum chosen-plaintext attacks* (qCPA). In such an attack, the quantum adversary has the ability to query the cryptographic functionality in superposition. The vulnerable cryptosystems include the Even-Mansour block cipher, the three-round Feistel network, the Encrypted-CBC-MAC, and many others.

In this article, we study simple algebraic adaptations of such schemes that replace $(\mathbb{Z}/2)^n$ addition with operations over alternate finite groups—such as $\mathbb{Z}/2^n$—and provide evidence that these adaptations are qCPA-secure. These adaptations furthermore retain the classical security properties and basic structural features enjoyed by the original schemes.

We establish security by treating the (quantum) hardness of the well-studied *Hidden Shift problem* as a cryptographic assumption. We observe that this problem has a number of attractive features in this cryptographic context, including random self-reducibility, hardness amplification, and—in many cases of interest—a reduction from the "search version" to the "decisional version." We then establish, under this assumption, the qCPA-security of several such Hidden Shift adaptations of symmetric-key constructions. We show that a Hidden Shift version of the Even-Mansour block cipher yields a quantum-secure pseudorandom function, and that a Hidden Shift version of the Encrypted CBC-MAC yields a collision-resistant hash function. Finally, we observe that such adaptations frustrate the direct Simon's algorithm-based attacks in more general circumstances, e.g., Feistel networks and slide attacks.

1 Introduction

The discovery of efficient quantum algorithms for algebraic problems with long-standing roles in cryptography, like factoring and discrete logarithm [30], has led to a systematic re-evaluation of cryptography in the presence of quantum attacks. Such attacks can, for example, recover private keys directly from public

© International Association for Cryptologic Research 2017
J.-S. Coron and J.B. Nielsen (Eds.): EUROCRYPT 2017, Part III, LNCS 10212, pp. 65–93, 2017.
DOI: 10.1007/978-3-319-56617-7_3

keys for many public-key cryptosystems of interest. A 2010 article of Kuwakado and Morii [18] identified a new family of quantum attacks on certain generic constructions of private-key cryptosystems. While the attacks rely on similar quantum algorithmic tools (that is, algorithms for the hidden subgroup problem), they qualitatively differ in several other respects. Perhaps most notably, they break reductions which are information-theoretically secure[1] in the classical setting. On the other hand, these attacks require a powerful "quantum CPA" setting which permits the quantum adversary to make queries—in superposition—to the relevant cryptosystem.

These quantum chosen-plaintext attacks (qCPA) have been generalized and expanded to apply to a large family of classical symmetric-key constructions, including Feistel networks, Even-Mansour ciphers, Encrypted-CBC-MACs, tweakable block ciphers, and others [14,18,19,29]. A unifying feature of all these new attacks, however, is an application of Simon's algorithm for recovering "hidden shifts" in the group $(\mathbb{Z}/2)^n$. Specifically, the attacks exploit an internal application of addition (mod 2) to construct an instance of a hidden shift problem—solving the hidden shift problem then breaks the cryptographic construction. As an illustrative example, consider two (independent) uniformly random permutations $P, Q : \{0,1\}^n \to \{0,1\}^n$ and a uniformly random element z of $\{0,1\}^n$. It is easy to see that no classical algorithm can distinguish the function $(x,y) \mapsto (P(x), Q(y))$ from the function $(x,y) \mapsto (P(x), P(y \oplus z))$ with a polynomial number of queries; this observation directly motivates the classical Even-Mansour block-cipher construction. On the other hand, an efficient quantum algorithm with oracle access to $(x,y) \mapsto (P(x), P(y \oplus z))$ can apply Simon's algorithm to recover the "hidden shift" z efficiently; this clearly allows the algorithm to distinguish the two cases above.

While these attacks threaten many classical private-key constructions, they depend on an apparent peculiarity of the group $(\mathbb{Z}/2)^n$—the HIDDEN SHIFT problem over $(\mathbb{Z}/2)^n$ admits an efficient quantum algorithm. In contrast, HIDDEN SHIFT problems in general have resisted over 20 years of persistent attention from the quantum algorithms community. Indeed, aside from Simon's polynomial-time algorithm for hidden shifts over $(\mathbb{Z}/2)^n$, generalizations to certain groups of constant exponent [10], and Kuperberg's $2^{O(\sqrt{\log N})}$ algorithm for hidden shifts over $\mathbb{Z}/N$ [16], very little is known. This dearth of progress is not for lack of motivation. In fact, it is well-known that efficient quantum algorithms for HIDDEN SHIFT over $\mathbb{Z}/N$ would (via a well-known reduction from the HIDDEN SUBGROUP PROBLEM on D_N) yield efficient quantum attacks on important public-key cryptosystems [26,27], including prime candidates for quantum security and the eventual replacement of RSA in Internet cryptography [3]. Likewise, efficient algorithms for the symmetric group would yield polynomial-time quantum algorithms for GRAPH ISOMORPHISM, a longstanding challenge in the area.

---

[1] The adversary is permitted to query the oracle a polynomial number of times, but may perform arbitrarily complex computations between queries.

On the other hand, $(\mathbb{Z}/2)^n$ group structure is rather incidental to the security of typical symmetric-key constructions. For example, the classical Even-Mansour construction defines a block cipher $E_{k_1,k_2}(m)$ by the rule

$$E_{k_1,k_2}(m) = P(m \oplus k_1) \oplus k_2,$$

where P is a public random permutation and the secret key (k_1, k_2) is given by a pair of independent elements drawn uniformly from $(\mathbb{Z}/2)^n$. The security proofs, however, make no particular assumptions about group structure, and apply if the $\oplus$ operation is replaced with an alternative group operation, e.g., $+$ modulo N or multiplication in $\mathbb{F}_{2^n}$.

This state of affairs suggests the possibility of ruling out quantum attacks by the simple expedient of adapting the underlying group in the construction. Moreover, the apparently singular features of $(\mathbb{Z}/2)^n$ in the quantum setting suggest that quite mild adaptations may be sufficient. As mentioned above, many classical security proofs are unaffected by this substitution; our primary goal is to add security against quantum adversaries. *Our approach is to reduce well-studied Hidden Shift problems directly to the security of these symmetric-key cryptosystems. Thus, efficient quantum chosen-plaintext attacks on these systems would resolve long-standing open questions in quantum complexity theory.*

1.1 Contributions

Hidden Shift as a Cryptographic Primitive. We propose the intractability of the HIDDEN SHIFT problem as a fundamental assumption for establishing quantum security of cryptographic schemes. In the general problem, we are given two functions on some finite group G, and a promise that one is a shift of the other; our task is to identify the shift. Our assumptions have the following form:

Assumption 1 (The $\mathcal{G}$-Hidden Shift Assumption, informal). *Let* $\mathcal{G} = \{G_i \mid i \in I\}$ *be a family of finite groups indexed by a set* $I \subset \{0,1\}^*$. *For all polynomial-time quantum algorithms* $\mathcal{A}$,

$$\mathbb{E}_f\left[\min_{s \in G_i} \Pr\left[\mathcal{A}^{f,f_s}(i) = s\right]\right] \leq \mathrm{negl}(|i|),$$

where $f_s(x) = f(sx)$, *the expectation is taken over random choice of the function* f, *the minimum is taken over all shifts* $s \in G_i$, *and the probability is taken over internal randomness and measurements of* $\mathcal{A}$.

This assumption asserts that there is no quantum algorithm for HIDDEN SHIFT (over $\mathcal{G}$) in the *worst-case* over s, when function values are chosen randomly. Note that the typical formulation in the quantum computing literature is worst case over s and f; on the other hand, known algorithmic approaches are invariant under arbitrary relabeling of the value space of f. The "random-valued" case thus seems satisfactory for our cryptographic purposes. (In fact, our results can alternatively depend on the case where f is injective, rather than random.)

See Sect. 3 below for further discussion and precise versions of Assumption 1. In general, formulating such an assumption requires attention to the encoding of the group. However, we will focus entirely on groups with conventional encodings which directly provide for efficient group operations, inversion, generation of random elements, etc. Specifically, we focus on the two following particular variants:

Assumption 2 (The 2^n-Cyclic Hidden Shift Assumption). *This is the Hidden Shift Assumption with the group family $C_2 = \{\mathbb{Z}/2^n \mid n \geq 0\}$ where the index consists of the number n written in unary.*

Assumption 3 (The Symmetric Hidden Shift Assumption). *This is the Hidden Shift Assumption with the group family $S = \{S_n \mid n \geq 0\}$ where S_n denotes the symmetric group on n symbols and the index consists of the number n written in unary.*

In both cases the size of the group is exponential in the length of the index.

We remark that the HIDDEN SHIFT problem has polynomial quantum query complexity [7]—thus one cannot hope that HIDDEN-SHIFT-based schemes possess information-theoretic security in the quantum setting (as they do in the classical setting); this motivates introduction of HIDDEN SHIFT intractability assumptions.

To explore the hardness of HIDDEN SHIFT problems against quantum polynomial-time (QPT) algorithms, we describe several reductions. First, we prove that HIDDEN SHIFT is equivalent to a randomized version of the problem where the shift s is random (RANDOM HIDDEN SHIFT), and provide an amplification theorem which is useful in establishing security of schemes based on Assumption 1.

Proposition (Amplification, informal). *Assume there exists a QPT algorithm which solves RANDOM HIDDEN SHIFT for an inverse-polynomial fraction of inputs. Then there exists a QPT algorithm for solving both HIDDEN SHIFT and RANDOM HIDDEN SHIFT for all but a negligible fraction of inputs.*

We then show that, for many group families, HIDDEN SHIFT over the relevant groups is equivalent to a *decisional* version of the problem. In the decisional version, we are guaranteed that the two functions are either (i) both random and independent, or (ii) one is random and the other is a shift; the goal is to decide which is the case.

Theorem (Search and decision are equivalent, informal). *Let G be the group family C_2 or the group family S (or a group family with an efficient subgroup series). Then there exists a QPT algorithm for RANDOM HIDDEN SHIFT (with at most inverse-poly error) over G if and only if there exists a QPT algorithm for DECISIONAL RANDOM HIDDEN SHIFT (with at most inverse-poly error) over G.*

Finally, we provide some evidence that HIDDEN SHIFT over the family C_2 is as hard as HIDDEN SHIFT over general cyclic groups. Specifically, we show that efficient algorithms for an approximate version of HIDDEN SHIFT over C_2 give rise to efficient algorithms for the same problem over C, the family of all cyclic groups.

We also briefly discuss the connections between HIDDEN SHIFT, the assumptions above, and assumptions underlying certain candidates for quantum-secure public-key cryptography [5,26]. For completeness, we recall known connections to the HIDDEN SUBGROUP PROBLEM. Both the HIDDEN SHIFT and HIDDEN SUBGROUP PROBLEM families have received significant attention from the quantum algorithms community, and are believed to be quantumly hard with the exception of particular families of groups [5,12,21,22,26].

Quantum-Secure Symmetric-Key Cryptographic Schemes. With the above results in hand, we describe a generic method for using Assumption 1 to "adapt" classically-secure schemes in order to remove vulnerabilities to quantum chosen-plaintext attacks. The adaptation is simple: replace the underlying $(\mathbb{Z}/2)^n$ structure of the scheme with that of either C_2 or S. This amounts to replacing bitwise XOR with a new group operation. In the case of C_2, the adaptation is particularly simple and efficient.

While our basic approach presumably applies in broad generality, we focus on three emblematic examples: the Even-Mansour construction—both as a PRF and as a block cipher—and the CBC-MAC construction. We focus throughout on the group families C_2 and S, though we also discuss some potential advantages of other choices (see Sect. 3.2). Finally, we discuss related quantum attacks on cryptographic constructions, including the 3-round Feistel cipher and quantum slide attacks [14]. We remark that the Feistel cipher over groups other than $(\mathbb{Z}/2)^n$ has been considered before, in a purely classical setting [24].

Hidden Shift Even-Mansour. Following the prescription above, we define group variants of the Even-Mansour cipher. We give a reduction from the worst-case HIDDEN SHIFT problem to the natural *distinguishability* problem (i.e., distinguishing an Even-Mansour cipher from a random permutation). Thus, under the Hidden Shift Assumption, the Even-Mansour construction is a quantum-query-secure pseudorandom function (qPRF). In particular, key-recovery is computationally infeasible, even for a quantum adversary. We also provide (weaker) reductions between HIDDEN SHIFT and the problem of breaking Even-Mansour in the more challenging case where the adversary is provided access to both the public permutation and its inverse (and likewise for the encryption map). In any case, these adaptations frustrate the "Simon algorithm key recovery attack" [14,19], as this would now require a subroutine for HIDDEN SHIFT in the relevant group family. Moreover, one can also apply standard results (see, e.g., [13]) to show that, over some groups, all bits of the key are as hard as the entire key (and hence, by our reductions, as hard as HIDDEN SHIFT). We remark that considering $\mathbb{Z}/2^n$ structure to define an adaptation of Even-Mansour has been considered before in the context of classical slide attacks [6].

Hidden Shift CBC-MAC. Following our generic method for transforming schemes, we define group variants of the Encrypted-CBC-MAC. We establish that this primitive is collision-free against quantum adversaries. Specifically, we show that any efficient quantum algorithm which discovers collisions in the Hidden-Shift Encrypted-CBC-MAC with non-negligible probability would yield an efficient worst-case quantum algorithm for HIDDEN SHIFT over the relevant group family. As with Even-Mansour, this adaptation also immediately frustrates the Simon's algorithm collision-finding attacks [14,29].

Feistel Ciphers, Slide Attacks. We also define group variants of the well-known Feistel cipher for constructing pseudorandom permutations from pseudorandom functions. Our group variants frustrate Simon-style attacks [18]; a subroutine for the more general HIDDEN SHIFT problem is now required. Finally, we also address the exponential quantum speedup of certain classical slide attacks, as described in [14]. We show how one can once again use HIDDEN SHIFT to secure schemes vulnerable to these "quantum slide attacks."

2   Preliminaries

Notation; Remarks on Finite Groups. For a finite group G and an element $s \in G$, let $L_s : G \to G$ denote the permutation given by left multiplication by s, so $L_s : x \mapsto s \cdot x$. We discuss a number of constructions in the paper requiring computation in finite groups and assume, throughout, that elements of the group in question have an encoding that efficiently permits such natural operations as product, inverse, selection of uniformly random group elements, etc. As our discussion focuses either on specific groups—such as $(\mathbb{Z}/2)^n$ or $\mathbb{Z}/N$—where such encoding issues are straightforward or, alternatively, *generic* groups in which we assume such features by fiat, we routinely ignore these issues of encoding.

Classical and Quantum Algorithms. Throughout we use the abbreviation PPT for "probabilistic polynomial time," referring to an efficient classical algorithm, and QPT for "quantum polynomial time," referring to an efficient quantum algorithm. Our convention is to denote algorithms of either kind with calligraphic letters, e.g., $\mathcal{A}$ will typically denote an algorithm which models an adversary. If f is a function, the notation $\mathcal{A}^f$ stands for an algorithm (either classical or quantum) with oracle access to the function f. A classical oracle is simply the black-box gate $x \mapsto f(x)$; a quantum oracle is the unitary black-box gate $|x\rangle|y\rangle \mapsto |x\rangle|y \oplus f(x)\rangle$. Unless stated otherwise, oracle QPT algorithms are assumed to have quantum oracle access.

Quantum-Secure Pseudorandomness. We now set down a way of quantifying the ability of a QPT adversary to distinguish between families of functions. Fix a function family $\mathcal{F} \subset \{h : \{0,1\}^m \to \{0,1\}^\ell\}$, a function $f : \{0,1\}^n \times \{0,1\}^m \to \{0,1\}^\ell$, and define $f_k := f(k, \cdot)$. We say that f is an *indexed subfamily* of

$\mathcal{F}$ if $f_k \in \mathcal{F}$ for every $k \in \{0,1\}^n$. We will generally assume that m and ℓ are polynomial functions of n and treat n to be the complexity (or security) parameter.

Definition 1. *Let $\mathcal{F}$ be a function family, f an indexed subfamily, and $\mathcal{D}$ an oracle QPT algorithm. The distinguishing advantage of $\mathcal{D}$ is the quantity*

$$\mathbf{Adv}_{\mathcal{F},f}^{\mathcal{D}} := \left| \Pr_{k \in_R \{0,1\}^n} \left[\mathcal{D}^{f_k}(1^n) = 1 \right] - \Pr_{g \in_R \mathcal{F}} \left[\mathcal{D}^g(1^n) = 1 \right] \right|.$$

Next, we define efficient indexed function families which are pseudorandom against QPT adversaries. We emphasize that these function families are computed by deterministic classical algorithms.

Definition 2. *Let $\mathcal{F}_n$ be the family of all functions from $m(n)$ bits to $\ell(n)$ bits, and f a efficiently computable, indexed subfamily of $\bigcup_n \mathcal{F}_n$ (so that $f_k \in \mathcal{F}_n$ for $|k| = n$). We say that f is a quantum-secure pseudorandom function (qPRF) if $\mathbf{Adv}_{\mathcal{F}_n,f}^{\mathcal{D}} \leq \mathrm{negl}(n)$ for all QPT $\mathcal{D}$.*

It is known how to construct qPRFs from standard assumptions (i.e., existence of quantum-secure one-way functions) [33].

The pseudorandom function property is not enough in certain applications, e.g., in constructing block ciphers. It is then often useful to add the property that each function in the family is a permutation, which can be inverted efficiently (provided the index is known).

Definition 3. *Let $\mathcal{P}$ be the family of all permutations, and f an efficiently computable, indexed subfamily of $\mathcal{P}$. We say that f is a quantum-secure pseudorandom permutation (qPRP) if (i) f is a qPRF, (ii) each f_k is a permutation, and (iii) there is an efficient algorithm which, given k, computes the inverse f_k^{-1} of f_k.*

A recent result shows how to construct qPRPs from one-way functions [32]. Finding simpler constructions is an open problem. Two simple constructions which are known to work classically, Even-Mansour and the 3-round Feistel, are both broken by a simple attack based on Simon's algorithm for HIDDEN SHIFT on $(\mathbb{Z}/2)^n$. As we discuss in detail later, we conjecture that the adaptations of these constructions to other group families are qPRPs.

We will also make frequent use of a result of Zhandry (Theorem 3.1 in [34]) which states that $2k$-wise independent functions are indistinguishable from random to quantum adversaries making no more than k queries.

Theorem 1. *Let $\mathcal{H}$ be a $2k$-wise independent family of functions with domain $\mathcal{X}$ and range $\mathcal{Y}$. Let $\mathcal{D}$ be a quantum algorithm making no more than k oracle queries. Then*

$$\Pr_{h \in_R \mathcal{H}} \left[\mathcal{D}^h(1^n) = 1 \right] = \Pr_{g \in_R \mathcal{Y}^{\mathcal{X}}} \left[\mathcal{D}^g(1^n) = 1 \right].$$

Collision-Freeness. We will also need a (standard) definition of collision-resistance against efficient quantum adversaries with oracle access.

Definition 4. *Let $f : \{0,1\}^* \times \{0,1\}^* \to \{0,1\}^*$ be an efficiently-computable function family defined for all (k,x) for which $|x| = m(|k|)$ (for a polynomial m). We say that f is collision-resistant if for all QPT $\mathcal{A}$,*

$$\Pr_{k \in_R \{0,1\}^n} \left[\mathcal{A}^{f_k}(1^n) = (x,y) \wedge f_k(x) = f_k(y) \wedge x \neq y \right] \leq \mathrm{negl}(n).$$

3   Hidden Shift as a Cryptographic Primitive

We begin by discussing a few versions of the basic oracle promise problem related to finding hidden shifts of functions on groups. In the problems below, the relevant functions are given to the algorithm via black-box oracle access and we are interested in the setting where the complexity of the algorithm (both number of queries and running time) scales in $\mathrm{poly}(\log|G|)$.

3.1   Hidden Shift Problems

Basic Definitions. We begin with the HIDDEN SHIFT problem. As traditionally formulated in the quantum computing literature, the problem is the following:

Problem 1 (The traditional Hidden Shift problem). Let G be a group and V a set. Given oracle access to an injective function $f : G \to V$ and an unknown shift $g = f \circ L_s$ of f, find s.

It is convenient for us to parameterize this definition in terms of a specific group family and fix the range of the oracles f and g. This yields our basic asymptotic definition for the problem.

Problem 2 (HIDDEN SHIFT (HS)). Let $\mathcal{G} = \{G_i \mid i \in I\}$ be a family of groups with index set $I \subset \{0,1\}^*$ and let $\ell : \mathbb{N} \to \mathbb{N}$ be a polynomial. Then the HIDDEN SHIFT problem over $\mathcal{G}$ (with length parameter ℓ) is the following: given an index i and oracle access to a pair of functions $f, g : G_i \to \{0,1\}^{\ell(|i|)}$ where $g(x) = f(sx)$, determine $s \in G_i$. We assume, throughout, that $2^{\ell(|i|)} \gg |G_i|$.

This generic formulation is more precise, but technically still awkward for cryptographic purposes as it permits oracle access to completely arbitrary functions f. To avoid this technical irritation, we focus on the performance of HIDDEN SHIFT algorithms over specific classes of functions f. Specifically, we either assume f is random or that it is injective. When a HIDDEN SHIFT algorithm is applied to solve problems in a typical computational setting, the actual functions f, g are injective and given by efficient computations. We remark that established algorithmic practice in this area ignores the actual function values altogether, merely relying on the structure of the level sets of the function

$$\Phi(x,b) = \begin{cases} f(x) & \text{if } b = 0, \\ g(x) & \text{if } b = 1. \end{cases}$$

In particular, such structural conditions of f appear to be irrelevant to the success of current quantum-algorithmic techniques for the problem. This motivates the following notion of "success" for an algorithm.

Definition 5 (Completeness). *Let $\mathcal{A}$ be an algorithm for the* HIDDEN SHIFT *problem on $\mathcal{G}$ with length parameter ℓ. Let f be a function defined on all pairs (i, x) where $x \in G_i$ so that $f(i, x) \in \{0, 1\}^{\ell(|i|)}$. Then we define the completeness of $\mathcal{A}$ relative to f to be the quantity*

$$1 - \epsilon_f(i) \triangleq \min_{s \in G_i} \Pr[\mathcal{A}^{f, f_s}(i) = s].$$

The completeness of $\mathcal{A}$ *relative to random functions is the average*

$$1 - \epsilon_R(i) \triangleq \mathop{\mathbb{E}}_{f} \left[\min_{s \in G_i} \Pr[\mathcal{A}^{f, f_s}(i) = s] \right] = \mathop{\mathbb{E}}_{f}[1 - \epsilon_f(i)],$$

where $f(i, x)$ is drawn uniformly at random. Note that these notions are worst-case in s, the shift.

Note that this definition does not specify how the algorithm should behave on instances that are not hidden shifts. For simplicity, we assume that the algorithm returns a value for s in any case, with no particular guarantee on s in the case when the functions are not shifts of each other.

Our basic hardness assumption is the following:

Assumption 4 (The $\mathcal{G}$-Hidden Shift Assumption; randomized). *Let $\mathcal{G} = \{G_i \mid i \in I\}$ be a family of finite groups indexed by a set $I \subset \{0, 1\}^*$ and $\ell : \mathbb{N} \to \mathbb{N}$ be a length parameter. Then for all efficient algorithms $\mathcal{A}$, $1 - \epsilon_R(i) = \mathrm{negl}(|i|)$.*

For completeness, we also record a version of the assumption for injective f. In practice, our cryptographic constructions will rely only on the randomized version.

Assumption 5 (The $\mathcal{G}$-Hidden Shift Assumption; injective). *Let $\mathcal{G} = \{G_i \mid i \in I\}$ be a family of finite groups indexed by a set $I \subset \{0, 1\}^*$ and $\ell : \mathbb{N} \to \mathbb{N}$ be a length parameter. Then for all efficient algorithms $\mathcal{A}$ there exists an injective f (satisfying the criteria of Definition 5 above), so that $1 - \epsilon_f(i) = \mathrm{negl}(|i|)$.*

In preparation for establishing results on security amplification, we define two additional variants of the HIDDEN SHIFT problem: a variant where both the function and the shift are randomized, and a decisional variant. Our general approach for constructing security proofs will be to reduce one of these variants to the problem of breaking the relevant cryptographic scheme. As we will later show, an efficient solution to either variant implies an efficient solution to both, which in turn results in a violation of Assumption 4 above.

*Problem 3 (*RANDOM HIDDEN SHIFT *(RHS))*. Let $\mathcal{G} = \{G_i \mid i \in I\}$ be a family of finite groups indexed by a set $I \subset \{0,1\}^*$ and $\ell : \mathbb{N} \to \mathbb{N}$ be a length parameter. Then the RANDOM HIDDEN SHIFT problem over $\mathcal{G}$ is the HIDDEN SHIFT problem where the input function $f(i, x)$ is drawn uniformly and the shift s is drawn (independently and uniformly) from G_i.

We define the completeness $1 - \epsilon(i)$ for a RANDOM HIDDEN SHIFT algorithm $\mathcal{A}$ analogously to Definition 5. Observe that a small error is unavoidable for any algorithm, as there exist pairs of functions for which s is not uniquely defined. We will also need a decisional version of the problem, defined as follows.

*Problem 4 (*DECISIONAL RANDOM HIDDEN SHIFT *(DRHS))*. Let $\mathcal{G} = \{G_i \mid i \in I\}$ be a family of finite groups indexed by a set $I \subset \{0,1\}^*$ and $\ell : \mathbb{N} \to \mathbb{N}$ be a length parameter. The DECISIONAL RANDOM HIDDEN SHIFT problem is the following: Given i and oracle access to two functions $f, g : G_i \to \{0,1\}^{\ell(|i|)}$ with the promise that either (i) both f and g are drawn independently at random, or (ii) f is random and $g = f \circ L_s$ for some $s \in G$, decide which is the case.

We say that an algorithm for DRHS has completeness $1 - \epsilon(i)$ and soundness $\delta(i)$ if the algorithm errs with probability no more than $\epsilon(i)$ in the case that the functions are shifts and errs with probability no more than $\delta(i)$ in the case that the functions are drawn independently.

Next, we briefly recall the definition of the (closely-related) HIDDEN SUBGROUP PROBLEM. The problem is primarily relevant in our context because of its historical significance (and relationship to HIDDEN SHIFT); we will not use it directly in any security reductions.

*Problem 5 (*HIDDEN SUBGROUP PROBLEM *(HSP))*. Let G be a group and S a set. Given a function $f : G \to S$, and a promise that there exists $H \leq G$ such that f is constant and distinct on the right cosets of H, output a complete set of generators for H.

Some further details, including explicit reductions between HS and HSP, are given in Appendix A.

Of interest are both classical and quantum algorithms for solving the various versions of HS and HSP. The relevant metrics for such algorithms are the query complexity (i.e., the number of times that the functions are queried, classically or quantumly) as well as their time and space complexity. An algorithm is said to be efficient if all three are polynomial in $\log |G|$.

Hardness Results. Next, we establish several reductions between these problems. Roughly, these results show that the average-case and decisional versions of the problem are as hard as the worst-case version.

Self-reducibility and Amplification. First, we show that (i) both HS and RHS are random self-reducible, and (ii) an efficient solution to RHS implies an efficient solution to HS.

Proposition 1. *Let* $\mathcal{G} = \{G_i \mid i \in I\}$ *be a family of finite groups indexed by a set* $I \subset \{0,1\}^*$ *and* $\ell : \mathbb{N} \to \mathbb{N}$ *be a length parameter. Assume there exists a QPT* $\mathcal{A}$ *which solves* RANDOM HIDDEN SHIFT *over* $\mathcal{G}$ *(with parameter* $\ell(|i|)$*) with inverse-polynomial completeness. Then there exists a QPT* $\mathcal{A}'$ *which satisfies all of the following:*

1. $\mathcal{A}'$ *solves* HIDDEN SHIFT *with random* f *with completeness* $1 - \text{negl}(|i|)$;
2. $\mathcal{A}'$ *solves* HIDDEN SHIFT *for any injective* f *with completeness* $1 - \text{negl}(|i|)$;
3. $\mathcal{A}'$ *solves* RANDOM HIDDEN SHIFT *with completeness* $1 - \text{negl}(|i|)$.

Proof. We are given oracles f, g and a promise that $g = f \circ L_s$. For a particular choice of n, there is an explicit (polynomial-size) bound k on the running time of $\mathcal{A}$. Let $\mathcal{H}$ be a $2k$-wise independent function family which maps the range of f to itself. The algorithm $\mathcal{A}'$ will repeatedly execute the following subroutine. First, an element $h \in \mathcal{H}$ and an element $t \in G_i$ are selected independently and uniformly at random. Then $\mathcal{A}$ is executed with oracles

$$f' := h \circ f \quad \text{and} \quad g' := h \circ g \circ L_t.$$

It's easy to see that $g' = f' \circ L_{st}$. If $\mathcal{A}$ outputs a group element r, $\mathcal{A}'$ checks if $g'(x) = f'(rx)$ at a polynomial number of random values x. If the check succeeds, $\mathcal{A}'$ outputs rt^{-1} and terminates. If the check fails (or if $\mathcal{A}$ outputs garbage), we say that the subroutine fails. The subroutine is repeated m times, each time with a fresh h and t.

Continuing with our fixed choice of f and g, we now argue that $\mathcal{A}$ (when used as above) cannot distinguish between (f', g') and the case where f' is uniformly random, and g' is a uniformly random shift of f'. First, the fact that the shift is randomized is clear. Second, if f is injective, then f' is simply h with permuted inputs, and is thus indistinguishable from random (by the $2k$-wise independence of h and Theorem 1). Third, if f is random, then it is indistinguishable from injective (by the collision bound of [35]), and we may thus apply the same argument as in the injective case.

It now follows that, with inverse-polynomial probability ϵ (over the choice of h and t), the instance (f', g') is indistinguishable from an instance (φ, φ_{st}) on which the subroutine succeeds with inverse-polynomial probability δ. After m repetitions of the subroutine, $\mathcal{A}'$ will correctly compute the shift $r = st$ with probability at least $(1 - \epsilon\delta)^m \approx e^{-\epsilon\delta m}$, as desired. $\square$

Decision Versus Search. Next, we consider the relationship between searching for shifts (given the promise that one exists), and deciding if a shift exists or not. Roughly speaking, we establish that the two problems are equivalent for most group families of interest. We begin with a straightforward reduction from DRHS to RHS.

Proposition 2. *If there exists a QPT algorithm for* RANDOM HIDDEN SHIFT *on* $\mathcal{G}$ *with completeness* $1 - \epsilon(i)$, *then there exists a QPT algorithm for* DECISIONAL RANDOM HIDDEN SHIFT *on* $\mathcal{G}$ *with completeness* $1 - \epsilon(i)$, *and negligible soundness error.*

Proof. Let $\ell(\cdot)$ be the relevant length parameter. Consider an RHS algorithm for G with completeness $1 - \epsilon$ and the following adaptation to DRHS.

– Run the RHS algorithm.
– When the algorithm returns a purported shift s, check s for veracity with a polynomial number of (classical) oracle queries to f and g (ensuring that $g(x_i) = f(sx_i)$ for $k(n)$ distinct samples $x_1, \ldots, x_{k(n)}$).

Observe that if f and g are indeed hidden shifts, this procedure will determine that with probability $1 - \epsilon$. When f and g are unrelated random functions, the "testing" portion of the algorithm will erroneously succeed with probability no more than $|G| \cdot 2^{-k \cdot \ell(|i|)}$. Thus, under the assumption that $|G| \geq k(n)$, the resulting DRHS algorithm has completeness $1 - \epsilon$ and soundness $|G| \cdot 2^{-k \cdot \ell(|i|)}$. For any nontrivial length function ℓ, this soundness can be driven exponentially close to zero by choosing $k = \log |G| + k'$.   □

On the other hand, we are only aware of reductions from RHS to DRHS under the additional assumption that G has a "dense" tower of subgroups. In that case, an algorithmic approach of Fenner and Zhang [9] can be adapted to provide a reduction. Both S_n and $\mathbb{Z}/2^n$ have such towers.

Proposition 3. *Let $\mathcal{G}$ be either the group family $\{\mathbb{Z}/2^n\}$, or the group family $\{S_n\}$. If there exists a QPT algorithm for* DECISIONAL RANDOM HIDDEN SHIFT *on $\mathcal{G}$ with at most inverse-polynomial completeness and soundness errors, then there exists a QPT algorithm for* RANDOM HIDDEN SHIFT *on $\mathcal{G}$ with negligible completeness error.*

Proof. The proof adapts techniques of [9] to our probabilistic setting, and relies on the fact that these group families have an efficient subgroup tower. Specifically, each G_i possesses a subgroup series $\{1\} = G^{(0)} < G^{(1)} < G^{(2)} < \cdots < G^{(s)} = G_i$ for which (i) uniformly random sampling and membership in $G^{(t)}$ can be performed efficiently for all t, and (ii) for all t, there is an efficient algorithm for producing a left transversal of $G^{(t-1)}$ in $G^{(t)}$. For $\mathbb{Z}/2^n$, the subgroup series is $\{1\} < \mathbb{Z}/2 < \mathbb{Z}/2^2 < \mathbb{Z}/2^3 < \cdots$. For S_n (i.e., the group of permutations of n letters), the subgroup series is $\{1\} < S_1 < S_2 < S_3 < \cdots$, where each step of the series adds a new letter. We remark that such series can be efficiently computed for general permutation groups using a *strong generating set*, which can be efficiently computed from a presentation of the group in terms of generating permutations [11].

We recursively define a RHS algorithm by considering the case of a group G with a subgroup H of polynomial index with a known left transversal $A = \{a_1, \ldots, a_k\}$ (so that G is the disjoint union of the $a_i H$). Assume that the DRHS algorithm for H has soundness δ_H and completeness $1 - \epsilon_H$. In this case, the algorithm (for G) may proceed as follows:

1. For each $\alpha \in A$, run the DRHS algorithm on the two functions f and $\breve{g} : x \mapsto g(\alpha x)$ restricted to the subgroup H.

2. If exactly one of these recursive calls reports that the function f and $x \mapsto g(\alpha x)$ are hidden shifts, recursively apply the RHS algorithm to recover the hidden shift s' (so that $f(x) = g(\alpha s' x)$ for $x \in H$). Return the shift $s = \alpha s'$.
3. Otherwise assert that the functions are unrelated random functions.

In the case that f and g are independent random functions, the algorithm above errs with probability no more than $[G : H]\delta_H$.

Consider instead the case that $f : G \to S$ is a random function and $g(x) = f(sx)$ for an element $s \in G$. Observe that if $s^{-1} \in \alpha_i H$, so that $s^{-1} = \alpha_i h_s$ for an element $h_s \in H$, we have $g(\alpha_i h_s x) = f(x)$. It follows that f and $\breve{g} : x \mapsto g(\alpha_i x)$ are shifts of each other; in particular, this is true when restricted to the subgroup H. Moreover, the hidden shift s can be determined directly from the hidden shift between f and $\breve{g}$. Note that, as above, the probability that any of the recursive calls to DRHS are answered incorrectly is no more than $[G : H]\delta_H + \epsilon_H$.

It remains to analyze the completeness of the resulting recursive RHS algorithm: in the case of the subgroup chain above, let γ_t denote the completeness of the resulting RHS algorithm on $G^{(t+1)}$ and note that

$$\gamma_{t+1} \leq [G^{(t+1)} : G^{(t)}]\delta_{G^{(t)}} + \epsilon_{G^{(t)}} + \gamma_t$$

and thus that the resulting error on G is no more than

$$\sum_t [G^{(t+1)} : G^{(t)}]\delta_{G^{(t)}} + \sum_t \epsilon_{G^{(t)}}. \tag{3.1}$$

As mentioned above, both the group families $\{\mathbb{Z}/2^n \mid n \geq 0\}$ and $\{S_n \mid n \geq 0\}$ satisfy this subgroup chain property.   □

Remark. Note that the groups $\mathbb{Z}/N$ for general N are not treated by the results above; indeed, when N is prime, there is no nontrivial tower of subgroups. (Such groups do have other relevant self-reducibility and amplification properties [13].) We remark, however, that a generalization of the HIDDEN SHIFT problem which permits *approximate equality* results in a tight relationship between HIDDEN SHIFT problems for different cyclic groups. In particular, consider the δ-APPROXIMATE HIDDEN SHIFT problem given by two functions $f, g : G \to S$ with the promise that there exists an element $s \in G$ so that $\Pr_x[g(x) = f(sx)] \geq 1 - \delta$ (where x is chosen uniformly in G); the problem is to identify an element $s' \in G$ with this property. Note that s' may not be unique in this case.

In particular, consider an instance $f, g : \mathbb{Z}/n \to V$ of a HIDDEN SHIFT problem on a cyclic group $\mathbb{Z}/n$. We wish to "lift" this instance to a group $\mathbb{Z}/m$ for $m \gg n$ in such a way that a solution to the $\mathbb{Z}/m$ instance yields a solution to the $\mathbb{Z}/n$ instance. For a function $\phi : \mathbb{Z}/n \to V$, define the function $\hat{\phi} : \mathbb{Z}/m \to V$ by the rule $\hat{\phi}(x) = \phi(x \bmod n)$. Note, then, that $\Pr_x[\hat{f}(x) = \hat{g}(\hat{s} + x)] \geq 1 - n/m$ for the shift $\hat{s} = s$; moreover, recovering *any* shift for the $\mathbb{Z}/m$ problem which achieves equality with probability near $1 - n/m$ yields a solution to the $\mathbb{Z}/n$ problem (by taking the answer modulo n, perhaps after correcting for the $m \bmod n$ overhang at the end of the $\mathbb{Z}/m$ oracle). Note that this function is not injective.

We remark that the HIDDEN SHIFT problem for non-injective Boolean functions (i.e., with range $\mathbb{Z}/2$) sometimes admits efficient algorithms (see, e.g., [23,28]). Whether these techniques can be extended to the general setting above is an interesting open problem.

3.2 Selecting Hard Groups

Efficiently Solvable Cases. For some choices of underlying group G, some of the above problems admit polynomial-time algorithms. A notable case is the HIDDEN SUBGROUP PROBLEM on $G = \mathbb{Z}$, which can be solved efficiently by Shor's algorithm [30]. The HSP with arbitrary abelian G also admits a polynomial-time algorithm [15]. The earliest and simplest example was Simon's algorithm [31], which efficiently solves the HSP in the case $G = (\mathbb{Z}/2)^n$ and $H = \{1, s\}$ for unknown s, with only $O(n)$ queries to the oracle. Due to the fact that $(\mathbb{Z}/2)^n \rtimes \mathbb{Z}/2 \cong (\mathbb{Z}/2)^{n+1}$, Simon's algorithm also solves the HIDDEN SHIFT problem on $(\mathbb{Z}/2)^n$. Additionally, Friedl et al. [10] have given efficient (or quasi-polynomial) algorithms for hidden shifts over solvable groups of constant exponent; for example, their techniques yield efficient algorithms for the groups $(\mathbb{Z}/p)^n$ (for constant p) and $(S_4)^n$.

Cyclic Groups. In contrast with the HIDDEN SUBGROUP PROBLEM, the general abelian HIDDEN SHIFT is believed to be hard. The only nontrivial algorithm known is due to Kuperberg, who gave a subexponential-time algorithm for the HSP on dihedral groups [16]. He also gave a generalization to the abelian HIDDEN SHIFT problem, as follows.

Theorem 2 (Theorem 7.1 in [16]). *The abelian* HIDDEN SHIFT *problem has a quantum algorithm with time and query complexity* $2^{O((\log |G|)^{1/2})}$, *uniformly for all finitely-generated abelian groups.*

Regev and Kuperberg later improved the above algorithm (so it uses polynomial quantum space, and gains various knobs for tuning complexity parameters), but the time and query complexity remains the same [17,25].

There is also evidence connecting HSP on the dihedral group D_N (and hence also HS on $\mathbb{Z}/N$) to other hard problems. Regev showed that, if there exists an efficient quantum algorithm for the dihedral HSP which uses coset sampling (the only nontrivial technique known), then there's an efficient quantum algorithm for poly(n)-unique-SVP [26]. This problem, in turn, is the basis of several lattice-based cryptosystems. However, due to the costs incurred in the reduction, Kuperberg's algorithm only yields exponential-time attacks. An efficient solution to HS on $\mathbb{Z}/N$ could also be used to break a certain isogeny-based cryptosystem [4].

We will focus particularly on the case $\mathbb{Z}/2^n$. This is the simplest group for which all of our constructions and results apply. Moreover, basic computational tasks (encoding/decoding group elements as bitstrings, sampling uniformly random group elements, performing basic group operations, etc.) all have straightforward and extremely efficient implementations over $\mathbb{Z}/2^n$. The existence of a quantum attack with complexity $2^{O(\sqrt{n})}$ in this case will only become practically relevant in the very long term, when the costs of quantum and classical computations become somewhat comparable. If such attacks are truly a concern, then there are other natural group choices, as we discuss below.

Permutation Groups. In the search for quantum algorithms for HSP and HS, arguably the most-studied group family is the family of symmetric groups S_n. It is well-known that an algorithm for HSP over $S_n \wr (\mathbb{Z}/2)$ would yield a polynomial-time quantum algorithm for Graph Isomorphism. As discussed in Appendix A, this is precisely the case of HSP relevant to the HIDDEN SHIFT problem over S_n.

For these groups, the efforts of the quantum algorithms community have so far amounted only to negative results. First, it was shown that the standard Shor-type approach of computing with individual "coset states" cannot succeed [22]. In fact, entangled measurements over $\Omega(n \log n)$ coset states are needed [12], matching the information-theoretic upper bound [7]. Finally, the only nontrivial technique for performing entangled measurements over multiple registers, the so-called Kuperberg sieve, is doomed to fail as well [21].

While encoding, decoding, and computing over the symmetric groups is more complicated and less efficient than the cyclic case, it is a well-understood subject (see, e.g., [11]). When discussing these groups below, we will assume (without explicit mention) an efficient solution to these problems.

Matrix Groups. Another relevant family of groups are the matrix groups $\mathsf{GL}_2(\mathbb{F}_q)$ and $\mathsf{SL}_2(\mathbb{F}_q)$ over finite fields. These nonabelian groups exhibit many structural features which are similar to the symmetric groups, such as high-dimensional irreducible representations. Many of the negative results concerning the symmetric groups also carry over to matrix groups [12,21].

Efficient encoding, decoding, and computation over finite fields $\mathbb{F}_q$ is standard. Given these ingredients, extending to matrix groups is not complicated. In the case of $\mathsf{GL}_2(\mathbb{F}_q)$, we can encode an arbitrary pair (not both zero) $(a, c) \in \mathbb{F}_q^2$ in the first column, and any pair (b, d) which is not a multiple of (a, c) in the second column. For $\mathsf{SL}_2(\mathbb{F}_q)$, we simply have the additional constraint that d is fixed to $a^{-1}(1 + bc)$ by the choices of a, b, c.

Product Groups. Arguably the simplest group family for which the negative results of [12] apply, are certain n-fold product groups. These are groups of the form G^n where G is a fixed, constant-size group (e.g., S_5). This opens up the possibility of simply replacing the XOR operation (i.e., $\mathbb{Z}/2$ addition) with composition in some other constant-size group (e.g., S_5), and retaining the same n-fold product structure.

Some care is needed, however, because there do exist nontrivial algorithms in this case. When the base group G is solvable, then there are efficient algorithms for both HSP and HIDDEN SHIFT (see Theorem 4.17 in [10]). It is important to note that this efficient algorithm applies even to some groups (e.g., $(S_4)^n$) for which the negative results of [12] also apply. Nevertheless, solvability seems crucial for [10], and choosing $G = S_5$ for the base group gives a family for which no nontrivial HIDDEN SHIFT algorithms are known. We remark that there is however a $2^{O(\sqrt{n \log n})}$-time algorithm for order-2 HIDDEN SUBGROUP PROBLEMS on G^n based on Kuperberg's sieve [1]; this suggests the possibility of subexponential (i.e., $2^{O(n^\delta)}$ for $\delta < 1$) algorithms for HIDDEN SHIFT over these groups.

4   Hidden Shift Even-Mansour Ciphers

We now address the question of repairing classical symmetric-key schemes which are vulnerable to Simon's algorithm. We begin with the simplest construction, the so-called Even-Mansour cipher [8].

4.1   Generalizing the Even-Mansour Scheme

The Standard Scheme. The Even-Mansour construction turns a publicly known, random permutation $P : \{0,1\}^n \rightarrow \{0,1\}^n$ into a keyed, pseudorandom permutation

$$E_{k_1,k_2}^P : \{0,1\}^n \longrightarrow \{0,1\}^n$$

$$x \longmapsto P(x \oplus k_1) \oplus k_2$$

where $k_1, k_2 \in \{0,1\}^n$, and $\oplus$ denotes bitwise XOR. This scheme is relevant in two settings:

1. simply as a source of pseudorandomness; in this setting, oracle access to P is provided to all parties.
2. as a block cipher; now oracle access to both P and P^{-1} is provided to all parties. Access to P^{-1} is required for decryption. One can then ask if E^P is a PRP (adversary gets access to E^P), or a strong PRP (adversary gets access to both E^P and its inverse).

In all of these settings, Even-Mansour is known to be information-theoretically secure against classical adversaries making at most polynomially-many queries [8].

Quantum Chosen Plaintext Attacks on the Standard Scheme. The proofs of classical security of Even-Mansour carry over immediately to the setting of quantum adversaries with only classical access to the relevant oracles. However, if an adversary is granted quantum oracle access to the P and E^P oracles, but no access at all to the inverse oracles, then Even-Mansour is easily broken. This attack was first described in [19]; a complete analysis is given in [14]. The attack is simple: First, one uses the quantum oracles for P and E^P to create a quantum oracle for $P \oplus E^P$, i.e., the function

$$f(x) = P(x) \oplus P(x \oplus k_1) \oplus k_2.$$

One then runs Simon's algorithm [31] on the function f. The claim is that, with high probability, Simon's algorithm will output k_1. To see this, note that f satisfies half of Simon's promise, namely $f(x \oplus k_1) = f(x)$. Moreover, if it is classically secure, then it *almost* satisfies the entire promise. More precisely, for any fixed P and random pair (x, y), either the probability of a collision $f(x) = f(y)$ is low enough for Simon's algorithm to succeed, or there are so many collisions that there exists a classical attack [14]. Once we have recovered k_1, we also immediately recover k_2 with a classical query, since $k_2 = E^P_{k_1, k_2}(x) \oplus P(x \oplus k_1)$ for any x.

Hidden Shift Even-Mansour. To address the above attack, we propose simple variants of the Even-Mansour scheme. The construction generalizes the standard Even-Mansour scheme in the manner described in Sect. 1. Each variant is parameterized by a family of exponentially-large finite groups G. The general construction is straightforward to describe. We begin with a public permutation $P : G \to G$, and from it construct a family of keyed permutations

$$E^P_{k_1, k_2}(x) = P(x \cdot k_1) \cdot k_2,$$

where k_1, k_2 are now uniformly random elements of G, and $\cdot$ denotes composition in G. The formal definition, as a block cipher, follows.

Scheme 1 (Hidden Shift Even-Mansour block cipher). *Let $\mathcal{G}$ be a family of finite, exponentially large groups, satisfying the efficient encoding conditions given in Sect. 3.2. The scheme consists of three polynomial-time algorithms, parameterized by a permutation P of the elements of a group G in $\mathcal{G}$:*

- KeyGen $: \mathbb{N} \to G \times G$; on input $|G|$, outputs $(k_1, k_2) \in_R G \times G$;
- $\mathsf{Enc}^P_{k_1, k_2} : G \to G$; defined by $m \mapsto P(m \cdot k_1) \cdot k_2$;
- $\mathsf{Dec}^P_{k_1, k_2} : G \to G$; defined by $c \mapsto P^{-1}(c \cdot k_2^{-1}) \cdot k_1^{-1}$.

For simplicity of notation, we set $E^P_{k_1, k_2} := \mathsf{Enc}^P_{k_1, k_2}$. Note that $\mathsf{Dec}^P_{k_1, k_2} = \left(E^P_{k_1, k_2}\right)^{-1}$. Correctness of the scheme is immediate; in the next section, we present several arguments for its security in various settings. All of these arguments are based on the conjectured hardness of certain HIDDEN SHIFT problems over $\mathcal{G}$.

4.2   Security Reductions

We consider two settings. In the first, the adversary is given oracle access to the permutation P, and then asked to distinguish the Even-Mansour cipher $E^P_{k_1, k_2}$ from a random permutation unrelated to P. In the second setting, the adversary is given oracle access to P, P^{-1}, as well as $E^P_{k_1, k_2}$ and its inverse; the goal in this case is to recover the key (k_1, k_2) (or some part thereof).

Distinguishability from Random. We begin with the first setting described above. We fix a group G, and let $\mathcal{P}_G$ denote the family of all permutations of G. Select a uniformly random $P \in \mathcal{P}_G$. The encryption map for the Hidden Shift Even-Mansour scheme over G can be written as

$$E_{k_1,k_2}^P = L_{k_2} \circ P \circ L_{k_1}.$$

If we have oracle access to P, then this is clearly an efficiently computable sub-family of $\mathcal{P}_G$, indexed by key-pairs. For pseudorandomness, the relevant problem is then to distinguish E^P from a random permutation which is unrelated to the oracle P.

Problem 6 (Even-Mansour Distinguishability (EMD)). Given oracle access to permutations $P, Q \in \mathcal{P}_G$ and a promise that either (i) both P and Q are random, or (ii) P is random and $Q = E_{k_1,k_2}^P$ for random k_1, k_2, decide which is the case.

It is straightforward to connect this problem to the decisional version of RANDOM HIDDEN SHIFT, as follows.

Proposition 4. *If there exists a QPT $\mathcal{D}$ for EMD on $\mathcal{G}$, then there exists a QPT algorithm for the DRHS problem on $\mathcal{G}$, with soundness and completeness at most negligibly different from those of $\mathcal{D}$.*

Proof. Let f, g be the two oracle functions for the DRHS problem over G. We know that f is a random function from G to G, and we must decide if g is also random, or simply a shift of f. We sample t_1, t_2 uniformly at random from G, and provide $\mathcal{D}$ with oracles f (in place of P), and $g' := L_{t_2} \circ g \circ L_{t_1}$ (in place of E^P). We then simply output what $\mathcal{D}$ outputs. Note that (f, g) are uniformly random permutations if and only if (f, g') are. In addition, $g = f \circ L_s$ if and only if $g' = L_{t_2} \circ f \circ L_{st_1}$. It follows that the input distribution to $\mathcal{D}$ is as in EMD, modulo the fact that the oracles in DRHS are random functions rather than random permutations. The error resulting from this is at most negligible, by the collision-finding bound of Zhandry [35]. □

Next, we want to amplify the DRHS distinguisher, and then apply the reduction from HIDDEN SHIFT given in Proposition 3. Combining this with Proposition 4, we arrive at a complete security reduction.

Theorem 3. *Let $\mathcal{G}$ be either the $\mathbb{Z}/2^n$ group family or the S_n group family. Under Assumption 4, the Hidden Shift Even-Mansour cipher over $\mathcal{G}$ is a quantum-secure pseudorandom function.*

Proof. Let $\mathcal{G}$ be either the $\mathbb{Z}/2^n$ group family, or the S_n group family. If the Even-Mansour cipher over $\mathcal{G}$ is not a qPRP, then by Definition 3, there exists an algorithm $\mathcal{D}_{\mathsf{EMD}}$ for the EMD problem with total (i.e., completeness plus soundness) error at most $1 - 1/s(n)$ for some polynomial s. To give the adversary as much freedom as possible, we assume that the probability of selecting the public permutation P is taken into account here; that is, $\mathcal{D}_{\mathsf{EMD}}$ need only succeed with inverse-polynomial probability over the choices of permutation P, keys k_1, k_2, and its internal randomness.

By Proposition 4, we then also have a DRHS algorithm $\mathcal{D}_{\mathsf{DRHS}}$ with error at most $1 - 1/s(n)$ (up to negligible terms). We can amplify this algorithm by means of a $2k$-wise independent hash function family $\mathcal{H}$, where k is an upper bound on the running time of $\mathcal{D}_{\mathsf{DRHS}}$ (for the given input size n and required error bound $1/s(n)$). Given functions f, g for the DRHS problem on G, we select a random function $h \in \mathcal{H}$ and a random group element $t \in G$. We then call $\mathcal{D}_{\mathsf{DRHS}}$ with oracles

$$ f' := h \circ f \qquad \text{and} \qquad g_t' := h \circ g \circ L_t $$

Note that, to any efficient quantum algorithm, (i) f and g are random if and only if f' and g_t' are, and (ii) $g(x) = f(sx)$ if and only if $g_t'(x) = f'(stx)$. We know that $\mathcal{D}_{\mathsf{DRHS}}$ will succeed with probability $1 - 1/s(n)$, except the probability is now taken over the choice of t and h (rather than f and g). We repeat this process with different random choices of h and t. A straightforward application of a standard Chernoff bound shows that, after $O(p(n))$ runs, we will correctly distinguish with $1 - \mathrm{negl}(n)$ probability.

Finally, we apply Proposition 3, to get an algorithm for RANDOM HIDDEN SHIFT with negligible error; by Proposition 1, we get an equally strong algorithm for HIDDEN SHIFT.   □

Key Recovery Attacks. We now consider partial or complete key recovery attacks, in the setting where the adversary also gets oracle access to the inverses of P and $E^P_{k_1,k_2}$. Note that, for the Even-Mansour cipher on any group G, knowing the first key k_1 suffices to produce the second key k_2, since

$$ k_2 = P(x \cdot k_1)^{-1} E^P_{k_1,k_2}(x) $$

for every $x \in G$.

We remark that giving security reductions is now complicated by the fact that RANDOM HIDDEN SHIFT and its variants all become trivial if we are granted even a partial ability to invert f or g; querying $f^{-1} \circ g$ on any input x produces $x \cdot s^{-1}$, which immediately yields the shift s. However, we can still give a nontrivial reduction, as follows.

Theorem 4. *Consider the Even-Mansour cipher over $G \times G$, for any group G. Suppose there exists a QPT algorithm which, when granted oracle access to P, $E^{P_{k_1,k_2}}$, and their inverses, outputs k_1, k_2. Then there exists an efficient quantum algorithm for the* HIDDEN SHIFT *problem over G.*

Proof. We are given oracle access to functions $f, g : G \to G$ and a promise that there exists $s \in G$ such that $f(x) = g(x \cdot s)$ for all $x \in G$. We define the following oracles, which can be constructed from access to f and g. First, we have permutations $P_f, P_g : G \times G \to G \times G$ defined by

$$ P_f(x, y) = (x, y \cdot f(x)) \qquad \text{and} \qquad P_g(x, y) = (x, y \cdot g(x)). $$

Now we sample keys $k_1 = (x_1, y_1), k_2 = (x_2, y_2)$ from $G \times G$ and define the function $E := E_{k_1,k_2}^{P_f}$. To the key-recovery adversary $\mathcal{A}$ for Even-Mansour over $G \times G$, we provide the oracles E and E^{-1} for the encryption/decryption oracles, and the oracles P_g and P_g^{-1} for the public permutation oracles.

To see that we can recover the shift s from the output of $\mathcal{A}$, we rewrite E in terms of g, as follows:

$$\begin{aligned}
E(x, y) &= P_f(xx_1, yy_1) \cdot (x_2, y_2) \\
&= (xx_1, yy_1 f(xx_1)) \cdot (x_2, y_2) \\
&= (xx_1 s, yy_1 f(xx_1)) \cdot (s^{-1}x_2, y_2) \\
&= (xx_1 s, yy_1 g(xx_1 s)) \cdot (s^{-1}x_2, y_2) \\
&= P_g(xx_1 s, yy_1) \cdot (s^{-1}x_2, y_2).
\end{aligned}$$

After complete key recovery, $\mathcal{A}$ will output $(x_1 s, y_1)$ and $(s^{-1}x_2, y_2)$, from which we easily deduce s.   □

Remark. The reduction above focuses on the problem of recovering the entire key. Note that for certain groups, e.g., $\mathbb{Z}/p$ for prime p, predicting any bit of the key with inverse-polynomial advantage is sufficient to recover the entire key (see Håstad and Nåslund [13]). In such cases we may conclude that predicting individual bits of the key is difficult.

5 Hidden Shift CBC-MACs

5.1 Generalizing the Encrypted-CBC-MAC Scheme

The Standard Scheme. The standard Encrypted-CBC-MAC construction requires a pseudorandom permutation $E_k : \{0, 1\}^n \rightarrow \{0, 1\}^n$. A message m is subdivided into blocks $m = m_1 || m_2 || \cdots || m_l$, each of length n. The tag is then computed by repeatedly encrypting-and-XORing the message blocks, terminating with one additional round of encryption with a different key. Specifically, we set

$$\text{CBC-MAC}_{k,k'} := E_{k'}(E_k(m_l \oplus E_k(\cdots E_k(m_2 \oplus E_k(m_1)) \cdots))).$$

This yields a secure MAC for variable-length messages.

Quantum Chosen Plaintext Attacks on the Standard Scheme. If we are granted quantum CPA access to $\text{CBC-MAC}_{k,k'}$, then there is a $(\mathbb{Z}/2)^n$-hidden-shift attack, described below. This attack was described in [14]; another version of the attack appears in [29]. Consider messages consisting of two blocks, and fix the first block to be one of two distinct values $\alpha_0 \neq \alpha_1$. We use the oracle for $\text{CBC-MAC}_{k,k'}$ to construct an oracle for the function

$$f(b, x) := \text{CBC-MAC}_{k,k'}(\alpha_b || x) = E_{k'}(E_k(x \oplus E_k(\alpha_b))).$$

Note that f satisfies Simon's promise, since

$$f(b \oplus 1, x \oplus E_k(\alpha_0) \oplus E_k(\alpha_1)) = f(b, x)$$

for all b, x. We can thus run Simon's algorithm to recover the string $s_k = E_k(\alpha_0) \oplus E_k(\alpha_1)$. Knowledge of s_k enables us to find an exponential number of collisions, since

$$\text{CBC-MAC}_{k,k'}(\alpha_0 || x) = \text{CBC-MAC}_{k,k'}(\alpha_1 || x \oplus E_k(\alpha_0) \oplus E_k(\alpha_1)).$$

In particular, this CBC-MAC does not satisfy the Boneh-Zhandry notion of a secure MAC in the quantum world [2].

Hidden Shift CBC-MAC. We propose generalizing the Encrypted-CBC-MAC construction above, by allowing the bitwise XOR operation to be replaced by composition in some exponentially-large family of finite groups G. Each message block is then identified with an element of G, and we view the pseudorandom permutation E_k as a permutation of the group elements of G. We then define

$$\text{CBC-MAC}_{k,k'}^G : G^* \longrightarrow G$$
$$(m_1, \ldots, m_l) \longmapsto E_{k'}(E_k(m_l \cdot E_k(\cdots E_k(m_2 \cdot E_k(m_1)) \cdots))),$$

where $\cdot$ denotes the group operation in G.

Scheme 2 (Hidden Shift Encrypted-CBC-MAC). *Let G be a family of finite, exponentially large groups satisfying the efficient encoding conditions given in Sect. 3.2. Let $E_k : G \to G$ be a quantum-secure pseudorandom permutation. The scheme consists of three polynomial-time algorithms:*

- KeyGen; *on input $|G|$, outputs two keys k, k' using key generation for E;*
- Mac$_{k,k'}$: $m \longmapsto E_{k'}(E_k(m_l \cdot E_k(\cdots E_k(m_2 \cdot E_k(m_1)) \cdots)$;
- Ver$_{k,k'}$: $(m, t) \mapsto$ *accept if* Mac$_{k,k'}(m) = t$, *and* reject *otherwise.*

We consider the security of this scheme in the next section.

5.2 Security Reduction

We now give a reduction from the RANDOM HIDDEN SHIFT problem to collision-finding in the above CBC-MAC.

Theorem 5. *Let $\mathcal{G}$ be either the $\mathbb{Z}/2^n$ group family or the S_n group family. Under Assumption 4, the Hidden-Shift CBC-MAC over $\mathcal{G}$ is a collision-resistant function.*

Proof. For simplicity, we assume that the collision-finding adversary finds collisions between equal-length messages. This is of course trivially true, for example, if the MAC is used only for messages of some a priori fixed length.

Suppose we are given an instance of the HIDDEN SHIFT problem, i.e., a pair of functions F_0, F_1 with the promise that F_0 is random and F_1 is a shift of F_0.

We have at our disposal a QPT $\mathcal{A}$ which finds collisions in the Hidden Shift Encrypted-CBC-MAC. We assume without loss of generality that, whenever $\mathcal{A}$ outputs a collision (c, c'), there is no pair of prefixes of (c, c') that also give a valid collision; indeed, we can easily build an $\mathcal{A}'$ which, whenever such prefixes exist, simply outputs the prefix collision instead.

We assume for the moment that the number of message blocks in c and c' is the same number t. Since the number of blocks and the running time of $\mathcal{A}$ are polynomial, we can simply guess t, and we will guess correctly with inverse-polynomial probability. We run $\mathcal{A}$ with a modified oracle $\mathcal{O}$ which "inserts" our hidden shift problem at stage t. This is defined as follows.

Let m be our input message, and l the number of blocks. If $l < t$, we simply output the usual Encrypted-CBC-MAC of m. If $l \geq t$, we first perform $t - 1$ rounds of the CBC procedure, computing a function

$$h(m) := E_k(m_{t-1} \cdot E_k(\cdots E_k(m_2 \cdot E_k(m_1)) \cdots)).$$

Note that h only depends on the first $t-1$ blocks of m. Next, we choose a random bit b and compute $F_{b(m)}(m_t \cdot h(m))$. We then finish the rest of the rounds of the CBC procedure, outputting

$$O(m) := E'_k(E_k(m_l \cdot E_k(\cdots E_k(F_{b(m)}(m_t \cdot h(m))) \cdots).$$

It's not hard to see that the distribution that the adversary observes will be indistinguishable from the usual Encrypted-CBC-MAC. Suppose a collision (m, m') is output. We set $x_1 = m_1 || m_2 || \cdots || m_{t-1}$ and $x_2 = m'_1 || m'_2 || \cdots || m'_{t-1}$ and $y_1 = m_t$ and $y_2 = m'_t$. The collision then means that

$$F_{b(m)}(y_1 \cdot h(x_1)) = F_{b(m')}(y_2 \cdot h(x_2)).$$

Since $m \neq m'$, with probability $1/2$ we have $b(m) \neq b(m')$. We repeat $\mathcal{A}$ until we achieve inequality of these bits. We then have

$$F_0(y_1 \cdot h(x_1)) = F_1(y_2 \cdot h(x_2)) = F_0(y_2 \cdot h(x_2) \cdot s)$$

and so the shift is simply $s = y_2^{-1} h(x_2)^{-1} y_1 h(x_1)$. $\square$

6   Thwarting the Simon Attack on Other Schemes

It is reasonable to conjecture that our transformation secures (classically secure) symmetric-key schemes against quantum CPA, generically. So far, we have only been able to give complete security reductions in the cases of the Even-Mansour cipher and the Encrypted-CBC-MAC. For the case of all other schemes vulnerable to the Simon algorithm attacks of [14, 18, 19], we can only say that the attack is thwarted by passing from $(\mathbb{Z}/2)^n$ to $\mathbb{Z}/2^n$ or S_n. We now briefly outline two cases of particular note. For further details, see Appendix B.

The first case is the Feistel network construction, which transforms random functions into pseudorandom permutations. While the three-round Feistel cipher is known to be classically secure [20], no security proof is known in the quantum CPA case, for any number of rounds. In [18], a quantum chosen-plaintext attack is given for the three-round Feistel cipher, again based on Simon's algorithm. The attack is based on the observation that, if one fixes the first half of the input to one of two fixed values $\alpha_0 \neq \alpha_1$, then the output contains one of two functions $f_{\alpha_0}, f_{\alpha_1}$, which are $(\mathbb{Z}/2)^n$-shifts of each other. However, if we instead replace each bitwise XOR in the Feistel construction with addition modulo $\mathbb{Z}/2^n$, the two functions become $\mathbb{Z}/2^n$-shifts, and the attack now requires a cyclic HIDDEN SHIFT subroutine.

The second case is what [14] refer to as the "quantum slide attack," which uses Simon's algorithm to give a linear-time quantum chosen-plaintext attack, an exponential speedup over classical slide attacks. The attack works against ciphers $E_{k,t}(x) := k \oplus (R_k)^t(x)$ which consist of t rounds of a function $R_k(x) := R(x \oplus k)$. In the attack, one simply observes that $E_{k,t}(R(x))$ is a shift of $R(E_{k,t}(x))$ by the key k, and then applies Simon's algorithm. To defeat this attack, we simply work over $\mathbb{Z}/2^n$, setting $E_{k,t}(x) := k + (R_k)^t(x)$ and $R_k(x) := R(x + k)$. It's easy to see that the same attack now requires a HIDDEN SHIFT subroutine for $\mathbb{Z}/2^n$.

Acknowledgments. G.A. would like to thank Tommaso Gagliardoni and Christian Majenz for helpful discussions. G.A. acknowledges financial support from the European Research Council (ERC Grant Agreement 337603), the Danish Council for Independent Research (Sapere Aude) and VILLUM FONDEN via the QMATH Centre of Excellence (Grant 10059). A.R. acknowledges support from NSF grant IIS-1407205.

A   Hidden Subgroups and Hidden Shifts

Basic Definitions. We now briefly discuss the HIDDEN SUBGROUP PROBLEM, which is closely related to HIDDEN SHIFT.

Problem 7 (HIDDEN SUBGROUP PROBLEM *(HSP))*. Let G be a group and S a set. Given a function $f : G \to S$, and a promise that there exists $H \leq G$ such that f is constant and distinct on the right cosets of H, output a complete set of generators for H.

Another, equivalent formulation of the HSP promise on f is that for $x \neq y$, $f(x) = f(y)$ iff $x = h \cdot y$ for $h \in H$. As before, one can also consider decision versions of HSP (e.g., where one has to decide if f hides a trivial or nontrivial subgroup) and promise versions where the function f is a random function satisfying the constraint that $f(hx) = x$ for all $x \in G$ and $h \in H$. This last variant is important for our purposes so we separately define it.

Problem 8 (RANDOM HIDDEN SUBGROUP PROBLEM *(RHSP))*. Let G be a group and S a set. Given a function $f : G \to S$ chosen uniformly among all functions for which $f(x) = f(hx)$ for all $x \in G$ and $h \in H$, output a complete set of generators for H.

Some Reductions. Traditionally, the Hidden Subgroup problem (HSP) has played a prominent role in the literature, as it offers a simple framework to which many other problems can be directly reduced. Indeed, there is a general reduction from HS to HSP.

A Canonical Reduction from HS to HSP. Consider an instance of a HIDDEN SHIFT problem over G given by the functions $f_0, f_1 : G \to S$ such that $f_0(x) = f_1(x \cdot s)$. Recall that the *wreath product* $K = G \wr \mathbb{Z}/2$ is the semi-direct product $(G \times G) \rtimes \mathbb{Z}/2$, where the action of the nontrivial element of $\mathbb{Z}/2$ on $G \times G$ is the swap $(a, b) \mapsto (b, a)$. Now define the function

$$\varphi : (G \times G) \rtimes \mathbb{Z}/2 \longrightarrow S \times S$$
$$((x, y), b) \longmapsto (f_b(x), f_{b \oplus 1}(y)).$$

One then easily checks that the function φ is constant and distinct on the cosets of the order-two subgroup of K generated by $((s, s^{-1}), 1)$.

We remark that the reduction above can significantly "complicate" the underlying group. In particular, note that $A \rtimes \mathbb{Z}/2$ is always non-abelian (unless the action of $1 \in \mathbb{Z}/2 = \{0, 1\}$ on A is trivial). Note that this reduction does not yield a reduction from RHS to RHSP as the resulting HSP instance is not uniformly random in the fully random case.

The Special Case of $(\mathbb{Z}/2)^n$; Reductions from RHS to RHSP. On $(\mathbb{Z}/2)^n$, the HIDDEN SHIFT problem can be reduced to the RANDOM HIDDEN SUBGROUP PROBLEM on $(\mathbb{Z}/2)^n$ via a special reduction that exploits $\mathbb{Z}/2$ structure. Specifically, for a pair of injective functions $f_0, f_1 : (\mathbb{Z}/2)^n \to S$ (for which $f_0(x) = f_1(x \oplus s)$,) construct the function $f : (\mathbb{Z}/2)^{n+1} \to S$ so that

$$g(bx) = f_b(x), \qquad \text{for } b \in \mathbb{Z}/2 \text{ and } x \in (\mathbb{Z}/2)^n.$$

Then observe that g hides the subgroup generated by $1s$.

Note, furthermore, that if the f_i are (independent) random functions, then the same can be said of g; likewise, if f_1 is a shift of the random function f_0, the function g is precisely a random function subject to the constraint that $g(x) = g(x \oplus 1s)$; thus this reduces RHS to RHSP. In this RHS setting, it is possible to develop an alternate reduction that more closely resembles the attacks we discussed above. Specifically, given the functions $f_0, f_1 : (\mathbb{Z}/2)^n \to (\mathbb{Z}/2)^n$ (so that $S = (\mathbb{Z}/2)^n$), consider the oracle $g = f_0 \oplus f_1$. When $f_1(x) = f_0(x \oplus s)$, note that this oracle satisfies the symmetry condition

$$[f_0 \oplus f_1](x \oplus s) = f_0(x \oplus s) \oplus f_1(x \oplus s)$$
$$= f_1(x \oplus s \oplus s) \oplus f_0(x) = f_0(x) \oplus f_1(x) = [f_0 \oplus f_1](x)$$

so that $g = f_0 \oplus f_1$ is a random function subject to the constraint that $g(x) = g(x \oplus s)$, as desired. If the functions f_i are independent, the function g has the uniform distribution. Thus this reduces RHS to RHSP.

B   Other Hidden Shift Constructions

B.1   Feistel Ciphers

The Standard Scheme. The Feistel cipher is a method for turning random functions into pseudorandom permutations. The core ingredient is a one-round Feistel cipher, which, for a function $f : \{0,1\}^n \to \{0,1\}^n$, is given by

$$F_f : \{0,1\}^{2n} \longrightarrow \{0,1\}^{2n}$$
$$x\|y \longmapsto y \oplus f(x)\|x.$$

The function f is called the "round function." The multi-round version of the Feistel cipher is defined by concatenating multiple one-round ciphers, each with a different choice of round function. Of particular interest is the three-round cipher, defined by

$$F_{R_1,R_2,R_3}(x\|y) = F_{R_3}(F_{R_2}(F_{R_1}(x\|y))).$$

A well-known result of Luby and Rackoff says that, if the R_j are random and independent, then F_{R_1,R_2,R_3} is indistinguishable from a random permutation [20].

Quantum Chosen Plaintext Attacks on the Standard Scheme. Suppose we are given quantum oracle access to a function F, and promised that F is either a random permutation, or that $F := F_{R_1,R_2,R_3}$ for some unknown, random functions R_j. The following attack was first shown in [18]; a thorough analysis appears in [14]. We first fix two n-bit strings $\alpha_0 \neq \alpha_1$. We then use the oracle for F to build oracles f_0, f_1 defined by

$$f_b(y) := F(\alpha_b\|y)\big|_{n+1}^{2n} \oplus \alpha_b.$$

Here $s|_j^k := s_j s_{j+1} \cdots s_k$. We then run Simon's algorithm to see if there's a shift between f_0 and f_1. If a shift is produced, we output "Feistel." Otherwise we output "random."

To see why the attack is successful, first note that if F is a random permutation, then the f_b are random functions. On the other hand, if $F = F_{R_1,R_2,R_3}$, then one easily checks that $f_b(y) = R_2(y \oplus R_1(\alpha_b))$. We then have

$$f_1(y) = f_0(y \oplus (R_1(\alpha_0) \oplus R_1(\alpha_1)))$$

for all y. Since the R_j are random, one can check that there are not too many other collisions [14]. It follows that Simon's algorithm will output $R_1(\alpha_0) \oplus R_1(\alpha_1)$ with high probability.

Hidden Shift Feistel Cipher. For simplicity, we will work over the group $\mathbb{Z}/2^n$. Our construction generalizes to other group families in a straightforward way.

Given a function $f : \mathbb{Z}/2^n \to \mathbb{Z}/2^n$, we define the one-round Feistel cipher using round function f to be

$$F_f : \mathbb{Z}/2^n \times \mathbb{Z}/2^n \longrightarrow \mathbb{Z}/2^n \times \mathbb{Z}/2^n$$
$$(x, y) \longmapsto (y + f(x), x).$$

Since $\mathbb{Z}/2^n \times \mathbb{Z}/2^n \cong \mathbb{Z}/2^{2n}$, we can then view F_f as a permutation on $2n$-bit strings. For multi-round ciphers, we define composition as before. In particular, given three functions R_1, R_2, R_3 on G, we get the three-round Feistel cipher

$$F_{R_1, R_2, R_3} : \mathbb{Z}/2^n \times \mathbb{Z}/2^n \longrightarrow \mathbb{Z}/2^n \times \mathbb{Z}/2^n$$
$$(x, y) \longmapsto F_{R_3}(F_{R_2}(F_{R_1}(x, y))).$$

Next, we check that the attack of [18] now appears to require a subroutine for the Hidden Shift problem over $\mathbb{Z}/2^n$, contrary to our Cyclic Hidden Shift Assumption (Assumption 2). We are given a function F on $\mathbb{Z}/2^{2n}$ with the promise that F is either random or a Feistel cipher F_{R_1, R_2, R_3} as above. Proceeding precisely as before, we pick two elements $\alpha_0 \neq \alpha_1$ of $\mathbb{Z}/2^n$ and build two functions

$$f_b(y) := F(\alpha_b, y)|_{n+1}^{2n} - \alpha_b.$$

If F is random, then clearly so are f_0 and f_1. But if $F = F_{R_1, R_2, R_3}$ then one easily checks that

$$f_b(y) = R_2(y + R_1(\alpha_b)),$$

from which it follows that

$$f_1(y) = R_2(y + R_1(\alpha_1)) = R_2(y + R_1(\alpha_0) - R_1(\alpha_0) + R_1(\alpha_1)) = f_0(y + s)$$

where we set $s := R_1(\alpha_1) - R_1(\alpha_0)$. We are thus presented with a HIDDEN SHIFT problem over the group $\mathbb{Z}/2^n$, which is hard according to Assumption 2. The only known subroutine (analogous to Simon) that one could apply here would be Kuperberg's algorithm, which would find s in time $2^{\Theta(\sqrt{n})}$ [16]. Defining the Feistel network over other groups (such as S_n) would frustrate all nontrivial quantum-algorithmic approaches, including the Kuperberg approach [21].

B.2   Protecting Against Quantum Slide Attacks

Quantum Slide Attack. Classically, slide attacks are a class of subexponential-time attacks against ciphers which encrypt simply by repeatedly applying some function R_k, with a fixed key k. Kaplan et al. [14] showed how Simon's algorithm can be used to give a polynomial-time "quantum slide attack" against ciphers of the form

$$E_{k,t} := k \oplus R_k^t(x) = k \oplus (R_k \circ R_k \cdots \circ R_k)(x),$$

where $R_k(x) = R(x \oplus k)$, and R is a known permutation. As usual, the attack requires quantum CPA access to $E_{k,t}$. The attack follows directly from the observation that the functions

$$f_b(x) = \begin{cases} E_{k,t}(R(x)) \oplus x & \text{if } b = 0, \\ R(E_{k,t}(x)) \oplus x & \text{if } b = 1. \end{cases}$$

are shifts of each other by the key k, i.e., $f_0(x \oplus k) = f_1(x)$ for all x. This means we can extract k with Simon's algorithm.

Eliminating the Attack via Hidden Shifts. Following our established pattern, we adapt schemes $E_{k,t}$ to use modular addition over $\mathbb{Z}/2^n$ instead of bitwise XOR. Given a permutation R of $\{0,1\}^n$, we now set

$$R_k(x) := R(x + k) \qquad \text{and} \qquad E_{k,t} := k + R_k^t(x).$$

Proceeding with the attack as before, we now define

$$f_b(x) = \begin{cases} E_{k,t}(R(x)) - x & \text{if } b = 0, \\ R(E_{k,t}(x)) - x & \text{if } b = 1. \end{cases}$$

We then check that

$$\begin{aligned} f_0(x + k) &= E_{k,t}(R(x + k)) - (x + k) \\ &= k + R_k^t(R(x + k)) - x - k \\ &= R(E_{k,t}(x)) - x \\ &= f_1(x). \end{aligned}$$

Continuing as in the Simon attack would now require a solution to the HIDDEN SHIFT problem over $\mathbb{Z}/2^n$.

References

1. Alagic, G., Moore, C., Russell, A.: Quantum algorithms for Simon's problem over general groups. In: Proceedings of the Eighteenth Annual ACM-SIAM Symposium on Discrete Algorithms (SODA), pp. 1217–1224. ACM Press (2007)
2. Boneh, D., Zhandry, M.: Quantum-secure message authentication codes. In: Johansson, T., Nguyen, P.Q. (eds.) EUROCRYPT 2013. LNCS, vol. 7881, pp. 592–608. Springer, Heidelberg (2013). doi:10.1007/978-3-642-38348-9_35
3. Chen, L., Jordan, S., Liu, Y.-K., Moody, D., Peralta, R., Perlner, R., Smith-Tone, D.: Report on post-quantum cryptography. Technical report, National Institute of Standards and Technology (2016). http://nvlpubs.nist.gov/nistpubs/ir/2016/NIST.IR.8105.pdf
4. Childs, A.M., Jao, D., Soukharev, V.: Constructing elliptic curve isogenies in quantum subexponential time. J. Math. Cryptol. **8**(1), 1–29 (2014)
5. Dinh, H., Moore, C., Russell, A.: Limitations of single coset states and quantum algorithms for code equivalence. Quantum Inf. Comput. **15**(3–4), 260–294 (2015)

6. Dunkelman, O., Keller, N., Shamir, A.: Slidex attacks on the Even-Mansour encryption scheme. J. Cryptol. **28**(1), 1–28 (2015). doi:10.1007/s00145-013-9164-7. ISSN 1432-1378

7. Ettinger, M., Høyer, P., Knill, E.: The quantum query complexity of the hidden subgroup problem is polynomial. Inf. Process. Lett. **91**(1), 43–48 (2004). doi:10.1016/j.ipl.2004.01.024

8. Even, S., Mansour, Y.: A construction of a cipher from a single pseudorandom permutation. J. Cryptol. **10**(3), 151–161 (1997). doi:10.1007/s001459900025

9. Fenner, S., Zhang, Y.: On the complexity of the hidden subgroup problem. Int. J. Found. Comput. Sci. **24**(8), 1221–1234 (2013)

10. Friedl, K., Ivanyos, G., Magniez, F., Santha, M., Sen, P.: Hidden translation and translating coset in quantum computing. SIAM J. Comput. **43**(1), 1–24 (2014). doi:10.1137/130907203

11. Furst, M., Hopcroft, J., Luks, E.: Polynomial-time algorithms for permutation groups. In: Proceedings of the 21st Annual Symposium on Foundations of Computer Science, FOCS 1980, Washington, DC, USA, pp. 36–41. IEEE Computer Society (1980). doi:10.1109/SFCS.1980.34

12. Hallgren, S., Moore, C., Rötteler, M., Russell, A., Sen, P.: Limitations of quantum coset states for graph isomorphism. J. ACM **57**(6), 34:1–34:33 (2010). doi:10.1145/1857914.1857918

13. Håstad, J., Näslund, M.: The security of all RSA and discrete log bits. J. ACM **51**(2), 187–230 (2004). doi:10.1145/972639.972642

14. Kaplan, M., Leurent, G., Leverrier, A., Naya-Plasencia, M.: Breaking symmetric cryptosystems using quantum period finding. In: Robshaw, M., Katz, J. (eds.) CRYPTO 2016. LNCS, vol. 9815, pp. 207–237. Springer, Heidelberg (2016). doi:10.1007/978-3-662-53008-5_8

15. Kitaev, A.Y.: Quantum measurements and the abelian stabilizer problem. Technical report, November 1995. arXiv:quant-ph/9511026

16. Kuperberg, G.: A subexponential-time quantum algorithm for the dihedral hidden subgroup problem. SIAM J. Comput. **35**(1), 170–188 (2005). doi:10.1137/S0097539703436345

17. Kuperberg, G.: Another subexponential-time quantum algorithm for the dihedral hidden subgroup problem. Technical report, December 2011. arXiv:quant-ph/1112.3333

18. Kuwakado, H., Morii, M.: Quantum distinguisher between the 3-round Feistel cipher and the random permutation. In: 2010 IEEE International Symposium on Information Theory Proceedings (ISIT), pp. 2682–2685, June 2010. doi:10.1109/ISIT.2010.5513654

19. Kuwakado, H., Morii, M.: Security on the quantum-type Even-Mansour cipher. In: Proceedings of the International Symposium on Information Theory and Its Applications (ISITA), pp. 312–316. IEEE Computer Society (2012)

20. Luby, M., Rackoff, C.: How to construct pseudorandom permutations from pseudorandom functions. SIAM J. Comput. **17**(2), 337–386 (1988)

21. Moore, C., Russell, A., Śniady, P.: On the impossibility of a quantum sieve algorithm for graph isomorphism. In: Proceedings of the Thirty-Ninth Annual ACM Symposium on Theory of Computing, STOC 2007, pp. 536–545. ACM, New York (2007). doi:10.1145/1250790.1250868

22. Moore, C., Russell, A., Schulman, L.J.: The symmetric group defies strong fourier sampling. SIAM J. Comput. **37**, 1842–1864 (2008). doi:10.1137/050644896

23. Ozols, M., Roetteler, M., Roland, J.: Quantum rejection sampling. ACM Trans. Comput. Theory **5**(3), 11:1–11:33 (2013). doi:10.1145/2493252.2493256

24. Patel, S., Ramzan, Z., Sundaram, G.S.: Luby-Racko. Ciphers: why XOR is not so exclusive. In: Nyberg, K., Heys, H. (eds.) SAC 2002. LNCS, vol. 2595, pp. 271–290. Springer, Heidelberg (2003). doi:10.1007/3-540-36492-7_18
25. Regev, O.: A subexponential time algorithm for the dihedral hidden subgroup problem with polynomial space. Technical report, June 2004. arXiv:quant-ph/0406151
26. Regev, O.: Quantum computation and lattice problems. SIAM J. Comput. $33(3)$, 738–760 (2004). doi:10.1137/S0097539703440678
27. Regev, O.: On lattices, learning with errors, random linear codes, and cryptography. In: Proceedings of the Thirty-Seventh Annual ACM Symposium on Theory of Computing, STOC 2005, pp. 84–93. ACM, New York (2005). doi:10.1145/1060590.1060603
28. Roetteler, M.: Quantum algorithms for abelian difference sets and applications to dihedral hidden subgroups. In: 11th Conference on the Theory of Quantum Computation, Communication and Cryptography, TQC 2016, Berlin, 27–29 September 2016
29. Santoli, T., Schaffner, C.: Using Simon's algorithm to attack symmetric-key cryptographic primitives. Quantum Inf. Comput. $17(1\&2)$, 65–78 (2017)
30. Shor, P.W.: Algorithms for quantum computation: discrete logarithms and factoring. In: Proceedings of the 35th Annual Symposium on Foundations of Computer Science, FOCS 1994, Washington, DC, USA, pp. 124–134. IEEE Computer Society (1994). doi:10.1109/SFCS.1994.365700. ISBN 0-8186-6580-7
31. Simon, D.R.: On the power of quantum computation. SIAM J. Comput. $26(5)$, 1474–1483 (1997). doi:10.1137/S0097539796298637
32. Zhandry, M.: A note on quantum-secure PRPs, November 2016. arXiv preprint: arXiv:1607.07759
33. Zhandry, M.: How to construct quantum random functions. In: Proceedings of the IEEE 53rd Annual Symposium on Foundations of Computer Science, FOCS 2012, Washington, DC, USA, pp. 679–687. IEEE Computer Society (2012). doi:10.1109/FOCS.2012.37. ISBN 978-0-7695-4874-6
34. Zhandry, M.: Secure identity-based encryption in the quantum random Oracle model. In: Safavi-Naini, R., Canetti, R. (eds.) CRYPTO 2012. LNCS, vol. 7417, pp. 758–775. Springer, Heidelberg (2012). doi:10.1007/978-3-042-32009-5_44
35. Zhandry, M.: A note on the quantum collision and set equality problems. Quantum Info. Comput. $15(7–8)$, 557–567 (2015)

Boolean Searchable Symmetric Encryption with Worst-Case Sub-linear Complexity

Seny Kamara$^{(\boxtimes)}$ and Tarik Moataz

Brown University, Providence, USA
{seny,tarik_moataz}@brown.edu

Abstract. Recent work on searchable symmetric encryption (SSE) has focused on increasing its expressiveness. A notable example is the OXT construction (Cash et al., *CRYPTO '13*) which is the first SSE scheme to support conjunctive keyword queries with sub-linear search complexity. While OXT efficiently supports disjunctive and boolean queries that can be expressed in searchable normal form, it can only handle *arbitrary* disjunctive and boolean queries in linear time. This motivates the problem of designing expressive SSE schemes with *worst-case* sub-linear search; that is, schemes that remain highly efficient for *any* keyword query.

In this work, we address this problem and propose non-interactive highly efficient SSE schemes that handle *arbitrary* disjunctive and boolean queries with worst-case sub-linear search and optimal communication complexity. Our main construction, called IEX, makes black-box use of an underlying single keyword SSE scheme which we can instantiate in various ways. Our first instantiation, IEX-2Lev, makes use of the recent 2Lev construction (Cash et al., *NDSS '14*) and is optimized for search at the expense of storage overhead. Our second instantiation, IEX-ZMF, relies on a new single keyword SSE scheme we introduce called ZMF and is optimized for storage overhead at the expense of efficiency (while still achieving asymptotically sub-linear search). Our ZMF construction is the first adaptively-secure highly compact SSE scheme and may be of independent interest. At a very high level, it can be viewed as an encrypted version of a new Bloom filter variant we refer to as a Matryoshka filter. In addition, we show how to extend IEX to be dynamic and forward-secure. To evaluate the practicality of our schemes, we designed and implemented a new encrypted search framework called *Clusion*. Our experimental results demonstrate the practicality of IEX and of its instantiations with respect to either search (for IEX-2Lev) and storage overhead (for IEX-ZMF).

1 Introduction

A structured encryption (STE) scheme encrypts a data structure in such a way that it can be privately queried. An STE scheme is secure if it reveals nothing

---

S. Kamara—Work done in part at Microsoft Research.

T. Moataz—Work done in part at IMT Atlantique and Colorado State.

J.-S. Coron and J.B. Nielsen (Eds.): EUROCRYPT 2017, Part III, LNCS 10212, pp. 94–124, 2017.
DOI: 10.1007/978-3-319-56617-7_4

about the structure and query beyond a well-specified and "reasonable" leakage profile [13,15]. STE schemes come in two forms: response-revealing and response-hiding. The former reveals the query response in plaintext whereas the latter does not. An important special case of STE is searchable symmetric encryption (SSE) which encrypts search structures such as inverted indexes [10,11,13,15,23,24] or search trees [19,23]. Another example is graph encryption which encrypts various kinds of graphs [13,27]. STE has received a lot of attention from Academia and Industry due to: (1) its potential applications to cloud storage and database security; and (2) the fact that, among a host of different encrypted search solutions (e.g., property-preserving encryption, fully-homomorphic encryption, oblivious RAM, functional encryption) it seems to provide the best tradeoffs between security and efficiency.

In recent years, much of the work on STE has focused on supporting more complex structures and queries. A notable example in the setting of SSE is the work of Cash et al. which proposed the first SSE scheme to support conjunctive queries in sub-linear time [11]. Their scheme, OXT, is also shown to support disjunctive and even boolean queries. Faber et al. later showed how to extend OXT to achieve even more complex queries including range, substring, wildcard and phrase queries. Another example is the BlindSeer project from Pappas et al. [30] and Fisch et al. [17] which present a solution that supports boolean and range queries as well as stemming in sub-linear time.

Naive Solutions. Any boolean query $\phi(w_1, \ldots, w_q)$, where $w_1, \ldots, w_q$ are keywords and ϕ is a boolean formula, can be handled using a single-keyword SSE scheme in a naive way. In the case of response-revealing schemes it suffices to search for each keyword and have the server take the intersection and unions of the result sets appropriately. The issue with this approach, of course, is that the server learns more information than necessary: namely, it learns the result sets $\mathsf{DB}(w_1)$, ..., $\mathsf{DB}(w_q)$ whereas it should only learn the set $\mathsf{DB}(\phi(w_1, \ldots, w_q))$. For response-hiding schemes, one can search for each keyword and compute the intersections and unions at the client. The problem with this approach is that the parties communicate more information than necessary: namely, the server sends elements within the intersections of the result sets multiple times. With this in mind, any boolean SSE solution should improve on one of the naive approaches depending on whether it is response-hiding or response-revealing.

Worst-Case Sub-linear Search Complexity. While OXT achieves sub-linear search complexity for conjunctive queries, its extension to disjunctive and arbitrary boolean queries does not. More precisely, OXT remains sub-linear only for queries in *searchable normal form* (SNF) which have the form $w_1 \wedge \phi(w_2, \ldots, w_q)$, where w_1 through w_q are keywords and ϕ is an arbitrary boolean formula. For non-SNF queries, OXT requires linear time in the number of documents. This motivates the following natural question: *can we design SSE schemes that support arbitrary disjunctive and arbitrary boolean queries with sub-linear search complexity?* In other words, can we design solutions for these queries that are efficient even in the worst-case?

1.1   Our Contributions and Techniques

In this work, we address this problem and propose efficient disjunctive and boolean SSE schemes with worst-case sub-linear search complexity and optimal communication overhead. Our schemes are non-interactive and, as far as we know, the first to achieve optimal communication complexity. To do this we make several contributions which we summarize below

Worst-Case Disjunctive Search. Our first solution, which we call IEX, is a worst-case sub-linear disjunctive SSE scheme. While it leaks more than the naive response-hiding solution, we stress that it achieves *optimal* communication complexity which, for response-hiding schemes, is the main tradeoff we seek. In addition, it leaks *less* than OXT (when used for disjunctive queries) while achieving worst-case efficiency.

The underlying idea behind IEX's design is best expressed in set-theoretic terms where we view the result of a disjunctive query $w_1 \vee \cdots \vee w_q$ as the union of the results of each individual term. More precisely, if we denote by $\mathsf{DB}(w)$ the set of document identifiers that contain the query w, then $\mathsf{DB}(w_1 \vee \cdots \vee w_q) = \mathsf{DB}(w_1) \cup \cdots \cup \mathsf{DB}(w_q)$. Using the naive response-hiding approach, one could use a single-keyword response-hiding scheme to query each keyword and compute the union at the client but, as discussed above, this would incur poor communication complexity. Our approach is different and, intuitively speaking, makes use of the *inclusion-exclusion* principle as follows. Consider a three-term query $w_1 \vee w_2 \vee w_3$. Instead of searching for $\mathsf{DB}(w_1)$, $\mathsf{DB}(w_2)$, $\mathsf{DB}(w_3)$ and computing the union, we compute $\mathsf{DB}(w_1)$ and remove from it

$$\mathsf{DB}(w_1) \cap \mathsf{DB}(w_2) \quad \text{and} \quad \mathsf{DB}(w_1) \cap \mathsf{DB}(w_3).$$

We then compute $\mathsf{DB}(w_2)$ and remove from it $\mathsf{DB}(w_2) \cap \mathsf{DB}(w_3)$. Finally, we take the union of the remaining sets and add $\mathsf{DB}(w_3)$. It follows by the inclusion-exclusion principle that this results in exactly $\mathsf{DB}(w_1) \cup \mathsf{DB}(w_2) \cup \mathsf{DB}(w_3)$. If we could somehow support the intersection and removal operations at the server, then we could achieve optimal communication complexity. Note that this high-level approach is "purely disjunctive" in the sense that it does not rely on transforming the query into another form as done in OXT. The avoidance of SNF in particular is what enables us to achieve worst-case efficiency.

We stress that the intuition provided thus far is only a very high-level conceptual explanation of our approach and cannot be translated directly to work on encrypted data. The challenge is that no SSE scheme we are aware of directly supports the kind of set operations needed to implement this idea. Therefore, a major part of our contribution is in designing and analyzing such a scheme.

Boolean Search. While IEX is naturally disjunctive, we show that it also supports boolean queries. Similarly to the disjunctive case, we explain our high-level approach in set-theoretic terms. First, recall that any boolean query can be written in conjunctive normal form (CNF) so it has the form $\Delta_1 \wedge \cdots \wedge \Delta_\ell$, where each $\Delta_i = w_{i,1} \vee \cdots \vee w_{i,q}$ is a disjunction. Given a response-hiding *disjunctive*-search scheme like IEX, a naive approach for CNF queries is to execute disjunctive searches for each disjunction $\Delta_1, \ldots, \Delta_\ell$ and have the client perform the intersection of the results. This approach is problematic, however, because it requires more communication than necessary. To avoid this we take the following alternative approach. We note that the result $\mathsf{DB}(\Delta_1 \wedge \cdots \wedge \Delta_\ell)$ is a subset of $\mathsf{DB}(\Delta_1)$ and that it can be computed by progressively keeping only the identifiers in $\mathsf{DB}(\Delta_1)$ that are also included in $\mathsf{DB}(\Delta_2)$ through $\mathsf{DB}(\Delta_\ell)$. Again, we stress that this description is only a high-level conceptual explanation of our approach and requires more work to instantiate over encrypted data.

The IEX Structure. As mentioned above, a major challenge in this work is the design of an encrypted structure that supports the set-theoretic operations needed to implement the strategies discussed above. To achieve this, IEX makes use of a more complex structure than the traditional encrypted inverted index. In particular, IEX combines several instantiations of two kinds of structures: dictionaries and multi-maps. A dictionary (i.e., a key-value store) maps labels to values whereas a multi-map (i.e., an inverted index) maps labels to tuples of values. More precisely, the IEX design consists of an encrypted *global* multi-map that maps every keyword w to its document identifiers $\mathsf{DB}(w)$ and an encrypted dictionary that maps every keyword to a *local* multi-map for w. The local multi-map of a keyword w maps all the keywords v that co-occur with w to the identifiers of the documents that contain both v and w. At a high-level, with the encrypted global multi-map we can recover $\mathsf{DB}(w_1)$. With the encrypted dictionary, we can recover the encrypted local multi-map for keywords w_2 through w_ℓ. And, finally, by querying the (encrypted) local multi-map of a keyword w with a keyword v, we can recover the identifiers of the documents that contain both w and v. With these basic operations, we can then execute a full disjunctive query as discussed above.

Instantiations. IEX is an abstract construction that makes black-box use of encrypted multi-maps and dictionaries which, in turn, can be instantiated with several concrete constructions, e.g., [10,13,15,23].[1] While its asymptotic complexity is not affected by how the building blocks are instantiated, its concrete efficiency is so we consider this choice carefully—especially how the local multi-maps are instantiated. We consider two instantiations. The first, IEX-2Lev, uses the 2Lev construction of Cash et al. [10] to encrypt the multi-maps (local and global). This particular instantiation is very efficient with respect to search time but produces large encrypted structures (e.g., 9.8 GB for datasets of 34M keyword/id pairs).

To address this we propose a second instantiation called IEX-ZMF which trades off efficiency for compactness. In fact, we show that IEX-ZMF is an order of magnitude more compact than IEX-2Lev (e.g., producing 0.9 GB EDBs for datasets with 34M keyword/id pairs). This compactness is achieved by

---

[1] Other constructions such as [11,24,28,32] could also be used but these are either dynamic or conjunctive which is not needed for the IEX.

encrypting IEX's local multi-maps with a new construction called ZMF which may be of independent interest and that we detail below.[2]

The ZMF Scheme. ZMF is a multi-map encryption scheme that is inspired by and has similarities to the classic Z-IDX construction of Goh [19]. Its core design as well as its security are very different, however. While Z-IDX produces a collection of non-adaptively-secure *fixed*-size encrypted Bloom filters, ZMF produces a collection of *adaptively*-secure *variable*-sized encrypted Bloom filters. In addition, the hash functions used for each filter can all be derived from a fixed set of hash functions (even though the filters store a different number of elements). This last property is non-standard but is crucial for our approach to be practical as it allows us to generate constant-size tokens that can be used with every filter in the collection. We refer to such collections of Bloom filters as *matryoshka filters* and, as far as we know, they have not been considered in the past. As we detail in Sect. 7, encrypting matryoshka filters with adaptive security is quite challenging. For this, we rely on the random oracle model and on a non-standard use of online ciphers [4] which are streaming block ciphers in the sense that every ciphertext block depends only on the previous plaintext blocks. Note that like Z-IDX, ZMF has linear search time but we use it in our IEX construction only to encrypt the *local* multi-maps which guarantees that IEX-ZMF is still sub-linear.

Dynamism and Forward-Security. We extend IEX to be dynamic resulting in a new scheme DIEX. An important security property for dynamic SSE schemes is *forward security* which guarantees that updates to an encrypted structure cannot be correlated with previous queries. Forward security was introduced by Stefanov, Papamanthou and Shi [32] and recent work of Zhang, Katz and Papamanthou [34] has shown that it mitigates certain injection attacks on SSE schemes. One advantage of our DIEX construction is that it naturally inherits the forward-security of its underlying encrypted multi-maps and dictionaries. That is, if the underlying structures are forward-secure then so is DIEX.

Reduced Leakage. As we mentioned above, IEX leaks more than the naive response-hiding solution *while achieving optimal communication complexity*. We stress, however, that it leaks less than the naive response-revealing solution and than OXT. As an example, consider that if OXT is used to search for two conjunctions $\mathbf{w} = w_1 \wedge w_2$ and $\mathbf{w}' = w_3 \wedge w_2$ which share a common term, the server can recover the results for $\mathbf{w}'' = w_1 \wedge w_2 \wedge w_3$. In the case of disjunctions, OXT's leakage is equivalent to the naive response-revealing solution.

Experiments. To evaluate the efficiency of IEX and its instantiations we designed and built a new encrypted search framework called *Clusion* [22]. It is written in Java and leverages the Apache Lucene search library [1]. It also includes a Hadoop-based distributed parser and indexer we implemented to handle massive datasets. Our experiments show that IEX—specifically our IEX-2Lev

---

[2] Multi-map encryption schemes are equivalent to SSE schemes so ZMF is an *adaptively*-secure compact SSE scheme with linear-time search.

instantiation—is very efficient and even achieves faster search times than those reported for a C++ implementation of OXT [11] on a comparable system. For example, for conjunctive, disjunctive and boolean queries with selectivity on the order of thousands, IEX-2Lev takes 12, 14.8 and 23.7 ms, respectively. For the same conjunctive query, OXT is reported to take 200 ms on a comparable system. Clearly, a C/C++ implementation of IEX would perform even better.

We also implemented IEX-ZMF to evaluate its efficiency and compactness. In our experiments, it produced EDBs of size 198 MB and 0.9 GB from datasets with 1.5M and 34M keyword/id pairs, respectively. This is highly compact in comparison to IEX-2Lev which produced 1.6 GB, 9.8 GB EDBs for 1.5M and 34M keyword/id pairs, respectively. We also evaluated the efficiency of IEX-ZMF and, as expected, its performance for setup, search and token size are worse than IEX-2Lev. For example, for a dataset with 34M keyword/id pairs, EDB setup takes 7.58 h to process compared to 31 mins for IEX-2Lev.

On a boolean query of the form $(w \vee x) \wedge (y \vee z)$, where the disjunctions had selectivity 2 K and 10 K, respectively, IEX-ZMF took 1610 ms whereas IEX-2Lev took only 23.7 ms. As expected due to its high degree of compactness, IEX-ZMF is slower than IEX-2Lev (this is the exact tradeoff we seek).

2   Related Work

SSE was first considered by Song, Wagner and Perrig [31]. Curtmola, Garay, Kamara and Ostrovsky [15] introduced the notion of adaptive-security for SSE and presented the first constructions that achieved optimal search time with a space-efficient index. STE was introduced by Chase and Kamara [13] who proposed constructions for two-dimensional arrays, graphs and web graphs.

In [19], Goh introduced the Z-IDX construction which has linear search complexity and produces highly compact indexes due to its use of Bloom filters. Here, we extract a general transformation implicitly used in the Z-IDX construction and use it in part to construct our ZMF scheme. Kamara, Papamanthou and Roeder gave the first optimal-time dynamic SSE scheme [24]. Cash et al. [11] proposed OXT; the first optimal-time conjunctive keyword search scheme. Faber et al. [16] extend OXT to handle range, substring, wildcard and phrase queries. Pappas et al. [30] and Fisch et al. [17] present solutions based on garbled circuits and Bloom filters that can support boolean formulas, ranges and stemming. In [30], the authors show how to build the first worst-case sub-linear time boolean encrypted search solution. Like Goh's Z-IDX construction and our ZMF scheme, the solution makes use of Bloom filters. In addition, it is the first adaptively-secure construction based on Bloom filters. For a disjunctive query $\mathbf{w}$, the scheme has search complexity $O(\log(n) \cdot C \cdot \mathsf{DB}(\mathbf{w}))$, where n is the number of documents and C is the cost of a 2-party secure function evaluation of a function that takes as input a Bloom filter of size $O(\#\mathrm{W})$ (i.e., the number of unique keywords in DB) and a q-term disjunctive query. We note that unlike IEX and OXT, it does not achieve optimal communication complexity. Also, while its search is sub-linear it involves multiple rounds of interactions.

Ishai, Kushilevitz, Lu and Ostrovsky propose a two-server SSE scheme that hides the access pattern and supports various complex queries including ranges, stemming and substring [21]. Cash et al. [10] design several I/O-efficient SSE schemes including the 2Lev construction which we use in one of our IEX instantiations. Kurosawa and Ohtaki [26] designed the first UC secure SSE scheme. Kurosawa [25] designed a linear-time construction that handles arbitrary boolean queries while not disclosing the structure of the boolean query itself. Forward Secrecy was first considered by Stefanov, Papamanthou and Shi [32]. In [9], Bost introduced an efficient forward secure construction. In [12], Cash and Tessaro give lower bounds on the locality of SSE by showing tradeoffs between locality, space overhead and read efficiency. Recently, Asharov, Naor, Segev and Shahaf gave SSE constructions with optimal locality, optimal space overhead and nearly-optimal read efficiency [3]. Encrypted search can also be achieved with other primitives like property-preserving encryption [5,6], functional encryption [7,8,29], oblivious RAM [20], full-homomorphic encryption [18] and multi-party computation [33].

Online ciphers were introduced by Bellare, Boldyreva et al. [4], where they propose several schemes including the HCB1 construction which we make use of in our ZMF implementation. More efficient constructions were later proposed by Andreeva et al. [2].

3   Preliminaries

Notation. The set of all binary strings of length n is denoted as $\{0,1\}^n$, and the set of all finite binary strings as $\{0,1\}^*$. $[n]$ is the set of integers $\{1,\ldots,n\}$, and $2^{[n]}$ is the corresponding power set. We write $x \leftarrow \chi$ to represent an element x being sampled from a distribution χ, and $x \xleftarrow{\$} X$ to represent an element x being sampled uniformly at random from a set X. The output x of an algorithm $\mathcal{A}$ is denoted by $x \leftarrow \mathcal{A}$. Given a sequence $\mathbf{v}$ of n elements, we refer to its ith element as v_i or $\mathbf{v}[i]$. If S is a set then $\#S$ refers to its cardinality. If s is a string then $|s|$ refers to its bit length and s_i to its ith bit. $s^{|n}$ denotes the string s padded with $n - |s|$ 0's and $s_{|n}$ represents the first n bits of s. Given strings s and r, we refer to their concatenation as either $\langle s, r \rangle$ or $s\|r$. For an n-bit string s and for all nonnegative d, we denote by $s^{\|d}$ the string $\langle s_1^{|d}, \cdots, s_n^{|d} \rangle$. In this work, padding takes precedence over truncation; that is, $s_{|p}^{\|d} = (s^{\|d})_{|p}$.

Data Types. An *abstract data type* is a collection of objects together with a set of operations defined on those objects. Examples include sets, dictionaries (also known as key-value stores or associative arrays) and graphs. The operations associated with an abstract data type fall into one of two categories: query operations, which return information about the objects; and update operations, which modify the objects. If the abstract data type supports only query operations it is *static*, otherwise it is *dynamic*.

Data Structures. A *data structure* for a given data type is a representation in some computational model[3] of an object of the given type. Typically, the representation is optimized to support the type's query operation as efficiently as possible. For data types that support multiple queries, the representation is often optimized to efficiently support as many queries as possible. As a concrete example, the dictionary *type* can be represented using various data structures depending on which queries one wants to support efficiently. Hash tables support Get and Put in expected $O(1)$ time whereas balanced binary search trees support both operations in worst-case $\log(n)$ time. For ease of understanding and to match colloquial usage, we will sometimes blur the distinction between data types and structures. So, for example, when referring to a *dictionary structure* or a *multi-map structure* what we are referring to is an unspecified instantiation of the dictionary or multi-map data type.

Basic Structures. We make use of several basic data types including arrays, dictionaries and multi-maps which we recall here. An array A of capacity n stores n items at locations 1 through n and supports read and write operations. We write $v = A[i]$ to denote reading the item at location i and $A[i] = v$ the operation of storing an item at location i. A dictionary DX of capacity n is a collection of n label/value pairs $\{(\ell_i, v_i)\}_{i \leq n}$ and supports Get and Put operations. We write $v_i = DX[\ell_i]$ to denote getting the value associated with label ℓ_i and $DX[\ell_i] = v_i$ to denote the operation of associating the value v_i in DX with label ℓ_i. A multi-map MM with capacity n is a collection of n label/tuple pairs $\{(\ell_i, V_i)_i\}_{i \leq n}$ that supports Get and Put operations. Similarly to dictionaries, we write $V_i = MM[\ell_i]$ to denote getting the tuple associated with label ℓ_i and $MM[\ell_i] = V_i$ to denote operation of associating the tuple V_i to label ℓ_i. We sometimes write $MM^{-1}[v]$ to refer to the set of labels in MM associated with tuples that include the value v. Multi-maps are the abstract data type instantiated by an inverted index. In the encrypted search literature multi-maps are sometimes referred to as indexes, databases or tuple-sets (T-sets) [10,11].

Document Collections. A document collection is a set of documents $\mathbf{D} = (D_1, \ldots, D_n)$, each document consisting of a set of keywords from some universe W. We assume the universe of keywords is totally ordered (e.g., using lexicographic order) and denote by W[i] the ith keyword in W. We assume every document has an identifier that is independent of its contents and denote it $\mathsf{id}(D_i)$. We assume the existence of an efficient indexing algorithm that takes as input a data collection $\mathbf{D}$ and outputs a multi-map that maps every keyword w in W to the identifiers of the documents that contain w. In previous work, this multi-map is referred to as an inverted index or as a database. For consistency, we refer to any multi-map derived in this way from a document collection as a database and denote it DB. Given a keyword w, we denote by $\mathsf{co}_{DB}(w) \subseteq W$ the set of keywords in W that co-occur with w; that is, the keywords that are contained in documents that contain w. When DB is clear from the context we omit DB and write only $\mathsf{co}(w)$.

---

[3] In this work, the underlying model will always be the word RAM.

3.1   Cryptographic Primitives

Basic Cryptographic Primitives. A private-key encryption scheme is a set
of three polynomial-time algorithms $\mathsf{SKE} = (\mathsf{Gen}, \mathsf{Enc}, \mathsf{Dec})$ such that Gen is a
probabilistic algorithm that takes a security parameter k and returns a secret key
K; Enc is a probabilistic algorithm takes a key K and a message m and returns
a ciphertext c; Dec is a deterministic algorithm that takes a key K and a ciphertext c and returns m if K was the key under which c was produced. Informally, a
private-key encryption scheme is secure against chosen-plaintext attacks (CPA)
if the ciphertexts it outputs do not reveal any partial information about the
plaintext even to an adversary that can adaptively query an encryption oracle.
We say a scheme is random-ciphertext-secure against chosen-plaintext attacks
(RCPA) if the ciphertexts it outputs are computationally indistinguishable from
random even to an adversary that can adaptively query an encryption oracle.[4]
In addition to encryption schemes, we also make use of pseudo-random functions
(PRF) and permutations (PRP), which are polynomial-time computable functions that cannot be distinguished from random functions by any probabilistic
polynomial-time adversary.

Online Ciphers. An online cipher (OC) is a block cipher that can encrypt data
streams. In particular, with an OC the encryption of the ith block in a stream
depends only on the 1st through ith message blocks. OCs were introduced by
Bellare, Boldyreva, Knudsen and Namprempre [4]. More formally, we say that a
cipher $\mathsf{OC} : \{0,1\}^k \times \{0,1\}^{n \times B} \rightarrow \{0,1\}^{n \times B}$, where $B > 1$ is the block length, is
B-online if there exists a function $X : \{0,1\}^k \times \{0,1\}^{n \times B} \rightarrow \{0,1\}^B$ such that
for any $\mathbf{m} \in \{0,1\}^{n \times B}$,

$$\mathsf{OC}_K(\mathbf{m}) = \mathsf{OC}_K^1(\mathbf{m}) \| \ldots \| \mathsf{OC}_K^n(\mathbf{m}),$$

where $\mathsf{OC}_K^i(\mathbf{m}) = X(K, m_1, \ldots, m_i)$ for all $i \in [n]$ and where m_i is the ith block
of $\mathbf{m}$. OCs cannot be pseudo-random permutations (see [4] for a simple distinguisher) but can satisfy the weaker requirement of being computationally indistinguishable from a random *online* permutation. An online permutation is simply
a permutation on a domain $\{0,1\}^{n \times B}$ whose ith block depends only on the first i
blocks of its input. We denote by $\mathsf{OPerm}_{n,B}$ the set of all online permutations over
$\{0,1\}^{n \times B}$. Security for an online cipher $\mathsf{OC} : \{0,1\}^k \times \{0,1\}^{n \times B} \rightarrow \{0,1\}^{n \times B}$
then holds if for all PPT adversaries $\mathcal{A}$,

$$\left| \Pr\left[\mathcal{A}^{\mathsf{OC}_K(\cdot)} = 1 : K \xleftarrow{\$} \{0,1\}^k \right] - \Pr\left[\mathcal{A}^{f(\cdot)} = 1 : f \xleftarrow{\$} \mathsf{OPerm}_{n,B} \right] \right| \leq \mathsf{negl}(k).$$

4   Definitions

Structured encryption schemes encrypt data structures in such a way that they
can be privately queried. There are several natural forms of structured encryption. The original definition of [13] considered schemes that encrypt both a

---

[4] RCPA-secure encryption can be instantiated practically using either the standard
PRF-based private-key encryption scheme or, e.g., AES in counter mode.

structure and a set of associated data items (e.g., documents, emails, user profiles etc.). In [14], the authors also describe *structure-only* schemes which only encrypt structures. Another distinction can be made between *interactive* and *non-interactive* schemes. Interactive schemes produce encrypted structures that are queried through an interactive two-party protocol, whereas non-interactive schemes produce structures that can be queried by sending a single message, i.e., the token. One can also distinguish between *response-hiding* and *response-revealing* schemes: the former reveal the response to queries whereas the latter do not.

STE schemes are used as follows. During a setup phase, the client constructs an encrypted data structure EDS under a key K. The client then sends EDS to the server. During the query phase, the client constructs and sends a token tk generated from its query q and the key K. The server then uses the token tk to query EDS. If the scheme is response-revealing, it recovers a response r. On the other hand, if the scheme is response-hiding it recovers a message that it returns to the client who in turn decrypts it with a resolving algorithm.

Definition 1 (Structured encryption). *A single-round response-hiding structured encryption scheme* $\Sigma_{\mathscr{T}} = ($Setup, Token, Query, Resolve$)$ *for data type* $\mathscr{T}$ *consists of four polynomial-time algorithms that work as follows:*

- $(K, EDS) \leftarrow $Setup$(1^k, DS)$: *is a probabilistic algorithm that takes as input a security parameter* 1^k *and a structure* DS *of type* $\mathscr{T}$ *and outputs a secret key* K *and an encrypted structure* EDS.
- tk $\leftarrow$ Token(K, q): *is a (possibly) probabilistic algorithm that takes as input a secret key* K *and a query* q *and returns a token* tk.
- $c \leftarrow$ Query$($EDS, tk$)$: *is a (possibly) probabilistic algorithm that takes as input an encrypted structure* EDS *and a token* tk *and outputs a message* c.
- $r \leftarrow$ Resolve(K, c): *is a deterministic algorithm that takes as input a secret key* K *and a message* c *and outputs a response* r.

We say that a structured encryption scheme Σ *is correct if for all* $k \in \mathbb{N}$, *for all* poly(k)-*size structures* DS *of type* $\mathscr{T}$, *for all* (K, EDS) *output by* Setup$(1^k, DS)$ *and all sequences of* $m = $poly$(k)$ *queries* $q_1, \ldots, q_m$, *for all tokens* tk$_i$ *output by* Token(K, q_i), *for all messages* c *output by* Query$($EDS, tk$_i)$, Resolve(K, c) *returns the correct response with all but negligible probability. The syntax of a response-revealing STE scheme can be recovered by omitting the* Resolve *algorithm and having* Query *output the response.*

Security. The standard notion of security for STE guarantees that an encrypted structure reveals no information about its underlying structure beyond the setup leakage $\mathcal{L}_\mathsf{S}$, and that the query algorithm reveals no information about the structure and the queries beyond the query leakage $\mathcal{L}_\mathsf{Q}$. If this holds for non-adaptively chosen operations then this is referred to as non-adaptive security. If, on the other hand, the operations are chosen adaptively, this leads to the stronger notion of adaptive security [15]. This notion of security was first formalized by Curtmola *et al.* in the context of searchable encryption [15] and later generalized to structured encryption in [13].

Definition 2 (Adaptive security [13,15]). *Let $\Sigma_{\mathcal{T}} = (\mathsf{Setup}, \mathsf{Token}, \mathsf{Query})$ be a structured encryption scheme for type $\mathcal{T}$ and consider the following probabilistic experiments where $\mathcal{A}$ is a stateful adversary, $\mathcal{S}$ is a stateful simulator, $\mathcal{L}_S$ and $\mathcal{L}_Q$ are leakage profiles and $z \in \{0,1\}^*$:*

Real$_{\Sigma,\mathcal{A}}(k)$*: given z the adversary $\mathcal{A}$ outputs a structure DS of type $\mathcal{T}$ and receives EDS from the challenger, where $(K, \mathsf{EDS}) \leftarrow \mathsf{Setup}(1^k, \mathsf{DS})$. The adversary then adaptively chooses a polynomial number of queries $q_1, \ldots, q_m$. For all $i \in [m]$, the adversary receives $\mathsf{tk}_i \leftarrow \mathsf{Token}(K, q_i)$. Finally, $\mathcal{A}$ outputs a bit b that is output by the experiment.*

Ideal$_{\Sigma,\mathcal{A},\mathcal{S}}(k)$*: given z the adversary $\mathcal{A}$ generates a structure DS of type $\mathcal{T}$ which it sends to the challenger. Given z and leakage $\mathcal{L}_S(\mathsf{DS})$ from the challenger, the simulator $\mathcal{S}$ returns an encrypted data structure EDS to $\mathcal{A}$. The adversary then adaptively chooses a polynomial number of operations $q_1, \ldots, q_m$. For all $i \in [m]$, the simulator receives query leakage $\mathcal{L}_Q(\mathsf{DS}, q_i)$ and returns a token tk_i to $\mathcal{A}$. Finally, $\mathcal{A}$ outputs a bit b that is output by the experiment.*

We say that Σ is adaptively $(\mathcal{L}_S, \mathcal{L}_Q)$-secure if for all PPT *adversaries $\mathcal{A}$, there exists a* PPT *simulator $\mathcal{S}$ such that for all $z \in \{0,1\}^*$,*

$$\left| \Pr\left[\mathbf{Real}_{\Sigma,\mathcal{A}}(k) = 1 \right] - \Pr\left[\mathbf{Ideal}_{\Sigma,\mathcal{A},\mathcal{S}}(k) = 1 \right] \right| \leq \mathsf{negl}(k).$$

5   IEX: A Worst-Case Sub-linear Disjunctive SSE Scheme

Our main construction, IEX, makes black-box use of a dictionary encryption scheme $\Sigma_{\mathsf{DX}} = (\mathsf{Setup}, \mathsf{Token}, \mathsf{Get})$, a multi-map encryption scheme $\Sigma_{\mathsf{MM}} = (\mathsf{Setup}, \mathsf{Token}, \mathsf{Get})$, a pseudo-random function F, and of a private-key encryption scheme $\mathsf{SKE} = (\mathsf{Gen}, \mathsf{Enc}, \mathsf{Dec})$. The details of the scheme are provided in Fig. 1. At a high-level, it works as follows.

Setup. The Setup algorithm takes as input a security parameter k and an index DB. It makes use of two data structures: a dictionary DX and a *global* multi-map MM_g. MM_g maps every keyword in $w \in \mathsf{W}$ to an encryption of the identifiers in $\mathsf{DB}(w)$. We refer to these encryptions as *tags* and they are computed by evaluating SKE.Enc using as coins the evaluation of F on keyword w and the identifier. The global multi-map MM_g is then encrypted using Σ_{MM}, resulting in EMM_g.

For each keyword $w \in \mathsf{W}$, the algorithm creates a *local* multi-map MM_w, that maps the keywords $v \in \mathsf{co}(w)$ to tags of identifiers in $\mathsf{DB}(v) \cap \mathsf{DB}(w)$. Intuitively, the purpose of the local multi-map MM_w is to quickly find out which documents contain both w and v, for any $v \neq w$. The local multi-maps MM_w are then encrypted with Σ_{MM}. This results in encrypted multi-maps EMM_w which are then stored in the dictionary DX such that $\mathsf{DX}[w] = \mathsf{EMM}_w$. In other words, it stores label/value pairs (w, MM_w) in DX. Finally, DX is encrypted with Σ_{DX}, resulting in an encrypted dictionary EDX. The output of Setup includes the encrypted structures $(\mathsf{EDX}, \mathsf{EMM}_g)$ as well as their keys.

There are several optimizations possible for Setup that we omit in our formal description for ease of exposition. The first is that the encrypted local multi-maps can be stored "by reference" in the encrypted dictionary EDX instead of "by value". More precisely, instead of storing the actual encrypted local multi-maps EMM_w in EDX one can just store a *pointer* to them. Another optimization is that, depending on how Σ_{MM} is designed, the keys for the local encrypted multi-maps could all be generated from a single key using a PRF (with a counter). This would reduce the size of K. This optimization can be easily applied to most known encrypted multi-map schemes including the ones from [10,11,13,15,23].

Token. The Token algorithm takes as input a key and a vector of keywords $\mathbf{w} = (w_1, \ldots, w_q)$. For all $i \in [q-1]$ it creates a "sub-token" $\mathbf{tk}_i = (\mathsf{dtk}_i, \mathsf{gtk}_i, \mathsf{ltk}_{i+1}, \ldots, \mathsf{ltk}_q)$ composed of a dictionary token dtk_i, a global token gtk_i for w_i and, for all keywords w_{i+1} through w_q in the disjunction, a local token ltk_j for w_j, with $i+1 \leq j \leq q$. Intuitively, the global token will allow the server to query the encrypted global multi-map EMM_g to recover tags of the ids in $\mathsf{DB}(w_i)$. The dictionary token for w_i will then allow the server to query the encrypted dictionary EDX to recover w_i's local multi-map EMM_i. Finally, the local tokens will allow the server to query w_i's encrypted local multi-map EMM_i to recover the tags of the ids of the documents that contain both w_i and w_{i+1}, w_i and w_{i+2}, etc. As we will see next, this information will be enough for the server to find the relevant documents. For the last keyword w_q in the disjunction, the algorithm only needs to create a global token.

Search. The Search algorithm takes as input $\mathsf{EDB} = (\mathsf{EDX}, \mathsf{EMM}_g)$ and a token $\mathsf{tk} = (\mathbf{tk}_1, \ldots, \mathbf{tk}_{q-1}, \mathsf{gtk}_q)$. For each sub-token $\mathbf{tk}_i = (\mathsf{dtk}_i, \mathsf{gtk}_i, \mathsf{ltk}_{i+1}, \ldots, \mathsf{ltk}_q)$, the server does the following. It first uses gtk_i to query the global multi-map EMM_g and recover a set of identifier tags T_i for $\mathsf{DB}(w_i)$. It then uses dtk_i to query the encrypted dictionary EDX to recover the local multi-map EMM_i for w_i and uses ltk_{i+1} to query EMM_i to recover the tags T' for identifiers of the documents that contain both w_i and w_{i+1}; that is, the tags for the set $I' = \mathsf{DB}(w_i) \cap \mathsf{DB}(w_{i+1})$. The server then removes T_i' from T_i. It then repeats this process for all local tokens ltk_{i+2} to ltk_q. Once it finishes processing all local tokens in $\mathbf{tk}_i$, it holds the set of tags for the set

$$\mathsf{DB}(w_i) \setminus \bigcup_{j=i}^{q-1} \Big(\mathsf{DB}(w_i) \cap \mathsf{DB}(w_{j+1}) \Big). \tag{1}$$

Once it finishes processing all the sub-tokens, the server holds tags T_1 through T_{q-1}. For gtk_q, the server just queries the global multi-map to recover T_q. Finally, it outputs the set

$$T = \bigcup_{i=1}^{q} T_i. \tag{2}$$

Let F be a pseudo-random function, $\mathsf{SKE} = (\mathsf{Gen}, \mathsf{Enc}, \mathsf{Dec})$ be a private-key encryption scheme, $\Sigma_{\mathsf{DX}} = (\mathsf{Setup}, \mathsf{Token}, \mathsf{Get})$ be a dictionary encryption scheme and $\Sigma_{\mathsf{MM}} = (\mathsf{Setup}, \mathsf{Token}, \mathsf{Get})$ be a multi-map encryption scheme. Consider the disjunctive SSE scheme $\mathsf{IEX} = (\mathsf{Setup}, \mathsf{Token}, \mathsf{Search})$ defined as follows:

- $\mathsf{Setup}(1^k, \mathsf{DB})$:
    1. sample $K_1, K_2 \overset{\$}{\leftarrow} \{0,1\}^k$;
    2. initialize a dictionary DX and a multi-map MM_g;
    3. for all $w \in \mathsf{W}$,
        (a) for all $\mathsf{id} \in \mathsf{DB}(w)$, let $\mathsf{tag}_{\mathsf{id}} := \mathsf{Enc}_{K_1}\big(\mathsf{id}; F_{K_2}(\mathsf{id} \| w)\big)$;
        (b) set $\mathsf{MM}_g[w] := \big(\mathsf{tag}_{\mathsf{id}}\big)_{\mathsf{id} \in \mathsf{DB}(w)}$;
        (c) initialize a multi-map MM_w of size $\#\mathsf{co}(w)$;
        (d) for all $v \in \mathsf{co}(w)$,
            i. for all $\mathsf{id} \in \mathsf{DB}(v) \cap \mathsf{DB}(w)$, let $\mathsf{tag}_{\mathsf{id}} := \mathsf{Enc}_{K_1}\big(\mathsf{id}; F_{K_2}(\mathsf{id} \| w)\big)$;
            ii. set $\mathsf{MM}_w[v] := \big(\mathsf{tag}_{\mathsf{id}}\big)_{\mathsf{id} \in \mathsf{DB}(v) \cap \mathsf{DB}(w)}$;
        (e) compute $(K_w, \mathsf{EMM}_w) \leftarrow \Sigma_{\mathsf{MM}}.\mathsf{Setup}(1^k, \mathsf{MM}_w)$;
        (f) set $\mathsf{DX}[w] := \mathsf{EMM}_w$;
    4. compute $(K_g, \mathsf{EMM}_g) \leftarrow \Sigma_{\mathsf{MM}}.\mathsf{Setup}(1^k, \mathsf{MM}_g)$;
    5. compute $(K_d, \mathsf{EDX}) \leftarrow \Sigma_{\mathsf{DX}}.\mathsf{Setup}(1^k, \mathsf{DX})$;
    6. set $K = \big(K_g, K_d, \{K_w\}_{w \in \mathsf{W}}\big)$ and $\mathsf{EDB} = (\mathsf{EMM}_g, \mathsf{EDX})$;
    7. output (K, EDB).
- $\mathsf{Token}(K, \mathbf{w})$:
    1. parse $\mathbf{w}$ as $(w_1, \ldots, w_q)$;
    2. for all $i \in [q-1]$,
        (a) compute $\mathsf{gtk}_i \leftarrow \Sigma_{\mathsf{MM}}.\mathsf{Token}(K_g, w_i)$;
        (b) compute $\mathsf{dtk}_i \leftarrow \Sigma_{\mathsf{DX}}.\mathsf{Token}(K_d, w_i)$;
        (c) for all $i+1 \leq j \leq \#\mathbf{w}$, compute $\mathsf{ltk}_j \leftarrow \Sigma_{\mathsf{MM}}.\mathsf{Token}(K_{w_i}, w_j)$;
        (d) set $\mathbf{tk}_i = \big(\mathsf{dtk}_i, \mathsf{gtk}_i, \mathsf{ltk}_{i+1}, \ldots, \mathsf{ltk}_{\#\mathbf{w}}\big)$;
    3. compute $\mathsf{gtk}_q \leftarrow \Sigma_{\mathsf{MM}}.\mathsf{Token}(K_g, w_q)$;
    4. output $\mathsf{tk} = \big(\mathbf{tk}_1, \ldots, \mathbf{tk}_{q-1}, \mathsf{gtk}_q\big)$.
- $\mathsf{Search}(\mathsf{EDB}, \mathsf{tk})$:
    1. parse EDB as $(\mathsf{EMM}_g, \mathsf{EDX})$;
    2. parse tk as $(\mathbf{tk}_1, \ldots, \mathbf{tk}_{q-1}, \mathsf{gtk}_q)$;
    3. for all $i \in [q-1]$,
        (a) parse $\mathbf{tk}_i$ as $\big(\mathsf{dtk}_i, \mathsf{gtk}_i, \mathsf{ltk}_{i+1}, \ldots, \mathsf{ltk}_q\big)$;
        (b) compute $T_i \leftarrow \Sigma_{\mathsf{MM}}.\mathsf{Get}(\mathsf{EMM}_g, \mathsf{gtk}_i)$;
        (c) compute $\mathsf{EMM}_i \leftarrow \Sigma_{\mathsf{DX}}.\mathsf{Get}(\mathsf{EDX}, \mathsf{dtk}_i)$;
        (d) for all $i+1 \leq j \leq q$,
            i. compute $T' \leftarrow \Sigma_{\mathsf{MM}}.\mathsf{Get}(\mathsf{EMM}_i, \mathsf{ltk}_j)$;
            ii. set $T_i = T_i \setminus T'$;
    4. compute $T_q \leftarrow \Sigma_{\mathsf{MM}}.\mathsf{Get}(\mathsf{EMM}_g, \mathsf{gtk}_q)$;
    5. output $\bigcup_{i \in [q]} T_i$;

Fig. 1. Our disjunctive SSE scheme IEX.

5.1   Correctness and Efficiency

We now analyze the correctness and efficiency of our construction. The correctness of IEX follows from Eqs. (1) and (2) and from the inclusion-exclusion principle. Given a disjunctive query $\mathbf{w} = (w_1, \ldots, w_q)$, by Eq. (2), IEX.Search(EDB, Token$(K, \mathbf{w})$) will output

$$
\begin{aligned}
T &= \bigcup_{i=1}^{q} T_i \\
&= \left(\bigcup_{i=1}^{q-1} T_i \right) \bigcup T_q \\
&= \left(\bigcup_{i=1}^{q-1} \left(\mathsf{DB}(w_i) \setminus \bigcup_{j=i}^{q-1} \left(\mathsf{DB}(w_i) \bigcap \mathsf{DB}(w_{j+1}) \right) \right) \right) \bigcup \mathsf{DB}(w_q) \\
&= \left(\bigcup_{i=1}^{q-2} \left(\mathsf{DB}(w_i) \setminus \bigcup_{j=i}^{q-1} \left(\mathsf{DB}(w_i) \bigcap \mathsf{DB}(w_{j+1}) \right) \right) \right) \\
&\qquad \bigcup \underbrace{\left(\mathsf{DB}(w_{q-1}) \setminus \left(\mathsf{DB}(w_{q-1}) \bigcap \mathsf{DB}(w_q) \right) \right) \bigcup \mathsf{DB}(w_q)}_{U}
\end{aligned}
$$

$$(3)$$

where the first and third equalities hold by Eqs. (2) and (1), respectively. Note, however, that U equals $\mathsf{DB}(w_{q-1}) \bigcup \mathsf{DB}(w_q)$:

$$
\begin{aligned}
U &= \mathsf{DB}(w_{q-1}) \bigcap \left(\overline{\mathsf{DB}(w_{q-1})} \bigcup \overline{\mathsf{DB}(w_q)} \right) \bigcup \mathsf{DB}(w_q) \\
&= \left(\mathsf{DB}(w_{q-1}) \bigcap \overline{\mathsf{DB}(w_{q-1})} \right) \bigcup \left(\mathsf{DB}(w_{q-1}) \bigcap \overline{\mathsf{DB}(w_q)} \right) \bigcup \mathsf{DB}(w_q) \\
&= \left(\mathsf{DB}(w_{q-1}) \bigcap \overline{\mathsf{DB}(w_q)} \right) \bigcup \mathsf{DB}(w_q) \\
&= \left(\mathsf{DB}(w_{q-1}) \bigcup \mathsf{DB}(w_q) \right) \bigcap \left(\overline{\mathsf{DB}(w_q)} \bigcup \mathsf{DB}(w_q) \right) \\
&= \mathsf{DB}(w_{q-1}) \bigcup \mathsf{DB}(w_q)
\end{aligned}
$$

Repeating the same argument for $q-2$, $q-3$ and so on and plugging into Eq. (3), we get that $T = \bigcup_{i=1}^{q} \mathsf{DB}(w_i)$.

Efficiency. The search complexity of IEX is $O(q^2 \cdot M)$, where $M = \max_{i \in [q]} \#\mathsf{DB}(w_i)$ and q is the number of terms in the disjunction. Tokens are of size $O(q)$. We also note that unlike BXT and OXT [11], IEX tokens are *selectivity-independent* in the sense that they do not depend on the size of the result. The IEX storage complexity is,

$$
O\left(\mathsf{strg}\left(\sum_w \#\mathsf{DB}(w) \right) + \sum_w \mathsf{strg}\left(\sum_{v \in \mathsf{co}(w)} \#\mathsf{DB}(v) \cap \mathsf{DB}(w) \right) \right),
$$

where strg is the storage complexity of the underlying encrypted multi-map encryption scheme Σ_{MM}.

A Storage Optimization. As we can see, the storage complexity of IEX can be large, especially if the underlying encrypted multi-maps are. This is indeed the case when they are instantiated with standard sub-linear constructions. We observe, however, that we can tradeoff storage complexity (and setup time) for the communication complexity of search as follows. When constructing a local multi-map EMM_w for a keyword w, we normally insert tags for the identifiers in $DB(w) \cap DB(v)$ for all $v \in co(w)$. This is not necessary for correctness, however, so we can omit some of the co-occurring keywords from w's local multi-map. The tradeoff is that this will increase the communication complexity of IEX's search operation and, in particular, make it non-optimal.

To do this, we suggest using the following approach to decide whether to add a keyword $v \in co(w)$ or not. Let $p < 1$ be a *filtering* parameter and let

$$T_{w,v} \stackrel{def}{=} \frac{\#DB(v) \cap DB(w)}{\max(\#DB(w), \#DB(v))}.$$

If $T_{w,v} > p$, then add v to EMM_w otherwise do not. With this filtering in place, the storage complexity of IEX is now

$$O\left(strg\left(\sum_w \#DB(w) \right) + \sum_w strg\left(\sum_{\substack{v \in co(w) \\ T_{w,v} > p}} \#DB(v) \cap DB(w) \right) \right).$$

In our experiments we set $p = 0.2$.

Remark. We note that when all the terms of the disjunctive query have selectivity $O(n)$, IEX has linear search complexity. This is, however, the best one can do. On the other hand, the communication complexity of IEX remains optimal independently of the selectivity of the terms. This similarly applies to OXT but not to BlindSeer since it induces a logarithmic (multiplicative) overhead.

5.2   Security

The setup leakage of IEX consists of the setup leakage of its underlying building blocks. In particular, this includes the setup leakage of the encrypted global multi-map and of the encrypted dictionary. Assuming the use of standard optimal-time multi-map and dictionary encryption schemes [10,13,15,23], this reveals the size of the database DB as well as the total size of the local multi-maps stored in the dictionary. The query leakage of IEX for a query **w** includes, for each keyword $w_i \in W$, the query leakage of the encrypted dictionary and of the encrypted global multi-map. It also includes the query leakage of every queried local multi-map as well as their setup leakage. Again if instantiated with standard constructions, this will consist of the search and access patterns which, respectively, capture whether or not the same query has been searched for and (in our case) the tags. Finally, the query leakage also includes the number of documents containing $\mathsf{DB}(w_i) \bigcap \mathsf{DB}(w_{j+1})$, for all $j \geq i$ and $i \in [q-1]$.

We now give a precise description of IEX's leakage profile and show that it is adaptively-secure with respect to it. Its setup leakage is

$$\mathcal{L}_S^{iex}(\mathsf{DB}) = \left(\mathcal{L}_S^{dx}(\mathsf{DX}), \mathcal{L}_S^{mm}(\mathsf{MM}_g) \right),$$

where $\mathcal{L}_S^{dx}(\mathsf{DX})$ and $\mathcal{L}_S^{mm}(\mathsf{MM}_g)$ are the setup leakages of the underlying dictionary and multi-map encryption schemes, respectively. Its query leakage is

$$\mathcal{L}_Q^{iex}(\mathsf{DB}, \mathbf{w}) = \left(\left(\mathcal{L}_Q^{dx}(\mathsf{DX}, w_i), \mathcal{L}_S^{mm}(\mathsf{MM}_i), \right. \right.$$
$$\left. \mathcal{L}_Q^{mm}(\mathsf{MM}_g, w_i), \dots, \mathcal{L}_Q^{mm}(\mathsf{MM}_i, w_q), \mathsf{TagPat}_i(\mathsf{DB}, \mathbf{w}) \right)_{i \in [q-1]},$$
$$\left. \mathcal{L}_Q^{mm}(\mathsf{MM}_g, w_q), \mathsf{TagPat}_q(\mathsf{DB}, \mathbf{w}) \right),$$

where, for all $i \in [q]$,

$$\mathsf{TagPat}_i(\mathsf{DB}, \mathbf{w}) = \left(\left(f_i(\mathsf{id}) \right)_{\mathsf{id} \in \mathsf{DB}(w_i) \cap \mathsf{DB}(w_{i+1})}, \dots, \left(f_i(\mathsf{id}) \right)_{\mathsf{id} \in \mathsf{DB}(w_i) \cap \mathsf{DB}(w_q)} \right),$$

and f_i is a random function from $\{0,1\}^{|\mathsf{id}| + \log \#W}$ to $\{0,1\}^k$.

Theorem 1. *If Σ_{DX} is adaptively $(\mathcal{L}_S^{dx}, \mathcal{L}_Q^{dx})$-secure, Σ_{MM} is adaptively $(\mathcal{L}_S^{mm}, \mathcal{L}_Q^{mm})$-secure, SKE is RCPA-secure and F is pseudo-random, then IEX is $(\mathcal{L}_S^{iex}, \mathcal{L}_Q^{iex})$-secure.*

The proof of Theorem 1 is deferred to the full version of the paper.

6 Boolean Queries with IEX

While IEX is naturally disjunctive, it can also support boolean queries. The boolean variant is similar to IEX in that it uses the same encrypted structures (i.e., it has the same Setup algorithm) but different Token and Search algorithms. We refer to the boolean variant of IEX as BIEX. We now provide an overview of how BIEX works.

Overview of BIEX. Recall that any query can be written in conjunctive normal form (CNF) so it has the form $\Delta_1 \wedge \cdots \wedge \Delta_\ell$, where each $\Delta_i = w_{i,1} \vee \cdots \vee w_{i,q}$ is a disjunction. Note that the result $\mathsf{DB}(\Delta_1 \wedge \cdots \wedge \Delta_\ell)$ is the intersection of $\mathsf{DB}(\Delta_1)$ through $\mathsf{DB}(\Delta_\ell)$. But this intersection does not have to be computed "directly" by executing a naive intersection operation. A better alternative (from a leakage point of view) is to compute the intersection by starting with $\mathsf{DB}(w_1)$, keeping only the subset of identifiers of $\mathsf{DB}(w_1)$ that are also in $\mathsf{DB}(w_2)$, then keeping only the subset of identifiers that are also in $DB(w_3)$ and so on. This alternative approach requires only information about $DB(w_1)$ and the progressive subsets. Moreover, it uses operations that are already supported by the IEX structures.

How we do this exactly, is best explained through a concrete example. Suppose we have a CNF query with $\Delta_1 = w_1 \vee w_2$ and $\Delta_2 = w_3$. The first step would be to perform a disjunctive query for Δ_1, resulting in tags for the identifiers in $DB(\Delta_1)$. In the second step, we want to filter out and keep the tags of identifiers in $DB(\Delta_1) \cap DB(w_3)$. To find these tags, it suffices to query the local multi-maps of w_1 and w_2 on w_3. In the first case, EMM_{w_1} will return tags for $DB(w_1) \cap DB(w_3)$ and in the second case EMM_{w_2} will return tags for $DB(w_2) \cap DB(w_3)$. Finally, we take the union of both of these intersections and perform a final intersection with $DB(\Delta_1)$. The final result equals $DB(\Delta_1) \cap DB(w_3)$. Figure 2 describes this process in more detail and for arbitrary boolean queries.

Correctness. To show correctness we need to show that, given a boolean query in CNF form $\Delta_1 \wedge \cdots \wedge \Delta_\ell$ such that $\Delta_i = w_{i,1} \vee \cdots \vee w_{i,q}$ (for simplicity we assume the disjunctions all have q terms), BIEX.Search outputs

$$\bigcap_{i \in [\ell]} \bigcup_{j \in [q]} DB(w_{i,j}). \qquad (4)$$

Looking at the description of BIEX.Search in Fig. 2, one can see that every time Step 4(d)i is invoked it outputs

$$\bigcup_{j \in [q]} DB(w_{i,t}) \cap DB(w_{1,j}),$$

for all $t \in [q]$ and $i \in [\ell]$. Note that this stems from the fact that $\Sigma_{MM}.Get(EMM_j, ltk_{t,i,j})$ outputs $DB(w_{i,t}) \cap DB(w_{1,j})$ for every $j \in [q]$.

Also, based on the correctness of IEX we know that the search for the first disjunction will output $\bigcup_{j \in [q]} DB(w_{1,j})$ (with no redundant identifiers). So we have the final result of the query

$$I_\ell = \underbrace{\bigcup_{j \in [q]} DB(w_{1,j}) \bigcap \Big(\bigcup_{j,l \in [q]} (DB(w_{2,j}) \cap DB(w_{1,l})) \Big) \bigcap \cdots \Big(\bigcup_{j,l \in [q]} (DB(w_{\ell,j}) \cap DB(w_{1,l})) \Big)}_{\ell-1 \text{ terms}} \qquad (5)$$

On the other hand, note that for all $i \in [\ell]$ we have by Morgan's laws that

$$\bigcup_{j \in [q]} DB(w_{1,j}) \bigcap \bigcup_{j \in [q]} DB(w_{i,j}) = \bigcup_{j \in [q]} DB(w_{1,j}) \bigcap \bigcup_{j \in [q]} DB(w_{1,j}) \bigcap \bigcup_{j \in [q]} DB(w_{i,j})$$

$$= \bigcup_{j \in [q]} DB(w_{1,j}) \bigcap \Big(\bigcup_{j,l \in [q]} (DB(w_{i,j}) \cap DB(w_{1,l})) \Big)$$

That is, we can recursively apply the above result on Eq. (5) for all $l \in [\ell]$ to obtain Eq. (4).

Efficiency. The storage complexity of BIEX is the same as IEX. Its search complexity is

$$O\left(q^2 \cdot \left(\max_{w \in \Delta_1} \#DB(w) + \ell \cdot \#DB(\Delta_1)\right)\right).$$

The term $q^2 \cdot \max_{w \in \Delta_1} \#DB(w)$ is the time to search for the first disjunction and the second term $q^2 \cdot \ell \cdot \#DB(\Delta_1)$ is the total number of local multi-map queries.

We can clearly see from the search complexity of BIEX that we can achieve better efficiency if the selectivity of the first disjunction is as small as possible. In practice, therefore, the first disjunction should be the one with the smallest selectivity; similarly to how the first keyword is chosen in OXT. Note that if the first disjunction in the CNF form of the boolean query matches the entire database then the search complexity of BIEX will be linear while the optimal complexity might be sub-linear (the communication complexity of BIEX will remain optimal, however). It is not obvious to us how to improve this without pre-computing every possible query as it seems almost inherent to the query itself. With this in mind, it follows that BIEX has a sub-linear worst-case search complexity when the first disjunction's selectivity is sub-linear.

The communication complexity of BIEX is optimal since the final set I_ℓ does not contain any redundant identifiers. Finally, note that it is non-interactive and token size is independent of the query's selectivity.

Security. The setup leakage of BIEX is the same as IEX's. Its query leakage includes the query leakage of IEX on the first disjunction and the query leakage of the encrypted local multi-maps when queried on all the terms of disjunctions $\Delta_2, \ldots, \Delta_\ell$. Finally, it also includes the number of documents that match the terms of the first disjunction and the terms of remaining disjunctions.

We now give a precise description of the leakage profile of BIEX and show that it is adaptively-secure with respect to it. The setup leakage is

$$\mathcal{L}_S^{iexb}(DB) = \mathcal{L}_S^{iex}(DB),$$

where $\mathcal{L}_S^{iex}(DB)$ is the setup leakages of IEX. Given a CNF query $\Delta_1 \wedge \cdots \wedge \Delta_\ell$, the query leakage is

$$\mathcal{L}_Q^{iexb}\left(DB, \bigwedge_{i=1}^{\ell} \Delta_i\right) = \left(\mathcal{L}_Q^{iex}(DB, \Delta_1), \left(\mathcal{L}_Q^{mm}(MM_i, w_{l,1}), \cdots, \mathcal{L}_Q^{mm}(MM_i, w_{l,q}),\right.\right.$$

$$\left.\left.\mathsf{TagPat}_{i,l}\left(DB, \bigwedge_{i=1}^{\ell} \Delta_i\right)\right)_{\substack{i \in [q] \\ l \in [2, \cdots, \ell]}}\right).$$

where,

$$\mathsf{TagPat}_{i,l}\left(DB, \bigwedge_{i=1}^{\ell} \Delta_i\right) = \left(\left(f_i(id)\right)_{DB(w_{1,i}) \cap DB(w_{l,1})}, \cdots, \left(f_i(id)\right)_{DB(w_{1,i}) \cap DB(w_{l,q})}\right)$$

and f_i is a random function from $\{0,1\}^{n+\log \#W}$ to $\{0,1\}^k$.

Let IEX = (Setup, Token, Search) be the IEX scheme described in Figure 1 and let Σ_{DX} = (Setup, Token, Get) and Σ_{MM} = (Setup, Token, Get) be its underlying dictionary and multi-map encryption schemes, respectively. Consider the boolean SSE encryption scheme BIEX = (Setup, Token, Search) defined as follows:

- Setup(1^k, DB): output $(K, \mathsf{EDB}) \leftarrow$ IEX.Setup(1^k, DB).
- Token($K, \mathbf{w}$):
  1. parse K as $(K_g, K_d, \{K_w\}_{w \in \mathsf{W}})$;
  2. parse $\mathbf{w}$ as $\left(\Delta_1 \bigwedge \cdots \bigwedge \Delta_\ell \right)$ where for all $i \in [\ell]$, $\Delta_i = \left(w_{i,1} \bigvee \cdots \bigvee w_{i,d} \right)$;
  3. compute $\mathbf{tk}_1 \leftarrow$ IEX.Token$_K(\Delta_1)$;
  4. for all $2 \le i \le \ell$ and all $j \in [q]$,
     (a) for all $1 \le s \le q$, compute $\mathsf{ltk}_{s,i,j} \leftarrow \Sigma_{\mathsf{MM}}.\mathsf{Token}(K_{w_{1,s}}, w_{i,j})$;
     (b) set $\mathbf{tk}_{i,j} = (\mathsf{ltk}_{1,i,j}, \ldots, \mathsf{ltk}_{q,i,j})$;
     (c) set $\mathbf{tk}_i = (\mathbf{tk}_{i,1}, \ldots, \mathbf{tk}_{i,q})$;
  5. output tk = $(\mathbf{tk}_1, \ldots, \mathbf{tk}_\ell)$.
- Search(EDB, tk):
  1. parse EDB as $(\mathsf{EMM}_g, \mathsf{EDX})$;
  2. parse tk as $(\mathbf{tk}_1, \ldots, \mathbf{tk}_\ell)$;
  3. compute $I_1 \leftarrow$ IEX.Search(EDB, $\mathbf{tk}_1$);
  4. for all $2 \le i \le \ell$,
     - instantiate an empty set I_i;
     - parse $\mathbf{tk}_i = (\mathbf{tk}_{i,1}, \ldots, \mathbf{tk}_{i,q})$;
     - for $j \in [q]$,
       (a) get dtk_j from $\mathbf{tk}_1$;
       (b) compute $\mathsf{EMM}_j \leftarrow \Sigma_{\mathsf{DX}}.\mathsf{Get}(\mathsf{EDX}, \mathsf{dtk}_j)$;
       (c) parse $\mathbf{tk}_{i,j} = (\mathsf{ltk}_{1,i,j}, \ldots, \mathsf{ltk}_{q,i,j})$;
       (d) for $s \in [q]$,
           i. compute $I' \leftarrow \Sigma_{\mathsf{MM}}.\mathsf{Get}(\mathsf{EMM}_j, \mathsf{ltk}_{s,i,j})$;
           ii. compute $I_i = I_i \bigcup \left(I_{i-1} \bigcap I' \right)$;
  5. output I_ℓ;

Fig. 2. The scheme BIEX.

Theorem 2. *If Σ_{DX} is adaptively $\left(\mathcal{L}_S^{\mathsf{dx}}, \mathcal{L}_Q^{\mathsf{dx}} \right)$-semantically secure and Σ_{MM} is adaptively $\left(\mathcal{L}_S^{\mathsf{mm}}, \mathcal{L}_Q^{\mathsf{mm}} \right)$-secure, then BIEX is adaptively $\left(\mathcal{L}_S^{\mathsf{iexb}}, \mathcal{L}_Q^{\mathsf{iexb}} \right)$-secure.*

The proof of Theorem 2 is similar (at a high-level) to the proof of Theorem 1.

7   ZMF: A Compact and Adaptively-Secure SSE Scheme

The main limitation of IEX is its storage complexity of

$$O\left(\mathsf{strg}_g(\mathsf{MM}_g) + \sum_w \mathsf{strg}_\ell(\mathsf{MM}_w) \right),$$

where strg_g and strg_ℓ are the storage complexity of the global and local EMMs, respectively. If the latter are instantiated with standard sub-linear-time constructions such as [10,11,15,24], we have

$$O\left(\sum_w \#\mathsf{DB}(w) + \sum_w \sum_{v \in \mathsf{co}(w)} \#\mathsf{DB}(v) \cap \mathsf{DB}(w)\right), \qquad (6)$$

which does not compare favorably to standard single-keyword search solutions which require only $O(\sum_w \#\mathsf{DB}(w))$, or to the OXT construction of [11] which requires

$$O\left(\sum_w \#\mathsf{DB}(w) + \log\left(\frac{1}{\varepsilon}\right) \cdot \sum_w \#\mathsf{DB}(w)\right)$$

when XSet is instantiated with a Bloom filter with a false positive rate of ε. In particular, note that the second term in the asymptotic expression above hides a constant of 1, which makes OXT reasonably compact.

Our Approach. The main storage inefficiency in IEX comes from the local EMMs which contribute the second term in Eq. (6). Ideally, we could improve things if we could use more compact local EMMs. Unfortunately, all known sub-linear constructions require $O(\sum_w \#\mathsf{DB}(w))$ storage. We observe, however, that for local EMMs sub-linear search is not necessary since in practice the number of label/tuple pairs they store is small in comparison to the total number of documents n. So, for our purposes, a linear-time construction would work as long as it was compact. In [19], Goh proposed a very compact construction called Z-IDX based on Bloom filters. Specifically, it needs only

$$O\left(\log\left(\frac{1}{\varepsilon}\right) \cdot \sum_{v \in \mathbf{V}} \#\mathsf{MM}^{-1}[v]\right)$$

bits of storage, where $\mathbf{V}$ is the value space of the multi-map and ε is the false positive rate. If we could encrypt the local EMMs of IEX with Z-IDX, the former's storage would be

$$O\left(\sum_w \#\mathsf{DB}(w) + \log\left(\frac{1}{\varepsilon}\right) \cdot \sum_w \#\mathsf{co}(w)\right),$$

which is much more competitive with OXT (note that the second term here also has a constant of 1). Unfortunately, this approach does not work because Z-IDX is not adaptively secure. Nevertheless, we show how to construct a highly compact scheme that is. In the following, we first recall how Z-IDX works.

Goh's ZIDX Scheme. Like any SSE scheme, Z-IDX can be viewed as a STE scheme and, in particular, as a multi-map encryption scheme. Conceptually, we observe that Z-IDX can be abstracted into two parts: (1) a compiler that transforms an underlying set encryption scheme into a multi-map encryption

Let $\Sigma_{\mathsf{SET}} = (\mathsf{Gen}, \mathsf{Enc}, \mathsf{Token}, \mathsf{Test})$ be a multi-structure set encryption scheme and consider the multi-map encryption scheme $\Sigma_{\mathsf{MM}} = (\mathsf{Setup}, \mathsf{Token}, \mathsf{Get})$ defined as follows:

- $\mathsf{Setup}(1^k, \mathsf{MM})$:
    1. compute $K \leftarrow \mathsf{Gen}(1^k)$;
    2. let $\mathbf{V}$ be the range of MM;
    3. for all $v \in \mathbf{V}$,
        (a) let $S_v = \mathsf{MM}^{-1}(v)$;
        (b) compute $\mathsf{ESET}_v \leftarrow \Sigma_{\mathsf{SET}}.\mathsf{Enc}(K, S_v)$;
    4. output $\mathsf{EMM} = (\mathsf{ESET}_v)_{v \in \mathbf{V}}$.
- $\mathsf{Token}(K, \ell)$: output $\mathsf{tk} \leftarrow \Sigma_{\mathsf{SET}}.\mathsf{Token}(K, \ell)$
- $\mathsf{Get}(\mathsf{EMM}, \mathsf{tk})$:
    1. let $I = \emptyset$;
    2. for all $v \in \mathbf{V}$,
        (a) if $\Sigma_{\mathsf{SET}}.\mathsf{Test}(\mathsf{ESET}_v, \mathsf{tk})$ outputs 1, set $I = I \cup \{v\}$;
    3. output I.

Fig. 3. The Z-IDX transformation.

scheme; and (2) a concrete set encryption scheme based on Bloom filters and PRFs. We refer to the former as the Z-IDX transformation and describe it in detail in Fig. 3. Given a set encryption scheme Σ_{SET}, it produces a multi-map encryption scheme Σ_{MM} that works as follows. The $\Sigma_{\mathsf{MM}}.\mathsf{Setup}$ algorithm takes as input a multi-map MM that maps labels to tuples of values from $\mathbf{V}$. It creates $\#\mathbf{V}$ sets $(S_v)_{v \in \mathbf{V}}$ such that S_v holds the labels in MM that map to v. It then encrypts each set S_v with Σ_{SET} resulting in an encrypted set ESET_v. The encrypted multi-map EMM is simply the collection of encrypted sets $(\mathsf{ESET}_v)_{v \in \mathbf{V}}$. A Σ_{MM} token for a label ℓ is a Σ_{SET} token for ℓ and $\Sigma_{\mathsf{MM}}.\mathsf{Get}$ uses the token to test each set in $\mathsf{EMM} = (\mathsf{ESET}_v)_{v \in \mathbf{V}}$ and outputs v if the test succeeds.

Note that for Σ_{MM} to work, Σ_{SET} must satisfy a stronger STE form than what is described in Definition 1. In particular, it must be what we call *multi-structure* in the sense that the tokens produced with a key K can be used to query all the structures encrypted under K. We provide formal syntax and security definitions of multi-structure STE schemes in the full version of the paper. The main difference between standard and multi-structure STE schemes are that in the latter the Setup algorithm is replaced with a key generation algorithm $\mathsf{Gen}(1^k)$ that takes as input a security parameter and outputs a secret key K; and an encryption algorithm $\mathsf{Enc}(K, \mathsf{DS})$ that takes as input a secret key K and a data structure DS and outputs an encrypted structure EDS.

Adaptive Security. From our abstract perspective, the reason Z-IDX is not adaptively-secure is because the Z-IDX transformation is (implicitly) applied to a set encryption scheme that is not adaptively-secure. We show in Theorem 3 below, however, that if the transformation is applied to an adaptively-secure set encryption scheme then the result is adaptively-secure as well. More precisely, we show that if the set encryption scheme is adaptively $(\mathcal{L}_S^{set}, \mathcal{L}_Q^{set})$-secure then the Z-IDX transformation yields a multi-map encryption scheme with the following leakage profile:

$$\mathcal{L}_S^{mm}(MM) = \left(\left(\mathcal{L}_S^{set}(MM^{-1}[v]) \right)_{v \in \mathbf{V}}, \#\mathbf{V} \right),$$

and

$$\mathcal{L}_Q^{mm}(MM, q) = \mathcal{L}_Q^{set} \left(\left(MM^{-1}[v] \right)_{v \in \mathbf{V}}, q \right).$$

Theorem 3. *If Σ_{SET} is adaptively $(\mathcal{L}_S^{set}, \mathcal{L}_Q^{set})$-secure then the scheme Σ_{MM} that results from applying the Z-IDX transformation to it is adaptively $(\mathcal{L}_S^{mm}, \mathcal{L}_Q^{mm})$-secure.*

Due to space constraints, the proof of Theorem 3 will appear in the full version of the paper.

7.1   An Adaptively-Secure and Multi-structure Set Encryption Scheme

In this Section, we construct an adaptively-secure, highly-compact and multi-structure set encryption scheme. Then, by applying the Z-IDX transformation to it we get an adaptively-secure and highly-compact multi-map encryption scheme which we then use in IEX.

Adaptive Security. The main difficulty in designing adaptively-secure encrypted structures is supporting *equivocation* during simulation. Roughly speaking, the issue is that during the **Ideal**(k) experiment the simulator first needs to simulate an encrypted structure for the adversary and later needs to be able to simulate tokens that work correctly with the simulated structure produced in the first step. The challenge in supporting equivocation is that at the time the encrypted structure is simulated, the simulator has no information about the adversary's queries so it is not clear how to simulate the structure in a way that will work correctly at query time. So to handle equivocation, the construction needs to be carefully designed and, typically, needs expensive cryptographic primitives. Fortunately, as first shown by Chase and Kamara [13], in the setting of symmetric STE, equivocation can be achieved very efficiently based only on XOR and PRF operations.

Our Base Scheme. One possible way to design an encrypted Bloom filter is as follows. Let $\mathbf{U}$ be a universe of elements. Given a set $S \subseteq \mathbf{U}$, insert the value $F_K(a)$, for all $a \in S$, in a standard Bloom filter, where F is a pseudo-random function. The token for an element $a \in \mathbf{U}$ is $tk = F_K(a)$ and the Bloom filter can be queried by doing a standard Bloom filter test on tk.

The main problems with this construction are that: (1) it reveals information about the size of S; and (2) it is not adaptively-secure. To achieve adaptive security, we can encrypt the Bloom filter by XORing each of its bits with a pad generated from another pseudo-random function G. This encryption step both hides the size of S and allows for equivocation. Now the token tk for an element $a \in \mathbf{U}$ includes $F_{K_1}(a)$ and the pads for locations $H_1(F_{K_1}(a))$ through $H_\lambda(F_{K_1}(a))$, where $(H_1, \ldots, H_\lambda)$ are the hash functions used for the Bloom filter.

For this to work, however, the pads have to be designed carefully. More precisely, correctness requires that the pads only depend on the locations that they mask otherwise two (or more) elements a_1 and a_2 that collide under one of the hash functions will produce different masks for the same location. To get such *location-dependent* pads we compute them as $G_{K_2}(\ell)$, where ℓ is the ℓth bit of the filter. Now, a token for element a is set to

$$\mathsf{tk} = \left(F_{K_1}(a), G_{K_2}\Big(H_1\big(F_{K_1}(a)\big) \Big), \ldots, G_{K_2}\Big(H_\lambda\big(F_{K_1}(a)\big) \Big) \right).$$

The base construction described so far is *compact* and *adaptively-secure* but not multi-structure.

Reusability. Recall that a multi-structure STE scheme can produce *multiple* encrypted structures $(\mathsf{EDS}_1, \ldots, \mathsf{EDS}_n)$ under a single key K in such a way that a *single* (constant-size) token tk can be used to query all the structures generated under key K. So to make our base scheme multi-structure, the pads have to be filter-dependent in addition to being location-dependent so that different pads are used for different filters *even if they mask the same location*. We do this by setting the pads to be the output of a random oracle applied to the pair $(G_{K_2}(\ell), \mathsf{id})$ where id is the identifier of the filter. The purpose of the random oracle here is twofold. First, it enables the extraction of n (random) pads from pairs $(G_{K_2}(\ell), \mathsf{id}_1)$ through $(G_{K_2}(\ell), \mathsf{id}_n)$ without relying on n secret keys. This, in turn, means the tokens can be of size independent of n. Second, it allows the simulator to equivocate on the pads while, again, keeping the tokens independent of n.

While the base scheme is now compact, adaptively-secure and multi-structure, it produces very large tokens. The problem is that if two sets S_1 and S_2 have different sizes, then the parameters of their Bloom filters (i.e., the array sizes, number of hash functions and hash function ranges) have to be different. The consequence is that in our encrypted set scheme, we will need different sets of hash functions for *each* filter which, in turn, means the tokens will have to include multiple pads for *every filter*.

Matryoshka Filters. We solve this problem as follows. Instead of encrypting a set of standard Bloom filters as in our base construction, we encrypt a new filter-based structure we refer to as *matryoshka filters* (MF).[5] MFs are essentially a set of *nested* Bloom filters of varying sizes whose hash functions are all derived from a fixed set of hash functions. More precisely, consider a sequence of sets $S_1, \ldots, S_n \subseteq \mathbf{U}$ not necessarily of the same size. We assume for simplicity that the sets have size a multiple of 2. For some false negative rate $2^{-\lambda}$, choose λ

---

[5] The term matryoshka here refers to Russian nested dolls which are called matryoshka dolls.

independent and ideal random hash functions $(H_1, \ldots, H_\lambda)$ from $\mathbf{U}$ to $[(\lambda/\ln 2) \cdot \max_i \#S_i]$. We refer to these functions as the *maximal* hash functions and to their associated filter as the *maximal* filter. For every set S_i, construct a Bloom filter of size $[\lceil(\lambda/\ln 2) \cdot \#S_i\rceil]$ with hash functions $(H_1^i, \ldots, H_h^i)$ where, for all $j \in [\lambda]$, $H_j^i(a) = H_j(a)_{\|p_i}$ with $p_i = \lceil \log((\lambda/\ln 2) \cdot \#S_i) \rceil$. We refer to these hash functions as the *derived* functions and to their associated filters as the *derived* filters. Note that if the maximal hash functions are ideal random functions then so are the derived functions so the standard Bloom filter analysis holds.

Encrypting Matryoshka Filters. As mentioned above, our final solution consists of adapting our base scheme to encrypt matryoshka filters instead of standard Bloom filters. In other words, we XOR each bit of each matryoshka filter with location- and filter-dependent pads. The main difference with the base scheme is that here the pads also need to be nested; that is, given a pad for the maximal filter we need to be able to construct the pads for the derived filters. To support this, we make use of the properties of online ciphers; namely, that given an n-bit string s and a B-online cipher OC, the following equality holds:

$$\mathsf{OC}_K\left(s_{|p \times B}^{\|B}\right) = \mathsf{OC}_K\left(s^{\|B}\right)_{|p \times B}, \tag{7}$$

where $p < n$. This can be derived as follows. From the correctness property of online ciphers, we have

$$\mathsf{OC}_K\left(s_{|p \times B}^{\|B}\right) = \mathsf{OC}_K^1\left(s_{|p \times B}^{\|B}\right) \| \cdots \| \mathsf{OC}_K^p\left(s_{|p \times B}^{\|B}\right)$$

$$= X\left(K, s_{|B}^{\|B}\right) \| \cdots \| X\left(K, s_{|p \times B}^{\|B}\right),$$

and

$$\mathsf{OC}_K\left(s^{\|B}\right) = \mathsf{OC}_K^1\left(s^{\|B}\right) \| \cdots \| \mathsf{OC}_K^n\left(s^{\|B}\right)$$

$$= X\left(K, s_{|B}^{\|B}\right) \| \cdots \| X\left(K, s_{|n \times B}^{\|B}\right),$$

for some function X. It follows then that

$$\mathsf{OC}_K\left(s^{\|B}\right)_{|p \times B} = X\left(K, s_{|B}^{\|B}\right) \| \cdots \| X\left(K, s_{|p \times B}^{\|B}\right)$$

$$= \mathsf{OC}_K\left(s_{|p \times B}^{\|B}\right).$$

Now, to encrypt the ℓth bit of a matryoshka filter, we use a pad constructed as

$$\mathsf{R}\left(\mathsf{OC}_K\left(\ell_{|p \times B}^{\|B}\right), \mathsf{id}(S)\right),$$

where R is a random oracle and $\mathrm{id}(S)$ is the identifier of the filter. Note that the pad is both filter- and location-dependent. In addition, if the server is provided the value $\mathrm{OC}_K(\ell^{\|B})$ for the maximal filter, it follows by Eq. (7) that it can derive the above pad as

$$\mathsf{R}\Big(\mathrm{OC}_K\big(\ell^{\|B}\big)_{|p\times B}, \mathrm{id}(S)\Big).$$

The detailed description of our set encryption scheme is given in Fig. 4. In the Theorem below we show that it is adaptively-secure with the following leakage profile:

$$\mathcal{L}_\mathsf{S}^\mathrm{est}(S) = \#S \quad \text{and} \quad \mathcal{L}_\mathsf{Q}^\mathrm{set}\Big(S_1,\ldots,S_n,q\Big) = \Big(b_1,\ldots,b_n,\mathsf{SP}(q)\Big),$$

where SP is the *search pattern*; that is, if and when two queries are the same. More formally, if t queries have been made, $\mathsf{SP}(q)$ outputs a t-bit string with a 1 at location i if q is equal to the ith query.

Theorem 4. *If OC is secure, then the multi-structure set encryption scheme described in Fig. 4 is adaptively $(\mathcal{L}_\mathsf{S}^\mathrm{set}, \mathcal{L}_\mathsf{Q}^\mathrm{set})$-secure in the random oracle model.*

The proof of Theorem 4 is in the full version of the paper.

7.2   The ZMF Multi-map Encryption Scheme

By applying the Z-IDX transformation to our multi-structure set encryption scheme from Fig. 4, we get a new adaptively-secure multi-map encryption scheme we call ZMF. We state its security formally in the following Corollary of Theorems 3 and 4. Its leakage profile is,

$$\mathcal{L}_\mathsf{S}^\mathrm{zmf}(\mathsf{MM}) = \Big(\big(\#\mathsf{MM}^{-1}[v]\big)_{v\in\mathbf{V}}, \#\mathbf{V}\Big) \quad \text{and} \quad \mathcal{L}_\mathsf{Q}^\mathrm{zmf}(\mathsf{MM},q) = \Big(b_1,\ldots,b_{\#\mathbf{V}},\mathsf{SP}(q)\Big)$$

where b_i is 1 if $q \in \mathsf{MM}^{-1}[v_i]$ and 0 otherwise, and v_i is the ith value in $\mathbf{V}$.

Corollary 1. *The ZMF multi-map encryption scheme which results from applying the Z-IDX transformation to the set encryption scheme of Fig. 4 is $(\mathcal{L}_\mathsf{S}^\mathrm{zmf}, \mathcal{L}_\mathsf{Q}^\mathrm{zmf})$-adaptively secure.*

8   DIEX: A Dynamic SSE Scheme

We describe our dynamic SSE construction DIEX. As far as we know, it is the first adaptively-secure dynamic SSE scheme that is forward-secure and supports Boolean search queries in sub-linear time. In particular, it supports the addition, deletion and editing of files. In the full version of the paper, we recall the syntax and security definitions for dynamic STE.

Let F be a pseudo-random function family, $\mathcal{H} : \{0,1\}^* \to [\sigma]$ be a family of hash functions modeled as random oracles where σ is a public upper bound, and $\mathsf{R} : \{0,1\}^* \to \{0,1\}$ be a random oracle. Let $\mathsf{OC} : \{0,1\}^{g(k)} \times \{0,1\}^{\gamma \times B} \to \{0,1\}^{\gamma \times B}$ a B-online cipher with $\gamma = \log \sigma$ blocks. Let $\varepsilon \in [0,1]$ be a false positive rate that is hardcoded in each algorithm. Set $\lambda = \log(1/\varepsilon)$ and set $\mathsf{H}_1, \ldots, \mathsf{H}_\lambda \leftarrow \mathcal{H}$. Consider the set encryption scheme $\Sigma = (\mathsf{Gen}, \mathsf{Enc}, \mathsf{Token}, \mathsf{Test})$ defined as follows:

- $\mathsf{Gen}(1^k)$:
    1. sample $K_1 \overset{\$}{\leftarrow} \{0,1\}^k$ and $K_2 \overset{\$}{\leftarrow} \{0,1\}^{g(k)}$;
    2. output $K = (K_1, K_2)$;
- $\mathsf{Enc}(K, S)$:
    1. let A be a binary array of size $m = \lceil \lambda \cdot \#S / \ln 2 \rceil$ initialized to all 0's;
    2. for all items $a \in S$ and all $i \in [\lambda]$,
        (a) compute $T = F_{K_1}(a)$;
        (b) compute $\ell = \mathsf{H}_i(T)$;
        (c) compute $s = \mathsf{OC}_{K_2}(\ell_{|\log m \times B}^{\|B})$;
        (d) set $\mathsf{A}[\ell_{|\log m}] = 1 \oplus \mathsf{R}(s, \mathsf{id}(S))$;
    3. for all $i \in [m]$ such that $\mathsf{A}[i] = 0$,
        (a) compute $s = \mathsf{OC}_{K_2}(i_{|\log m \times B}^{\|B})$;
        (b) set $\mathsf{A}[i] = 0 \oplus \mathsf{R}(s, \mathsf{id}(S))$;
    4. set $\mathsf{ESET} = \mathsf{A}$;
    5. output ESET.
- $\mathsf{Token}(K, a)$:
    1. compute $T = F_{K_1}(a)$;
    2. for all $i \in [\lambda]$,
        (a) compute $\ell = \mathsf{H}_i(T)$;
        (b) compute $s_i = \mathsf{OC}_{K_2}(\ell^{\|B})$;
    3. output $\mathsf{tk} = (T, s_1, \ldots, s_\lambda)$.
- $\mathsf{Test}(\mathsf{ESET}, \mathsf{tk})$:
    1. parse tk as $(T, s_1, \ldots, s_\lambda)$;
    2. parse ESET as A;
    3. set $m = |\mathsf{A}|$;
    4. for all $i \in [\lambda]$,
        (a) compute $b_i = \mathsf{A}[\mathsf{H}_i(T)_{|\log m}] \oplus \mathsf{R}\left((s_i)_{|\log m \times B}, \mathsf{id}(\mathsf{ESET}) \right)$;
    5. if, for all $i \in [\lambda]$, $b_i = 1$ output 1, otherwise output 0.

Fig. 4. An adaptively-secure multi-structure set encryption scheme.

Overview. As a starting point, we describe a dynamic version of IEX that is *not* forward-secure. For this, we make two changes to our static construction. First, we replace the encrypted dictionary EDX and the global encrypted multi-map EMM_g with a dynamic encrypted dictionary EDX^+ and a dynamic global encrypted multi-map EMM_g^+. The encrypted local multi-maps remain static.

Second, we require that these new structures be *response-hiding*. We provide a high level description of our construction which is described in detail in Fig. 5.

The DIEX.Setup algorithm is the same as the IEX.Setup with the exception that it uses a dynamic encrypted dictionary and a dynamic encrypted multi-map and outputs state information st. The DIEX.Token^{sr} algorithm is similar to IEX.Token with the exception that it is stateful. Here, the state is just used to generate tokens for the underlying dynamic dictionary and global multi-map encrypted structures. The DIEX.Token^{up} algorithm works as follows. It takes as inputs the key K, the state st and an update $u = (\mathsf{op}, \mathsf{id}, \mathsf{W_{id}})$ that consists of an operation $\mathsf{op} \in \{\mathsf{edit}^+, \mathsf{edit}^-\}$, the document identifier id being edited and a set of keywords $\mathsf{W_{id}}$ to add or delete based on op. We have the following cases:

– if $u = (\mathsf{edit}^+, \mathsf{id}, \mathsf{W_{id}})$, the client will update the global multi-map EMM_g with pairs $(w, \mathsf{tag_{id}})$ for all $w \in \mathsf{W_{id}}$. Here, $\mathsf{tag_{id}} := \mathsf{Enc}_{K_1}(\mathsf{id}; F_{K_2}(\mathsf{id}\|w))$ as in IEX. This is done by generating update tokens $(\mathsf{utk}_g^w)_{w \in \mathsf{W_{id}}}$ for EMM_g using $\Sigma_{\mathsf{MM}}.\mathsf{Token}^{up}$. For all $w \in \mathsf{W_{id}}$, the client generates a new local multi-map MM_w that maps all $v \in \mathsf{W_{id}} \setminus \{w\}$ to $\mathsf{tag_{id}}$. It encrypts all these local multi-maps $(\mathsf{MM}_w)_{w \in \mathsf{W_{id}}}$ with $\Sigma_{\mathsf{DX}}.\mathsf{Setup}$, resulting in $(\mathsf{EMM}_w)_{w \in \mathsf{W_{id}}}$ and creates update tokens $(\mathsf{utk}_d^w)_{w \in \mathsf{W_{id}}}$ for EDX. The algorithm outputs an update token

$$\mathsf{utk} = \left(\mathsf{op}, (\mathsf{utk}_d^w)_{w \in \mathsf{W_{id}}}, (\mathsf{utk}_g^w)_{w \in \mathsf{W_{id}}} \right),$$

and $st = (st_d, st_g)$, where the former is the state maintained by Σ_{DX} and the latter is the state maintained by Σ_{MM}.

– if $u = (\mathsf{edit}^-, \mathsf{id}, \mathsf{W_{id}})$, the client only updates EMM_g. Specifically, it removes all pairs $(w, \mathsf{tag_{id}})$ for $w \in \mathsf{W_{id}}$. This can be done by computing tags as above and generating update tokens $(\mathsf{utk}_g^w)_{w \in \mathsf{W_{id}}}$ using $\Sigma_{\mathsf{MM}}.\mathsf{Token}^{up}$. The algorithm outputs the update token

$$\mathsf{utk} = \left(\mathsf{op}, (\mathsf{utk}_g^w)_{w \in \mathsf{W_{id}}} \right),$$

and $st = st_g$ where st_g is the state maintained by Σ_{MM}.

The Update algorithm takes as input EDB and an update token utk and outputs EDB′. If $\mathsf{op} = \mathsf{edit}^+$, it uses the sub-tokens in utk to update EMM_g and EDX. If $\mathsf{op} = \mathsf{edit}^-$, it only updates EMM_g. The Search algorithm is the same as IEX.Search. Recall that we do not update the local multi-maps already in EDX. This is not necessary to for correctness because, during search, the server will take the intersection of the tags returned from the global multi-map EMM_g and from the appropriate local multi-maps. However, because EMM_g is properly updated, the intersection operation will filter out the old/stale tags from the local multi-map.

Forward Security. We note that DIEX is forward secure if its underlying structures are. Specifically, if Σ_{MM} and Σ_{DX} are forward secure then so is DIEX. This is easy to see from the fact the DIEX tokens only consist of Σ_{DX} and Σ_{MM}

Let Σ_{DX}^+ = (Setup, Token^{sr}, Get, Token^{up}, Update) and Σ_{MM}^+ = (Setup, Token^{sr}, Get, Token^{up}, Update) be dynamic dictionary and multi-map encryption schemes, respectively. Let IEX^+ = (Setup$^+$, Token$^+$, Search$^+$) be the IEX scheme described in Fig. 1 with Σ_{MM} and Σ_{DX} replaced with Σ_{MM}^+ and Σ_{DX}^+, respectively, and let Σ_{MM} = (Setup, Token, Query) be the static multi-map encryption scheme used to encrypt the local multi-maps. Consider the dynamic disjunctive SSE scheme DIEX = (Setup, Token^{sr}, Search, Token^{up}, Update) defined as follows:

- Setup(1^k, DB): output $(K, st, \mathsf{EDB}) \leftarrow \mathsf{IEX}^+.\mathsf{Setup}(1^k, \mathsf{DB})$;
- Token$^{\mathsf{sr}}(K, \mathbf{w})$: output tk $\leftarrow \mathsf{IEX}^+.\mathsf{Token}(K, st, \mathbf{w})$;
- Token$^{\mathsf{up}}(K, st, u)$
  1. parse u as $(\mathsf{op}, \mathsf{id}, \mathsf{W}_{\mathsf{id}})$ and st as (st_g, st_d)
  2. if $\mathsf{op} = \mathsf{edit}^+$,
     (a) for all $w \in \mathsf{W}_{\mathsf{id}}$,
        i. let $\mathsf{tag}_{\mathsf{id}} := \mathsf{Enc}_{K_1}(\mathsf{id}; F_{K_2}(\mathsf{id}\|w))$;
        ii. compute $(\mathsf{utk}_g^w, st_g) \leftarrow \Sigma_{\mathsf{MM}}^+.\mathsf{Token}^{\mathsf{up}}(K, st_g, (\mathsf{op}, w, \mathsf{tag}_{\mathsf{id}}))$;
        iii. initialize a multi-map MM_w of size $\#\mathsf{W}_{\mathsf{id}}$;
        iv. for all $v \in \mathsf{W}_{\mathsf{id}} \setminus \{w\}$, set $\mathsf{MM}_w[v] = \mathsf{tag}_{\mathsf{id}}$;
        v. compute $(K_w, \mathsf{EMM}_w) \leftarrow \Sigma_{\mathsf{MM}}.\mathsf{Setup}(1^k, \mathsf{MM}_w)$;
        vi. compute $(\mathsf{utk}_d^w, st_d) \leftarrow \Sigma_{\mathsf{MM}}^+.\mathsf{Token}^{\mathsf{up}}(K, st_d, (\mathsf{op}, w, \mathsf{EMM}_w))$;
     (b) output $\mathsf{utk} = (\mathsf{op}, (\mathsf{utk}_d^w)_{w \in \mathsf{W}_{\mathsf{id}}}, (\mathsf{utk}_g^w)_{w \in \mathsf{W}_{\mathsf{id}}})$;
  3. if $\mathsf{op} = \mathsf{edit}^-$,
     (a) for all $w \in \mathsf{W}_{id}$
        i. let $\mathsf{tag}_{\mathsf{id}} := \mathsf{Enc}_{K_1}(\mathsf{id}; F_{K_2}(\mathsf{id}\|w))$;
        ii. compute $(\mathsf{utk}_g^w, st_g) \leftarrow \Sigma_{\mathsf{MM}}^+.\mathsf{Token}^{\mathsf{up}}(K, st_g, (\mathsf{op}, \mathsf{W}_{\mathsf{id}}, \mathsf{tag}_{\mathsf{id}}))$;
     (b) output $\mathsf{utk} = (\mathsf{op}, (\mathsf{utk}_g^w)_{w \in \mathsf{W}_{\mathsf{id}}})$ and the updated state $st = (st_g, st_d)$;
- Update(EDB, utk)
  1. parse utk as $(\mathsf{op}, (\mathsf{utk}_i)_{i \in [\#\mathsf{tk}]})$ and $\mathsf{EDB} = (\mathsf{EDX}, \mathsf{EMM}_g)$;
  2. if $\mathsf{op} = \mathsf{edit}^-$, then for all $i \in [\#\mathsf{utk}]$ compute $\mathsf{EMM}_g \leftarrow \Sigma_{\mathsf{MM}}^+.\mathsf{Update}(\mathsf{EMM}_g, \mathsf{utk}_i, \mathsf{op})$;
  3. if $\mathsf{op} = \mathsf{edit}^+$, then for all $i \in [\#\mathsf{utk}/2]$, compute $\mathsf{EMM}_g \leftarrow \Sigma_{\mathsf{MM}}^+.\mathsf{Update}(\mathsf{EMM}_g, \mathsf{utk}_i, \mathsf{op})$ and $\mathsf{EDX} \leftarrow \Sigma_{\mathsf{DX}}^+.\mathsf{Update}(\mathsf{EDX}, \mathsf{utk}_{i+\#\mathsf{utk}/2+1}, \mathsf{op})$;
  4. output $\mathsf{EDB} = (\mathsf{EDX}, \mathsf{EMM}_g)$;
- Search(EDB, utk): output $\bigcup_{i \in [q]} T_i \leftarrow \mathsf{IEX}^+.\mathsf{Search}(\mathsf{EDB}, \mathsf{tk})$.

Fig. 5. The scheme DIEX.

tokens so if the former can be simulated from the security parameter, then the latter can. Due to space constraints, the definition of forward security is differed to the full version of the paper. As a possible instantiation of a forward secure multi-map and dictionary encryption scheme, one can use the Sophos scheme of Bost [9].

Efficiency. The efficiency of DIEX depends on the underlying multi-map and dictionary encryption schemes. Using optimal constructions, the search complexity of DIEX is the same as IEX; that is, $O(q^2 \cdot M)$, where $M = \max_{i \in [q]} \#\mathsf{DB}(w_i)$ and q is the number of terms in the disjunction.

Security. We show that DIEX is adaptively secure with respect to the following well-defined leakage profile. The setup and query leakages are the same as IEX so we only describe the update leakage. For an update $u = (\mathsf{edit}^+, \mathsf{id}, \mathsf{W_{id}})$,

$$\mathcal{L}_{\mathsf{U}}\Big(\mathsf{DB}, u\Big) = \Big(\mathcal{L}_{\mathsf{U}}^{\mathsf{mm}}(\mathsf{MM}_g, (\mathsf{op}, w, \mathsf{id})), \mathcal{L}_{\mathsf{U}}^{\mathsf{dx}}(\mathsf{DX}, (\mathsf{op}, w, \mathsf{id})), \mathcal{L}_{\mathsf{S}}^{\mathsf{mm}}(\mathsf{MM}_w)\Big)_{w \in \mathsf{W_{id}}}.$$

If $u = (\mathsf{edit}^-, \mathsf{id}, \mathsf{W_{id}})$: $\mathcal{L}_{\mathsf{U}}^{\mathsf{diex}}\Big(\mathsf{DB}, u\Big) = \Big(\mathcal{L}_{\mathsf{U}}^{\mathsf{mm}}(\mathsf{MM}_g, (\mathsf{op}, w, \mathsf{id}))\Big)_{w \in \mathsf{W_{id}}}.$

Theorem 5. *If Σ_{DX} is adaptively $\left(\mathcal{L}_{\mathsf{S}}^{\mathsf{dx}}, \mathcal{L}_{\mathsf{Q}}^{\mathsf{dx}}, \mathcal{L}_{\mathsf{U}}^{\mathsf{dx}}\right)$-semantically secure and Σ_{MM} is adaptively $\left(\mathcal{L}_{\mathsf{S}}^{\mathsf{mm}}, \mathcal{L}_{\mathsf{Q}}^{\mathsf{mm}}, \mathcal{L}_{\mathsf{U}}^{\mathsf{mm}}\right)$-secure, then DIEX is adaptively $\left(\mathcal{L}_{\mathsf{S}}^{\mathsf{diex}}, \mathcal{L}_{\mathsf{Q}}^{\mathsf{diex}}, \mathcal{L}_{\mathsf{U}}^{\mathsf{diex}}\right)$-secure.*

We defer the proof to the final version of this work.

9   Empirical Evaluation

To evaluate the practicality of our schemes, we designed and built an open source encrypted search framework called Clusion [22]. Due to space limitations, however, we defer the empirical analysis of our constructions to the full version of this work. There, we evaluate IEX-2Lev and IEX-ZMF which are instantiations of IEX with 2Lev and ZMF, respectively. We also evaluate our Boolean scheme BIEX. Our experiments report setup time, search time, storage and token size for all our constructions.

References

1. Apache lucene. http://lucene.apache.org
2. Andreeva, E., Bogdanov, A., Luykx, A., Mennink, B., Tischhauser, E., Yasuda, K.: Parallelizable and authenticated online ciphers. In: Sako, K., Sarkar, P. (eds.) ASIACRYPT 2013. LNCS, vol. 8269, pp. 424–443. Springer, Heidelberg (2013). doi:10.1007/978-3-642-42033-7_22
3. Asharov, G., Naor, M., Segev, G., Shahaf, I.: Searchable symmetric encryption: optimal locality in linear space via two-dimensional balanced allocations. In: STOC (2016)
4. Bellare, M., Boldyreva, A., Knudsen, L.R., Namprempre, C.: On-line ciphers and the hash-CBC constructions. IACR Cryptology ePrint Archive 2007:197 (2007)
5. Bellare, M., Boldyreva, A., O'Neill, A.: Deterministic and efficiently searchable encryption. In: Menezes, A. (ed.) CRYPTO 2007. LNCS, vol. 4622, pp. 535–552. Springer, Heidelberg (2007). doi:10.1007/978-3-540-74143-5_30
6. Boldyreva, A., Chenette, N., Lee, Y., O'Neill, A.: Order-preserving symmetric encryption. In: Joux, A. (ed.) EUROCRYPT 2009. LNCS, vol. 5479, pp. 224–241. Springer, Heidelberg (2009). doi:10.1007/978-3-642-01001-9_13

7. Boneh, D., Crescenzo, G., Ostrovsky, R., Persiano, G.: Public key encryption with keyword search. In: Cachin, C., Camenisch, J.L. (eds.) EUROCRYPT 2004. LNCS, vol. 3027, pp. 506–522. Springer, Heidelberg (2004). doi:10.1007/978-3-540-24676-3_30

8. Boneh, D., Sahai, A., Waters, B.: Functional encryption: definitions and challenges. In: Ishai, Y. (ed.) TCC 2011. LNCS, vol. 6597, pp. 253–273. Springer, Heidelberg (2011). doi:10.1007/978-3-642-19571-6_16

9. Bost, R.: Sophos - forward secure searchable encryption. In: ACM CCS (2016)

10. Cash, D., Jaeger, J., Jarecki, S., Jutla, C., Krawczyk, H., Rosu, M., Steiner, M.: Dynamic searchable encryption in very-large databases: data structures and implementation. In: NDSS (2014)

11. Cash, D., Jarecki, S., Jutla, C., Krawczyk, H., Roşu, M.-C., Steiner, M.: Highly-scalable searchable symmetric encryption with support for boolean queries. In: Canetti, R., Garay, J.A. (eds.) CRYPTO 2013. LNCS, vol. 8042, pp. 353–373. Springer, Heidelberg (2013). doi:10.1007/978-3-642-40041-4_20

12. Cash, D., Tessaro, S.: The locality of searchable symmetric encryption. In: Nguyen, P.Q., Oswald, E. (eds.) EUROCRYPT 2014. LNCS, vol. 8441, pp. 351–368. Springer, Heidelberg (2014). doi:10.1007/978-3-642-55220-5_20

13. Chase, M., Kamara, S.: Structured encryption and controlled disclosure. In: Abe, M. (ed.) ASIACRYPT 2010. LNCS, vol. 6477, pp. 577–594. Springer, Heidelberg (2010). doi:10.1007/978-3-642-17373-8_33

14. Chase, M., Kamara, S.: Structured encryption and controlled disclosure. Technical report 2011/010.pdf, IACR Cryptology ePrint Archive (2010)

15. Curtmola, R., Garay, J., Kamara, S., Ostrovsky, R.: Searchable symmetric encryption: improved definitions and efficient constructions. In: ACM CCS (2006)

16. Faber, S., Jarecki, S., Krawczyk, H., Nguyen, Q., Rosu, M., Steiner, M.: Rich queries on encrypted data: beyond exact matches. In: Pernul, G., Ryan, P.Y.A., Weippl, E. (eds.) ESORICS 2015. LNCS, vol. 9327, pp. 123–145. Springer, Heidelberg (2015). doi:10.1007/978-3-319-24177-7_7

17. Fisch, B.A., Vo, B., Krell, F., Kumarasubramanian, A., Kolesnikov, V., Malkin, T., Bellovin, S.M.: Malicious-client security in blind seer: a scalable private DBMS. In: IEEE S&P (2015)

18. Gentry, C.: Fully homomorphic encryption using ideal lattices. In: STOC (2009)

19. Goh, E.-J.: Secure indexes. Technical report 2003/216, IACR ePrint Cryptography Archive (2003). http://eprint.iacr.org/2003/216

20. Goldreich, O., Ostrovsky, R.: Software protection and simulation on oblivious RAMs. J. ACM **43**(3), 431–473 (1996)

21. Ishai, Y., Kushilevitz, E., Lu, S., Ostrovsky, R.: Private large-scale databases with distributed searchable symmetric encryption. In: Sako, K. (ed.) CT-RSA 2016. LNCS, vol. 9610, pp. 90–107. Springer, Heidelberg (2016). doi:10.1007/978-3-319-29485-8_6

22. Kamara, S., Moataz, T.: Clusion. https://github.com/orochi89/Clusion

23. Kamara, S., Papamanthou, C.: Parallel and dynamic searchable symmetric encryption. In: FC (2013)

24. Kamara, S., Papamanthou, C., Roeder, T.: Dynamic searchable symmetric encryption. In: ACM CCS (2012)

25. Kurosawa, K.: Garbled searchable symmetric encryption. In: Christin, N., Safavi-Naini, R. (eds.) FC 2014. LNCS, vol. 8437, pp. 234–251. Springer, Heidelberg (2014). doi:10.1007/978-3-662-45472-5_15

26. Kurosawa, K., Ohtaki, Y.: UC-secure searchable symmetric encryption. In: Keromytis, A.D. (ed.) FC 2012. LNCS, vol. 7397, pp. 285–298. Springer, Heidelberg (2012). doi:10.1007/978-3-642-32946-3_21

27. Meng, X., Kamara, S., Nissim, K., Kollios, G.: GRECS: graph encryption for approximate shortest distance queries. In: ACM CCS (2015)

28. Naveed, M., Prabhakaran, M., Gunter, C.: Dynamic searchable encryption via blind storage. In: IEEE S&P (2014)

29. O'Neill, A.: Definitional issues in functional encryption, Cryptology ePrint Archive, report 2010/556 (2010)

30. Pappas, V., Krell, F., Vo, B., Kolesnikov, V., Malkin, T., Choi, S.-G., George, W., Keromytis, A., Bellovin, S.: Blind seer: a scalable private DBMS. In: IEEE S&P (2014)

31. Song, D., Wagner, D., Perrig, A.: Practical techniques for searching on encrypted data. In: IEEE S&P (2000)

32. Stefanov, E., Papamanthou, C., Shi, E.: Practical dynamic searchable encryption with small leakage. In: NDSS (2014)

33. Yao, A.: Protocols for secure computations. In: FOCS (1982)

34. Zhang, Y., Katz, J., Papamanthou, C.: All your queries are belong to us: the power of file-injection attacks on searchable encryption. In: USENIX (2016)

Obfuscation I

Patchable Indistinguishability Obfuscation: $i\mathcal{O}$ for Evolving Software

Prabhanjan Ananth[1(✉)], Abhishek Jain[2], and Amit Sahai[1]

[1] Center for Encrypted Functionalities and Department of Computer Science,
UCLA, Los Angeles, USA
{prabhanjan,sahai}@cs.ucla.edu
[2] Johns Hopkins University, Baltimore, USA
abhishek@cs.jhu.edu

Abstract. In this work, we introduce *patchable indistinguishability obfuscation*: our notion adapts the notion of indistinguishability obfuscation ($i\mathcal{O}$) to a very general setting where obfuscated software evolves over time. We model this broadly by considering software patches P as arbitrary Turing Machines that take as input the description of a Turing Machine M, and output a new Turing Machine description $M' = P(M)$. Thus, a short patch P can cause changes everywhere in the description of M and can even cause the description length of the machine to increase by an arbitrary polynomial amount. We further consider *multi-program* patchable indistinguishability obfuscation where a patch is applied not just to a single machine M, but to an unbounded set of machines $M_1, \ldots, M_n$ to yield $P(M_1), \ldots, P(M_n)$.

We consider both single-program and multi-program patchable indistinguishability obfuscation in a setting where there are an unbounded number of patches that can be *adaptively* chosen by an adversary. We show that sub-exponentially secure $i\mathcal{O}$ for circuits and sub-exponentially secure re-randomizable encryption schemes (Re-randomizable encryption schemes can be instantiated under standard assumptions such as

The full version of this paper can be found in [6].

Work done in part while visiting the Simons Institute for Theoretical Computer Science, supported by the Simons Foundation and by the DIMACS/Simons Collaboration in Cryptography through NSF grant #CNS-1523467.

P. Ananth—This work was partially supported by grant #360584 from the Simons Foundation and the grants listed under Amit Sahai.

A. Jain—Supported in part by a DARPA/ARL Safeware Grant W911NF-15-C-0213 and NSF CNS-1414023.

A. Sahai—Research supported in part from a DARPA/ARL SAFEWARE award, NSF Frontier Award 1413955, NSF grants 1619348, 1228984, 1136174, and 1065276, BSF grant 2012378, a Xerox Faculty Research Award, a Google Faculty Research Award, an equipment grant from Intel, and an Okawa Foundation Research Grant. This material is based upon work supported by the Defense Advanced Research Projects Agency through the ARL under Contract W911NF-15-C-0205. The views expressed are those of the authors and do not reflect the official policy or position of the Department of Defense, the National Science Foundation, or the U.S. Government.

© International Association for Cryptologic Research 2017
J.-S. Coron and J.B. Nielsen (Eds.): EUROCRYPT 2017, Part III, LNCS 10212, pp. 127–155, 2017.
DOI: 10.1007/978-3-319-56617-7_5

DDH, LWE.) imply single-program patchable indistinguishability obfuscation; and we show that sub-exponentially secure $i\mathcal{O}$ for circuits and sub-exponentially secure DDH imply multi-program patchable indistinguishability obfuscation.

At the our heart of results is a new notion of *splittable $i\mathcal{O}$* that allows us to transform any $i\mathcal{O}$ scheme into a patchable one. Finally, we exhibit some simple applications of patchable indistinguishability obfuscation, to demonstrate how these concepts can be applied.

1 Introduction

Program obfuscation is the process of making a program "unintelligible" to any polynomial-time entity while preserving its functionality. A formal study of program obfuscation was initiated more than a decade ago in the works of [10,41]. In the recent years, this research area has seen renewed activity with the emergence of candidate constructions [30] for a type of general-purpose program obfuscation called indistinguishability obfuscation. This notion has proven to be both extremely useful and the most plausible of existing notions of program obfuscation.

A major limitation of existing notions of program obfuscation is that they only consider "static" programs that do not change with time. In reality, however, programs are rarely changeless. We typically alter programs over time, with *patches* (a.k.a updates) causing the programs to grow and vary, in response to demands for greater or new functionality. Can program obfuscation be adapted to deal with this reality? Specifically, can we obfuscate programs that *evolve* over time? The central intellectual and theoretical focus of this work is to answer this question.

Obfuscation for Evolving Software. A trivial solution to obfuscating evolving software would be to simply apply the obfuscator afresh to each updated version of a particular program. For example, to modify an obfuscation of a program M, the obfuscator may simply release a fresh obfuscation of M' where M' is the patched version of M. Note, however, that in this solution, the total communication complexity is at least $|M| + |M'|$. In particular, this is the case *even if the difference between the programs M and M' can be described in the form of a small patch P*. In contrast, if M was not obfuscated, then we could modify it by simply communicating the patch P to a user, yielding a total communication complexity of only $|M| + |P|$. Our goal is to develop a mechanism for program obfuscation that approximately preserves this communication complexity.

A bit more precisely, we define a notion of *patchable* obfuscation where, informally, there are four algorithms:

- $\mathsf{Obf}(M; r)$ taking as input a program M, and outputting an obfuscated program $\langle M \rangle$, using randomness r.
- $\mathsf{GenPatch}(P; r, r')$ taking as input a patch P, and outputting an encoded patch $\langle P \rangle$, using a combination of the original randomness r and new randomness r'.

- AppPatch $(\langle M \rangle, \langle P \rangle)$ taking as input an obfuscated program $\langle M \rangle$ and a patch encoding $\langle P \rangle$, and outputting an obfuscated patched program $\langle M' = P(M) \rangle$.
- Eval $(\langle M \rangle, x)$, taking as input an obfuscated program $\langle M \rangle$ and an input x, and outputting the value $y = M(x)$.

The key efficiency requirement is that the size of a patch encoding should not depend on the size of the original program M. Specifically, we want that $|\langle P \rangle| = \mathrm{poly}(|P|, \lambda)$, where λ is the security parameter.

Beyond this basic efficiency requirement, we also discuss some other important considerations w.r.t. patchable obfuscation.

I. NO RESTRICTION ON PATCHES: An important consideration for patchable obfuscation is the class of patches that we wish to allow. Clearly, the larger the class of patches that we can support, the larger the potential application pool.

To maximize the applicability of our notion, we allow for *arbitrary* patches. Specifically, we model a patch P as a Turing machine that takes as input a program M (also modeled as a TM) and outputs a new program M'. We allow for the unpatched program to *grow* in size after patching. That is, M' may be arbitrarily bigger than M.

II. MULTIPLE PATCHES: Another consideration is the number of patches that we wish to allow. In reality, it may be difficult to anticipate in advance how many times a program may need to be patched. Thus, we allow for an *unlimited* number of patches.

Specifically, we consider two modes of patching:

- Sequential patching: Here, given an obfuscated program $\langle M_0 \rangle$ and a sequence of patch encodings $\langle P_1 \rangle, \ldots, \langle P_n \rangle$, one can apply the patches one-by-one, *in order*, to obtain $\langle M_1 \rangle, \ldots, \langle M_n \rangle$ s.t. $M_i = P_i(M_{i-1})$.
- Parallel patching: Here, given an obfuscated program $\langle M_0 \rangle$ and a sequence of patch encodings $\langle P_1 \rangle, \ldots, \langle P_n \rangle$, one can apply each patch to $\langle M \rangle$, *in parallel*, to obtain $\langle M_1 \rangle, \ldots, \langle M_n \rangle$ s.t. $M_i = P_i(M_0)$.

While sequential patching seems to better capture patching of programs in reality, as we discuss later, parallel patching also enables interesting applications of patchable obfuscation. Thus, we consider both patching modes in this work.

III. SUPPORT FOR MULTIPLE PROGRAMS: So far, we have only discussed patching for a single obfuscated program. Now consider the case where an authority wishes to patch *multiple* obfuscated programs $\langle M_1 \rangle, \ldots, \langle M_n \rangle$. Such a situation often arises in practice where, for example, the programs $M_1, \ldots, M_n$ may correspond to different copies of the same core program M that are individualized to different users.

One approach to address this scenario would be to release a separate patch for every obfuscated program. In this case, however, the communication complexity grows linearly with the number of obfuscated programs and may quickly become prohibitive. Instead, we would like to build patchable obfuscation where the obfuscator can release *one* patch that can be applied to all of the obfuscated programs. We refer to this notion as *multi-program patchable obfuscation*.

How to Define Security? Of course, we must define security for patchable obfuscation. The natural direction is to start with a "base" notion of obfuscation (without patching) and extend it to the setting of patching. Our goal in this work is to obtain general positive results for patchable obfuscation. With this viewpoint, we identify indistinguishability obfuscation ($i\mathcal{O}$) [10] as a natural choice for the base notion. Indeed, over the last few years, several general-purpose candidate constructions, (for example: [9,22,30]) for $i\mathcal{O}$ have been proposed, and no impossibility results are known. Furthermore, it was shown by [39] that $i\mathcal{O}$ is, in fact, "best-possible" obfuscation. $i\mathcal{O}$ has already enabled a long sequence of exciting applications (see e.g., [18,28,30,51]) and its patchable analogue can be expected to find even more applications. Finally, we stress that while the security of $i\mathcal{O}$ remains an area of intense study, there are several known $i\mathcal{O}$ candidates and even *universal* $i\mathcal{O}$ candidates under well-studied assumptions [3].

In contrast, powerful (base) notions such as virtual black-box obfuscation [10] and differing-inputs obfuscation [1,10,20] have been shown to be impossible to realize for general functions [10,13,15,31,36]. This, in turn, means that patchable analogues of these notions are also impossible, in general. The notion of virtual grey-box obfuscation [14,16] is impossible for general Turing Machines but seems to circumvent general impossibility results for circuits; however, it has found rather limited applicability so far.

In light of the above, in this work, we focus on patching in the context of $i\mathcal{O}$. We do believe that the study of patchable obfuscation for other base obfuscation notions (e.g., obfuscation in weaker adversarial models such as virtual black-box obfuscation in hardware token model [34,38,40] or generic model [9,22]) is interesting, and we leave this study to future work. We remark that many of the ideas that we develop in this work should be more widely applicable to other notions of obfuscation, and are not intrinsically tied to $i\mathcal{O}$. As such, we envision these ideas to be portable to other notions of patchable obfuscation.

Patchable Indistinguishability Obfuscation. We develop a notion of *patchable indistinguishability obfuscation* (*pa-i$\mathcal{O}$*) that naturally extends the standard notion of $i\mathcal{O}$ to the setting of patching. Let us explain our notion for the single-program case, for sequential and parallel patches.

- *Sequential patches*: Recall that $i\mathcal{O}$ security dictates that given two equivalent programs M_0 and M_1, obfuscations of M_0 and M_1 are computationally indistinguishable. In single-program *pa-i$\mathcal{O}$* for sequential patches, we require that given two equivalent programs M_0^0 and M_1^0 and a sequence of patch pairs $(P_0^1, P_1^1), \ldots, (P_0^n, P_1^n)$ such that for every "level" $i \in [n]$, the patched programs $M_0^i = P_0^i(M_0^{i-1})$ and $M_1^i = P_1^i(M_1^{i-1})$ are also equivalent, it should be hard to distinguish the tuples $(\langle M_0^0 \rangle, \{\langle P_0^i \rangle\}_{i=1}^n)$ and $(\langle M_1^0 \rangle, \{\langle P_1^i \rangle\}_{i=1}^n)$. Intuitively, the equivalence requirement at every patch level i rules out the trivial attack of using a splitting input for the patched programs M_0^i and M_1^i to distinguish the tuples.
- *Parallel patches*: Single-program *pa-i$\mathcal{O}$* for parallel patches is defined similarly to above, except that here we require equivalence for the patched programs $M_0^i = P_0^i(M_0^0)$ and $M_1^i = P_1^i(M_1^0)$ at every (parallel) "branch" $i \in [n]$.

A few remarks are in order: **(1)** It is easy to see that these definitions ensure *patch hiding*, which is crucial for some of the applications discussed later. **(2)** Our definitions naturally extend to multi-program *pa-iO* where we start with multiple pairs of programs and equivalence is required for every pair at every level/branch. **(3)** We, in fact, consider *adaptive* security, where the adversary can make the patch queries in an adaptive fashion. See Sect. 2 for further details.

Implications of *pa-iO*. We view *pa-iO* as a powerful primitive that is likely to have several applications in the future. To see the power of *pa-iO*, it is instructive to first compare it with *iO*. While *iO* exists if **P=NP**,[1] we show that multi-program *pa-iO* for parallel patches implies secret-key functional encryption (FE) [19,49,50]. The construction is remarkably simple: let $M_{f,x}$ be an input-less machine that simply outputs $f(x)$. We construct an FE scheme as follows:

- A secret key for a function f is computed as $\langle M_{f,\perp} \rangle$, i.e., an obfuscation of $M_{f,x}$ where $x = \perp$.
- Encryption of a message m corresponds to generating an encoding $\langle P_m \rangle$ for a patch P_m that modifies $M_{f,\perp}$ to $M_{f,m}$.
- Decryption simply corresponds to applying the patch encoding $\langle P_m \rangle$ on $\langle M_{f,\perp} \rangle$ to obtain $\langle M_{f,m} \rangle$ and then evaluating it to obtain $f(m)$.

Correctness and security of the construction follow in a straightforward manner from the correctness and security of *pa-iO*.[2] As we discuss later, the above basic idea can, in fact, be easily extended to multi-input functional encryption [35], yielding new results.

Alternate Viewpoint: Obfuscation with Private Homomorphism. Another way of looking at our notion of *pa-iO* is as a form of *iO* that supports a kind of semi-private *homomorphism*: the generation of the patch encoding is private – requiring secret information that was used to obfuscate the original program – although the application of the patch encoding is public. Note that unlike encryption, for the security of obfuscation it is critical that this homomorphism is semi-private – if an adversary was allowed to use public information to arbitrarily modify the program underlying an obfuscation, this would trivially allow the adversary to break the security of the original obfuscated program. On the other hand, our notion of *pa-iO* and the notion of fully homomorphic encryption [33] share a similarity in that they both require a form of compactness for the notions to be non-trivial.

---

[1] Assuming **NP** $\neq$ **co-RP**, it was shown that *iO* implies one-way functions [43,48].

[2] An observant reader may notice that in the above construction, it is not important whether the size of a patch encoding depends on the size of an unpatched machine $M_{f,\perp}$ or not. However, it is important that the size of the patch encoding is independent of the number of obfuscated machines that it can be applied to – a property guaranteed by multi-program *pa-iO*.

1.1   Our Results

We state our results below.

I. Patchable Indistinguishability Obfuscation. In this work, we formalize the notion of patchable indistinguishability obfuscation. We focus on the setting where programs to be obfuscated and patched are described as Turing Machines.

Multi-program pa-iO: Our main result is a construction of a multi-program *pa-iO* scheme from sub-exponentially secure *iO* and sub-exponentially secure DDH.

Theorem 1 (Multi-program *pa-iO*: Sequential patches). *Assuming the existence of sub-exponentially secure iO for circuits, sub-exponentially secure DDH, there exists an adaptively secure multi-program pa-iO scheme with unbounded sequential patches, for Turing Machines where the running time of the patch generation algorithm for a patch P is bounded by $poly(\lambda, |P|, \ell)$, where λ is a security parameter and ℓ is a bound on the input size to the patched program.*

Note that the runtime efficiency of the patch generation algorithm in the above theorem implies the necessary size efficiency for a patch encoding, namely, the size of the encoding of a patch P is bounded by $poly(\lambda, |P|, \ell)$.

Single-Program pa-iO: We obtain the above result in two steps. Our first, and key step is to construct a single-program *pa-iO* scheme for TMs which achieves the desired size efficiency for patches but requires a large state (proportional to the size of the TM being updated) as well as a large patch generation time.

Theorem 2 (Single-program *pa-iO*: Sequential patches). *Assuming the existence of sub-exponentially secure iO for circuits and sub-exponentially secure re-randomizable encryption schemes, there exists an adaptively secure single-program pa-iO scheme with unbounded sequential patches, for Turing Machines where the size of the obfuscation of a patch P is bounded by $poly(\lambda, |P|, \ell)$, where λ is a security parameter and ℓ is a bound on the input size to the patched program.*

Main Tool: Splittable iO: The main tool in our construction of single-program is an intermediate notion between *iO* and patchable *iO*, that we refer to as *splittable iO*. Very roughly, splittable *iO* allows us to reduce the problem of building patchable *iO* to the problem of building a patchable "encoding" scheme, a seemingly simpler problem. Very roughly, an obfuscation of M w.r.t. splittable *iO* consists of two parts: an encoding of M w.r.t. a patchable encoding scheme, and some auxiliary information z computed on the encoding as well as the secret key used to encode M. We place suitable efficiency and security requirements on the auxiliary information so as to allow us to transfer the patching property of the encoding scheme to the setting of *iO*. We refer the reader to the technical overview section for further details on this notion.

From Single-Program to Multi-program pa-iO: Next, we devise a generic transformation from any such single-program *pa-iO* scheme to a multi-program *pa-iO* scheme with the aforementioned efficient patch generation property.

Theorem 3 (Single-program to Multi-program $pa\text{-}i\mathcal{O}$). *Assuming the existence of a succinct garbled TM scheme with persistent memory and a compact secret-key functional encryption scheme for general circuits, there exists a general transformation from any single-program $pa\text{-}i\mathcal{O}$ scheme to a multi-program $pa\text{-}i\mathcal{O}$ scheme for TMs with efficient patch generation.*

In particular, when the underlying primitives are all adaptively secure, then the resulting multi-program $pa\text{-}i\mathcal{O}$ scheme is also adaptively secure. An adaptively secure succinct garbled TM scheme with persistent memory is known from the works of [2,24] based on sub-exponentially secure $i\mathcal{O}$ and DDH assumption, while a compact secret-key functional encryption scheme is known from $i\mathcal{O}$ for general circuits.

For the theorems above, we stress that we place no restrictions on the patches. A patch P can be an arbitrary Turing Machine that takes the original program description M as input, and outputs an arbitrary Turing Machine description $M' = P(M)$ that can differ in arbitrary ways from M. In particular, the description size of $P(M)$ can be any unbounded polynomial in the security parameter, and thus the program size can grow by arbitrary polynomial factors. Furthermore any unbounded polynomial number of patches can be applied sequentially, and the adversary can specify these patches adaptively given all obfuscated programs and patches constructed earlier.

Parallel Patching: We can obtain a similar result for multi-program $pa\text{-}i\mathcal{O}$ in the context of parallel patches. This result follows the same approach as the case of sequential patches. The first step is to obtain single-program $pa\text{-}i\mathcal{O}$ scheme with unbounded parallel patches and the second step is to obtain multi-program $pa\text{-}i\mathcal{O}$ from single-program $pa\text{-}i\mathcal{O}$. The construction of single-program $pa\text{-}i\mathcal{O}$ with parallel patches will be identical to the one in the sequential patch setting. The transformation from single-program $pa\text{-}i\mathcal{O}$ to multi-program $pa\text{-}i\mathcal{O}$ is, however, different from the sequential setting to enable this transformation. Instead of using garbled TM scheme with persistent memory, we instead employ functional encryption for TMs [7,37] scheme. Since the techniques employed in the parallel patch setting are similar to the sequential patch setting, we omit the transformation. We have the following theorem.

Theorem 4 (Multi-program $pa\text{-}i\mathcal{O}$: Parallel patches). *Assuming the existence of sub-exponentially secure $i\mathcal{O}$ for circuits, sub-exponentially secure DDH, there exists an adaptively secure multi-program $pa\text{-}i\mathcal{O}$ scheme with unbounded parallel patches, for Turing Machines where the running time of the patch generation algorithm for a patch P is bounded by $poly(\lambda, |P|, \ell)$, where λ is a security parameter and ℓ is a bound on the input size to the patched program.*

II. Applications of $pa\text{-}i\mathcal{O}$. We view $pa\text{-}i\mathcal{O}$, and especially multi-program $pa\text{-}i\mathcal{O}$ as a powerful primitive that is likely to have several applications in the future. As initial evidence of this, we demonstrate implications of $pa\text{-}i\mathcal{O}$ to functional encryption and $i\mathcal{O}$ for TMs. In our eyes, the main appeal of these implications is their remarkable *simplicity* that highlights the potential of *pa-iO* as a replacement for *iO* in cryptographic applications.

Multi-input FE for Unbounded Arity Functions: We first show that multi-program *pa-iO* for parallel updates implies secret-key multi-input functional encryption (MIFE) [4,21,35] for unbounded arity functions. This implication follows from a straightforward extension of the *pa-iO* to (single-input) FE implication discussed earlier.

Theorem 5 (Unbounded-Arity MIFE). *Adaptively secure multi-program pa-iO with unbounded parallel updates implies secret-key MIFE for unbounded arity functions with security against pre-ciphertext key queries.*

Combining the above with Theorem 4, we obtain secret-key MIFE for unbounded arity functions from sub-exponentially secure *iO* for circuits, sub-exponentially secure DDH. Previously, this result was only known [8] from a knowledge assumption, namely public-coin differing-input obfuscation [42] and one-way functions.

FE for TMs with Unbounded Length Inputs: The following implication follows as a simple corollary of Theorem 5.

Theorem 6 (Unbounded-Input FE). *Adaptively secure multi-program pa-iO implies secret-key functional encryption for TMs with unbounded input length with security against pre-ciphertext key queries.*

A construction of FE for TMs with unbounded input was recently given by [7] based on *iO*. We emphasize that our construction from multi-program *pa-iO* is extremely simple, in contrast to the involved construction of [7].

We now discuss implications of *pa-iO* to *iO* for TMs. We first recall that all recent progress on achieving *iO* for TMs/RAMs [17,25–27,44] from *iO* for circuits has required a polynomial bound ℓ to be placed on the input length to the obfuscated Turing Machine. We share this need for a polynomial bound ℓ on the input size, and the size of our obfuscated patches do grow with this bound. Indeed, if we could remove this restriction, then we would show how to bootstrap *iO* for circuits to *iO* for Turing Machines without any input length restriction from *iO* for circuits – this remains a major open question. Achieving *iO* for Turing Machines without any input length restriction currently requires strong assumption such as output-compressing randomized encodings [45] or knowledge-type assumptions such as public-coin *diO* [1,20,42]. We do not know how to achieve these objects using only *iO* for circuits.

iO for TMs with Unbounded Length Inputs: So far, in our definition of *pa-iO*, we have only considered "single-use" patches. More accurately, in our definition of single-program (resp., multi-program) *pa-iO* for sequential patching, the ith patch P_i can only be applied to the updated machine (resp., machines) at level $i - 1$. As we discuss now, such "single-use" patches are, in fact, inherent given the current state of art in *iO* for TMs.

In particular, is not difficult to see that single-program $pa\text{-}i\mathcal{O}$ with *reusable* patches (i.e., where a patch P is not tied to any "level" and can be applied an arbitrary number of times, to any machine) in fact, implies $i\mathcal{O}$ for TMs with unbounded length inputs. The construction is extremely simple: let M_x be a family of (input-less) machines parameterized by strings x of arbitrary length, where every machine simply outputs $M(x)$. Obfuscation of a TM M consists of an obfuscation of a machine $M_\perp$ w.r.t. the $pa\text{-}i\mathcal{O}$ scheme along with encodings of two reusable patches P_0 and P_1. Patch P_0 is such that it updates any machine M_x to $M_{x\|0}$ while P_1 updates any machine M_x to $M_{x\|1}$.

To evaluate the above obfuscation on any input $x = x_1, \ldots, x_\ell$ for an arbitrary ℓ, a user can transform obfuscation of $M_\perp$ to M_x by applying the patches $P_{x_1}, \ldots, P_{x_n}$ and then execute M_x to obtain $M(x)$. The correctness of the construction is easy to verify.

While we do not consider security for reusable patches in this work, we view the above as a potential new template for building $i\mathcal{O}$ for TMs with unbounded length inputs.

1.2 Technical Overview

We now give an overview of the main technical ideas in our constructions. We start by building a general template for building $pa\text{-}i\mathcal{O}$, and then discuss our ideas for implementing this template.

1.2.1 A Template for $pa\text{-}i\mathcal{O}$

In this section, we devise a general template for building $pa\text{-}i\mathcal{O}$ starting from any non-patchable obfuscation scheme. We keep the discussion in this section to a high-level, focusing on issues directly related to *patching*, and largely ignoring implementation issues that may arise due to the specific properties of the underlying non-patchable obfuscation scheme. For simplicity, in this section, we advise the reader to think of the non-patchable obfuscation scheme as general-purpose virtual-black-box obfuscation. Later, in Sect. 1.2, we discuss the additional challenges that arise in implementing our template when the non-patchable obfuscation scheme is $i\mathcal{O}$, and our solutions for the same.

Let us start with the weaker goal of building single-program $pa\text{-}i\mathcal{O}$ where the authority issues a single obfuscated program that can then be patched multiple times, in a sequential order. Our initial idea towards achieving this goal is to identify an encoding scheme that supports patching and then combine it with a non-patchable obfuscation scheme to build a $pa\text{-}i\mathcal{O}$ scheme. Intuitively, we say that an encoding scheme is patchable if given an encoding of a machine M and an encoding of a patch P, it is possible to derive an encoding of $M' = P(M)$. The hope here is that the patching property of the encoding scheme can be translated into patching property for obfuscation.

A natural candidate for a patchable encoding scheme is *fully homomorphic encryption* (FHE). Indeed, given an encryption (i.e., encoding) of a machine M and an encryption of a patch P, one can obtain an encryption of the patched machine $M' = P(M)$ by homomorphically evaluating the function $f(M, P) = P(M)$. Starting with FHE and any non-patchable obfuscation scheme, we can build an initial template for $pa\text{-}i\mathcal{O}$ as follows: to obfuscate M, first encrypt M using FHE and then provide an obfuscation of the FHE decryption circuit that has the FHE decryption key hardcoded into it. Evaluation on an input x can be done as follows: first use FHE evaluation to transform encryption of M into an encryption of $M(x)$, and then use the obfuscated decryption circuit to obtain $M(x)$. To patch the obfuscated program, we can simply patch the encryption of M in the manner as described above.

While this solution seems to offer the functionality of patching, it does not offer any security. Specifically, in the above template, an adversary can choose an arbitrary patch P^* on its own and then use FHE evaluation of the function $f_{P^*}(M) = P^*(M)$ to transform encryption of M into an encryption of $P^*(M)$. If this patch P^* is such that for two equivalent machines M_0 and M_1, $P^*(M_0)$ and $P^*(M_1)$ are not equivalent, then the adversary can easily break the security of $pa\text{-}i\mathcal{O}$. Indeed, the security of $pa\text{-}i\mathcal{O}$ prevents an adversary from creating patches on its own, while the above template does not place this restriction in any way. In particular, we need to crucially use the fact that patch generation is a secret key operation.

Towards that end, we modify the above template such that an evaluator can only apply *authenticated* patches. The obfuscation of M consists of an FHE encryption of M as before but the obfuscated FHE decryption circuit now takes as input old encryption $Enc(M)$, updated encryption $Enc(M')$, encrypted patch $Enc(P)$, a signature σ on $Enc(P)$ and an input x. It checks if the signature is valid and also if $Enc(M')$ is obtained by updating $Enc(M)$ using P. If the check passes, then it decrypts $Enc(M')$ and evaluates M' on x. During the patching phase, the authority sends both $Enc(P)$ and the signature σ. This signature now prevents a user from applying "invalid" patches to the obfuscation; however, we note that in the context of $i\mathcal{O}$, this authentication will need to be done in a much more careful manner, as we elaborate below.

Enforcing Ordered Executions of Patches. While the above template does not seem to suffer from any immediate issues when we consider a single patch, unfortunately, its security breaks down when we consider the setting of multiple patches. Indeed, in the above template, given (say) two patch encodings $(Enc(P_1), \sigma_1)$, $(Enc(P_2), \sigma_2)$, an adversary may first apply the second patch and then the first patch, which may break the equivalence requirement on the patched machines in the security definition of $pa\text{-}i\mathcal{O}$. In fact, an adversary can also repeatedly apply the same patch multiple times in the above template, which may also break the equivalence requirement on the patched machines in the security definition of $pa\text{-}i\mathcal{O}$. Indeed, the definition of $pa\text{-}i\mathcal{O}$ requires that the patch encodings can only be applied *in order*, namely, the ith patch encoding can only be applied to the $(i-1)^{th}$ patched obfuscation, *once*.

Towards this, we introduce a mechanism to force a user to apply the patches in order. We begin by observing that instead of authenticating the encrypted patch in the above template, if we instead authenticate the encrypted patched machine, then we can enforce ordered executions of patches. That is, suppose we want to update the machine M using patch P, the authority first computes $Enc(P)$ and then updates $Enc(M)$ using $Enc(P)$ to obtain $Enc(M')$. It then signs $Enc(M')$ and sends the signature[3] σ and the encrypted patch $Enc(P)$ to the user. The user now updates $Enc(M)$ using $Enc(P)$ to obtain $Enc(M')$. To evaluate the patched obfuscation on an input x, it inputs $(Enc(M'), \sigma, x)$ to the obfuscated FHE decryption circuit that first checks for validity of the signature and then decrypts $Enc(M')$ followed by computation of $M'(x)$, as before. Crucially, by shifting the authentication to the updated encrypted machine instead of encrypted patch, we are now able to prevent the "out-of-order patching" attacks (as well as "repeated patching" attacks) by an adversary discussed above.

A disadvantage of the above solution is that it requires the authority to maintain large state. In particular, at any time, the authority must remember the last patched machine M_{i-1} in order to generate a valid encoding for the ith patch P_i. Furthermore, the patch encoding generation time now depends on the size of the machine M_{i-1}. While this loss in efficiency may be acceptable for the setting of single-program $pa\text{-}i\mathcal{O}$, it unfortunately becomes a significant barrier for the setting of multi-program $pa\text{-}i\mathcal{O}$. Indeed, in the multi-program setting, the number of obfuscated programs are not a priori bounded; as such, if we were to extend the above template to this case, then the authority's state size becomes unbounded! (This is because the authority would need to maintain a separate state for every obfuscated program.)

Compressing the State of Authority. In order to resolve this issue we introduce the next idea: "delegating" the state of the authority to the user. That is, the authority now maintains the state at the user's end. Implementing this idea introduces several issues: not only should the state be encrypted at the user's end but it should also be possible to repeatedly update and also compute on this (updated) encrypted state. To address these issues, we turn to a cryptographic primitive called garbled RAMs with persistent memory. This notion allows for encoding a database and repeatedly update this encoding and compute on the updated encodings. The updating and computation operations are enabled by using encodings of RAM programs which are issued by the authority. Using this primitive, we propose a solution template.

- To obfuscate M, the authority computes: (i) $Enc(M)$ and a signature upon it. (ii) An obfuscation of the FHE decryption circuit (as before) that takes an input x, $Enc(M)$ and a signature σ, and outputs $M(x)$ if the signature is valid. (iii) A database encoding $\widetilde{Enc(M)}$ of $Enc(M)$. It then sends $\widetilde{Enc(M)}$, $Enc(M)$, σ and the obfuscated decryption circuit to the user.
- To evaluate the obfuscation on an input x, the user inputs $(x, Enc(M), \sigma)$ to the obfuscated decryption circuit to recover the output $M(x)$.

---

[3] For this discussion, let us assume that we have a signature scheme where the size of the signature is independent of the length of the message. We will revisit this later when we discuss implementation issues.

– To compute a patch encoding of P, the authority first computes $Enc(P)$ (as before) and then computes a garbled RAM encoding $\widetilde{T}$ of a RAM machine T that has $Enc(P)$ hardcoded in it. The machine T uses FHE evaluation over $Enc(M)$ (in the database encoding) and $Enc(P)$ to compute $Enc(M')$ and additionally computes signature σ' over $Enc(M)'$. It outputs σ' in the clear. The user, upon receiving the patch encoding, first computes $\widetilde{Enc(M')}$ using $Enc(P)$. It then updates the database encoding $\widetilde{Enc(M)}$ using $\widetilde{T}$. The result is an updated database encoding $\widetilde{Enc(M')}$ and the signature σ' on $Enc(M')$. The user can now evaluate the updated machine on any input in the same manner as before.

Some remarks are in order: first, from an efficiency viewpoint, we need the garbled RAM scheme to be *succinct* where the size of RAM machine encoding is independent of its running time. This is because we are applying the above idea on a single-program *pa-iO* scheme where the patch generation time depends on the size of the machine being updated. Second, in order to argue security in the setting of adaptively chosen patches, we need the garbled RAM scheme to satisfy adaptive security as well. Such a garbled RAM scheme (with persistent memory) was recently constructed in the independent works of [2, 24].

Finally, we note that while the above idea successfully compresses the state size of the authority, it still does not suffice for the multi-program setting. This is because in the above solution, when extended to the multi-program case, the authority would need to maintain some small state, namely, the garbling key, for *every* obfuscated machine, which still leads to a state of unbounded size. We address this problem by developing a generic transformation from any single-program *pa-iO* scheme with small state (or alternatively, a *stateless* scheme) into a multi-program *pa-iO* scheme by using a compact secret-key functional encryption scheme for general circuits. We defer the discussion of this transformation to the next section.

1.2.2 Implementation

Issues Related to Indistinguishability Obfuscation. While the above template seems promising, several issues arise when we have to implement it only assuming indistinguishability obfuscation *for circuits*. For starters, the above template requires an obfuscation scheme for Turing machines with unbounded length inputs. This is because, the size of the encrypted machine M can grow arbitrarily over a sequence of updates and thus the input to the obfuscated circuit cannot be a priori bounded. We currently know how to realize this only based on strong knowledge-type assumptions [1, 20, 42]. Another technical issue is that standard signature schemes are not "compatible" with iO and more generally, using iO restricts the type of cryptographic primitives that we can use. These challenges were encountered in many recent works [17, 26, 44] whose main goal was reducing the problem of constructing iO for Turing machines, where the length of inputs to be evaluated are a priori bounded, to the problem of constructing iO for circuits. We build upon the primitives and notions introduced in the work of [44] to address these challenges. We recall the Turing machine randomized encodings[4] construction by [44].

The core idea in the randomized encodings construction of Koppula et al. [44] is to leverage an obfuscated circuit to perform step-by-step computation of the machine M that is encoded. In more detail, a randomized encoding of (M, x) consists of: (a) input tape initialized with an encoding of M and, (b) an obfuscated circuit C_x that performs "step-by-step" computation of a machine $U_x(\cdot)$. Here, $U_x(\cdot)$ is a universal TM that takes as input machine M and outputs $M(x)$. By step-by-step computation, we mean that the circuit C_x takes as input time step i, encoded symbol and partial information about the current state in an encrypted form and produces a new encoded symbol and state, again in encrypted form, by executing the transition function of U_x. This enables the size of the circuit C_x to be independent of the length of M.

To see how the randomized encodings construction might be useful to our setting, note that we could potentially encode the machine M using a patchable encoding scheme that will allow us to patch M. Furthermore, we can allow the machine size to arbitrarily grow, over a sequence of updates, since the size of the circuit C_x is independent of the machine size M. However, the main issue is that their approach is tied to just a single computation $M(x)$ whereas we require that M be reused on multiple inputs. They propose an approach to achieve reusability by using another layer of obfuscation, with M hardwired in it, that produces fresh encodings of M for every computation. This is highly problematic for us, since patching M would now correspond to patching the underlying obfuscated circuit.

We need to make the randomized encodings construction of KLW reusable while preserving the underlying encoding of M. A recent work of Ananth et al. [5], proposed in a different context of building iO with constant overhead, achieves this goal. In more detail, they showed how to achieve iO for TMs, with a priori bound in the input length, such that an obfuscation of M proceeds in two phases: (a) M is encoded using a suitable encoding scheme and, (b) an obfuscation of a circuit that takes as input x and produces an encoding of x. The evaluation of the obfuscation on an input x proceeds by first obtaining an encoding of x (using the obfuscated circuit) and then decoding this using the encoding of M to recover $M(x)$.

While their work offers a starting point for building patchable iO, we still need to address several issues that specifically arise in the context of patching. For instance, their work only considers the setting when the adversary is given one obfuscated machine whereas in our setting she also receives additionally, patches that share some common randomness with the obfuscated machine. We need to argue that the security holds even with this additional information. Instead of directly digging into the details of [5] to apply it in the context of patching, we undertake a more modular approach. First, we propose an intermediate primitive

---

[4] A randomized encoding of (M, x) satisfies two properties: (a) it only reveals $M(x)$ and, (b) the size of the encoding is polynomial only in the length of M, x and security parameter.

called *splittable iO* and show that it suffices for building single-program patchable iO. We then show that splittable iO can be implemented assuming only iO for circuits by using the framework of [5]. We describe this primitive in detail next.

Splittable iO: Intermediate Notion Between iO and Patchable iO.
A splittable iO scheme is a strengthening of iO and is associated with respect to a patchable encoding scheme. A patchable encoding scheme consists of algorithms: Setup, Encode and Decode. Setup generates a secret key sk that will be used by Encode procedure to obtain an encoding of M, $\mathcal{E}_{sk}(M)$. Decode recovers the Turing machine M from the encoding $\mathcal{E}_{sk}(M)$ using the secret key sk. Additionally, it is associated with two algorithms: patch generation algorithm, used to generate secure patches and patch application algorithm, that enables applying secure patches on encodings of TMs. The security property requires that the encodings and patches hide the underlying TMs and patches, respectively.

We start with a oversimplified template of splittable iO and make suitable modifications later. An obfuscation of M, with respect to splittable iO, consists of two parts: $(\mathcal{E}_{sk}(M), aux_M)$, where (i) $\mathcal{E}_{sk}(M)$ is a patchable encoding of M computed using secret key sk, (ii) aux_M computed as a function of an additional PPT algorithm AuxGen, on $(sk, \mathcal{E}_{sk}(M))$.

Armed with the notion of splittable iO, we show how to construct single-program patchable iO. At first glance, it seems that splittable iO already allows for patching: indeed, since M is encoded with respect to a patchable encoding scheme, we can use the patching algorithm to update this encoding. However, this does not work because the obfuscation also contains aux_M that is tied to encoding of M. Indeed, this is necessary for the security of obfuscation to hold. So if the encoding of M is updated, it is necessary to also update aux_M. A naive way of achieving this is to issue a fresh aux_M every time the encoding is patched. That is, initially the user is issued an encoding of M, $\mathcal{E}_{sk}(M)$ and auxiliary information aux_M. During the patching phase, a secure version of patch P with respect to the patchable encoding scheme is issued. Along with this, a fresh $aux_{M'}$ is issued, which is generated by first patching $\mathcal{E}_{sk}(M)$ using $\widetilde{P}$, secure patch of P, and then executing AuxGen on input $(sk, \mathcal{E}_{sk}(M'))$.

However this raises the question of efficiency: the patch size now grows with the size of $aux_{M'}$. This can be taken care of imposing an efficiency constraint on splittable iO: we require that the size of aux be a polynomial in security parameter and specifically, independent of the size of the machine obfuscated. The next issue is correctness: why should the patched obfuscated machine be correct? for instance: AuxGen could abort on input patched encodings. To take care of this issue, we impose an additional property on splittable iO: the correctness of the obfuscated machine should hold irrespective of whether fresh encodings or patched encodings of the machine are fed to AuxGen.

Finally, we move on to proving the security of patchable iO. A first attempt is to use the security of the underlying patchable encoding scheme to argue this. However, it is unclear why the security of encoding scheme is guaranteed at all given that aux contains information about the secret key of the encoding scheme. If we additionally impose aux to hide the secret key, we can then hope to invoke the security of patchable encoding scheme to argue the security of patchable iO. A natural approach of formalizing this is to use a simulation-based argument – there exists a simulator that can simulate the aux even without knowing the secret key. But this would mean that aux will not able to decode any information about the encoding of M. In order to maintain correctness of the obfuscation of M, we need to hardwire all possible outputs which is clearly infeasible. Instead we use an indistinguishability-based definition: instead of having one encoding of M, we will consider a pair of encodings of M. That is, obfuscation of M consists of $(\mathcal{E}_{sk_0}(M), \mathcal{E}_{sk_1}(M))$, computed with respect to secret keys sk_0, sk_1. In addition, it consists of aux generated using $\mathsf{AuxGen}(sk_0, \mathcal{E}_{sk_0}(M), \mathcal{E}_{sk_1}(M))$. Now, we impose a security property that says that aux generated using sk_0 is computationally indistinguishable from aux generated using sk_1.

We summarize the (informal) definition of splittable iO below. The formal definition can be found in Sect. 3.2. In addition to the properties of any iO scheme, a splittable iO scheme has the following properties.

1. *Splittable Property:* An obfuscation of M can be performed in two steps: the first step is encoding M twice using two secret keys sk_0 and sk_1 of a patchable encoding scheme. The second step is generation of aux by computing AuxGen on input $(sk_0, \mathcal{E}_{sk_0}(M), \mathcal{E}_{sk_1}(M))$, where $\mathcal{E}_{sk_0}(M)$ and $\mathcal{E}_{sk_1}(M)$ are two encodings of M and sk_0 is the secret key used to encode $\mathcal{E}_{sk_0}(M)$.
2. *Correctness of* AuxGen: The correctness of obfuscation of M holds irrespective of whether AuxGen is executed on fresh encodings of M or whether it is executed on encodings of M obtained as a result of patching. This will be used to argue the correctness of the resulting patchable iO scheme.
3. *Efficiency of* aux: We require that the size of aux is a polynomial in λ and in particular, independent of the size of the machine obfuscated. This will be used to argue the patch size efficiency of patchable iO.
4. *Indistinguishability of* aux: We require that it is computationally hard to distinguish aux generated using secret key sk_0 from aux generated using sk_1. This property will be helpful to argue security of patchable iO.

Going from Single-Program to Multi-program Patchable Obfuscation. In the solution sketched above, every time the authority has to generate a patch, she has to spend time proportional to the size of the obfuscated machine. In particular, recall that one of the steps in the generation of secure patch is computing aux_M: this step involves first patching the old encoding $\mathcal{E}_{sk}(M)$ and then executing AuxGen. We will use the trick described earlier to solve the problem: we delegate the state of the authority as well as the computation of the secure patches to the user. This can be implemented by using a suitable garbling scheme that works in the persistent memory setting. Once this mechanism is implemented, the authority is only required to store the garbling key.

While this is a viable solution in the single-program setting, this is undesirable when the authority is issuing multiple obfuscated programs. She has to store the garbling keys corresponding to all the machines in this case. The storage space of the authority thus puts a bound on the number of obfuscated machines it can issue.

To overcome this difficulty, we employ another idea for delegating responsibility to the user! The garbling key of every user is maintained at her own storage space in an encrypted form. The computation of the garbled program encodings are then delegated to every user. This mechanism is implemented by using a functional encryption scheme. Every user along with the obfuscated machine, garbled encoding of state, also contains an FE encryption of the garbling key. During the patching phase, the authority sends a FE key containing patch P, that takes as input a garbling key and produces a garbled encoding of P with respect to this garbling key. To carry this out, we only require a *secret-key* FE scheme for *circuits*.

Putting it Together: A Framework for (Multi-program) Patchable Obfuscation. Putting all the components together, we construct a multi-program patchable iO in the following steps:

1. The first step involves formalizing the notion of splittable iO. This is shown in Sect. 3.
2. Next, we show how to obtain single-program patchable iO from splittable iO. This is shown in Sect. 4. The resulting single-program patchable iO scheme is statefull, i.e., the authority is required to maintain a large state.
3. We show how to overcome this problem by giving a transformation from any statefull to a stateless single-program patchable iO scheme. This is presented in the full version.
4. In the next step, we give a transformation from single-program to multi-program patchable iO. This is presented in the full version.
5. In the last step, we instantiate splittable iO using the framework of [5]. This is presented in the full version.

1.3 Related Work: Incremental Cryptography

The area of incremental cryptography was pioneered by Bellare et al. [11]. Subsequently, this concept of incremental updates has been studied for various standard primitives such as encryption schemes, signature schemes and so on [12,23,29,46,47]. We remark that none of these works handled the setting of arbitrary updates.

In a concurrent and independent work, [32] consider a related notion called *incremental obfuscation*. In incremental obfuscation, individual bits of an existing obfuscated program can be updated one-by-one. While their work shares much in spirit with our work, there are several important differences that we describe below.

Our work focuses on support for arbitrary, adaptively chosen patches that may potentially increase the size of the program(s) being patched, and we consider both single-program and multi-program setting. In contrast, their work considers the single-program setting where bit-wise, non-adaptively chosen patches can be applied such that the size of the circuit being patched remains unchanged. Our main efficiency requirement is that the size of the secure patches (or more strongly, the time to generate the secure patches) is independent of the size of the program. In contrast, their work considers the stronger runtime efficiency requirement where the time to apply the secure patch is also independent of the size of the circuit.

2 Patchable $i\mathcal{O}$: Definitions and Implications

In this section, we present the formal definitions of patchable indistinguishability obfuscation ($pa\text{-}i\mathcal{O}$) in the single program and multi program setting.

2.1 Definition: Single-Program $pa\text{-}i\mathcal{O}$

In this section, we present a formal definition of single-program patchable indistinguishability obfuscation, denoted as $pa\text{-}i\mathcal{O}_{\mathsf{sp}}$. We start by presenting the syntax, and then proceed to give a security definition for sequential updates.

Syntax. A $pa\text{-}i\mathcal{O}_{\mathsf{sp}}$ scheme, defined for a class of Turing machines $\mathcal{M}$ with an associated family of patches $\mathcal{P}$ and update algorithm Update, consists of a tuple of probabilistic polynomial-time algorithms $pa\text{-}i\mathcal{O}_{\mathsf{sp}} =$ (Setup, Obf, GenPatch, AppPatch, Eval) which are defined below.

- **Setup**, $\mathsf{Setup}(1^\lambda)$: It takes as input the security parameter λ and outputs the secret key SK.
- **Obfuscate**, $\mathsf{Obf}(\mathsf{SK}, M)$: It takes as input the secret key SK and a TM $M \in \mathcal{M}$. It outputs an obfuscated TM $\langle M \rangle$ along with state st.
- **(Stateful) Patch Generation**, $\mathsf{GenPatch}(\mathsf{SK}, P, \mathsf{st})$: It takes as input the secret key SK, a description of a patch $P \in \mathcal{P}$, and state st. It outputs a patch encoding $\langle P \rangle$ along with the updated state st$'$.
- **Applying Patch**, $\mathsf{AppPatch}\big(\langle M \rangle, \langle P \rangle\big)$: It takes as input an obfuscated TM $\langle M \rangle$ and a patch encoding $\langle P \rangle$. It outputs an updated obfuscation $\langle M' \rangle$.
- **Evaluation**, $\mathsf{Eval}\big(\langle M \rangle, x\big)$: It takes as input an obfuscated TM $\langle M \rangle$ and an input x. It outputs a value y.

Efficiency. We define two efficiency properties:

- *Patch Size Efficiency:* For every patch $P \in \mathcal{P}$, we require that the size of the patch encoding $|\langle P \rangle|$ is a fixed polynomial in $(|P|, \lambda)$, where $(\langle P \rangle, \mathsf{st}') \leftarrow \mathsf{GenPatch}(\mathsf{SK}, P, \mathsf{st})$.
- *Patch Generation Efficiency:* For every patch $P \in \mathcal{P}$, we require that the running time of $\mathsf{GenPatch}(\mathsf{SK}, P, \mathsf{st})$ to be a fixed polynomial in $(|P|, \lambda)$. The length of st could depend on the size of the obfuscated machine its associated with and we require that the running time of GenPatch to be independent of $|\mathsf{st}|$.

It is easy to see that the second property implies the first property. Our first construction of $pa\text{-}i\mathcal{O}_{\mathsf{sp}}$ only satisfies the first property. In the full version, we describe a modified construction that also achieves the second property.

Correctness for Sequential Patches. At a high level, the correctness property states that executing Update on a TM M and a patch P is equivalent to executing AppPatch on the obfuscation of M and a secure patch of P. In fact we require that this holds even if there are multiple patches that are applied sequentially.

For any TM $M_0 \in \mathcal{M}$, $L > 0$, sequence of patches $P_1, \ldots, P_L \in \mathcal{P}$, consider two processes:

- **Obfuscate-then-Update**: Compute the following: (a) $\mathsf{SK} \leftarrow \mathsf{Setup}(1^\lambda)$,
  (b) $\left(\langle M_0 \rangle, \mathsf{st}_0 \right) \leftarrow \mathsf{Obf}(\mathsf{SK}, M_0)$, (c) $\left(\langle P_i \rangle, \mathsf{st}_i \right) \leftarrow \mathsf{GenPatch}(\mathsf{SK}, P_i, \mathsf{st}_{i-1})$,
  (d) $\langle M_i \rangle \leftarrow \mathsf{AppPatch}\left(\langle M_{i-1} \rangle, \langle P_i \rangle \right)$.
- **Update**: $M_i \leftarrow \mathsf{Update}(M_{i-1}, P_i)$.

We require that for all $x \in \{0,1\}^*$, every $i \in [L]$, $\mathsf{Eval}\left(\langle M_i \rangle, x \right) = M_i(x)$.

Remark 1. For the case of parallel patching, we require that $\langle M_i \rangle \leftarrow \mathsf{AppPatch}\left(\langle M_0 \rangle, \langle P_i \rangle \right)$ is a valid obfuscation of machine M_i. We emphasize that for the case of parallel patching, the patches are applied only on the original machine.

Adaptive Security for Sequential Patches. We next give an indistinguishability (IND)-style definition for modeling the security of an $pa\text{-}i\mathcal{O}_{\mathsf{sp}}$ scheme for the case of sequential patches. In an IND-security definition, we consider a security game between the challenger and the adversary. In this game, the adversary sends two machines (M_0^0, M_1^0) to the challenger and in response receives an obfuscation $\langle M_b^0 \rangle$, where b is the challenge bit chosen randomly by the challenger. Then the adversary submits patch queries, adaptively, to the challenger in a series of phases. In each phase, the adversary chooses a pair of patches (P_0^i, P_1^i) and in return gets the patch encoding $\langle P_b^i \rangle$. The patch queries of the adversary are restricted in the following manner: suppose $\left((P_0^1, P_1^1), \ldots, (P_0^L, P_1^L) \right)$ is a sequence of adaptive patch queries made by the adversary. We require that the machine M_0^i is functionally equivalent with M_1^i, for every $i \in [L]$, where $M_0^i \leftarrow \mathsf{Update}(M_0^{i-1}, P_0^i)$ (resp., $M_1^i \leftarrow \mathsf{Update}(M_1^{i-1}, P_1^i)$). At the end of the game, the adversary attempts to guess the bit b. If the adversary's guess is the same as b only with probability negligibly close to $1/2$, then we say that the scheme is secure. Henceforth, we use the term *adaptive security* to refer to this notion. We proceed to formally defining this notion.

The experiment for the adaptive security definition is formulated below. Let $\mathcal{A}$ be any PPT adversary.

$\underline{\mathsf{Expt}_{\mathcal{A}}^{pa\text{-}i\mathcal{O}_{\mathsf{sp}}}(1^\lambda, b)}$:

1. $\mathcal{A}$ sends (M_0^0, M_1^0) to the challenger.
2. Challenger executes the setup algorithm to obtain $\mathsf{SK} \leftarrow \mathsf{Setup}(1^\lambda)$. It then sends $\langle M_b^0 \rangle \leftarrow \mathsf{Obf}(\mathsf{SK}, M_b^0)$ to $\mathcal{A}$.

3. Repeat the following steps for $i \in \{1, \ldots, L\}$, where L is chosen by $\mathcal{A}$.
   - $\mathcal{A}$ sends (P_0^i, P_1^i) to the challenger.
   - Challenger checks if $M_0^i \equiv M_1^i$, where $M_0^i \leftarrow \mathsf{Update}(M_0^{i-1}, P_0^i)$ and $M_1^i \leftarrow \mathsf{Update}(M_1^{i-1}, P_1^i)$.
   - Challenger computes $\langle P_b^i \rangle \leftarrow \mathsf{GenPatch}(\mathsf{SK}, P_b^i)$ and sends $\langle P_b^i \rangle$ to $\mathcal{A}$.
4. $\mathcal{A}$ outputs the bit b'.

Definition 1 (Adaptive Security). *A single-program patchable indistinguishability obfuscation scheme pa-$i\mathcal{O}_{\mathsf{sp}}$ is said to be adaptively secure against sequential updates if for any PPT adversary $\mathcal{A}$, there exists a negligible function $\mathsf{negl}(\cdot)$ s.t.*

$$\left| \Pr\left[1 \leftarrow \mathsf{Expt}_{\mathcal{A}}^{pa\text{-}i\mathcal{O}_{\mathsf{sp}}}(1^\lambda, 1) \right] - \Pr\left[1 \leftarrow \mathsf{Expt}_{\mathcal{A}}^{pa\text{-}i\mathcal{O}_{\mathsf{sp}}}(1^\lambda, 0) \right] \right| \leq \mathsf{negl}(\lambda)$$

Remark 2. For the case of parallel patching, the same security is defined with the only difference being that it is required that the machine M_0^i is functionally equivalent to M_1^i, where M_b^i is obtained by patching M_b^0 (the original machine) using P_i.

2.2   Definition: Multi-program *pa-i$\mathcal{O}$*

We now present a formal definition of multi-program *pa-i$\mathcal{O}$*, denoted as *pa-i$\mathcal{O}_{\mathsf{mp}}$*. Informally speaking, *pa-i$\mathcal{O}_{\mathsf{mp}}$* allows an authority to obfuscate an arbitrary number of programs in such a way that it is possible to later issue a patch encoding that can be used to update all the obfuscated programs at once. The authority who issues the obfuscated programs stores just a "short" information about all the obfuscated programs issued that enables it to produce a single patch that can act on all these programs. In particular, the size of the storage space of the authority is independent of the joint size of all these programs.[5] This is in contrast to the single-program setting described above, where the authority maintains state and this state can be as big as the program whose obfuscation is issued. There is another difference between both the settings: in the single-program setting, if we were to relax the size of the secure patch to be proportional to the size of the updated program then achieving a feasibility result is straightforward – the secure patch will just be the obfuscation of the updated program. Hence the primary goal is to reduce the size of the patch. However, in the multi-program setting, even if we relax the size of the secure patch to be proportional to the size of any of the updated programs, achieving a feasibility result is already non-trivial. As mentioned earlier, the authority does not have enough space to store all the updated programs and hence the above naïve solution, of sending a fresh obfuscation of the updated program, does not work. As we will see later we not only give a feasibility result in this setting but we also

---

[5] The reason why the authority can't store all the programs is because it is a machine that has a priori bounded memory and yet has the capability to produce an unbounded number of obfuscated programs.

achieve a solution with optimal efficiency where the size of the secure patches depend only on the size of their original patches and in particular, independent of the size of any obfuscated programs issued.

Syntax. A *pa-*$i\mathcal{O}_{\mathsf{mp}}$ scheme, defined for a class of Turing machines $\mathcal{M}$ and a family of patches $\mathcal{P}$, consists of a tuple of probabilistic polynomial-time algorithms *pa-*$i\mathcal{O}_{\mathsf{mp}}$ = (Setup, Obf, GenPatch, AppPatch, Eval) which are defined below. We denote the update algorithm associated with $(\mathcal{M}, \mathcal{P})$ to be Update.

- **Setup**, Setup(1^{λ}): It takes as input the security parameter λ and outputs the secret key SK.
- **Obfuscate**, Obf(SK, M): It takes as input the secret key SK and a TM $M \in \mathcal{M}$ id. It outputs an obfuscated TM $\langle M \rangle$.
- **(Stateless) Patch Generation**, GenPatch(SK, P): It takes as input the secret key SK and a description of a patch $P \in \mathcal{P}$. It outputs a patch encoding $\langle P \rangle$.
- **Applying Patch**, AppPatch$\big(\langle M \rangle, \langle P \rangle \big)$: It takes as input an obfuscated TM $\langle M \rangle$ and a patch encoding $\langle P \rangle$. It outputs an updated obfuscation $\langle M' \rangle$.
- **Evaluation**, Eval$\big(\langle M \rangle, x \big)$: It takes as input an obfuscated TM $\langle M \rangle$ and an input x. It outputs a value y.

Efficiency. Similar to *pa-*$i\mathcal{O}_{\mathsf{sp}}$, we define two efficiency properties for *pa-*$i\mathcal{O}_{\mathsf{mp}}$:

- *Patch Size Efficiency:* For every patch $P \in \mathcal{P}$, we require that the size of the patch encoding $|\langle P \rangle|$ is a fixed polynomial in $(|P|, \lambda)$, where $(\langle P \rangle, \mathsf{st}') \leftarrow$ GenPatch(SK, P, st).
- *Patch Generation Efficiency:* For every patch $P \in \mathcal{P}$, we require that the running time of GenPatch(SK, P,) to be a fixed polynomial in $(|P|, \lambda)$.

It is easy to see that the second property implies the first property. Our construction of *pa-*$i\mathcal{O}_{\mathsf{mp}}$ presented in the full version achieves both of the properties.

Correctness for Sequential Patches. For every $Q, L > 0$, any sequence of TMs $M_{0^1}, \ldots, M_0^Q \in \mathcal{M}$, sequence of patches $P_1, \ldots, P_L \in \mathcal{P}$, consider the following two processes. For every $j \in \{1, \ldots, Q\}, i \in \{1, \ldots, L\}$, we have:

- **Obfuscate-then-Update**: Compute the following: (a) SK $\leftarrow$ Setup(1^{λ}), (b) $\langle M_0^j \rangle \leftarrow$ Obf(SK, M_0^j), (c) $\langle P_i \rangle \leftarrow$ GenPatch(SK, P_i), (d) $\langle M_i^j \rangle \leftarrow$ AppPatch$\big(\langle M_{i-1}^j \rangle, \langle P_i \rangle \big)$.
- **Update**: $M_i^j \leftarrow$ Update(M_{i-1}^j, P_i).

We require that $\forall x \in \{0, 1\}^*, \forall j \in [Q], \forall i \in [L]$, we have Eval$\big(\langle M_i^j \rangle, x \big) = M_i^j(x)$.

Adaptive Security for Sequential Patches. We next give indistinguishability (IND)-style definitions for modeling the security of a patchable obfuscation scheme. As in the case of single-program patchable obfuscation, the definition is based on a game between the challenger and the adversary. The adversary makes TM queries and patch queries to the challenger. One important distinction is that in this setting, the adversary can make multiple TM queries whereas in the case of single-program obfuscation, it makes just one TM query. We describe the experiment below.

$\underline{\mathsf{Expt}_{\mathcal{A}}^{pa\text{-}i\mathcal{O}_{mp}}(1^\lambda, b)}$:

1. $\mathcal{A}$ submits a sequence of TM pairs $\left((M_{0,0}^1, M_{0,1}^1), \ldots, (M_{0,0}^Q, M_{0,1}^Q)\right)$.
2. Challenger executes the setup algorithm to obtain $\mathsf{SK} \leftarrow \mathsf{Setup}(1^\lambda)$. For every $j \in [Q]$, it computes $\langle M_{0,b}^j \rangle \leftarrow \mathsf{Obf}(\mathsf{SK}, M_{0,b}^j)$ and sends $\left\{\langle M_{0,b}^j \rangle\right\}_{j \in [Q]}$ to the adversary.
3. Repeat the following steps for $i \in \{1, \ldots, L\}$, where $L(\lambda)$ is chosen by $\mathcal{A}$:
   - $\mathcal{A}$ sends (P_0^i, P_1^i) to the challenger.
   - Challenger computes $\langle P_b^i \rangle \leftarrow \mathsf{GenPatch}(\mathsf{SK}, P_b^i)$. It sends $\langle P_b^i \rangle$ to $\mathcal{A}$.
4. For every $i \in \{1, \ldots, L\}$, every $j \in \{1, \ldots, Q\}$, the challenger checks if $M_{i,0}^j \equiv M_{i,1}^j$, where $M_{i,0}^j \leftarrow \mathsf{Update}(M_{i-1,0}^j, P_0^i)$ and $M_{i,1}^j \leftarrow \mathsf{Update}(M_{i-1,1}^j, P_1^i)$. If check fails then the challenger aborts the experiment.
5. $\mathcal{A}$ outputs the bit b'.

Definition 2. (Adaptive security). *A multi-program patchable obfuscation scheme pa-$i\mathcal{O}_{mp}$ is said to be adaptively secure if for any PPT adversary $\mathcal{A}$, there exists a negligible function $\mathsf{negl}(\cdot)$ s.t.*

$$\left| \Pr\left[0 \leftarrow \mathsf{Expt}_{\mathcal{A}}^{pa\text{-}i\mathcal{O}_{mp}}(1^\lambda, 0)\right] - \Pr\left[0 \leftarrow \mathsf{Expt}_{\mathcal{A}}^{pa\text{-}i\mathcal{O}_{mp}}(1^\lambda, 1)\right] \right| \leq \mathsf{negl}(\lambda)$$

Remark 3. For the case of parallel patching, the correctness and security can be similarly defined.

3   Splittable iO

We describe the notion of splittable iO next. This notion will be associated with a patchable encoding scheme. We define patchable encoding scheme first.

3.1   Patchable Encoding Scheme

A patchable encoding scheme is an encoding scheme associated with a class of Turing machines. This scheme allows for updating an encoding of a machine M using an encoding of a patch P to obtain an encoding of another machine M', where $M' \leftarrow \mathsf{Update}(M, P)$. The secret key, used in the computation of the encodings, is generated using algorithm Gen. Turing machines are encoded using

Encode and the patches are encoded using GenPatch. Algorithm AppPatch is used to apply update the encoding of machine M using encoding of patch P. Finally, Decode is used to decode an encoding of M using the secret key produced by Gen.

Syntax. A patchable encoding scheme is described by the algorithms UE = (Gen, Encode, GenPatch, AppPatch, Decode) which are defined below. We denote by $\mathcal{M}$, the class of Turing machines it is associated with. We further denote the update algorithm associated with $\mathcal{M}$ to be Update.

- $sk \leftarrow$ Gen(1^λ): On input λ, it produces the secret key sk.
- $\mathcal{E}_{sk}(M) \leftarrow$ Encode(sk, M): On input secret key sk, Turing machine M, it produces an encoding of M, namely $\mathcal{E}_{sk}(M)$, with respect to sk.
- $\widetilde{P} \leftarrow$ GenPatch(sk, P): On input secret key sk, patch P, it produces a secure patch $\widetilde{P}$.
- $\mathcal{E}_{sk}(M') \leftarrow$ AppPatch $\left(\mathcal{E}_{sk}(M), \widetilde{P} \right)$: On input encoding $\mathcal{E}_{sk}(M)$, secure patch $\widetilde{P}$, it produces the updated encoding $\mathcal{E}_{sk}(M)$.
- $M \leftarrow$ Decode($sk, \mathcal{E}_{sk}(M)$): On input secret key sk, machine encoding $\mathcal{E}_{sk}(M)$, it produces the machine M.

Efficiency. We require that the size of the secure patches is a (a priori fixed) polynomial in the security parameter and the size of the underlying patch. That is, $|\widetilde{P}| = \text{poly}(\lambda, |P|)$, where $\widetilde{P} \leftarrow$ GenPatch(sk, P).

Correctness of Sequential Updating. Consider $M \in \mathcal{M}$ and a sequence of patches $P_1, \ldots, P_L$. We consider the following two processes:

- **Encode-then-Update:** Compute the following: (a) $sk \leftarrow$ Gen(1^λ); (b) $\mathcal{E}_{sk}(M_1) \leftarrow$ Encode(sk, M); (c) For every $i \in [L]$, $\widetilde{P}_i \leftarrow$ GenPatch(sk, P_i); (d) $\mathcal{E}_{sk}(M_{i+1}) \leftarrow$ AppPatch $\left(\mathcal{E}_{sk}(M_i), \widetilde{P}_i \right)$.
- **Update:** For every $i \in [L]$, $M_{i+1} \leftarrow$ Update(M_i, P_i) with $M_1 = M$.

We require that Decode($sk, \mathcal{E}_{sk}(M_L)$) $= M_L$.

Security. We require any patchable encoding scheme to satisfy the following.

Definition 3. *A patchable encoding scheme,* UE = (Gen, Encode, GenPatch, AppPatch, Decode) *is said to be* **secure** *if the following holds: Consider the game between a challenger and an adversary. The adversary submits machines* $(M_0^1, M_1^1) \ldots, (M_0^Q, M_1^Q) \in \mathcal{M}$ *to the challenger. In return, the adversary receives* $\{\mathcal{E}_{sk}(M_b^j)\}_{j \in [Q]}$, *where* $b \in \{0, 1\}$ *is picked at random. The adversary can then make patch queries* (P_0^i, P_1^i), *for every* $i \in [L]$, *adaptively. In return it receives* $\widetilde{P_b^i}$. *The probability that the adversary outputs* b *is negligibly close to* $1/2$.

We can correspondingly define an encoding scheme supporting parallel patches.

In the full version, we present an instantiation of the above primitive using fully homomorphic encryption.

3.2 Definition of Splittable iO

We define the notion of splittable iO next. A splittable iO is an indistinguishability obfuscation scheme, satisfying additional properties. The model of computation is Turing machines and we work in succinct iO setting [17, 26, 44]. Although the algorithms associated with succinct iO take the input length bound as input, we omit this in the description below. For simplicity, set the input length bound to be λ. Our results can easily be extended to the case when the input bound is an arbitrary polynomial in λ and our parameter sizes would blow accordingly.

Firstly, we require that the obfuscation of M proceeds in two steps: in the first step, M is encoded (twice) using the underlying patchable encoding scheme UE. This is done by generating the setup of UE twice and encoding M using both these secret keys sk_0 and sk_1. Call the two encodings $\mathcal{E}_{sk_0}(M)$ and $\mathcal{E}_{sk_1}(M)$. The second step involves generation of auxiliary information as a function of the encodings $\mathcal{E}_{sk_0}(M)$ and $\mathcal{E}_{sk_1}(M)$ and one of the secret keys. This is enabled via an additional algorithm AuxGen. This requirement on the structure of the obfuscate algorithm is termed as *splittable property*. The second property we require is *correctness of* AuxGen – this says that the correctness of the obfuscated machine should not be affected by whether the two encodings (part of the obfuscated machine) fed to AuxGen are freshly computed or if they are obtained as a result of patching. The third property, which is *efficiency of aux*, states that the auxiliary information produced by AuxGen should be a fixed polynomial in λ. Finally, we have the *indistinguishability of aux* property that states that the auxiliary information obtained by AuxGen on input two encodings $\mathcal{E}_{sk_0}(M)$ and $\mathcal{E}_{sk_1}(M)$ and secret key sk_0 is indistinguishability the output of AuxGen on input $\mathcal{E}_{sk_0}(M)$, $\mathcal{E}_{sk_1}(M)$ and secret key sk_1.

Definition 4 (Splittable iO). *A splittable iO scheme, denoted by* siO = (Obf, Eval) *for a class of Turing machines* $\mathcal{M}$, *is an indistinguishability obfuscation scheme that is associated with a patchable encoding scheme* UE = (Gen, Encode, GenPatch, AppPatch, Decode) *and satisfies the following properties:*

- **Splittable Property:** Obf *consists of* Gen, Encode *and an additional PPT algorithm* AuxGen. *On input* $(1^\lambda, M)$ *it proceeds in the following three phases:*
  1. Encoding of M *using* UE: *(a)* $sk_0 \leftarrow$ Gen(1^λ); $sk_1 \leftarrow$ Gen(1^λ).
     (b) $\mathcal{E}_{sk_0}(M) \leftarrow$ Encode(sk_0, M); $\mathcal{E}_{sk_1}(M) \leftarrow$ Encode(sk_1, M)
  2. Generation of aux: $aux \leftarrow$ AuxGen $(sk_0, \mathcal{E}_{sk_0}(M), \mathcal{E}_{sk_1}(M))$
     Output $\langle M \rangle = (\mathcal{E}_{sk_0}(M), \mathcal{E}_{sk_1}(M), aux)$. *The secret state associated with this execution is set to be* (sk_0, sk_1).

- **Correctness of** AuxGen: *Let* $M \in \mathcal{M}$ *and let* $P_1, \ldots, P_L$ *be a sequence of patches. Let* M_i *be the* i^{th} *updated machine,* $M_i \leftarrow$ Update(M_{i-1}, P), *for every* $i \in [L]$, *where* $M_0 = M$.
  Consider the following process:
  - *Let* sk_0, sk_1 *be such that* $sk_0 \leftarrow$ UE.Gen(1^λ), $sk_1 \leftarrow$ UE.Gen(1^λ).
  - *Let* $\mathcal{E}_{sk_0}(M) \leftarrow$ UE.Encode(sk_0, M) *and* $\mathcal{E}_{sk_1}(M) \leftarrow$ UE.Encode(sk_0, M).

- Consider the ith updated encodings, $\mathcal{E}_{sk_0}(M_i) \leftarrow$ UE.AppPatch $(\mathcal{E}_{sk_0}(M_{i-1}), \mathsf{UE.GenPatch}(sk_0, P_i))$ and $\mathcal{E}_{sk_1}(M_i) \leftarrow$ UE.AppPatch $(\mathcal{E}_{sk_1}(M_{i-1}), \mathsf{UE.GenPatch}(sk_1, P_i))$.
- Let $aux \leftarrow$ AuxGen$(sk_0, \mathcal{E}_{sk_0}(M_L), \mathcal{E}_{sk_1}(M_L))$ and set $\langle M_L \rangle = (\mathcal{E}_{sk_0}(M_L), \mathcal{E}_{sk_1}(M_L), aux)$.

  For every x, we have Eval$(\langle M_L \rangle, x) = M_L(x)$.

- **Efficiency of aux:** There exists a polynomial p such that the following holds. Let $(\mathcal{E}_{sk_0}(M), \mathcal{E}_{sk_1}(M), aux) \leftarrow$ Obf$(1^\lambda, M)$ for $M \in \mathcal{M}$. Then, $|aux| = p(\lambda)$.

- **Indistinguishability of aux:** Consider $M_0, M_1 \in \mathcal{M}$ such that $M_0(x) = M_1(x)$ for every $x \in \{0,1\}^*$. Suppose E_0, E_1, sk_0, sk_1 are such that $M_0 \leftarrow$ Decode(sk_0, E_0) and $M_1 \leftarrow$ Decode(sk_1, E_1). We have,

$$\{E_0, E_1, sk_0, sk_1, aux_0\} \approx_c \{E_0, E_1, sk_0, sk_1, aux_1\},$$

where $aux_b \leftarrow$ AuxGen(sk_b, E_0, E_1) for $b \in \{0,1\}$.

An instantiation of splittable iO is presented in the full version.

We note that the above definition can be extended to the parallel patches setting if the underlying patchable encoding scheme supports parallel patches.

4   Splittable iO to Single-Program $pa\text{-}i\mathcal{O}$

We give a generic transformation from splittable iO to single-program patchable iO.

Construction. The main tool we use in our construction is a splittable iO scheme siO = (siO.Obf, siO.Eval) associated with the updatable encoding scheme UE = (Gen, Encode, GenPatch, AppPatch, Decode). We construct a single-program patchable obfuscation scheme $pa\text{-}i\mathcal{O}$ below.

Setup, Setup(1^λ): It outputs SK = $\perp$.

Obfuscate, Obf(SK, M): It takes as input the secret key SK = $\perp$ and a TM $M \in \mathcal{M}$. The obfuscation of M is essentially the obfuscation of M with respect to siO. That is, it executes the obfuscate algorithm of siO on M; $(\mathcal{E}_{sk_0}(M), \mathcal{E}_{sk_1}(M), aux) \leftarrow$ siO.Obf$(1^\lambda, M)$. Denote $(\mathcal{E}_{sk_0}(M), \mathcal{E}_{sk_1}(M), aux)$ by $\langle M \rangle$. Let the state associated with this execution be (sk_0, sk_1) (refer to Splittable Property in Definition 4).

It outputs the obfuscated TM $\langle M \rangle$. The state is set to be st = $(sk_0, sk_1, \mathcal{E}_{sk_0}(M), \mathcal{E}_{sk_1}(M))$. That is, the state consists of the two secret keys and the patchable encodings of M with respect to sk_0 and sk_1.

Secure Patch Generation, GenPatch(SK, P, st): It takes as input the secret key SK = $\perp$, a description of a patch $P \in \mathcal{P}$ and state st = $(sk_0, sk_1, \mathcal{E}_{sk_0}(M), \mathcal{E}_{sk_1}(M))$. Then,

- It computes the secure patches, $\widetilde{P}^0 \leftarrow$ UE.GenPatch(sk_0, P) and $\widetilde{P}^1 \leftarrow$ UE.GenPatch(sk_1, P).

- It applies the secure patches on the encodings, $\mathcal{E}_{sk_0}(M') \leftarrow$ UE.AppPatch $(\mathcal{E}_{sk_0}(M), \widetilde{P}^0)$ and $\mathcal{E}_{sk_1}(M') \leftarrow$ UE.AppPatch$(\mathcal{E}_{sk_1}(M), \widetilde{P}^1)$.
- It then executes AuxGen algorithm of siO. It computes $aux' \leftarrow$ AuxGen$(sk_0, \mathcal{E}_{sk_0}(M_0), \mathcal{E}_{sk_1}(M_1))$.

It outputs a secure patch $\langle P \rangle = (\widetilde{P}^0, \widetilde{P}^1, aux')$. It updates the state to be $\mathsf{st}' = (sk_0, sk_1, \mathcal{E}_{sk_0}(M'), \mathcal{E}_{sk_1}(M'))$.

Note: It suffices to just include the encodings $(\widetilde{P}^0, \widetilde{P}^1)$ (and not the updated encodings $\mathcal{E}_{sk_0}(M'), \mathcal{E}_{sk_1}(M'))$ as part of secure patch because anyone having the original pair of encodings $(\mathcal{E}_{sk_0}(M), \mathcal{E}_{sk_1}(M))$ can now recompute the $(\mathcal{E}_{sk_0}(M'), \mathcal{E}_{sk_1}(M'))$ by using just $(\widetilde{P}^0, \widetilde{P}^1)$.

Applying Patch, AppPatch$(\langle M \rangle, \langle P \rangle)$: It takes as input an obfuscated TM $\langle M \rangle = (\mathcal{E}_{sk_0}(M), \mathcal{E}_{sk_1}(M), aux)$ and a secure patch $\langle P \rangle = (\widetilde{P}^0, \widetilde{P}^1, aux')$.

- It applies the secure patches on the encodings, $\mathcal{E}_{sk_0}(M') \leftarrow$ UE.AppPatch $(\mathcal{E}_{sk_0}(M), \widetilde{P}^0)$ and $\mathcal{E}_{sk_1}(M') \leftarrow$ UE.AppPatch$(\mathcal{E}_{sk_1}(M), \widetilde{P}^1)$.
- It replaces aux with aux' which is sent as part of the patch.

It outputs an updated obfuscation $\langle M' \rangle = (\mathcal{E}_{sk_0}(M'), \mathcal{E}_{sk_1}(M'), aux')$.

Evaluation, Eval$(\langle M \rangle, x)$: It takes as input an obfuscated TM $\langle M \rangle$ and an input x. It executes the evaluation algorithm of siO; $y \leftarrow$ siO.Eval$(\langle M \rangle, x)$. Output y.

Efficiency. We claim that the size of the secure patch solely depends on the size of the patch and the security parameter. In particular, it is independent of the size of the machine.

Consider a patch P. Let the output of GenPatch(SK, P, st) be $\langle P \rangle = (\widetilde{P}^0, \widetilde{P}^1, aux')$. From the efficiency of the underlying patchable encoding scheme, $|(\widetilde{P}^0, \widetilde{P}^1)| = \mathrm{poly}(\lambda, |P|)$. From the efficiency of the underlying spittable iO scheme, $|aux'| = \mathrm{poly}(\lambda)$.

Remark 4. The secure patch generation time in the above scheme is proportional to the size of the obfuscated machine. This is in general undesirable and we show how to deal with this issue in the full version.

Correctness of Sequential Updating. Consider a TM $M_0 \in \mathcal{M}$ and a sequence of patches $P_1, \ldots, P_L \in \mathcal{P}$. Consider the following two processes generated using the above scheme. For every $i \in \{1, \ldots, L\}$, we have:

- **Obfuscate-then-Update:** Compute the following: (a) SK $\leftarrow$ Setup(1^λ), (b) $(\langle M_0 \rangle, \mathsf{st}_0) \leftarrow$ Obf(SK, M_0), (c) $(\langle P_i \rangle, \mathsf{st}_i) \leftarrow$ GenPatch$(SK, P_i, \mathsf{st}_{i-1})$, (d) $\langle M_i \rangle \leftarrow$ AppPatch$\big(\langle M_{i-1} \rangle, \langle P_i \rangle\big)$.
- **Update:** $M_i \leftarrow$ Update(M_{i-1}, P_i).

We have the following claim.

Claim. For every x, we have $\mathsf{Eval}(\langle M_L \rangle, x) = M_L(x)$.

Proof. Let $\langle M_0 \rangle = (E_0^0, E_1^0, aux^0)$, $\mathsf{st} = (sk_0, sk_1, E_0^0, E_1^0)$ and $\langle M_L \rangle = (E_0^L, E_1^L, aux^L)$. Note that E_0 is the output of an execution of $\mathsf{Encode}(sk_0, M_0)$ and aux^0 is the output of $\mathsf{AuxGen}(sk_0, E_0^0, E_1^0)$. From the correctness of patchable encoding scheme, we have $\mathsf{Decode}(\mathsf{SK}_0, E_0^L) = M_L$. Using this fact along with the correctness of AuxGen property of siO, we get that the output of $\mathsf{Eval}(\langle M_L \rangle, x)$ to be $M_L(x)$.

Security of Sequential Updating. We prove,

Theorem 7. *pa-iO satisfies security of sequential updating property.*

A formal proof for the above theorem can be found in the full version.

References

1. Ananth, P., Boneh, D., Garg, S., Sahai, A., Zhandry, M.: Differing-inputs obfuscation and applications. IACR Cryptology ePrint Archive 2013:689 (2013)
2. Ananth, P., Chen, Y.-C., Chung, K.-M., Lin, H., Lin, W.-K.: Delegating RAM computations with adaptive soundness and privacy. In: Hirt, M., Smith, A. (eds.) TCC 2016. LNCS, vol. 9986, pp. 3–30. Springer, Heidelberg (2016). doi:10.1007/978-3-662-53644-5_1
3. Ananth, P., Jain, A., Naor, M., Sahai, A., Yogev, E.: Universal obfuscation and witness encryption: boosting correctness and combining security. In: CRYPTO (2016)
4. Ananth, P., Jain, A.: Indistinguishability obfuscation from compact functional encryption. In: CRYPTO (2015)
5. Ananth, P., Jain, A., Sahai, A.: Indistinguishability obfuscation with constant size overhead. Cryptology ePrint Archive, report 2015/1023 (2015)
6. Ananth, P., Jain, A., Sahai, A.: Patchable obfuscation. Cryptology ePrint Archive, report 2015/1084 (2015). http://eprint.iacr.org/2015/1084
7. Ananth, P., Sahai, A.: Functional encryption for turing machines. In: Kushilevitz, E., Malkin, T. (eds.) TCC 2016. LNCS, vol. 9562, pp. 125–153. Springer, Heidelberg (2016). doi:10.1007/978-3-662-49096-9_6
8. Badrinarayanan, S., Gupta, D., Jain, A., Sahai, A.: Multi-input functional encryption for unbounded arity functions. In: Iwata, T., Cheon, J.H. (eds.) ASIACRYPT 2015. LNCS, vol. 9452, pp. 27–51. Springer, Heidelberg (2015). doi:10.1007/978-3-662-48797-6_2
9. Barak, B., Garg, S., Kalai, Y.T., Paneth, O., Sahai, A.: Protecting obfuscation against algebraic attacks. In: Nguyen, P.Q., Oswald, E. (eds.) EUROCRYPT 2014. LNCS, vol. 8441, pp. 221–238. Springer, Heidelberg (2014). doi:10.1007/978-3-642-55220-5_13
10. Barak, B., Goldreich, O., Impagliazzo, R., Rudich, S., Sahai, A., Vadhan, S.P., Yang, K.: On the (im)possibility of obfuscating programs. J. ACM **59**(2), 6 (2012)
11. Bellare, M., Goldreich, O., Goldwasser, S.: Incremental cryptography: the case of hashing and signing. In: Desmedt, Y.G. (ed.) CRYPTO 1994. LNCS, vol. 839, pp. 216–233. Springer, Heidelberg (1994). doi:10.1007/3-540-48658-5_22

12. Bellare, M., Goldreich, O., Goldwasser, S.: Incremental cryptography and application to virus protection. In: Proceedings of the Twenty-Seventh Annual ACM Symposium on Theory of Computing, pp. 45–56. ACM (1995)
13. Bellare, M., Stepanovs, I., Waters, B.: New negative results on differing-inputs obfuscation. IACR Cryptology ePrint Archive 2016:162 (2016)
14. Bitansky, N., Canetti, R.: On strong simulation and composable point obfuscation. In: Rabin, T. (ed.) CRYPTO 2010. LNCS, vol. 6223, pp. 520–537. Springer, Heidelberg (2010). doi:10.1007/978-3-642-14623-7_28
15. Bitansky, N., Canetti, R., Cohn, H., Goldwasser, S., Kalai, Y.T., Paneth, O., Rosen, A.: The impossibility of obfuscation with auxiliary input or a universal simulator. In: Garay, J.A., Gennaro, R. (eds.) CRYPTO 2014. LNCS, vol. 8617, pp. 71–89. Springer, Heidelberg (2014). doi:10.1007/978-3-662-44381-1_5
16. Bitansky, N., Canetti, R., Kalai, Y.T., Paneth, O.: On virtual grey box obfuscation for general circuits. In: Garay, J.A., Gennaro, R. (eds.) CRYPTO 2014. LNCS, vol. 8617, pp. 108–125. Springer, Heidelberg (2014). doi:10.1007/978-3-662-44381-1_7
17. Bitansky, N., Garg, S., Lin, H., Pass, R., Telang, S.: Succinct randomized encodings and their applications. In: STOC (2015)
18. Bitansky, N., Goldwasser, S., Jain, A., Paneth, O., Vaikuntanathan, V., Waters, B.: Time-lock puzzles from randomized encodings. In: ITCS (2016)
19. Boneh, D., Sahai, A., Waters, B.: Functional encryption: definitions and challenges. In: Ishai, Y. (ed.) TCC 2011. LNCS, vol. 6597, pp. 253–273. Springer, Heidelberg (2011). doi:10.1007/978-3-642-19571-6_16
20. Boyle, E., Chung, K.-M., Pass, R.: On extractability obfuscation. In: Lindell, Y. (ed.) TCC 2014. LNCS, vol. 8349, pp. 52–73. Springer, Heidelberg (2014). doi:10.1007/978-3-642-54242-8_3
21. Brakerski, Z., Komargodski, I., Segev, G.: From single-input to multi-input functional encryption in the private-key setting. In: EUROCRYPT (2016)
22. Brakerski, Z., Rothblum, G.N.: Virtual black-box obfuscation for all circuits via generic graded encoding. In: Lindell, Y. (ed.) TCC 2014. LNCS, vol. 8349, pp. 1–25. Springer, Heidelberg (2014). doi:10.1007/978-3-642-54242-8_1
23. Buonanno, E., Katz, J., Yung, M.: Incremental unforgeable encryption. In: Matsui, M. (ed.) FSE 2001. LNCS, vol. 2355, pp. 109–124. Springer, Heidelberg (2002). doi:10.1007/3-540-45473-X_9
24. Canetti, R., Chen, Y., Holmgren, J., Raykova, M.: Succinct adaptive garbled RAM. In: TCC (2016-B)
25. Canetti, R., Holmgren, J.: Fully succinct garbled RAM. In: ITCS (2016)
26. Canetti, R., Holmgren, J., Jain, A., Vaikuntanathan, V.: Indistinguishability obfuscation of iterated circuits and RAM programs. In: STOC (2015)
27. Chen, Y.-C., Chow, S.S.M., Chung, K.-M., Lai, R.W.F., Lin, W.-K., Zhou, H.-S.: Computation-trace indistinguishability obfuscation and its applications. In: ITCS (2016)
28. Cohen, A., Holmgren, J., Nishimaki, R., Vaikuntanathan, V., Wichs, D.: Watermarking cryptographic capabilities. In: STOC (2016)
29. Fischlin, M.: Incremental cryptography and memory checkers. In: Fumy, W. (ed.) EUROCRYPT 1997. LNCS, vol. 1233, pp. 393–408. Springer, Heidelberg (1997). doi:10.1007/3-540-69053-0_27
30. Garg, S., Gentry, C., Halevi, S., Raykova, M., Sahai, A., Waters, B.: Candidate indistinguishability obfuscation and functional encryption for all circuits. In: 54th Annual IEEE Symposium on Foundations of Computer Science, FOCS 2013, Berkeley, CA, USA, 26–29 October 2013, pp. 40–49. IEEE Computer Society (2013)

31. Garg, S., Gentry, C., Halevi, S., Wichs, D.: On the implausibility of differing-inputs obfuscation and extractable witness encryption with auxiliary input. In: Garay, J.A., Gennaro, R. (eds.) CRYPTO 2014. LNCS, vol. 8616, pp. 518–535. Springer, Heidelberg (2014). doi:10.1007/978-3-662-44371-2_29
32. Garg, S., Pandey, O.: Incremental program obfuscation. Cryptology ePrint Archive, report 2015/997 (2015). http://eprint.iacr.org/
33. Gentry, C.: Fully homomorphic encryption using ideal lattices. In: Mitzenmacher, M. (ed.) Proceedings of the 41st Annual ACM Symposium on Theory of Computing, STOC, Bethesda, MD, USA, 31 May–2 June 2009, pp. 169–178. ACM (2009)
34. Goldreich, O., Ostrovsky, R.: Software protection and simulation on oblivious RAMs. J. ACM **43**(3), 431–473 (1996)
35. Goldwasser, S., Gordon, S.D., Goyal, V., Jain, A., Katz, J., Liu, F.-H., Sahai, A., Shi, E., Zhou, H.-S.: Multi-input functional encryption. In: Nguyen, P.Q., Oswald, E. (eds.) EUROCRYPT 2014. LNCS, vol. 8441, pp. 578–602. Springer, Heidelberg (2014). doi:10.1007/978-3-642-55220-5_32
36. Goldwasser, S., Kalai, Y.T.: On the impossibility of obfuscation with auxiliary input. In: Proceedings of 46th Annual IEEE Symposium on Foundations of Computer Science (FOCS 2005), Pittsburgh, PA, USA, 23–25 October 2005, pp. 553–562 (2005)
37. Goldwasser, S., Kalai, Y.T., Popa, R.A., Vaikuntanathan, V., Zeldovich, N.: How to run turing machines on encrypted data. In: Canetti, R., Garay, J.A. (eds.) CRYPTO 2013. LNCS, vol. 8043, pp. 536–553. Springer, Heidelberg (2013). doi:10.1007/978-3-642-40084-1_30
38. Goldwasser, S., Kalai, Y.T., Rothblum, G.N.: One-time programs. In: Wagner, D. (ed.) CRYPTO 2008. LNCS, vol. 5157, pp. 39–56. Springer, Heidelberg (2008). doi:10.1007/978-3-540-85174-5_3
39. Goldwasser, S., Rothblum, G.N.: On best-possible obfuscation. In: Vadhan, S.P. (ed.) TCC 2007. LNCS, vol. 4392, pp. 194–213. Springer, Heidelberg (2007). doi:10.1007/978-3-540-70936-7_11
40. Goyal, V., Ishai, Y., Sahai, A., Venkatesan, R., Wadia, A.: Founding cryptography on tamper-proof hardware tokens. In: Micciancio, D. (ed.) TCC 2010. LNCS, vol. 5978, pp. 308–326. Springer, Heidelberg (2010). doi:10.1007/978-3-642-11799-2_19
41. Hada, S.: Zero-knowledge and code obfuscation. In: Okamoto, T. (ed.) ASIACRYPT 2000. LNCS, vol. 1976, pp. 443–457. Springer, Heidelberg (2000). doi:10.1007/3-540-44448-3_34
42. Ishai, Y., Pandey, O., Sahai, A.: Public-coin differing-inputs obfuscation and its applications. In: Dodis, Y., Nielsen, J.B. (eds.) TCC 2015. LNCS, vol. 9015, pp. 668–697. Springer, Heidelberg (2015). doi:10.1007/978-3-662-46497-7_26
43. Komargodski, I., Moran, T., Naor, M., Pass, R., Rosen, A., Yogev, E.: One-way functions and (im)perfect obfuscation. In: 55th IEEE Annual Symposium on Foundations of Computer Science, FOCS, Philadelphia, PA, USA, 18–21 October 2014, pp. 374–383 (2014)
44. Koppula, V., Lewko, A.B., Waters, B.: Indistinguishability obfuscation for turing machines with unbounded memory. In: STOC (2015)
45. Lin, H., Pass, R., Seth, K., Telang, S.: Output-compressing randomized encodings and applications. In: Kushilevitz, E., Malkin, T. (eds.) TCC 2016. LNCS, vol. 9562, pp. 96–124. Springer, Heidelberg (2016). doi:10.1007/978-3-662-49096-9_5
46. Micciancio, D.: Oblivious data structures: applications to cryptography. In: Proceedings of the Twenty-Ninth Annual ACM Symposium on Theory of Computing, pp. 456–464. ACM (1997)

47. Mironov, I., Pandey, O., Reingold, O., Segev, G.: Incremental deterministic public-key encryption. In: Pointcheval, D., Johansson, T. (eds.) EUROCRYPT 2012. LNCS, vol. 7237, pp. 628–644. Springer, Heidelberg (2012). doi:10.1007/978-3-642-29011-4_37

48. Moran, T., Rosen, A.: There is no indistinguishability obfuscation in pessiland. IACR Cryptology ePrint Archive 2013:643 (2013)

49. O'Neill, A.: Definitional issues in functional encryption. IACR Cryptology ePrint Archive 2010:556 (2010)

50. Sahai, A., Waters, B.: Fuzzy identity-based encryption. In: Cramer, R. (ed.) EUROCRYPT 2005. LNCS, vol. 3494, pp. 457–473. Springer, Heidelberg (2005). doi:10.1007/11426639_27

51. Sahai, A., Waters, B.: How to use indistinguishability obfuscation: deniable encryption, and more. In: Symposium on Theory of Computing, STOC 2014, New York, NY, USA, 31 May–03 June 2014, pp. 475–484 (2014)

Breaking the Sub-Exponential Barrier in Obfustopia

Sanjam Garg[1], Omkant Pandey[2], Akshayaram Srinivasan[1(✉)],
and Mark Zhandry[3]

[1] University of California, Berkeley, USA
{sanjamg,akshayaram}@berkeley.edu
[2] Stony Brook University, Stony Brook, USA
omkant@gmail.com
[3] Princeton University, Princeton, USA
mzhandry@princeton.edu

Abstract. Indistinguishability obfuscation ($i\mathcal{O}$) has emerged as a surprisingly powerful notion. Almost all known cryptographic primitives can be constructed from general purpose $i\mathcal{O}$ and other minimalistic assumptions such as one-way functions. A major challenge in this direction of research is to develop novel techniques for using $i\mathcal{O}$ since $i\mathcal{O}$ by itself offers virtually no protection for secret information in the underlying programs. When dealing with complex situations, often these techniques have to consider an exponential number of hybrids (usually one per input) in the security proof. This results in a *sub-exponential* loss in the security reduction. Unfortunately, this scenario is becoming more and more common and appears to be a fundamental barrier to many current techniques.

A parallel research challenge is building obfuscation from simpler assumptions. Unfortunately, it appears that such a construction would likely incur an exponential loss in the security reduction. Thus, achieving any application of $i\mathcal{O}$ from simpler assumptions would also require a sub-exponential loss, *even if the $i\mathcal{O}$-to-application security proof incurred a polynomial loss*. Functional encryption ($\mathcal{FE}$) is known to be equivalent to $i\mathcal{O}$ up to a sub-exponential loss in the $\mathcal{FE}$-to-$i\mathcal{O}$ security reduction; yet, unlike $i\mathcal{O}$, $\mathcal{FE}$ can be achieved from simpler assumptions (namely, specific multilinear map assumptions) with only a polynomial loss.

In the interest of basing applications on weaker assumptions, we therefore argue for using $\mathcal{FE}$ as the starting point, rather than $i\mathcal{O}$, and restricting to reductions with only a polynomial loss. By significantly expanding on ideas developed by Garg, Pandey, and Srinivasan (CRYPTO 2016), we achieve the following early results in this line of study:

– We construct *universal samplers* based only on polynomially-secure public-key $\mathcal{FE}$. As an application of this result, we construct a *non-interactive multiparty key exchange* (NIKE) protocol for an unbounded number of users without a trusted setup. Prior to this work, such constructions were only known from indistinguishability obfuscation.

© International Association for Cryptologic Research 2017
J.-S. Coron and J.B. Nielsen (Eds.): EUROCRYPT 2017, Part III, LNCS 10212, pp. 156–181, 2017.
DOI: 10.1007/978-3-319-56617-7_6

- We also construct trapdoor one-way permutations (OWP) based on polynomially-secure public-key $\mathcal{FE}$. This improves upon the recent result of Bitansky, Paneth, and Wichs (TCC 2016) which requires $i\mathcal{O}$ of *sub-exponential strength*. We proceed in two steps, first giving a construction requiring $i\mathcal{O}$ of *polynomial strength*, and then specializing the $\mathcal{FE}$-to-$i\mathcal{O}$ conversion to our specific application.

Many of the techniques that have been developed for using $i\mathcal{O}$, including many of those based on the "punctured programming" approach, become inapplicable when we insist on polynomial reductions to $\mathcal{FE}$. As such, our results above require many new ideas that will likely be useful for future works on basing security on $\mathcal{FE}$.

1   Introduction

Indistinguishability obfuscation ($i\mathcal{O}$) [5,16] has emerged as a powerful cryptographic primitive in the past few years. It has proven sufficient to construct a plethora of cryptographic primitives, many of them for the first time,[4,8,10,12,30]. Recently, $i\mathcal{O}$ also proved instrumental in proving the hardness of complexity class PPAD [7].

A major challenge in this direction of research stems from the fact that $i\mathcal{O}$ by itself is "too weak" to work with. The standard security of $i\mathcal{O}$ may not even hide any secrets present in the underlying programs. Therefore, the crucial part of most $i\mathcal{O}$-based constructions lies in developing novel techniques for using $i\mathcal{O}$ to obfuscate "programs with secrets".

Despite its enormous power, we only know of a limited set of techniques for working with $i\mathcal{O}$. In complex situations, these techniques often run into what we call the *sub-exponential barrier*. More specifically, the security proof of many $i\mathcal{O}$-based constructions end up considering an exponential number of hybrid experiments in order to make just one change in the underlying obfuscation. The goal is usually to eliminate all "troublesome" inputs, one at a time, that may be affected by the change. There are often exponentially many such inputs, resulting in a sub-exponential loss in the security reduction.

To make matters worse, a sub-exponential loss seems inherent to achieving $i\mathcal{O}$ from "simple" assumptions, such as those based on multilinear maps[1]. Indeed, all known security proofs for $i\mathcal{O}$ relative to "simple" assumptions[2] iterate over all (exponentially-many) inputs anyway, and there are reasons to believe that this loss may be necessary [18][3]. Indeed, any reduction from $i\mathcal{O}$ to a simple assumption would need to work for equivalent programs, but should fail for inequivalent programs (since inequivalent programs can be distinguished). Thus,

---

[1] Here, we do not define "simple". However, one can consider various notions of "simplicity" or "niceness" for assumptions, such as falsifiable assumptions [28] or complexity assumptions [22].

[2] Here, we exclude über assumptions such as semantically secure graded encodings [29], which encompass exponentially many distinct complexity assumptions.

[3] We stress that this argument has not yet been formalized.

such a reduction would seemingly need to decide if two programs compute equivalent functions; assuming $\mathcal{P} \neq \mathcal{NP}$, this in general cannot be done in polynomial time. This exponential loss would then carry over to any application of $i\mathcal{O}$, even if the $i\mathcal{O}$-to-application security reduction only incurred a polynomial loss. On the other hand, this exponential loss does *not* seem inherent to the vast majority of $i\mathcal{O}$ applications. This leaves us in an undesirable situation where the only way we know to instantiate an application from "simple" assumptions requires sub-exponential hardness assumptions, even though sub-exponential hardness is not inherent to the application.

One application for which an exponential loss does not appear inherent is Functional encryption ($\mathcal{FE}$), and indeed starting from the work of Garg et al. [17], it has been shown in [26,27] how to build $\mathcal{FE}$ from progressively simpler assumptions on multilinear maps with only a polynomial loss. Therefore, to bypass the difficulties above, we ask the following:

> Can applications of $i\mathcal{O}$ be based instead on $\mathcal{FE}$ with a polynomial security reduction?

There are two results that give us hope in this endeavor. First, it is known that $\mathcal{FE}$ is actually *equivalent* to $i\mathcal{O}$, except that *the $\mathcal{FE}$-to-$i\mathcal{O}$ reduction* [3,9] *incurs an exponential loss*. This hints at the possibility that, perhaps, specializing the $\mathcal{FE}$-to-$i\mathcal{O}$-to-application reduction to particular applications can aleviate the need for sub-exponential hardness.

Second and very recently, Garg et al. [19] took upon the issue of sub-exponential loss in $i\mathcal{O}$-based constructions in the context of PPAD hardness. They developed techniques to eliminate the sub-exponential loss in the work of Bitansky et al. [7] and reduced the hardness of PPAD to the hardness of standard, *polynomially-secure $i\mathcal{O}$* (and injective one-way functions). More importantly for us, they also presented a new reduction which bases the hardness of PPAD on standard polynomially-secure *functional encryption*, thus giving essentially the first non-trivial instance of using $\mathcal{FE}$ to build applications with only a polynomial loss.

This Work. Our goal is to develop techniques to break the sub-exponential barrier in cryptographic constructions based on $i\mathcal{O}$ and $\mathcal{FE}$. Towards this goal, we build upon and significantly extend the techniques in [19]. Our techniques are applicable, roughly, to any $i\mathcal{O}$ setting where the computation is changed on just a polynomial number of points; on all other points, the exact same circuit is used to compute the outputs. Notice that for such settings there exists an efficient procedure for checking functional equivalence. This enables us to argue indistinguishability based only on polynomial hardness assumptions. As it turns out, for many applications of $i\mathcal{O}$, the hybrid arguments involve circuits with the above specified structure. In this work, we focus on two such applications: *trapdoor permutations* and *universal samplers*.

We start with the construction of trapdoor permutations of Bitanksy et al. [8] based on sub-exponentially secure $i\mathcal{O}$. We improve their work by constructing trapdoor permutations based only on *polynomially-secure iO* (and one-way permutations). We further extend our results and obtain a construction based on standard, polynomial hard, functional encryption (instead of iO). Together with the result of [17,26,27], this gives us trapdoor permutations based on simple polynomial-hard assumptions on multilinear maps.

We then consider *universal samplers*, a notion put forward by Hofheinz et al. [23]. It allows for a single trusted setup which can be used to sample common parameters for *any* protocol. Hofheinz et al. construct universal samplers from iO. They also show how to use them to construct *multi-party non-interactive key-exchange* (NIKE) and broadcast encryption.

We consider the task of constructing universal samplers from the weaker notion of only polynomially-secure functional encryption. As noted earlier, we cannot use the generic reduction of [3,9] between $\mathcal{FE}$ and iO since it incurs sub-exponential loss. Intuitively, a fresh approach that is not powerful enough to imply iO is essential to obtaining a polynomial-time reduction for this task.

We present a new construction of universal samplers directly from $\mathcal{FE}$. We also consider the task of constructing multiparty NIKE for an unbounded number of users based on $\mathcal{FE}$. As detailed later, this turns out to be non-trivial even given the work of Hofheinz et al. This is because the definitions presented in [23] are not completely suitable to deal with an unbounded number of users. To support unbounded number of users, we devise a new security notion for universal samplers called *interactive simulation*. We present a construction of universal samplers based on $\mathcal{FE}$ that achieves this notion and gives us multiparty NIKE for unbounded number of users.

Remark 1. Our construction of TDP from FE is weaker in comparison to our construction from iO (and the construction of Bitansky et al. in [8]). In particular, given the random coins used to sample the function and the trapdoor, the output of the sampler is no longer pseudorandom. This property is important for some applications of TDPs like the construction of OT.

An Overview of Our Approach. In the following sections, we present a detailed overview of our approach of constructing Universal Samplers and NIKE for unbounded number of parties. Our techniques used for constructing trapdoor permutations are closely related to the techniques developed in proving PPAD-hardness of Garg et al. [19]. However, constructing trapdoor permutations poses additional challenges, namely the design of an efficient sampling algorithm that samples a domain element. Solving this problem requires development of new techniques and we elaborate them in the full version [20].

1.1 Universal Samplers and Multiparty Non-interactive Key Exchange from $\mathcal{FE}$

Multiparty Non-Interactive Key Exchange (multiparty NIKE) was one of the early applications of multilinear maps and iO. In multiparty NIKE, n parties simultaneously post a single message to a public bulletin board. Then they each read off the contents of the board, and are then able to derive a shared key K which is hidden to any adversary that does not engage in the protocol, but is able to see the contents of the public bulletin board.

Boneh and Silverberg [11] show that multilinear maps imply multiparty NIKE. However, (1) their protocol requires an a priori bound on the number of users n, and (2) due to limitations with current multilinear map candidates [13,15], the protocol requires a trusted setup. The party that runs the trusted setup can also learn the shared key k, even if that party does not engage in the protocol.

Boneh and Zhandry [12] show how to use $i\mathcal{O}$ to remove the trusted setup. Later, Ananth et al. [1] shows how to remove the bound on the number of users by using the very strong *differing inputs* obfuscation. Khurana et al. [25] further modify the Boneh-Zhandry protocol to get unbounded users with just $i\mathcal{O}$. In [12,25], $i\mathcal{O}$ is invoked on programs for which are guaranteed to be equivalent; however it is computationally infeasible to actually verify this equivalence. Thus, following the arguments of [18], it would appear that any reduction to a few simple assumptions, no matter how specialized to the particular programs being obfuscated, would need to incur an exponential loss. Hence, these approaches do not seem suitable to achieving secure multiparty NIKE from polynomially secure $\mathcal{FE}$.

Universal Samplers. Instead, we follow an alternate approach given by Hofheinz et al. [23] using universal samplers. A universal sampler is an algorithm that takes as input the description of a sampling procedure (say, the sampling procedure for the common parameters of some protocol) and outputs a sample from that procedure (a set of parameters for that protocol). The algorithm is deterministic, so that anyone running the protocol on a given sampling procedure gets the same sampled parameters. Yet the generated parameters should be "as good as" a freshly generated set of parameters. Therefore, the only set of common parameters needed for all protocols is just a single universal sampler. When a group of users wish to engage in a protocol involving a trusted setup, they can each feed the setup procedure of that protocol into the universal sampler, and use the output as the common parameters.

Unfortunately, defining a satisfactory notion of "as good as" above is non-trivial. Hofheinz et al. give two definitions: a static definition which only remains secure for a bounded number of generated parameters, as well as an adaptive definition that is inherently tied to the random oracle model, but allows for an unbounded number of generated parameters. They show how to use the stronger definitions to realize primitives such as adaptively secure multiparty non-interactive key exchange (NIKE) and broadcast encryption.

In this work, we focus on the standard model, and here we review the static standard-model security definition for universal samplers. Fix some bound k on the number of generated parameters. Intuitively, the k-time static security definition says that up to k freshly generated parameters $s_1, \ldots, s_k$ for sampling algorithms $C_1, \ldots, C_k$ can be embedded into the universal sampler without detection. Thus, if the sampler is used on any of the sampling algorithms C_i, the generated output will be the fresh sample s_i. Formally, there is a simulator Sim that takes as input up to k sampler/sample pairs (C_i, s_i), and outputs a simulated universal sampler Sampler, such that $\mathsf{Sampler}(C_i) = s_i$. As long as the s_i are fresh samples from C_i, the simulated universal sampler will be indistinguishable from a honestly generated sampler.

Fortunately for us, the $i\mathcal{O}$-based construction of [23] only invokes $i\mathcal{O}$ on programs for which it is trivial to verify equivalence. Thus, there seems hope that universal samplers can be based on simple assumptions without an exponential loss. In particular, there is hope to base universal samplers on the polynomial hardness of functional encryption.

Application to Multiparty NIKE. From the static definition above, it is straightforward to obtain a statically secure multiparty NIKE protocol analogous to the adaptive protocol of Hofheinz et al. [23]. Each party simply publishes a public key pk_i for a public key encryption scheme, and keeps the corresponding secret key sk_i hidden. Then to generate the shared group key, all parties run Sampler on the sampler $C_{\mathsf{pk}_1,\dots,\mathsf{pk}_n}$. Here, $C_{\mathsf{pk}_1,\dots,\mathsf{pk}_n}$ is the randomized procedure that generates a random string K, and encrypts K under each of the public keys $\mathsf{pk}_1,\dots,\mathsf{pk}_n$, resulting in n ciphertexts $c_1,\dots,c_n$ which it outputs. Then party i decrypts c_i using sk_i. The result is that all parties in the protocol learn K.

Meanwhile, an eavesdropper who does not know any of the secret keys will only have the public keys, the sampler, and thus the ciphertexts c_i outputted by the sampler. The proof that the eavesdropper will not learn K is as follows. First, we consider a hybrid experiment where K is generated uniformly at random, and the universal sampler is simulated on sampler $C_{\mathsf{pk}_1,\dots,\mathsf{pk}_n}$, and sample $s = (c_1,\dots,c_n)$, where c_i are fresh encryptions of K under each of the public keys pk_i. 1-time static security of the universal sampler implies that this hybrid is indistinguishable to the adversary from the real world. Next, we change each of the c_i to encrypt 0. Here, indistinguishability follows from the security of the public key encryption scheme. In this final hybrid, the view of the adversary is independent of the shared secret key K, and security follows.

Unbounded Multiparty NIKE. One limitation of the protocol above is that the number of users must be a priori bounded. There are several reasons for this, the most notable being that in order to simulate, the universal sampler must be as large as the sample $s = (c_1,\dots,c_n)$, which grows with n. Thus, once the universal sampler is published, the number of users is capped. Unfortunately, the only prior protocols for achieving an unbounded number of users, [1,25], seems inherently tied to the Boneh-Zhandry approach, and it is not clear that their techniques can be adapted to universal samplers.

In order to get around this issue, we change the sampling procedure $C_{\mathsf{pk}_1,\dots,\mathsf{pk}_n}$ fed into the universal sampler. Instead, we feed in circuits of the form $D_{\mathsf{pk},\mathsf{pk}'}$, which generate a new secret and public key $(\mathsf{sk}'', \mathsf{pk}'')$, encrypt sk'' under both pk and pk', and output both encryptions as well as the new public key pk''. A group of users with public keys $\mathsf{pk}_1,\dots,\mathsf{pk}_n$ then generates the shared key in an iterative fashion as follows. Run the universal sampler on $D_{\mathsf{pk}_1,\mathsf{pk}_2}$, obtaining a new public key pk_3', as well as encryptions of the corresponding secret key sk_3' under both $\mathsf{pk}_1, \mathsf{pk}_2$. Notice that users 1 and 2 can both recover sk_3' using their secret keys. Then run the universal sampler on $D_{\mathsf{pk}_3, \mathsf{pk}_3'}$, obtaining a new public key pk_4' and encryptions of the corresponding secret key sk_4'. Notice that user 3 can recover sk_4' by decrypting the appropriate ciphertext using sk_3, and users 1 and 2 can recover sk_4' by decrypting the other ciphertext using sk_3'. Continue in this way until public key pk_{n+1}' is generated, and all users 1 through n recover the corresponding secret key sk_{n+1}'. Set sk_{n+1}' to be the shared secret key.

For security, since an eavesdropper does not know any of the secret keys and the ciphertexts are "as good as" fresh ciphertexts, he should not be able to decrypt any of the ciphertexts in the procedure above. However, turning this intuition into a security proof using the static notion of security is problematic. The straightforward approach requires constructing a simulated Sampler where the outputs on each of the circuits $D_{\mathsf{pk}_i, \mathsf{pk}_i'}$ are fresh samples. Then, each of the ciphertexts in the samples are replaced with encryptions of 0 (instead of the correct secret decryption key). However, as there are n such circuits, a standard incompressibility argument shows that Sampler must grow linearly in n. Thus again, once the universal sampler is published, the number of users is capped.

Simulating at Fewer Points. To get around this issue, we devise a sequence of hybrids where in each hybrid, we only need replace $\log n$ outputs of the sampler with fresh samples. The core idea is the following. Say that a circuit $D_{\mathsf{pk}_i, \mathsf{pk}_i'}$ has been "treated" if the public key pk_{i+1}' outputted by the universal sampler is freshly sampled and the corresponding ciphertexts are changed to encrypt 0 (instead of the secret key sk_{i+1}'). We observe that to switch circuit $D_{\mathsf{pk}_i, \mathsf{pk}_i'}$ from untreated to treated, circuit $D_{\mathsf{pk}_{i-1}, \mathsf{pk}_{i-1}'}$ needs to currently be treated so that the view of the adversary is independent of the secret key sk_i'. However the status of all the other circuits is irrelevant. Moreover, once we have treated $D_{\mathsf{pk}_i, \mathsf{pk}_i'}$, we can potentially "untreat" $D_{\mathsf{pk}_{i-1}, \mathsf{pk}_{i-1}'}$ and reset its ciphertexts to the correct values, assuming $D_{\mathsf{pk}_{i-2}, \mathsf{pk}_{i-2}'}$ is currently treated. Our goal is to start from no treated circuits, and arrive at a hybrid where $D_{\mathsf{pk}_n, \mathsf{pk}_n'}$ is treated, which implies that the view of the adversary is independent of the shared secret sk_{n+1}.

This gives rise to an interesting algorithmic problem. The goal is to get a pebble at position n, where the only valid moves are (1) placing or removing a pebble at position 1, or (2) placing or removing a pebble at position i provided there is currently a pebble at position $i-1$. We desire to get a pebble at position n while minimizing the number of pebbles used at any time. The trivial solution is to place a pebble at 1, then 2, and so on, requiring n pebbles. We show a pebbling scheme that gets a pebble to position n using only $\approx \log n$ pebbles by removing certain pebbles as we go. Interestingly, the pebbling scheme is exactly same as the one used in [6] in the context of reversible computation. The pebbling scheme is also efficient: the number of moves is polynomial in n.

Using our pebbling algorithm, we derive a sequence of hybrids corresponding to each move in the algorithm. Thus we show that the number of circuits that need simulating can be taken to be $\approx \log n$.

A New Universal Sampler Definition. Unfortunately, we run into a problem when trying to base security on the basic static sampler definition of Hofheinz et al. [23]. The issue stems from the fact that the simulator in the static definition requires knowing all of the circuits $D_{\mathsf{pk}_i,\mathsf{pk}'_i}$ up front. However, in our pebbling approach, some of the pk'_i (and thus the $D_{\mathsf{pk}_i,\mathsf{pk}'_i}$) are determined by the sampler Sampler - namely, all the pk'_i for which $D_{\mathsf{pk}_{i-1},\mathsf{pk}'_{i-1}}$ is "untreated". Thus we encounter a circularity where we need to know Sampler to compute the circuit $D_{\mathsf{pk}_i,\mathsf{pk}'_i}$, but we need $D_{\mathsf{pk}_i,\mathsf{pk}'_i}$ in order to simulate the Sampler.

To get around this issue, we devise a new security notion for universal samplers that allows for *interactive simulation.* That is, *before* the simulator outputs Sampler, we are allowed to query it on various inputs, learning what the output of the sampler will be on that input (called as the *read* query). Moreover, we are allowed to feed circuit/sample pairs (C, s) (called as *write* query) interactively, potentially *after* seeing some of the sample outputs, and the simulator will guarantee that the simulated Sampler will output s on C. For security, we require that for a statically chosen query index i^* and a circuit C^* the simulator's outputs in the following two cases are computationally indistinguishable:

1. i^{*th} query is a read query on C^*.
2. i^{*th} query is a write query on (C^*, s^*) where s^* is fresh sample from C^*.

This new definition allows us to avoid the circularity above and complete the security proof for our NIKE protocol.

Construction. Before we describe our construction of universal samplers from $\mathcal{FE}$, we first describe a construction from $i\mathcal{O}$ that satisfies the above definition of interactive simulation.

The universal sampler is an obfuscation of a circuit that has a puncturable PRF key K hardwired in its description and on input C outputs $C(; \mathsf{PRF}_K(C))$ i.e. It uses the PRF key to generate the random coins. This is precisely the same construction as given by Hofheinz et al. [23] for the static security case. To prove that this construction satisfies the stronger definition of interactive simulation we construct a simulator that works as follows. It first samples a fresh PRF key K' and answers the read queries using it. At the end of the simulation, it outputs an obfuscation of a circuit that has the PRF key K' as well as (C_i, s_i) for every write query made by the adversary hardwired in its description. When run on input C where C is one of the write queries, it outputs the corresponding s. On other inputs, it outputs $C(; \mathsf{PRF}_{K'}(C))$.

The security is shown via a hybrid argument. The initial hybrid corresponds to the output of the simulator when the challenge query (made at index i^*) is a write query on (C_{i^*}, s_{i^*}) where s_{i^*} is a fresh random sample from C_{i^*}. We first change the obfuscated circuit to have the PRF key K' punctured at C_{i^*}. This is possible since the circuit does not use K' to compute the output on C_{i^*}. Relying on the security of puncturable PRF, we change s_{i^*} from $C_{i^*}(; r)$ where r is random string to $C_{i^*}(; \mathsf{PRF}_{K'}(C_{i^*}))$. We then unpuncture the key K' and finally remove C_{i^*}, s_{i^*} from the hardwired list.

We adapt the above construction from $i\mathcal{O}$ to the $\mathcal{FE}$ setting using techniques from [9,19]. Recall that the "obfuscated" universal sampler consists of $\ell + 1$ (ℓ is the maximum size of the input circuit) function keys (where each function key computes a bit extension function) along with an initial ciphertext c_ϕ that encrypts the empty string ϕ and a prefix constrained PRF key K^4. These bit extension functions form a natural binary tree structure and "parsing" an input circuit C corresponds to traveling along the path from the root to the leaf labeled C. Each node x along the path from the root to C contains the key K prefix constrained at x. The prefix constrained PRF key appearing at the leaf C is precisely equal to the PRF value at C and we use this to generate a "pseudorandom" sample from C.

We are now ready to describe the construction of our simulator. As in the $i\mathcal{O}$ case, the simulator samples a random prefix constrained PRF key K' and uses it to answer the read queries made by the adversary. Recall that for every write query (C_i, s_i) the adversary makes, the simulator must ensure that the sampler on C_i outputs s_i. The simulator accomplishes this by "tunneling" the underlying binary tree along path C_i. To give a bit more details, the simulator "forces" the function keys at every level i to output a precomputed value say V_i (instead of the bit-extension) if the input to the function matches with a prefix of C_i. At the leaf level, if the input matches C_i then the function outputs s_i. Illustration of "tunneling" is given in Fig. 1. We now explain how this "tunneling" is done.

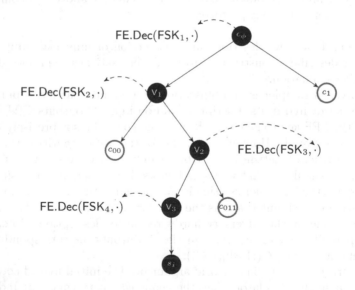

Fig. 1. Illustration of "tunneling" on $C_i = 010$ and $\kappa = 3$.

---

[4] [19] used the term prefix-punctured PRF to denote the same primitive. We use the term prefix constrained PRF as we feel that this name is more appropriate. This was also suggested by an anonymous Eurocrypt reviewer.

At a high level, the "tunneling" is achieved by triggering a hidden "trapdoor" thread in the function keys using techniques illustrated in [2, 19]. This technique proceeds by first encrypting a set of precomputed values under a symmetric key sk and hardwires them in the description of bit-extension function in each level. The symmetric key sk is encrypted in the initial ciphertext c_ϕ along with the empty string and the prefix constrained PRF key. The trapdoor thread (that is triggered only along the write query paths) uses this secret key sk to decrypt the hardcoded ciphertext and outputs the appropriate pre-computed value.

To complete the security proof, we want to show that we can indistinguishably "tunnel" the binary tree along a new path C_i^* and output s_i^* which is a fresh random sample from C_i^* at the leaf. Recall that in the construction of Garg et al. in [19] a single secret key sk is used to for computing the encryptions of pre-computed values along multiple paths. But having a single secret key does not allow us to "tunnel" along a new path C_i^* as this secret key already appears in the initial ciphertext c_ϕ. Hence, we cannot rely on the semantic security of symmetric key encryption to augument the pre-computed values to include values along the new path C_i^*. In order to get around this issue, we use multiple secret keys: one for each write query[5] which enables us to "tunnel" along a new path C_i^*.

2   Preliminaries

κ denotes the security parameter. A function $\mu(\cdot) : \mathbb{N} \to \mathbb{R}^+$ is said to be negligible if for all polynomials $\mathsf{poly}(\cdot)$, $\mu(k) < \frac{1}{\mathsf{poly}(k)}$ for large enough k. We will use PPT to denote Probabilistic Polynomial Time algorithm. We denote $[k]$ to be the set $\{1, \cdots, k\}$. We will use $\mathsf{negl}(\cdot)$ to denote an unspecified negligible function and $\mathsf{poly}(\cdot)$ to denote an unspecified polynomial. We denote the identity polynomial by $I(\cdot)$ i.e. $I(x) = x$. All adversarial functions are modeled as polynomial sized circuits. We assume that all cryptographic algorithms take the security parameter in unary as input and would not explicitly mention it in all cases. We assume without loss of generality that the length of the random tape used by all cryptographic algorithms is κ.

A binary string $x \in \{0, 1\}^k$ is represented as $x_1 \cdots x_k$. x_1 is the most significant (or the highest order bit) and x_k is the least significant (or the lowest order bit). The i-bit prefix $x_1 \cdots x_i$ of the binary string x is denoted by $x_{[i]}$. We denote $|x|$ to be the length of the binary string $x \in \{0, 1\}^*$. We use $x\|y$ to denote concatenation of binary strings x and y. We say that a binary string y is a prefix of x if and only if there exists a string $z \in \{0, 1\}^*$ such that $x = y\|z$.

We assume the reader's familiarity with standard cryptographic primitives like injective pseudorandom generator, puncturable pseudorandom functions, indistinguishability obfuscation, functional encryption, symmetric and public

---

[5] In the security definition, the number of write queries that an adversary could make is apriori bounded. On the otherhand, the adversary could make an unbounded number of read queries. Thus, we can fix the number of secret keys to be sampled at the time of setup.

key encryption. Below, we give the definition of Prefix Constrained Pseudorandom Function [19].

Prefix Constrained Pseudorandom Function. A PCPRF is a tuple of algorithms (KeyGen$_{\mathcal{PCPRF}}$, PrefixCons) with the following syntax. KeyGen$_{\mathcal{PCPRF}}$ takes the security parameter (encoded in unary) and descriptions of two polynomials p_{in} and p_{out} as input and outputs a PCPRF key $S \in \{0,1\}^\kappa$. PrefixCons is a deterministic algorithm and has two modes of operation:

1. **Normal Mode:** In the normal mode, PrefixCons takes a PCPRF key S and a string $y \in \cup_{k=0}^{p_{in}(\kappa)} \{0,1\}^k$ and outputs a prefix constrained key $S_y \in \{0,1\}^\kappa$ if $|y| < p_{in}(\kappa)$; else outputs $S_y \in \{0,1\}^{p_{out}(\kappa)}$. We assume that S_y contains implicit information about $|y|$.
2. **Repeated Constraining Mode:** In the repeated constraining mode, PrefixCons takes a prefix constrained key S_y and a string $z \in \cup_{k=0}^{p_{in}(\kappa)} \{0,1\}^k$ as input and works as follows. If $|y| + |z| > p_{in}(\kappa)$, it outputs $\bot$; else if $|y| + |z| < p_{in}(\kappa)$, it outputs the prefix constrained key $S_{y\|z} \in \{0,1\}^\kappa$; else it outputs $S_{y\|z} \in \{0,1\}^{p_{out}(\kappa)}$.

Henceforth, unless it is not directly evident from the context, we will not explicitly mention if PrefixCons is in the normal mode or in the repeated constraining mode. We note that there is no explicit evaluation procedure for PCPRF and the output of PCPRF on an input $x \in \{0,1\}^{p_{in}(\kappa)}$ is given by PrefixCons$(S, x) \in \{0,1\}^{p_{out}(\kappa)}$.

We now give the formal definition of PCPRF.

Definition 1. *A prefix constrained pseudorandom function $\mathcal{PCPRF}$ is a tuple of PPT algorithms (KeyGen$_{\mathcal{PCPRF}}$, PrefixCons) satisfying the following properties:*

- ***Functionality is preserved under repeated constraining:** For all κ, polynomials $p_{in}(\cdot), p_{out}(\cdot)$ and for all $x \in \cup_{k \in [p_{in}(\kappa)]} \{0,1\}^k$, $y, z \in \{0,1\}^*$ s.t. $x = y\|z$,*

$$\Pr[\mathsf{PrefixCons}(\mathsf{PrefixCons}(S, y), z) = \mathsf{PrefixCons}(S, x)] = 1$$

  where $S \leftarrow \mathsf{KeyGen}_{\mathcal{PCPRF}}(1^\kappa, p_{in}(\cdot), p_{out}(\cdot))$.
- ***Pseudorandomness at constrained prefix:** For all κ, polynomials $p_{in}(\cdot), p_{out}(\cdot)$, for all $x \in \cup_{k \in [p_{in}(\kappa)]} \{0,1\}^k$, and for all poly sized adversaries $\mathcal{A}$*

$$|\Pr[\mathcal{A}(\mathsf{PrefixCons}(S, x), \mathsf{Keys}) = 1] - \Pr[\mathcal{A}(U_\ell, \mathsf{Keys}) = 1]| \leq \mathsf{negl}(\kappa)$$

  where $S \leftarrow \mathsf{KeyGen}_{\mathcal{PCPRF}}(1^\kappa, p_{in}(\cdot), p_{out}(\cdot))$, $\ell = |\mathsf{PrefixCons}(S, x)|$ and $\mathsf{Keys} = \{\mathsf{PrefixCons}(S, x_{[i-1]}\|(1 - x_i))\}_{i \in [|x|]}$.

The above properties are satisfied by the construction of the pseudorandom function in [21].

Notation. For a key S_i (indexed by i), we will use $S_{i,y}$ to denote PrefixCons(S_i, y).

3   TDP from IO in Poly Loss

We consider trapdoor permutation with pseudorandom sampling which is a weakened notion than the traditional uniform sampling. We refer the reader to [8] for a formal definition.

3.1   Construction of Trapdoor Permutations

In this section, we give a construction of trapdoor permutations and prove the one-wayness assuming the existence polynomially hard $i\mathcal{O}$, puncturable pseudorandom function $\mathcal{PRF}$ and injective $\mathcal{PRG}$ (used only in the proof).

Theorem 1. *Assuming the existence of one-way permutations and indistinguishablity obfuscation against polytime adversaries there exists a trapdoor permutation with pseudorandom sampling.*

Our Construction. Our construction uses the following primitives:

1. An indistinguishability Obfuscator $i\mathcal{O}$.
2. A puncturable pseudorandom function $\mathcal{PRF} = (\mathsf{KeyGen}_{\mathcal{PRF}}, \mathsf{PRF}, \mathsf{Punc})$.
3. A length doubling pseudorandom generator $\mathsf{PRG} : \{0,1\}^{\kappa/2} \to \{0,1\}^{\kappa}$.
4. Additionally, in the proof of security, we use a length doubling injective pseudorandom generator $\mathsf{InjPRG} : [2^{\kappa/4}] \to [2^{\kappa/2}]$.

The formal description of our construction appears in Fig. 2.

Due to lack of space we give the proof of security in the full version of our paper [20].

- $\mathsf{KeyGen}(1^{\kappa})$:
  1. Sample $\{S_i\}_{i \in [\kappa]} \leftarrow \mathsf{KeyGen}_{\mathcal{PRF}}(1^{\kappa})$. For all $i \in [\kappa]$, S_i is a seed for a PRF mapping i bits to κ bits. That is, $\mathsf{PRF}_{S_i} : \{0,1\}^i \to \{0,1\}^{\kappa}$.
  2. The public key is given by $i\mathcal{O}(F_{S_1,\cdots,S_{\kappa}})$ where $F_{S_1,\cdots,S_{\kappa}}$ is described in Figure 3 and the secret key is given by $S_1, \cdots, S_{\kappa}$.
- TDP_{PK} : Run the obfuscated circuit $i\mathcal{O}(F_{S_1,\cdots,S_{\kappa}})$ on the given input $(x, \sigma_1, \cdots, \sigma_{\kappa})$.
- TDP^{-1}_{SK}: The Inverter $I_{S_1,\cdots,S_{\kappa}}$ is described in Figure 3.
- $\mathsf{SampGen}(SK)$: The sampler is given by $i\mathcal{O}(X_{S_1,\cdots,S_{\kappa}})$ where $X_{S_1,\cdots,S_{\kappa}}$ is described in Figure 3.
- Samp: Run the circuit $i\mathcal{O}(X_{S_1,\cdots,S_{\kappa}})$ on the given randomness r.

Fig. 2. Construction of trapdoor permutation

$$F_{S_1,\cdots,S_\kappa}$$

Input: $(i, \sigma_1, \cdots, \sigma_\kappa)$
Constants: $S_1, \cdots, S_\kappa$

1. For all $j \in [\kappa]$, check if $\sigma_j = \mathsf{PRF}_{S_j}(i_{[j]})$.
2. If any of the above checks fail, output $\perp$.
3. Else, for all $j \in [\kappa]$ compute $\sigma'_j = \mathsf{PRF}_{S_j}((i+1)_{[j]})$ where $i+1$ is computed modulo 2^κ.
4. Output $(i+1, \sigma'_1, \cdots, \sigma'_\kappa)$.

Padding: The circuit would be padded to size $p(\kappa)$ where $p(\cdot)$ is a polynomial that would be specified later.

$$X_{S_1,\cdots,S_\kappa}$$

Input: $r \in \{0,1\}^{\kappa/2}$
Constants: $S_1, \cdots, S_\kappa$

1. Compute $i = \mathsf{PRG}(r)$.
2. For every $j \in [\kappa]$, compute $\sigma_j = \mathsf{PRF}_{S_j}(i_{[j]})$.
3. Output $(i, \sigma_1, \sigma_2, \cdots, \sigma_\kappa)$.

Padding: The circuit would be padded to size $q(\kappa)$ where $q(\cdot)$ is a polynomial that would be specified later.

$$I_{S_1,\cdots,S_\kappa}$$

Input: $(i, \sigma_1, \cdots, \sigma_\kappa)$
Constants: $S_1, \cdots, S_\kappa$

1. Check whether for all $j \in [\kappa]$, $\sigma_j = \mathsf{PRF}_{S_j}(i_{[j]})$.
2. If any of the checks fail, output $\perp$.
3. Else, for all $j \in [\kappa]$ compute $\sigma'_j = \mathsf{PRF}_{S_j}((i-1)_{[j]})$ where $i-1$ is computed modulo 2^κ.
4. Output $(i-1, \sigma'_1, \sigma'_2, \cdots, \sigma'_\kappa)$.

Fig. 3. Public key, sampler and the inverter for the trapdoor permutations

4   Trapdoor Permutation from FE

We start by defining a weaker (with respect to pseudorandom sampling) notion of trapdoor permutation.

Definition 2. *An efficiently computable family of functions:*

$$\mathcal{TDP} = \{\mathsf{TDP}_{PK} : D_{PK} \to D_{PK} \text{ and } PK \in \{0,1\}^{\mathsf{poly}(\kappa)}\}$$

---

1. $(PK, SK) \leftarrow$ KeyGen(1^κ).
2. Samp $\leftarrow$ SampGen(SK)
3. **if**$(b = 0)$, $x \xleftarrow{\$} D_{PK}$.
4. **else**, $x \leftarrow$ Samp.
5. Output $\mathcal{A}(PK, \mathsf{Samp}, x)$

Fig. 4. Exp$_{\mathcal{A},b,\mathsf{wPRS}}$

over the domain D_{PK} with associated (probabilistic) (KeyGen, SampGen) algorithms is a weakly samplable trapdoor permutation if it satisfies:

- **Trapdoor Invertibility:** For any $(PK, SK) \leftarrow$ KeyGen(1^κ), TDP$_{PK}$ is a permutation over D_{PK}. For any $y \in D_{PK}$, $TDP_{SK}^{-1}(y)$ is efficiently computable given the trapdoor SK.
- **Weak Pseudorandom Sampling:** For any $(PK, SK) \leftarrow$ KeyGen(1^κ) and Samp $\leftarrow$ SampGen(SK), Samp$(\cdot)$ samples pseudo random points in the domain D_{PK}. Formally, for any polysized distinguisher $\mathcal{A}$,

$$\left| \Pr\left[\mathsf{Exp}_{\mathcal{A},0,\mathsf{wPRS}} = 1\right] - \Pr\left[\mathsf{Exp}_{\mathcal{A},1,\mathsf{wPRS}} = 1\right] \right| \leq \mathsf{negl}(\kappa)$$

where Exp$_{\mathcal{A},b,\mathsf{wPRS}}$ is described in Fig. 4.
- **One-wayness:** For all poly sized adversaries $\mathcal{A}$,

$$\Pr\left[\mathcal{A}(PK, \mathsf{Samp}, TDP_{PK}(x)) = x \,\middle|\, \begin{array}{l} (PK, SK) \leftarrow \mathsf{KeyGen}(1^\kappa) \\ \mathsf{Samp} \leftarrow \mathsf{SampGen}(SK) \\ x \leftarrow \mathsf{Samp} \end{array} \right] \leq \mathsf{negl}(\kappa)$$

Remark 2. The requirement of pseudorandom sampling considered in Bitanksy et. al.'s work [8] is stronger than the one considered here in sense that they require the pseudorandomness property to hold even when given the random coins used by KeyGen and the SampGen algorithms. We do not achieve the stronger notion in this work. In particular, given the random coins used in SampGen the sampler's output is no longer pseudorandom. Therefore, our trapdoor permutations can be only used in applications where an honest party runs the KeyGen and SampGen algorithm. It cannot be used for example to achieve receiver privacy in EGL Oblivious Transfer protocol [14].

In this section, we construct trapdoor permutation satisfying the Definition 2 from polynomially hard public key functional encryption, prefix puncturable pseudorandom function, left half injective pseudorandom generator, strong randomness extractor and public key encryption with random public keys.

Theorem 2. *Assuming the existence of one-way permutations, single-key, selective secure, public key functional encryption and public key encryption with (pseudo) random public keys, there exists a weakly samplable trapdoor permutation.*

We now recall the special key structure [19] which forms a crucial part of our construction of trapdoor permutation.

Notation. We treat $1^i + 1$ as 0^i and $\phi + 1$ as ϕ. Let LeftInjPRG be a left half injective pseudorandom generator. Let τ be the size of public key output by PK.KeyGen(1^κ). Below, for every $i \in [\kappa + \tau]$, $S_i \leftarrow \mathsf{KeyGen}_{\mathcal{PCPRF}}(1^\kappa.C_i(\cdot), I(\cdot))$ where $C_i(\kappa) = i$ and $I(\kappa) = \kappa$. Recall $S_{i,x}$ denotes a prefix constrained PRF key S_i constrained at a prefix x.

Special Key Structure.

$$\mathsf{U}_x = \bigcup_{i \in [\tau+\kappa]} \mathsf{U}_x^i \qquad \mathsf{U}_x^i = \begin{cases} \{S_{i,x_{[i]}}\} & \text{if } |x| > i \\ \{S_{i,x}\} & \text{otherwise} \end{cases}$$

$$\mathsf{V}_x = \bigcup_{i \in [\tau+\kappa]} \mathsf{V}_x^i \qquad \mathsf{V}_x^i = \begin{cases} \{S_{i,x_{[i]}}, S_{i,x_{[i]}+1}\} & \text{if } |x| > i \text{ and } x = x_{[i]}\|1^{|x|-i} \\ \{S_{i,x}, S_{i,(x+1)\|0^{i-|x|}}\} & \text{if } |x| \leq i \\ \emptyset & \text{if } |x| > i \text{ and } x \neq x_{[i]}\|1^{|x|-i} \end{cases}$$

$$\mathsf{W}_x = \bigcup_{i \in [\tau+\kappa]} \mathsf{W}_x^i \qquad \mathsf{W}_x^i = \begin{cases} \{\mathsf{LeftInjPRG}_0(S_{i,x_{[i]}})\} & \text{if } |x| \geq i \\ \emptyset & \text{otherwise} \end{cases}$$

For the empty string $x = \phi$, these sets can be initialized as follows.

$$\mathsf{U}_\phi = \bigcup_{i \in [\tau+\kappa]} \mathsf{U}_\phi^i \qquad \mathsf{U}_\phi^i = \{S_i\}$$

$$\mathsf{V}_\phi = \bigcup_{i \in [\tau+\kappa]} \mathsf{V}_\phi^i \qquad \mathsf{V}_\phi^i = \{S_i\}$$

$$\mathsf{W}_\phi = \bigcup_{i \in [\tau+\kappa]} \mathsf{W}_\phi^i \qquad \mathsf{W}_\phi^i = \emptyset$$

Jumping ahead, the set of keys in U_x would be used by the sampler to generate the set of associated signatures on the sampled point. The set W_x (called as the vestigial set in [19]) is used to check the validity of input i.e. checking whether the input belongs to the domain. The set V_x is used to generate the associated signatures on the "next" point as defined by the permutation.

Our Construction. The construction of weakly samplable trapdoor permutation uses the following primitives:

1. A single-key, selective secure public key functional encryption scheme $\mathcal{FE}$.
2. A prefix constrained pseudorandom function $\mathcal{PCPRF}$.
3. An injective length doubling pseudorandom generator InjPRG : $\{0,1\}^{\kappa/8} \rightarrow \{0,1\}^{\kappa/4}$
4. A length doubling Left half injective pseudorandom generator LeftInjPRG : $\{0,1\}^\kappa \rightarrow \{0,1\}^{2\kappa}$

In the construction, we denote $\mathsf{SK.Enc}_{sk_1,\ldots,sk_n}(m)$ to be $\mathsf{SK.Enc}_{sk_n}$ ($\mathsf{SK.Enc}_{sk_{n-1}}(\cdots \mathsf{SK.Enc}_{sk_1}(m)))$. The formal description our construction appears in Fig. 5.

Setting rand($\cdot$) We set rand(κ) to be the maximum number of random bits needed to generate $\tau + \kappa$ encryptions under $\gamma_1, \cdots, \gamma_\kappa$ as well as $\tau + \kappa + 1$ encryptions under the public keys pk.

Due to shortage of space, we defer the proof of Theorem 2 to the full version of the paper [20].

5   Universal Samplers

Intuitively, a universal sampler, defined by Hofheinz et al. [23] is a box that takes as input the description of a sampling procedure, and outputs a fresh-looking sample according to the sampling procedure. The difficulty is that we want the box to be public code, and that every user, when they run the sampler on a particular procedure, gets the same result. Moreover, we want the sample to appear as if it were a fresh random sample.

5.1   Definition

A *Universal Sampler* consists of an algorithm Setup that takes as input a security parameter κ (encoded in unary) and a size bound $\ell(\cdot)$, random tape size $r(\cdot)$ and an output size $t(\cdot)$. It outputs a program Sampler. Sampler takes as input a circuit of size at most $\ell(\kappa)$, uses $r(\kappa)$ bits of randomness and outputs an $t(\kappa)$-bit string.

Intuitively, Sampler(C) will be a pseudorandom sample from C: Sampler(C) = $C(s)$ for some s pseudorandomly chosen based on C. We will actually not formalize a standalone correctness requirement, but instead correctness will follow from our security notion.

For security, we ask that the sample output by Sampler(C) actually looks like a fresh random sample from C. Unfortunately, formalizing this requirement is tricky. Hofheinz et al. [23] defined two notions: the first is a "static" and "bounded" security notion, while the second stronger notion is "adaptive" and "unbounded". The latter definition requires random oracles, so it is unfortunately uninstantiable in the standard model. We will provide a third definition which strikes some middle ground between the two, and is still instantiable in the standard model.

Definition 3. *A Universal Sampler given by* Setup *is n-time statically secure with interactive simulation if there exists an efficient randomized simulator* Sim *such that the following hold.*

- Sim *takes as input κ (encoded in unary) and three polynomials $\ell(\cdot), r(\cdot), t(\cdot)$ (for ease of notation, we denote $\ell = \ell(\kappa)$, $r = r(\kappa)$ and $t = t(\kappa)$), and ultimately will output a simulated sampler* Sampler. *However, before doing so,* Sim *provides the following interface for additional input:*
  - **Read** *queries: here the user submits an input circuit C of size at most ℓ, that uses r bits of randomness and has output length t.* Sim *will respond with a sample s that will ultimately be the output of the simulated sampler on C.* Sim *supports an unbounded number of* **Read** *queries.*

- KeyGen(1^κ):
  1. For each $i \in [\tau + \kappa]$, sample $S_i \leftarrow \mathsf{KeyGen}_{\mathcal{PCPRF}}(1^\kappa, C_i(\cdot), I(\cdot))$ where $C_i(\kappa) = i$ and $I(\kappa) = \kappa$. Sample $\widetilde{K} \leftarrow \mathsf{KeyGen}_{\mathcal{PCPRF}}(1^\kappa, \mathsf{quad}(\cdot), \mathsf{rand}(\cdot))$ where $\mathsf{quad}(\kappa) = 2(\kappa + \tau) + 1$. For every $i \in [\tau + \kappa]$, initialize $\mathsf{V}_\phi^i := S_i$, $\mathsf{V}_\phi = \bigcup_{i \in [\tau + \kappa]} \mathsf{V}_\phi^i$ and $\mathsf{W}_\phi = \emptyset$.
  2. Let $\mathsf{Ext}_w : \{0,1\}^{\tau + \kappa} \to \{0,1\}^{\kappa/8}$ be a $(\kappa/4, \mathsf{negl}(\kappa))$ strong randomness extractor with seed length $q(\kappa)$. Sample a seed $w \xleftarrow{\$} \{0,1\}^{q(\kappa)}$ for the extractor Ext.
  3. Sample $(\mathsf{PK}_i^1, \mathsf{MSK}_i^1) \leftarrow \mathsf{FE.Setup}(1^\kappa)$ for all $i \in [\tau + \kappa + 1]$.
  4. Sample $sk_1 \leftarrow \mathsf{SK.KeyGen}(1^\kappa)$ where $|sk_1| = p(\kappa)$ and let $\Pi_1 \leftarrow \mathsf{SK.Enc}_{sk_1}(\pi_1)$ and $\Lambda_1 \leftarrow \mathsf{SK.Enc}_{sk_1}(\lambda_1)$ where $\pi_1 = 0^{\ell_1(\kappa)}$ and $\lambda_1 = 0^{\ell_1'(\kappa)}$. Here, $\ell_1(\cdot)$ and $\ell_1'(\cdot)$ are appropriate length functions specified later.
  5. Sample $v \xleftarrow{\$} \{0,1\}^{\kappa/4}$.
  6. For each $i \in [\tau + \kappa]$, generate $\mathsf{FSK}_i^1 \leftarrow \mathsf{FE.KeyGen}(\mathsf{MSK}_i^1, F_{i,\mathsf{PK}_{i+1}^1, \Pi_1}^1)$ and $\mathsf{FSK}_{\tau + \kappa + 1}^1 \leftarrow \mathsf{FE.KeyGen}(\mathsf{MSK}_{\tau + \kappa + 1}^1, G_{v, \Lambda_1, w}^1)$, where $F_{i,\mathsf{PK}_{i+1}^1, \Pi_1}^1$ and $G_{v, \Lambda_1, w}^1$ are circuits described in Figure 6.
  7. Let $c_\phi^1 = \mathsf{FE.Enc}_{\mathsf{PK}_1}(\phi, \mathsf{V}_\phi, \mathsf{W}_\phi, \widetilde{K}_\phi, 0^{p(\kappa)}, 0)$.
  8. The Public Key PK is given by $(\{\mathsf{FSK}_i^1\}_{i \in [\tau + \kappa + 1]}, c_\phi^1)$ and the secret key SK is given by $(S_1, \cdots, S_{\tau + \kappa})$.
- TDP_{PK}: The evaluation algorithm takes as input $(x, \sigma_1, \ldots, \sigma_{\tau + \kappa})$ and outputs $(x + 1, \sigma_1', \ldots, \sigma_{\tau + \kappa}')$ if the associated signatures $\sigma_1, \ldots, \sigma_{\tau + \kappa}$ are valid. It proceeds as follows:
  1. For $i \in [\tau + \kappa]$, compute $c_{x_{[i-1]}\|0}^1, c_{x_{[i-1]}\|1}^1 := \mathsf{FE.Dec}(\mathsf{FSK}_i^1, c_{x_{[i-1]}}^1)$.
  2. Obtain $d_x = ((\psi_1, \ldots, \psi_{\tau + \kappa}), (\beta_j, \ldots, \beta_{\tau + \kappa}))$ as output of $\mathsf{FE.Dec}(\mathsf{FSK}_{\tau + \kappa + 1}^1, c_x^1)$. Here, $\mathsf{j} = \mathsf{f(x)}$ where $\mathsf{f(x)}$ is the smallest k that $f(x)$ is the smallest k such that $x = x_{[k]}\|1^{\tau + \kappa - k}$.
  3. Output $\perp$ if $\mathsf{LeftInjPRG}_0(\sigma_i) \neq \psi_i$ for any $i \in [\tau + \kappa]$.
  4. For each $i \in [j - 1]$, set $\sigma_i' = \sigma_i$.
  5. For each $i \in \{j, \ldots, \tau + \kappa\}$, set $\gamma_i = \mathsf{LeftInjPRG}_1(\sigma_i)$ and σ_i' as $\mathsf{SK.Dec}_{\gamma_j, \ldots, \gamma_{\tau + \kappa}}(\beta_i)$, iteratively decrypting β_i encrypted under $\gamma_j, \ldots, \gamma_{\tau + \kappa}$.
  6. Output $(x + 1, \sigma_1', \cdots, \sigma_{\tau + \kappa}')$.

Fig. 5. Construction of TDP from $\mathcal{FE}$

- **Set** *queries: here the user submits in input circuit C of size at most ℓ, that uses r bits of randomness with output length t, as well as a sample s of length t. Sim will record (C, s), and set the output of the simulated sampler on C to be s. Sim supports up to n **Set** queries. We require that there is no overlap between circuits C in **Read** and **Set** queries, and that all **Set** queries are for distinct circuits.*
- **Finish** *query: here, the user submits nothing, and Sim closes its interfaces, terminates, and outputs a sampler* Sampler.

- TDP^{-1}_{SK} : The inversion algorithm on input $(x, \sigma_1, \cdots, \sigma_{\tau+\kappa})$ checks for all $i \in [\tau + \kappa]$ if $\sigma_i = S_{i,x_{[i]}}$ and if so it outputs $(x - 1, \sigma'_1, \cdots, \sigma'_{\tau+\kappa})$ where $x - 1$ is computed modulo $2^{\tau+\kappa}$ and for all $i \in [\tau + \kappa]$ $\sigma'_i = S_{i,(x-1)_{[i]}}$.
- $\mathsf{SampGen}(SK):$
  1. Choose $\overline{K} \leftarrow \mathsf{KeyGen}_{\mathcal{PCPRF}}(1^\kappa, 2v(\cdot) + 1, \mathsf{rand}(\cdot))$ and $K \leftarrow \mathsf{KeyGen}_{\mathcal{PCPRF}}(1^\kappa, v(\cdot), I(\cdot))$ where $v(\kappa) = \tau$. Initialize $\mathsf{U}^i_\phi := S_i$ and $\mathsf{U}_\phi = \bigcup_{i \in [\tau+\kappa]} \mathsf{U}^i_\phi$.
  2. For every $i \in [\tau + 1]$, choose $(\mathsf{PK}^2_i, \mathsf{MSK}^2_i) \leftarrow \mathsf{FE.Setup}(1^\kappa)$.
  3. Sample $sk_2 \leftarrow \mathsf{SK.KeyGen}(1^\kappa)$ where $|sk_2| = p(\kappa)$ and set $\Pi_2 \leftarrow \mathsf{SK.Enc}_{sk_2}(\pi_2)$ and $\Lambda_2 \leftarrow \mathsf{SK.Enc}_{sk_2}(\lambda_2)$ where $\pi_2 = 0^{\ell_2(\kappa)}$ and $\lambda_2 = 0^{\ell'_2(\kappa)}$. Here $\ell_2(\cdot)$ and $\ell'_2(\cdot)$ are appropriate length functions specified later.
  4. For each $i \in [\tau]$, generate $\mathsf{FSK}^2_i \leftarrow \mathsf{FE.KeyGen}(\mathsf{MSK}^2_i, F^2_{i,\mathsf{PK}^2_{i+1},\Pi_2})$ and $\mathsf{FSK}^2_{\tau+1} \leftarrow \mathsf{FE.KeyGen}(\mathsf{MSK}^2_{\tau+1}, G^2_{\Lambda_2})$ where $F^2_{i,\mathsf{PK}^2_{i+1},\Pi_2}, G^2_{\Lambda_2}$ are described in Figure 7.
  5. Let $c^2_\phi \leftarrow \mathsf{FE.Enc}_{\mathsf{PK}^2_1}(\phi, \mathsf{U}_\phi, K, \overline{K}, 0^{p(\kappa)}, 0)$.
  6. The sampler circuit has $\{\mathsf{FSK}^2_i\}_{i \in [\tau+1]}$ and c^2_ϕ hardwired in its description and works as described below.
- Samp: The sampler takes pk where $(pk, sk) \leftarrow \mathsf{PK.KeyGen}(1^\kappa)$. It proceeds as follows:
  1. For $i \in [\tau]$, compute $c^2_{pk_{[i-1]}\|0}, c^2_{pk_{[i-1]}\|1} := \mathsf{FE.Dec}(\mathsf{FSK}^2_i, c^2_{pk_{[i-1]}})$.
  2. Obtain $(pk, h_{pk}) = (pk, (pk, \rho, \rho_1, \cdots, \rho_{\tau+\kappa}))$ as output of $\mathsf{FE.Dec}(\mathsf{FSK}^2_{\tau+1}, c^2_{pk})$.
  3. Compute $K_{pk} := \mathsf{PK.Dec}_{sk}(\rho)$ and $\sigma_i := \mathsf{PK.Dec}_{sk}(\rho_i)$ for all $i \in [\tau + \kappa]$
  4. Output $(pk\|K_{pk}, \sigma_1, \cdots, \sigma_{\tau+\kappa})$.

Fig. 5. (*continued*)

Sim *must be capable of taking the queries above in any order.*

- **Correctness.** Sampler *is consistent with any queries made. That is, if a* **Read** *query was made on* C *and the response was* s, *then* Sampler$(C) = s$. *Similarly, if a* **Set** *query was made on* (C, s), *then* Sampler$(C) = s$.
- **Indistinguishability from honest generation.** *Roughly, this requirement says that in the absence of any* **Write** *queries, and honest and simulated sampler are indistinguishable. More precisely, the advantage of any polynomial-time algorithm A is negligible in the following experiment:*
  - *The challenger flips a random bit* b. *If* $b = 0$, *the challenger runs* Sampler $\leftarrow$ Setup$(1^\kappa, \ell, r, t)$. *If* $b = 1$, *the challenger initiates* Sim$(1^\kappa, \ell, r, t)$.
  - *A is allowed to make* **Read** *queries on arbitrary circuits* C *of size at most* ℓ, *using* r *bits of randomness and output length* t. *If* $b = 0$, *the challenger runs* $s \leftarrow$ Sampler(C) *and responds with* s. *If* $b = 1$, *the challenger forwards* C *to* Sim *as a* **Read** *query, and when* Sim *responds with* s, *the challenger forwards* s *to A.*

$$F^1_{i,\mathsf{PK}^1_{i+1},\Pi_1}$$

Hardcoded Values: $i, \mathsf{PK}^1_{i+1}, \Pi_1$.
Input: $(x \in \{0,1\}^{i-1}, \mathsf{V}_x, \mathsf{W}_x, \widetilde{K}_x, sk, \mathsf{mode})$

1. If ($\mathsf{mode} = 0$)
    (a) Output $\mathsf{FE.Enc}_{\mathsf{PK}^1_{i+1}}(x\|0, \mathsf{V}_{x\|0}, \mathsf{W}_{x\|0}, \widetilde{K}_{x\|0}, sk, \mathsf{mode}; \widetilde{K}'_{x\|0})$ and
    $\mathsf{FE.Enc}_{\mathsf{PK}^1_{i+1}}(x\|1, \mathsf{V}_{x\|1}, \mathsf{W}_{x\|1}, \widetilde{K}_{x\|1}, sk, \mathsf{mode}; \widetilde{K}'_{x\|1})$, where for $b \in \{0,1\}$,
    $\widetilde{K}_{x\|b} = \mathsf{PrefixCons}(\widetilde{K}_x, b\|0)$ and $\widetilde{K}'_{x\|b} = \mathsf{PrefixCons}(\widetilde{K}_x, b\|1)$ and
    $(\mathsf{V}_{x\|0}, \mathsf{W}_{x\|0})$, $(\mathsf{V}_{x\|1}, \mathsf{W}_{x\|1})$ are computed using the efficient procedure
    from the Computability Lemma [20]
2. Else, compute $\pi_1 \leftarrow \mathsf{SK.Dec}_{sk_1}(\Pi_1)$ and parse π_1 as a set of tuples of the form
    (z, c^1_z). Recover $(x\|0, c^1_{x\|0})$ and $(x\|1, c^1_{x\|1})$ from π_1. Output $c^1_{x\|0}$ and $c^1_{x\|1}$.

$$G^1_{v, \Lambda_1, w}$$

Hardcoded Values: v, Λ_1, w
Input: $x \in \{0,1\}^{\tau+\kappa}, \mathsf{V}_x, \mathsf{W}_x, \widetilde{K}_x, sk_1, \mathsf{mode}$

1. If ($\mathsf{InjPRG}(\mathsf{Ext}_w(x)) = v$) then output $\perp$.
2. If $\mathsf{mode} = 0$, (Below, $j = \mathsf{f}(x)$ where $\mathsf{f}(x)$ is the smallest k that $f(x)$ is the
    smallest j such that $x = x_{[j]}\|1^{\tau+\kappa-j}$.)
    (a) For each $i \in [\tau+\kappa]$, set $\psi_i = \mathsf{LeftInjPRG}_0(S_{i,x_{[i]}})$ (obtained from W^i_x for
    $i \leq j$ and from V^i_x for $i > j$).
    (b) For each $i \in \{j, \ldots, \tau+\kappa\}$ set $\gamma_i = \mathsf{LeftInjPRG}_1(S_{i,x_{[i]}})$ and $\beta_i =$
    $\mathsf{SK.Enc}_{\gamma_j, \cdots, \gamma_{\tau+\kappa}}(S_{i,x_{[i]}+1})$, encrypting $S_{i,x_{[i]}+1}$ under $\gamma_j, \ldots \gamma_{\tau+\kappa}$ using
    $\mathsf{PrefixCons}(\widetilde{K}_x, 0)$ as the random tape. In the above, $S_{i,x_{[i]}}$ and $S_{i,x_{[i]}+1}$
    are obtained from V^i_x for all $i \in [j, \tau+\kappa]$.
    (c) Output $((\psi_1, \ldots, \psi_{\tau+\kappa}), (\beta_j, \ldots, \beta_{\tau+\kappa}))$
3. Else, recover (x, d_x) from $\mathsf{SK.Dec}_{sk_1}(\Lambda_1)$ and output d_x.

Fig. 6. Circuits for simulating public key.

- *Finally, A sends a* **Finish** *query. If $b = 0$, the challenger then sends*
  Sampler *to A. If $b = 1$, the challenger sends a* **Finish** *query to* Sim, *gets*
  Sampler *from* Sim, *and forwards* Sampler *to A.*
- *A then tries to guess b. The advantage of A is the advantage A has in*
  guessing b.
- **Pseudorandomness of samples.** *Roughly, this requirement says that, in the*
  simulated sampler, if an additional **Set** *query is performed on (C, s) where s*
  is a fresh sample from C, then the simulated sampler is indistinguishable from

$$F^2_{i,\mathsf{PK}^2_{i+1},\Pi_2}$$

Hardcoded Values: $i, \mathsf{PK}^2_{i+1}, \Pi_2$.
Input: $(x \in \{0,1\}^{i-1}, \mathsf{U}_x, K_x, \overline{K}_x, sk_2, \mathsf{mode})$

1. If $(\mathsf{mode} = 0)$,
   (a) Output $\mathsf{FE.Enc}_{\mathsf{PK}^2_{i+1}}(x\|0, \mathsf{U}_{x\|0}, K_{x\|0}, \overline{K}_{x\|0}, sk, \mathsf{mode}; K'_{x\|0})$ and
   $\mathsf{FE.Enc}_{\mathsf{PK}^2_{i+1}}(x\|1, \mathsf{U}_{x\|1}, K_{x\|1}, \overline{K}_{x\|1}, sk, \mathsf{mode}; K'_{x\|1})$, where for $b \in \{0,1\}$,
   $\overline{K}_{x\|b} = \mathsf{PrefixCons}(\overline{K}_x, b\|0)$ and $K'_{x\|b} = \mathsf{PrefixCons}(\overline{K}_x, b\|1)$ and
   $\mathsf{U}_{x\|0}$ and $\mathsf{U}_{x\|1}$ are computed as described in Computability Lemma
   [20].
2. Else recover $(x\|0, c^2_{x\|0})$ and $(x\|1, c^2_{x\|1})$ from $\mathsf{SK.Dec}_{sk_2}(\Pi_2)$ and output $c^2_{x\|0}$
   and $c^2_{x\|1}$.

$$G^2_{\Lambda_2}$$

Hardcoded Values: Λ_2
Input: $pk \in \{0,1\}^{\kappa}, \mathsf{U}_{pk}, K_{pk}, \overline{K}_{pk}, sk_2, \mathsf{mode}$

1. If $\mathsf{mode} = 0$,
   (a) For all $i \in [\tau + \kappa]$, compute $\sigma_i := S_{i,(pk\|K_{pk})_{[i]}}$ from U_{pk}.
   (b) Compute $\rho \leftarrow \mathsf{PK.Enc}_{pk}(K_{pk})$ and $\rho_i \leftarrow \mathsf{PK.Enc}_{pk}(\sigma_i)$ for all $i \in [\tau + \kappa]$
   using $\mathsf{PrefixCons}(\overline{K}_{pk}, 0)$ as the random tape.
   (c) Output $(pk, (pk, \rho, \rho_1, \cdots, \rho_{\tau+\kappa}))$.
2. Else, recover (pk, h_{pk}) from $\mathsf{SK.Dec}_{sk_2}(\Lambda_2)$ and output h_{pk}.

Fig. 7. Circuits for simulating sampler

the case where the **Set** query was not performed. More precisely, the advantage
of any polynomial-time algorithm B is negligible in the following experiment:
- The challenger flips a random bit b. It then initiates $\mathsf{Sim}(1^{\kappa}, \ell, r, t)$.
- B first makes a **Challenge** query on circuit C^* of size at most ℓ, using
  r bits of randomness and output length t, as well as an integer i^*.
- B is allowed to make arbitrary **Read** and **Set** queries, as long as the
  number of **Set** queries is at most $n-1$, and the queries are all on distinct
  circuits that are different from C^*. The **Read** and **Set** queries can occur
  in any order; the only restriction is that the **Challenge** query comes
  before all **Read** and **Set** queries.
- After $i^* - 1$ **Read** and **Set** queries, the challenger does the following:
  * If $b = 0$, the challenger makes a **Read** query to Sim, and forwards
    the response s^* to B.

∗ If $b = 1$, the challenger computes a fresh random sample $s^* \leftarrow C^*(r)$, and makes a **Set** query to Sim on (C^*, s^*). Then it gives s^* to B.

Thus the i^*th query made to Sim is on circuit C^*, and the only difference between $b = 0$ and $b = 1$ is whether the output of the simulated sampler will be a pseudorandom sample or a fresh random sample from C^*.

- B is allowed to continue making arbitrary **Read** and **Set** queries, as long as the number of **Set** queries is at most $n - 1$ and the queries are all on distinct circuits that are different from C^*.
- Finally B makes a **Finish** query, at which point the challenger makes a **Finish** query to Sim. It obtained a simulated sampler Sampler, which it then gives to B.
- B then tries to guess b. The advantage of B is the advantage B has in guessing b.

5.2 Construction from FE

In this section, we will construct Universal Samplers that satisfies Definition 3 from polynomially hard, compact Functional Encryption and Prefix Constrained Pseudorandom Function (which is implied by Functional Encryption).

Theorem 3. *Assuming the existence of selective secure, single key, compact public key functional encryption there exists an Universal Sampler scheme satisfying Definition 3.*

Our Construction. The formal description our construction appears in Fig. 8.

Due to lack of space, we give the proof of security in the full version of the paper [20].

6 Multiparty Non-interactive Key Exchange

In this section, we build multiparty non-interactive key exchange for an unbounded number of users. Moreover, in constrast to the original multilinear map protocols [15], our protocol has no trusted setup.

6.1 Definition

A multiparty key exchange protocol consists of:

- Publish(κ) takes as input the security parameter and outputs a user secret sv and public value pv. pv is posted to the bulletin board.
- KeyGen($\{pv_j\}_{j \in S}, sv_i, i$) takes as input the public values of a set S of users, plus one of the user's secrets sv_i. It outputs a group key $k \in \mathcal{K}$.

Setup

- **Input:** 1^κ and three polynomials $\ell(\cdot), r(\cdot), t(\cdot)$.
- **Sampled Ingredients:**
  1. Sample $S \leftarrow \mathsf{KeyGen}_{\mathcal{PCPRF}}(1^\kappa, \ell(\cdot), r(\cdot))$ and $K \leftarrow \mathsf{KeyGen}_{\mathcal{PCPRF}}(1^\kappa, \mathsf{rand}(\cdot), I(\cdot))$ where $\mathsf{rand}(\kappa) = 2\ell(\kappa)$ and $I(\kappa) = \kappa$. For ease of notation, we denote $\ell = \ell(\kappa)$ and $r = r(\kappa)$.
  2. For every $i \in [\ell+1]$, sample $(\mathsf{PK}_i, \mathsf{MSK}_i) \leftarrow \mathsf{FE.Setup}(1^\kappa)$.
  3. For every $j \in [n]$, sample $sk_j \leftarrow \mathsf{SK.KeyGen}(1^\kappa)$. Let $|sk_j| = p(\kappa)$. For $i \in [\ell+1]$ and $j \in [n]$, let $\Pi_i^j \leftarrow \mathsf{SK.Enc}_{sk_j}(\pi_i^j)$ where $\pi_i^j = 0^{\mathsf{len}(\kappa)}$. Here $\mathsf{len}(\cdot)$ is an appropriate length function that would be specified later. For all $i \in [\ell+1]$, let $\Pi_i = \{\Pi_i^j\}_{j \in [n]}$.
- **Functional encryption ciphertext and keys to simulate obfuscation of Setup:**
  1. For each $i \in [\ell]$, generate $\mathsf{FSK}_i \leftarrow \mathsf{FE.KeyGen}(\mathsf{MSK}_i, F_{i, PK_{i+1}, \Pi_i})$ and $\mathsf{FSK}_{\ell+1} \leftarrow \mathsf{FE.KeyGen}(\mathsf{MSK}_{\ell+1}, G_{\Pi_{\ell+1}})$, where F_{i, PK_{i+1}, Π_i} and $G_{\Pi_{\ell+1}}$ are circuits described in Figure 9.
  2. For every $j \in [n]$, $Z_j = (j, \perp)$. Let $Z := \{Z_j\}_{j \in [n]}$.
  3. Let $c_\phi = \mathsf{FE.Enc}_{PK_1}(\phi, S, K, Z, 0)$.
  4. Output $(c_\phi, \{\mathsf{FSK}_i\}_{i \in [\ell+1]})$ as the sampler.

Evaluating the Sampler

- **Input:** Circuit C of size ℓ (padded with dummy symbols if its size is less than ℓ) using r bits of randomness and output length t and the sampler given by $(c_\phi, \{\mathsf{FSK}_i\}_{i \in [\ell+1]})$.
- **Evaluation:**
  1. For $i \in [\ell]$, compute $c_{C_{[i-1]}\|0}, c_{C_{[i-1]}\|1} := \mathsf{FE.Dec}(\mathsf{FSK}_i, c_{C_{[i-1]}})$.
  2. Compute d_C as output of $\mathsf{FE.Dec}(\mathsf{FSK}_{\ell+1}, c_C)$.
  3. Output d_C.

Fig. 8. Setup and evaluating the sampler

For correctness, we require that all users generate the same key:

$$\mathsf{KeyGen}(\{\mathsf{pv}_j\}_{j \in S}, \mathsf{sv}_i, i) = \mathsf{KeyGen}(\{\mathsf{pv}_j\}_{j \in S}, \mathsf{sv}_{i'}, i')$$

for all $(\mathsf{sv}_j, \mathsf{pv}_j) \leftarrow \mathsf{Publish}(\kappa)$ and $i, i' \in S$. For security, we have the following:

Definition 4. *A non-interactive multiparty key exchange protocol is statically secure if the following distributions are indistinguishable for any polynomial-sized set S:*

$$\{\mathsf{pv}_j\}_{j \in S}, k \text{ where } (\mathsf{sv}_j, \mathsf{pv}_j) \leftarrow \mathsf{Publish}(\kappa) \forall j \in S, k \leftarrow \mathsf{KeyGen}(\{\mathsf{pv}_j\}_{j \in S}, s_1, 1) \text{ and}$$

$$\{\mathsf{pv}_j\}_{j \in S}, k \text{ where } (\mathsf{sv}_j, \mathsf{pv}_j) \leftarrow \mathsf{Publish}(\kappa) \forall j \in G, k \leftarrow \mathcal{K}$$

$$F_{i,PK_{i+1},\Pi_i}$$

Hardcoded Values: i, PK_{i+1}, Π_i.
Input: $C \in \{0,1\}^{i-1}$, S_C, K_C, Z, mode

1. If (mode $= 0$),
   (a) Output $\mathsf{FE.Enc}_{PK_{i+1}}(C\|0, S_{C\|0}, K_{C\|0}, Z, \mathsf{mode}; K'_{C\|0})$ and
       $\mathsf{FE.Enc}_{PK_{i+1}}(C\|1, S_{C\|1}, K_{C\|1}, Z, \mathsf{mode}; K'_{C\|1})$, where for $b \in \{0,1\}$,
       $K_{C\|b} = \mathsf{PrefixCons}(K_C, b\|0)$ and $K'_{C\|b} = \mathsf{PrefixCons}(K_C, b\|1)$ and
       $S_{C\|b} := \mathsf{PrefixCons}(S_C, b)$.
2. Else,
   (a) Let j^* be the minimum value of $j \in [n]$ such that $Z_{j+1} = (j+1, \perp)$.
   (b) Let $\pi_i^{j^*} \leftarrow \mathsf{SK.Dec}_{sk_{j^*}}(\Pi_i^{j^*})$ where $\pi_i^{j^*}$ is a collection of elements of
       the form $(C', \cdot, \cdot)$ for $C' \in \{0,1\}^{i-1}$. Recover $(C, (C\|b, c_{C\|b}), (C\|(1-b), c_{C\|(1-b)}))$ (if there are more than one value of $(C, \cdot, \cdot)$, select the lexicographically first such value) from $\pi_i^{j^*}$ and output $(c_{C\|0}, c_{C\|1})$.

$$G_{\Pi_{\ell+1}}$$

Hardcoded Values: $\Pi_{\ell+1}$
Input: $C \in \{0,1\}^{\ell}, S_C, K_C, sk, \mathsf{mode}$

1. If mode $= 0$, output $C(S_C)$.
2. Else, let j^* be the minimum value of $j \in [n]$ such that $Z_{j+1} = (j+1, \perp)$.
   Recover (C, d_C) from $\mathsf{SK.Dec}_{sk_{j^*}}(\Pi_i^{j^*})$ and output d_C.

Fig. 9. Circuits for simulating public key.

Notice that our syntax does not allow a trusted setup, as the original constructions based on multilinear maps [11,13,15] require. Boneh and Zhandry [12] give the first multiparty key exchange protocol without trusted setup, based on obfuscation. A construction of obfuscation from a finite set of assumptions with polynomial security appears implausible due to an argument of [18]. Notice as well that our syntax does not allow the key generation to depend on the number of users who wish to share a group key. To date, prior key exchange protocols satisfying this property relied on strong knowledge variants of obfuscation [1]. Recently Khurana, Rao and Sahai in [25] constructed a key exchange protocol supporting unbounded number of users based on indistinguishability obfuscation and a tool called as *somewhere statistically binding hash functions* [24]. Here, we get an unbounded protocol based on *functiona encryption* only, and without using complexity leveraging.

6.2 Construction

Our construction will use the universal samplers built in Sect. 5, as well as any public key encryption scheme.

- Publish(κ). Run (sk, pk) $\leftarrow$ PK.KeyGen(κ). Also run the universal sampler setup algorithm Sampler $\leftarrow$ Setup(κ, ℓ, t) where output size ℓ and circuit size bound t will be decided later. Output pv $=$ (pk, Sampler) as the public value and keep sv $=$ sk as the secret value.
- KeyGen($\{(\mathsf{pk}_j, \mathsf{Sampler}_j)\}_{j \in S}, \mathsf{sk}_i, i$). Interpret S as the set $[1, n]$ for $n = |S|$, choosing some canonical ordering for the users in S (say, the lexicographic order of their public values). Define Sampler $=$ Sampler$_1$.

  Define $C_{\mathsf{pk},\mathsf{pk}'}$ for two public keys pk, pk$'$ to be the circuit that samples a random (sk$''$, pk$''$) $\leftarrow$ PK.KeyGen(κ), then encrypts sk$''$ under both pk and pk$'$, obtaining encryptions c and c' respectively, and then outputs (pk$''$, c, c'). Let $D_{\mathsf{pk},\mathsf{pk}'}$ be a similar circuit that samples a uniformly random string sk$''$ in the key space of $\mathcal{PKE}$, encrypts sk$''$ to get c, c' as before, and outputs $(0, c, c')$ where 0 is a string of zeros with the same length as a public key for $\mathcal{PKE}$. Let ℓ the length of (pk$''$, c, c') and let t be the size of $C_{\mathsf{pk},\mathsf{pk}'}$ (which we will assume is at least as large as $D_{\mathsf{pk},\mathsf{pk}'}$).

  Next, define $\mathsf{pk}'_2 = \mathsf{pk}_1$, and recursively define $(\mathsf{pk}'_{j+1}, c_j, c'_j) = $ Sampler($C_{\mathsf{pk}_j, \mathsf{pk}'_j}$) for $j = 2, \ldots, n-1$. Define sk'_{j+1} to be the secret key corresponding to pk'_{j+1}, which is also the secret key encrypted in c_j, c'_j. Finally, define $(0, c_n, c'_n) = $ Sampler($D_{\mathsf{pk}_n, \mathsf{pk}'_n}$), and define sk'_{n+1} to be the secret key encrypted in c_n, c'_n.

  First, it is straightforward that given $\{\mathsf{pk}_j\}_{j \in [n]}$ and Sampler, it is possible to compute $\mathsf{pk}'_j, c_j, c'_j$ for all $k \in [2, n]$. Thus anyone, including an eavesdropper, can compute these values.

  Next, we claim that if additionally given secret keys sk_j or sk'_j, it is possible to compute sk'_{j+1}. Indeed, sk'_{j+1} can be computed by decrypting c_j (using sk_j) or decrypting c'_j (using sk'_j). By iterating, it is possible to compute sk'_k for every $k > j$. This implies that all users in $[n]$ can compute sk_{n+1}.

Security. We now argue that any eavesdropper cannot learn any information about sk. Our theorem is the following:

Theorem 4. *If $\mathcal{PKE}$ is a secure public key encryption scheme and Setup is a m-time statically secure universal sampler with interactive simulation, the the construction above is a statically secure NIKE for up to 2^m users. In particular, by setting $m = \kappa$, the scheme is secure for an unbounded number of users.*

Due to lack of space, we give the proof of Theorem 4 in the full version of the paper [20].

Acknowledgments. Research supported in part from DARPA/ARL SAFEWARE Award W911NF15C0210, DARPA/ARO Award W911NF15C0226, AFOSR Award

FA9550-15-1-0274, NSF CRII Award 1464397, AFOSR YIP Award and research grants by the Okawa Foundation and Visa Inc. The views expressed are those of the author and do not reflect the official policy or position of the funding agencies.

References

1. Ananth, P., Boneh, D., Garg, S., Sahai, A., Zhandry, M.: Differing-inputs obfuscation and applications. Cryptology ePrint Archive, Report 2013/689 (2013). http://eprint.iacr.org/2013/689
2. Ananth, P., Brakerski, Z., Segev, G., Vaikuntanathan, V.: From selective to adaptive security in functional encryption. In: Gennaro, R., Robshaw, M. (eds.) CRYPTO 2015. LNCS, vol. 9216, pp. 657–677. Springer, Heidelberg (2015). doi:10.1007/978-3-662-48000-7_32
3. Ananth, P., Jain, A.: Indistinguishability obfuscation from compact functional encryption. In: Gennaro, R., Robshaw, M. (eds.) CRYPTO 2015. LNCS, vol. 9215, pp. 308–326. Springer, Heidelberg (2015). doi:10.1007/978-3-662-47989-6_15
4. Ananth, P., Sahai, A.: Functional encryption for turing machines. In: Kushilevitz, E., Malkin, T. (eds.) TCC 2016. LNCS, vol. 9562, pp. 125–153. Springer, Heidelberg (2016). doi:10.1007/978-3-662-49096-9_6
5. Barak, B., Goldreich, O., Impagliazzo, R., Rudich, S., Sahai, A., Vadhan, S.P., Yang, K.: On the (im)possibility of obfuscating programs. J. ACM 59(2), 6 (2012)
6. Bennett, C.H.: Time/space trade-offs for reversible computation. SIAM J. Comput. 18(4), 766–776 (1989)
7. Bitansky, N., Paneth, O., Rosen, A.: On the cryptographic hardness of finding a nash equilibrium. In: FOCS (2015)
8. Bitansky, N., Paneth, O., Wichs, D.: Perfect structure on the edge of chaos. In: Kushilevitz, E., Malkin, T. (eds.) TCC 2016. LNCS, vol. 9562, pp. 474–502. Springer, Heidelberg (2016). doi:10.1007/978-3-662-49096-9_20
9. Bitansky, N., Vaikuntanathan, V.: Indistinguishability obfuscation from functional encryption. In: FOCS (2015)
10. Boneh, D., Lewi, K., Raykova, M., Sahai, A., Zhandry, M., Zimmerman, J.: Semantically secure order-revealing encryption: multi-input functional encryption without obfuscation. In: Oswald, E., Fischlin, M. (eds.) EUROCRYPT 2015. LNCS, vol. 9057, pp. 563–594. Springer, Heidelberg (2015). doi:10.1007/978-3-662-46803-6_19
11. Boneh, D., Silverberg, A.: Applications of multilinear forms to cryptography. Cryptology ePrint Archive, Report 2002/080 (2002). http://eprint.iacr.org/2002/080
12. Boneh, D., Zhandry, M.: Multiparty key exchange, efficient traitor tracing, and more from indistinguishability obfuscation. In: Garay, J.A., Gennaro, R. (eds.) CRYPTO 2014. LNCS, vol. 8616, pp. 480–499. Springer, Heidelberg (2014). doi:10.1007/978-3-662-44371-2_27
13. Coron, J.-S., Lepoint, T., Tibouchi, M.: Practical multilinear maps over the integers. In: Canetti, R., Garay, J.A. (eds.) CRYPTO 2013. LNCS, vol. 8042, pp. 476–493. Springer, Heidelberg (2013). doi:10.1007/978-3-642-40041-4_26
14. Even, S., Goldreich, O., Lempel, A.: A randomized protocol for signing contracts. Commun. ACM 28(6), 637–647 (1985)
15. Garg, S., Gentry, C., Halevi, S.: Candidate multilinear maps from ideal lattices. In: Johansson, T., Nguyen, P.Q. (eds.) EUROCRYPT 2013. LNCS, vol. 7881, pp. 1–17. Springer, Heidelberg (2013). doi:10.1007/978-3-642-38348-9_1

16. Garg, S., Gentry, C., Halevi, S., Raykova, M., Sahai, A., Waters, B.: Candidate indistinguishability obfuscation and functional encryption for all circuits. In: 54th FOCS, Berkeley, CA, USA, 26–29 October 2013, pp. 40–49. IEEE Computer Society Press (2013)

17. Garg, S., Gentry, C., Halevi, S., Zhandry, M.: Fully secure functional encryption from multilinear maps. In: TCC (2016)

18. Garg, S., Gentry, C., Sahai, A., Waters, B.: Witness encryption and its applications. In: Boneh, D., Roughgarden, T., Feigenbaum, J. (eds.) 45th ACM STOC, pp. 467–476, Palo Alto, CA, USA, 1–4 June 2013. ACM Press (2013)

19. Garg, S., Pandey, O., Srinivasan, A.: Revisiting the cryptographic hardness of finding a nash equilibrium. In: Robshaw, M., Katz, J. (eds.) CRYPTO 2016. LNCS, vol. 9815, pp. 579–604. Springer, Heidelberg (2016). doi:10.1007/978-3-662-53008-5_20

20. Garg, S., Pandey, O., Srinivasan, A., Zhandry, M.: Breaking the sub-exponential barrier in obfustopia. IACR Cryptology ePrint Archive, 2016:102 (2016)

21. Goldreich, O., Goldwasser, S., Micali, S.: How to construct random functions. J. ACM **33**(4), 792–807 (1986)

22. Goldwasser, S., Kalai, Y.T.: Cryptographic assumptions: a position paper. Cryptology ePrint Archive, Report 2015/907 (2015). http://eprint.iacr.org/2015/907

23. Hofheinz, D., Jager, T., Khurana, D., Sahai, A., Waters, B., Zhandry, M.: How to generate and use universal samplers. In: Cheon, J.H., Takagi, T. (eds.) ASIACRYPT 2016. LNCS, vol. 10032, pp. 715–744. Springer, Heidelberg (2016). doi:10.1007/978-3-662-53890-6_24. http://eprint.iacr.org/2014/507

24. Hubacek, P., Wichs, D.: On the communication complexity of secure function evaluation with long output. In: Roughgarden, T. (ed.) ITCS 2015, pp. 163–172, Rehovot, Israel, 11–13 January 2015. ACM (2015)

25. Khurana, D., Rao, V., Sahai, A.: Multi-party key exchange for unbounded parties from indistinguishability obfuscation. In: Iwata, T., Cheon, J.H. (eds.) ASIACRYPT 2015. LNCS, vol. 9452, pp. 52–75. Springer, Heidelberg (2015). doi:10.1007/978-3-662-48797-6_3

26. Lin, H.: Indistinguishability obfuscation from DDH on 5-linear maps and locality-5 PRGs. IACR Cryptology ePrint Archive, 2016:1096 (2016)

27. Lin, H., Vaikuntanathan, V.: Indistinguishability obfuscation from DDH-like assumptions on constant-degree graded encodings. In: IEEE 57th Annual Symposium on Foundations of Computer Science, FOCS 2016, Hyatt Regency, New Brunswick, New Jersey, USA, 9–11 October 2016, pp. 11–20 (2016)

28. Naor, M.: On cryptographic assumptions and challenges. In: Boneh, D. (ed.) CRYPTO 2003. LNCS, vol. 2729, pp. 96–109. Springer, Heidelberg (2003). doi:10.1007/978-3-540-45146-4_6

29. Pass, R., Seth, K., Telang, S.: Indistinguishability obfuscation from semantically-secure multilinear encodings. In: Garay, J.A., Gennaro, R. (eds.) CRYPTO 2014. LNCS, vol. 8616, pp. 500–517. Springer, Heidelberg (2014). doi:10.1007/978-3-662-44371-2_28

30. Sahai, A., Waters, B.: How to use indistinguishability obfuscation: deniable encryption, and more. In: Symposium on Theory of Computing, STOC 2014, New York, NY, USA, 31 May–03 June 2014, pp. 475–484 (2014)

Symmetric Cryptanalysis II

New Impossible Differential Search Tool from Design and Cryptanalysis Aspects
Revealing Structural Properties of Several Ciphers

Yu Sasaki[✉] and Yosuke Todo

NTT Secure Platform Laboratories,
3-9-11 Midori-cho, Musashino-shi, Tokyo 180-8585, Japan
{sasaki.yu,todo.yosuke}@lab.ntt.co.jp

Abstract. In this paper, a new tool searching for impossible differentials is presented. Our tool can detect any contradiction between input and output differences. It can also take into account the property inside the S-box when its size is small e.g. 4 bits. This is natural for ciphers with bit-wise diffusion like PRESENT, while finding such impossible differentials for ciphers with word-wise diffusion is novel. In addition, several techniques are proposed to evaluate 8-bit S-box. The tool improves the number of rounds of impossible differentials from the previous best results for Midori128, LILLIPUT, and Minalpher. The tool also finds new impossible differentials for ARIA and MIBS. We manually verify the impossibility of the searched results, which reveals new structural properties of those designs. The tool can be implemented by slightly modifying the previous differential search tool using Mixed Integer Linear Programming (MILP). This motivates us to discuss the usage of our tool particular for the design process. With this tool, the maximum number of rounds of impossible differentials can be proven under reasonable assumptions and the tool is applied to various concrete designs.

Keywords: Symmetric-key · Impossible differential · Mixed integer linear programming · Midori · LILLIPUT · Minalpher · ARIA · MIBS

1 Introduction

Designing symmetric-key primitives becomes more and more complicated to simultaneously satisfy various goals such as security against many notions, efficiency in high-end software, low-implementation cost in hardware, and so on.

A popular design approach is *substitution-permutation network (SPN)*, in which a state is composed of small words, and is updated by iteratively applying a round function consisting of a non-linear layer and a linear layer. In the non-linear layer, the state is updated by looking up a word-wise precomputed table called S-box. In the linear layer, the state is mixed with some linear operations.

A lot of designs were proposed in the last decade. It is now necessary for the community to carefully but quickly evaluate their security. Automated evaluation tools are useful to evaluate various designs in short term. Regarding the

© International Association for Cryptologic Research 2017
J.-S. Coron and J.B. Nielsen (Eds.): EUROCRYPT 2017, Part III, LNCS 10212, pp. 185–215, 2017.
DOI: 10.1007/978-3-319-56617-7_7

differential cryptanalysis and linear cryptanalysis, automated tools have been well-developed. In particular, evaluating the lower bound of the number of active S-boxes with mixed-integer-linear programming (MILP) is becoming popular in the design of SPN primitives [1]. Meanwhile, automated tools for other cryptanalytic approaches are not as sophisticated as differential and linear cryptanalysis.

Impossible differential cryptanalysis [2,3] is one of the most major and effective cryptanalytic approaches. In short, for a target keyed cipher E_K, it exploits a pair of input and output differences (Δ_i, Δ_o) that cannot be connected for any K. Namely, two input values x, x' satisfying $x \oplus x' = \Delta_i$ never satisfy $E_K(x) \oplus E_K(x') = \Delta_o$.

Such (Δ_i, Δ_o) are detected by the *miss-in-the-middle* approach [4]. The first automated search attempt was done in [3] with a technique called *shrink*. It shrinks the word size to 3 bits and finds impossible differentials of the global structure of the cipher by exhaustively testing all possible differences and values. The shrink technique is useful when the cipher consists of small number of words with big word size, e.g. 4 words of 32 bits in Skipjack, while the recent design trend is using many words with small word size, e.g. 16 words of 8 bits in AES.

Kim et al. [5] presented the automated tool called $\mathcal{U}$-method. Suppose that one wants to examine if (Δ_i, Δ_o) is impossible. First it propagates Δ_i in forwards (with F) by r_f rounds, and checks if the difference of each word is known active, active, inactive, or unknown. Then, it propagates Δ_o in backwards (with F^{-1}) by r_b rounds and checks the same information. Finally, it finds contradiction in the middle, detecting that (Δ_i, Δ_o) is impossible for $r_f + r_b$ rounds.

Several researches extended the $\mathcal{U}$-method, e.g. UID-method by Luo et al. [6,7] or some extension by Wu and Wang [8]. Those detect more complicated contradiction than the $\mathcal{U}$-method. Although some advancement was made, usability of the previous tools is limited as explained below.

- To be as generic as possible, the recent tools consider complicated differential impact through the linear layer, which requires more sensitive implementation. Even with this effort, only particular contradictions can be analyzed.
- Most of the previous tools cannot take into account differential property inside the S-box. Several analysis against a particular S-box in a particular primitive may analyze its differential property [9,10], however such an analysis cannot be extended to a generic tool.
- Most of the previous tools for impossible differential cryptanalysis cannot be used to evaluate other cryptanalytic approaches, e.g. differential and linear cryptanalysis. Derbez and Fouque proposed a tool for the meet-in-the-middle attack that can also be used for impossible differential cryptanalysis [11]. However, it cannot find better impossible pairs compared to [5,6,8].

Our Contributions. In this paper, we propose a new automated tool to find impossible differentials. Our tool is based on the previous MILP-based tools for (standard) differential cryptanalysis, which models S-boxes in bitwise [12–14].

In the differential search with MILP, the attacker describes possible differential propagation patterns in a round function by using linear inequalities. Then, the attacker runs a solver for MILP, which returns the minimum number of active S-boxes under the given propagation patterns. In this research, to examine the impossibility of (Δ_i, Δ_o), we simply add constraints to fix the input and output differences to (Δ_i, Δ_o). Due to the added constraints, the lower bound of the number of active S-boxes usually increases. In some case, (Δ_i, Δ_o) cannot be satisfied, thus the MILP solver returns an error code implying that no solution exists. In other words, Δ_i and Δ_o are impossible pairs.

We then iterate this test to examine multiple pairs of (Δ_i, Δ_o) e.g. all pairs with 1 active word both in input and output. We note that, for all existing ciphers, the longest impossible differentials have only 1 active word in both input and output. Thus, it is reasonable to conjecture that if such impossible differentials do not exist, any impossible differentials do not.

Our tool leads to stronger cryptanalytic results than the previous tools owing to the following advantages.

Analyzing inside S-boxes: The previous differential-bound search using MILP [12] can model the possible differential propagation patterns in the differential distribution table (DDT) of the S-box. Our tool inherits this advantage. Thus impossible differentials taking into account DDT can be found.

Arbitrary Contradiction: The MILP solver automatically judges whether or not the solution exists. Thus, the attacker does not have to predict the mechanism of contradiction in advance, which significantly increases the versatility of the tool.

Multi-purpose Tool: We convert the previous MILP-based differential search into impossible differential search by just adding constraints to fix input and output differences. Thus only with a single tool, security against differential cryptanalysis and impossible differential cryptanalysis can be evaluated. This feature is especially useful for future primitive designers who need to evaluate both cryptanalyses.

Arbitrary S-box Mode: MILP requires too many inequalities to represent differential propagations in DDT of 8-bit S-boxes. Thus, the tool is infeasible for 8-bit S-boxes in a straightforward manner. Here, we introduce an *arbitrary S-box* where impossible differentials for the arbitrary S-box are always valid for arbitrary S-box choice. The arbitrary S-box can be described efficiently, which enables us to evaluate 8-bit S-boxes. We note that previous work on MILP based tool aimed to model DDT precisely. One can see the catchphrase "MILP whose feasible region is exactly the set of all valid differential" in [13,15], while modeling 8-bit S-box precisely is infeasible. Our approach is opposite of previous work, which describes DDT only roughly but can be executed in practice.

Quick Search for Truncated Impossible Differential: A single pair of input and output differences can be impossible for more rounds than truncated differentials. Meanwhile, evaluating all the pairs is infeasible and the search range is often limited to single-active word. Here we present a technique to make the tool more efficient only by aiming truncated impossible differentials, which can be implemented only by changing the constraints of input and output differences.

Table 1. Application results. 'KR' denotes 'key recovery.'

Target	Ref	#Rounds		Search mode	Goal	Remarks
		Prev.	Ours			
Midori128	[16]	6	7	Specific S-box	Characteristic	
LILLIPUT	[17]	8	9	Specific S-box	Characteristic	
Minalpher	[18]	6.5	7.5	Arbitrary S-box	Truncated	Large state
ARIA	[19]	4	4	Arbitrary S-box	Truncated	8-bit S-box, improve KR
MIBS	[20]	8	8	Specific S-box	Characteristic	New impossible differentials

Note that running time of our tool for a single pair of input and output differences is significantly shorter than the differential search. This can be explained that the solver can stop only by detecting one characteristic. In the previous differential bound search, the bottleneck of the tool is increasing the lower bound. Finding some upper bound (some solution of the system) is usually fast.

We apply the proposed tool to various designs. The results improving the existing impossible differentials are summarized in Table 1. Although one of the advantages of the tool is that the attacker can detect impossible differentials without analyzing contradicting reasons, we manually analyze why the detected (Δ_i, Δ_o) is impossible. The manual verification not only demonstrates the correctness of the tool, but also reveals the structural properties of the target designs that have not been known before. We believe that the contradicting reasons analyzed in this paper for Midori128, LILLIPUT, and Minalpher lead to new understanding about their designs.

Our automated tool is useful to test many design choices during the design process of new primitives. Thus, we also discuss the usage of the tool for the design. For example, when the tool finds several impossible pairs of (Δ_i, Δ_o), the designers may want to patch the design to avoid such (Δ_i, Δ_o). By using the arbitrary S-box mode, we can easily check whether (Δ_i, Δ_o) is dependent on the S-box. If it is dependent on the S-box, it may be prevented by replacing the S-box. If it is independent, it needs to modify the linear layer to prevent it.

Moreover, because it catches any contradiction, the tool provides a certain level of provable security about the existence of impossible differentials with reasonable assumptions and reasonable search range. In details, provable security can be discussed when a single word is active in the input and output differences, and we can set two-level of the assumption; (1) S-box is public and each subkey is chosen independently and uniformly at random and (2) keyed S-box is used and for each key the S-box is chosen uniformly at random. We apply the tool to various designs to prove the maximum number of rounds of impossible differentials. Finally, we propose an *optimal pick technique* which dramatically reduces the execution time only when the tool is used for obtaining the proof.

Paper Outline. Notations and related work are introduced in Sect. 2. Framework of our tool is introduced in Sect. 3. Application on various designs improving previous impossible differentials are shown in Sect. 4. A technique to reduce the search complexity is explained in Sect. 5. Advantages of our tool in the design process are explained in Sect. 6. Our research is partially overlapped with [21]. The relationship between [21] and this paper is explained in Appendix A.

2   Related Work

2.1   Terminologies in Impossible Differential Cryptanalysis

- We call a pair of input and output differences (Δ_i, Δ_o) that cannot be connected an *impossible differential characteristic* or *impossible characteristic*.
- We call a pair of a closed set of input differences and a closed set of output differences in which any pair cannot be connected as a *truncated impossible differential*.
- When we do not distinguish the above two, we call it *impossible differential*.

2.2   Differential Search with Mixed Integer Linear Programming

Here we explain an automated tool for differential cryptanalysis, not impossible differential cryptanalysis, which will be a base of our tool.

Mouha et al. [1] showed that the problem to search for the minimum number of active S-boxes can be modeled with mixed integer linear programming (MILP). The approach is now very popular for designing a new primitive. For example, resistance against differential and linear cryptanalysis of Skinny [22] recently proposed at CRYPTO 2016 was evaluated by MILP.

The approach by Mouha et al. [1] is effective for evaluating word-oriented ciphers, while several ciphers are not word-oriented. For example, PRESENT [23] applies 4-bit S-box, then the bit-permutation moves four bits from a single S-box to four different S-boxes. In order to apply MILP to such a structure, Sun et al. [12] developed a method to model all possible differential propagations bit by bit even for the S-box.

Modeling Differential Propagations with MILP. We explain how to model valid differential propagations of PRESENT in bitwise. Note that one round of PRESENT consists of subkey addition, S-box applications, and bit-permutation.

At first, binary variables to represent whether the bits are active or inactive are defined for all rounds; $x_0, x_1, \ldots, x_{63}$ are for 64 bits in the plaintext, $x_{64}, x_{65}, \ldots, x_{127}$ are for 64 bits after round 1, $x_{128}, x_{129}, \ldots, x_{191}$ are for 64 bits after round 2, and so on. Each variable takes '1' if the bit has the difference, and takes '0' otherwise. Then, the constraint to ensure at least 1 active bit is added, which can be written as '$x_0 + x_1 + \cdots + x_{63} \geq 1$.' Finally, constraints to be valid differential propagations are added. Here, the bit-permutation only changes the order of variables and subkey addition can be ignored because it does not change the difference. The following denotes the variables involved in the first round, in which a 64-bit plaintext difference $x_0, \ldots, x_{63}$ are updated to $x_{64}, \ldots, x_{127}$.

$$\begin{bmatrix} x_0, x_1, x_2, x_3 \\ x_4, x_5, x_6, x_7 \\ x_8, x_9, x_{10}, x_{11} \\ x_{12}, x_{13}, x_{14}, x_{15} \\ \cdots \\ x_{60}, x_{61}, x_{62}, x_{63} \end{bmatrix} \xrightarrow{\text{S-box}} \begin{bmatrix} x_{64}, x_{68}, x_{72}, x_{76} \\ x_{80}, x_{84}, x_{88}, x_{92} \\ x_{96}, x_{100}, x_{104}, x_{108} \\ x_{112}, x_{116}, x_{120}, x_{124} \\ \cdots \\ x_{115}, x_{119}, x_{123}, x_{127} \end{bmatrix} \xrightarrow{\text{BitPerm}} \begin{bmatrix} x_{64}, x_{65}, x_{66}, x_{67} \\ x_{68}, x_{69}, x_{70}, x_{71} \\ x_{72}, x_{73}, x_{74}, x_{75} \\ x_{76}, x_{77}, x_{78}, x_{79} \\ \cdots \\ x_{124}, x_{125}, x_{126}, x_{127} \end{bmatrix}$$

The most difficult part is describing all possible propagation patterns for 16 S-boxes, e.g. $x_0, x_1, x_2, x_3 \longrightarrow x_{64}, x_{68}, x_{72}, x_{76}$, with a system of linear inequalities. Sun et al. [12] showed two approaches to solve the problem.

Fact 1. *Linear inequalities to constrain input and output variables of the S-box only to valid patterns can be generated by using either the computation tool called SageMath or several logical operations.*

How to use SageMath is well explained in [12] and more details of logical computations can be seen in [14]. We rely on Fact 1 about the description of S-box, and the choice of SageMath and logical operations does not impact to our tool. Meanwhile, the following limitation of those approaches should be noted.

Fact 2. *Both of SageMath and the logical operations can be used only when the S-box size is small.*

In our computational environment, both methods are feasible for S-boxes of size five bits or less. No method is known to model bigger S-box, e.g. 8-bit S-box.

MILP returns a solution of the system optimizing a given objective function. In differential cryptanalysis, the attacker's goal is minimizing the number of active S-boxes, which can be defined as "Minimize $\sum_i (x_{4i} \vee x_{4i+1} \vee x_{4i+2} \vee x_{4i+3})$."

The system can be solved by the MILP solver to find the optimal solution. We use Gurobi Optimizer [24] as the MILP solver.

3   Composite Framework for Differential and Impossible Differential Searches

We begin with explaining the basic concept of our impossible differential search tool, which has been independently discovered by Cui et al. and their paper was posted on Cryptology ePrint Archive prior to our paper [21]. Comparison between [21] and this work will be explained in Appendix A.

The tool adds several constraints to the previous differential bound search for fixing an input and output difference to a specific pair (Δ_i, Δ_o). Due to those additional constraints, the MILP solver may not be able to find the solution, thus returns some error code indicating that *the system is infeasible*, which tells that (Δ_i, Δ_o) is an impossible differential characteristic.

Algorithm 1. Generating System of Inequalities in Previous Differential Search

---

Require: number of rounds r, system of inequalities for S-boxes and linear layer
Ensure: system of inequalities
  1: <u>Write an objective function.</u>$_A$
  2: <u>Write constraints ensuring at least 1 active bit in input.</u>$_B$
  3: **for** round $= 1$ to r **do**
  4:     <u>Write constraints for the S-boxes.</u>$_C$
  5:     Write constraints for the linear layer.
  6: **end for**

Example 1. *Let $p_0, p_1, \ldots, p_{b-1}$ and $c_0, c_1, \ldots, c_{b-1}$ be variables that represent active/inactive of plaintext bits and ciphertext bits, respectively, where b is the block size. To test if $(\Delta_i, \Delta_o) = (\texttt{0x1}, \texttt{0x1})$ is impossible, the MILP solver should run with the following constraints added.*

$$p_0 = 1, p_1 = 0, c_2 = 0, \ldots, p_{b-1} = 0,$$
$$c_0 = 1, c_1 = 0, c_2 = 0, \ldots, c_{b-1} = 0.$$

We then iterate this test to examine multiple pairs of (Δ_i, Δ_o) e.g. all pairs with 1 active word both in input and output.

3.1 Composite Framework

A remarkable advantage of our tool is that users can switch differential-bound search and impossible-differential search very easily. This helps primitive designers, generally required to evaluate the resistance against both of differential and impossible differential cryptanalyses. Here we introduce our framework to generate system of inequalities depending on the target to evaluate.

Most of the symmetric-key primitives can be described as an iteration of the round function consisting of the non-linear and linear layers. We explain our tool by following this structure. Our tool focuses on the primitive whose non-linear layer is the parallel application of S-boxes. The tool relies on the previous MILP-based differential search that models differential propagations through S-box in bitwise [12–14]. Here, we recall how a system of inequalities is generated.

First, the number of rounds, r, is fixed. Then, an objective function, e.g. minimizing the number of active S-boxes, is defined. It also constrains the system so that at least one S-box is activated. The remaining is writing constraints for the valid differential propagations through the S-boxes and linear layer for r rounds, which can be done with [12–14]. The procedure is summarized in Algorithm 1. Underlines in Algorithm 1 will be later referred by Algorithm 2.

We slightly modify Algorithm 1 so that impossible differentials can be evaluated with several techniques. The goal of the tool can be either the differential bound (DB) or the impossibility of the given input and output differences (ID), which can be specified in the parameter "GOAL". For converting DB to ID, the users need to modify only two parts; make the objective function empty and specify input and output differences.

Algorithm 2. Generating System of Inequalities in Composite Framework

---

Require: number of rounds r, system of inequalities for S-boxes and linear layer, GOAL $\in$ {DB, ID}, MODE $\in$ {SPECIFIC, ARBITRARY}, and OBJECT $\in$ {TRUNCATED,CHARACTERISTIC}

Ensure: system of inequalities

```
    /* Lines 1–5 correspond to __A in Alg. 1. */
 1: if GOAL = DB then
 2:     Write an objective function.
 3: else if GOAL = ID then
 4:     Leave an objective function empty.
 5: end if

    /* Lines 6–14 correspond to __B in Alg. 1. */
 6: if GOAL = DB then
 7:     Write constraints ensuring at least 1 active bit in input.
 8: else if GOAL = ID then
 9:     if OBJECT = CHARACTERISTIC then
10:         Fully specify active or inactive for each input and output bit.
11:     else if OBJECT = TRUNCATED then
12:         Specify input and output difference in a truncated level.
13:     end if
14: end if

15: for round = 1 to r do

    /* Lines 16–20 correspond to __C in Alg. 1. */
16:     if TARGET = ID and MODE = ARBITRARY then
17:         Write constraints for the differentially ideal S-box.
18:     else
19:         Write constraints for the S-boxes as in specification.
20:     end if

21:     Write constraints for the linear layer.
22: end for
```

For impossible differentials, the users can further choose several search modes specified in the parameter "MODE". To be more precise, the S-boxes can be fixed to particular ones (SPECIFIC) or can be treated as general ones (ARBITRARY).

The users can also choose which of truncated differential (TRUNCATED) or a single impossible differential characteristic (CHARACTERISTIC) is searched as a parameter "OBJECT".

The updated framework to generate the system of inequalities for each setting is given in Algorithm 2. Note that the basic idea in [21] corresponds to "GOAL = ID", "MODE = SPECIFIC", and "OBJECT = CHARACTERISTIC." In the following sections, we will discuss the purpose of each search mode.

Hereafter, we explain details of impossible differential search ("GOAL = ID"). We first explain how to search impossible differential characteristics ("OBJECT = CHARACTERISTIC") with the specific S-box mode and the arbitrary S-box mode in Sects. 3.2 and 3.3, respectively. We then explain the case of truncated impossible differential ("OBJECT = TRUNCATED") in Sect. 3.4.

3.2 Specific S-Box Mode for Impossible Characteristic

In the specific S-box mode, the users derive the differential distribution table (DDT) from the actual S-boxes, and construct the MILP model to describe all valid differential propagations by using the existing method [12–14]. Then differences in all input and output bits are constrained to the target pair. The analysis is iterated for various input and output differences chosen from a reasonable subset, i.e. only one word is active.

The specific S-box mode can maximize the number of rounds of impossible differentials. Thus the attackers may prefer to choose this mode.

Impact of Key Schedule. The tool does not take into account the key schedule, thus we need a careful discussion about the impact of its omission.

The search by MILP describes a system of inequalities for the entire rounds by iterating a system of one-round differential propagation. Thus all valid propagations for one round are also valid in the evaluation of multiple rounds independently of the propagation in neighboring rounds and subkey values. This is true only if all subkeys are independent and chosen uniformly at random, which is not true in practical designs with a particular key schedule.

In summary, what the MILP simulates is the worst-case scenario (for the attackers). Namely, even if some differential propagations cannot occur for multiple rounds, the tool regards it possible, which leads to the following observation.

Observation 1. *Impossible differential characteristics found in the specific S-box mode are always impossible independently of the choice of key schedule.*

3.3 Arbitrary S-Box Mode for Impossible Characteristic

In the arbitrary S-box mode, we assume an imaginary S-box in which any non-zero input difference can be propagated to any non-zero output difference. Then, a set of valid differential propagations of any bijective S-box can be a subset of the one in the arbitrary S-box.

Valid differential propagations of the n-bit arbitrary S-box can be described only by $2n$ inequalities. Let $i_0, i_1, \ldots, i_{n-1}$ and $o_0, o_1, \ldots, o_{n-1}$ be binary variables to represent whether input and output bits are active or inactive, respectively. We write the constraints such that if input (resp. output) is 0, each output bit (resp. input bit) is 0, namely

$$i_0 + i_1 + \cdots + i_{n-1} - o_0 \geq 0, \qquad o_0 + o_1 + \cdots + o_{n-1} - i_0 \geq 0,$$
$$i_0 + i_1 + \cdots + i_{n-1} - o_1 \geq 0, \qquad o_0 + o_1 + \cdots + o_{n-1} - i_1 \geq 0,$$
$$\cdots \qquad\qquad\qquad \cdots$$
$$i_0 + i_1 + \cdots + i_{n-1} - o_{n-1} \geq 0, \qquad o_0 + o_1 + \cdots + o_{n-1} - i_{n-1} \geq 0.$$

The advantage of the arbitrary S-box compared to the specific S-box is efficiency owing to a small number of constrains to describe differential propagations. The arbitrary S-box mode is useful in the following two cases.

8-bit S-boxes: There is no known method to describe differential propagations of 8-bit S-boxes in MILP. Here by using the arbitrary S-box, the tool can be applied to 8-bit S-boxes.

Large Block Size: Even if the S-box size is small, say 4 bits, it is computationally hard to evaluate a large block size, say 256 bits. Again the arbitrary S-box enables analysis.

Note that, differently from the specific S-box mode, the analysis can no longer exploit properties inside the S-box. However, the analysis can still exploit another advantage that the tool catches any contradiction, and this advantage is often big enough to find new impossible differential characteristics. Actually, we found new characteristics of ARIA (8-bit S-boxes) [19] and of Minalpher (4-bit S-box, 256-bit block) [18], which will be explained in Sect. 4.

Similarly to Sect. 3.2, MILP simulates the worst-case scenario. Namely, even if some differential propagations cannot occur for some specific S-box, the tool regards it possible.

Observation 2. *Impossible differential characteristics found in the arbitrary S-box mode are always impossible independently of the choice of S-box and key schedule.*

3.4   Searching for Truncated Impossible Differential

The tool for a single characteristic can be extended to truncated differentials by simply running the tool for multiple pairs of input and output differences. However, this approach easily becomes computational infeasible when the number of active words is more than 1. Actually, searching for two active words is already too heavy. Let n and c be the number of S-boxes per round and the size of each S-box, respectively. Then, the number of pairs of input and output differences with 1-active word is $\left(n \cdot (2^c - 1)\right)^2$, which is $O(n^2 \cdot 2^{2c})$, while one with two active words is $\left(\binom{n}{2} \cdot (2^{2c} - 1)\right)^2$, which is $O(n^4 \cdot 2^{4c})$. Generally for d input active words and d' output active words, the number of pairs to test is given by

$$O(n^{d+d'} \cdot 2^{(d+d')c}). \tag{1}$$

With $n = 16$ and $c = 4$, which is a popular choice for lightweight ciphers, we need to evaluate 2^{16} pairs for single-active word ($d = d' = 1$) while 2^{32} pairs for 2-active words ($d = d' = 2$).

Here, we show a technique to make the tool more efficient only by aiming truncated impossible differentials in both of the specific S-box and the arbitrary S-box modes. Let $i_0, i_1, \ldots, i_{n-1}$ and $o_0, o_1, \ldots, o_{n-1}$ be variables to represent whether n input and n output bits in the truncated position are active or inactive.

Then, we write the following constraint (along with constraints fixing the other bits to 0):

$$i_0 + i_1 + \cdots + i_{n-1} \geq 1, \qquad o_0 + o_1 + \cdots + o_{n-1} \geq 1.$$

Note that if there exists at least one solution satisfying the constraints, the tool returns that the system is feasible. Hence, the truncated impossible differential search is less accurate than the impossible characteristic search, while execution time is significantly reduced. Compared to Eq. (1), 1 inequality is enough for each active word position. Thus the number of pairs to test is given by

$$O(n^{d+d'}), \tag{2}$$

which enables to evaluate multiple active words differences. Actually, we searched for truncated impossible differentials on ARIA [19] with this technique. Then, we found new truncated impossible differentials with $d = 2$ active input words and $d' = 5$ output active words, which will be explained in Sect. 4.3.

4 Applications from Cryptanalysis Aspect

4.1 Midori128

Midori is a low energy block cipher designed by Banik et al. in 2015 [16]. Midori provides two different block lengths; Midori64 and Midori128 have 64-bit and 128-bit block lengths, respectively. Both ciphers accept 128-bit secret key.

Specification. Midori128 uses the SPN structure with AES-like state. The state is arranged in a 4×4 matrix as

$$S = \begin{pmatrix} s_0 & s_4 & s_8 & s_{12} \\ s_1 & s_5 & s_9 & s_{13} \\ s_2 & s_6 & s_{10} & s_{14} \\ s_3 & s_7 & s_{11} & s_{15} \end{pmatrix}. \tag{3}$$

The bit length of every cell s_i is 8 bits.

The round function consists of SubCell, ShuffleCell, MixColumn, and KeyAdd. In SubCell, 8-bit S-boxes SSb$_0$, SSb$_1$, SSb$_2$, and SSb$_3$ are used and $s_i \leftarrow \text{SSb}_{i \bmod 4}(s_i)$ where $0 \leq i \leq 15$. Four 8-bit S-boxes SSb$_i$ are constructed by 4-bit S-box Sb$_1$, where Sb$_1$ is defined as follows.

x	0 1 2 3 4 5 6 7 8 9 A B C D E F
Sb$_1(x)$	1 0 5 3 E 2 F 7 D A 9 B C 8 4 6

Then, SSb$_i$ are constructed as SSb$_i = p_i^{-1} \circ (\text{Sb}_1 \| \text{Sb}_1) \circ p_i$, where two Sb$_1$ are applied to top and bottom halves in (Sb$_1 \| \text{Sb}_1$). Note that SSb$_i$ is involution, and we later show that impossible differentials are improved by exploiting this property. Figure 1 shows the specification

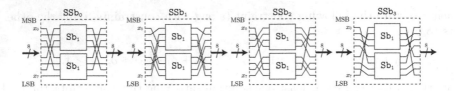

Fig. 1. SSb_0, SSb_1, SSb_2, and SSb_3

of SSb_i. In ShuffleCell, each cell is permuted as $(s_0, s_1, \ldots, s_{15})$ ←
$(s_0, s_{10}, s_5, s_{15}, s_{14}, s_4, s_{11}, s_1, s_9, s_3, s_{12}, s_6, s_7, s_{13}, s_2, s_8)$. In MixColumns, the
following multiplication

$$\begin{pmatrix} s_i \\ s_{i+1} \\ s_{i+2} \\ s_{i+3} \end{pmatrix} = \begin{pmatrix} 0 & 1 & 1 & 1 \\ 1 & 0 & 1 & 1 \\ 1 & 1 & 0 & 1 \\ 1 & 1 & 1 & 0 \end{pmatrix} \begin{pmatrix} s_i \\ s_{i+1} \\ s_{i+2} \\ s_{i+3} \end{pmatrix}$$

is applied for $i = 0, 4, 8, 12$. In KeyAdd, the i-th n-bit round key is XORed with
a state. The number of rounds of Midori128 is 20. Moreover, only SubCell is
applied in the final round function.

Previous Cryptanalysis. Several third-party cryptanalyses have been proposed, and the full-round Midori64 was broken by the invariant subspace
attack [25] and nonlinear invariant attack [26] under the weak-key setting. On
the other hand, there are no cryptanalysis against full-round Midori128. Regarding the impossible differential attack on Midori128, the designers found 6-round
impossible differentials such that only one cell is active in the input and output [16]. Then, Zhen et al. found 6-round impossible differentials that are advantageous for the key recovery but the number of rounds is not increased [27].

Configurations for the Tool. The block size of Midori128 is 128 bits and
the S-boxes size is 8 bits. However, since the 8-bit S-boxes are represented as
concatenation of two 4-bit S-boxes, we can regard that there are thirty-two 4-bit
S-boxes in each round. The search space for impossible differential characteristics
is large, hence we run our tool in the arbitrary S-box mode.

When the arbitrary S-box mode is chosen for Midori, it is sufficient to evaluate
truncated impossible differentials rather than impossible differential characteristics. This is because, for any choice of the differential value of the active nibble
in the plaintext, the set of possible output differences of the active S-box in the
first round is identical. In other words, when (Δ_i, Δ_o) is an impossible differential characteristic, for any other 1-nibble difference Δ_i' in the same active nibble
position, (Δ_i', Δ_o) becomes impossible.

We limit the input and output differences to 1 active nibble. The number of
such input differences is 32, and we have the same number of output differences.
In the end, we run MILP for $32 * 32 = 1024$ pairs of input and output differences.

Table 2. 7-round truncated impossible differentials against Midori128

ID	ΔP	ΔC	Remarks
001T	$(0\alpha_1 00, 0000, 0000, 0000)$	$(0\beta_1 00, 0000, 0000, 0000)$	Manually verified
002T	$(0\beta_1 00, 0000, 0000, 0000)$	$(0\alpha_1 00, 0000, 0000, 0000)$	Manually verified
003T	$(0000, \alpha_0 000, 0000, 0000)$	$(0000, \beta_0 000, 0000, 0000)$	
004T	$(0000, \beta_0 000, 0000, 0000)$	$(0000, \alpha_0 000, 0000, 0000)$	
005T	$(0000, 0\alpha_1 00, 0000, 0000)$	$(0000, 0\beta_1 00, 0000, 0000)$	
006T	$(0000, 0\beta_1 00, 0000, 0000)$	$(0000, 0\alpha_1 00, 0000, 0000)$	
007T	$(0000, 0000, \alpha_0 000, 0000)$	$(0000, 0000, \beta_0 000, 0000)$	
008T	$(0000, 0000, \beta_0 000, 0000)$	$(0000, 0000, \alpha_0 000, 0000)$	
009T	$(0000, 0000, 0\alpha_1 00, 0000)$	$(0000, 0000, 0\beta_1 00, 0000)$	
010T	$(0000, 0000, 0\beta_1 00, 0000)$	$(0000, 0000, 0\alpha_1 00, 0000)$	
011T	$(0000, 0000, 0000, \alpha_0 000)$	$(0000, 0000, 0000, \beta_0 000)$	
012T	$(0000, 0000, 0000, \beta_0 000)$	$(0000, 0000, 0000, \alpha_0 000)$	

List of 7-Round Truncated Impossible Differentials. We ran our tool with the above configuration. The tool required about 0.03 seconds per pair and it took about 0.5 min to test 1024 pairs.

As a result, our tool found 12 truncated impossible differentials for 7 rounds, which improves the previous best result by 1 round. We list 12 truncated impossible differentials in Table 2. Note that α_i is active in 4 bits where the active bits go to top four bits after p_i is applied, while β_i is active in 4 bits where the active bits go to bottom four bits after p_i is applied. Every truncated impossible differential consists of $15^2 = 225$ impossible differential characteristics.

Manual Verification of ID001T and ID002T. Although one of the major advantages of the tool is that the attacker does not have to analyze the reason of contradiction, we would like to verify the reason. The analysis reveals a new structural property of Midori128 exploiting the involution of SSb_i, which seems to be useful for future analysis. We first prove ID001T.

Theorem 1. *The input difference* $(0\alpha_1 00, 0000, 0000, 0000)$ *cannot propagate to the output difference* $(0\beta_1 00, 0000, 0000, 0000)$ *after 7 rounds of Midori128, where only top four bits of* $p_1(\alpha_1)$ *and bottom four bits of* $p_1(\beta_1)$ *are active.*

Proof. In Fig. 2, the input difference is propagated in forwards by 3.5 rounds, and the output difference is propagated in backwards by 3 rounds.

Let us focus on the forward propagation. From the definition, the differential form of α_1 is $(*, *, 0, 0, 0, 0, *, *)$ thus $p_1(\alpha_1) = (*, *, *, *, 0, 0, 0, 0)$, where $*$ and 0 are active and inactive, respectively. In $\mathsf{SubCell}$ in the first round, $\mathsf{SSb}_1(\alpha_1) = p_1^{-1} \circ (\mathsf{Sb}_1 \| \mathsf{Sb}_1) \circ p_1(\alpha_1)$ is computed. $(\mathsf{Sb}_1 \| \mathsf{Sb}_1)$ preserves that only top 4 bits are active, and active bit positions go back to α_i after the application of p_1^{-1}. The position of the active byte moves from s_1 to s_7 by $\mathsf{ShuffleCell}$, then is

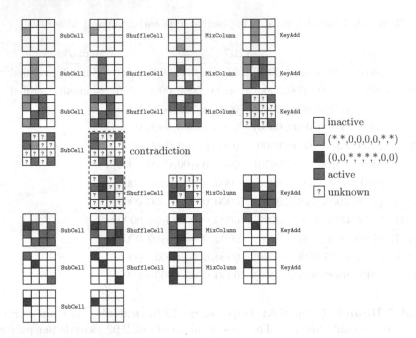

Fig. 2. 7-round truncated impossible differential of Midori128; ID001T

diffused to s_4, s_5, and s_6 by MixColumns. S-boxes are applied in the second round again, but SSb_0 and SSb_2 do not preserve the form of α_1 due to the different bit permutations p_0 and p_2. Therefore, only s_5 preserves the differential form of α_1. Similar analysis is continued during the 3.5-round forward propagation.

The differential form of β_1 is $(0, 0, *, *, *, *, 0, 0)$. With the same reason as α_1, the differential form of β_1 is preserved after the computation of $SSb_1^{-1}(\beta_1)$, and 1 byte preserves the difference β_1 after 3 round decryption.

On one hand, from the forward 3.5-round propagation, only top half of $p_1(s_5)$ is active and bottom half is inactive. On the other hand, from the 3-round backward propagation, only bottom half of $p_1(s_5)$ is active and top half is inactive. This is a contradiction, therefore ID001T is manually verified.   □

ID002T can be proved by exchanging the position of α_1 and β_1 of ID001T. Note that all impossible differentials found by our tool have the similar structure. Therefore, we expect that ID003T–ID012T can be verified similarly.

4.2   LILLIPUT

LILLIPUT is a lightweight block cipher designed by Berger et al. in 2015 [17] in which the block size and the key size are 64 bits and 80 bits, respectively. LILLIPUT adopts an extended generalized Feistel network (EGFN) [28].

Specification. A 64-bit plaintext is loaded to a 64-bit state X^0, which is divided into sixteen 4-bit nibbles, $X_{15}^0 \| X_{14}^0 \| \cdots \| X_0^0$. The round function, RF, takes as

Table 3. S-box in LILLIPUT (hex)

x	0	1	2	3	4	5	6	7	8	9	A	B	C	D	E	F
$S(x)$	4	8	7	1	9	3	2	E	0	B	6	F	A	5	D	C

Table 4. Nibble permutation (decimal)

x	0	1	2	3	4	5	6	7	8	9	10	11	12	13	14	15
$\pi(x)$	13	9	14	8	10	11	12	15	4	5	3	1	2	6	0	7

input a previous state X^j and a 32-bit subkey $SK^j \triangleq SK_7^j \| SK_6^j \| \cdots SK_0^j$ and updates the state to X^{j+1} with three operations $\mathcal{F}$, $\mathcal{L}$, and $\mathcal{P}$.

Non-linear layer $\mathcal{F}$: Copy the right half of the state, XOR the subkey, apply an S-box to each nibble, finally XOR the results to the left half of the state. Namely, $X_{8+i}^j \leftarrow X_{8+i}^j \oplus S(X_{7-i}^j \oplus SK_i^j)$, $i = 0, 1, \ldots, 7$, where $S(\cdot)$ is a 4-bit S-box defined in Table 3.

Linear layer $\mathcal{L}$: Update the left half of the state with several XORs.

$$X_{15}^j \leftarrow X_{15}^j \oplus X_7^j \oplus X_6^j \oplus X_5^j \oplus X_4^j \oplus X_3^j \oplus X_2^j \oplus X_1^j,$$

$$X_{15-i}^j \leftarrow X_{15-i}^j \oplus X_7^j \text{ for } i = 1, 2, \ldots, 6.$$

Permutation layer $\mathcal{P}$: Permute nibble positions with π defined in Table 4.

The round function is iterated 30 times in which the permutation π is omitted in the last round. Because we are discussing distinguishers in which several rounds will be added for the key recovery, we do not omit the last permutation. The illustration of the round function can be seen in Fig. 3.

Previous Impossible Differential. The designers searched for truncated impossible differentials with $\mathcal{U}$-method [5] and found two 8-round truncated impossible differentials, e.g. the input difference $(0, 0, 0, 0, 0, 0, 0, \alpha, 0, 0, 0, 0, 0, 0, 0, 0)$ is incompatible with the output difference $(0, 0, 0, 0, \beta, 0, 0, 0, 0, 0, 0, 0, 0, 0, 0, 0)$. We stress that the designers searched for them independently of the S-box choice.

Configurations for the Tool. Because both of the block size and the S-box size are small in LILLIPUT, we run our tool in the specific S-box mode to maximize the number of rounds of the distinguisher. In our experiment, we limited the input and output differences to only 1 active nibble.

Considering the Feistel network, having an active nibble in the left half of the input and in the right half of the output can maximize the number of rounds. The number of such input differences is $8 * 15 = 120$, where 8 is for the active nibble position and 15 is for non-zero difference in the active nibble. The number of output differences is the same. In the end, we run MILP for $120 * 120 = 14400$ pairs of input and output differences.

List of 9-Round Impossible Differential Characteristics. We ran our tool with the above configuration. The tool required about 0.2 seconds per pair and it took about 1 h to test 14400 pairs.

Table 5. 9-round impossible differential characteristics against LILLIPUT

ID	$(\Delta L^0, \Delta R^0)$	$(\Delta L^9, \Delta R^9)$	Remarks
001–015	$(0000000\alpha, 00000000)$	$(00000000, 00000\alpha00)$	Manually verified
016–030	$(000000\alpha0, 00000000)$	$(00000000, 00\alpha00000)$	
031–045	$(000000\alpha0, 00000000)$	$(00000000, 0000000\alpha)$	
...	...	...	...
181–195	$(000\alpha0000, 00000000)$	$(00000000, 0000000\alpha)$	
196	$(00000020, 00000000)$	$(00000000, 00000200)$	Manually verified
197	$(00000030, 00000000)$	$(00000000, 00000300)$	Manually verified
198	$(00000080, 00000000)$	$(00000000, 00000800)$	Manually verified
199	$(00000090, 00000000)$	$(00000000, 00000900)$	Manually verified
200	$(000000e0, 00000000)$	$(00000000, 00000e00)$	Manually verified
201	$(000000f0, 00000000)$	$(00000000, 00000f00)$	Manually verified
202	$(00007000, 00000000)$	$(00000000, 00000700)$	
203	$(0000e000, 00000000)$	$(00000000, 00000e00)$	
204–216	$(000\beta0000, 00000000)$	$(00000000, 000000\beta0)$	Manually verified
217	$(00010000, 00000000)$	$(00000000, 00000050)$	

As a result, we found 217 impossible differential characteristics for 9 rounds, which improves the previous best result by 1 round. We list a part of 217 impossible characteristics in Table 5. Note that α in the impossible characteristics with ID 001 to 195 can be any non-zero value but must be the same between input and output. β in ID 204 to 216 can be 1,2,3,4,5,6,7,8,9,10,11,14, or 15.

Manual Verification of ID196 to ID201. Because some of detected impossible characteristics exploit the property of DDT, the analysis is completely different from the previous truncated impossible differentials. Verifying ID001–ID015 is relatively simple (but cannot be detected by the previous tools), which actually does not use the property inside the S-box.[1] Due to the page limitation, we omit the proof of ID001–ID015. We expect ID016–ID195 can be proven similarly.

ID196–ID201 essentially exploit the differential property of the S-box. Here, we explain the details of the contradicting reasons of ID196–ID201.

Theorem 2. *The input difference* $(000000\alpha0, 00000000)$ *cannot propagate to the output difference* $(00000000, 00000\alpha00)$ *after 9 rounds of* LILLIPUT, *where* $\alpha \in \{2, 3, 8, 9, e, f\}$.

Proof. In Fig. 3, the input (resp. output) difference is propagated in forwards (resp. backwards) by 4 rounds. We first focus on the forward propagation.

---

[1] We realized this fact only after we finished manual verification. The tool outputs a list of 217 pairs, and at that time we had no clue about the contradicting reason.

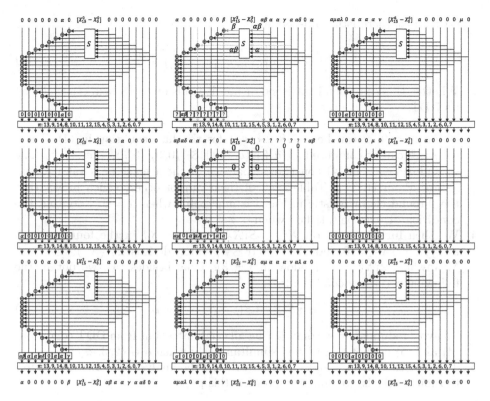

Fig. 3. 9-round impossible differential characteristic of LILLIPUT; ID196–201 (Color figure online)

- In the second round, we denote by β the output difference of the active S-box. Note that β may or may not be equal to α.
- In the third round, we further introduce γ and δ for the output difference from the S-boxes. In Fig. 3, we denote by $\alpha\beta$ and $\alpha\delta$ abbreviations of $\alpha \oplus \beta$ and $\alpha \oplus \delta$ respectively. Note that $\alpha \oplus \beta$ and $\alpha \oplus \delta$ may or may not be non-zero.
- In the forth round, difference is unknown in many nibbles, denoted by '?'.

We do the same for the last 4 rounds and detect the contradiction in the middle.

1. We focus on $X_8^4 \oplus S(X_7^4) = X_4^5$ in the fifth round, in which $\Delta X_8^4 = \Delta X_4^5 = \alpha$, which eventually leads to $\Delta X_7^4 = 0$ (red lines in Fig. 3).
2. We then focus on $X_{11}^4 \oplus S(X_4^4) \oplus X_7^4 = X_1^5$, in which $\Delta X_{11}^4 = \Delta X_1^5 = \alpha$ and $\Delta X_7^4 = 0$. Hence, $\Delta X_4^4 = 0$. Similarly, $\Delta X_2^4 = 0$ (blue in Fig. 3).
3. We focus on $X_8^3 \oplus S(X_7^3) = X_4^4$ in the fourth round, in which $\Delta X_8^3 = \beta$ and $\Delta X_4^4 = 0$. Hence $\Delta S(X_7^3)$ must be β while $\Delta X_7^3 = \alpha \oplus \beta$ (green in Fig. 3). Considering that β is originally defined as an output difference of the S-box whose input difference is α, we have the following necessary condition for this

9-round characteristic to be possible.

$$\exists \beta, x, y : \begin{cases} S(x) \oplus S(x \oplus \alpha) = \beta \\ S(y) \oplus S(y \oplus \alpha \oplus \beta) = \beta \end{cases} \tag{4}$$

Whether this condition is satisfied or not depends on the S-box, especially on its DDT.

When $\alpha = 9$, β can be $3, 7, 8, 9, c, e, f$ for the first equation in (4). Then, $(\alpha \oplus \beta, \beta)$ can be computed as $(a, 3), (e, 7), (1, 8), (0, 9), (5, c), (7, e), (6, f)$. The second equation in (4) constrains that one of them must be a valid propagation. From DDT in Table 5, all of them cannot occur, which proves that the 9-round characteristic in Fig. 3 is impossible when $\alpha = 9$. Note that the condition (4) can be satisfied when $\alpha \neq 0, 9$.

4. We then further focus on $X_{12}^3 \oplus S(X_3^3) \oplus X_7^3 = X_2^4$ in the fourth round. $\Delta X_{12}^3 = \Delta X_2^4 = 0$ and $\Delta X_7^3 = \alpha \oplus \beta$, which derives $\Delta S(X_3^3) = \alpha \oplus \beta$. Meanwhile, $\Delta X_3^3 = \alpha$ (yellow in Fig. 3). Thus besides (4), we obtain the following necessary condition.

$$\exists z : S(z) \oplus S(z \oplus \alpha) = \alpha \oplus \beta \tag{5}$$

To avoid redundancy, we omit listing all candidates, but from DDT conditions (4) and (5) cannot be satisfied simultaneously when $\alpha \in \{2, 3, e, f\}$.

5. To prove the case $\alpha = 8$, we further proceed the analysis. Because it requires too much details, we omit the proof in this paper.

With the above argument, Theorem 2 is proven. $\square$

Remarks. We would like to emphasize once again that the advantage of our tool is that we can obtain a list of all impossible differential characteristics without considering the contradicting reason. We also manually verified ID204 to ID216, while we could not catch the contradicting reason for ID202, ID203, and ID217 by hand. In particular, ID217 is the only pair that the difference of active nibbles in the input and output are different. We leave their verification open.

4.3   ARIA

ARIA is a 128-bit block cipher and provides three secret-key lengths: 128, 192, and 256 bits [19]. ARIA is standardized by Korean Agency for Technology and Standards (KATS) and is described by RFC5794 and RFC6209. ARIA uses Substitution-Permutation Network (SPN) structure, and the state is represented by 16 bytes. The round function consists of Substitution layer SL and Diffusion layer DL. We refer to [19] for its detailed specification.

Previous Cryptanalysis. Wu et al. proposed a truncated impossible differential on 4.5-round ARIA $((DL \circ SL)^4 \circ DL)$ as

$$(0, 0, 0, a, a, 0, a, 0, a, a, 0, 0, 0, a, a, 0) \xrightarrow{4.5R} (0, h, 0, 0, 0, 0, 0, 0, h, h, h, 0, 0, 0, h, 0),$$

where a and h denote any non-zero difference. Based on it, they attacked 6-round ARIA $(SL \circ (DL \circ SL)^5)$ [29]. Then, Li et al. showed new truncated impossible differentials on 4.5-round ARIA and the data-time tradeoff for the attack on 6-round ARIA [30]. One of Li's truncated impossible differentials improved Wu's by reducing the number of active output bytes to 4 from 5, implying that the number of involved subkeys is less, and the time complexity is improved. However, the data complexity is greater than the time complexity. The total complexity is not very improved. Another Li's truncated impossible differential is

$$(0, b, 0, a \oplus b, a \oplus b, 0, a, 0, a, a \oplus b, b, 0, 0, a, a \oplus b, b)$$
$$\xrightarrow{4.5R} (0, h, 0, 0, 0, h, 0, 0, 0, 0, 0, 0, h, 0, h, 0),$$

where a, b, and h denote any non-zero difference. This contributes to reducing the data complexity because the number of independent non-zero differences increases. Unfortunately, the number of involved subkeys increases to 14, and the time complexity is greater than the data complexity. In the end, the total complexity is not very improved.

Configurations for the Tool. Since the S-boxes size of ARIA is 8 bits, we run our tool in the arbitrary S-box mode. Similar to Midori128, we only execute truncated impossible differential search. Our goal is to improve Li's truncated impossible differentials. Namely, we search for 4.5-round truncated impossible differentials, where input and output differences take 3 independent differences and the number of involved subkey is reduced from 14. To search such truncated impossible differentials efficiently, our tool searches for truncated impossible differentials for 3.5 rounds $(SL \circ (DL \circ SL)^3)$, where every active byte can take any difference. Then, found truncated differentials are trivially extended to 4.5 rounds by applying DL to the beginning and end. Finally, we evaluate the number of input and output differences.

4.5-Round Truncated Impossible Differentials. We ran our tool with the above configuration. As a result, we found a truncated impossible differential as

$$(a, 0, 0, 0, 0, 0, 0, a, 0, a, 0, a, a, 0, a, 0) \xrightarrow{DL} (0, a, 0, 0, a, 0, 0, 0, 0, 0, 0, 0, 0, 0, 0, 0)$$
$$\xrightarrow{3.5R} (h, g, 0, 0, 0, 0, 0, h \oplus g, 0, h \oplus g, 0, g, 0, 0, 0, 0)$$
$$\xrightarrow{DL} (h \oplus g, 0, 0, 0, h \oplus g, h, h, 0, 0, h, 0, 0, 0, g, 0, g)$$

where a, h, and g are non-zero differences. The number of involved subkeys is 13, and it decreases by one byte from that of Li's truncated impossible differentials. It implies that we can improve the time complexity of their key recovery attack.

4.4 Minalpher

Minalpher is an authenticated encryption scheme designed by Sasaki et al. in 2015 [18]. Minalpher uses 256-bit core permutation called Minalpher-P, which

Table 6. 7.5-round truncated impossible differentials of Minalpher-P

ID	ΔP	ΔC	Remarks
0001T	$A[0][0]$	$A[0][2]$	Manually verified
0002T	$A[0][0]$	$A[0][3]$	
0003T	$A[0][0]$	$A[0][4]$	
0004T	$A[0][0]$	$A[0][5]$	
$\vdots$	$\vdots$	$\vdots$	
1152T	$B[3][7]$	$B[3][7]$	

is based on Substitution-Permutation Network (SPN) structure using 4-bit S-boxes. We refer to [18] for its detailed specification.

Previous Cryptanalysis. The designers found 6.5-round truncated impossible differentials by using the $\mathcal{U}$-method by Kim et al. These are the longest impossible differentials discovered by the $\mathcal{U}$-method.

Configurations for the Tool. While the S-boxes size is 4 bits, the block size, i.e., 256 bits, is very large. Therefore, we run our tool in the arbitrary S-box mode aiming truncated impossible differentials with 1 active nibble in the input and output differences. The number of such differences is 64 for both of input and output. In the end, we run MILP for $64 * 64 = 4096$ pairs.

List of 7.5-Round Truncated Impossible Differentials. The tool required about a few seconds per pair. As a result, our experiment found 1152 truncated impossible differentials for 7.5 rounds, which improves the previous best truncated impossible differentials by 1 round. Table 6 shows several examples. Column ΔP shows the position of the active nibble in plaintext, and column ΔC shows the position of the active nibble in ciphertext. Every truncated impossible differential consists of $15^2 = 225$ impossible differential characteristics.

4.5   MIBS

MIBS is a lightweight block cipher designed by Izadi et al. in 2009 [20]. The block length is 64, and it provides two key lengths: 64- and 80-bit secret key. We refer to [20] for its detailed specification.

Previous Cryptanalysis. Bay et al. found two 8-round truncated impossible differentials [31]. Then, Wu and Wang found four additional 8-round truncated impossible differentials [8].

Table 7. 8-round impossible differential characteristics against MIBS

ID	ΔP	ΔC	Remarks
001T	$(00000000, 000000\alpha0)$	$(0000\beta000, 00000000)$	Bay
002T	$(00000000, 0000\alpha000)$	$(000000\beta0, 00000000)$	Wu
003T	$(00000000, 00\alpha00000)$	$(0000000\beta, 00000000)$	Bay
004T	$(00000000, 0000000\alpha)$	$(00\beta00000, 00000000)$	Wu
005T	$(00000000, 00\alpha00000)$	$(0000\beta000, 00000000)$	Wu
006T	$(00000000, 0000\alpha000)$	$(00\beta00000, 00000000)$	Wu
001–120	$(00000000, 000\gamma0000)$	$(00000\epsilon00, 00000000)$	
121–240	$(00000000, 00000\epsilon00)$	$(000\gamma0000, 00000000)$	

Configurations for the Tool. The block size of MIBS is 64 bits and the S-boxes size is 4 bits. Therefore, we run our tool in the specific S-box mode to maximize the number of rounds of the distinguisher. In our experiment, we limited the input and output differences to only 1 active nibble.

Considering the Feistel network, the number of differences we need to test is exactly the same as the case of LILLIPUT in Sect. 4.2. Thus we run MILP for $120 \times 120 = 14400$ pairs of input and output differences.

List of 8-Round Impossible Differential Characteristics. The tool required about 7.7 seconds per pair using single core and it took about 30 h to test 14400 pairs.

Our tool found six 8-round truncated impossible differentials, which are the same as results by Wu's method. However, our method additionally found 2×120 impossible differential characteristics, which are not nibble-oriented truncated impossible differentials. We list all impossible differentials in Table 7, where α and β are any non-zero value. ID001–ID240 are impossible differential characteristics that our tool newly found. If the differences (γ, ϵ) takes differences that are shown by x in Table 8, the pairs of input and output differences is impossible differential characteristics.

5   Differential Possibility Equivalence Technique

In Sect. 4.1, we searched for all truncated impossible differentials with one active nibble. However, since ShuffleCell and MixColumn in Midori128 are byte-wise operations, we should search for all impossible characteristics with one active byte if possible. Moreover, the search in Sect 4.1 never exploited the property of Sb_1 because the tool was run in the arbitrary S-box mode. This section explains how to run the tool in the specific S-box mode in a feasible time.

As described in Sect. 3.4, the number of all pairs with d input active words and d' output active words is $O(n^{d+d'}2^{(d+d')c})$, where n and c are the number of S-boxes per round and the size of each S-box, respectively. If we want to evaluate

Table 8. Pairs of impossible differences found by our tool for MIBS

γ \ ϵ	1	2	3	4	5	6	7	8	9	a	b	c	d	e	f
1	x	x	x	0	x	x	0	0	0	x	0	0	x	0	x
2	0	x	0	x	x	x	0	x	x	0	0	0	x	x	0
3	x	0	x	x	0	0	0	x	x	x	0	0	0	x	x
4	x	x	0	x	0	0	0	0	0	x	x	x	x	x	0
5	x	0	0	0	x	x	0	x	x	x	x	x	0	0	0
6	x	0	x	x	0	x	x	0	x	0	x	0	x	0	0
7	0	0	0	0	0	0	x	x	x	0	x	x	0	x	x
8	x	x	x	0	x	0	x	x	0	0	x	0	0	x	0
9	0	x	x	0	0	x	x	0	x	x	0	x	0	x	0
a	0	x	0	x	x	0	x	0	x	x	x	0	0	0	x
b	x	0	0	0	x	0	x	0	x	0	0	x	x	x	x
c	0	0	x	x	x	0	x	x	0	x	0	x	x	0	0
d	0	0	x	x	x	x	0	0	0	0	x	x	0	x	x
e	0	x	x	0	0	0	0	x	x	0	x	x	x	0	x
f	x	x	0	x	0	x	x	x	0	0	0	x	0	0	x

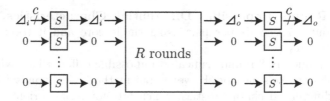

Fig. 4. Differential possibility equivalence technique

all impossible differential characteristics on Midori128 with 1 byte active, our tool has to execute $16^2 \times 255^2 \approx 2^{24}$ MILP instances. Then, it takes about 200 days to complete all instances even if an MILP instance is solved within one second. Therefore, we need an efficient method to evaluate all instances.

5.1 Procedure of Differential Possibility Equivalence Technique

The *differential possibility equivalence technique* reduces the number of MILP instances that our tool has to solve.[2] Figure 4 shows the outline of the technique.

---

[2] The motivation of the differential possibility equivalence technique is quite different from truncated impossible differential. The truncated impossible search overlooks impossible characteristics only with one possible characteristic in the truncated set. When the number of impossible characteristics is small, truncated impossible differential search is not useful.

Table 9. 7-round impossible differential characteristics against Midori128

ID	ΔP		ΔC	
	Position	Value	Position	Value
001	s_1	0x04	s_8	0x43
002	s_1	0x0C	s_8	0x43

Assuming that we search for impossible differential characteristics in which the first words of plaintexts and ciphertexts are active, we want to evaluate $(2^c - 1)^2$ pairs of input and output differences. First, we solve one MILP instance and obtain that $(\Delta_i \rightarrow \Delta'_i \rightarrow \Delta'_o \rightarrow \Delta_o)$ is possible differential characteristic for one tuple of $(\Delta_i, \Delta'_i, \Delta'_o, \Delta_o)$. Next, we evaluate a set $\mathcal{I}$ whose elements are all Δ such that $\Delta \rightarrow \Delta'_i$ is possible. Similarly, we evaluate a set $\mathcal{O}$ whose elements are all Δ such that $\Delta'_o \rightarrow \Delta$ is possible. Then, pairs in $(\mathcal{I} \times \mathcal{O})$ are possible characteristics via (Δ'_i, Δ'_o), we thus do not need to evaluate them using MILP. We note that some MILP solvers have API for programming languages, e.g. Gurobi Optimizer supports API for C-language. Thus, adding such auxiliary codes is easily done. Since the numbers of elements in $\mathcal{I}$ and $\mathcal{O}$ are $2^{c/2}$ on average, we can efficiently reduce the number of MILP instances that our tool has to solve.

We estimate the effectiveness of differential possibility equivalence technique.

Theorem 3. *Let n and c be the number of S-boxes per round and the size of each S-box, respectively. Our tool aims to find impossible differential with d input active words and d' output active words. Then, the number of trials that we have to solve MILP instances is $2^{d+d'} ((d + d') \log_e(2^c - 1) + O(1))$ on average.*

Due to the page limitation, we omit the proof of Theorem 3. Accurately, we can more efficiently collect N input and output differences than the estimation by Theorem 3 because every trial can always choose a pair without duplication. On the other hand, this error is not serious because N' differences are evaluated in the same time in one trial.

We searched for impossible differential characteristics with one active byte on Midori128 by using the differential possibility equivalence technique. In our experiment, this technique reduces the number of MILP instances that our tool has to solve from 2^{24} to $546865 \approx 2^{19}$. As a result, we found two new impossible differential characteristics, which are shown in Table 9. The total time that our tool evaluates all impossible differential characteristics with one active byte is about 24 days in single core.

6 Applications from Design Aspect

6.1 Design Tool Using Specific S-Box Mode

Let us discuss using the tool for the design process of new primitives. Attack tools can always be used to evaluate how many rounds are attacked after the design is completed. Here we want to discuss a more interactive process. In many

SPN-based designs, the designers evaluate many candidates with MILP and pick up the best choice. For example, the designers of Midori chose an almost-MDS matrix for `MixColumn`, and tested all parameters for `ShuffleCell`. Similarly, the designers of Skinny tested all light non-MDS matrices for `MixColumns` and the designers of Minalpher tested all parameters of a `ShiftRows`-like operation.

To run our tool in the specific S-box mode, S-boxes must be fixed in advance. This situation occurs when the choice of S-boxes has a high priority in the design. For example, Midori [16] chose the S-box with the lowest depth, and FIDES [32] and PICARO [33] chose the S-box that can be masked easily.

In our tool, all the components but for key schedule are simulated. Therefore, when we assume that subkeys are XORed to all words of the state before S-boxes, the tool can provide a certain level of proof, which is detailed below.

Observation 3. *Suppose that the tool does not find any impossible differential characteristic for r rounds after testing all paired input and output differences in a certain subset in the specific S-box mode. Then, the number of rounds of the longest impossible differential satisfying those input and output differences is at most $r-1$ by assuming that all subkeys are independent and chosen uniformly at random.*

Proof. Suppose that the tool can find specific differential propagations for given input and output differences. We now assume subkeys are XORed to all words of the state. Therefore, the output difference of any S-box are computed as

$$\Delta_o = S(x \oplus sk) \oplus S(x \oplus sk \oplus \Delta_i),$$

where Δ_i and Δ_o denote the input and output difference, respectively. The tool does not evaluate the value of x, but we now assume that all subkeys are independent and chosen uniformly at random. Since Δ_o can take all possible output differences in DDT, the differential propagations that the tool finds are always valid in this assumption.   □

If we can verify that all input and output differences with one active word are possible in the specific S-box mode, we say that the cipher is secure against impossible differential with one active word under the *subkey uniform assumption*.

Remarks About Proof in [21]. Cui et al. claimed that the tool can be used to prove the longest impossible differentials under the condition that input and output differences belong to the tested subset. After evaluating several ciphers, they claimed that *"we proof that the longest impossible differentials for LBlock, TWINE and Piccolo ciphers are really 14, 14 and 7 rounds respectively."* Unfortunately, Cui et al. are misinterpreting what the tool does.

In the evaluation with MILP, all valid propagations for one round are also valid in the evaluation of multiple rounds irrespectively of the propagation in neighboring rounds and subkey values. This is true only if all subkeys are independent and chosen uniformly at random. Therefore, even if no impossible differential is found for r rounds by MILP, it cannot ensure the non-existence for r rounds for real ciphers with particular key schedule.

6.2 Design Tool Using Arbitrary S-Box Mode

The arbitrary S-box mode is also useful for the design tool. When we run our tool in the specific S-box mode for the design tool, S-boxes must be fixed in advance. Meanwhile, if the choice of the linear layer has a higher priority, we would like to recommend the arbitrary S-box mode. The arbitrary S-box mode have two advantages: it can be executed before S-boxes are not specified and is generally more efficient than the specific S-box mode. In addition, the arbitrary S-box mode leads to several benefit to the designers.

Evaluating Linear Layer: The designers often test many choices of the S-boxes and of the linear layer. Because exhaustively testing all combinations is infeasible, the designers need to evaluate them independently. The arbitrary S-box mode finds impossible differential characteristics that are independent from the choice of the S-box, which makes possible to evaluate the security of the linear layer. In addition, the arbitrary S-box mode enables the designers to proceed the design of S-boxes and the design linear layer in parallel, which can shorten the design period.

Distinguishing Contradicting Reasoning: When impossible differentials are found for some rounds, the designers may prefer to patch the design or choose other design candidates. Then it is convenient for the designer to know whether the detected differentials can be prevented by changing S-boxes or not. In the arbitrary S-box mode, the contradiction is clearly caused by the linear layer.

Actually, impossible differential characteristics ID001–ID195 of LILLIPUT can be found by both the specific and arbitrary S-box modes, but the others ID196–ID217 can be found only by the specific S-box mode. Thus, we can immediately know ID001–ID195 are impossible differential characteristics independent of the choice of the S-box and cannot be prevented by replacing the S-box.

Similarly to Sect. 6.1, the fact that no impossible differential is found gives a certain level of security proof as follows.

Observation 4. *Suppose that the tool does not find any impossible differential characteristic for r rounds after testing all paired input and output differences in a certain subset in the arbitrary S-box mode. Then, the number of rounds of the longest impossible differential satisfying those input and output differences is at most r − 1 by assuming that all S-boxes are keyed bijective S-boxes that are independent and chosen uniformly at random.*

If we can verify that all pairs of input and output differences with one active word are possible in the arbitrary S-box mode, we say that the cipher is secure against impossible differential with one active word under the *keyed (uniform) bijective S-boxes assumption.*

6.3 Optimal Pick Technique; Application to MIBS

When ciphers have heavy diffusion layer, MILP solver requires too much time to verify whether or not a given pair of input and output differences is possible. For example, suppose that we evaluate resistance of MIBS against 9-round impossible differential. As discussed in Sect. 4.5, we need to test 14400 pairs of input and output differences. However, the tool could not finish the evaluation of 1 pair even after a couple of hours. Proving the security of 9-round MIBS with the direct application of our tool is infeasible.

Optimal Pick Technique. We propose an *optimal pick technique*, which dramatically reduces the computation time to prove the resistance against impossible differentials, i.e. to prove the existence of differential characteristic. Suppose that we are given a pair of input and output differences. The optimal pick technique well works when there are many differential characteristics satisfying a pair of given input and output differences. The intuition of this technique is as follows. We partially constrain the difference of the state in a middle round as well as the input and output differences. Suppose that our aim is to prove the resistance against r-rounds impossible differentials, and we expect that the proof is possible. Let X_{i-1} be a difference of the input of the i-th round. Our tool constrains a pair of input and output differences (X_0, X_r), and additional b bits of $X_{\lceil r/2 \rceil}$, where b is heuristically chosen. In our experiments, these additional constraints often reduce the execution time of the MILP solver. To prove the resistance against impossible differential, it is sufficient to find only one characteristic satisfying the constraint. Therefore, if the solver takes too long for a choice of constrained b bits, we give up searching for the b bits, and test another b bits by expecting that the new b bits are easy to compute.

In application to 9-round MIBS, for pairs of input and output differences (X_0, X_9) we used the optimal pick technique with the following strategy.

- Four nibbles in X_4 are additionally constrained ($b = 16$).
- For all 2^{16} choices of additional constraints, we evaluate whether or not it is possible to satisfy (X_0, X_4, X_9). If the execution time reaches 10 s, we stop the evaluation and proceed the next additional constraints.
- Once we find an additional constraint X_4 satisfying the input and output differences (X_0, X_9), we return that the pair (X_0, X_9) is possible.

The second strategy is the essence of the optimal pick technique. The execution time of the MILP solver becomes too long for some choice of X_4, and the second strategy allows us to escape from the unlucky choice. As a result, we successfully proved that there is no 9-round impossible differential characteristics with one active nibble under the subkey uniform assumption. Note that the optimal pick technique only can be used for the proving approach, i.e. it cannot be used to find impossible differential characteristics because we terminate the MILP search when the execution time reaches 10 s.

6.4   List of Evaluated Designs

We proved the maximal number of rounds of impossible differential characteristics for many designs. Besides the already discussed five designs, we evaluated SIMON [34], TWINE [35], LBlock [36], Piccolo [37], RECTANGLE [10],

Table 10. Provable security against impossible differentials

Target	#Rounds	Assumption	Remarks
Midori128	8	Subkey uniform	1 active byte
	8	Keyed bijective 4-bit S-boxes	1 active byte
	7	Keyed bijective 8-bit S-boxes	1 active byte
LILLIPUT	10	Subkey uniform	1 active nibble
Minalpher	9.5	Keyed bijective S-boxes	1 active nibble
ARIA	5	Keyed bijective S-boxes	1 active byte
MIBS	9	Subkey uniform	1 active nibble
SIMON	12	Subkey uniform	1 active bit
TWINE	15	Subkey uniform	1 active nibble
LBlock	15	Subkey uniform	1 active nibble
Piccolo	8	Subkey uniform	1 active nibble
RECTANGLE	9	Subkey uniform	1 active nibble
Skinny-64	12	Subkey uniform	1 active nibble
Midori64	7	Subkey uniform	1 active nibble
CLEFIA	10	Keyed bijective 8-bit S-boxes	1 active byte

Skinny [22], Midori64 [16], and CLEFIA [38] as shown in Table 10. We confirmed that there are no impossible differential characteristics within the parameters of input and output differences in Remark column. For example, if we regard SSb_i in Midori128 as keyed 8-bit bijective S-boxes, we proved that there are no 7-round impossible differential characteristics with 1 active byte. However, if we exploit the structure of SSb_i and regard Sb_1 as keyed 4-bit bijective S-boxes, 7-round impossible characteristics can be found as explained in Sect. 4.1. In such an assumption, we proved that there are no 8-round impossible differential characteristics with 1 active byte. Moreover, 8 rounds are also secure in the subkey uniform assumption.

A   Relationship Between [21] and This Paper

Cui et al. [21] have recently posted their work to Cryptology ePrint Archive (received by ePrint Archive at 11 July 2016) presenting that impossible differentials can be searched with MILP. Although we have independently reached the same idea and used it to evaluate a lot of designs, the work by Cui et al. became the first article to report the impossible differential search tool based on MILP.

Though the basic idea of the tool is the same, two papers extend the basic idea to quite different directions. The main focus of [21] seems to be the extension to the ARX structure and zero-correlation cryptanalysis, which is not covered by our work. Meanwhile, we are focusing on the impossible differential cryptanalysis much deeper, and trying to extend the structure that can be evaluated by the tool. Therefore, we obtained new results even for 8-bit S-boxes, in which [21] left application to 8-bit S-box open.

Another difference is enthusiasm for the application to practical designs. Considering the number of applications, [21] seems to focus on the theoretical aspects, while we are trying to evaluate more and more targets and the usage of the tool for designing new primitives is another main focus.

Advantages of [21] Over Our Work.

- By converting differential evaluation to linear evaluation, the tool is extended to zero-correlation approximations.
- By borrowing the idea by Fu et al. about MILP on the ARX structure [39], the tool is extended to the impossible differentials for the ARX structure.
- By applying the basic idea to PRESENT, new impossible differentials are recovered while the number of attacked rounds is not improved.
- By applying the extended tool to ARX, new impossible differentials and new zero-correlation approximations are discovered against HIGHT, which improves the previous best results by 1 round.

Advantages of Our Work Over [21].

- The arbitrary S-box mode to apply the tool to 8-bit S-box.
- Focusing on the property of the tool that it can catch any contradiction, which leads to find improvement of impossible differential using 8-bit S-box.
- More applications are examined and we improved the previous best results in several applications.
- Analyzing the contradicting reasons for the detected pairs and revealed the new structural properties that may be used in future analysis.
- More precise arguments for provable security.
- The differential possibility equivalence technique for the efficient search.
- The optimal pick technique for the efficient proof.

References

1. Mouha, N., Wang, Q., Gu, D., Preneel, B.: Differential and linear cryptanalysis using mixed-integer linear programming. In: Wu, C.-K., Yung, M., Lin, D. (eds.) Inscrypt 2011. LNCS, vol. 7537, pp. 57–76. Springer, Heidelberg (2012). doi:10. 1007/978-3-642-34704-7_5
2. Knudsen, L.: DEAL - a 128-bit block cipher. Technical report no. 151, Department of Informatics, University of Bergen, Norway (1998)
3. Biham, E., Biryukov, A., Shamir, A.: Cryptanalysis of skipjack reduced to 31 rounds using impossible differentials. In: Stern, J. (ed.) EUROCRYPT 1999. LNCS, vol. 1592, pp. 12–23. Springer, Heidelberg (1999). doi:10.1007/3-540-48910-X_2
4. Biryukov, A.: Miss-in-the-middle attack. In: van Tilborg, H.C.A. (ed.) Encyclopedia of Cryptography and Security. Springer, Heidelberg (2005)

5. Kim, J., Hong, S., Sung, J., Lee, S., Lim, J., Sung, S.: Impossible differential cryptanalysis for block cipher structures. In: Johansson, T., Maitra, S. (eds.) INDOCRYPT 2003. LNCS, vol. 2904, pp. 82–96. Springer, Heidelberg (2003). doi:10.1007/978-3-540-24582-7_6

6. Luo, Y., Wu, Z., Lai, X., Gong, G.: A unified method for finding impossible differentials of block cipher structures. Cryptology ePrint Archive, report 2009/627 (2009)

7. Luo, Y., Lai, X., Wu, Z., Gong, G.: A unified method for finding impossible differentials of block cipher structures. Inf. Sci. **263**, 211–220 (2014)

8. Wu, S., Wang, M.: Automatic search of truncated impossible differentials for word-oriented block ciphers. In: Galbraith, S., Nandi, M. (eds.) INDOCRYPT 2012. LNCS, vol. 7668, pp. 283–302. Springer, Heidelberg (2012). doi:10.1007/978-3-642-34931-7_17

9. Tezcan, C.: Improbable differential attacks on present using undisturbed bits. J. Comput. Appl. Math. **259**, 503–511 (2014)

10. Zhang, W., Bao, Z., Lin, D., Rijmen, V., Yang, B., Verbauwhede, I.: RECTANGLE: a bit-slice lightweight block cipher suitable for multiple platforms. Cryptology ePrint Archive, report 2014/084 (2014). http://eprint.iacr.org/2014/084

11. Derbez, P., Fouque, P.-A.: Automatic search of meet-in-the-middle and impossible differential attacks. In: Robshaw, M., Katz, J. (eds.) CRYPTO 2016. LNCS, vol. 9815, pp. 157–184. Springer, Heidelberg (2016). doi:10.1007/978-3-662-53008-5_6

12. Sun, S., Hu, L., Wang, P., Qiao, K., Ma, X., Song, L.: Automatic security evaluation and (related-key) differential characteristic search: application to SIMON, PRESENT, LBlock, DES(L) and other bit-oriented block ciphers. In: Sarkar, P., Iwata, T. (eds.) ASIACRYPT 2014. LNCS, vol. 8873, pp. 158–178. Springer, Heidelberg (2014). doi:10.1007/978-3-662-45611-8_9

13. Sun, S., Hu, L., Wang, M., Wang, P., Qiao, K., Ma, X., Shi, D., Song, L., Fu, K.: Towards finding the best characteristics of some bit-oriented block ciphers and automatic enumeration of (related-key) differential and linear characteristics with predefined properties. IACR Cryptology ePrint Archive 2014/747 (2014)

14. Sasaki, Y., Todo, Y.: New differential bounds and division property of LILLIPUT: block cipher with extended generalized Feistel network. In: Avanzi, R., Heys, H. (eds.) SAC 2016. LNCS. Springer, Cham (2016)

15. Sun, S., Hu, L., Wang, M., Wang, P., Qiao, K., Ma, X., Shi, D., Song, L., Fu, K.: Constructing mixed-integer programming models whose feasible region is exactly the set of all valid differential characteristics of SIMON. Cryptology ePrint Archive, report 2015/122 (2015). http://eprint.iacr.org/2015/122

16. Banik, S., Bogdanov, A., Isobe, T., Shibutani, K., Hiwatari, H., Akishita, T., Regazzoni, F.: Midori: a block cipher for low energy. In: Iwata, T., Cheon, J.H. (eds.) ASIACRYPT 2015. LNCS, vol. 9453, pp. 411–436. Springer, Heidelberg (2015). doi:10.1007/978-3-662-48800-3_17

17. Berger, T.P., Francq, J., Minier, M., Thomas, G.: Extended generalized Feistel networks using matrix representation to propose a new lightweight block cipher: LILLIPUT. IEEE Trans. Comput. **65**, 2074–2089 (2015)

18. Sasaki, Y., Todo, Y., Aoki, K., Naito, Y., Sugawara, T., Murakami, Y., Matsui, M.: Minalpher v1.1. Submitted to CAESAR (2015)

19. Kwon, D., et al.: New block cipher: ARIA. In: Lim, J.-I., Lee, D.-H. (eds.) ICISC 2003. LNCS, vol. 2971, pp. 432–445. Springer, Heidelberg (2004). doi:10.1007/978-3-540-24691-6_32

20. Izadi, M., Sadeghiyan, B., Sadeghian, S.S., Khanooki, H.A.: MIBS: a new lightweight block cipher. In: Garay, J.A., Miyaji, A., Otsuka, A. (eds.) CANS 2009. LNCS, vol. 5888, pp. 334–348. Springer, Heidelberg (2009). doi:10.1007/978-3-642-10433-6_22

21. Cui, T., Jia, K., Fu, K., Chen, S., Wang, M.: New automatic search tool for impossible differentials and zero-correlation linear approximations. Cryptology ePrint Archive, report 2016/689 (2016). http://eprint.iacr.org/2016/689

22. Beierle, C., et al.: The SKINNY family of block ciphers and its low-latency variant MANTIS. In: Robshaw, M., Katz, J. (eds.) CRYPTO 2016. LNCS, vol. 9815, pp. 123–153. Springer, Heidelberg (2016). doi:10.1007/978-3-662-53008-5_5

23. Bogdanov, A., Knudsen, L.R., Leander, G., Paar, C., Poschmann, A., Robshaw, M.J.B., Seurin, Y., Vikkelsoe, C.: PRESENT: an ultra-lightweight block cipher. In: Paillier, P., Verbauwhede, I. (eds.) CHES 2007. LNCS, vol. 4727, pp. 450–466. Springer, Heidelberg (2007). doi:10.1007/978-3-540-74735-2_31

24. Gurobi Optimization, Inc.: Gurobi optimizer 6.5 (2015). http://www.gurobi.com/

25. Guo, J., Jean, J., Nikolić, I., Qiao, K., Sasaki, Y., Sim, S.M.: Invariant subspace attack against full Midori64. Cryptology ePrint Archive, report 2015/1189 (2015). http://eprint.iacr.org/2015/1189

26. Todo, Y., Leander, G., Sasaki, Y.: Nonlinear invariant attack - practical attack on full SCREAM, iSCREAM, and Midori64. Cryptology ePrint Archive, report 2016/732 (2016). http://eprint.iacr.org/2016/732

27. Zhan, C., Xiaoyun, W.: Impossible differential cryptanalysis of Midori. Cryptology ePrint Archive, report 2016/535 (2016). http://eprint.iacr.org/2016/535

28. Berger, T.P., Minier, M., Thomas, G.: Extended generalized feistel networks using matrix representation. In: Lange, T., Lauter, K., Lisoněk, P. (eds.) SAC 2013. LNCS, vol. 8282, pp. 289–305. Springer, Heidelberg (2014). doi:10.1007/978-3-662-43414-7_15

29. Wu, W., Zhang, W., Feng, D.: Impossible differential cryptanalysis of reduced-round ARIA and Camellia. J. Comput. Sci. Technol. **22**(3), 449–456 (2007)

30. Li, R., Sun, B., Zhang, P., Li, C.: New impossible differential cryptanalysis of ARIA. Cryptology ePrint Archive, report 2008/227 (2008). http://eprint.iacr.org/2008/227

31. Bay, A., Nakahara Jr., J., Vaudenay, S.: Cryptanalysis of reduced-round MIBS block cipher. In: Heng, S.-H., Wright, R.N., Goi, B.-M. (eds.) CANS 2010. LNCS, vol. 6467, pp. 1–19. Springer, Heidelberg (2010). doi:10.1007/978-3-642-17619-7_1

32. Bilgin, B., Bogdanov, A., Knežević, M., Mendel, F., Wang, Q.: FIDES: lightweight authenticated cipher with side-channel resistance for constrained hardware. In: Bertoni, G., Coron, J.-S. (eds.) CHES 2013. LNCS, vol. 8086, pp. 142–158. Springer, Heidelberg (2013). doi:10.1007/978-3-642-40349-1_9

33. Piret, G., Roche, T., Carlet, C.: PICARO – a block cipher allowing efficient higher-order side-channel resistance. In: Bao, F., Samarati, P., Zhou, J. (eds.) ACNS 2012. LNCS, vol. 7341, pp. 311–328. Springer, Heidelberg (2012). doi:10.1007/978-3-642-31284-7_19

34. Beaulieu, R., Shors, D., Smith, J., Treatman-Clark, S., Weeks, B., Wingers, L.: The SIMON and SPECK families of lightweight block ciphers. Cryptology ePrint Archive, report 2013/404 (2013). http://eprint.iacr.org/2013/404

35. Suzaki, T., Minematsu, K., Morioka, S., Kobayashi, E.: TWINE: a lightweight block cipher for multiple platforms. In: Knudsen, L.R., Wu, H. (eds.) SAC 2012. LNCS, vol. 7707, pp. 339–354. Springer, Heidelberg (2013). doi:10.1007/978-3-642-35999-6_22

36. Wu, W., Zhang, L.: LBlock: a lightweight block cipher. In: Lopez, J., Tsudik, G. (eds.) ACNS 2011. LNCS, vol. 6715, pp. 327–344. Springer, Heidelberg (2011). doi:10.1007/978-3-642-21554-4_19

37. Shibutani, K., Isobe, T., Hiwatari, H., Mitsuda, A., Akishita, T., Shirai, T.: Piccolo: an ultra-lightweight blockcipher. In: Preneel, B., Takagi, T. (eds.) CHES 2011. LNCS, vol. 6917, pp. 342–357. Springer, Heidelberg (2011). doi:10.1007/978-3-642-23951-9_23

38. Shirai, T., Shibutani, K., Akishita, T., Moriai, S., Iwata, T.: The 128-bit blockcipher CLEFIA (extended abstract). In: Biryukov, A. (ed.) FSE 2007. LNCS, vol. 4593, pp. 181–195. Springer, Heidelberg (2007). doi:10.1007/978-3-540-74619-5_12

39. Fu, K., Wang, M., Guo, Y., Sun, S., Hu, L.: MILP-based automatic search algorithms for differential and linear trails for speck. In: Peyrin, T. (ed.) FSE 2016. LNCS, vol. 9783, pp. 268–288. Springer, Heidelberg (2016). doi:10.1007/978-3-662-52993-5_14

New Collision Attacks on Round-Reduced Keccak

Kexin Qiao[1,3,4], Ling Song[1,2,3(✉)], Meicheng Liu[1], and Jian Guo[2]

[1] State Key Laboratory of Information Security,
Institute of Information Engineering, Chinese Academy of Sciences, Beijing, China
{qiaokexin,songling,liumeicheng}@iie.ac.cn
[2] Nanyang Technological University, Singapore, Singapore
guojian@ntu.edu.sg
[3] Data Assurance and Communication Research Center,
Chinese Academy of Sciences, Beijing, China
[4] University of Chinese Academy of Sciences, Beijing, China

Abstract. In this paper, we focus on collision attacks against Keccak hash function family and some of its variants. Following the framework developed by Dinur *et al.* at FSE 2012 where 4-round collisions were found by combining 3-round differential trails and 1-round connectors, we extend the connectors one round further hence achieve collision attacks for up to 5 rounds. The extension is possible thanks to the large degree of freedom of the wide internal state. By linearization of all S-boxes of the first round, the problem of finding solutions of 2-round connectors are converted to that of solving a system of linear equations. However, due to the quick freedom reduction from the linearization, the system has solution only when the 3-round differential trails satisfy some additional conditions. We develop a dedicated differential trail search strategy and find such special differentials indeed exist. As a result, the first practical collision attack against 5-round SHAKE128 and two 5-round instances of the Keccak collision challenges are found with real examples. We also give the first results against 5-round Keccak-224 and 6-round Keccak collision challenges. It is remarked that the work here is still far from threatening the security of the full 24-round Keccak family.

Keywords: Keccak · SHA-3 · Hash function · Linearization · Differential

1 Introduction

The Keccak [3,5] family of hash functions has attracted intensive cryptanalysis since its submission to the SHA-3 competition in 2008 [1,9–11,13,14,16,17,19]. In 2012, the National Institute of Standards and Technology of the U.S. selected Keccak as the winner of the SHA-3 competition. The SHA-3 family consists of four

---

K. Qiao and M. Liu—This work was done while the authors were visiting Nanyang Technological University.

J.-S. Coron and J.B. Nielsen (Eds.): EUROCRYPT 2017, Part III, LNCS 10212, pp. 216–243, 2017.
DOI: 10.1007/978-3-319-56617-7_8

cryptographic hash functions of fixed digest sizes and two eXtendable-Output Functions (XOFs) named SHAKE128 and SHAKE256, each of which is based on an instance of the KECCAK algorithms [18]. KECCAK$[r, c, d]$ applies sponge construction with bitrate r and capacity c to generate d bit digests from arbitrary length messages where $d = 224, 256, 384, 512$ in the official SHA-3 versions and $d = 160, 80$ in the KECCAK Crunchy Crypto Collision and Pre-image Contest [2]. Depending on the size of the internal state in $r + c$ bits from the set $\{200, 400, 800, 1600\}$, each of the challenge versions contains 4 variants. SHAKE128 and SHAKE256 generate digests that can be extended to any desired length. The suffixes "128" and "256" indicate the security strengths against generic attacks that these two functions support.

In this paper, we focus on collision attacks against the KECCAK family, *i.e.*, to find two different messages such that their hash digests are the same. The best previous practical collision attacks on KECCAK family are on KECCAK-224 and KECCAK-256 reduced to 4 rounds found by Dinur *et al.* [10] in 2012 and later furnished in the journal version [12]. After this, theoretical results improved to 5-round KECCAK-256 [11]. However, the number of practically attacked rounds remains at 4. To promote cryptanalysis against KECCAK, the KECCAK design team proposed smaller variants in the KECCAK challenge [2] with 160 digest size for collision attack and 80 digest size for preimage attack with each of the 4 sizes of internal states reduced to from 1 to 12 rounds. The ideal security levels of both are set to be 2^{80} unit computations for collision and preimages, respectively. This is a level much lower than that of the main 4 instances of SHA-3, but still beyond the reach of current computation resource one may have. The current best solutions of collision challenges are instances reduced up to 4 rounds by Dinur *et al.* [10] and Mendel *et al.* [17]. Theoretical results were found by Dinur *et al.* [11] against KECCAK-256 with complexities 2^{115} using generalized internal differentials. To the best of our knowledge, this remains as the only results on collision attack against KECCAK reduced to 5 or more rounds up to date.

Our Contribution. We develop an algebraic and differential hybrid method to launch collision attacks on KECCAK family and practically find collisions of 5-round SHAKE128 and two 5-round instances of the KECCAK collision challenges. Theoretical results, with complexities below the birthday bound, against 5-round KECCAK-224 and 6-round KECCAK collision challenges are also achieved.

These results follow a crucial observation that, the KECCAK S-box can be re-expressed as linear transformations, when the input is restricted to some affine subspaces. It was already noted by Daemen *et al.* [3,8] and Dinur *et al.* [10] that when the input and output differences are fixed, the solution set of the KECCAK S-box contains affine subspaces of dimension up to 3. In this paper, we show the maximum subspaces allowing linearization of S-box is of dimension 2. Furthermore, all affine subspaces of dimension up to 2 allow S-box linearization, and for those of dimension 3, six 2-dimensional affine subspaces out of it could allow the linearization. With this property in mind, we enforce linearization of all S-boxes in the first round, under which the first round function of the KECCAK permutation is transformed into a linear one. Combining with an inversion method of the S-box layer of the second round, we convert the problem of finding two-round connectors into that of solving a system of linear equations. Solving the equation once will produce sufficiently many solutions so that at least one of them will follow the differential trails in the following 3 rounds or more.

A side effect of linearization of all S-boxes is quick reduction of freedom degrees, which in turn decides the existence of such two-round connectors. To solve this problem, we aim to find differential trails, which impose least possible conditions to the two-round connectors. We design a dedicated search strategy to find suitable differential trails of up to 4 rounds. Implementation confirmed the correctness of this idea, and found real examples of collisions for 5-round SHAKE128 and two instances of challenge versions.

We list our results together with the related previous work in Table 1. Note, the algorithm for building 2-round connectors is heuristic and there is no theoretical bound for the solving time. However, it applies to our attacked instances within practical time, so we indicate the real time cost instead of complexities here. Experiments are run on a server with 32 cores of AMD processors.

Table 1. Collision attack results and comparison

Target $[r, c, d]$	n_r	Searching complexity	Searching time	Solving time[2]	Reference
SHAKE128	5	2^{39}	30 m	25 m	Sect. 6.2
KECCAK[1440,160,160]	5	2^{40}	2.48 h	9.6 s	Sect. 6.1
	6	$2^{70.24}$	N.A.[1]	1 h	Sect. 6.4
KECCAK[640,160,160]	5	2^{35}	2.67 h	30 m	Sect. 6.3
KECCAK-224	4	2^{24}	2–3 m		[10]
		2^{12}	0.3 s	2 m 15 s	Sect. 6.6
	5	2^{101}	N.A.	N.A.	Sect. 6.5
KECCAK-256	4	2^{24}	15–30 m		[10]
		2^{12}	0.28 s	7 m	Sect. 6.6

[1] N.A.: Not Available.
[2] There is no theoretical estimate for the solving time of the heuristic algorithms used here.

Organization. The rest of the paper is organized as follows. In Sects. 2 and 3, notations and a brief description of KECCAK family are given. In Sect. 4, we give a detailed description of the algebraic methods to achieve 2-round connectors. In Sect. 5, we give the dedicated search strategies for differential trails. Then the experimental results are presented in Sect. 6. We conclude the paper in Sect. 7.

2   Notations

We summarize the majority of notations to be used in this paper here.

c	Capacity of a sponge function
r	Rate of a sponge function
b	Width of a KECCAK permutation in bits, $b = r + c$
d	Length of the digest of a hash function
n_r	Number of rounds
$\theta, \rho, \pi, \chi, \iota$	The five step mappings that comprise a round, a subscript i denotes the operation at i-th round, e.g., θ_i denotes the θ layer at i-th round for $i = 0, 1, 2, \cdots$
L	composition of θ, ρ, π
L^{-1}	Inverse of L
RC	Round constant for a round of KECCAK-f permutation
$S(\cdot)$	5-bit S-box operating on each row of KECCAK state
$R^i(\cdot)$	KECCAK permutation reduced to the first i rounds
δ_{in}	5-bit input difference of an S-box,
δ_{out}	5-bit output difference of an S-box
DDT	Differential distribution table
α_i	Input difference of the i-th round function, $i = 0, 1, 2, \cdots$
β_i	Input difference of χ in the i-th round, $i = 0, 1, 2, \cdots$
w_i	Weight of the i-th round, $i = 0, 1, 2, \cdots$
$m_1 \| m_2$	Concatenation of strings m_1 and m_2
x	Bit value vector before χ in the first round
y	Bit value vector after the first round
z	Bit value vector before χ in the second round

3   Description of KECCAK

In this section, we give a brief description of KECCAK family of hash functions. KECCAK family applies sponge construction which processes messages in two phases—absorbing phase and squeezing phase, as shown in Fig. 1. The message is firstly padded by appending a bit string of 10*1, where 0* represents a shortest string of 0's so that the length of padded message is multiple of r. We denote the original message by M and the padded message by $\overline{M} = M \| 10^*1$. The b-bit internal state is initialized to be all 0's. In absorbing phase, the padded message is split into blocks of r-bits and each message block is XORed into the first r bits of current internal state, followed by the application of the fixed permutation to the entire b-bit state. This is repeated until all message blocks are processed. In the squeezing phase, the first r bits of the state are returned as output, then the permutation is applied and another r bits are outputted until all d output bits are produced. When restricted to the case of $b = 1600$ and $c = 2d$, the four official instances of KECCAK family are denoted by KECCAK-224/256/384/512 respectively for $d = 224, 256, 384, 512$. SHAKE128 and SHAKE256 are defined from two instances of KECCAK with the capacity c being 256 and 512, respectively, and the additional appending of a four-bit suffix 1111 to the original message M before applying the KECCAK padding. Without further specifications, we presume the digest sizes are 256 and 512 for SHAKE128 and SHAKE256, respectively. We use $\overline{M}$ to denote the padded message for SHAKE as well. Instances of KECCAK challenges will be denoted as KECCAK$[r, c, n_r, d]$, where the parameters are explicitly indicated for the rate, capacity, number of rounds the permutation is reduced to, and bit size of the digest, respectively.

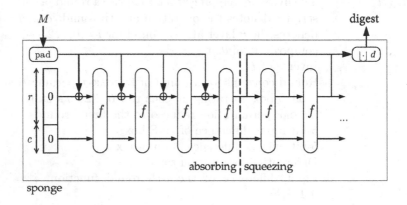

Fig. 1. Sponge construction [4].

The KECCAK permutation function in SHA-3 consists of 24 rounds of five layers operating on the 1600-bit state that can be represented as 5×5 64-bit lanes. In general 2^l is used to denote the bit length of lanes. If A denotes a 5-by-5-by-2^l array of bits that represents the state, then its indices are the integer triples (i, j, k) for which $0 \le i < 5, 0 \le j < 5$, and $0 \le k < 2^l$. The bit that corresponds to (i, j, k) is denoted by $A[i, j, k]$. Names for single-dimensional sub-arrays and two-dimensional ones are defined by the KECCAK designers: $A[\cdot][j][k]$ is called a row, $A[i][\cdot][k]$ is a column, and $A[i][j][\cdot]$ is a lane; $A[i][\cdot][\cdot]$ is called a sheet, $A[\cdot][j][\cdot]$ is a plane, and $A[\cdot][\cdot][k]$ is a slice.

The five layers in each round of the permutation are given below:

$\theta: A[i][j][k] \leftarrow A[i][j][k] + \sum_{j'=0}^{4} A[i-1][j'][k] + \sum_{j'=0}^{4} A[i+1][j'][k-1]$

$\rho: A[i][j][k] \leftarrow A[i][j][k + T(i,j)]$, where $T(i,j)$ is a predefined constant

$\pi: A[i][j][k] \leftarrow A[i'][j'][k]$, where $\begin{pmatrix} i \\ j \end{pmatrix} = \begin{pmatrix} 0 & 1 \\ 2 & 3 \end{pmatrix} \cdot \begin{pmatrix} i' \\ j' \end{pmatrix}$.

$\chi: A[i][j][k] \leftarrow A[i][j][k] + ((\neg A[i+1][j][k]) \wedge A[i+2][j][k])$,

$\iota: A \leftarrow A + RC[i_r]$, where $RC[i_r]$ is the round constants.

It is interesting to note that the size of permutation can be reduced to one of $\{25, 50, 100, 200, 400, 800\}$ by choosing $2^l = 1, 2, 4, 8, 16, 32$, respectively for the size of the lanes. In such cases, the round functions are defined exactly in the same way except the rotation constants of the ρ operation are now in modulo the respective 2^l instead of 64 in the original 1600-bit full permutation. These size-reduced permutations are not used in the SHA-3 instances, but in the Keccak challenges.

The first three layers are linear mappings and we denote their composition by $L \triangleq \theta \circ \rho \circ \pi$. The only non-linear layer of the permutation is χ, which can be seen as a S-box layer that applies 5-bit substitution to 320 rows of the state. We use $S(x)$ to denote the substitution of a 5-bit input value x. The difference distribution table of the S-box is denoted by DDT, where $DDT(\delta_{in}, \delta_{out})$ represents the size of the set $\{x : S(x) + S(x + \delta_{in}) = \delta_{out}\}$. We denote the Keccak permutation reduced to the first i rounds as R^i (note the round functions are identical up to a difference of constant addition in ι and we will omit ι as it has little impact on our differential collision attack), $i.e.$, $R^i(\overline{M})$ is the state after i rounds processing of the padded message $\overline{M}$.

4   Overview of Our Collision Attack

In this section, we give an overview of our collision attacks, followed by the details of the algebraic methods to achieve two-round connectors. Without further specification, we assume in this paper the length of the messages used are of one block after padding. To fulfil the Keccak padding rule, one needs to fix the last bit of the padded message to be "1", hence the first $r - 1$ bits of the state are under the full control of the attacker through the message bits, and the last c bits of the state are fixed to zeros as in the IV specified by Keccak. When applied to SHAKE, there are $r - 6$ free bits under control, by setting the last 6 bits of the padded message to be all 1's so it is compatible with the specific SHAKE padding rule.

Following the framework by Dinur $et\ al.$ [10], as well as many other collision attacks utilizing differential trails, our collision attacks consist of two parts, $i.e.$, a high probability differential trail and a connector linking the differential trail with the initial value, as depicted in Fig. 2. Let ΔS_I and ΔS_O denote the input and output differences of the differential trail, respectively. Dinur $et\ al.$ explored a method, which they call "target difference algorithm", to find message pairs (M, M') such that the output difference after one round permutation is ΔS_I, formally $R^1(\overline{M}\|0^c) + R^1(\overline{M'}\|0^c) = \Delta S_I$. In what follows, we show an algebraic method to extend this connector to two rounds, $i.e.$, a new target difference algorithm to find (M, M') such that $R^2(\overline{M}\|0^c) + R^2(\overline{M'}\|0^c) = \Delta S_I$. The differential trail is then fulfilled probabilistically with many such message pairs, collision can be produced if the first d bits of ΔS_O are 0. As we are aiming at low complexity attacks, finding solutions of connectors should be practical so that this part will not dominate the overall complexities of collision finding. Details of the differential trail search will be discussed in Sect. 5.

Overall, the constraints of the two-round connectors are that the last $c + 1$ (or $c + 6$) bits of the initial state are fixed, and the output difference after two rounds is given and fixed (this is determined by the differential trail to be used), we are to utilize the degree of freedom from the first $r - 1$ (or $r - 6$) bits of the initial state to find solutions efficiently. We will start with some observations on the KECCAK S-box, then move to the details of solution finding algorithm.

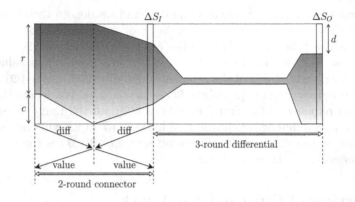

Fig. 2. Overview of 5-round collision attack

4.1 S-Box Linearization and Affine Subspaces

The key observation is that internal state is much larger than the digest size, providing large number of freedom degrees to attackers. One can choose some subsets of the available spaces with special properties to achieve fast enumerations. In case of KECCAK, we are to choose the subsets which are *linear* with respective to the S-box, *i.e.*, the expression of S-box can be re-written as linear transformation when the input is restricted to such subsets. It is obvious to note the S-box is non-linear when the entire 2^5 input space is considered. However, affine subspaces of size up to 4, as to be shown below, could be found so that the S-box can be linearized. Note that the S-box is the only nonlinear part of the KECCAK round function. Hence, the entire round function becomes linear when restricted to such subspaces. Formally, we define

Definition 1 (Linearizable affine subspace). *Linearizable affine subspaces are affine input subspaces on which S-box substitution is equivalent to a linear transformation. If V is a linearizable affine subspace of an S-box operation $S(\cdot)$, $\forall x \in V, S(x) = A \cdot x + b$, where A is a matrix and b is a constant vector.*

For example, when input is restricted to the subset $\{00000, 00001, 00100, 00101\}$ ($\{00, 01, 04, 05\}$ in hex), the corresponding output set of the KECCAK S-box is $\{00000, 01001, 00101, 01100\}$($\{00, 09, 05, 0C\}$ in hex), and the expression of the S-box can be re-written as linear transformation:

$$y = \begin{pmatrix} 1 & 0 & 1 & 0 & 0 \\ 0 & 1 & 0 & 0 & 0 \\ 0 & 0 & 1 & 0 & 0 \\ 1 & 0 & 0 & 1 & 0 \\ 0 & 0 & 0 & 0 & 1 \end{pmatrix} \cdot x \tag{1}$$

where x and y are bit vector representation of input and output values of the Keccak S-box with the last bit on top. By rotation symmetry, four more linearizable affine subspaces can be deduced from one.

Exhaustive search for the linearizable affine subspaces of the Keccak S-box shows:

Observation 1. *Out of the entire 5-dimensional input space,*

a. *there are totally 80 2-dimensional linearizable affine subspaces, as listed in Table 5 in Appendix A.*
b. *there does not exist any linearizable affine subspace with dimension 3 or more.*

For completeness, any 1-dimensional subspace is automatically linearizable affine subspace.

Since the affine subspaces are to be used together with differential trails, we are interested in those linearizable affine subspaces with fixed input and output differences, which is more relevant with the differential distribution table (DDT) of S-boxes. Referring to the DDT of Keccak S-box postponed to Appendix B, we observe:

Observation 2. *Given a 5-bit input difference δ_{in} and a 5-bit output difference δ_{out} such that $\mathrm{DDT}(\delta_{in}, \delta_{out}) \neq 0$, denote the value solution set $V = \{x : \mathrm{S}(x) + \mathrm{S}(x + \delta_{in}) = \delta_{out}\}$ and $\mathrm{S}(V) = \{\mathrm{S}(x) : x \in V\}$, we have*

a. *if $\mathrm{DDT}(\delta_{in}, \delta_{out}) = 2$ or 4, then V is a linearizable affine subspace.*
b. *if $\mathrm{DDT}(\delta_{in}, \delta_{out}) = 8$, then there are six 2-dimensional subsets $W_i \subset V, i = 0, 1, \cdots, 5$ such that $W_i(i = 0, 1, \cdots, 5)$ are linearizable affine subspaces.*

It is interesting to note the 2-dimensional linearizable affine subspaces obtained from analysis of DDT cover all the 80 cases in Observation 1. It is already noted in [15] there is one-to-one correspondence between linearizable affine subspaces and entries with value 2 or 4 in DDT. As for the DDT entries of value 8, we will leave the 6 choices of 2-dimensional linearizable affine subspaces for later usage. As an example, the 3-dimensional affine subspace corresponding to $\mathrm{DDT}(01, 01)$, *i.e.*, with both input and output differences being 01, is $\{10, 11, 14, 15, 18, 19, 1C, 1D\}$ and the six 2-dimensional linearizable affine subspaces from it are

$$\begin{aligned} &\{10, 11, 14, 15\}, \\ &\{10, 11, 18, 19\}, \\ &\{10, 11, 1C, 1D\}, \\ &\{14, 15, 18, 19\}, \\ &\{14, 15, 1C, 1D\}, \\ &\{18, 19, 1C, 1D\}. \end{aligned} \tag{2}$$

When projected to the whole KECCAK state, direct product of affine subspaces of each individual S-box form affine subspaces of the entire state with larger dimensions. In other words, when all the S-boxes in the round function are linearized, the entire round function becomes linear. This will be the way we are to handle the S-box layer of the first round of the 2-round connector.

4.2   A Connector Covering Two Rounds

The core idea of our two-round connector is to convert the problem to solving a system of linear equations. Two rounds of KECCAK permutation can be expressed as $\chi_1 \circ L_1 \circ \chi_0 \circ L_0$ (omitting the ι). With the χ_0 layer linearized by the techniques discussed above, $i.e.$, given the input and output differences of χ, the first three operations $L_1 \circ \chi_0 \circ L_0$ become linear. We will give details of the method how input and output differences of χ_0 are selected later. Now, we show how the χ_1 can be inverted by adding more constraints of linear equations. In our attack setting, the output difference of χ_1 is given as ΔS_I—input difference of the 3-round differential trail. It is not necessary that all S-boxes of the χ_1 layer are active, $i.e.$, with a non-zero difference. Here only active S-boxes are concerned, and each of them is inverted by randomly choosing an input difference with non-zero number of solutions, we call it *compatible* input difference. Formally, given the output difference δ_{out}, compatible input differences are $\{\delta_{in} : \text{DDT}(\delta_{in}, \delta_{out}) \neq 0\}$. As noted previously [3,8,10], for any pair of $(\delta_{in}, \delta_{out})$, the solution set $V = \{x : \text{S}(x) + \text{S}(x + \delta_{in}) = \delta_{out}\}$ forms an affine subspace. In other words, V can be deduced from the set $\{0,1\}^5$ by setting up i constraints that turn to be binary linear equations, when the size of solution set V is 2^{5-i}. For example, corresponding to $\text{DDT}(03, 02)$ is the 2-dimensional affine subspace $\{14, 17, 1C, 1F\}$ which can be formulated by the following three linear equations:

$$\begin{pmatrix} 0\,0\,1\,0\,0 \\ 1\,1\,0\,0\,0 \\ 0\,0\,0\,0\,1 \end{pmatrix} \cdot x = \begin{pmatrix} 1 \\ 0 \\ 1 \end{pmatrix}. \tag{3}$$

It is important to note, under the i linear constraints or set V, there is a bijective relation between δ_{in} and δ_{out}, $i.e.$, given one the other can be deduced deterministically. Hence, each active S-box in χ_1 layer is inverted by a choice of compatible input difference together with the corresponding i linear constraints on the input values. Once input difference and linear constraints for all active S-boxes of χ_1 are enforced and fulfilled, solutions of 2-round connector are found. Note a compatible input difference of χ_1 is a choice of β_1, and α_1 can be uniquely determined by the relation $\alpha_1 = L^{-1}(\beta_1)$. In the remaining part of this subsection, more details on implementation of this idea are given.

As depicted in Fig. 3, the variables of our equation system are the bit values before the first χ layer denoted by vector x. y and z are bit vectors of intermediate values for further interpretation where y represents the output after the first χ layer and z the bits before the second χ layer. The main task is to derive all constraints on differences and affine subspaces to that on the variables x. Now, suppose β_1 and β_0 (details will be given in Sect. 4.4) are fixed, and ΔS_I (aka. α_2) is given, we show how the system of equations could be set up. With the input difference β_1 and output difference ΔS_I of χ_1, all the linear constraints on the input affine subspaces of the active S-boxes can be derived and stored as

$$G \cdot z = m,$$

where G is a block-diagonal matrix in which each diagonal block together with corresponding constants in m formulates the constraints of one active S-box. Similar procedure is done for input affine subspaces of the first round, except that the input is restricted to linearizable affine subspaces for all S-boxes regardless whether or not the S-box is active so that χ_0 layer can be replaced by a linear transformation χ_L. We denote the constraints by

$$A \cdot x = t. \tag{4}$$

Then x and y can be linked by

$$\chi_L \cdot x + \chi_C = y,$$

where χ_C denotes the constant offsets for the affine subspaces. Furthermore, the two equation systems can be linked by

$$L \cdot (y + RC[0]) = z,$$

where $RC[0]$ denotes the round constant of the first round. Note, only active S-boxes of the second round are concerned, *i.e.*, only part of bits of z are known, hence the same applies to y, and we use y' to denote the known bits of y for later. Overall, the constraints on z can be derived to that on x as

$$G \cdot L \cdot (\chi_L \cdot x + \chi_C + RC[0]) = m. \tag{5}$$

Note an additional constraint x needs to fulfil is that the last $c + 1$ (or $c + 6$) bits of initial state are pre-fixed, which can be derived as

$$L^{-1}(x) = CA, \tag{6}$$

where CA denotes the preset values for bits of the inner state and padding bits. We use E_M to denote the equation systems (4), (5) and (6), solutions fulfilling E_M will be solutions of 2-round connectors.

Algorithm for Building Two-Round Connectors. We use *the basic linearization procedure* to generate the equations for confining x to a smaller subspace suitable for linearization of the first χ layer and use *the main linearization procedure* to generate the final equations to bypass the second χ layer. One of the inputs of the basic procedure is the equation system E_M on x values, other inputs include the input and output differences of the first S-box layer β_0, α_1 and y'.

The Basic Linearization Procedure.

Inputs: $E_M, \beta_0, \alpha_1, y'$.
Outputs: updated E_M, χ_L, χ_C.

1. Initialize a matrix χ_L and a vector χ_C.
2. Iterate on each bit of y', calculate the index of the bit in S-box level, say the j-th bit of the i-th S-box in the first round. Then for the i-th S-box in the first round:
   (a) If the i-th S-box has not been processed in this procedure before, then:
       (i) If it is non-active, randomly choose a linearizable 2-dimensional subspace, check whether the 3 equations specifying this 2-dimensional affine subspace is consistent with the current E_M.
           If so, add them to E_M and update χ_L and χ_C with the j-th line of the matrix which specifies the affine linear transformation. Continue to next bit of y' in step 2.
           Otherwise, try another linearizable 2-dimensional subspace. If all linearizable 2-dimensional subspaces have been tried and no consistent equations exist, output "No Solution in basic procedure".
       (ii) Otherwise it is active: find its input and output differences from β_0 and α_1, i.e., $\delta_{in}, \delta_{out}$.
           Case 1. When $\text{DDT}(\delta_{in}, \delta_{out}) = 8$, randomly choose one of the six linearizable 2-dimensional subspaces and the corresponding equation to specialize this 2-dimensional subspace (the other two of the three equations to formulate the 2-dimentional subspace have already been indicated in E_M after choosing β_0 procedure).
           If current E_M is consistent with this linear equation, add it to E_M and update χ_L and χ_C with the j-th line of the matrix which specifies the linear map from the 2-dimensional subspace to the output 2-dimensional subspace of S-box. Continue to next bit of y' in step 2.
           Otherwise, try another randomly chosen 2-dimensional linearizable subspace. If all six 2-dimensional linearizable subspaces have been chosen and no consistent equation exist, output "No Solution in basic procedure".
           Case 2. When $\text{DDT}(\delta_{in}, \delta_{out}) = 2$ or 4, update χ_L and χ_C with the j-th line of the matrix which specifies the affine linear transformation of the input 1 or 2-dimensional subspace to the output 1 or 2-dimensional subspace of S-box. Continue to next bit of y' in step 2.
   (b) Otherwise, if the i-th S-box has already been processed in this procedure: update χ_L and χ_C with the j-th line of the matrix which specifies the affine linear transformation of the predefined linearizable subspace to the output subspace of S-box.
3. Output the current equations system E_M as well as χ_L and χ_C such that $\chi_L \cdot x + \chi_C = y'$.

The inputs to the Main procedure are $\beta_0, \alpha_1, \beta_1, \alpha_2(\Delta S_I)$ and E_M we get after choosing β_0.
The Main Linearization Procedure.

Input: $E_M, \beta_0, \alpha_1, \beta_1, \alpha_2$.
Output: Updated E_M.

1. Using β_1 and α_2, initialize a coefficient matrix G and a constant vector m that specify the linear equations to constrain the input bits of the second S-box layer for deriving the equation $G \cdot z = m$.
2. Derive the L into the matrix format for $L \cdot (y + RC[0]) = z$.
3. Initialize a counter to 0.
4. Execute the basic linear procedure with indexes of know bits y' in y and E_M, β_0 and α_1. If the procedure succeeds, it will return the matrix specifying the linearization of the first S-box layer such that $\chi_L \cdot x + \chi_C = y'$, then continue to Step 6. Otherwise, go to step 5.
5. Increment the counter. If the counter's value is equal to a preset threshold $T1$, output "Failed". Otherwise, go to step 4.
6. Test whether the equation system (5) is consistent with E_M. If so, add the new system to E_M and output final E_M. Otherwise, go to step 5.

Note that the algorithms do not succeed all the time. To overcome this problem, from the input difference ΔS_I of a 3-round differential trail, we repeat random picks of compatible input differences β_1 until the main procedure succeeds. As the number of active S-boxes in α_2 is large enough (range from tens to hundreds in our experiments), there are enough different cases for β_1 resulting in high final success probability. An interesting point is that the invertion from α_2 to β_1 does not need to maintain high probability because this transition is covered in our two-round connector. Besides, the unconstrained number of active S-boxes of an input difference allows more freedom in searching of the most suitable three round differential trails. We will describe the searching strategies in Sect. 5. Finally, exhaustive search of solution for the following 3-round differential trails can be performed from the solution space of E_M.

4.3   Analysis of Degree of Freedom

The degree of freedom of solution space of final E_M is a key factor on success of our method. A solution space with degree of freedom larger than the weight of the 3-round differential trail is possible to suggest a message pair with collision digest. After the linearization of the first round, the degree of freedom is $\sum_{i=0}^{\frac{b}{5}-1} \mathrm{DF}_i^{(1)}$ in which $\mathrm{DF}_i^{(1)}$ is the degree of freedom of 5-bit input space of the i-th S-box in the first round. The value is assigned for $\mathrm{DF}_i^{(1)}$ according to rules in Table 2.

The constraints on the initial state reduce $(c + p)$ degree of freedom where c is the capacity and p is due to the padding rule. We have $p = 1$ for KECCAK and $p = 6$ for SHAKE. Another decrease on degree of freedom is due to the constraints on the input values of the S-box layer in the second round. The definition of $\mathrm{DF}_i^{(2)}$, the degree of freedom of 5-bit input values to S-boxes in the second round, is

$$
\mathrm{DF}_i^{(2)} = \begin{cases} 1, & \mathrm{DDT}(\delta_{in}, \delta_{out}) = 2, \\ 2, & \mathrm{DDT}(\delta_{in}, \delta_{out}) = 4, \\ 3, & \mathrm{DDT}(\delta_{in}, \delta_{out}) = 8, \\ 5, & \mathrm{DDT}(\delta_{in}, \delta_{out}) = 0, \end{cases} \tag{7}
$$

Table 2. Rules for value assignment for DF_i^1.

$DF_i^{(1)}$ *	Non-active	$DDT(\delta_{in}, \delta_{out}) = 2$	$DDT(\delta_{in}, \delta_{out}) = 4$	$DDT(\delta_{in}, \delta_{out}) = 8$
Involved in y'	2	1	2	2
Not involved in y'	5	1	2	3

* The value of $DF_i^{(1)}$ is based on whether the i-th S-box is involved in y' and the value $DDT(\delta_{in}, \delta_{out})$ where δ_{in} and δ_{out} are the input and output differences of the i-th S-box in the first round.

where δ_{in} and δ_{out} are the input and output differences of the i-th S-box in the second round. For the i-th S-box in the second round, we add $(5 - DF_i^{(2)})$ equations to E_M and suppose to deduce the degree of freedom by this amount.

The degree of freedom of the final E_M is estimated as

$$DF = \sum_{i=0}^{\frac{b}{5}-1} DF_i^{(1)} - (c+p) - \sum_{i=0}^{\frac{b}{5}-1} (5 - DF_i^{(2)}). \tag{8}$$

Large DF benefits our search for collisions in rounds beyond the second round.

4.4   How to Choose β_0

So far we have not given details on how β_0 can be selected. We follow Dinur et al.'s work [10] in a more general way to uniquely determine β_0, the difference before χ layer in the first round. The algorithm is called "target difference algorithm" and consists of difference phase and value phase.

Given ΔS_I, we have randomly chosen a compatible input difference β_1. We then build two equation systems E_Δ and E_M accordingly. E_Δ is on differences of the message pairs and E_M is on values of one message. The initialization of E_Δ should abide by (1) the constraints implied by padding rules that the last $c+1$ difference bits of initial state equal to 0, and (2) the input difference bits of nonactive S-boxes in the first round equal to 0. The initialization of E_M should abide by the padding rules that the last $c + p$ value bits equal to $1^p\|0^c$. We set $p = 1$ for KECCAK and $p = 6$ for SHAKE. These rules are easy to be implemented as the variable vector x is an invertible linear mapping of the initial vector. Therefore, in the initialization period, we equate the corresponding bits to their enforced values in E_Δ and E_M.

For E_Δ, we add additional equations to enforce that α_1 is possibly deduced from β_0. Though the obvious way is to equate the 5 input difference bits to a specific value for each active S-box in β_0, this will restricts the solution space significantly. As suggested in [10], we chose one of the 2-dimensional affine subsets of input differences instead of a specific value for each active S-box. This is based on the fact that given any nonzero 5-bit output difference to a KECCAK S-box, the set of possible input differences contains at least five 2-dimensional affine subspaces. After a consistent E_Δ system has been constructed, the solution space is an affine subspace of candidates for β_0. Then we continue to maintain E_Δ by iteratively add the additional 2 equations to uniquely specify each 5-bit input difference for the active S-boxes. For all active S-boxes, once the specific input differences is determined, we add equations to E_M system to enforce every active 5-bit of x (input bits to active S-box) to an affine subspace corresponding to the uniquely determined δ_{in} and δ_{out}. In this way, we always find a compatible β_0 from α_1 fulfilling the constraints from the $c + p$ bits of padding and pre-set bits of capacities.

5 Search for Differential Trails

In this section, we elaborate on our searching algorithms for finding differential trails of KECCAK. Our ideas greatly benefit from previous works of searching differential trails for KECCAK [9, 14, 19]. We start by recalling several properties of the operations in the round function, followed by our considerations in finding differential trails. Then, we describe our searching algorithms which provide differential trails for practical collision attacks against KECCAK[1440, 160, 5, 160], 5-round SHAKE128 and KECCAK[640, 160, 5, 160] respectively, and trails for theoretical collision attack against 5-round KECCAK-224 and KECCAK[1440, 160, 6, 160].

5.1 Properties of θ, ρ, π, ι and χ

θ, ρ, π, ι are linear operations while χ acts as the parallel application of 5-bit nonlinear S-boxes on the rows of the state. Since ι adds a round constant and has no essential effect on difference, we ignore it in this section. Additionally, ρ and π do not change the number of active bits in a differential trail, but only positions. Therefore, θ and χ are the crucial parts for differential analysis.

To describe the properties of θ, we take definitions from [3]. The *column parity* (or parity for short) $P(A)$ of a value (or difference) A is defined as the parity of the columns of A, i.e. $P(A)[i][k] = \Sigma_j A[i][j][k]$. A column is even, if its parity is 0, otherwise it is odd. A state is in CP-kernel if all its columns are even.

θ adds a pattern to the state, and this pattern is called the θ-effect. The θ-effect of a state A is $E(A)[i][k] = P(A)[i-1][k] + P(A)[i+1][k-1]$. So θ depends only on column parities. The θ-gap is defined as the Hamming weight of the θ-effect divided by two. Note that if the θ-gap is g, after applying θ there are $10g$ bits flipped. Given a state A in CP-kernel, the θ-gap is zero and hence the Hamming weight of A remains after θ. Another interesting property is that θ^{-1} diffuses much faster than θ. More exactly, a single bit difference can be propagated to about half state bits through θ^{-1}.

Given an input difference to χ, all possible output differences occur with the same probability. On the contrary, given an output difference to χ, it is not the same case, but the highest probability of all possible input differences is determined. Moreover, for one-bit differences, each S-box of χ acts as identity with probability 2^{-2}.

5.2   Representation of Trails and Their Weights

As in previous sections, we denote the differences before and after i-th round by α_i and α_{i+1}, respectively. Let $\beta_i = L(\alpha_i)$. Therefore an n-round differential trail starting from 0-th round is of the following form

$$\alpha_0 \xrightarrow{L} \beta_0 \xrightarrow{\chi} \alpha_1 \xrightarrow{L} \cdots \alpha_{n-1} \xrightarrow{L} \beta_{n-1} \xrightarrow{\chi} \alpha_n.$$

For the sake of simplicity, a trail can also be represented with only β_i's or α_i's.

The weight of a differential $\beta \rightarrow \alpha$ over a function f with domain $\{0,1\}^b$ is defined as

$$w(\beta \rightarrow \alpha) = b - \log_2 |\{x : f(x) \oplus f(x \oplus \beta) = \alpha\}|.$$

In other words, the weight of a differential $\beta \rightarrow \alpha$ is equal to $-\log_2 \Pr(\beta \rightarrow \alpha)$. If $\Pr(\beta \rightarrow \alpha) > 0$, we say α and β are compatible, otherwise the weight of $\beta \rightarrow \alpha$ is undefined.

We denote the weight of i-th round differential by w_i where i starts from 0, and thus the weight of a trail is the sum of the weights of round differentials that constitute the trail. In addition, we use $\#AS(\alpha)$ to represent the number of active S-boxes in a state difference α. According to the properties of χ, given β_i the weight of $(\beta_i \rightarrow \alpha_{i+1})$ is determined; also, given β_i the minimum reverse weight of $(\beta_{i-1} \rightarrow L^{-1}(\beta_i))$ is fixed.

As in [3], $n-1$ consecutive β_i's, say $(\beta_1, \cdots, \beta_{n-1})$ is called an n-round **trail core** which defines a set of n-round trails $\alpha_0 \xrightarrow{L} \beta_0 \xrightarrow{\chi} \alpha_1 \xrightarrow{L} \beta_1 \cdots \xrightarrow{L} \beta_{n-1} \xrightarrow{\chi} \alpha_n$ where the first round is of the minimal weight determined by $\alpha_1 = L^{-1}(\beta_1)$, and α_n is compatible with β_{n-1}. The first step of mount collision attacks against n-round KECCAK is to find good $(n-1)$-round trail cores.

5.3   Requirements for Differential Trails

Good trail cores are those satisfying all the requirements which we will explain as follows. The first requirement is that the difference of the output is zero, i.e. $\alpha_{n_r}^d = 0$ (we denote output digest difference after n_r rounds with $\alpha_{n_r}^d$). The second requirement relates to the freedom degree budget.

With the definition of weight, Eq. (8) can be represented in an alternative way

$$\text{DF} = \sum_{i=0}^{b/5-1} \text{DF}_i^{(1)} - (c+p) - w_1. \tag{9}$$

The first term of the formula depends on the number of S-boxes that need to be linearized and its corresponding DDT entry as depicted in Table 2. Empirically, when all S-boxes are active and linearized in the first round it is more possible to get a consistent equation system. Therefore, we heuristically set $\frac{b}{5} \times 2$ as a threshold for the first term in (9), and denote a threshold of the first two terms in (9) for further search conditions by

$$\text{TDF} = \frac{b}{5} \times 2 - (c+p).$$

To mount collision attacks against KECCAK$[r, c, n_r, d]$ with methods described in Sect. 4, it is necessary that

$$\mathrm{TDF} > w_1 + \cdots + w_{n_r-2} + w^d_{n_r-1} \tag{10}$$

where $w^d_{n_r-1}$ is the part of w_{n_r-1} that relates to the digest[1]. The trail searching phase is performed to provide ΔS_I for the connector building algorithm. However, the sufficient conditions for a good trail core is restrained by solving results of the connector, i.e. the number of freedom degrees of the solution space of E_M. So we take (10) as a heuristic condition for searching good trail cores which are promising for collision attacks.

Thirdly, the collision attack should be practical. Note that after we obtain a subspace of message pairs making it sure to bypass the first two rounds, the complexity for searching a collision is $2^{w_2+\cdots+w^d_{n_r-1}}$. To make our attacks practical, we restrict $w_2 + \cdots + w^d_{n_r-1}$ to be small enough, say 48.

We summarize the requirements for differential trails as follows and list TDFs for different versions of KECCAK$[r, c, n_r, d]$ in Table 3.

(1) $\alpha^d_{n_r} = 0$, i.e. the difference of output must be zero.
(2) $\mathrm{TDF} > w_1 + \cdots + w^d_{n_r-1}$, i.e. the degree of freedom must be sufficient;
(3) $w_2 + \cdots + w^d_{n_r-1} \le 48$, the complexity for finding a collision should be low.

Table 3. TDFs of different versions of KECCAK$[r, c, n_r, d]$.

KECCAK$[r, c, n_r, d]$	TDF	Remarks
KECCAK[1440, 160, 5, 160]	479	Challenge
KECCAK[1344, 256, 5, 256]	378	SHAKE128
KECCAK[640, 160, 5, 160]	159	Challenge
KECCAK[1440, 160, 6, 160]	479	Challenge
KECCAK[1152, 448, 5, 224]	191	KECCAK-224

* 'Challenge' means that version is included in Keccak Crunchy Crypto Collision and Pre-image Contest [2].

5.4   Searching Strategies

Searching From Light β_3's. Our initial goal is to find collisions for 5-round KECCAK. To facilitate a 5-round collision of KECCAK, we need to find 4-round differential trails satisfying the three requirements mentioned previously. However it is difficult to meet all of them simultaneously even though each of them can be fulfilled easily.

---

[1] Suppose all the equations are independent. In later parts we will show these equations are not necessarily independent.

We explain as follows. Since we aim for practical attacks, $w_2 + w_3 + w_4^d$ must be small enough, say 48. That is to say, the last three rounds of the trail must be light and sparse. When we restrict a 3-round trail to be lightweight and extend it backwards for one round, we almost always unfortunately get a heavy state α_2 (usually $\#AS(\alpha_2) > 120$) whose weight may exceeds the TDF. We take KECCAK-224 as an example. The TDF of KECCAK-224 is 191, which indicates $\#AS(\alpha_2) < 92$ as the least weight for an S-box is 2. For a lightweight 3-round trail, it satisfies Requirement (1) occasionally. The greater d is, the less trails satisfy Requirement (1).

With these requirements in mind, we search for 4-round differential trail cores from light middle state differences β_3's. From light β_3's we search forwards and backwards, and check whether Requirement (1) and (2) are satisfied respectively; once these two requirements are satisfied, we compute the weight $w_2 + w_3 + w_4^d$ for brute force, hoping it is small enough for practical attacks.

α_3, α_4 **in CP-kernel.** The designers of KECCAK show in [3] that it is not possible to construct 3-round low weight differential trails which stay in CP-kernel. However, 2-round differential trails in CP-kernel are possible, as studied in [9,14,19].

We restrict α_3 in CP-kernel. If $\rho^{-1} \circ \pi^{-1}(\beta_3)$ is outside the CP-kernel and sparse, say 8 active bits, the active bits of $\alpha_3 = L^{-1}(\beta_3)$ will increase due to the strong diffusion of θ^{-1} and the sparseness of β_3. When $\#AS(\alpha_3) > 10$, the complexity for searching backwards for one β_3 is greater than $2^{31.7}$ which is too time-consuming. We had better also confine α_4 to the CP-kernel. If not, the requirement $\alpha_{n_r}^d = 0$ may not be satisfied. As can be seen from the lightest 3-round trail for KECCAK-f[1600] [14], even though the θ-gap is only one, after θ the difference bits are diffused among the state making a 224-bit collision impossible (a 160-bit collision is still possible). So our starting point is special β_3's which makes sure $\alpha_3 = L^{-1}(\beta_3)$ lies in CP-kernel, and for which there exists a compatible α_4 in CP-kernel. Fortunately, such kind of β_3's can be obtained with KeccakTools [6].

Steps for Searching 4-Round Differential Trails. We sketch below our steps for finding 4-round differential trail cores for KECCAK and provide a description in more detail in Appendix C. To mount collision attacks on 6-round KECCAK, 5-round differential trail cores are needed. In this case, we just extend our forward extension for one more round.

1. Using KeccakTools, find special β_3's with a low Hamming weight, say 8.
2. For every β_3 obtained, traverse all possible α_4 using a tree structure, compute $\beta_4 = L(\alpha_4)$ and test whether there exists a compatible α_5 where $\alpha_5^d = 0$. If so, keep this β_3 and record its forward extension, otherwise discard it.
3. For remaining β_3's, also using a tree structure traverse all possible β_2 which is compatible with $L^{-1}(\beta_3)$'s, compute $\#AS(\alpha_2)$ from β_2. If $\#AS(\alpha_2)$ is small enough, say below 110, check whether this trail core $(\beta_2, \beta_3, \beta_4)$ under consideration is sufficient for collision attacks.

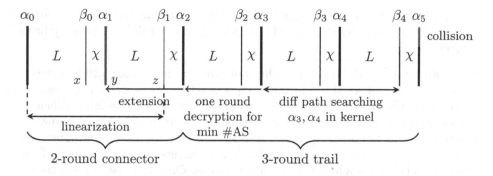

Fig. 3. Collision attacks on 5-round Keccak.

5.5 Searching Results

Some of the best differential trail cores we obtained are listed in Table 4. As can be seen that Trail cores No. 1–3 are all suitable for collision attacks against Keccak[1440,160,5,160], and Trail cores No. 1 and 2 for SHAKE128. Trail core No. 4 is sufficiently good for collision attacks against Keccak[640, 160, 5, 160]. However, to mount collision attacks on Keccak-224, all the first three trail cores are not good enough. Fortunately, a doubled version of Trail core No. 4 can make our two-round attack possible because $85 \times 2 = 170 < 191$. For Keccak[1440, 160, 6, 160], we also find a trail core ripe for collision attacks except that Requirement (3) is not satisfied. Details of these differential trail cores are provided in Appendix D.

Table 4. Differential trail cores for Keccak$[r, c, n_r, d]$.

No	$r + c$	$\#AS(\alpha_2\text{-}\beta_2\text{-}\beta_3\text{-}\beta_4^d)$	$w_1\text{-}w_2\text{-}w_3\text{-}w_4^d$	d
1	1600	102-8-8-2	240-19-16-4	256
2	1600	88-8-7-0	195-21-15-0	256
3	1600	85-9-10-2	190-25-20-3	224
4	800	38-8-8-0	85-20-16-0	160
No.	$r + c$	$\#AS(\alpha_2\text{-}\beta_2\text{-}\beta_3\text{-}\beta_4\text{-}\beta_5^d)$	$w_1\text{-}w_2\text{-}w_3\text{-}w_4\text{-}w_5^d$	d
5	1600	145-6-6-10-14	340-15-12-22-23	160

6 Experiments and Results

In this section, we employ 4-round (5-round) trail cores to mount collision attacks against 5-round (6-round) Keccak$[r, c, n_r, d]$. Our attack consists of two main stages:

– *Connecting stage.* Find a subspace of messages bypassing the first two rounds.
– *Brute-force searching stage.* Find a colliding pair from this subspace by brute force.

In the first stage, with α_2 fixed by the trail core, we choose compatible β_1 where $\alpha_1 = L^{-1}(\beta_1)$ and all the S-boxes in α_1 are active. In order to save freedom degrees, we also restrict that $\beta_1 \rightarrow \alpha_2$ should be of least weight. When β_1 is chosen, we run the two-round connector. If a certain number of failures is reached, we select another β_1 until a solution is found, i.e. a subspace of message pairs definitely reaching to α_2 is obtained. If the number freedom degrees of this subspace is large enough, the first stage succeeds. Once the first stage succeeds, we move on to the second stage for finding a colliding message pair.

6.1 Collision Attack of KECCAK[1440,160,5,160]

We apply Trail core No. 2 to the collision attack of 5-round KECCAK[1440, 160, 5, 160]. In this case, we choose compatible β_1s randomly. After solving the two-round problem in 9.6 s, the degree of freedom is 162, which is enough for collision search of the remaining 3 rounds with probability 2^{-40}. The searching time for the collision is 2.48 h. We give one example of collisions in Table 10, with which we solve a challenge of Keccak Crunchy Crypto Collision and Pre-image Contest [2].

6.2 Collision Attack of 5-Round SHAKE128

We apply Trail core No. 1 to the collision attack of 5-round SHAKE128[2]. As the capacity of SHAKE128 is much larger than that of KECCAK[1440, 160, 5, 160], which means about 100 more freedom degrees are needed, we just choose compatible β_1s where $\beta_1 \rightarrow \alpha_2$ is of least weight. We also follow this rule in later collision attacks. After solving the two-round problem with 25 min, the degree of freedom is 94 and the search for 3-round collision with probability 2^{-39} costs half an hour. We give an instance of collision in Table 11.

6.3 Collision Attack of KECCAK[640,160,5,160]

We apply Trail core No. 5 to the collision attack of KECCAK[640, 160, 5, 160]. The methods used in this case are similar to those of 5-round SHAKE128. The first stage succeeds in 30 min. The second stage takes 2 h 40 min to find a collision which happens with probability 2^{-35}. An example of collision is provide in Table 12, with which we solve another challenge of Keccak Crunchy Crypto Collision and Pre-image Contest [2].

---

[2] We also utilized Trail core No. 2, but Trail core No. 1 produces a colliding pair in a relatively shorter time.

6.4   Collision Attack of KECCAK[1440,160,6,160]

We found four trail cores for which there exist zero 160-bit output differences. The one with the best probability is Trail core No. 5 which is displayed in Table 9. From β_4 there are 24 trails to zero α_6^d. Taking all these trails into consideration, we get a complexity of $2^{67.24}$–$2^{70.24}$ for the second stage. If we let $\#AS(\alpha_2)$ (w_2) be the smallest, the complexity for the second stage is $2^{70.24}$ ($2^{67.24}$). In the experiments, we let $\#AS(\alpha_2)$ be the smallest. In one hour our two-round algorithm returns a subspace of messages with freedom degree 135, and in 20 min we get a message pair shown in Table 13 that follows the first four rounds of the differential trail, which demonstrates that in time complexity of $2^{70.24}$ a collision for 6-round KECCAK[1440, 160, 6, 160] will be found with great confidence.

6.5   Collision Attack of 5-Round Keccak-224

For the collision attack of 5-round KECCAK-224, all the 4-round trail cores we found for KECCAK-f[1600] are not good enough, i.e. the weight of the trail cores exceeds TDF too much and even $w_2 >$ TDF. However, our two-round connector is still likely to work. For one hand, from Trail core No. 4 for KECCAK-f[800] we can construct a 4-round trail core for KECCAK-f[1600] with weight pattern (170-40-32-0) which makes our two-round connector possible. From the other, as the capacity increases, it is probable that equations added in connecting phase are not always mutually independent, which means the assumption of freedom degrees of our connector may be less than TDF. The applicability of our connector in this case is verified with experiments. With Trail core No. 4, the two round connector returns a subspace of messages of freedom degree 11 and 2 or 3 for Trail core No. 3. Since the message subspaces derived are too small to mount collision attacks against 5-round KECCAK-224, we turn to two-block messages. Once we get c bits from the first block, we set corresponding c bit constants in E_M to the value we obtained and then solve the system to find a subspace of messages for the second block. Now the attack proceeds in the following way.

– *Connecting stage.*
  • Use the two-round connector to find a message subspace with freedom degree s as large as possible, hoping that $t = (c+p)+\text{rank}(E_M \cap E_{(c,p)}) - \text{rank}(E_M)$ is as small as possible.
– *Brute-force searching stage.*
  • Choose the first message randomly and compute the c-bit value for the second block. Replace the corresponding c bit constants in E_M and check whether it is still consistent. If it is consistent, we obtain another subspace with size 2^s.
  • Search for collision with the subspace.
  • Repeat until we find a two-block collision.

In our experiment, using Trail core No. 3 the connector returns a message subspace with freedom degree $s = 2$, and $t = 55$. Then the complexity for find a two-block collision is $2^{55+(48-2)} = 2^{101}$.

6.6   Re-launch 4-Round Collision Attacks of Keccak-224 and Keccak-256

Though the 4-round collisions of KECCAK-224 and KECCAK-256 have already been found [10], we use our method to optimize the complexity. We start from the same 2-round differential trail in Dinur *et al.*'s work [10] and build a two-round connector. The time spent on building and solving the two-round connectors is 2 min 15 s for KECCAK-224 and 7 min for KECCAK-256. Then the complexity for brute forth searching is reduced to 2^{12} and cost 0.325 s and 0.28 s respectively which outperforms 2^{24} on-line complexity in [10]. Besides, it is pointed out in [10] that even though they got subsets with more than 2^{30} message pairs from their target difference algorithm, they were not able to find collisions within some of these subsets. The reason was suspected to be the incomplete diffusion within the first two rounds and the closely related message pairs within a subset. While in our algorithm, we did not encounter such a problem. In other words, we always find collisions from the subsets deduced from the two-round connector. Thus once we succeed in the 2-round connector building phase with a large enough subset, we never need to repeat it.

7   Conclusion

In conclusion, we observed that the KECCAK S-box can be re-expressed as linear transformations under some restricted input subspaces. With this property, we linearized all S-boxes of the first round, and extended the existing connector by one round. Implementations confirmed our idea, and found us real examples of 5-round SHAKE128, and two instances of KECCAK challenges. Theoretical results on 5-round KECCAK-224 and a 6-round KECCAK challenge version are projected.

It is noted that the algorithm for solving the two-round connectors are heuristic, further work includes finding the theoretical bounds of this algorithm and factors deciding the complexities for possible improvements. Note, any relaxation on the restrictions of ΔS_I might lead us to better differential trails in the searching phase.

Acknowledgement. The authors would like to thank anonymous reviewers and Joan Daemen for their helpful comments and suggestions. The work of this paper was supported by the National Key Basic Research Program of China (2013CB834203) and the National Natural Science Foundation of China (Grants 61472417, 61472415, 61402469, and 61672516).

A   2-Dimentional Linearizable Affine Subspaces of KECCAK S-box

There are totally 80 2-dimensional linearizable affine subspaces for KECCAK S-box as listed in Table 5.

Table 5. Linearizable affine subspaces of KECCAK S-box

{0, 1, 4, 5}	{2, 3, 6, 7}	{0, 1, 8, 9}	{4, 5, 8, 9}	{0, 2, 8, A}
{1, 2, 9, A}	{0, 3, 8, B}	{1, 3, 9, B}	{2, 3, A, B}	{6, 7, A, B}
{0, 1, C, D}	{4, 5, C, D}	{8, 9, C, D}	{4, 6, C, E}	{5, 6, D, E}
{4, 7, C, F}	{5, 7, D, F}	{2, 3, E, F}	{6, 7, E, F}	{A, B, E, F}
{0, 2, 10, 12}	{8, A, 10, 12}	{1, 3, 11, 13}	{9, B, 11, 13}	{0, 4, 10, 14}
{1, 5, 10, 14}	{2, 4, 12, 14}	{0, 4, 11, 15}	{1, 5, 11, 15}	{3, 5, 13, 15}
{10, 11, 14, 15}	{0, 6, 10, 16}	{2, 6, 12, 16}	{3, 7, 12, 16}	{4, 6, 14, 16}
{C, E, 14, 16}	{1, 7, 11, 17}	{2, 6, 13, 17}	{3, 7, 13, 17}	{5, 7, 15, 17}
{D, F, 15, 17}	{12, 13, 16, 17}	{10, 11, 18, 19}	{14, 15, 18, 19}	{0, 2, 18, 1A}
{8, A, 18, 1A}	{10, 12, 18, 1A}	{11, 12, 19, 1A}	{10, 13, 18, 1B}	{1, 3, 19, 1B}
{9, B, 19, 1B}	{11, 13, 19, 1B}	{12, 13, 1A, 1B}	{16, 17, 1A, 1B}	{8, C, 18, 1C}
{9, D, 18, 1C}	{A, C, 1A, 1C}	{8, C, 19, 1D}	{9, D, 19, 1D}	{B, D, 1B, 1D}
{10, 11, 1C, 1D}	{14, 15, 1C, 1D}	{18, 19, 1C, 1D}	{8, E, 18, 1E}	{A, E, 1A, 1E}
{B, F, 1A, 1E}	{4, 6, 1C, 1E}	{C, E, 1C, 1E}	{14, 16, 1C, 1E}	{15, 16, 1D, 1E}
{9, F, 19, 1F}	{A, E, 1B, 1F}	{B, F, 1B, 1F}	{14, 17, 1C, 1F}	{5, 7, 1D, 1F}
{D, F, 1D, 1F}	{15, 17, 1D, 1F}	{12, 13, 1E, 1F}	{16, 17, 1E, 1F}	{1A, 1B, 1E, 1F}

B Differential Distribution Table of KECCAK S-box [14]

C Steps for Finding Differential Trials

In this section, we describe more at length about the steps for finding differential trails.

1. Generate β_3s each of which makes sure $\alpha_3 = L^{-1}(\beta_3)$ lies in CP-kernel, and for each β_3 there exists a compatible α_4 in CP-kernel using TrailCoreInKernelAtC of KeccakTools [6] where the parameter $aMaxWeight$ is set to be 52. As a result, 503 such β_3s are obtained.

2. For every β_3 obtained, if $\#AS(\beta_3) < 16$ we traverse all possible α_4 using a tree structure, compute $\beta_4 = L(\alpha_4)$ and test whether there exists a compatible α_5 where $\alpha_5^d = 0$. If so, keep this β_3 and record its forward extension, otherwise discard it. For $d = 224$, 351 β_3s are left while 495 β_3s are left for $d = 160$.

3. For remaining β_3s, we also use a tree structure to search backwards. For each β_3 compute $\alpha_3 = L^{-1}(\beta_3)$. If $\#AS(\alpha_3) < 10$, traverse all possible β_2s which is compatible with α_3, compute $\#AS(\alpha_2)$ from β_2. If $\#AS(\alpha_2)$ is small enough, say below 110, check whether this trail core $(\beta_2, \beta_3, \beta_4)$ under consideration is sufficient for collision attacks.

Parameters and conditions in our algorithm can be changed and we just set them as described for the sake of practicality. For example, in the third step if $\#AS(\alpha_3) \geqslant 10$, it costs too much time for one β_3 ($\geqslant 2^{31.7}$). For each β_3, the backward search costs more time than the forward search because of the property of χ. Since the corresponding α_3 has the same number of active bits with β_3, the numbers of active S-boxes are the same for both extension. Further, active bits are rather sparse in both α_3 and β_3, and the active S-boxes in them

Table 6. The differential distribution table of the χ when viewed as S-box. The first bit of a row is viewed as the least significant bit. Given input difference Δ_{in} and output difference Δ_{out} the number in the table shows the size of the solution set $\{v \mid \chi(v) + \chi(v + \Delta_{in}) = \Delta_{out}\}$. Differences are in hex number.

$\Delta_{in}\backslash\Delta_{out}$	00	01	02	03	04	05	06	07	08	09	0A	0B	0C	0D	0E	0F	10	11	12	13	14	15	16	17	18	19	1A	1B	1C	1D	1E	1F
00	32	-	-	-	-	-	-	-	-	-	-	-	-	-	-	-	-	-	-	-	-	-	-	-	-	-	-	-	-	-	-	-
01	-	8	-	-	-	-	-	-	-	8	-	-	-	-	-	-	-	8	-	-	-	-	-	-	-	8	-	-	-	-	-	-
02	-	-	8	8	-	-	-	-	-	-	-	-	-	-	-	-	-	-	8	8	-	-	-	-	-	-	-	-	-	-	-	-
03	-	-	4	4	-	-	-	-	-	-	4	4	-	-	-	-	-	-	4	4	-	-	-	-	-	-	4	4	-	-	-	-
04	-	-	-	-	8	8	8	8	-	-	-	-	-	-	-	-	-	-	-	-	-	-	-	-	-	-	-	-	-	-	-	-
05	-	-	-	-	4	-	4	-	-	-	-	-	4	-	4	-	-	-	-	-	-	4	-	4	-	-	-	-	-	4	-	4
06	-	-	-	-	4	4	4	4	-	-	-	-	-	-	-	-	-	-	-	-	4	4	4	4	-	-	-	-	-	-	-	-
07	-	-	-	-	2	2	2	2	-	-	-	-	2	2	2	2	-	-	-	-	2	2	2	2	-	-	-	-	2	2	2	2
08	-	-	-	-	-	-	-	-	8	-	8	-	8	-	8	-	-	-	-	-	-	-	-	-	-	-	-	-	-	-	-	-
09	-	4	-	4	-	-	-	-	-	-	-	-	-	4	-	4	-	4	-	4	-	-	-	-	-	-	-	-	-	4	-	4
0A	-	-	-	-	-	-	-	-	4	-	-	4	4	-	-	4	-	-	-	-	-	-	-	-	4	-	-	4	4	-	-	4
0B	-	4	4	-	-	-	-	-	-	-	-	-	-	4	4	-	-	4	4	-	-	-	-	-	-	-	-	-	-	4	4	-
0C	-	-	-	-	-	-	-	-	4	4	4	4	4	4	4	4	-	-	-	-	-	-	-	-	-	-	-	-	-	-	-	-
0D	-	-	-	-	4	-	4	-	4	-	4	-	-	-	-	-	-	-	-	-	-	4	-	4	-	4	-	4	-	-	-	-
0E	-	-	-	-	-	-	-	-	2	2	2	2	2	2	2	2	-	-	-	-	-	-	-	-	2	2	2	2	2	2	2	2
0F	-	-	-	-	2	2	2	2	2	2	2	2	-	-	-	-	-	-	-	-	2	2	2	2	2	2	2	2	-	-	-	-
10	-	-	-	-	-	-	-	-	-	-	-	-	-	-	-	-	8	-	-	-	8	-	-	-	8	-	-	-	8	-	-	-
11	-	4	-	-	-	4	-	-	-	4	-	-	-	4	-	-	-	4	-	-	-	4	-	-	-	4	-	-	-	4	-	-
12	-	-	4	4	-	-	4	4	-	-	-	-	-	-	-	-	-	-	-	-	-	-	-	-	-	-	4	4	-	-	4	4
13	-	-	2	2	-	-	2	2	-	-	2	2	-	-	2	2	-	-	2	2	-	-	2	2	-	-	2	2	-	-	2	2
14	-	-	-	-	-	-	-	-	-	-	-	-	-	-	-	-	4	4	-	-	-	-	4	4	4	4	-	-	-	-	4	4
15	-	4	-	-	-	-	-	4	-	4	-	-	-	-	-	4	4	-	-	-	-	-	4	-	4	-	-	-	-	-	4	-
16	-	-	4	4	4	4	-	-	-	-	-	-	-	-	-	-	-	-	-	-	-	-	-	-	-	-	4	4	4	4	-	-
17	-	-	2	2	2	2	-	-	-	-	2	2	2	2	-	-	-	-	2	2	2	2	-	-	-	-	2	2	2	2	-	-
18	-	-	-	-	-	-	-	-	-	-	-	-	-	-	-	-	4	-	4	-	4	-	4	-	4	-	4	-	4	-	4	-
19	-	2	-	2	-	2	-	2	-	2	-	2	-	2	-	2	-	2	-	2	-	2	-	2	-	2	-	2	-	2	-	2
1A	-	-	-	-	-	-	-	-	4	-	-	4	4	-	-	4	4	-	-	4	4	-	-	4	-	-	-	-	-	-	-	-
1B	-	2	2	-	-	2	2	-	-	2	2	-	-	2	2	-	-	2	2	-	-	2	2	-	-	2	2	-	-	2	2	-
1C	-	-	-	-	-	-	-	-	-	-	-	-	-	-	-	-	2	2	2	2	2	2	2	2	2	2	2	2	2	2	2	2
1D	-	2	-	2	-	2	-	2	-	2	-	2	-	2	-	2	2	-	2	-	2	-	2	-	2	-	2	-	2	-	2	-
1E	-	-	-	-	-	-	-	-	2	2	2	2	2	2	2	2	2	2	2	2	2	2	2	2	-	-	-	-	-	-	-	-
1F	-	2	2	-	2	-	-	2	2	-	-	2	-	2	2	-	2	-	-	2	-	2	2	-	-	2	2	-	2	-	-	2

are almost all with one active bit. Note that given a one-bit input difference of an S-box, there are only 4 possible output difference, while there are 9 possible input differences given an one-bit output difference. This is the reason why the backward search is more time-consuming.

To find a 5-round trail core for KECCAK[1440, 160, 6, 160], we adapt the second step as follows.

2. We first extend forwards from β_3 for one round using KeccakFTrailExtension of KeccakTools [6] with weight up to 36. Then for β_4 of each trail core, if $\#AS(\beta_4) < 16$, traverse all possible α_5, compute $\beta_5 = L(\alpha_5)$ and test whether there exists a compatible α_6 where $\alpha_6^d = 0$. If so, we keep the β_3 and record the three-round trail core $(\beta_3, \beta_4, \beta_5)$, otherwise we discard the β_3.

In the end four trail cores remain. In order to reduce the weight, we take multiple trails from β_4 to α_5 into consideration and the trail core in Table 9 is the best among the four trail cores.

D   Differential Trails

In this section, we give details of differential trails of KECCAK mentioned in Sect. 5. Actually we present trail cores. For example, a 4-round tail core $(\beta_2, \beta_3, \beta_4)$ consisting three state differences represents a set of 4-round differential trails

$$\alpha_1 \xrightarrow{L} \beta_1 \xrightarrow{\chi} \alpha_2 \xrightarrow{L} \beta_2 \xrightarrow{\chi} \alpha_3 \xrightarrow{L} \beta_3 \xrightarrow{\chi} \alpha_4 \xrightarrow{L} \beta_4 \xrightarrow{\chi} \alpha_5$$

where α_5 is compatible with β_4 and $\beta_1 \rightarrow \alpha_2$ is of the least weight determined by β_2. In our collision attacks of 5-round KECCAK, 4-round trail cores are needed. In this section, we not only present trail cores used in collision attacks, but also two 5-round trail cores we found.

Each state difference is represented with a matrix of 5×5 lanes, ordered from left to right, where '|' is used as a separator between lanes; each lane is given in hexadecimal using the little-endian format and '0' is replaced with '-' (Tables 7, 8 and 9).

Table 7. Trail core No. 1–3 used in the collision attacks

```
Trail core No. 1, used in the collision attack of 5-round SHAKE128
     ----------------|--------------24|--------------2-|----------------|----------------
     ----------------|-------------4-|----------------|----------------|----------4-----
β₂   ----------------|------4--------|--------------2-|----------------|----------------
     ----------------|---------------|----------------|----------------|----------------
     --------4-------|------4--------|--------4-------|--------4------|--------4-------
     ----------------|----4-----------|---1------------|----------------|-----1----------
     ----------------|----4-----------|----------------|----------------|----------------
β₃   --------------8-|----------------|----------------|----------------|----------------
     ----------------|----------------|---1------------|----------------|----------------
     --------------8-|----------------|----------------|----------------|-----------1----
     ----------------|----------------|----------------|----------------|2-4-------------
     ----------------|--------------8-|----------4-|----------8-----|----------------
β₄   ----------------|----------------|--8-------------|8---------------|----------2-----
     ------4-8--------|----------4----|-1--------------|----------------|---1-----
     ----4-----------|----------------|--------4-------|----------------|---1-----------
Trail core No. 2, used in the collision attack of KECCAK[1440, 160, 5, 160]
     ----------------|----------------|----------------|----------------|----------------
     ------1--------|--------------8---|--------------1--|--------1----1--|----------------
β₂   --------1-------|----------------|-1--------------|-1-------------|-1--------------
     ----------------|--------------8---|----------------|----------------|-1--------------
     ----------------|----------------|--------------1--|--------------1--|----------------
     ----------------|8---------------|----------------|----------------|----------------
     ----------------|----------------|--------8-------|----------------|-1|-------------
β₃   ----------------|----------------|----------------|----------------|-1|-------------
     ---------------|--------------1|----------------|----------------|---------------1
     ----------------|8---------------|--------8------|----------------|----------------
     ----------------|----------------|----------------|----------------|----------------
     ----------------|----------------|----------------|----2-----------|--------1-------
β₄   ---------------1|------2---------|----------------|----------------|----------------
     ----------------|----------------|----------4--|----------------|----------------
     ----------------|--8-------------|----------------|----------------|--------------2
Trail core No. 3, used in the collision attack of 5-round KECCAK-224
     ----------------|---4------------|---4-------1---|---4------------|----4-----------
     ----4-------|----------4-----|----------------|--------4-------|----------4-----
β₂   ----------------|----------------|----------------|----------------|----------------
     ---------1---|----------4-----|-------------1---|---4------------|----------------
     --------4-------|----4-----------|--------4------|--------4-------|--------4-------
     ----------------|--------------4-|----------------|----------------8|----------------
     --------4--|----------------|----------------|---------8-----8|----------------
β₃   ----------------|----------------|----------------|----------------|---1------------
     ----------------|----4-----------|----------------|--------8-----|----------------
     --------4--|----------------|----------------|----------------|---1------------
     ----------------|----------------|----------------|---1------------|4---------------
     ------8--------|----------------|----------------|----8-----------|----------------
β₄   ---------------8|----------------|----------------|----------------|--------1-------
     ----------------|----4-----------|----------------|----------------|----------------
     --------------|-4---------4---|---------8----|----------------|----------------
```

E　Collisions for Keccak$[r, c, n_r, d]$

In this section, we give instances of collisions against KECCAK$[1440, 160, 5, 160]$, 5-round SHAKE128 and KECCAK$[640, 160, 5, 160]$ respectively. Note that we denote two colliding messages with M_1, M_2. For 5-round KECCAK-224 and KECCAK $[1440, 160, 6, 160]$, we are unable to find collisions for them because of the limitation of computation power. However, we can demonstrate the soundness of our method by providing instances of massage pairs that follow first 4 rounds of the trail of KECCAK$[1440, 160, 6, 160]$ and first 2 rounds of the trail of 5-round KECCAK-224.

Table 8. Trail core No. 4, used in the collision attack of KECCAK$[640, 160, 5, 160]$ and its doubled version can be used in the collision attack of KECCAK-224. β_4 has two choices.

```
                                  Trail core No. 4
       --------|-----8--|----A--|----8--|----8--|          --------|--------|--------|--------|--------
       --------|--------|--------|--------|--------| 1------4|-4------|--------|--------|--------
β₂  2-------|2-------|--------|2-------|--------| β₄⁽¹⁾ --------|------4-|-2------|--------|-2------
    2-------|--------|----8--|----8--|-1------|          ------2|----8--|--------|--------|4-------
       --------|--------|----2--|--------|-1------|          4-------|--------|--------|--------|--------
       --------|--------|--------|------1|-----4-|          --------|--------|--------|--------|--------
       ----8-|--------|------1|--------|-----4-| 1-------|-4------|--------|--------|--------
β₃  --------|--------|--------|------1|--------| β₄⁽²⁾ --------|------4-|-2------|--------|-2------
       --------|--------|--------|--------|--------|          ------2|----8--|--------|--------|4-------
       ------8-|--------|--------|----4-|--------|          4-------|2-------|--------|--------|--------
```

Table 9. Trail cores No. 5, used in the collision attack of KECCAK$[1440, 160, 6, 160]$.

```
        ----------------|--------------|--------------|--------------|--------------
        ----------------1-|---2-----------|--------------|---2----------|--------------1-
β₂   ----------------1|--------------|---1---------|---2----------|--------------
        ----------------1-|--------------|---1---------|--------------|--------------
        ----------------|--------------|--------------|--------------|--------------
        ----------------|--------------|--------------|--------------|--------------
        ----------------|--------------|--------------|---2---------|--------------
β₃   ----------------|--------------|------4--|--------------|--------------
        ----------------|-----1---------|------4--|--------------|--------------
        ----------------|-----1---------|--------------|---2----------|--------------
        ----------------|--------------|---2----------|--------------|--------------
        ----------------|--------------|-------2--|--------------2----|----4---------
β₄   ----------------|--8-----------|--------------|--------------|--------------
        ----------------|--------------|--------------|-----2-------|-----2--------
        ----------------|-----1---------|--------------|--8----------|----4---------
        ----------------|-4-----4-------|--------8--28-22|-C--4--4-----4--|-1-8--88-------2
β₅   -----2-----2--4|42--22--------8-|--------------|-8-----8--------|---2---A--88----
        --8---------8---|-4--14-11--8----|8-----4-----4---|-8----22--------|--------------
        --11--------4--1|--------------|2--------1-----1|--28-22--------8|----28--4-----2-
        ---C--14-11-----|----1---2---1-1-|1--------4--1--1|--------------|-12---4----1---
```

Table 10. Collision for KECCAK[1440, 160, 5, 160]

M_1	C09C5501A913CC3C │ 7406D907E6569334 │ 89182C870A0387A0 │ 980A9D8F82C40A90 │ 9306194AEBBC1C17 6D7DFE249ED35BB5 │ 35C1981BFF84755C │ 37E7FA11AAD390EB │ 19485675C7530B8E │ 042893444D9EC364 6D317B9B40DE874C │ E2EC2A3613678DDA │ 3939A7F72AC29BF6 │ 4FABBC80AE5192EA │ AB50ABCBCC7E5CC7 0F152006D01F65AC │ AEC5B4B7EEC068E1 │ 58E287388571520A │ 569ED102CB7D2EFA │ 4AC1C2A0645D5B2C 4C323DADBB2DAFC4 │ 36F6BEEB558F2B22 │ 0000000089F71BE8 │ 0000000000000000 │ 0000000000000000
M_2	D634EAE0EF26F002 │ 90371C35BB5CFABC │ 7396C3D058D2F577 │ 78CDF403D882B742 │ 22ECA6BCBFC9501F 2352A9667EB05FCD │ 4CA3FD90EFB8A2D3 │ 8DDFF276C0B60599 │ 4B4CCD54AD6B2646 │ A490FAFA55BF4E37 234734EA58D9191D │ 3C580CA9664107ED │ 29E6AEB01815FB08 │ 8FB33829BABDF8C2 │ 48A21B6E764A7987 D9FA24DCB0331C80 │ 9272D67CEF52F8E3 │ 0C82810B4BE7307A │ CF164B325F4DEEBE │ BA41517B4D315C3C 99CD68FF39016FC4 │ AB018238479D9A8D │ 00000000E3233895 │ 0000000000000000 │ 0000000000000000
digest	A6E173DCDFC3E8EF │ 8242EAEA1EE736D5 │ E33875A0

Table 11. Collision for 5-round SHAKE128

M_1	0A3E44EBE62104A0 │ 1E8617C352E80FBC │ B69A38114369962F │ 1237F5EEA8045DAB │ D4144AC64E22044C 1240A93D79FCCB2E │ C8C63A830CBACFFC │ B36B34C0E1719824 │ F94803ECC5586680 │ ACF133FE29839CAD CA5F88F260DEFAA7 │ 972FE7E882A4AB03 │ D11344BE12431A54 │ 814488EBAE68F93B │ 56D10CF0251FAED3 77A665FCC5F52D9F │ D50EF69FAB128ACC │ 87F3F1816E740894 │ 770D4D55489234B0 │ 737134B1243F3A3D FE0E2AE7F23D8E40 │ 0000000000000000 │ 0000000000000000 │ 0000000000000000 │ 0000000000000000
M_2	0CDDD5D25A8BD7CA │ F71D259EE445A4D8 │ EB84F177D51C9D45 │ 0A70C1FC50024C24 │ 7108096E3F024F63 3B8D1EEBC5E9150E │ EADCC9FB19824E75 │ B8A97CB74697BAAD │ 5988E2CD64063AD1 │ FB55123185E2E4A4 E74FF74033CA1486 │ 915F016B41BDAF6B │ 145441AAC9EFA342 │ D9A609CF15E6C626 │ 5609C4F58F5DEE0A AB4E178C43BA8687 │ 3774B01D78F2ABE4 │ AA35E3D371664594 │ A26EAD50F73069A7 │ DE4E25F8A0F8E928 FE431BE34F8371D8 │ 0000000000000000 │ 0000000000000000 │ 0000000000000000 │ 0000000000000000
digest	9D2E953AD7C6A939 │ 326F59A68A6016EF │ A71EAFEE371700D7 │ 3C463D5D098D9B76

Table 12. Collision for KECCAK[640, 160, 5, 160].

M_1	297DB73F │ CE5FB46D │ 63EFD5AB │ AB75DBB2 │ 020119E7 06927773 │ A645A6A4 │ 68E6E3F8 │ 15282462 │ 633AAB83 96C7A5FB │ 5E4CBEB5 │ 92614C96 │ DD9647DA │ D4B0094F 4C68376F │ D3B63751 │ 6286AB56 │ DE577A52 │ 9003EA0F 00000000 │ 00000000 │ 00000000 │ 00000000 │ 00000000	M_2	5B150BCB │ C0F3F2FC │ 5907B5A5 │ 22736DC3 │ 914CF0C5 87477D63 │ A675A649 │ 8BBEA96F │ 52EB8AE3 │ 19402D41 D9FB4CC3 │ 669FD630 │ D8C9FC71 │ 57558554 │ 0662F64A 64B4B5C5 │ 7F12BF56 │ 2BEADBF0 │ F6207B10 │ F2FD9787 00000000 │ 00000000 │ 00000000 │ 00000000 │ 00000000
digest	F90B5ABA │ 7430682D │ 85668C62 │ 66E1B0AD │ B052AC35		

Table 13. Collision attack of KECCAK[1440, 160, 6, 160]. From a message space of freedom degree 135 which definitely leads to β_2 of Trail core No. 5, we present a pair of messages following the first 4 rounds. To find a collision of KECCAK[1440, 160, 6, 160], it costs a time complexity of $2^{70.24}$

M_1	F33E499A09AD9B73 │ CA5358DF1D89473D │ 1B984D8B14D538AB │ F7B7ADD4FBE9425A │ B9B58D552AB12786 CEA1C136D6ADE06C │ 7CCD4A72E45C8222 │ 3337867744F8E6B1 │ 416948B8E8F30A01 │ E9BA1151837A99C0 CB2F620C029A0E29 │ 6BDA24629364CE16 │ C1B6C702D518B1C0 │ DF3D3F6121C87C5E │ 1A5154511DCF5069 97CF66E84A7DF86F │ E6D669B526963387 │ EC88B00FD1D1328A │ 7DB7FCD1A05744B2 │ 288722B23E653CF2 051F63BA5C5EC16E │ C7F1EF8734BF8A4FA │ 000000008CE14DA4 │ 0000000000000000 │ 0000000000000000
M_2	B3411F19C5C972E6 │ C6CF4F990A6FAD57 │ 354F4EA8568AB4A4 │ E4A48E7A516C6A62 │ 92B3E65F1A4114A9 28D238874AD48D50 │ 2F4B715A451EFF5E │ 516C7EC96CCF73BB │ 24638F92E701C38B │ D83A34C323CBB335 6F9D34ADABA26565 │ 48276ACC061BF678 │ 37B2B688A051EDB2 │ E93C28A0A17F2CEA │ 5A976AD0DAF3CB8D 5235F8FF84041376 │ 95F8173A97D0D448 │ D3D8B045A3008325 │ 28C4FD73A1542DB7 │ B8AB1796BACD1E17 C8ECEF0F993328A3 │ C2F7160C897CD9A4 │ 00000000C6BAF84B │ 0000000000000000 │ 0000000000000000

Table 14. Collision attack of 5-round KECCAK-224. From a message space of freedom degree 2 which definitely leads to β_2 of Trail core No. 3, we present a pair of messages following the first 2 rounds. To find a collision of 5-round KECCAK-224, it costs a time complexity of 2^{101}.

M_1	8E85F0BC15BBA27B	776FF9140B9AED24	52C6D4A9251C9886	D74E6FC4FB7EE6CB	40C5BF16312FEA11
	0618F45C4F30B4EA	FEA4F176FFB65180	8B03CFC2E0C168A8	E47CFF2303F924D6	280AC9CC77707399
	790244BCD16F3621	4125A834D1FEB877	5DA576EB0306BE03	5498B00302BECD5C	F13E10DD2A230829
	26AADDAF76496EA0	E3EC0DC10D9FC852	9CDF3DE3421ACD7B	0000000000000000	0000000000000000
	0000000000000000	0000000000000000	0000000000000000	0000000000000000	0000000000000000
M_2	D756F690FEDC7326	248DA8A7D0F3432D	276828C7C3A3728A	CFAC96E3956C5F35	AC2B2D679F6F1745
	933191AEDE3EB500	6562F098D4099896	D24C31AF425CE2E7	ADC9058BB5E4FEFC	06D4880875530CB0
	C2C4BCB373FFAD24	C2E9DCD965CAA725	06B1F37EE7F51056	4D43F63490D82FBC	18FA91952DC4DB40
	14F2289283846D81	73FE0284B87D9815	91590D20B7E9251F	0000000000000000	0000000000000000
	0000000000000000	0000000000000000	0000000000000000	0000000000000000	0000000000000000

References

1. Aumasson, J.P., Meier, W.: Zero-sum distinguishers for reduced Keccak-f and for the core functions of Luffa and Hamsi. rump session of Cryptographic Hardware and Embedded Systems-CHES 2009, p. 67 (2009)
2. Bertoni, G., Daemen, J., Peeters, M., Van Assche, G.: Keccak Crunchy Crypto Collision and Pre-image Contest. http://keccak.noekeon.org/crunchy_contest.html
3. Bertoni, G., Daemen, J., Peeters, M., Van Assche, G.: The Keccak reference, version 3.0 (2011). http://keccak.noekeon.org/Keccak-reference-3.0.pdf
4. Bertoni, G., Daemen, J., Peeters, M., Van Assche, G.: Cryptographic Sponge functions. Submission to NIST (Round 3) (2011)
5. Bertoni, G., Daemen, J., Peeters, M., Van Assche, G.: The Keccak SHA-3 submission. Submiss. NIST (Round 3) 6(7), 16 (2011)
6. Bertoni, G., Daemen, J., Peeters, M., Van Assche, G.: Keccaktools (2015). http://keccak.noekeon.org/
7. Canteaut, A. (ed.): FSE 2012. LNCS, vol. 7549. Springer, Heidelberg (2012)
8. Daemen, J.: Cipher and hash function design strategies based on linear and differential cryptanalysis. Ph.D. thesis, Doctoral dissertation, KU Leuven, March 1995
9. Daemen, J., Assche, G.: Differential propagation analysis of Keccak. In: Canteaut, A. (ed.) FSE 2012. LNCS, vol. 7549, pp. 422–441. Springer, Heidelberg (2012). doi:10.1007/978-3-642-34047-5_24. [7]
10. Dinur, I., Dunkelman, O., Shamir, A.: New attacks on Keccak-224 and Keccak-256. In: Canteaut, A. (ed.) FSE 2012. LNCS, vol. 7549, pp. 442–461. Springer, Heidelberg (2012). doi:10.1007/978-3-642-34047-5_25. [7]
11. Dinur, I., Dunkelman, O., Shamir, A.: Collision attacks on up to 5 rounds of SHA-3 using generalized internal differentials. In: Moriai, S. (ed.) FSE 2013. LNCS, vol. 8424, pp. 219–240. Springer, Heidelberg (2014)
12. Dinur, I., Dunkelman, O., Shamir, A.: Improved practical attacks on round-reduced Keccak. J. Cryptol. 27(2), 183–209 (2014)
13. Dinur, I., Morawiecki, P., Pieprzyk, J., Srebrny, M., Straus, M.: Cube attacks and cube-attack-like cryptanalysis on the round-reduced Keccak sponge function. In: Oswald, E., Fischlin, M. (eds.) EUROCRYPT 2015. LNCS, vol. 9056, pp. 733–761. Springer, Heidelberg (2015). doi:10.1007/978-3-662-46800-5_28
14. Duc, A., Guo, J., Peyrin, T., Wei, L.: Unaligned rebound attack: application to Keccak. In: Canteaut, A. (ed.) FSE 2012. LNCS, vol. 7549, pp. 402–421. Springer, Heidelberg (2012). doi:10.1007/978-3-642-34047-5_23

15. Guo, J., Jean, J., Nikolic, I., Qiao, K., Sasaki, Y., Sim, S.M.: Invariant subspace attack against Midori64 and the resistance criteria for S-box designs. IACR Trans. Symmetric Cryptol. 1(1) (2017, to appear)
16. Jean, J., Nikolić, I.: Internal differential boomerangs: practical analysis of the round-reduced Keccak-f permutation. In: Leander, G. (ed.) FSE 2015. LNCS, vol. 9054, pp. 537–556. Springer, Heidelberg (2015). doi:10.1007/978-3-662-48116-5_26
17. Mendel, F., Nad, T., Schläffer, M.: Finding SHA-2 characteristics: searching through a minefield of contradictions. In: Lee, D.H., Wang, X. (eds.) ASIACRYPT 2011. LNCS, vol. 7073, pp. 288–307. Springer, Heidelberg (2011)
18. National Institute of Standards and Technology: SHA-3 STANDARD: PERMUTATION-BASED HASH AND EXTENDABLE-OUTPUT FUNCTIONS. Federal Information Processing Standards (FIPS) Publication Series (2015)
19. Naya-Plasencia, M., Röck, A., Meier, W.: Practical analysis of reduced-round Keccak. In: Bernstein, D.J., Chatterjee, S. (eds.) INDOCRYPT 2011. LNCS, vol. 7107, pp. 236–254. Springer, Heidelberg (2011)

Obfuscation II

Lattice-Based SNARGs and Their Application to More Efficient Obfuscation

Dan Boneh[1,2], Yuval Ishai[1,3,4], Amit Sahai[1,4], and David J. Wu[1,2(✉)]

[1] Center for Encrypted Functionalities, Los Angeles, USA
[2] Stanford University, Stanford, USA
{dabo,dwu4}@cs.stanford.edu
[3] Technion, Haifa, Israel
yuvali@cs.technion.ac.il
[4] UCLA, Los Angeles, USA
sahai@cs.ucla.edu

Abstract. Succinct non-interactive arguments (SNARGs) enable verifying NP computations with substantially lower complexity than that required for classical NP verification. In this work, we give the first lattice-based SNARG candidate with quasi-optimal succinctness (where the argument size is quasilinear in the security parameter). Further extension of our methods yields the first SNARG (from any assumption) that is quasi-optimal in terms of *both* prover overhead (polylogarithmic in the security parameter) as well as succinctness. Moreover, because our constructions are lattice-based, they plausibly resist quantum attacks. Central to our construction is a new notion of *linear-only vector encryption* which is a generalization of the notion of linear-only encryption introduced by Bitansky et al. (TCC 2013). We conjecture that variants of Regev encryption satisfy our new linear-only definition. Then, together with new information-theoretic approaches for building statistically-sound linear PCPs over small finite fields, we obtain the first quasi-optimal SNARGs.

We then show a surprising connection between our new lattice-based SNARGs and the concrete efficiency of program obfuscation. All existing obfuscation candidates currently rely on multilinear maps. Among the constructions that make black-box use of the multilinear map, obfuscating a circuit of even moderate depth (say, 100) requires a multilinear map with multilinearity degree in excess of 2^{100}. In this work, we show that an ideal obfuscation of both the decryption function in a fully homomorphic encryption scheme and a variant of the verification algorithm of our new lattice-based SNARG yields a general-purpose obfuscator for all circuits. Finally, we give some concrete estimates needed to obfuscate this "obfuscation-complete" primitive. We estimate that at 80-bits of security, a (black-box) multilinear map with $\approx 2^{12}$ levels of multilinearity suffices. This is over 2^{80} times more efficient than existing candidates, and thus, represents an important milestone towards implementable program obfuscation for all circuits.

The full version of this paper is available from https://crypto.stanford.edu/people/dwu4/snargs.html.

J.-S. Coron and J.B. Nielsen (Eds.): EUROCRYPT 2017, Part III, LNCS 10212, pp. 247–277, 2017.
DOI: 10.1007/978-3-319-56617-7_9

1   Introduction

Interactive proofs systems [49] are fundamental to modern cryptography and complexity theory. In this work, we consider computationally sound proof systems for NP languages, also known as *argument systems*. An argument system is *succinct* if its communication complexity is *polylogarithmic* in the running time of the NP verifier for the language. Notably, the size of the argument is polylogarithmic in the size of the NP witness.

Kilian [53] gave the first succinct four-round interactive argument system for NP based on collision-resistant hash functions and probabilistically-checkable proofs (PCPs). Subsequently, Micali [63] showed how to convert Killian's four-round argument into a single-round argument for NP by applying the Fiat-Shamir heuristic [38]. Micali's "computationally-sound proofs" (CS proofs) is the first candidate construction of a *succinct non-interactive argument* (i.e., a "SNARG" [46]) in the random oracle model. In the standard model, single-round argument systems are impossible for sufficiently hard languages, so we consider the weaker goal of two-message succinct argument systems where the verifier's initial message is generated independently of the statement being proven. This message is often referred to as the *common reference string* (CRS).

In this work, we are interested in minimizing the prover complexity and proof length of SNARGs. Concretely, for a security parameter λ, we measure the asymptotic cost of achieving soundness against provers of circuit size 2^λ with $\mathrm{negl}(\lambda)$ error. We say that a SNARG has *quasi-optimal succinctness* if its proof length is $\widetilde{O}(\lambda)$ and that it is *quasi-optimal* if in addition, the SNARG prover's running time is larger than that of a classical prover by only a polylogarithmic factor (in λ and the running time). In this paper, we construct the first SNARG that is quasi-optimal in this sense. The soundness of our SNARG is based on a new plausible intractability assumption, which is in the spirit of assumptions on which previous SNARGs were based (see Sect. 1.2). Moreover, based on a stronger variant of the assumption, we get a SNARK [15] (i.e., a SNARG of knowledge) with similar complexity (see Remark 4.9). All previous SNARGs, including heuristic ones, were suboptimal in at least one of the two measures by a factor of $\Omega(\lambda)$. For a detailed comparison with previous approaches, see Table 1.

We give two SNARG constructions: one with quasi-optimal succinctness based on standard lattices, and another that is quasi-optimal based on ideal lattices over polynomial rings. Because all of our SNARGs are lattice-based, they plausibly resist known quantum attacks. All existing SNARGs with quasi-optimal succinctness rely, at the minimum, on number-theoretic assumptions such as the hardness of discrete log. Thus, they are vulnerable to quantum attacks [72,73].

Application to Efficient Obfuscation. Independently of their asymptotic efficiency, our SNARGs can also be used to significantly improve the *concrete* efficiency of program obfuscation. Program obfuscation is the task of making code unintelligible such that the obfuscated program reveals nothing more about the implementation details beyond its functionality. The theory of program obfuscation was first formalized by Barak et al. [12]. In their work, they introduced the natural notion of virtual black-box (VBB) obfuscation, and moreover, showed that VBB obfuscation for all circuits is impossible in the standard model. In the same work, Barak et al. also introduced the weaker notion of *indistinguishability obfuscation* ($i\mathcal{O}$); subsequently, Garg et al. [41] gave the first candidate construction of $i\mathcal{O}$ for general circuits based on multilinear maps [24,33,39,43].

Since the breakthrough result of Garg et al., there has been a flurry of works showcasing the power of $i\mathcal{O}$ [19,25,40,41,69]. However, in spite of the numerous constructions and optimizations that have been developed in the last few years [5,7,10,11,30,74], concrete instantiations of program obfuscation remain purely theoretical. Even obfuscating a relatively simple function such as the AES block cipher requires multilinear maps capable of supporting unimaginable levels of multilinearity ($\gg 2^{100}$ [74]). In this work, we show that our new lattice-based SNARG constructions can be combined with existing lattice-based fully homomorphic encryption schemes (FHE) to obtain an "obfuscation-complete" primitive[1] with significantly better concrete efficiency. Targeting 80 bits of security, we show that we can instantiate our obfuscation-complete primitive over a composite-order multilinear map supporting $\approx 2^{12}$ levels of multilinearity. The number of multilinear map encodings in the description of the obfuscated program is $\approx 2^{44}$. While the levels of multilinearity required is still beyond what we can efficiently realize using existing composite-order multilinear map candidates [33], future multilinear map candidates with better efficiency as well as further optimizations to the components that underlie our transformation will bring our constructions closer to reality. Concretely, our results are many orders of magnitude more efficient than existing constructions (that make black-box use of the underlying multilinear map), and thus, represent an important stepping stone towards implementable obfuscation.

Non-black-box Alternatives. Nearly all obfuscation constructions [5,7,10,11,30, 74] rely on the underlying multilinear map as a black-box. Recently, several works [4,57–59] gave the first candidate constructions of $i\mathcal{O}$ based on *constant-degree* multilinear maps (by going through the functional encryption route introduced in [3,20]). Even more impressively, the most recent constructions by Lin [58] as well as Ananth and Sahai [4] only require a degree-5 multilinear map, which is certainly implementable [56]. However, this reduction in multilinearity comes at the cost of a *non-black-box* construction. Notably, their construction requires a gate-by-gate transformation to be applied to a Boolean circuit

---

[1] An "obfuscation-complete" primitive is a function whose ideal obfuscation (e.g., using tamper-proof hardware) can be used for obfuscating arbitrary functions. While we do not provide a provably secure instantiation of this primitive using $i\mathcal{O}$, it can be heuristically instantiated using existing $i\mathcal{O}$ candidates. Moreover, our obfuscation-complete primitive has the appealing property that it needs to be invoked exactly *once* regardless of the function being obfuscated. This is in contrast to alternative constructions [6,50] where the obfuscated primitive needs to be invoked for each gate in the circuit or each step of a Turing machine evaluation.

description of the encoding function of the underlying multilinear map. While further investigation of non-black-box approaches is certainly warranted, due to the complexity of existing multilinear map constructions [33,39], this approach faces major hurdles with regards to implementability. In this work, we focus on constructions that use the multilinear map in a black-box manner.

1.1 Background

Constructing SNARGs. Gentry and Wichs [46] showed that no SNARG (for a sufficiently difficult language) can be proven secure under any "falsifiable" assumption [65]. Consequently, all existing SNARG constructions for NP in the standard model (with a CRS) have relied on non-falsifiable assumptions such as knowledge-of-exponent assumptions [14,35,42,51,61,64], extractable collision-resistant hashing [15,36], homomorphic encryption with a homomorphism extraction property [17] and linear-only encryption [18].

Designated-Verifier Arguments. Typically, in a non-interactive argument system, the arguments can be verified by anyone. Such systems are said to be "publicly verifiable." In some applications (notably, bootstrapping certain types of obfuscation), it suffices to consider a relaxation where the setup algorithm for the argument system also outputs a *secret* verification state which is needed for proof verification. Soundness holds provided that the prover does not know the secret verification state. These systems are said to be *designated verifier*. A key question that arises in the design and analysis of designated verifier arguments is whether the same common reference string can be reused for multiple proofs. Formally, this "multi-theorem" setting is captured by requiring soundness to hold even against a prover that makes adaptive queries to a proof verification oracle. If the prover can choose its queries in a way that induces noticeable correlations between the outputs of the verification oracle and the secret verification state, then the adversary can potentially compromise the soundness of the scheme. Thus, special care is needed to construct designated-verifier argument systems in the multi-theorem setting.

SNARGs from Linear-Only Encryption. Bitansky et al. [18] introduced a generic compiler for building SNARGs in the "preprocessing" model based on a notion called "linear-only" encryption. In the preprocessing model, the setup algorithm that constructs the CRS can run in time that depends polynomially on a time bound T of the computations that will be verified. The resulting scheme can then be used to verify computations that run in time at most T. The compiler of [18] can be decomposed into an information-theoretic transformation and a cryptographic transformation, which we outline here:

- First, they restrict the interactive proof model to only consider "affine-bounded" provers. An affine-bounded prover is only able to compute affine functions (over a ring) of the verifier's queries.[2] Bitansky et al. give several constructions of succinct two-message interactive proofs in this restricted model by applying a generic transformation to existing "linear PCP" constructions.

– Next, they introduce a new cryptographic primitive called linear-only encryption, which is a (public-key) encryption scheme that *only* supports linear homomorphisms on ciphertexts. Bitansky et al. show that combining a linear-only encryption scheme with the affine-restricted interactive proofs from the previous step suffices to construct a designated-verifier SNARG in the preprocessing model. The construction is quite natural: the CRS for the SNARG system is a linear-only encryption of what would be the verifier's first message. The prover then homomorphically computes its response to the verifier's encrypted queries. The linear-only property of the encryption scheme constrains the prover to only using affine strategies. This ensures soundness for the SNARG. To check a proof, the verifier decrypts the prover's responses and applies the decision algorithm for the underlying two-message proof system. Bitansky et al. give several candidate instantiations for their linear-only encryption scheme based on Paillier encryption [66] as well as bilinear maps [22,52].

Linear PCPs. Like [18], our SNARG constructions rely on linear PCPs (LPCPs). A LPCP of length m over a finite field $\mathbb{F}$ is an oracle computing a linear function $\boldsymbol{\pi} : \mathbb{F}^m \to \mathbb{F}$. On any query $\mathbf{q} \in \mathbb{F}^m$, the LPCP oracle responds with $\mathbf{q}^\top \boldsymbol{\pi}$. More generally, if ℓ queries are made to the LPCP oracle, the ℓ queries can be packed into the columns of a query matrix $\mathbf{Q} \subset \mathbb{F}^{m \times \ell}$. The response of the LPCP oracle can then be written as $\mathbf{Q}^\top \boldsymbol{\pi}$. We provide more details in Sect. 3.

1.2 Our Results: New Constructions of Preprocessing SNARGs

In this section, we summarize our main results on constructing preprocessing SNARGs based on a more advanced form of linear-only encryption. Our results extend the framework introduced by Bitansky et al. [18].

New Compiler for Preprocessing SNARGs. The preprocessing SNARGs we construct in this work enjoy several advantages over those of [18]. We enumerate some of them below:

– **Direct construction of SNARGs from linear PCPs.** Our compiler gives a *direct* compilation from linear PCPs over a finite field $\mathbb{F}$ into a preprocessing SNARG. In contrast, the compiler in [18] first constructs a two-message linear interactive proof from a linear PCP by introducing an additional linear consistency check. The additional consistency check not only increases the communication complexity of their construction, but also introduces a

---

[2] Bitansky et al. [18] refer to this as "linear-only," even though the prover is allowed to compute affine functions. To be consistent with their naming conventions, we will primarily write "linear-only" to refer to "affine-only".

soundness error $O(1/|\mathbb{F}|)$. As a result, their construction only provides soundness when working over a large field (that is, when $|\mathbb{F}|$ is super-polynomial in the security parameter). By using a direct compilation of linear PCPs into SNARGs, we avoid both of these problems. Our construction does not require any additional consistency checks and moreover, it preserves the soundness of the underlying linear PCP. Thus, as long as the underlying linear PCP is statistically sound, applying our compiler yields a computationally sound argument (even if $|\mathbb{F}|$ is small).

- **Constructing linear PCPs with strong soundness.** As noted in the previous section, constructing multi-theorem designated-verifier SNARGs can be quite challenging. In [18], this is handled at the information-theoretic level (by constructing interactive proof systems satisfying a notion of "strong" or "reusable" soundness) and at the cryptographic level (by introducing strengthened definitions of linear-only encryption). A key limitation in their approach is that the information-theoretic construction of two-round interactive proof systems again requires LPCPs over super-polynomial-sized fields. This is a significant barrier to applying their compiler to natural LPCP constructions over small finite fields (which are critical to our approach for bootstrapping obfuscation). In this work, we show how to apply soundness amplification to standard LPCPs with constant soundness error against linearly-bounded provers (and which do not necessarily satisfy strong soundness) to obtain strong, statistically-sound LPCPs against affine-bounded provers. Coupled with our direct compilation of LPCPs to preprocessing SNARGs, we obtain multi-theorem designated-verifier SNARGs.

We describe our construction of strong statistically sound LPCPs against affine provers from LPCPs with constant soundness error against linear provers in Sect. 3. Applying our transformation to linear PCPs based on the Walsh-Hadamard code [9] as well as those based on quadratic-span programs (QSPs) [42], we obtain two LPCPs with strong statistical soundness against affine provers over polynomial-size fields.

From Linear PCPs to Preprocessing SNARGs. The primary tool we use construction of preprocessing SNARGs from linear PCPs is a new cryptographic primitive we call linear-only *vector encryption*. A vector encryption scheme is an encryption scheme where the plaintexts are vectors of ring (or field) elements. Next, we extend the notion of linear-only encryption [18] to the context of vector encryption. We say that a vector encryption scheme is linear-only if the only homomorphisms it supports is addition (and scalar multiplication) of vectors.

Our new notion of linear-only vector encryption gives an immediate method of compiling an ℓ-query linear PCP (over a finite field $\mathbb{F}$) into a designated-verifier SNARG. The construction works as follows. In a ℓ-query linear PCP over $\mathbb{F}$, the verifier's query can be written as a matrix $\mathbf{Q} \in \mathbb{F}^{m \times \ell}$ where m is the query length of the LPCP. The LPCP oracle's response is $\mathbf{Q}^\top \boldsymbol{\pi}$ where $\boldsymbol{\pi} \in \mathbb{F}^m$ is the proof. To compile this LPCP into a preprocessing SNARG, we use a linear-only vector encryption scheme with plaintext space $\mathbb{F}^\ell$. The setup algorithm takes the verifier's query matrix $\mathbf{Q}$ (which is *independent* of the statement being proved) and encrypts each row of $\mathbf{Q}$ using the vector encryption scheme. The key observation is that the product $\mathbf{Q}^\top \boldsymbol{\pi}$ is a linear combination of the rows of $\mathbf{Q}$. Thus, the prover can homomorphically compute an encryption of $\mathbf{Q}^\top \boldsymbol{\pi}$. To check the proof, the verifier decrypts to obtain the prover's responses and then invokes the decision algorithm for the underlying LPCP. Soundness is ensured by the linear-only property of the underlying vector encryption scheme. The advantage of linear-only vector encryption (as opposed to standard linear-only encryption) is that the prover is constrained to evaluating a single linear function on *all* of the query vectors simultaneously. This insight enables us to remove the extra consistency check introduced in [18], and thus, avoids the soundness penalty $O(1/|\mathbb{F}|)$ incurred by the consistency check.[3] Consequently, we can instantiate our transformation with statistically-sound linear PCPs over *any* finite field $\mathbb{F}$. We describe our construction in Sect. 4.

New Lattice-Based SNARG Candidates. We then conjecture that the Regev-based [68] encryption scheme of Peikert et al. [67] is a secret-key linear-only vector encryption scheme over $\mathbb{Z}_p$ where p is a prime whose bit-length is polynomial in the security parameter λ. Then, applying our generic compiler from LPCPs to SNARGs (Construction 4.5) to our new LPCP constructions over polynomial-size fields $\mathbb{Z}_p$, we obtain a lattice-based construction of a designated-verifier SNARG (for Boolean circuit satisfiability) in the preprocessing model.[4] Specifically, starting with a QSP-based LPCP [42], we obtain a SNARG with quasi-optimal succinctness. As discussed above, this is the first such SNARG that can plausibly resist quantum attacks. We note here that a direct instantiation of the construction in [18] with a Regev-based candidate for linear-only encryption yields a SNARG that is suboptimal in *both* prover complexity and proof length (Remark 4.13). Thus, for Boolean circuit satisfiability, using lattice-based linear-only *vector encryption* provides some concrete advantages over vanilla linear-only encryption.

Quasi-Optimal SNARGs. In the full version of this paper, we further extend our techniques to obtain the first instantiation of a *quasi-optimal* SNARG for Boolean circuit satisfiability—that is, a SNARG where the prover complexity is $\widetilde{O}(s)$ and the argument size is $\widetilde{O}(\lambda)$, where s is the size of the Boolean circuit and λ is a security parameter guaranteeing soundness against 2^λ-size provers

---

[3] This is the main difference between our approach and that taken in [18]. By making the *stronger* assumption of linear-only *vector encryption*, we avoid the need for an extra consistency check, thus allowing for a *direct* compilation from linear PCPs to SNARGs. In contrast, [18] relies on the weaker assumption of linear-only encryption, but requires an extra step of first constructing a two-message linear interactive proof (incorporating the consistency check) from the linear PCP.

[4] While it would be preferable to obtain a construction based on the hardness of standard lattice assumptions like learning with errors (LWE) [68], the separation results of Gentry and Wichs [46] suggest that stronger, non-falsifiable assumptions may be necessary to construct SNARGs.

with negl(λ) error. All previous constructions with quasi-optimal succinctness (including our lattice-based candidate described above) achieved at best prover complexity $\tilde{O}(s\lambda)$. We refer to Table 1 for a detailed comparison. Our construction relies on a new information-theoretic construction of a linear PCP operating over rings. In conjunction with a linear-only vector encryption scheme where the underlying message space is a ring, we can apply our compiler to obtain a SNARG. To achieve quasi-optimality, we require that the ciphertext expansion factor of the underlying vector encryption scheme be polylogarithmic. Using Regev-based vector encryption based on the ring learning with errors (RLWE) problem [62] and conjecturing that it satisfies our linear-only requirements, we obtain the first quasi-optimal SNARG construction. We leave open the question of realizing a stronger notion of quasi-optimality, where the soundness error (against 2^λ-size provers) is $2^{-\lambda}$ rather than negl(λ).

1.3   Our Results: Concrete Efficiency of Bootstrapping Obfuscation

In spite of the numerous optimizations and simplifications that have been proposed for indistinguishability obfuscation ($i\mathcal{O}$) and VBB obfuscation (in a generic model), obfuscating even relatively simple functions like AES remains prohibitively expensive. In this section, we describe how the combination of our new lattice-based SNARG candidate and fully homomorphic encryption (FHE) allows us to obtain VBB obfuscation for all circuits (in a generic model) with concrete parameters that are significantly closer to being implementable. Our construction is over 2^{80} times more efficient than existing constructions.

Background. The earliest candidates of $i\mathcal{O}$ and VBB obfuscation operated on matrix branching programs [11,30,41], which together with multilinear maps [33,39,43], yielded obfuscation for NC^1 (via Barrington's theorem [13]).[5] The primary source of inefficiency in these branching-program-based obfuscation candidates is the enormous overhead incurred when converting NC^1 circuits to an equivalent branching program representation. While subsequent work [5,10] has provided significant asymptotic improvements for representing NC^1 circuits as matrix branching programs, the levels of multilinearity required to obfuscate a computation of depth d still grows *exponentially* in d. Thus, obfuscating even a simple function like AES, which has a circuit of relatively low depth (≈ 100), still requires a multilinear map capable of supporting $\gg 2^{100}$ levels of multilinearity and a similarly astronomical number of encodings. This is completely infeasible.

Zimmerman [74] as well as Applebaum and Brakerski [7] showed how to directly obfuscate circuits. While their constructions do not incur the exponential overhead of converting NC^1 circuits to matrix branching programs, due to the noise growth in existing multilinear map candidates, the level of multilinearity

---

[5] Garg et al. [41] as well as Brakerski and Rothblum [30] show how to combine obfuscation for NC^1 together with fully homomorphic encryption (FHE) and low-depth checkable proofs to bootstrap $i\mathcal{O}$ and VBB obfuscation from NC^1 to P/poly.

required again grows exponentially in the depth of the circuit d. However, the number of multilinear map encodings is substantially smaller with these candidates. In the case of VBB obfuscation of AES, Zimmerman estimates that the obfuscation would contain $\approx 2^{17}$ encodings of a multilinear map capable of supporting $\gg 2^{100}$ levels of multilinearity. Despite the more modest number of encodings required, the degree of multilinearity required remains prohibitively large.

Revisiting the Branching-Program Based Obfuscation. In this work, we revisit the branching-program-based constructions of obfuscation. However, rather than follow the traditional paradigm of taking a Boolean circuit, converting it to a matrix branching program via Barrington's theorem, and then obfuscating the resulting branching program, we take the more direct approach of using the matrix branching program to compute simple functions over $\mathbb{Z}_q$ (for polynomial-sized q). The key observation is that the additive group $\mathbb{Z}_q$ embeds into the symmetric group S_q of $q \times q$ permutation matrices. This technique was previously used by Alperin-Sheriff and Peikert [2] for improving the efficiency of bootstrapping for FHE. While the functions that can be evaluated in this way are limited, they are expressive enough to include both the decryption function for lattice-based FHE [2,27,28,31,37,45] and the verification algorithm of our new lattice-based SNARG. Using a variant of the bootstrapping theorem in [30], VBB obfuscation of these two functionalities suffice for VBB obfuscation of all circuits.

We remark here that Applebaum [6] described a simpler approach for bootstrapping VBB obfuscation of all circuits based on obfuscating a pseudorandom function (PRF) in conjunction with randomized encodings. While this approach is conceptually simpler, it is unclear whether this yields a scheme with concrete efficiency. One problem is that we currently do not have any candidate PRFs that are amenable to existing obfuscation candidates. Constructing an "obfuscation-friendly" PRF remains an important open problem. Perhaps more significantly, this approach requires invoking the obfuscated program multiple times (a constant number of times per gate in the circuit, or per step of the computation in the case of Turing machines [55]). In contrast, in this work, we focus on building an "obfuscation-complete" primitive such that a *single* call to the obfuscated program suffices for program evaluation.

Computing in $\mathbb{Z}_q$ via Matrix Branching Programs. By leveraging the power of bootstrapping, it suffices to obfuscate a program that performs FHE decryption and SNARG verification. Using FHE schemes based on standard lattices [2,27, 28,31,37,45] and our new lattice-based SNARG, both computations effectively reduce to computing rounded inner products over $\mathbb{Z}_q$—that is, functions where we first compute the inner product $\langle \mathbf{x}, \mathbf{y} \rangle$ of two vectors $\mathbf{x}$ and $\mathbf{y}$ in $\mathbb{Z}_p^\ell$ and then reduce the result modulo a smaller value p. In our setting, one of the vectors $\mathbf{y}$ is embedded within the obfuscated program. We briefly describe the technique here. Our presentation is adapted from [2], who use this technique to improve the efficiency of FHE bootstrapping.

The key idea is to embed the group $\mathbb{Z}_q$ in the symmetric group S_q. The embedding is quite straightforward. A group element $y \in \mathbb{Z}_q$ is represented by the basis vector $\mathbf{e}_y \in \{0, 1\}^q$ (i.e., the vector with a single 1 in the yth position). Addition by an element $x \in \mathbb{Z}_q$ corresponds to multiplying by a permutation matrix that implements a cyclic rotation by x positions. Specifically, to implement the function $f_x(y) = x + y$ where $x, y \in \mathbb{Z}_q$, we define the permutation matrix $\mathbf{B}_x \in \{0, 1\}^{q \times q}$ where $\mathbf{B}_x \mathbf{e}_y = \mathbf{e}_{x+y \bmod q}$ for all $y \in [q]$. Then, to compute f_x on an input y, we simply take the q-by-q permutation matrix $\mathbf{B}_x$ and multiply it with the basis vector $\mathbf{e}_y$ representing the input. Scalar multiplication can be implemented by repeated additions. Finally, modular reduction with respect to p can be implemented via multiplication by a p-by-q matrix where the ith row sums the entries of the q-dimensional indicator vector corresponding to those values in $\mathbb{Z}_q$ that reduce to i modulo p. As long as q is small, this method gives an efficient way to compute simple functions over $\mathbb{Z}_q$.

Optimizing the SNARG Construction. While computing a single rounded inner product suffices for FHE decryption, it is not sufficient for SNARG verification. We introduce a series of additional optimizations to make our SNARG verification algorithm more branching-program-friendly and minimize the concrete parameters needed to obfuscate the functionality. These optimizations are described in detail in the full version. We highlight the most significant ones here:

- **Modulus switching.** Recall that the SNARG verifier has to first decrypt a proof (encrypted under the linear-only vector encryption scheme) before applying the underlying LPCP decision procedure. While decryption in this case does consist of evaluating a rounded inner product, the size of the underlying field scales *quadratically* in the running time of the computation being verified.[6] As a result, the width of the branching programs needed to implement the SNARG verification scales quadratically in the running time of the computation, which can quickly grow out of hand. However, since the ciphertexts in question are essentially LWE ciphertexts, we can apply the modulus switching trick that has featured in many FHE constructions [28,31,37]. With modulus switching, after the prover homomorphically computes its response (a ciphertext vector over a large ring), the prover rescales each component of the ciphertext to be defined with respect to a much smaller modulus (one that grows *polylogarithmically* with the running time of the computation). The actual decryption then operates on the rescaled ciphertext, which can be implemented as a (relatively) small branching program.
- **Strengthening the linear-only assumption.** To further reduce the overhead of the SNARG verification, we also consider strengthened definitions of (secret-key) linear-only vector encryption. In particular, we conjecture that our candidate lattice-based vector encryption scheme only supports a *restricted* set of affine homomorphisms. This allows us to use LPCPs with simpler and more branching-program-friendly verification procedures. We introduce the definition and state our conjecture in the full version. We note that when considering the public-key notion of linear-only encryption [18], one *cannot* restrict the set

---

[6] This is fine from the SNARG perspective since the number of *bits* in the proof is still growing logarithmically in the running time of the computation.

of affine homomorphisms available to the adversary. By definition, the adversary can compute arbitrary linear functions on the ciphertexts, and moreover, it can also encrypt values of its choosing and linearly combine those values with the ciphertexts. This allows the adversary to realize arbitrary affine functions in the public-key setting. However, in the secret-key setting, the adversary does *not* have the flexibility of constructing arbitrary ciphertexts of its own, and so, it is plausible that the encryption scheme only permits more limited homomorphisms. Our techniques here are not specific to our particular SNARG instantiation, and thus, may be useful in optimizing other SNARG constructions (at the expense of making stronger linear-only assumptions).

- **Parallelization via CRT.** Unlike FHE decryption, the SNARG verification algorithm requires computing a matrix-vector product of the form $\mathbf{Ax}$, where the matrix $\mathbf{A} \in \mathbb{Z}_q^{m \times \ell}$ is embedded inside the program and $\mathbf{x} \in \mathbb{Z}_q^\ell$ is part of the input. The verification algorithm then applies an (independent) test to each of the components of $\mathbf{Ax}$. Verification succeeds if and only if each of the underlying tests pass. While a matrix-vector product can be computed by iterating the algorithm for computing an inner product m times and performing the m checks sequentially, this increases the length of the branching program by a factor of m. A key observation here is that since the components of $\mathbf{Ax}$ are processed independently of one another, this computation can be performed in parallel if we consider matrix branching programs over composite-order rings. Then, each of the rows of $\mathbf{A}$ can be embedded in the different sub-rings according to the Chinese Remainder Theorem (CRT). Assuming the underlying multilinear map is composite-order, this method can potentially yield a factor m reduction in the length of the branching program. Indeed, using the CLT multilinear map [33], the plaintext space naturally decomposes into sufficiently many sub-rings, thus allowing us to take advantage of parallelism with essentially no extra cost. A similar technique of leveraging CRT to parallelize computations was also used in [2] to improve the concrete efficiency of FHE bootstrapping.

A Concrete Obfuscation Construction. In the full version, we describe our methodology for instantiating the building blocks for our obfuscation-complete primitive (for VBB obfuscation). Our parameter estimates show that targeting $\lambda = 80$ bits of security, implementing FHE decryption together with SNARG verification can be done with a branching program (over composite-order rings[7] of length 4150 and size $\approx 2^{44}$. While publishing 2^{44} encodings of a multilinear map capable of supporting 4150 levels of multilinearity is likely beyond the scope of existing candidates, further optimizations to the underlying multilinear map as well as to the different components of our pipeline can lead to a realizable construction. Compared to previous candidates which require $\gg 2^{100}$ levels of multilinearity, our construction is over 2^{80} times more efficient.

---

[7] To minimize the degree of multilinearity required, we require a composite-order ring that splits into ≈ 200 sub-rings. Instantiating our construction with the composite-order CLT multilinear map [33], the plaintext ring already supports the requisite number of sub-rings, so using CRT for parallelization does not incur any overhead.

2 Preliminaries

We begin by defining the notation that we use throughout this paper. For an integer n, we write $[n]$ to denote the set of integers $\{1, \ldots, n\}$. For a positive integer p, we write $\mathbb{Z}_p$ to denote the ring of integers modulo p. We typically use bold uppercase letters (e.g., $\mathbf{A}$, $\mathbf{B}$) to denote matrices and bold lowercase letters (e.g., $\mathbf{u}, \mathbf{v}$) to denote vectors.

For a finite set S, we write $x \overset{\text{R}}{\leftarrow} S$ to denote that x is drawn uniformly at random from S. For a distribution $\mathcal{D}$, we write $x \leftarrow \mathcal{D}$ to denote a sample from $\mathcal{D}$. Unless otherwise noted, we write λ to denote a computational security parameter and κ to denote a statistical security parameter. We say a function $f(\lambda)$ is negligible in λ if $f(\lambda) = o(1/\lambda^c)$ for all $c \in \mathbb{N}$. We write $f(\lambda) = \text{negl}(\lambda)$ to denote that f is a negligible function in λ and $f(\lambda) = \text{poly}(\lambda)$ to denote that f is a polynomial in λ. We say an algorithm is efficient if it runs in probabilistic polynomial time. For two families of distributions $\mathcal{D}_1$ and $\mathcal{D}_2$, we write $\mathcal{D}_1 \overset{c}{\approx} \mathcal{D}_2$ if the two distributions are computationally indistinguishable (that is, if no efficient algorithm is able to distinguish $\mathcal{D}_1$ from $\mathcal{D}_2$, except with negligible probability). We will also use the Schwartz-Zippel lemma [71, 75]:

Lemma 2.1 (Schwartz-Zippel Lemma [71,75]**).** *Let p be a prime and let $f \in \mathbb{Z}_p[x_1, \ldots, x_n]$ be a multivariate polynomial of total degree d, not identically zero. Then,*

$$\Pr[\alpha_1, \ldots, \alpha_n \overset{\text{R}}{\leftarrow} \mathbb{Z}_p : f(\alpha_1, \ldots, \alpha_n) = 0] \leq \frac{d}{p}.$$

In the full version, we also review the standard definitions of succinct non-interactive arguments (SNARGs).

3 Linear PCPs

We begin by reviewing the definition of linear probabilistically checkable proofs (LPCPs). In an LPCP system for a binary relation $\mathcal{R}$ over a finite field $\mathbb{F}$, the proof consists of a vector $\boldsymbol{\pi} \in \mathbb{F}^m$ and the PCP oracle is restricted to computing a linear function on the verifier's query vector. Specifically, on input a query matrix $\mathbf{Q} \in \mathbb{F}^{m \times \ell}$, the PCP oracle responds with $\mathbf{y} = \mathbf{Q}^\top \boldsymbol{\pi} \in \mathbb{F}^\ell$. We now give a formal definition adapted from [18].

Definition 3.1 (Linear PCPs [18]**).** *Let $\mathcal{R}$ be a binary relation, $\mathbb{F}$ be a finite field, P_{LPCP} be a deterministic prover algorithm, and V_{LPCP} be a probabilistic oracle verification algorithm. Then, $(P_{\text{LPCP}}, V_{\text{LPCP}})$ is a ℓ-query linear PCP for $\mathcal{R}$ over $\mathbb{F}$ with soundness error ε and query length m if it satisfies the following requirements:*

- ***Syntax:** For a vector $\boldsymbol{\pi} \in \mathbb{F}^m$, the verification algorithm $V_{\text{LPCP}}^{\boldsymbol{\pi}} = (Q_{\text{LPCP}}, D_{\text{LPCP}})$ consists of an input-oblivious probabilistic query algorithm Q_{LPCP} and a deterministic decision algorithm D_{LPCP}. The query algorithm Q_{LPCP} generates a query matrix $\mathbf{Q} \in \mathbb{F}^{m \times \ell}$ (independently of the statement*

x) *and some state information* st. *The decision algorithm* D_{LPCP} *takes the statement* $\mathbf{x}$, *the state* st, *and the response vector* $\mathbf{y} = \mathbf{Q}^\top \boldsymbol{\pi} \in \mathbb{F}^\ell$ *and either "accepts" or "rejects."*

- **Completeness:** *For every* $(\mathbf{x}, \mathbf{w}) \in \mathcal{R}$, *the output of* $P_{\mathsf{LPCP}}(\mathbf{x}, \mathbf{w})$ *is a vector* $\boldsymbol{\pi} \in \mathbb{F}^m$ *such that* $V^{\boldsymbol{\pi}}_{\mathsf{LPCP}}(\mathbf{x})$ *accepts with probability 1.*
- **Soundness:** *For all* $\mathbf{x}$ *where* $(\mathbf{x}, \mathbf{w}) \notin \mathcal{R}$ *for all* $\mathbf{w}$ *and for all vectors* $\boldsymbol{\pi}^* \in \mathbb{F}^m$, *the probability that* $V^{\boldsymbol{\pi}^*}_{\mathsf{LPCP}}(\mathbf{x})$ *accepts is at most* ε.

We say that $(P_{\mathsf{LPCP}}, V_{\mathsf{LPCP}})$ *is statistically sound if* $\varepsilon(\kappa) = \mathrm{negl}(\kappa)$, *where* κ *is a statistical security parameter.*

Soundness Against Affine Provers. In Definition 3.1, we have only required soundness to hold against provers that employ a *linear* strategy, and not an *affine* strategy. Our construction of SNARGs (Sect. 4), will require the stronger property that soundness holds against provers using an affine strategy—that is, a strategy which can be described by a tuple $\boldsymbol{\varPi} = (\boldsymbol{\pi}, \mathbf{b})$ where $\boldsymbol{\pi} \in \mathbb{F}^m$ represents a linear function and $\mathbf{b} \in \mathbb{F}^\ell$ represents an affine shift. Then, on input a query matrix $\mathbf{Q} \in \mathbb{F}^{m \times \ell}$, the response vector is constructed by evaluating the affine relation $\mathbf{y} = \mathbf{Q}^\top \boldsymbol{\pi} + \mathbf{b}$. We now define this stronger notion of soundness against an affine prover.

Definition 3.2 (Soundness Against Affine Provers). *Let* $\mathcal{R}$ *be a relation and* $\mathbb{F}$ *be a finite field. A linear PCP* $(P_{\mathsf{LPCP}}, V_{\mathsf{LPCP}})$ *is a* ℓ*-query linear PCP for* $\mathcal{R}$ *over* $\mathbb{F}$ *with soundness error* ε *against affine provers if it satisfies the requirements in Definition 3.1 with the following modifications:*

- **Syntax:** *For any affine function* $\boldsymbol{\varPi} = (\boldsymbol{\pi}, \mathbf{b})$, *the verification algorithm* $V^{\boldsymbol{\varPi}}_{\mathsf{LPCP}}$ *is still specified by a tuple* $(Q_{\mathsf{LPCP}}, D_{\mathsf{LPCP}})$. *Algorithms* $Q_{\mathsf{LPCP}}, D_{\mathsf{LPCP}}$ *are the same as in Definition 3.1, except that the response vector* $\mathbf{y}$ *computed by the PCP oracle is an affine function* $\mathbf{y} = \mathbf{Q}^\top \boldsymbol{\pi} + \mathbf{b} \in \mathbb{F}^\ell$ *of the query matrix* $\mathbf{Q}$ *rather than a linear function.*
- **Soundness against affine provers:** *For all* $\mathbf{x}$ *where* $(\mathbf{x}, \mathbf{w}) \notin \mathcal{R}$ *for all* $\mathbf{w}$, *and for all affine functions* $\boldsymbol{\varPi}^* = (\boldsymbol{\pi}^*, \mathbf{b}^*)$ *where* $\boldsymbol{\pi}^* \in \mathbb{F}^m$ *and* $\mathbf{b}^* \in \mathbb{F}^\ell$, *the probability that* $V^{\boldsymbol{\varPi}^*}_{\mathsf{LPCP}}(\mathbf{x})$ *accepts is at most* ε.

Algebraic Complexity. There are many ways one can measure the complexity of a linear PCP system such as the number of queries or the number of field elements in the verifier's queries. Another important metric also considered in [18] is the *algebraic complexity* of the verifier. In particular, the verifier's query algorithm Q_{LPCP} and decision algorithm D_{LPCP} can both be viewed as multivariate polynomials (equivalently, arithmetic circuits) over the finite field $\mathbb{F}$. We say that the query algorithm Q_{LPCP} has degree d_Q if the output of Q_{LPCP} can be computed by a collection of multivariate polynomials of maximum degree d_Q in the verifier's choice of randomness. Similarly, we say that the decision algorithm D_{LPCP} has degree d_D if the output of D_{LPCP} can be computed by a multivariate polynomial of maximum degree d_D in the prover's response and the verification state.

Strong Soundness. In this work, we focus on constructing designated-verifier SNARGs. An important consideration that arises in the design of designated-verifier SNARGs is whether the same reference string σ can be reused across many proofs. This notion is formally captured by stipulating that the SNARG system remains sound even if the prover has access to a proof-verification oracle. While this property naturally follows from soundness if the SNARG system is publicly-verifiable, the same is not true in the designated-verifier setting. Specifically, in the designated-verifier setting, soundness is potentially compromised if the responses of the proof-verification oracle is correlated with the verifier's secrets. Thus, to construct a *multi-theorem* designated-verifier SNARG, we require linear PCPs with a stronger soundness property, which we state below.

Definition 3.3 Strong Soundness [18]). *A ℓ-query LPCP $(P_{\mathsf{LPCP}}, V_{\mathsf{LPCP}})$ with soundness error ε satisfies strong soundness if for every input $\mathbf{x}$ and every proof $\boldsymbol{\pi}^* \in \mathbb{F}^m$, either $V_{\mathsf{LPCP}}^{\boldsymbol{\pi}^*}(\mathbf{x})$ accepts with probability 1 or with probability at most ε.*

Roughly speaking, in an LPCP that satisfies strong soundness, *every* LPCP prover either causes the LPCP verifier to accept with probability 1 or with bounded probability. This prevents correlation attacks where a malicious prover is able to submit (potentially malformed) proofs to the verifier and seeing responses that are correlated with the verifier's secrets. We can define an analogous notion of strong soundness against affine provers.

3.1 Constructing Linear PCPs with Strong Soundness

A natural first question is whether linear PCPs with strong soundness against affine provers exist. Bitansky et al. [18] give two constructions of algebraic LPCPs for Boolean circuit satisfaction problems: one from the Hadamard-based PCP of Arora et al. [9], and another from the quadratic span programs (QSPs) of Gennaro et al. [42]. In both cases, the linear PCP is defined over a finite field $\mathbb{F}$ and the soundness error scales inversely with $|\mathbb{F}|$. Thus, the LPCP is statistically sound only if $|\mathbb{F}|$ is *superpolynomial* in the (statistical) security parameter. However, when we apply our LPCP-based SNARGs to bootstrap obfuscation, the size of the obfuscated program grows polynomially in $|\mathbb{F}|$, and so we require LPCPs with statistical soundness over small (polynomially-sized) fields.

In this section, we show that starting from any LPCP with constant soundness error against linear provers, we can generically obtain an LPCP that is statistically sound against affine provers. Our generic transformation consists of two steps. The first is a standard soundness amplification step where the verifier makes κ sets of independently generated queries (of the underlying LPCP scheme) to the PCP oracle, where κ is a statistical security parameter. The verifier accepts only if the prover's responses to all κ sets of queries are valid. Since the queries are independently generated, each of the κ sets of responses (for a false statement) is accepted with probability at most ε (where ε is proportional to $1/|\mathbb{F}|$). Thus, an honest verifier only accepts with probability at most $\varepsilon^{\kappa} = \mathrm{negl}(\kappa)$.

However, this basic construction does not achieve *strong* soundness against *affine* provers. For instance, a malicious LPCP prover using an affine strategy could selectively corrupt the responses to exactly one set of queries (by applying an affine shift to its response for a single set of queries). When this selective corruption is applied to a well-formed proof and the verifier's decision algorithm has low algebraic complexity, then the verifier will accept with some noticeable probability less than 1, which is sufficient to break strong soundness. To address this problem, the verifier first applies a (secret) random linear shift to its queries before submitting them to the PCP oracle. This ensures that any prover using an affine strategy with a non-zero offset will corrupt its responses to *every* set of queries, and the proof will be rejected with overwhelming probability. We now describe our generic construction in more detail.

Construction 3.4 (Statistically Sound Linear PCPs over Small Fields).
Fix a statistical security parameter κ. Let $\mathcal{R}$ be a binary relation, $\mathbb{F}$ be a finite field, and $\left(P_{\mathsf{LPCP}}^{(\mathrm{weak})}, V_{\mathsf{LPCP}}^{(\mathrm{weak})}\right)$ be an ℓ-query linear PCP for $\mathcal{R}$, where $V_{\mathsf{LPCP}}^{(\mathrm{weak})} = \left(Q_{\mathsf{LPCP}}^{(\mathrm{weak})}, D_{\mathsf{LPCP}}^{(\mathrm{weak})}\right)$. Define the $(\kappa\ell)$-query linear PCP $(P_{\mathsf{LPCP}}, V_{\mathsf{LPCP}})$ where $V_{\mathsf{LPCP}} = (Q_{\mathsf{LPCP}}, D_{\mathsf{LPCP}})$ as follows:

– **Prover's Algorithm** P_{LPCP}: On input $(\mathbf{x}, \mathbf{w})$, output $P_{\mathsf{LPCP}}^{(\mathrm{weak})}(\mathbf{x}, \mathbf{w})$.
– **Verifier's Query Algorithm** Q_{LPCP}: The query algorithm invokes $Q_{\mathsf{LPCP}}^{(\mathrm{weak})}$ a total of κ times to obtain (independent) query matrices $\mathbf{Q}_1, \ldots, \mathbf{Q}_\kappa \in \mathbb{F}^{m \times \ell}$ and state information $\mathsf{st}_1, \ldots, \mathsf{st}_\kappa$. It constructs the concatenated matrix $\mathbf{Q} = [\mathbf{Q}_1 | \mathbf{Q}_2 | \cdots | \mathbf{Q}_\kappa] \in \mathbb{F}^{m \times \kappa\ell}$. Finally, it chooses a random matrix $\mathbf{Y} \xleftarrow{\mathrm{R}} \mathbb{F}^{\kappa\ell \times \kappa\ell}$ and outputs the queries $\mathbf{Q}' = \mathbf{QY}$ and state $\mathsf{st} = (\mathsf{st}_1, \ldots, \mathsf{st}_\kappa, \mathbf{Y}')$ where $\mathbf{Y}' = (\mathbf{Y}^\top)^{-1}$.
– **Verifier's Decision Algorithm** D_{LPCP}: On input the statement $\mathbf{x}$, the prover's response vector $\mathbf{a}' \in \mathbb{F}^{\kappa\ell}$ and the state $\mathsf{st} = (\mathsf{st}_1, \ldots, \mathsf{st}_\kappa, \mathbf{Y}')$, the verifier's decision algorithm computes $\mathbf{a} = \mathbf{Y}'\mathbf{a}' \in \mathbb{F}^{\kappa\ell}$. Next, it writes $\mathbf{a}^\top = [\mathbf{a}_1^\top | \mathbf{a}_2^\top | \cdots | \mathbf{a}_\kappa^\top]$ where each $\mathbf{a}_i \in \mathbb{F}^\ell$ for $i \in [\kappa]$. Then, for each $i \in [\kappa]$, the verifier runs $D_{\mathsf{LPCP}}^{(\mathrm{weak})}(\mathbf{x}, \mathbf{a}_i, \mathsf{st}_i)$ and accepts if $D_{\mathsf{LPCP}}^{(\mathrm{weak})}$ accepts for all κ instances. It rejects otherwise.

Theorem 3.5. *Fix a statistical security parameter κ. Let $\mathcal{R}$ be a binary relation, $\mathbb{F}$ be a finite field, and $(P_{\mathsf{LPCP}}^{(\mathrm{weak})}, V_{\mathsf{LPCP}}^{(\mathrm{weak})})$ be a strongly-sound ℓ-query linear PCP for $\mathcal{R}$ with constant soundness error $\varepsilon \in [0, 1)$ against linear provers. If $|\mathbb{F}| > d_D$, where d_D is the degree of the verifier's decision algorithm $D_{\mathsf{LPCP}}^{(\mathrm{weak})}$, then the linear PCP $(P_{\mathsf{LPCP}}, V_{\mathsf{LPCP}})$ from Construction 3.4 is a $(\kappa\ell)$-query linear PCP for $\mathcal{R}$ with strong statistical soundness against affine provers.*

Proof. Completeness follows immediately from completeness of the underlying LPCP system, so it suffices to check that the linear PCP is statistically sound against affine provers. Take any statement $\mathbf{x}$, and consider an affine prover strategy $\boldsymbol{\Pi}^* = (\boldsymbol{\pi}^*, \mathbf{b}^*)$, where $\boldsymbol{\pi}^* \in \mathbb{F}^m$ and $\mathbf{b}^* \in \mathbb{F}^{\kappa\ell}$. We consider two cases:

- Suppose $\mathbf{b}^* \neq 0^{\kappa\ell}$. Then, the decision algorithm D_{LPCP} starts by computing

$$\mathbf{a} = \mathbf{Y}'\mathbf{a}' = \mathbf{Y}'(\mathbf{Y}^\top \mathbf{Q}^\top \boldsymbol{\pi}^* + \mathbf{b}^*) = \mathbf{Q}^\top \boldsymbol{\pi}^* + \mathbf{Y}'\mathbf{b}^* \in \mathbb{F}^{\kappa\ell}.$$

Next, the verifier invokes the decision algorithm $D_{\mathsf{LPCP}}^{(\mathrm{weak})}$ for the underlying LPCP on the components of $\mathbf{a}$. By assumption, $D_{\mathsf{LPCP}}^{(\mathrm{weak})}$ is a polynomial of maximum degree d_D in the components of the prover's response $\mathbf{a}$, and by extension, in the components of the matrix $\mathbf{Y}'$. Since $\mathbf{b}^*$ is non-zero, this is a non-zero polynomial in the $\mathbf{Y}'$. Since $\mathbf{Y}'$ is sampled uniformly at random (and independently of $\mathbf{Q}, \boldsymbol{\pi}^*, \mathbf{b}^*$), by the Schwartz-Zippel lemma, $D_{\mathsf{LPCP}}^{(\mathrm{weak})}(\mathbf{x}, \mathbf{a}_i, \mathsf{st}_i)$ accepts with probability at most $d_D / |\mathbb{F}|$ for each $i \in [\kappa]$. Thus, the verifier rejects with probability at least $1 - (d_D / |\mathbb{F}|)^\kappa = 1 - \mathsf{negl}(\kappa)$ since $|\mathbb{F}| > d_D$.
- Suppose $\mathbf{b}^* = 0^{\kappa\ell}$. Then, the prover's strategy is a linear function $\boldsymbol{\pi}^*$. Since the underlying PCP satisfies strong soundness against linear provers, it follows that $D_{\mathsf{LPCP}}^{(\mathrm{weak})}(\mathbf{a}_i, \mathsf{st}_i)$ either accepts with probability 1 or with probability at most ε. In the former case, D_{LPCP} also accepts with probability 1. In the latter case, because the verifier constructs the κ queries to the underlying LPCP independently, D_{LPCP} accepts with probability at most $\varepsilon^\kappa = \mathsf{negl}(\kappa)$. We conclude that the proof system $(P_{\mathsf{LPCP}}, V_{\mathsf{LPCP}})$ satisfies strong soundness against affine provers. $\qquad\square$

Remark 3.6 (Efficiency of Transformation). Construction 3.4 incurs a κ overhead in the number of queries made to the PCP oracle and a quadratic overhead in the algebraic complexity of the verifier's decision algorithm. Specifically, the degree of the verifier's decision algorithm in Construction 3.4 is d_D^2, where d_D is the degree of the verifier's decision algorithm in the underlying LPCP. The quadratic factor arises from undoing the linear shift in the prover's responses before applying the decision algorithm of the underlying LPCP. In many existing LPCP systems, the verifier's decision algorithm has low algebraic complexity (e.g., $d_D = 2$ for both the Hadamard-based LPCP [9] as well as the QSP-based LPCP [42]), so the verifier's algebraic complexity only increases modestly. However, the increase in degree means that we can no longer leverage pairing-based linear-only one-way encodings [18] to construct publicly-verifiable SNARGs (since these techniques only apply when the algebraic complexity of the verifier's decision algorithm is exactly 2). No such limitations apply in the designated-verifier setting.

Remark 3.7 (Comparison with [18, Lemma C.3]). Bitansky et al. [18, Lemma C.3] previously showed that any algebraic LPCP over a finite field $\mathbb{F}$ with soundness error ε is also strongly sound with soundness error $\varepsilon' = \max\left\{\varepsilon, \frac{d_Q d_D}{|\mathbb{F}|}\right\}$. For sufficiently large fields $\mathbb{F}$ (e.g., when $|\mathbb{F}|$ is superpolynomial), statistical soundness implies strong statistical soundness. However, when $|\mathbb{F}|$ is polynomial, then their lemma is insufficient to argue strong statistical soundness of the underlying LPCP. In contrast, using our construction (Construction 3.4), any LPCP with just constant soundness against linear provers can be used to construct an algebraic LPCP with strong statistical soundness against affine provers (at the cost of increasing the query complexity and the verifier's algebraic complexity).

Concrete Instantiations. Applying Construction 3.4 to the algebraic LPCPs for Boolean circuit satisfaction of Bitansky et al. [18], we obtain statistically sound LPCPs for Boolean circuit satisfaction over small finite fields. In the following, fix a (statistical) security parameter κ and let C be a Boolean circuit of size s.

- Starting from the Hadamard-based PCP of Arora et al. [9] over a finite field $\mathbb{F}$, there exists a 3-query LPCP with strong soundness error $2/|\mathbb{F}|$. The algebraic complexity of the decision algorithm for this PCP is $d_D = 2$. Applying Construction 3.4 and working over any finite field where $|\mathbb{F}| > 2$, we obtain a (3κ)-query LPCP with strong statistical soundness against affine provers and where queries have length $O(s^2)$.
- Starting from the quadratic span programs of Gennaro et al. [42], there exists a 3-query LPCP over any (sufficiently large) finite field $\mathbb{F}$ with strong soundness error $O(s/|\mathbb{F}|)$. The algebraic complexity of the decision algorithm for this PCP is $d_D = 2$. Applying Construction 3.4 and working over a sufficiently large finite field of size $|\mathbb{F}| = \widetilde{O}(s)$, we obtain a (3κ)-query LPCP with strong statistical soundness against affine provers where queries have length $O(s)$.

4   SNARGs from Linear-Only Vector Encryption

In this section, we introduce the notion of a linear-only vector encryption scheme. We then show how linear-only vector encryption can be directly combined with the linear PCPs from Sect. 3 to obtain multi-theorem designated-verifier pre-processing SNARGs in the standard model. We conclude by describing a candidate instantiation of our linear-only vector encryption scheme using the LWE-based encryption scheme of Peikert et al. [67]. In the full version of this paper, we also show how using linear-only vector encryption over polynomial rings, our techniques can be further extended to obtain the first quasi-optimal SNARG from any assumption (namely, a SNARG that is quasi-optimal in *both* the prover complexity and the proof length). Our notion of linear-only vector encryption is a direct generalization of the notion of linear-only encryption first introduced by Bitansky et al. [18].

4.1   Vector Encryption and Linear Targeted Malleability

A vector encryption scheme is an encryption scheme where the message space is a vector of ring elements. In this section, we take $\mathbb{Z}_p$ as the underlying ring and $\mathbb{Z}_p^\ell$ as the message space (for some dimension ℓ). In the full version, we also consider vector encryption schemes where the ring R is a polynomial ring and the message space is R^ℓ. We introduce the basic schema below:

Definition 4.1 (Vector Encryption Scheme over $\mathbb{Z}_p^\ell$). *A secret-key vector encryption scheme over $\mathbb{Z}_p^\ell$ consists of a tuple of algorithms $\Pi_{\mathsf{enc}} = (\mathsf{Setup},$ $\mathsf{Encrypt}, \mathsf{Decrypt})$ with the following properties:*

- Setup$(1^\lambda, 1^\ell) \to$ sk: *The setup algorithm takes as input the security parameter λ and the dimension ℓ of the message space and outputs the secret key* sk.
- Encrypt$(\text{sk}, \mathbf{v}) \to$ ct: *The encryption algorithm takes as input the secret key* sk *and a message vector $\mathbf{v} \in \mathbb{Z}_p^\ell$ and outputs a ciphertext* ct.
- Decrypt$(\text{sk}, \text{ct}) \to \mathbb{Z}_p^\ell \cup \{\bot\}$: *The decryption algorithm takes as input the secret key* sk *and a ciphertext* ct *and either outputs a message vector $\mathbf{v} \in \mathbb{Z}_p^\ell$ or a special symbol $\bot$ (to denote an invalid ciphertext).*

We can define the usual notions of correctness and semantic security [48] for a vector encryption scheme. Next, we say that a vector encryption scheme over $\mathbb{Z}_p^\ell$ is additively homomorphic if given encryptions ct_1, ct_2 of two vectors $\mathbf{v}_1, \mathbf{v}_2 \in \mathbb{Z}_p^\ell$, respectively, there is a public operation[8] that allows one to compute an encryption ct_{12} of the (component-wise) sum $\mathbf{v}_1 + \mathbf{v}_2 \in \mathbb{Z}_p^\ell$. Note that additively homomorphic vector encryption can be constructed directly from any additively homomorphic encryption scheme by simply encrypting each component of the vector separately. However, when leveraging vector encryption to build efficient SNARGs, we require that our encryption scheme satisfies a more restrictive homomorphism property. We define this now.

A vector encryption scheme satisfies *linear targeted malleability* [23] if the only homomorphic operations the adversary can perform on ciphertexts is evaluate affine functions on the underlying plaintext vectors. We now state our definition more precisely. Note that our definition is a vector generalization of the "weaker" notion of linear-only encryption introduced by Bitansky et al. [18]. This notion already suffices for constructing a designated-verifier SNARG.

Definition 4.2 (Linear Targeted Malleability [23, adapted]**).** *Fix a security parameter λ. A (secret-key) vector encryption scheme $\Pi_{\text{venc}} =$ (Setup, Encrypt, Decrypt) for a message space $\mathbb{Z}_p^\ell$ satisfies* linear targeted malleability *if for all efficient adversaries $\mathcal{A}$ and plaintext generation algorithms $\mathcal{M}$ (on input 1^ℓ, algorithm $\mathcal{M}$ outputs vectors in $\mathbb{Z}_p^\ell$), there exists a (possibly computationally unbounded) simulator $\mathcal{S}$ such that for any auxiliary input $z \in \{0,1\}^{\text{poly}(\lambda)}$, the following two distributions are computationally indistinguishable:*

Real Distribution:	Ideal Distribution:
1. sk $\leftarrow$ Setup$(1^\lambda, 1^\ell)$	*1.* $(s, \mathbf{v}_1, \ldots, \mathbf{v}_m) \leftarrow \mathcal{M}(1^\ell)$
2. $(s, \mathbf{v}_1, \ldots, \mathbf{v}_m) \leftarrow \mathcal{M}(1^\ell)$	*2.* $(\boldsymbol{\pi}, \mathbf{b}) \leftarrow \mathcal{S}(z)$ *where* $\boldsymbol{\pi} \in \mathbb{Z}_p^m$,
3. $\text{ct}_i \leftarrow$ Encrypt$(\text{sk}, \mathbf{v}_i)$ *for all* $i \in [m]$	$\quad\mathbf{b} \in \mathbb{Z}_p^\ell$
4. $\text{ct}' \leftarrow \mathcal{A}(\{\text{ct}_i\}_{i \in [m]} ; z)$ *where*	*3.* $\mathbf{v}' \leftarrow [\mathbf{v}_1 \vert \mathbf{v}_2 \vert \cdots \vert \mathbf{v}_m] \cdot \boldsymbol{\pi} + \mathbf{b}$
$\quad$ Decrypt$(\text{sk}, \text{ct}') \neq \bot$	*4. Output* $\left(\{\mathbf{v}_i\}_{i \in [m]}, s, \mathbf{v}_i'\right)$
5. Output	
$\quad\left(\{\mathbf{v}_i\}_{i \in [m]}, s, \text{Decrypt}(\text{sk}, \text{ct}')\right)$	

---

[8] In principle, homomorphic evaluation might require additional *public* parameters to be published by the setup algorithm. For simplicity of presentation, we will assume that no additional parameters are required, but all of our notions extend to the setting where the setup algorithm outputs a public evaluation key.

Remark 4.3 (Multiple Ciphertexts). Similar to [18,23], we can also define a variant of linear targeted malleability where the adversary is allowed to output multiple ciphertexts $ct'_1, \ldots, ct'_m$. In this case, the simulator should output an affine function $(\boldsymbol{\Pi}, \mathbf{B})$ where $\boldsymbol{\Pi} \in \mathbb{Z}_p^{m \times m}$ and $\mathbf{B} \in \mathbb{Z}_p^{\ell \times m}$ that "explains" the ciphertexts $ct'_1, \ldots, ct'_m$. However, the simple variant we have defined above where the adversary just outputs a single ciphertext is sufficient for our construction.

Remark 4.4 (Auxiliary Input Distributions). In Definition 4.2, the simulator is required to succeed for all auxiliary inputs $z \in \{0,1\}^{\mathrm{poly}(\lambda)}$. This requirement is quite strong since z can be used to encode difficult cryptographic problems that the simulator needs to solve in order to correctly simulate the output distribution [16]. However, many of these pathological auxiliary input distributions are not problematic for Definition 4.2, since the simulator is allowed to be *computationally unbounded*. In other cases where we require the simulator to be efficient (e.g., to obtain succinct arguments of knowledge via Remark 4.9), we note that Definition 4.2 can be relaxed to only consider "benign" auxiliary input distributions for which the definition plausibly holds. For instance, for the multi-theorem SNARK construction described in the full version, it suffices that the auxiliary information is a uniformly random string.

Construction 4.5 (SNARG from Linear-Only Vector Encryption). Fix a prime p (so the ring $\mathbb{Z}_p$ is a field), and let $\mathcal{C} = \{C_k\}_{k \in \mathbb{N}}$ be a family of arithmetic circuits over $\mathbb{Z}_p$.[9] Let $\mathcal{R}_{\mathcal{C}}$ be the relation associated with $\mathcal{C}$. Let $(P_{\mathsf{LPCP}}, V_{\mathsf{LPCP}})$ be an ℓ-query input-oblivious linear PCP for $\mathcal{C}$. Let $\Pi_{\mathsf{venc}} = (\mathsf{Setup}, \mathsf{Encrypt}, \mathsf{Decrypt})$ be a secret-key vector encryption scheme for $\mathbb{Z}_p^{\ell}$. Our single-theorem, designated-verifier SNARG $\Pi_{\mathsf{SNARG}} = (\mathsf{Setup}, \mathsf{Prove}, \mathsf{Verify})$ in the preprocessing model for $\mathcal{R}_{\mathcal{C}}$ is defined as follows:

- $\mathsf{Setup}(1^\lambda, 1^k) \to (\sigma, \tau)$: On input the security parameter λ and the circuit family parameter k, the setup algorithm first invokes the query algorithm Q_{LPCP} for the LPCP to obtain a query matrix $\mathbf{Q} \in \mathbb{Z}_p^{m \times \ell}$ and some state information st. Next, it generates a secret key for the vector encryption scheme $\mathsf{sk} \leftarrow \mathsf{Setup}(1^\lambda, 1^\ell)$. Then, it encrypts each row (an element of $\mathbb{Z}_p^\ell$) of the query matrix $\mathbf{Q}$. More specifically, for $i \in [m]$, let $\mathbf{q}_i \in \mathbb{Z}_p^\ell$ be the ith row of $\mathbf{Q}$. Then, the setup algorithm computes ciphertexts $ct_i \leftarrow \mathsf{Encrypt}(\mathsf{sk}, \mathbf{q}_i)$. Finally, the setup algorithm outputs the common reference string $\sigma = (ct_1, \ldots, ct_m)$ and the verification state $\tau = (\mathsf{sk}, \mathsf{st})$.
- $\mathsf{Prove}(\sigma, \mathbf{x}, \mathbf{w})$: On input a common reference string $\sigma = (ct_1, \ldots, ct_m)$, a statement $\mathbf{x}$, and a witness $\mathbf{w}$, the prover invokes the prover algorithm P_{LPCP} for the LPCP to obtain a vector $\boldsymbol{\pi} \in \mathbb{Z}_p^m$. Viewing $ct_1, \ldots, ct_m$ as vector encryptions of the rows of a query matrix $\mathbf{Q} \in \mathbb{Z}_p^{m \times \ell}$, the prover uses the linear homomorphic properties of Π_{venc} to homomorphically compute an encryption of the matrix vector product $\mathbf{Q}^\top \boldsymbol{\pi}$. In particular, the prover homomorphically

---

[9] While we describe a SNARG for arithmetic circuit satisfiability (over $\mathbb{Z}_p$), the problem of Boolean circuit satisfiability easily reduces to arithmetic circuit satisfiability with only constant overhead [18, Claim A.2].

computes the sum $\mathsf{ct}' = \sum_{i \in [m]} \pi_i \cdot \mathsf{ct}_i$. The prover outputs the ciphertext ct' as its proof.

- $\mathsf{Verify}(\tau, \mathbf{x}, \pi)$: On input the (secret) verification state $\tau = (\mathsf{sk}, \mathsf{st})$, the statement $\mathbf{x}$, and the proof $\pi = \mathsf{ct}'$, the verifier decrypts the proof ct' using the secret key sk to obtain the prover's responses $\mathbf{a} \leftarrow \mathsf{Decrypt}(\mathsf{sk}, \mathsf{ct}')$. If $\mathbf{a} = \bot$, the verifier stops and outputs 0. Otherwise, it invokes the verification decision algorithm D_{LPCP} on the statement $\mathbf{x}$, the responses $\mathbf{a}$, and the LPCP verification state st to decide whether the proof is valid or not. The verification algorithm echoes the output of the decision algorithm.

Theorem 4.6 [18, Lemma 6.3]. *Let* $(P_{\mathsf{LPCP}}, V_{\mathsf{LPCP}})$ *be a linear PCP that is statistically sound against affine provers, and let* $\Pi_{\mathsf{venc}} = (\mathsf{Setup}, \mathsf{Encrypt}, \mathsf{Decrypt})$ *be a vector encryption scheme with linear targeted malleability. Then, applying Construction 4.5 to* $(P_{\mathsf{LPCP}}, V_{\mathsf{LPCP}})$ *and* Π_{venc} *yields a (non-adaptive) designated-verifier SNARG in the preprocessing model.*

Proof. Our proof is similar to the proof of [18, Lemma 6.3]. Let P^* be a malicious prover that convinces the verifier of some false statement $\mathbf{x} \notin \mathcal{L}_C$ with non-negligible probability $\varepsilon(\lambda)$, where $\mathcal{L}_C$ is the language associated with C. Since Π_{enc} satisfies linear targeted malleability (Definition 4.2), there exists a simulator S such that the following distributions are computationally indistinguishable:

Real Distribution:	Ideal Distribution:
1. $\mathsf{sk} \leftarrow \mathsf{Setup}(1^\lambda, 1^\ell)$	1. $(\mathsf{st}, \mathbf{Q}) \leftarrow Q_{\mathsf{LPCP}}$ where $\mathbf{Q} \in \mathbb{Z}_p^{m \times \ell}$
2. $(\mathsf{st}, \mathbf{Q}) \leftarrow Q_{\mathsf{LPCP}}$ where $\mathbf{Q} \in \mathbb{Z}_p^{m \times \ell}$	2. $(\boldsymbol{\pi}, \mathbf{b}) \leftarrow S(\mathbf{x})$ where $\boldsymbol{\pi} \in \mathbb{Z}_p^m$ and
3. $\mathsf{ct}_i \leftarrow \mathsf{Encrypt}(\mathsf{sk}, \mathbf{q}_i)$ where $\mathbf{q}_i$ is the	$\quad \mathbf{b} \in \mathbb{Z}_p^\ell$
$\quad i^{\text{th}}$ row of $\mathbf{Q}$ for $i \in [m]$	3. $\hat{\mathbf{a}} \leftarrow \mathbf{Q}^\top \boldsymbol{\pi} + \mathbf{b}$
4. $\mathsf{ct}' \leftarrow P^*(\mathsf{ct}_1, \dots, \mathsf{ct}_q; \mathbf{x})$ such that	1. Output $(\mathbf{Q}, \mathsf{st}, \hat{\mathbf{a}})$
$\quad \mathsf{Decrypt}(\mathsf{sk}, \mathsf{ct}') \neq \bot$	
5. $\mathbf{a} \leftarrow \mathsf{Decrypt}(\mathsf{sk}, \mathsf{ct}') \in \mathbb{Z}_p^\ell$	
6. Output $(\mathbf{Q}, \mathsf{st}, \mathbf{a})$	

By assumption, P^* convinces an honest verifier with probability $\varepsilon = \varepsilon(\lambda)$, or equivalently, in the real distribution, $D_{\mathsf{LPCP}}(\mathbf{x}, \mathsf{st}, \mathbf{a}) = 1$ with probability at least ε. Since D_{LPCP} is efficiently computable, computational indistinguishability of the real and ideal experiments means that $D_{\mathsf{LPCP}}(\mathbf{x}, \mathsf{st}, \hat{\mathbf{a}}) = 1$ with probability at least $\varepsilon - \mathsf{negl}(\lambda)$. However, in the ideal distribution, the affine function $(\boldsymbol{\pi}, \mathbf{b})$ is generated *independently* of the verifier's queries $\mathbf{Q}$ and state st. By an averaging argument, this means that there must exist some affine function $(\boldsymbol{\pi}^*, \mathbf{b}^*)$ such that with probability at least $\varepsilon - \mathsf{negl}(\lambda)$ taken over the randomness of Q_{LPCP}, the verifier's decision algorithm D_{LPCP} on input $\mathbf{x} \notin \mathcal{L}_C$, st, and $\mathbf{Q}^\top \boldsymbol{\pi}^* + \mathbf{b}^*$ accepts. But this contradicts statistical soundness (against affine provers) of the underlying linear PCP. $\square$

Remark 4.7 (Adaptivity). In Theorem 4.6, we showed that instantiating Construction 4.5 with a vector encryption scheme with linear targeted malleability and a linear PCP yields a non-adaptive SNARG in the preprocessing model.

The same construction can be shown to satisfy *adaptive* soundness for proving efficiently decidable statements. As noted in [18, Remark 6.5], we can relax Definition 4.2 and allow the adversary to additionally output an arbitrary string in the real distribution which the simulator must produce in the ideal distribution. Invoking Construction 4.5 with an encryption scheme that satisfies this strengthened linear targeted malleability definition yields a SNARG with adaptive soundness for the case of verifying deterministic polynomial-time computations. Note that the proof system necessary to bootstrap obfuscation is used to verify correctness of a polynomial-time computation (i.e., FHE evaluation), so adaptivity for this restricted class of statements is sufficient for our primary application.

Remark 4.8 (Multi-theorem SNARGs). Our basic notion of linear targeted malleability for vector encryption only suffices to construct a single-theorem SNARG. While the same construction can be shown secure for an adversary that is allowed to make any constant number of queries to a proof verification oracle, we are not able to prove that the construction is secure against a prover who makes polynomially many queries to the proof verification oracle. In the full version, we present an analog of the strengthened version of linear-only encryption from [18, Appendix C] that suffices for constructing a multi-theorem SNARG. Combined with a linear PCP that is *strongly sound* against affine provers, Construction 4.5 can then be applied to obtain a multi-theorem, designated-verifier SNARG. This raises the question of whether the same construction using the weaker notion of linear targeted malleability also suffices when the underlying linear PCP satisfies strong soundness. While we do not know how to prove security from this weaker definition, we also do not know of any attacks. This is especially interesting because at the information-theoretic level, the underlying linear PCP satisfies *strong* soundness, which intuitively would suggest that the responses the malicious prover obtains from querying the proof verification oracle are *uncorrelated* with the verifier's state (strong soundness states that for any proof, either the verifier accepts with probability 1 or with negligible probability).

Remark 4.9 (Arguments of Knowledge). Theorem 4.6 shows that instantiating Construction 4.5 with a linear PCP with soundness against affine provers and a vector encryption scheme with linear targeted malleability suffices for a SNARG. In fact, the same construction yields a SNARK (that is, a succinct non-interactive argument *of knowledge*) if the soundness property of the underlying LPCP is replaced with a corresponding knowledge property,[10] and the vector encryption scheme satisfies a variant of linear targeted malleability (Definition 4.2) where the simulator is required to be *efficient* (i.e., polynomially-sized). For more details, we refer to [18, Lemma 6.3, Remark 6.4].

---

[10] Roughly, the knowledge property states that there exists an extractor such that for every affine strategy Π^* that convinces the verifier of some statement $\mathbf{x}$ with high probability, the extractor outputs a witness $\mathbf{w}$ such that $(\mathbf{x}, \mathbf{w}) \in \mathcal{R}$. The Hadamard LPCP from [9] also satisfies this stronger knowledge property.

4.2   A Candidate Linear-Only Vector Encryption Scheme

The core building block in our new SNARG construction is a *vector encryption* scheme for $\mathbb{Z}_p^{\ell}$ that plausible satisfies our notion of linear targeted malleability (Definition 4.2). In particular, we conjecture that the Regev-based encryption scheme [68] due to Peikert et al. [67, Sect. 7.2] satisfies our required properties. Before describing the scheme, we review some notation as well as the learning with errors (LWE) assumption which is essential (though not sufficient) for arguing security of the vector encryption scheme.

Notation. For $x \in \mathbb{Z}$ and a positive odd integer q, we write $[x]_q$ to denote the value $x \bmod q$, with values in the interval $(-q/2, q/2]$. For a lattice Λ and a positive real value $\sigma > 0$, we write $D_{\Lambda,\sigma}$ to denote the discrete Gaussian distribution over Λ with standard deviation σ. In particular, $D_{\Lambda,\sigma}$ assigns a probability proportional to $\exp(-\pi \|\mathbf{x}\|^2 / \sigma^2)$ to each element $\mathbf{x} \in \Lambda$.

Learning with Errors. The learning with errors problem [68] is parameterized by a dimension $n \geq 1$, an integer modulus $q \geq 2$ and an error distribution χ over the integers $\mathbb{Z}$. In this work, the noise distribution is always the discrete Gaussian distribution $\chi = D_{\mathbb{Z},\sigma}$. For $\mathbf{s} \in \mathbb{Z}_q^n$, the LWE distribution $A_{\mathbf{s},m,\chi}$ over $\mathbb{Z}_q^{m \times n} \times \mathbb{Z}_q^n$ is specified by choosing a uniformly random matrix $\mathbf{A} \xleftarrow{\text{R}} \mathbb{Z}_q^{m \times n}$ and error $\mathbf{e} \leftarrow \chi^n$ and outputting the pair $(\mathbf{A}, \mathbf{As} + \mathbf{e}) \in \mathbb{Z}_q^{m \times n} \times \mathbb{Z}_q^m$. The learning with errors assumption $\mathsf{LWE}_{n,q,\chi}$ (parameterized by parameters n, q, χ) states that for all $m = \mathrm{poly}(n)$, the LWE distribution $A_{\mathbf{s},m,\chi}$ for a randomly sampled $\mathbf{s} \xleftarrow{\text{R}} \mathbb{Z}_q^n$ is computationally indistinguishable from the uniform distribution over $\mathbb{Z}_q^{m \times n} \times \mathbb{Z}_q^m$.

The PVW Encryption Scheme. We now review the encryption scheme due to Peikert et al. [67, Sect. 7.2]. To slightly simplify the notation, we describe the scheme where the message is embedded in the least significant bits of the plaintext. Note that when the modulus q is odd, this choice of "most significant bit" and "least significant bit" encoding makes no difference and the encodings are completely interchangeable [1, Appendix A]. In our setting, it suffices to just consider the secret-key setting. Let $\mathbb{Z}_p^{\ell}$ be the plaintext space. The vector encryption scheme $\Pi_{\mathsf{venc}} = (\mathsf{Setup}, \mathsf{Encrypt}, \mathsf{Decrypt})$ in [67] is defined as follows:

- $\mathsf{Setup}(1^\lambda, 1^\ell)$: Choose $\bar{\mathbf{A}} \xleftarrow{\text{R}} \mathbb{Z}_q^{n \times m}$, $\bar{\mathbf{S}} \xleftarrow{\text{R}} \mathbb{Z}_q^{n \times \ell}$, and $\bar{\mathbf{E}} \leftarrow \chi^{\ell \times m}$, where $n = n(\lambda)$, $m = m(\lambda)$, and $q = q(\lambda)$ are polynomials in the security parameter. Define the matrices $\mathbf{A} \in \mathbb{Z}_q^{(n+\ell) \times m}$ and $\mathbf{S} \in \mathbb{Z}_q^{(n+\ell) \times \ell}$ as follows:

$$\mathbf{A} = \begin{bmatrix} \bar{\mathbf{A}} \\ \bar{\mathbf{S}}^\top \bar{\mathbf{A}} + p\bar{\mathbf{E}} \end{bmatrix} \qquad \mathbf{S} = \begin{bmatrix} -\bar{\mathbf{S}} \\ \mathbf{I}_\ell \end{bmatrix},$$

where $\mathbf{I}_\ell \in \mathbb{Z}_q^{\ell \times \ell}$ is the ℓ-by-ℓ identity matrix. Output the secret key $\mathsf{sk} = (\mathbf{A}, \mathbf{S})$.

- Encrypt(sk, $\mathbf{v}$): To encrypt a vector $\mathbf{v} \in \mathbb{Z}_p^\ell$, choose $\mathbf{r} \xleftarrow{\text{R}} \{0,1\}^m$ and output the ciphertext $\mathbf{c} \in \mathbb{Z}_q^{n+\ell}$ where

$$\mathbf{c} = \mathbf{Ar} + \begin{bmatrix} \mathbf{0}^n \\ \mathbf{v} \end{bmatrix}.$$

- Decrypt(sk, $\mathbf{c}$): Compute and output $[[\mathbf{S}^\top \mathbf{c}]_q]_p$.

Remark 4.10 (Low-Norm Secret Keys). For some of our applications (namely, those that leverage modulus switching), it is advantageous to sample the LWE secret $\mathbf{s} \in \mathbb{Z}_q^n$ from a low-norm distribution. Previously, Applebaum et al. [8] and Brakerski et al. [29] showed that the LWE variant where the secret key $\mathbf{s} \leftarrow \chi^n$ is sampled from the error distribution is still hard under the standard LWE assumption. In the same work, Brakerski et al. also showed that LWE instances with binary secrets (i.e., $\mathbf{s} \in \{0,1\}^n$) is as hard as standard LWE (with slightly larger parameters). Sampling the secret keys from a binary distribution has been used to achieve significant concrete performance gains in several implementations of lattice-based cryptosystems [37, 44].

Correctness. Correctness of the encryption scheme follows as in [67]. In the full version of this paper, we provide the concrete bounds on the parameters under which correctness holds. This analysis will prove useful for estimating the concrete parameters needed to instantiate our candidate obfuscation scheme in Sect. 5.

Additive Homomorphism. Like Regev encryption, the scheme is additively homomorphic and supports scalar multiplication. Since the error is additive, to compute a linear combination of ξ ciphertexts (where the coefficients for the linear combination are drawn from $\mathbb{Z}_p$), we need to scale the modulus q by a factor ξp for correctness to hold. In the full version, we show that this encryption scheme supports modulus switching, and thus, it is possible to work with a smaller modulus during decryption. However, this optimization is not necessary when using the vector encryption scheme to construct a SNARG (via Construction 4.5). It becomes important when we combine the SNARG with other tools to obtain more efficient bootstrapping of obfuscation for all circuits (Sect. 5).

Semantic Security. Security of this construction follows fairly naturally from the LWE assumption. We state the main theorem here, but refer readers to [67, Sect. 7.2.1] for the formal analysis.

Theorem 4.11 (Semantic Security [67]). *Fix a security parameter λ and let $n, q = \text{poly}(\lambda)$. Let $\chi = D_{\mathbb{Z}, \sigma}$ be a discrete Gaussian distribution with standard deviation $\sigma = \sigma(\lambda)$. Then, if $m \geq 3(n + \ell) \log q$, and assuming the $\text{LWE}_{n,q,\chi}$ assumption holds, then the vector encryption scheme Π_{venc} is semantically secure.*

4.3 Our Lattice-Based SNARG Candidate

We now state our concrete conjecture on the vector encryption scheme Π_{venc} from Sect. 4.2 that yields the first lattice-based candidate of a designated-verifier, preprocessing SNARG with quasi-optimal succinctness.

Conjecture 4.12. *The PVW vector encryption scheme Π_{venc} from Sect. 4.2 satisfies linear targeted malleability (Definition 4.2).*

Under Conjecture 4.12, we can apply Construction 4.5 in conjunction with algebraic LPCPs to obtain designated-verifier SNARGs in the preprocessing model (Theorem 4.6). To conclude, we give an asymptotic characterization of the complexity of our lattice-based SNARG system, and compare against existing SNARG candidates for Boolean circuit satisfiability. Let λ be a security parameter, and let C be a Boolean circuit of size $s = s(\lambda)$. We describe the parameters needed to achieve $2^{-\lambda}$ soundness against provers of size 2^{λ}.

- **Prover complexity.** In Construction 4.5, the prover performs m homomorphic operations on the encrypted vectors, where m is the length of the underlying linear PCP. When instantiating the vector encryption scheme Π_{venc} over the plaintext space $\mathbb{Z}_p^{\ell}$ where $p = \text{poly}(\lambda)$, the ciphertexts consist of vectors of dimension $O(\lambda + \ell)$ over a ring of size $q = \text{poly}(\lambda)$.[11] Homomorphic operations on ciphertexts corresponds to scalar multiplication (by values from $\mathbb{Z}_p$) and vector additions. Since all operations are performed over a *polynomial-sized* domain, all of the basic arithmetic operations can be performed in $\text{polylog}(\lambda)$ time. Thus, as long as the underlying LPCP operates over a polynomial-sized field, the prover's overhead is $\widetilde{O}(m(\lambda + \ell))$.
  If the underlying LPCP is instantiated with the Arora et al. [9] PCP based on the Walsh-Hadamard code, then $m = O(s^2)$ and $\ell = O(\lambda)$. The overall prover complexity in this case is thus $\widetilde{O}(\lambda s^2)$. If the underlying LPCP is instead instantiated with one based on the QSPs of Gennaro et al. [42], then $m = \widetilde{O}(s)$ and $\ell = O(\lambda)$. The overall prover complexity in this case is $\widetilde{O}(\lambda s)$.
- **Proof length.** Proofs in Construction 4.5 consist of a single ciphertext of the vector encryption scheme, which has length $\widetilde{O}(\lambda + \ell)$. Thus, both of our candidate instantiations of the LPCP (based on the Hadamard code and on QSPs) yield proofs of size $\widetilde{O}(\lambda)$.
- **Verifier complexity.** In Construction 4.5, the verifier first invokes the decryption algorithm of the underlying vector encryption scheme and then applies the verification procedure for the underlying linear PCP. Decryption consists of a rounded matrix-vector product over a polynomial-sized ring, which requires

---

[11] More precisely, the ciphertexts are actually vectors of dimension $n + \ell$, where n is the dimension of the lattice in the LWE problem. Currently, the most effective algorithms for solving LWE rely either on BKW-style [21,54] or BKZ-based attacks [32,70]. Based on our current understanding [26,32,54,60], the best-known algorithms for LWE all require time $2^{\Omega(n/\log^c n)}$ for some constant c. Thus, in terms of a concrete security parameter λ, we set the lattice dimension to be $n = \widetilde{O}(\lambda)$.

$\widetilde{O}(\lambda(\lambda+\ell))$ operations. In both of our candidate LPCP constructions, the verifier's decision algorithm runs in time $O(n)$, where n is the length of the statement. Moreover, the decision algorithm for the underlying LPCP is applied $O(\lambda)$ times for soundness amplification. Thus, the overall complexity of the verifier for both of our candidate instantiations is $\widetilde{O}(\lambda^2 + \lambda n)$.

Note that we can generically reduce the verifier complexity to $\widetilde{O}(\lambda^2 + n)$ by first applying a collision-resistant hash function to the statement and having the prover argue that it knows a preimage to the hash function and that the preimage is in the language. After applying this transformation, the length of the statement is simple the output length of of a collision-resistant hash function, namely $O(\lambda)$.

Remark 4.13 (Comparison with [18]). An alternative route to obtaining a lattice-based SNARG is to directly instantiate [18] with Regev-based encryption. However, to achieve soundness error $2^{-\lambda}$, Bitansky et al. [18] require a LPCP (and consequently, an additively homomorphic encryption) over a field of size 2^λ. Instantiating the construction in [18] with Regev-based encryption over a plaintext space of size 2^λ, the resulting SNARGs have length $\widetilde{O}(\lambda^2)$ and the prover complexity is $\widetilde{O}(s\lambda^2)$. Another possibility is to instantiate [18] with Regev-based encryption over a polynomial-size field (thus incurring $1/\mathrm{poly}(\lambda)$-soundness error) and perform parallel repetition at the SNARG level to amplify the soundness. But this method suffers from the same drawback as above. While each individual SNARG instance (over a polynomial-size field) is quasi-optimally succinct, the size of the overall proof is still $\widetilde{O}(\lambda^2)$ and the prover's complexity remains at $\widetilde{O}(s\lambda^2)$. This is a factor λ worse than using linear-only vector encryption over a polynomial-size field. We provide a concrete comparison in Table 1.

In Table 1, we compare our new lattice-based SNARG constructions to existing constructions for Boolean circuit satisfiability (the same results apply for arithmetic circuit satisfiability over polynomial-size fields). Amongst SNARGs with quasi-optimal succinctness (proof size $\widetilde{O}(\lambda)$), Construction 4.5 instantiated with a QSP-based LPCP achieves the same prover efficiency as the current state-of-the-art (GGPR [42] and BCIOP [18]). However, in contrast to current schemes, our construction is lattice-based, and thus, plausibly resists quantum attacks. One limitation is that our new constructions are designated-verifier, while existing constructions are publicly verifiable. We stress here though that a common limitation of designated-verifier SNARGs—that the common reference string cannot be reused for multiple proofs [15, 34, 47]—does not apply to our construction. As noted by [18], this limitation can be circumvented by SNARG constructions relying on algebraic PCPs such as ours. We show in the full version that a variant of our construction (with the same asymptotic complexity) gives a multi-theorem designated-verifier SNARG in the preprocessing model.

Remark 4.14 (Arithmetic Circuit Satisfiability over Large Fields). Construction 4.5 also applies to arithmetic circuit satisfiability over large finite fields (say,

Table 1. Asymptotic performance of different SNARG systems for Boolean circuit satisfiability. Here, s is the size of the circuit and λ is a security parameter guaranteeing $\mathrm{negl}(\lambda)$ soundness error against provers of size 2^{λ}. (Some of the schemes can achieve $2^{-\lambda}$ soundness error with the same complexity.) All of the schemes can be converted into an *argument of knowledge* (i.e., a SNARK)—in some cases, this requires a stronger cryptographic assumption.

Construction	Type*	Prover complexity	Proof size	Assumption
CS proofs [63]	PV	$\widetilde{O}(s+\lambda^2)$	$\widetilde{O}(\lambda^2)$	Random oracle
Groth [51]	PV	$\widetilde{O}(s^2\lambda + s\lambda^2)$	$\widetilde{O}(\lambda)$	Knowledge of exponent
GGPR [42]	PV	$\widetilde{O}(s\lambda)$	$\widetilde{O}(\lambda)$	
BCIOP [18]† (Paillier)	DV	$\widetilde{O}(s\lambda^3)$	$\widetilde{O}(\lambda^3)$	Linear-only encryption
BCIOP [18]† (Pairing)	PV	$\widetilde{O}(s\lambda)$	$\widetilde{O}(\lambda)$	
BCIOP [18]† (Regev)‡	DV	$\widetilde{O}(s\lambda^2)$	$\widetilde{O}(\lambda^2)$	
Construction 4.5§ (Hadamard LPCP)	DV	$\widetilde{O}(s^2\lambda)$	$\widetilde{O}(\lambda)$	Linear-only vector enc.
Construction 4.5§ (QSP-based LPCP)	DV	$\widetilde{O}(s\lambda)$	$\widetilde{O}(\lambda)$	
Construction 4.5 (RLWE-based)¶	DV	$\widetilde{O}(s)$	$\widetilde{O}(\lambda)$	Linear-only vector enc.

*We write "PV" to denote public verifiability and "DV" for designated verifiability.
†Instantiated using a LPCP based on QSPs.
‡Based on a direct instantiation of [18] using Regev-based encryption (Remark 4.13).
§Instantiated with the PVW [67] encryption scheme from Sect. 4.2.
¶Instantiated with the RLWE-based vector encryption scheme described in the full version. This construction is the first which is quasi-optimal with respect to *both* prover complexity and proof size.

$\mathbb{Z}_p$ where $p = 2^{\lambda}$). However, if the size of the plaintext space for the vector encryption scheme Π_{venc} from Sect. 4.2 is 2^{λ}, then the bit-length of the ciphertexts becomes $\widetilde{O}(\lambda^2)$ bits. Consequently, the proof system is no longer quasi-optimally succinct. In contrast, the QSP-based constructions [18, 42] remain quasi-optimally succinct for arithmetic circuit satisfiability over large fields.

Quasi-Optimal SNARG. In the full version of this paper, we also show how vector encryption over polynomial rings that satisfy linear targeted malleability can be leveraged to obtain the first SNARG construction that achieves quasi-optimal prover complexity as well as quasi-optimal succinctness. Our construction makes use of a new information-theoretic construction of LPCPs over rings.

5   Concrete Efficiency of Bootstrapping VBB Obfuscation

Due to space limitations, we defer our results on the concrete efficiency of bootstrapping obfuscation to the full version, and give an outline of our main results here. We start by describing how matrix branching programs can be used to perform simple computations over $\mathbb{Z}_q$. In particular, we show how we can implement

FHE decryption and SNARG verification as a matrix branching program. Then, we introduce a series of algorithmic as well as heuristic optimizations to improve the concrete efficiency of the candidate obfuscator. We conclude by giving an estimate of the parameters needed to instantiate our obfuscation candidate.

To summarize, after applying our optimizations, implementing FHE decryption together with SNARG verification can be done with a branching program (over composite-order rings) of length 4150 and size $\approx 2^{44}$ (at a security level of $\lambda = 80$). While publishing 2^{44} encodings of a multilinear map capable of supporting 4150 levels of multilinearity is likely beyond the scope of existing candidates, further optimizations to the underlying multilinear map as well as to the different components of our pipeline can plausibly lead to a realizable construction. Thus, our construction represents an important milestone towards the ultimate goal of implementable program obfuscation.

Acknowledgments. We thank the anonymous reviewers for helpful feedback on the presentation. D. Boneh and D.J. Wu are supported by NSF, DARPA, a grant from ONR, the Simons Foundation, and an NSF Graduate Research Fellowship. Y. Ishai and A. Sahai are supported in part from a DARPA/ARL SAFEWARE award, NSF Frontier Award 1413955, NSF grants 1619348, 1228984, 1136174, and 1065276, BSF grant 2012378, NSF-BSF grant 2015782, a Xerox Faculty Research Award, a Google Faculty Research Award, an equipment grant from Intel, and an Okawa Foundation Research Grant. Y. Ishai is additionally supported by ISF grant 1709/14. This material is based upon work supported by the Defense Advanced Research Projects Agency through the ARL under Contract W911NF-15-C-0205. The views expressed are those of the authors and do not reflect the official policy or position of the Department of Defense, the National Science Foundation, or the U.S. Government.

References

1. Alperin-Sheriff, J., Peikert, C.: Practical bootstrapping in quasilinear time. In: Canetti, R., Garay, J.A. (eds.) CRYPTO 2013. LNCS, vol. 8042, pp. 1–20. Springer, Heidelberg (2013). doi:10.1007/978-3-642-40041-4_1

2. Alperin-Sheriff, J., Peikert, C.: Faster bootstrapping with polynomial error. In: Garay, J.A., Gennaro, R. (eds.) CRYPTO 2014. LNCS, vol. 8616, pp. 297–314. Springer, Heidelberg (2014). doi:10.1007/978-3-662-44371-2_17

3. Ananth, P., Jain, A.: Indistinguishability obfuscation from compact functional encryption. In: Gennaro, R., Robshaw, M. (eds.) CRYPTO 2015. LNCS, vol. 9215, pp. 308–326. Springer, Heidelberg (2015). doi:10.1007/978-3-662-47989-6_15

4. Ananth, P., Sahai, A.: Projective arithmetic functional encryption and indistinguishability obfuscation from degree-5 multilinear maps. In: Coron, J.-S., Nielsen, J.B. (eds.) EUROCRYPT 2017, Part III. LNCS, vol. 10212, pp. 152–181. Springer, Heidelberg (2017)

5. Ananth, P.V., Gupta, D., Ishai, Y., Sahai, A.: Optimizing obfuscation: avoiding Barrington's theorem. In: ACM CCS (2014)

6. Applebaum, B.: Bootstrapping obfuscators via fast pseudorandom functions. In: Sarkar, P., Iwata, T. (eds.) ASIACRYPT 2014. LNCS, vol. 8874, pp. 162–172. Springer, Heidelberg (2014). doi:10.1007/978-3-662-45608-8_9

7. Applebaum, B., Brakerski, Z.: Obfuscating circuits via composite-order graded encoding. In: Dodis, Y., Nielsen, J.B. (eds.) TCC 2015. LNCS, vol. 9015, pp. 528–556. Springer, Heidelberg (2015). doi:10.1007/978-3-662-46497-7_21

8. Applebaum, B., Cash, D., Peikert, C., Sahai, A.: Fast cryptographic primitives and circular-secure encryption based on hard learning problems. In: Halevi, S. (ed.) CRYPTO 2009. LNCS, vol. 5677, pp. 595–618. Springer, Heidelberg (2009). doi:10.1007/978-3-642-03356-8_35

9. Arora, S., Lund, C., Motwani, R., Sudan, M., Szegedy, M.: Proof verification and the hardness of approximation problems. In: FOCS (1992)

10. Badrinarayanan, S., Miles, E., Sahai, A., Zhandry, M.: Post-zeroizing obfuscation: new mathematical tools, and the case of evasive circuits. In: Fischlin, M., Coron, J.-S. (eds.) EUROCRYPT 2016. LNCS, vol. 9666, pp. 764–791. Springer, Heidelberg (2016). doi:10.1007/978-3-662-49896-5_27

11. Barak, B., Garg, S., Kalai, Y.T., Paneth, O., Sahai, A.: Protecting obfuscation against algebraic attacks. In: Nguyen, P.Q., Oswald, E. (eds.) EUROCRYPT 2014. LNCS, vol. 8441, pp. 221–238. Springer, Heidelberg (2014). doi:10.1007/978-3-642-55220-5_13

12. Barak, B., Goldreich, O., Impagliazzo, R., Rudich, S., Sahai, A., Vadhan, S., Yang, K.: On the (im)possibility of obfuscating programs. In: Kilian, J. (ed.) CRYPTO 2001. LNCS, vol. 2139, pp. 1–18. Springer, Heidelberg (2001). doi:10.1007/3-540-44647-8_1

13. Barrington, D.A.M.: Bounded-width polynomial-size branching programs recognize exactly those languages in nc^1. In: STOC (1986)

14. Bellare, M., Palacio, A.: The knowledge-of-exponent assumptions and 3-round zero-knowledge protocols. In: Franklin, M. (ed.) CRYPTO 2004. LNCS, vol. 3152, pp. 273–289. Springer, Heidelberg (2004). doi:10.1007/978-3-540-28628-8_17

15. Bitansky, N., Canetti, R., Chiesa, A., Tromer, E.: From extractable collision resistance to succinct non-interactive arguments of knowledge, and back again. In: ITCS (2012)

16. Bitansky, N., Canetti, R., Paneth, O., Rosen, A.: On the existence of extractable one-way functions. In: STOC (2014)

17. Bitansky, N., Chiesa, A.: Succinct arguments from multi-prover interactive proofs and their efficiency benefits. In: Safavi-Naini, R., Canetti, R. (eds.) CRYPTO 2012. LNCS, vol. 7417, pp. 255–272. Springer, Heidelberg (2012). doi:10.1007/978-3-642-32009-5_16

18. Bitansky, N., Chiesa, A., Ishai, Y., Paneth, O., Ostrovsky, R.: Succinct non-interactive arguments via linear interactive proofs. In: Sahai, A. (ed.) TCC 2013. LNCS, vol. 7785, pp. 315–333. Springer, Heidelberg (2013). doi:10.1007/978-3-642-36594-2_18

19. Bitansky, N., Paneth, O., Wichs, D.: Perfect structure on the edge of chaos. In: Kushilevitz, E., Malkin, T. (eds.) TCC 2016. LNCS, vol. 9562, pp. 474–502. Springer, Heidelberg (2016). doi:10.1007/978-3-662-49096-9_20

20. Bitansky, N., Vaikuntanathan, V.: Indistinguishability obfuscation from functional encryption. In: FOCS (2015)

21. Blum, A., Kalai, A., Wasserman, H.: Noise-tolerant learning, the parity problem, and the statistical query model. In: STOC (2000)

22. Boneh, D., Franklin, M.K.: Identity-based encryption from the weil pairing. In: Kilian, J. (ed.) CRYPTO 2001. LNCS, vol. 2139, pp. 213–229. Springer, Heidelberg (2001). doi:10.1007/3-540-44647-8_13

23. Boneh, D., Segev, G., Waters, B.: Targeted malleability: homomorphic encryption for restricted computations. In: ITCS (2012)

24. Boneh, D., Silverberg, A.: Applications of multilinear forms to cryptography. Contemp. Math. **324**(1), 71–90 (2003)
25. Boneh, D., Zhandry, M.: Multiparty key exchange, efficient traitor tracing, and more from indistinguishability obfuscation. In: Garay, J.A., Gennaro, R. (eds.) CRYPTO 2014. LNCS, vol. 8616, pp. 480–499. Springer, Heidelberg (2014). doi:10.1007/978-3-662-44371-2_27
26. Bos, J., Costello, C., Ducas, L., Mironov, I., Naehrig, M., Nikolaenko, V., Raghunathan, A., Stebila, D.: Frodo: take off the ring! practical, quantum-secure key exchange from LWE. IACR Cryptology ePrint Archive 2016 (2016)
27. Brakerski, Z.: Fully homomorphic encryption without modulus switching from classical GapSVP. In: Safavi-Naini, R., Canetti, R. (eds.) CRYPTO 2012. LNCS, vol. 7417, pp. 868–886. Springer, Heidelberg (2012). doi:10.1007/978-3-642-32009-5_50
28. Brakerski, Z., Gentry, C., Vaikuntanathan, V.: (Leveled) fully homomorphic encryption without bootstrapping. In: ITCS (2012)
29. Brakerski, Z., Langlois, A., Peikert, C., Regev, O., Stehlé, D.: Classical hardness of learning with errors. In: STOC (2013)
30. Brakerski, Z., Rothblum, G.N.: Virtual black-box obfuscation for all circuits via generic graded encoding. In: Lindell, Y. (ed.) TCC 2014. LNCS, vol. 8349, pp. 1–25. Springer, Heidelberg (2014). doi:10.1007/978-3-642-54242-8_1
31. Brakerski, Z., Vaikuntanathan, V.: Efficient fully homomorphic encryption from (standard) LWE. In: FOCS (2011)
32. Chen, Y., Nguyen, P.Q.: BKZ 2.0: better lattice security estimates. In: Lee, D.H., Wang, X. (eds.) ASIACRYPT 2011. LNCS, vol. 7073, pp. 1–20. Springer, Heidelberg (2011). doi:10.1007/978-3-642-25385-0_1
33. Coron, J.-S., Lepoint, T., Tibouchi, M.: Practical multilinear maps over the integers. In: Canetti, R., Garay, J.A. (eds.) CRYPTO 2013. LNCS, vol. 8042, pp. 476–493. Springer, Heidelberg (2013). doi:10.1007/978-3-642-40041-4_26
34. Crescenzo, G., Lipmaa, H.: Succinct NP proofs from an extractability assumption. In: Beckmann, A., Dimitracopoulos, C., Löwe, B. (eds.) CiE 2008. LNCS, vol. 5028, pp. 175–185. Springer, Heidelberg (2008). doi:10.1007/978-3-540-69407-6_21
35. Damgård, I.: Towards practical public key systems secure against chosen ciphertext attacks. In: Feigenbaum, J. (ed.) CRYPTO 1991. LNCS, vol. 576, pp. 445–456. Springer, Heidelberg (1992). doi:10.1007/3-540-46766-1_36
36. Damgård, I., Faust, S., Hazay, C.: Secure two-party computation with low communication. In: Cramer, R. (ed.) TCC 2012. LNCS, vol. 7194, pp. 54–74. Springer, Heidelberg (2012). doi:10.1007/978-3-642-28914-9_4
37. Ducas, L., Micciancio, D.: FHEW: bootstrapping homomorphic encryption in less than a second. In: Oswald, E., Fischlin, M. (eds.) EUROCRYPT 2015. LNCS, vol. 9056, pp. 617–640. Springer, Heidelberg (2015). doi:10.1007/978-3-662-46800-5_24
38. Fiat, A., Shamir, A.: How to prove yourself: practical solutions to identification and signature problems. In: Odlyzko, A.M. (ed.) CRYPTO 1986. LNCS, vol. 263, pp. 186–194. Springer, Heidelberg (1987). doi:10.1007/3-540-47721-7_12
39. Garg, S., Gentry, C., Halevi, S.: Candidate multilinear maps from ideal lattices. In: Johansson, T., Nguyen, P.Q. (eds.) EUROCRYPT 2013. LNCS, vol. 7881, pp. 1–17. Springer, Heidelberg (2013). doi:10.1007/978-3-642-38348-9_1
40. Garg, S., Gentry, C., Halevi, S., Raykova, M.: Two-round secure MPC from indistinguishability obfuscation. In: Lindell, Y. (ed.) TCC 2014. LNCS, vol. 8349, pp. 74–94. Springer, Heidelberg (2014). doi:10.1007/978-3-642-54242-8_4
41. Garg, S., Gentry, C., Halevi, S., Raykova, M., Sahai, A., Waters, B.: Candidate indistinguishability obfuscation and functional encryption for all circuits. In: FOCS (2013)

42. Gennaro, R., Gentry, C., Parno, B., Raykova, M.: Quadratic span programs and succinct NIZKs without PCPs. In: Johansson, T., Nguyen, P.Q. (eds.) EUROCRYPT 2013. LNCS, vol. 7881, pp. 626–645. Springer, Heidelberg (2013). doi:10.1007/978-3-642-38348-9_37

43. Gentry, C., Gorbunov, S., Halevi, S.: Graph-induced multilinear maps from lattices. In: Dodis, Y., Nielsen, J.B. (eds.) TCC 2015. LNCS, vol. 9015, pp. 498–527. Springer, Heidelberg (2015). doi:10.1007/978-3-662-46497-7_20

44. Gentry, C., Halevi, S., Smart, N.P.: Fully homomorphic encryption with polylog overhead. In: Pointcheval, D., Johansson, T. (eds.) EUROCRYPT 2012. LNCS, vol. 7237, pp. 465–482. Springer, Heidelberg (2012). doi:10.1007/978-3-642-29011-4_28

45. Gentry, C., Sahai, A., Waters, B.: Homomorphic encryption from learning with errors: conceptually-simpler, asymptotically-faster, attribute-based. In: Canetti, R., Garay, J.A. (eds.) CRYPTO 2013. LNCS, vol. 8042, pp. 75–92. Springer, Heidelberg (2013). doi:10.1007/978-3-642-40041-4_5

46. Gentry, C., Wichs, D.: Separating succinct non-interactive arguments from all falsifiable assumptions. In: STOC (2011)

47. Goldwasser, S., Lin, H., Rubinstein, A.: Delegation of computation without rejection problem from designated verifier CS-proofs. IACR Cryptology ePrint Archive 2011 (2011)

48. Goldwasser, S., Micali, S.: Probabilistic encryption and how to play mental poker keeping secret all partial information. In: STOC (1982)

49. Goldwasser, S., Micali, S., Rackoff, C.: The knowledge complexity of interactive proof-systems (extended abstract). In: STOC (1985)

50. Goyal, V., Ishai, Y., Sahai, A., Venkatesan, R., Wadia, A.: Founding cryptography on tamper-proof hardware tokens. In: Micciancio, D. (ed.) TCC 2010. LNCS, vol. 5978, pp. 308–326. Springer, Heidelberg (2010). doi:10.1007/978-3-642-11799-2_19

51. Groth, J.: Short pairing-based non-interactive zero-knowledge arguments. In: Abe, M. (ed.) ASIACRYPT 2010. LNCS, vol. 6477, pp. 321–340. Springer, Heidelberg (2010). doi:10.1007/978-3-642-17373-8_19

52. Joux, A.: A one round protocol for tripartite Diffie–Hellman. In: Bosma, W. (ed.) ANTS 2000. LNCS, vol. 1838, pp. 385–393. Springer, Heidelberg (2000). doi:10.1007/10722028_23

53. Kilian, J.: A note on efficient zero-knowledge proofs and arguments (extended abstract). In: STOC (1992)

54. Kirchner, P., Fouque, P.-A.: An improved BKW algorithm for LWE with applications to cryptography and lattices. In: Gennaro, R., Robshaw, M. (eds.) CRYPTO 2015. LNCS, vol. 9215, pp. 43–62. Springer, Heidelberg (2015). doi:10.1007/978-3-662-47989-6_3

55. Koppula, V., Lewko, A.B., Waters, B.: Indistinguishability obfuscation for turing machines with unbounded memory. In: STOC (2015)

56. Lewi, K., Malozemoff, A.J., Apon, D., Carmer, B., Foltzer, A., Wagner, D., Archer, D.W., Boneh, D., Katz, J., Raykova, M.: 5Gen: a framework for prototyping applications using multilinear maps and matrix branching programs. In: ACM CCS (2016)

57. Lin, H.: Indistinguishability obfuscation from constant-degree graded encoding schemes. In: Fischlin, M., Coron, J.-S. (eds.) EUROCRYPT 2016. LNCS, vol. 9665, pp. 28–57. Springer, Heidelberg (2016). doi:10.1007/978-3-662-49890-3_2

58. Lin, H.: Indistinguishability obfuscation from DDH on 5-linear maps and locality-5 PRGs. IACR Cryptology ePrint Archive 2016 (2016)

59. Lin, R., Vaikuntanathan, V.: Indistinguishability obfuscation from DDH-like assumptions on constant-degree graded encodings. In: FOCS (2016)

60. Lindner, R., Peikert, C.: Better key sizes (and attacks) for LWE-based encryption. In: Kiayias, A. (ed.) CT-RSA 2011. LNCS, vol. 6558, pp. 319–339. Springer, Heidelberg (2011). doi:10.1007/978-3-642-19074-2_21

61. Lipmaa, H.: Progression-free sets and sublinear pairing-based non-interactive zero-knowledge arguments. In: Cramer, R. (ed.) TCC 2012. LNCS, vol. 7194, pp. 169–189. Springer, Heidelberg (2012). doi:10.1007/978-3-642-28914-9_10

62. Lyubashevsky, V., Peikert, C., Regev, O.: On ideal lattices and learning with errors over rings. In: Gilbert, H. (ed.) EUROCRYPT 2010. LNCS, vol. 6110, pp. 1–23. Springer, Heidelberg (2010). doi:10.1007/978-3-642-13190-5_1

63. Micali, S.: Computationally sound proofs. SIAM J. Comput. **30**(4), 1253–1298 (2000)

64. Mie, T.: Polylogarithmic two-round argument systems. J. Math. Cryptol. **2**(4), 343–363 (2008)

65. Naor, M.: On cryptographic assumptions and challenges. In: Boneh, D. (ed.) CRYPTO 2003. LNCS, vol. 2729, pp. 96–109. Springer, Heidelberg (2003). doi:10.1007/978-3-540-45146-4_6

66. Paillier, P.: Public-key cryptosystems based on composite degree residuosity classes. In: Stern, J. (ed.) EUROCRYPT 1999. LNCS, vol. 1592, pp. 223–238. Springer, Heidelberg (1999). doi:10.1007/3-540-48910-X_16

67. Peikert, C., Vaikuntanathan, V., Waters, B.: A framework for efficient and composable oblivious transfer. In: Wagner, D. (ed.) CRYPTO 2008. LNCS, vol. 5157, pp. 554–571. Springer, Heidelberg (2008). doi:10.1007/978-3-540-85174-5_31

68. Regev, O.: On lattices, learning with errors, random linear codes, and cryptography. In: STOC (2005)

69. Sahai, A., Waters, B.: How to use indistinguishability obfuscation: deniable encryption, and more. In: STOC (2014)

70. Schnorr, C., Euchner, M.: Lattice basis reduction: improved practical algorithms and solving subset sum problems. Math. Program. **66**, 181–199 (1994)

71. Schwartz, J.T.: Fast probabilistic algorithms for verification of polynomial identities. J. ACM **27**(4), 701–717 (1980)

72. Shor, P.W.: Algorithms for quantum computation: discrete logarithms and factoring. In: FOCS (1994)

73. Simon, D.R.: On the power of quantum computation. SIAM J. Comput. **26**(5), 1474–1483 (1997)

74. Zimmerman, J.: How to obfuscate programs directly. In: Oswald, E., Fischlin, M. (eds.) EUROCRYPT 2015. LNCS, vol. 9057, pp. 439–467. Springer, Heidelberg (2015). doi:10.1007/978-3-662-46803-6_15

75. Zippel, R.: Probabilistic algorithms for sparse polynomials. In: EUROSAM (1979)

Cryptanalyses of Candidate Branching Program Obfuscators

Yilei Chen[1]($\boxtimes$), Craig Gentry[2], and Shai Halevi[2]

[1] Boston University, Boston, MA, USA
chenyl@bu.edu
[2] IBM Research, Yorktown Heights, NY, USA
craigbgentry@gmail.com, shaih@alum.mit.edu

Abstract. We describe new cryptanalytic attacks on the candidate branching program obfuscator proposed by Garg, Gentry, Halevi, Raykova, Sahai and Waters (GGHRSW) using the GGH13 graded encoding, and its variant using the GGH15 graded encoding as specified by Gentry, Gorbunov and Halevi. All our attacks require very specific structure of the branching programs being obfuscated, which in particular must have some input-partitioning property. Common to all our attacks are techniques to extract information about the "multiplicative bundling" scalars that are used in the GGHRSW construction.

For GGHRSW over GGH13, we show how to recover the ideal generating the plaintext space when the branching program has input partitioning. Combined with the information that we extract about the "multiplicative bundling" scalars, we get a distinguishing attack by an extension of the annihilation attack of Miles, Sahai and Zhandry. Alternatively, once we have the ideal we can solve the principle-ideal problem (PIP) in classical subexponential time or quantum polynomial time, hence obtaining a total break.

For the variant over GGH15, we show how to use the left-kernel technique of Coron, Lee, Lepoint and Tibouchi to recover ratios of the bundling scalars. Once we have the ratios of the scalar products, we can use factoring and PIP solvers (in classical subexponential time or quantum polynomial time) to find the scalars themselves, then run mixed-input attacks to break the obfuscation.

Keywords: Cryptanalysis · Graded-encoding · Obfuscation

1 Introduction

General-purpose code obfuscation is an amazingly powerful technique, making it possible to hide secrets in arbitrary running software. The first plausible construction of a secure general-purpose obfuscation, described three years ago by Garg, Gentry, Halevi, Raykova, Sahai and Waters [22] (hereafter GGHRSW), opened up a new direction of research that transformed our thinking about what can and cannot be done in cryptography. The GGHRSW construction

© International Association for Cryptologic Research 2017
J.-S. Coron and J.B. Nielsen (Eds.): EUROCRYPT 2017, Part III, LNCS 10212, pp. 278–307, 2017.
DOI: 10.1007/978-3-319-56617-7_10

consists of a "core component" for obfuscating branching programs, and a bootstrapping procedure that uses the core component—in conjunction with homomorphic encryption and some proofs—to obfuscate arbitrary code (modeled as a binary circuit). Many different constructions were proposed since then e.g., [3,4,6–8,11,12,23,24,27,31,32,34,37,39], most of which only modify the "core component" for branching programs, then use the GGHRSW bootstrapping to obfuscate circuits.

All known obfuscation constructions rely crucially on the underlying tool of *graded encoding schemes*, for which there are (essentially) only three candidate constructions: one due to Garg, Gentry and Halevi [21] (GGH13), another due to Coron, Lepoint and Tibouchi [19] (CLT13), and the third due to Gentry, Gorbunov and Halevi [26] (GGH15). However, the security properties of these encoding schemes are poorly understood, and therefore the same holds for the obfuscation constructions that use them.

Known Attacks. The original publications of GGH13, CLT13 and GGH15 survey several number theoretical and algebraic attacks. For the GGH13 encoding scheme—that relies on the difficulty of the NTRU problem and the principle ideal problem (PIP) in certain number fields—we recently saw some advances in attacking these underlying problem [2,9,10,14,20], that may affect the choice of parameters.

The most serious attacks on all three encoding schemes are the so-called "zeroizing attacks": when encodings of zero are easy to find, some secrets can be extracted by linear algebraic techniques. The most devastating zeroizing attack is found by Cheon, Han, Lee, Ryu and Stehl?[13] against CLT13—when the encodings of zero form certain combinations, one can extract all the secret parameters. The attack is extended by Coron et al. [16, Sect. 3.4], breaking the GGHRSW branching-program obfuscator when instantiated using CLT13 encodings and used to obfuscate branching programs with certain input-partitioning features.

Applying zeroizing attacks to construction based on GGH13 and GGH15 appears somewhat harder, especially in the context of obfuscation. Nonetheless, Miles, Sahai and Zhandry recently introduced "annihilation attack" against many GGH13-based branching-program obfuscators, for specific types of branching programs [35]. Interestingly, these attacks do not apply to the GGHRSW construction, due to the presence of some random entries in the encoded matrices. Moreover, it was shown in [24] that such random entries (in conjunction with other techniques) provably eliminates all known variants of zeroizing attacks.

To the best of our knowledge, no polynomial time attacks (either classical or quantum) were known before the current work on the GGHRSW obfuscator using GGH13 encoding, nor were there any attacks on any GGH15-based branching-program obfuscators.

This Work. We describe new attacks on the GGHRSW branching-program obfuscator, when using GGH13 and GGH15 encodings. The attacks that we describe in this work require the underlying branching programs to satisfy some input-partitioning features, similar to the attack on the CLT variant [16,

Sect. 3.4]. Roughly, the indexes of the branching program can be partitioned into two or three consecutive intervals, each contains "sufficiently many" input bits that do not appear in the other intervals.

A common thread in our attacks is that they focus on the "multiplicative bundling" scalars that are used in the GGHRSW construction (as protection against "mixed-input attacks"). We show that some information about these scalars can be extracted using zeroizing techniques, if the underlying branching program satisfy certain input-partitioning features. We are not able to fully recover these scalars, and hence cannot quite mount mixed-input attacks, but we can still use the extracted information in weaker attacks.

For the GGH13-based candidates, we first apply a variant of the attacks due to Cheon et al. and Coron et al. [13,17] to recover a basis of the ideal $\langle g \rangle$ that defines the plaintext space, as well as some representatives of the scalars, then use the recovered information in a distinguishing attack, using an extension of the annihilation attack of Miles et al. [35]. Alternatively, once we have a basis for $\langle g \rangle$ we can solve PIP (in classical subexponential time or quantum polynomial time), resulting in a total break.

For the GGH15-based candidates, we recover some rational expressions in the bundling scalars using techniques from [17] (among others), then we can use factoring and PIP solvers (in classical subexponential time or quantum polynomial time) to recover the bundling scalars themselves from the rational expressions, then mount mixed-input attacks.

Applicability and Extensions of Our Attacks. We stress that all our attacks rely crucially on the input-partitioning of the branching program (in order to use the techniques of Cheon et al. or those of Coron et al.) In particular they do not seem to apply to "dual input" branching programs as used in many branching-program obfuscators. Also, our GGH13 attacks cannot be used against schemes that were proven secure in the "Weak Multilinear Map" model of Garg et al. [24], since our first step of recovering $\langle g \rangle$ fits in that model. However, some of our techniques do not seem to quite fit in that model (in particular Step II of the attack, see Sect. 3.2), so they should serve as a cautionary tale about relying too much on proofs of security in such idealized models. Also, the "immunization" techniques against GGH13 annihilation attack from [24] by themselves do not prevent our new attack if the branching programs are input-partitioning (see Sect. 3.5), it is only in combination with the "dual input" technique that they provide protection.

Finally, our techniques can potentially be combined with the recent techniques of Apon et al. and Coron et al. [5,18], to attack also some non-input-partitioned obfuscators. This seems a promising direction for future work.

2   Preliminaries

For a positive integer n, let $[n] = \{1, 2, \ldots, n\}$. Let Φ_n be the nth cyclotomic polynomial. The typical ring used in the paper $R := \mathbb{Z}[x]/\langle \Phi_n(x) \rangle$, and the fractional field of R_n: $K_n := \mathbb{Q}[x]/\langle \Phi_n(x) \rangle$. Below we denote matrices by boldface uppercase letter (e.g., $\mathbf{A}, \mathbf{B}, \ldots$).

2.1 Matrix Branching Programs

We consider oblivious matrix branching programs (as usual in the obfuscation literature). Such a branching program consists of a sequence of steps, where each step is associated with an index of some input bit and we have two matrices associated with each step. To evaluate such a branching program over some input string, we choose one of the two matrices from each step, depending on the value of the corresponding input bit, then multiply all these matrices in order, and compare the result to the identity matrix.

Definition 1. *A dimension-w, length-h branching program over ℓ-bit inputs consists of an index-to-input map and a sequence of pairs of 0–1 matrices,*

$$\mathcal{B} = \left\{ \iota : [h] \to [\ell], \{\mathbf{B}_{i,b} \in \{0,1\}^{w \times w}\}_{i \in [h], b \in \{0,1\}} \right\}.$$

This branching program is computing the function $f_\mathcal{B} : \{0,1\}^\ell \to \{0,1\}$, defined as

$$f_\mathcal{B}(x) = \begin{cases} 0 & \text{if } \prod_{i \in [h]} \mathbf{B}_{i,x_{\iota(i)}} = \mathbf{I} \\ 1 & \text{if } \prod_{i \in [h]} \mathbf{B}_{i,x_{\iota(i)}} \neq \mathbf{I} \end{cases}$$

where the matrix product is carried over some implicitly set ring that includes 0,1 (e.g., the ring R_n from above).

Input Partitioning. We say (somewhat informally) that a branching program $\mathcal{B}$ is input-partitioned if its set of steps can be partitioned into two or more consecutive intervals $[h] = \mathcal{H}_1 || \mathcal{H}_2 || \ldots$, such that for each interval there are "sufficiently many" input bits that control only steps in that interval and nowhere else. We sometime say that $\mathcal{B}$ is 2-partitioned or 3-partitioned if it can be broken to 2 or 3 intervals, respectively, and the number of bits that are unique to each interval will vary among the different attacks that we describe (and will typically be polylogarithmic).

When considering input-partitioned program $\mathcal{B}$, we will often consider its evaluation on inputs that differ in bits that only affect steps in one of the intervals. A simple (but important) observation that underlies most of our techniques is the following:

Lemma 1. *Let $\mathcal{B}$ be a branching program as per Definition 1 which is input-partitioned, $[h] = \mathcal{H}_1 || \mathcal{H}_2$, and let $x, x' \in \{0,1\}^\ell$ be two zeros of $f_\mathcal{B}$ that differ only in bits that are mapped to steps in $\mathcal{H}_1$. Namely, $f_\mathcal{B}(x) = f_\mathcal{B}(x') = 0$, and for all $i \notin \mathcal{H}_1$ we have $x_{\iota(i)} = x'_{\iota(i)}$. Then the product of the matrices corresponding to $\mathcal{H}_1$ yields the same result in the evaluation of $\mathcal{B}$ on x and x', that is $\prod_{i \in \mathcal{H}_1} \mathbf{B}_{i,x_{\iota(i)}} = \prod_{i \in \mathcal{H}_1} \mathbf{B}_{i,x'_{\iota(i)}}$.*

Similarly, if x, x' are two zeros of $f_\mathcal{B}$ that differ only in bits that are mapped to steps in $\mathcal{H}_2$, then $\prod_{i \in \mathcal{H}_2} \mathbf{B}_{i,x_{\iota(i)}} = \prod_{i \in \mathcal{H}_2} \mathbf{B}_{i,x'_{\iota(i)}}$.

Proof. For the first statement, denote $\mathbf{B} := \prod_{i \in \mathcal{H}_1} \mathbf{B}_{i,x_{\iota(i)}}$, $\mathbf{B}' := \prod_{i \in \mathcal{H}_1} \mathbf{B}_{i,x'_{\iota(i)}}$, and $\mathbf{C} := \prod_{i \in \mathcal{H}_2} \mathbf{B}_{i,x_{\iota(i)}} = \prod_{i \in \mathcal{H}_2} \mathbf{B}_{i,x'_{\iota(i)}}$, where the last equality follows since $x_{\iota(i)} = x'_{\iota(i)}$ whenever $i \in \mathcal{H}_2$. Since $f_{\mathcal{B}}(x) = f_{\mathcal{B}}(x') = 0$ then $\mathbf{B} \times \mathbf{C} = \mathbf{B}' \times \mathbf{C} = \mathbf{I}$, and as $\mathbf{B}, \mathbf{B}', \mathbf{C}$ are square matrices then $\mathbf{C}$ must be invertible and $\mathbf{B} = \mathbf{B}' = \mathbf{C}^{-1}$. The proof of the "similarly" statement is analogous.

2.2   Overview of the GGHRSW Branching-Program Obfuscator

We briefly review the candidate branching program obfuscator of Garg et al. [22] and its GGH15-based variant from [26, Sect. 5.2]. The GGHRSW branching-program obfuscator applies several different randomization steps to the underlying branching program, and then encodes the resulting randomized matrices, using either GGH13 or GGH15.

We defer the description of the GGH13 and GGH15 encoding schemes themselves to the corresponding attack sections, but just note that these schemes let us encode matrices in a way that allows checking whether certain degree-h polynomial expressions in these matrices evaluate to zero.

We also recall that these constructions are supposed to implement *indistinguishability obfuscation*. In the context of branching programs, this means that if two programs have the same length h and same input mapping function $\iota : [h] \rightarrow [\ell]$ and they compute the same function, then their obfuscations should be indistinguishable. Correspondingly when attacking these constructions we need to show two such equivalent programs for which we are able to distinguish the obfuscated versions.

Below we let $\mathcal{B} = \left\{ \iota : [h] \rightarrow [\ell], \{\mathbf{B}_{i,b} \in \{0,1\}^{w \times w}\}_{i \in [h], b \in \{0,1\}} \right\}$ be the branching program to be obfuscated. The obfuscation process consists of the following steps:

0. **Dummy branch.** The construction begins by introducing a "dummy branch", which is just a length-h branching program with the same input mapping function $\iota : [h] \rightarrow [\ell]$, but consisting of only identity matrices of the same dimension as the $\mathbf{B}_{i,b}$'s. (In particular the "dummy branch" computes the all-zero function.) We refer to the original branching program as the "functional branch", and apply the same randomization/encoding transformations to both branches.

1. **Random diagonal entries and bookends.** Next every matrix in each of the branches (all are $w \times w$ 0–1 matrices) is embedded inside a higher-dimension randomized matrix. Specifically, for each $i \in [h], b \in \{0,1\}$ we consider the matrices

$$\tilde{\mathbf{B}}_{i,b} := \begin{bmatrix} \mathbf{V}_{i,b} & \\ & \mathbf{B}_{i,b} \end{bmatrix} \text{ and } \tilde{\mathbf{B}}'_{i,b} := \begin{bmatrix} \mathbf{V}'_{i,b} & \\ & \mathbf{I} \end{bmatrix}, \tag{1}$$

where $\mathbf{V}_{i,b}$ and $\mathbf{V}'_{i,b}$ are "random diagonal matrices." In the GGHRSW construction from [22], these are $2(h + 3)$-by-$2(h + 3)$ diagonal matrices with the diagonal entries chosen uniformly at random from the plaintext space, whereas in the GGH15-based variant from [26] they are diagonal 2-by-2 matrices with "random small entries" that are drawn from some Gaussian distribution over R_n. Below we denote the dimension of these random matrices as $2m$-by-$2m$ (so we have $m = h + 3$ for the original GGHRSW and $m = 1$ for the GGH15-based variant). When the analysis requires fine grained structure of the padded matrices, we further split the notation for each m-by-m blocks and denote the whole as $\mathrm{diag}\,(\mathbf{U}_{i,b}, \mathbf{V}_{i,b})$ and $\mathrm{diag}\,(\mathbf{U}'_{i,b}, \mathbf{V}'_{i,b})$. The construction also chooses four "bookend" vectors $\mathbf{J}, \mathbf{J}', \mathbf{L}, \mathbf{L}' \in R^{2m+w}$, of the form:

$$\mathbf{J}, \mathbf{J}' \in \left[\, 0^m, \$^m, \$^w \,\right], \quad \mathbf{L}, \mathbf{L}' \in \left[\, \$^m, 0^m, \$^w \,\right]^T \tag{2}$$

where the $'s stand for uniformly random elements from the plaintext space for the original GGH13-based construction, and for "small random" elements drawn from some Gaussian distribution for the GGH15-based candidate. They satisfy $\mathbf{JL} = \mathbf{J}'\mathbf{L}'$.

2. **Killian-style randomization and bundling scalars.** Next the construction chooses invertible matrices $\{\mathbf{K}_i, \mathbf{K}'_i \in R_n^{(2m+w)\times(2m+w)}\}_{i\in[h]}$ and also scalars $\{\alpha_{i,b}, \alpha'_{i,b}\}_{i\in[h],b\in\{0,1\}}$. The scalars are chosen under the constraint that for any input bit $j \in [\ell]$, we have

$$\prod_{\iota(i)=j} \alpha_{i,0} = \prod_{\iota(i)=j} \alpha'_{i,0} \text{ and } \prod_{\iota(i)=j} \alpha_{i,1} = \prod_{\iota(i)=j} \alpha'_{i,1}.$$

Below we sometime use the notations $\beta_{j,b} := \prod_{\iota(i)=j} \alpha_{i,b}\left(= \prod_{\iota(i)=j} \alpha'_{i,b}\right)$. As before, here too the scalars and matrices are chosen at random from the plaintext space in the GGH13-based construction, and drawn from an appropriate Gaussian distribution with small parameters in the GGH15-based solution. Let us also denote $\mathbf{K}_0 = \mathbf{K}'_0 = \mathbf{I}$.

3. **Encoding.** Denote the randomized matrices by

$$\mathbf{S}_{i,b} := \alpha_{i,b}\mathbf{K}_{i-1}^{-1}\tilde{\mathbf{B}}_{i,b}\mathbf{K}_i \text{ and } \mathbf{S}'_{i,b} := \alpha'_{i,b}\mathbf{K}_{i-1}'^{-1}\tilde{\mathbf{B}}'_{i,b}\mathbf{K}'_i. \tag{3}$$

The obfuscation of the branching program $\mathcal{B}$ consists of encoding of all the matrices $\mathbf{S}_{i,b}$ and $\mathbf{S}'_{i,b}$ and also of the bookends $\mathbf{J}, \mathbf{J}', \mathbf{L}, \mathbf{L}'$.

To evaluate the obfuscated branching program on some input x, we use the operations and zero-test capabilities of the underlying encoding scheme to check that $\mathbf{J}\left(\prod_{i\in[h]} \mathbf{S}_{i,b}\right)\mathbf{L} - \mathbf{J}'\left(\prod_{i\in[h]} \mathbf{S}'_{i,b}\right)\mathbf{L}' = 0$.

Branching Program with Input Partitioning. Let $\mathcal{X}\|\mathcal{Y}\|\mathcal{Z} = [h]$ be a 3-partition of the branching program steps. In the attacks we use honest evaluation of the branching program on many inputs of the form $u^{(i,j,k)} = x^{(i)}y^{(j)}z^{(k)}$, where all the bits that only affect steps in $\mathcal{X}$ are in the $x^{(i)}$ part, all the bits that only affect steps in $\mathcal{Y}$ are in the $y^{(j)}$ part, all the bits that only affect steps in $\mathcal{Z}$ are in the $z^{(k)}$ part, and all the other bits are fixed. This notation *does not mean* that the bits of $x^{(i)}$, $y^{(j)}$, $z^{(k)}$ appear in this order in $u^{(i,j,k)}$, but it does mean that $u^{(i,j,k)}$ and $u^{(i',j,k)}$ can only differ in bits that affect steps in $\mathcal{X}$, and similarly $u^{(i,j,k)}$ and $u^{(i,j',k)}$ only differ in bits that affect steps in $\mathcal{Y}$ and $u^{(i,j,k)}$ and $u^{(i,j,k')}$ only differ in bits that affect steps in $\mathcal{Z}$.

For such an input $u = xyz$, we denote by $\mathbf{S}_x$ the plaintext product matrix of functional branch in the $\mathcal{X}$ interval, by $\mathbf{S}_y$ the plaintext product matrix of functional branch in the the $\mathcal{Y}$ interval, and by $\mathbf{S}_z$ the plaintext product matrix of the functional branch in the $\mathcal{Z}$ interval (including the bookends). We similarly denote by $\mathbf{S}'_x, \mathbf{S}'_y, \mathbf{S}'_z$, the plaintext product matrix of the dummy branch. Namely

$$
\begin{aligned}
\mathbf{S}_x &:= \mathbf{J} \cdot (\textstyle\prod_{i \in \mathcal{X}} \mathbf{S}_{i,u_{\iota(i)}}), \ \mathbf{S}_y := \textstyle\prod_{i \in \mathcal{Y}} \mathbf{S}_{i,u_{\iota(i)}}, \ \mathbf{S}_z := (\textstyle\prod_{i \in \mathcal{Z}} \mathbf{S}_{i,u_{\iota(i)}}) \cdot \mathbf{L}, \\
\mathbf{S}'_x &:= \mathbf{J}' \cdot (\textstyle\prod_{i \in \mathcal{X}} \mathbf{S}'_{i,u_{\iota(i)}}), \ \mathbf{S}'_y := \textstyle\prod_{i \in \mathcal{Y}} \mathbf{S}'_{i,u_{\iota(i)}}, \ \mathbf{S}'_z := (\textstyle\prod_{i \in \mathcal{Z}} \mathbf{S}'_{i,u_{\iota(i)}}) \cdot \mathbf{L}',
\end{aligned}
\tag{4}
$$

with products over the plaintext space. In some cases we only need 2-partition of the program, so we suppress the $\mathbf{S}_y$, $\mathbf{S}'_y$ parts.

When we have multiple inputs of the form $u^{(i,j,k)} = x^{(i)} y^{(j)} z^{(k)}$ that are all zeros of the function, then by Lemma 1 the parts of the plaintext matrices that come from the product of the branching program matrices must be the same for the different $x^{(i)}$'s (and similarly for the different $y^{(j)}$'s and $z^{(k)}$'s). We denote these matrices simply by $\mathbf{B}_x$, $\mathbf{B}_y$, and $\mathbf{B}_z$, independently of i, j, k. Namely we have:

$$
\begin{aligned}
\mathbf{S}_{x^{(i)}} &= \alpha_{x^{(i)}} \ \mathbf{J} \ \times \mathrm{diag}(\mathbf{U}_{x^{(i)}}, \mathbf{V}_{x^{(i)}}, \mathbf{B}_x) \times \mathbf{K}_y; \\
\mathbf{S}'_{x^{(i)}} &= \alpha'_{x^{(i)}} \ \mathbf{J}' \ \times \mathrm{diag}(\mathbf{U}'_{x^{(i)}}, \mathbf{V}'_{x^{(i)}}, \mathbf{I}) \times \mathbf{K}'_y \\
\mathbf{S}_{y^{(j)}} &= \alpha_{y^{(j)}} \mathbf{K}_y^{-1} \times \mathrm{diag}(\mathbf{U}_{y^{(j)}}, \mathbf{V}_{y^{(j)}}, \mathbf{B}_y) \times \mathbf{K}_z; \\
\mathbf{S}'_{y^{(j)}} &= \alpha'_{y^{(j)}} \mathbf{K}_y'^{-1} \times \mathrm{diag}(\mathbf{U}'_{y^{(j)}}, \mathbf{V}'_{y^{(j)}}, \mathbf{I}) \times \mathbf{K}'_z; \\
\mathbf{S}_{z^{(k)}} &= \alpha_{z^{(k)}} \mathbf{K}_z^{-1} \times \mathrm{diag}(\mathbf{U}_{z^{(k)}}, \mathbf{V}_{z^{(k)}}, \mathbf{B}_z) \times \mathbf{L}; \\
\mathbf{S}'_{z^{(k)}} &= \alpha'_{z^{(k)}} \mathbf{K}_z'^{-1} \times \mathrm{diag}(\mathbf{U}'_{z^{(k)}}, \mathbf{V}'_{z^{(k)}}, \mathbf{I}) \times \mathbf{L}'
\end{aligned}
\tag{5}
$$

where the scalars $\alpha_{x^{(i)}}$, $\alpha_{y^{(j)}}$, etc. are just the product of all the $\alpha_{i,b}$'s in the corresponding (partial) branch. Moreover, we observe that all the ratios of $\alpha_{x^{(i)}}/\alpha'_{x^{(i)}}, i = 1, 2, \ldots$ (and similarly for the $\alpha_{y^{(j)}}$ and $\alpha_{z^{(k)}}$) must also be equal.

Lemma 2. *With the notations above, we have* $\alpha'_{x^{(1)}}/\alpha_{x^{(1)}} = \alpha'_{x^{(2)}}/\alpha_{x^{(2)}} = \ldots$ *and similarly* $\alpha'_{y^{(1)}}/\alpha_{y^{(1)}} = \alpha'_{y^{(2)}}/\alpha_{y^{(2)}} = \ldots$ *and* $\alpha'_{z^{(1)}}/\alpha_{z^{(1)}} = \alpha'_{z^{(2)}}/\alpha_{z^{(2)}} = \ldots$.

Proof. To prove the statement for the $\alpha_{x^{(i)}}$'s consider an input bit $t \in [\ell]$ that affect some steps in $\mathcal{X}$. That bit either only affects steps in $\mathcal{X}$ or it affects steps in both $\mathcal{X}$ and in $\mathcal{Y}, \mathcal{Z}$. In the former case, by construction we have $\prod_{\iota(i')=t} \alpha_{i',b} = \prod_{\iota(i')=t} \alpha'_{i',b}$ (for $b = 0, 1$), so this input bit's contribution to the ratio $\alpha'_{x^{(i)}}/\alpha_{x^{(i)}}$ is 1 (for all i). In the latter case, this input bit has the same value (0 or 1) for all the inputs $x^{(i)}$, so it contributes the same factor to the ratio $\alpha'_{x^{(i)}}/\alpha_{x^{(i)}}$ for all i. The proof for the $\alpha_{y^{(j)}}$ and $\alpha_{z^{(k)}}$ is the same.

3 Cryptanalysis of the GGH13-Based Candidate

The GGH13 Encoding Scheme. The core secret parameter in the GGH13 encoding scheme is a small $g \in R_n$ (sampled from small Gaussian distribution), such that the inverse $g^{-1} \in K$ is also small. Let $\mathcal{I} = \langle g \rangle = gR_n$ be the ideal generated by g in R_n, the plaintext space of the GGH13 scheme is the quotient ring $R_n/\mathcal{I}$, and we typically choose g so that this plaintext space is isomorphic to some prime field $\mathbb{F}_p$. Other parameters of the scheme are an integer modulus $q \gg p$ and the multi-linearity degree k (which are public), and a random secret denominator $z \in R_n/qR_n$ (which is kept secret). Plaintext elements are encoded relative to levels between 0 and k.

The encoding of $s \in R_n/\mathcal{I}$ at level 0 is a short representative of the coset of the ideal shifted by s, i.e. $c \in s + \mathcal{I}$, $\|c\| \ll q$. To encode at level i, compute $c/z^i \pmod{q}$. (There is also an "asymmetric mode" of GGH13, in which there are many different denominators z_i.) The public zero-test parameter is $p_{zt} = \eta \cdot z^k/g$, with $\|\eta\| \le q^{1/2}$.[1] Additions and multiplications are simply adding and multiplying the encodings in R_n/qR_n, with the restrictions that correctness only holds when adding on the same level, or multiplying below the maximum level k. To zero-test, multiply the (potential) top-level encoding c/z^k by p_{zt} (modulo q). If c encodes zero then $c \in \mathcal{I}$, hence $c = c' \cdot g$, and therefore $c \cdot p_{zt} = \eta c'$, which is small since both η and c' are much smaller than q.

Attacking the GGH13-Based Obfuscator. When using GGH13 as the underlying encoding scheme in the GGHRSW obfuscator, we denote the encoding of the plaintext matrices $\mathbf{S}_{i,b}$, $\mathbf{S}'_{i,b}$ by

$$\mathbf{C}_{i,b} = (\mathbf{S}_{i,b} + g \cdot \mathbf{E}_{i,b})/z, \text{ and } \mathbf{C}'_{i,b} = (\mathbf{S}'_{i,b} + g \cdot \mathbf{E}'_{i,b})/z.$$

We also denote the encoding of the bookends by

$$\tilde{\mathbf{J}} = (\mathbf{J} + g \cdot \mathbf{E}_J)/z, \tilde{\mathbf{L}} = (\mathbf{L} + g \cdot \mathbf{E}_L)/z, \tilde{\mathbf{J}}' = (\mathbf{J}' + g \cdot \mathbf{E}'_{J'})/z, \text{ and } \tilde{\mathbf{L}}' = (\mathbf{L}' + g \cdot \mathbf{E}'_{L'})/z,$$

where all the calculations are modulo q.

We first recover the ideal $\langle g \rangle$ adapting the zeroing attack techniques of Cheon, Han, Lee, Ryu and Stehlé [13] and Coron, Lee, Lepoint and Tibouchi [17]. This part requires 2-partitioning of the branching program. Once we have a basis of $\langle g \rangle$, sub-exponential time classical algorithms [9] and polynomial-time quantum algorithms [10] are known to recover a short generator of $\langle g \rangle$ [20], thus breaking GGH13 completely [21, Sect. 6.3.3].

Alternatively, using a basis of $\langle g \rangle$ we can proceed with the zeroing attack modulo $\langle g \rangle$ to recover (some representation of) products of the bundling scalars. Then we can execute a simplified variant of the annihilation attack by Miles, Sahai and Zhandry [35]. This yields a classical polynomial time attack, and requires 3-partitioning of the branching program. We now proceed to describe the attack in more details.

Some More Notations. Consider a 3-partitioned branching program with the partitioning $\mathcal{X}||\mathcal{Y}||\mathcal{Z} = [h]$. We use the same notation as in Eq. (4) for the

---

[1] The scalar η is denoted h in [21], but we are already using h for the length of the branching program.

plaintext matrices, and also denote $\mathbf{C}_x, \mathbf{C}_y, \mathbf{C}_z$ and $\mathbf{C}'_x, \mathbf{C}'_y, \mathbf{C}'_z$ for the encoded matrices. Namely

$$\mathbf{C}_x := \tilde{\mathbf{J}} \cdot (\textstyle\prod_{i \in \mathcal{X}} \mathbf{C}_{i,u_{\iota(i)}}), \quad \mathbf{C}_y := \textstyle\prod_{i \in \mathcal{Y}} \mathbf{C}_{i,u_{\iota(i)}}, \quad \mathbf{C}_z := (\textstyle\prod_{i \in \mathcal{Z}} \mathbf{C}_{i,u_{\iota(i)}}) \cdot \tilde{\mathbf{L}},$$
$$\mathbf{C}'_x := \tilde{\mathbf{J}}' \cdot (\textstyle\prod_{i \in \mathcal{X}} \mathbf{C}'_{i,u_{\iota(i)}}), \quad \mathbf{C}'_y := \textstyle\prod_{i \in \mathcal{Y}} \mathbf{C}'_{i,u_{\iota(i)}}, \quad \mathbf{C}'_z := (\textstyle\prod_{i \in \mathcal{Z}} \mathbf{C}'_{i,u_{\iota(i)}}) \cdot \tilde{\mathbf{L}}',$$

with products over R_n/qR_n. As before, when we only need 2-partition we ignore the $\mathbf{C}_y$'s. With these notations, for any $u = xyz$ we can multiply, subtract, and zero-test to get

$$w := p_{zt}(\mathbf{C}_x\mathbf{C}_y\mathbf{C}_z - \mathbf{C}'_x\mathbf{C}'_y\mathbf{C}'_z) \tag{6}$$
$$= \frac{\eta}{g} \cdot [\mathbf{S}_x + g\mathbf{E}_x, -(\mathbf{S}'_x + g\mathbf{E}'_x)] \begin{bmatrix} \mathbf{S}_y + g\mathbf{E}_y, & 0 \\ 0, & \mathbf{S}'_y + g\mathbf{E}'_y \end{bmatrix} \begin{bmatrix} \mathbf{S}_z + g\mathbf{E}_z \\ \mathbf{S}'_z + g\mathbf{E}'_z \end{bmatrix} \pmod{q}$$

(or the without the middle matrix if we only use 2-partitioning). Moreover, if u is a zero of the function then the final zero-tested value is an encoding of zero, and hence Eq. 6 holds not only modulo q but also over the base ring R_n.

3.1 Step I: Recovering $\langle g \rangle$

Our first task is to recover (a basis for) the plaintext-space ideal $\mathcal{I} = \langle g \rangle$. To that end, we will construct two matrices $\mathbf{M}, \mathbf{N}$ which are both full rank over R_n (whp), but (after canceling some common factors) the determinant of $\mathbf{M}$ is divisible by a higher power of g than the determinant of $\mathbf{N}$. Computing $\mathbf{M} \times \mathbf{N}^{-1}$ over the field of fractions K_n and multiplying by the common denominator, we get an integral matrix whose determinant is divisible by g. Repeating this process many times and taking the common denominator of all the resulting determinants we obtain whp a basis for the ideal $\langle g \rangle$.

Let $\mathcal{X}||\mathcal{Z} = [h]$ be a 2-partition of the branching program steps, where we have sufficiently many input bits that only affect steps in the $\mathcal{X}$ interval and sufficiently many other input bits that only affect steps in the $\mathcal{Z}$ interval. (Denote these input bits by $J_x, J_z \subset [\ell]$, respectively.) Moreover, we can fix all the remaining input bits in such a way that for sufficiently many choices $x^{(i)} \in \{0,1\}^{|J_x|}$, $z^{(j)} \in \{0,1\}^{|J_z|}$ we get an input which is a zero of the function.

Finally, we assume that there are two distinguished input bits $j_1, j_2 \in J_x$ that we can set arbitrarily. Namely, for all the other choices of input bits as above, we can set these two bits to 00,01,10, and 11 and all four combinations will yield a zero of the function.

With these assumptions, let us denote by $w_{00}^{(i,j)}$ the zero-tested value which was obtained by honest evaluation of the obfuscated program on the input $x_{00}^{(i)} z^{(j)}$ with the two distinguished bits set to 00, and similarly $w_{01}^{(i,j)}$, $w_{10}^{(i,j)}$, $w_{11}^{(i,j)}$ with these bits set to 01, 10, 11, respectively. Note that:

- For every fixed i, j, the four inputs whose evaluation yields the scalars $w_{00}^{(i,j)}$, $w_{01}^{(i,j)}$, $w_{10}^{(i,j)}$, and $w_{11}^{(i,j)}$ differ only in the values of the distinguished input bit;

- For every $a \in \{00, 01, 10, 11\}$ and every fixed j, the inputs whose evaluation yields the different $\{w_a^{(i,j)}\}_i$ only differ in bits that affect the $\mathcal{X}$ interval of steps (but not the distinguished j_1, j_2); and
- For every $a \in \{00, 01, 10, 11\}$ and every fixed i, the inputs whose evaluation yields the different $\{w_a^{(i,j)}\}_j$ only differ in bits that affect the $\mathcal{Z}$ interval of steps.

Using Eq. (6), we have for all i, j and $a \in \{00, 01, 10, 11\}$,

$$w_a^{(i,j)} := p_{zt} \left(\mathbf{C}_{x_a^{(i)}} \mathbf{C}_{z^{(j)}} - \mathbf{C}'_{x_a^{(i)}} \mathbf{C}'_{z^{(j)}} \right) \tag{7}$$

$$= \frac{\eta}{g} \cdot \left[(\mathbf{S}_{x_a^{(i)}} + g\mathbf{E}_{x_a^{(i)}})(\mathbf{S}_{z^{(j)}} + g\mathbf{E}_{z^{(j)}}) - (\mathbf{S}'_{x_a^{(i)}} + g\mathbf{E}'_{x_a^{(i)}})(\mathbf{S}'_{z^{(j)}} + g\mathbf{E}'_{z^{(j)}}) \right]$$

$$= \frac{\eta}{g} \cdot \left[\mathbf{S}_{x_a^{(i)}} + g\mathbf{E}_{x_a^{(i)}}, (-\mathbf{S}'_{x_a^{(i)}} - g\mathbf{E}'_{x_a^{(i)}}) \right] \begin{bmatrix} \mathbf{S}_{z^{(j)}} + g\mathbf{E}_{z^{(j)}} \\ \mathbf{S}'_{z^{(j)}} + g\mathbf{E}'_{z^{(j)}} \end{bmatrix}$$

with Eq. (7) holding over the base ring R_n. Fixing $a \in \{00, 01, 10, 11\}$ and letting i, j range over sufficiently many inputs, we get the matrices

$$\mathbf{W}_a := [w_a^{(i,j)}]_{i,j} = \mathbf{X}_a \mathbf{Z}$$

$$:= \frac{\eta}{g} \left[\begin{array}{c} \cdots \\ \mathbf{S}_{x_a^{(i)}} + g\mathbf{E}_{x_a^{(i)}}, (-\mathbf{S}'_{x_a^{(i)}} - g\mathbf{E}'_{x_a^{(i)}}) \\ \cdots \end{array} \right] \left[\begin{array}{c} \cdots, \mathbf{S}_{z^{(j)}} + g\mathbf{E}_{z^{(j)}}, \cdots \\ \cdots, \mathbf{S}'_{z^{(j)}} + g\mathbf{E}'_{z^{(j)}}, \cdots \end{array} \right] \tag{8}$$

Specifically we choose as many different $x^{(i)}$'s and $z^{(j)}$'s to make $\mathbf{X}_a$ and $\mathbf{Z}$ square matrices (of dimension 2ρ, where $\rho = 2m + w$).

The two matrices $\mathbf{M}, \mathbf{N}$ that we consider in this part of the attack are

$$\mathbf{M} = \begin{bmatrix} \mathbf{W}_{00} & \mathbf{W}_{01} \\ \mathbf{W}_{10} & \mathbf{W}_{11} \end{bmatrix} = \frac{\eta}{g} \cdot \begin{bmatrix} \mathbf{X}_{00} & \mathbf{X}_{01} \\ \mathbf{X}_{10} & \mathbf{X}_{11} \end{bmatrix}_l \times \begin{bmatrix} \mathbf{Z} \\ & \mathbf{Z} \end{bmatrix},$$

$$\mathbf{N} = \begin{bmatrix} \mathbf{W}_{00} & 0 \\ 0 & \mathbf{W}_{11} \end{bmatrix} = \frac{\eta}{g} \cdot \begin{bmatrix} \mathbf{X}_{00} & 0 \\ 0 & \mathbf{X}_{11} \end{bmatrix} \times \begin{bmatrix} \mathbf{Z} \\ & \mathbf{Z} \end{bmatrix} \tag{9}$$

These matrices will have full rank over the base ring R_n whp due to the "random" error matrices $\mathbf{E}$ in the $\mathbf{X}$'s and $\mathbf{Z}$. However, we show now that whp, the determinant of $\mathbf{M}$ (after disregarding the common factor $\frac{\eta}{g}$) is divisible by a higher power of g than that of $\mathbf{N}$.

To see that, recall that the matrices $\mathbf{S}_x$ from Eq. (8) are the plaintext matrices of the GGHRSW constructions as per Eq. (5), and in particular they include the scalars $\beta_{j_1,b}, \beta_{j_2,b}$ for the two distinguished input bits j_1, j_2. To somewhat simplify notations we use below $\beta_b := \beta_{j_1,b}$ and $\beta'_b = \beta_{j_2,b}$. Specifically for any index i we have

$$\mathbf{S}_{x_{00}^{(i)}} = \beta_0 \beta'_0 \cdot \gamma^{(i)} \cdot \mathbf{J} \times \text{diag}(\mathbf{U}_{00}^{(i)}, \mathbf{V}_{00}^{(i)}, \mathbf{B}_x) \times \mathbf{K}_z$$

$$= \beta_0 \beta'_0 \cdot \gamma^{(i)} \cdot [0, \mathbf{v}_{00}^{(i)}, \mathbf{b}] \times \mathbf{K}_z \pmod{\mathcal{I}}$$

$$\mathbf{S}'_{x_{00}^{(i)}} = \delta \cdot \beta_0 \beta'_0 \cdot \gamma^{(i)} \cdot \mathbf{J}' \times \text{diag}(\mathbf{U}'^{(i)}_{00}, \mathbf{V}'^{(i)}_{00}, \mathbf{I}) \times \mathbf{K}'_z$$

$$= \delta \cdot \beta_0 \beta'_0 \cdot \gamma^{(i)} \cdot [0, \mathbf{v}'^{(i)}_{00}, \mathbf{b}'] \times \mathbf{K}'_z \pmod{\mathcal{I}}, \tag{10}$$

where above we used $\delta := \alpha_{x_{\sigma\tau}^{(i)}}/\alpha'_{x_{\sigma\tau}^{(i)}}$ (which by Lemma 2 is independent of i or the two bits σ, τ), and $\gamma^{(i)}$ is some scalar that depends on i but not on these two bits. We use similar notations for $\mathbf{S}_{x_{01}^{(i)}}, \mathbf{S}_{x_{10}^{(i)}}, \mathbf{S}_{x_{11}^{(i)}}$. For any two bits σ, τ, each row i of $\mathbf{X}_{\sigma\tau}$ (mod $\mathcal{I}$) has the form $[\mathbf{S}_{x_{\sigma\tau}^{(i)}} | - \mathbf{S}'_{x_{\sigma\tau}^{(i)}}]$, so we can write

$$\mathbf{X}_{\sigma\tau} = \left(\beta_\sigma \beta'_\tau \cdot \mathbf{X} + \mathbf{\Delta}_{\sigma\tau}\right) \times \mathrm{diag}(\mathbf{K}_z, \mathbf{K}'_z) \pmod{\mathcal{I}} \tag{11}$$

where $\mathrm{diag}(\mathbf{K}_z, \mathbf{K}'_z)$ is invertible, $\mathbf{X}$ is some fixed matrix independent of σ, τ, and where $\mathbf{\Delta}_{\sigma\tau}$ has only few non-zero columns (i.e., the ones corresponding to $\mathbf{v}_{\sigma\tau}^{(i)}$ and $\mathbf{v}'^{(i)}_{\sigma\tau}$ from Eq. (10)). Denoting by n the number of non-zero columns in the Δ's, we have (over $R_n/\mathcal{I}$)

$$\mathrm{rank}\begin{pmatrix} \beta_0\beta'_0\mathbf{X} + \Delta_{00} & \beta_0\beta'_1\mathbf{X} + \Delta_{01} \\ \beta_1\beta'_0\mathbf{X} + \Delta_{10} & \beta_1\beta'_1\mathbf{X} + \Delta_{11} \end{pmatrix} \leq 2n + \mathrm{rank}\begin{pmatrix} \beta_0\beta'_0\mathbf{X} & \beta_0\beta'_1\mathbf{X} \\ \beta_1\beta'_0\mathbf{X} & \beta_1\beta'_1\mathbf{X} \end{pmatrix}$$
$$= 2n + \mathrm{rank}(\mathbf{X}),$$

because $\beta_0\beta'_0 \cdot \beta_1\beta'_1 - \beta_0\beta'_1 \cdot \beta_1\beta'_0 = 0$. On the other hand, we have

$$\mathrm{rank}\begin{pmatrix} \beta_0\beta'_0\mathbf{X} + \Delta_{00} & 0 \\ 0 & \beta_1\beta'_1\mathbf{X} + \Delta_{11} \end{pmatrix} \stackrel{(whp)}{=} 2n + 2 \cdot \mathrm{rank}(\mathbf{X}).$$

Since it has lower rank modulo $\mathcal{I}$, then (at least heuristically[2]) the determinant of $\begin{bmatrix} \mathbf{X}_{00} & \mathbf{X}_{01} \\ \mathbf{X}_{10} & \mathbf{X}_{11} \end{bmatrix}$ is divisible by a higher power of g than that of $\begin{bmatrix} \mathbf{X}_{00} & 0 \\ 0 & \mathbf{X}_{11} \end{bmatrix}$.

Computing $\mathbf{M}\mathbf{N}^{-1}$ over K, the common factor η/g drops out, and we are left with a fractional matrix such that

$$\det(\mathbf{M}\mathbf{N}^{-1}) = \det\begin{pmatrix} \mathbf{X}_{00} & \mathbf{X}_{01} \\ \mathbf{X}_{10} & \mathbf{X}_{11} \end{pmatrix} / \det\begin{pmatrix} \mathbf{X}_{00} & 0 \\ 0 & \mathbf{X}_{11} \end{pmatrix} = \frac{\text{a multiple of } g}{\text{some denominator}},$$

where the denominator is not divisible by g. Multiplying by the denominator we thus get a multiple of g, as needed. Repeating this process several times with different distinguished indexes j_1, j_2, we can take the GCD and whp get a basis for some power $\mathcal{I}^t$ of the ideal $\mathcal{I}$.

Finally, when $\mathcal{I}$ is a prime ideal then it is easy to find $\mathcal{I}$ from $\mathcal{I}^t$: The norm of $\mathcal{I}^t$ is $\mathrm{norm}(\mathcal{I})^r$, and $p = \mathrm{norm}(\mathcal{I})$ is a prime integer, and we can find p from p^t (by exhaustive search over t). The Kummer-Dedekind theorem let us compute all the ideals of norm p in K, and one of these ideals is $\mathcal{I}$.

---

[2] Having $\mathrm{rank}(\mathbf{A}) > \mathrm{rank}(\mathbf{B})$ (mod g) does not always mean that $\det(\mathbf{B})$ is divisible by a higher power of g than $\det(\mathbf{A})$, since $\mathbf{A}$ could have one eigenvalue which is divisible by a high power of g, e.g., consider $\mathbf{A} = \begin{bmatrix} g^5 & 0 \\ 0 & 1 \end{bmatrix}$ and $\mathbf{B} = \begin{bmatrix} g & 0 \\ 0 & g \end{bmatrix}$. For our "random matrices", however, this is unlikely, as confirmed by our experiments.

3.2 Step II: Recovering Some Representatives of the Bundling Scalars

For this step we need the branching program to be 3-partitioned. Recall that Eq. (6) holds over R if the input $u = x^{(i)} y^{(b)} z^{(j)}$ is a zero of the function. Let i, j ranging over 2ρ inputs, and for $b \in \{0,1\}$, we get the matrices:

$$\mathbf{W}_b := \mathbf{X} \mathbf{Y}_b \mathbf{Z}$$

$$:= \frac{\eta}{g} \begin{bmatrix} \cdots \\ \mathbf{S}_{x^{(i)}} + g\mathbf{E}_{x^{(i)}}, -(\mathbf{S}'_{x^{(i)}} + g\mathbf{E}'_{x^{(i)}}) \\ \cdots \end{bmatrix} \tag{12}$$

$$+ \cdot \begin{bmatrix} \mathbf{S}_{y^{(b)}} + g\mathbf{E}_{y^{(b)}}, & 0 \\ 0, & \mathbf{S}'_{y^{(b)}} + g\mathbf{E}'_{y^{(b)}} \end{bmatrix} \cdot \begin{bmatrix} \cdots, \mathbf{S}_{z^{(j)}} + g\mathbf{E}_{z^{(j)}}, \cdots \\ \cdots, \mathbf{S}'_{z^{(j)}} + g\mathbf{E}'_{z^{(j)}}, \cdots \end{bmatrix}$$

where $\mathbf{X}, \mathbf{Y}_1, \mathbf{Y}_0, \mathbf{Z} \in R^{2\rho \times 2\rho}$ are full-rank w.h.p. due to the contribution of $\mathbf{E}$ terms from different paths.

We then compute the characteristic polynomial χ of $\mathbf{W}_1 \mathbf{W}_0^{-1} \in K^{2\rho \times 2\rho}$, which is equal to the characteristic polynomial of $\mathbf{Y}_1 \mathbf{Y}_0^{-1}$. Considering $\mathbf{Y}_1 \mathbf{Y}_0^{-1}$ modulo $\mathcal{I}$ we have:

$$\mathbf{Y}_1 \mathbf{Y}_0^{-1} = \begin{bmatrix} \mathbf{S}_{y^{(1)}} + g\mathbf{E}_{y^{(1)}}, & 0 \\ 0, & \mathbf{S}'_{y^{(1)}} + g\mathbf{E}'_{y^{(1)}} \end{bmatrix} \begin{bmatrix} \mathbf{S}_{y^{(0)}} + g\mathbf{E}_{y^{(0)}}, & 0 \\ 0, & \mathbf{S}'_{y^{(0)}} + g\mathbf{E}'_{y^{(0)}} \end{bmatrix}^{-1}$$

$$= \begin{bmatrix} \mathbf{S}_{y^{(1)}}, & 0 \\ 0, & \mathbf{S}'_{y^{(1)}} \end{bmatrix} \begin{bmatrix} \mathbf{S}_{y^{(0)}}, & 0 \\ 0, & \mathbf{S}'_{y^{(0)}} \end{bmatrix}^{-1} \pmod{\mathcal{I}} \tag{13}$$

Expanding the "functional term" of $\mathbf{Y}_1 \mathbf{Y}_0^{-1} \pmod{\mathcal{I}}$, i.e. $\mathbf{S}_{y^{(1)}} \mathbf{S}_{y^{(0)}}^{-1}$, we have:

$$\mathbf{S}_{y^{(1)}} \mathbf{S}_{y^{(0)}}^{-1} = \alpha_{y^{(1)}} \mathbf{K}_x^{-1} \begin{bmatrix} \mathbf{U}_{y^{(1)}}, & 0, & 0 \\ 0, & \mathbf{V}_{y^{(1)}}, & 0 \\ 0, & 0, & \mathbf{B}_{y^{(1)}} \end{bmatrix} \mathbf{K}_z$$

$$\times \left(\alpha_{y^{(0)}} \mathbf{K}_x^{-1} \begin{bmatrix} \mathbf{U}_{y^{(0)}}, & 0, & 0 \\ 0, & \mathbf{V}_{y^{(0)}}, & 0 \\ 0, & 0, & \mathbf{B}_{y^{(0)}} \end{bmatrix} \mathbf{K}_z \right)^{-1} \tag{14}$$

$$= \frac{\alpha_{y^{(1)}}}{\alpha_{y^{(0)}}} \cdot \mathbf{K}_x^{-1} \begin{bmatrix} \mathbf{U}_{y^{(1)}} \mathbf{U}_{y^{(0)}}^{-1}, & 0, & 0 \\ 0, & \mathbf{V}_{y^{(1)}} \mathbf{V}_{y^{(0)}}^{-1}, & 0 \\ 0, & 0, & \mathbf{B}_{y^{(1)}} \mathbf{B}_{y^{(0)}}^{-1} \end{bmatrix} \mathbf{K}_x$$

By Lemma 1, $\mathbf{B}_{y^{(1)}} \mathbf{B}_{y^{(0)}}^{-1} = \mathbf{I}^{w \times w}$, so $\alpha_{y^{(1)}} / \alpha_{y^{(0)}} \in K$ is an eigenvalue of $\mathbf{S}_{y^{(1)}} \mathbf{S}_{y^{(0)}}^{-1}$ with multiplicity at least the dimensions of the $\mathbf{B}$'s (i.e., at least w).[3] Similarly $\alpha'_{y^{(1)}} / \alpha'_{y^{(0)}}$ is an eigenvalue of $\mathbf{S}'_{y^{(1)}} \mathbf{S}'^{-1}_{y^{(0)}}$ of multiplicity at least w, and by Lemma 2 we have $\alpha'_{y^{(1)}} / \alpha'_{y^{(0)}} = \alpha_{y^{(1)}} / \alpha_{y^{(0)}}$. Hence $\alpha_{y^{(1)}} \alpha_{y^{(0)}}^{-1}$ is the eigenvalue

---

[3] We remark that this step of finding (the multiplicity of) an eigenvalue does not seem to fit in the "Weak Multilinear Map" model of Garg et al. [24].

of $\mathbf{Y}_1 \mathbf{Y}_0^{-1}$ (mod $\mathcal{I}$) of multiplicity at least $2w$. Given a basis of $\mathcal{I}$, we can solve the characteristic polynomial $\chi_{\mathbf{W}_1 \mathbf{W}_0^{-1}}$ (mod $\mathcal{I}$) and obtain eigenvalues in K. The eigenvalue of multiplicity $2w$ is $\alpha_{y^{(1)}} \alpha_{y^{(0)}}^{-1}$.

3.3   Step III: Annihilation Attack

The annihilation attacks described by Miles, Sahai and Zhandry [35] do not extend to break GGH13-based branching program obfuscators with the padded random diagonal entries. We show that with the knowledge of the ratios of scalars (even if their representations are big), this attack can be extended to handle the random diagonal entries. We begin with a brief overview of the attacks from [35].

Given many level-0 encodings $\{c_i = s_i + e_i \cdot g\}_i$, any degree-$d$ expression in them can be written as

$$c = r_0 + r_1 \cdot g^1 + r_2 \cdot g^2 + \ldots + r_d \cdot g^d \pmod{q}.$$

If that expression is encoded at level d, then multiplying it by the zero-test parameter yields $x = p_{zt} \cdot c/z^d = h(r_0 g^{-1} + r_1 + r_2 g + \ldots r_d g^{d-1}) \pmod{q}$ (which is small if $r_0 = 0$ and likely large when $r_0 \neq 0$).

An annihilation attack consists of collecting and zero-testing many encodings with $r_0 = 0$, getting the corresponding $x^{(1)}, x^{(2)}, \ldots$, then applying some carefully-selected polynomial to these $x^{(i)}$'es and examining the result. Specifically, Miles et al. observed that it is possible to check whether or not the terms that depends only on the r_1 values vanish in the resulting polynomial. They also observed that these r_1 values can be expressed as very structured expressions in the encoded secret and the error terms,

$$r_1 = e_1 s_2 \ldots s_d + s_1 e_2 s_3 \ldots s_d + \ldots + s_1 s_2 \ldots e_d.$$

Using these observation, Miles et al. described in [35] a particular polynomial in the $x^{(i)}$'s that can be used to distinguish the obfuscation of equivalent branching programs (under some contemporary obfuscators).

Introducing Our Running Example. To help describe our attack, we show below how it can be used to distinguish between GGHRSW obfuscation of two specific branching programs that compute the constant zero function. For this attack we need the branching programs to be 3-partitioned with intervals $\mathcal{X}||\mathcal{Y}||\mathcal{Z} = [h]$, and we need to have two distinguished input bit positions j_1, j_2 that only control steps in the $\mathcal{Y}$ interval but not $\mathcal{X}$ or $\mathcal{Z}$. In addition, we require that bit j_1 controls at least two steps (denoted u, w) in the $\mathcal{Y}$ interval, and that bit j_2 controls (at least) one step in between u and w (denoted v). That is, we need $u, v, w \in \mathcal{Y}$ with $u < v < w$, such that $\iota(u) = \iota(w) = j_1$, $\iota(v) = j_2$, and j_j does not control any steps before u or after w. As before, we shorten our notations somewhat and denote the relevant products of the bundling constants by

$$\beta_0 := \prod_{\iota(i)=j_1} \alpha_{i,0}, \quad \beta_1 := \prod_{\iota(i)=j_1} \alpha_{i,1}, \quad \beta_0' := \prod_{\iota(i)=j_2} \alpha_{i,0}, \quad \beta_1' := \prod_{\iota(i)=j_2} \alpha_{i,1}.$$

The two branching programs in our running example will have the identity matrix for both 0 and 1 in all the steps *except for the two steps u, w controlled by y_1*, and the zero matrices will be the identity also for these two steps. For the 1 matrices in these two steps, in one program they too will be the identity, and in the other program those two matrices are a permutation matrix and its inverse (denoted $\mathbf{P}, \mathbf{P}^{-1}$). The two programs $\mathcal{B}$ and $\mathcal{B}'$ are illustrated in Example 1.

Example 1. Two programs that compute the constant-zero function:

$$
\begin{array}{rrcccccc}
\mathcal{B} = & 0: & \mathbf{I} \ldots & \mathbf{I}\ \mathbf{I}\ \mathbf{I} & \mathbf{I} \ldots \mathbf{I} \\
& 1: & \mathbf{I} \ldots & \mathbf{I}\ \mathbf{I}\ \mathbf{I} & \mathbf{I} \ldots \mathbf{I} \\
\hline
\mathcal{B}' = & 0: & \mathbf{I} \ldots & \mathbf{I}\ \mathbf{I}\ \mathbf{I} & \mathbf{I} \ldots \mathbf{I} \\
& 1: & \mathbf{I} \ldots & \mathbf{I}\ \mathbf{P}\ \mathbf{I}\ \mathbf{P}^{-1} & \mathbf{I} \ldots \mathbf{I} \\
\hline
& \text{Steps}: & \mathcal{X} & u\ v\ w & \mathcal{Z} \\
& \text{input bits}: & * \ldots * & j_1\ j_2\ j_1 & * \ldots *
\end{array}
\tag{15}
$$

The Attack. Recall that the GGHRSW obfuscator embeds the branching-program matrices $\mathbf{B}_{i,b}$ (and the identity for the dummy branch) into higher-dimension randomized matrices

$$
\tilde{\mathbf{B}}_{i,b} := \begin{bmatrix} \mathbf{V}_{i,b} & \\ & \mathbf{B}_{i,b} \end{bmatrix} \text{ and } \tilde{\mathbf{B}}'_{i,b} := \begin{bmatrix} \mathbf{V}'_{i,b} & \\ & \mathbf{I} \end{bmatrix},
$$

where $\mathbf{V}_{i,b}, \mathbf{V}'_{i,b}$ are random diagonal matrices. The $\tilde{\mathbf{B}}$'s are multiplied by the bundling scalars and Kilian randomization matrices, and then encoded to get

$$
\mathbf{C}_{i,b} = \alpha_i^b \mathbf{K}_{i-1}^{-1} \tilde{\mathbf{B}}_{i,b} \mathbf{K}_i + g \cdot \mathbf{E}_{i,b} = \alpha_i^b \mathbf{K}_{i-1}^{-1} (\tilde{\mathbf{B}}_{i,b} + g \cdot \mathbf{F}_{i,b}) \mathbf{K}_i \pmod{q} \tag{16}
$$

where $\mathbf{K}_{i-1}^{-1}$ is the inverse of $\mathbf{K}_{i-1}$ *modulo* $\langle g \rangle$, and $\mathbf{F}_{i,b}$ is the matrix satisfying $\alpha_i^b \mathbf{K}_{i-1}^{-1} \mathbf{F} \mathbf{K}_i = \mathbf{E} \pmod{q}$. (We ignore the denominator z in these notations, since it gets canceled when we apply zero-test.)

From Step II above we can obtain (some representatives of) the ratios β_1/β_0 and β_1'/β_0'. Namely, we can compute four scalars $\nu_0, \nu_1, \gamma_{00}, \gamma_{11} \in R$ such that

$$
\frac{\nu_1}{\nu_0} = \frac{\beta_1'}{\beta_0'} \pmod{\mathcal{I}}, \text{ and } \frac{\gamma_{11}}{\gamma_{00}} = \frac{\beta_1 \beta_1'}{\beta_0 \beta_0'} \pmod{\mathcal{I}}. \tag{17}
$$

(Note that we chose notations that resemble their meaning: The scalars ν_0, ν_1 relate to the step v in the program, and γ_{00}, γ_{11} relate to the product of all relevant steps in the y interval.) Consider some values $x^{(i)} \in \{0, 1\}^{|J_x|}$ for the bits that control steps in the $\mathcal{X}$ interval, τ, σ for the two distinguished bits that control steps in the $\mathcal{Y}$ interval, and $z^{(j)} \in \{0, 1\}^{|J_z|}$ for the bits that control steps in the $\mathcal{Z}$ interval (all other bits are fixed). The resulting input is $u_{\sigma\tau}^{(i,j)} := x^{(i)} \sigma \tau z^{(j)}$, and it is a zero of the function. Also let $\mathrm{Eval}(u_{\sigma\tau}^{(i,j)})$ be the scalar obtained by evaluating the obfuscated branching program on this input:

$$
\mathsf{Eval}(u) := \frac{\eta}{g} \left(\mathbf{J}(\prod_{k \in [h]} \mathbf{C}_{k,u_{\iota(k)}})\mathbf{L} - \mathbf{J}'(\prod_{k \in [h]} \mathbf{C}'_{k,u_{\iota(k)}})\mathbf{L}' \right)
$$

$$
= \frac{\eta}{g} \left(\prod_{k \in [h]} \alpha_{k,u_{\iota(k)}} \mathbf{JL} - \prod_{k \in [h]} \alpha'_{k,u_{\iota(k)}} \mathbf{J}'\mathbf{L}' + g \cdot r_1(u) + g^2 \cdot r_2(u) + \ldots \right)
$$

(18)

where if u is a zero of the function then by construction we have

$$
\prod_{k \in [h]} \alpha_{k,u_{\iota(k)}} \mathbf{JL} - \prod_{k \in [h]} \alpha'_{k,u_{\iota(k)}} \mathbf{J}'\mathbf{L}' = 0.
$$

In our attack, we choose many different $x^{(i)}$'s and $z^{(j)}$'s and for each i, j we compute

$$
a_{i,j} := \quad \mathsf{Eval}(x^{(i)}11z^{(j)}) \cdot \gamma_{00} \cdot \nu_1\nu_0 - \mathsf{Eval}(x^{(i)}10z^{(j)}) \cdot \gamma_{00} \cdot \nu_1\nu_1
$$
$$
- \mathsf{Eval}(x^{(i)}01z^{(j)}) \cdot \gamma_{11} \cdot \nu_0\nu_0 + \mathsf{Eval}(x^{(i)}00z^{(j)}) \cdot \gamma_{11} \cdot \nu_0\nu_1,
$$

(19)

where all the operations are carried out in the base ring R. Using sufficiently many $x^{(i)}$'s and $z^{(j)}$'s we get a matrix $\mathbf{A} = [a_{i,j}]_{i,j}$, and we check if this matrix has full rank modulo $\mathcal{I}$. We guess that the branching program is $\mathcal{B}'$ if $\mathbf{A}$ has full rank, and otherwise we guess that it is $\mathcal{B}$.

3.4   Analysis

The Matrix H. We begin by considering the interval $\mathcal{Y}$ of the functional branch only. If $\mathcal{Y}$ consisted of only the steps u, v, w, then for any two bits $\sigma, \tau \in \{0,1\}$, the matrix that we get in the functional branch when evaluating on input with $u_{j_1} = \sigma$ and $u_{j_2} = \tau$ (namely $\mathbf{C}_{\sigma\tau} := \prod_{i \in \mathcal{Y}} \mathbf{C}_{i,u_{\iota(i)}}$) has the form

$$
\mathbf{C}_{\sigma\tau} = \beta_\sigma\beta'_\tau \cdot \mathbf{K}_{u-1}^{-1} \times \left(\overbrace{\tilde{\mathbf{B}}_{u,\sigma}\tilde{\mathbf{B}}_{v,\tau}\tilde{\mathbf{B}}_{w,\sigma}}^{:=\tilde{\mathbf{B}}_{\mathcal{Y}}^{\sigma\tau}} + g \cdot (\tilde{\mathbf{B}}_{u,\sigma}\tilde{\mathbf{B}}_{v,\tau}\overbrace{(\mathbf{F}_{w,\sigma} + \mathbf{E}'_v\tilde{\mathbf{B}}_{w,\sigma})}^{:=\tilde{\mathbf{F}}_{w,\sigma}} \right.
$$

$$
\left. + \tilde{\mathbf{B}}_{u,\sigma}\overbrace{(\mathbf{F}_{v,\tau} + \mathbf{E}'_u\tilde{\mathbf{B}}_{v,\tau})}^{:=\tilde{\mathbf{F}}_{v,\tau}}\tilde{\mathbf{B}}_{w,\sigma} + \overbrace{\mathbf{F}_{u,\sigma}}^{:=\tilde{\mathbf{F}}_{u,\sigma}} \tilde{\mathbf{B}}_{v,\tau}\tilde{\mathbf{B}}_{w,\sigma}) + g^2 \cdot \mathbf{E}_{\tau,\sigma} \right) \times \mathbf{K}_w
$$

(20)

$$
= \beta_\sigma\beta'_\tau \cdot \mathbf{K}_{u-1}^{-1} \times (\tilde{\mathbf{B}}_{\mathcal{Y}}^{\sigma\tau} + g \cdot \tilde{\mathbf{F}}_{\mathcal{Y}}^{\sigma\tau} + g^2 \cdot \mathbf{E}_{\mathcal{Y}}^{\tau\sigma}) \times \mathbf{K}_w
$$

with equality modulo q, where $\mathbf{K}, \mathbf{K}^{-1}$'es are the Kilian randomization matrices, and $\mathbf{E}_{\mathcal{Y}}^{\tau\sigma}$ is some error matrix. (In the last line we have $\tilde{\mathbf{F}}_{\mathcal{Y}}^{\sigma\tau}$ denoting the coefficient of g in the $\mathcal{Y}$ interval.) If there are more steps in the interval $\mathcal{Y}$ then we get the same form, except the matrices $\tilde{\mathbf{B}}, \tilde{\mathbf{F}}$ are not single-step matrices but rather a product of a few steps, and we have an extra scalar factor α' (independent of the bits σ, τ) that comes from the bundling factors in the fixed steps in $\mathcal{Y}$.

The coefficient of g in Eq. (20) is

$$\tilde{\mathbf{F}}_{\mathcal{Y}}^{\sigma\tau} := \tilde{\mathbf{B}}_{u,\sigma}\tilde{\mathbf{B}}_{v,\tau}\tilde{\mathbf{F}}_{w,\sigma} + \tilde{\mathbf{B}}_{u,\sigma}\tilde{\mathbf{F}}_{v,\tau}\tilde{\mathbf{B}}_{w,\sigma} + \tilde{\mathbf{F}}_{u,\sigma}\tilde{\mathbf{B}}_{v,\tau}\tilde{\mathbf{B}}_{w,\sigma}.$$

Let

$$\mathbf{H} := \tilde{\mathbf{F}}_{\mathcal{Y}}^{11} - \tilde{\mathbf{F}}_{\mathcal{Y}}^{10} - \tilde{\mathbf{F}}_{\mathcal{Y}}^{01} + \tilde{\mathbf{F}}_{\mathcal{Y}}^{00} \tag{21}$$
$$= \left(\tilde{\mathbf{B}}_{u,1}\tilde{\mathbf{B}}_{v,1}\tilde{\mathbf{F}}_{w,1} + \tilde{\mathbf{B}}_{u,1}\tilde{\mathbf{F}}_{v,1}\tilde{\mathbf{B}}_{w,1} + \tilde{\mathbf{F}}_{u,1}\tilde{\mathbf{B}}_{v,1}\tilde{\mathbf{B}}_{w,1}\right)$$
$$- \left(\tilde{\mathbf{B}}_{u,1}\tilde{\mathbf{B}}_{v,0}\tilde{\mathbf{F}}_{w,1} + \tilde{\mathbf{B}}_{u,1}\tilde{\mathbf{F}}_{v,0}\tilde{\mathbf{B}}_{w,1} + \tilde{\mathbf{F}}_{u,1}\tilde{\mathbf{B}}_{v,0}\tilde{\mathbf{B}}_{w,1}\right)$$
$$- \left(\tilde{\mathbf{B}}_{u,0}\tilde{\mathbf{B}}_{v,1}\tilde{\mathbf{F}}_{w,0} + \tilde{\mathbf{B}}_{u,0}\tilde{\mathbf{F}}_{v,1}\tilde{\mathbf{B}}_{w,0} + \tilde{\mathbf{F}}_{u,0}\tilde{\mathbf{B}}_{v,1}\tilde{\mathbf{B}}_{w,0}\right)$$
$$+ \left(\tilde{\mathbf{B}}_{u,0}\tilde{\mathbf{B}}_{v,0}\tilde{\mathbf{F}}_{w,0} + \tilde{\mathbf{B}}_{u,0}\tilde{\mathbf{F}}_{v,0}\tilde{\mathbf{B}}_{w,0} + \tilde{\mathbf{F}}_{u,0}\tilde{\mathbf{B}}_{v,0}\tilde{\mathbf{B}}_{w,0}\right).$$

The crux of the analysis is to show that $\mathbf{H}$ has a block of zeros when evaluating the program $\mathcal{B}$ (that has the identity matrices everywhere), but whp not when evaluating the branching program $\mathcal{B}'$ (that has $\mathbf{P}$ and $\mathbf{P}^{-1}$).

When evaluating $\mathcal{B}$, all the $\mathbf{B}_{i,b}$ matrices are the $w \times w$ identity $\mathbf{I}$, which are then embedded in the lower-right quadrant of the higher-dimension $\tilde{\mathbf{B}}_{i,b}$'s with the diagonal random $\mathbf{V}_i^b$'s in the upper-left quadrant. Below we also use the notation $\mathbf{V}_{ii'}^{\sigma\tau} := \mathbf{V}_i^{\sigma} \times \mathbf{V}_{i'}^{\tau}$ for the product of two of these diagonal matrices. We analyze separately the terms $\tilde{\mathbf{B}}\tilde{\mathbf{B}}\tilde{\mathbf{F}}$, $\tilde{\mathbf{B}}\tilde{\mathbf{F}}\tilde{\mathbf{B}}$, and $\tilde{\mathbf{F}}\tilde{\mathbf{B}}\tilde{\mathbf{B}}$, in order to establish that in this case the lower-right quadrant of $\mathbf{H}$ (that correspond to these identity matrices) is $\mathbf{0}$, i.e. $\mathbf{H} \in \begin{bmatrix} * & * \\ * & 0^{w\times w} \end{bmatrix}$.

(a) $\tilde{\mathbf{F}}\tilde{\mathbf{B}}\tilde{\mathbf{B}}$:

$$\tilde{\mathbf{F}}_u^1\tilde{\mathbf{B}}_v^1\tilde{\mathbf{B}}_w^1 - \tilde{\mathbf{F}}_u^1\tilde{\mathbf{B}}_v^0\tilde{\mathbf{B}}_w^1 - \tilde{\mathbf{F}}_u^0\tilde{\mathbf{B}}_v^1\tilde{\mathbf{B}}_w^0 + \tilde{\mathbf{F}}_u^0\tilde{\mathbf{B}}_v^0\tilde{\mathbf{B}}_w^0$$
$$= \tilde{\mathbf{F}}_u^1 \times \left(\begin{bmatrix} \mathbf{V}_{vw}^{11} & 0 \\ 0 & \mathbf{I} \end{bmatrix} - \begin{bmatrix} \mathbf{V}_{vw}^{01} & 0 \\ 0 & \mathbf{I} \end{bmatrix}\right) - \tilde{\mathbf{F}}_u^0 \times \left(\begin{bmatrix} \mathbf{V}_{vw}^{10} & 0 \\ 0 & \mathbf{I} \end{bmatrix} - \begin{bmatrix} \mathbf{V}_{vw}^{00} & 0 \\ 0 & \mathbf{I} \end{bmatrix}\right)$$
$$= \tilde{\mathbf{F}}_u^1 \times \begin{bmatrix} \mathbf{V}_{vw}^{11} - \mathbf{V}_{vw}^{01} & 0 \\ 0 & 0 \end{bmatrix} - \tilde{\mathbf{F}}_u^0 \times \begin{bmatrix} \mathbf{V}_{vw}^{10} - \mathbf{V}_{vw}^{00} & 0 \\ 0 & 0 \end{bmatrix} \tag{22}$$
$$\in \left[*^{(2m+w)\times 2m},\, 0^{(2m+w)\times w} \right]$$

(b) $\tilde{\mathbf{B}}\tilde{\mathbf{B}}\tilde{\mathbf{F}}$:

$$\tilde{\mathbf{B}}_u^1\tilde{\mathbf{B}}_v^1\tilde{\mathbf{F}}_w^1 - \tilde{\mathbf{B}}_u^1\tilde{\mathbf{B}}_v^0\tilde{\mathbf{F}}_w^1 - \tilde{\mathbf{B}}_u^0\tilde{\mathbf{B}}_v^1\tilde{\mathbf{F}}_w^0 + \tilde{\mathbf{B}}_u^0\tilde{\mathbf{B}}_v^0\tilde{\mathbf{F}}_w^0$$
$$= \begin{bmatrix} \mathbf{V}_{uv}^{11} - \mathbf{V}_{uv}^{10} & 0 \\ 0 & 0 \end{bmatrix} \times \tilde{\mathbf{F}}_w^1 - \begin{bmatrix} \mathbf{V}_{uv}^{01} - \mathbf{V}_{uv}^{00} & 0 \\ 0 & 0 \end{bmatrix} \times \tilde{\mathbf{F}}_w^0 \in \begin{bmatrix} *^{2m\times(2m+w)} \\ 0^{w\times(2m+w)} \end{bmatrix} \tag{23}$$

(c) The most interesting term is $\tilde{\mathbf{B}}\tilde{\mathbf{F}}\tilde{\mathbf{B}}$:

$$\tilde{\mathbf{B}}_u^1\tilde{\mathbf{F}}_v^1\tilde{\mathbf{B}}_w^1 - \tilde{\mathbf{B}}_u^1\tilde{\mathbf{F}}_v^0\tilde{\mathbf{B}}_w^1 - \tilde{\mathbf{B}}_u^0\tilde{\mathbf{F}}_v^1\tilde{\mathbf{B}}_w^0 + \tilde{\mathbf{B}}_u^0\tilde{\mathbf{F}}_v^0\tilde{\mathbf{B}}_w^0$$
$$= \begin{bmatrix} * & * \\ * \mathbf{I} \tilde{\mathbf{F}}_{v(LR)}^1 & \mathbf{I} \end{bmatrix} - \begin{bmatrix} * & * \\ * \mathbf{I} \tilde{\mathbf{F}}_{v(LR)}^0 & \mathbf{I} \end{bmatrix} - \begin{bmatrix} * & * \\ * \mathbf{I} \tilde{\mathbf{F}}_{v(LR)}^1 & \mathbf{I} \end{bmatrix} + \begin{bmatrix} * & * \\ * \mathbf{I} \tilde{\mathbf{F}}_{v(LR)}^0 & \mathbf{I} \end{bmatrix} \tag{24}$$
$$\in \begin{bmatrix} *^{2m\times 2m}, & *^{2m\times w} \\ *^{w\times 2m}, & 0^{w\times w} \end{bmatrix}$$

where the subscript $\tilde{\mathbf{F}}_{(LR)}$ denotes the lower-right quadrant (of dimension $w \times w$) in the corresponding matrix.

Adding Eqs. (22), (24) and (23), we get $\mathbf{H} \in \begin{bmatrix} * & * \\ * & 0^{w \times w} \end{bmatrix}$, as needed.

When evaluating $\mathcal{B}'$, the form of the terms $\tilde{\mathbf{B}}\tilde{\mathbf{F}}\tilde{\mathbf{B}}$ changes: Instead of Eq. (24), in the lower-right quadrant we now get $\mathbf{H}_{(LR)} = \tilde{\mathbf{F}}^1_{\mathbf{v}(LR)} - \tilde{\mathbf{F}}^0_{\mathbf{v}(LR)} - \mathbf{P}(\tilde{\mathbf{F}}^1_{\mathbf{v}(LR)} - \tilde{\mathbf{F}}^0_{\mathbf{v}(LR)})\mathbf{P}^{-1}$, which is unlikely to be the zero matrix.

The same analysis can be applied to the dummy branch, where we can define the matrix $\mathbf{H}'$ in the same way. In the dummy branch, however, the lower-right quadrant of $\mathbf{H}'$ is always zero, in both $\mathcal{B}$ and $\mathcal{B}'$ (since the dummy branch always consists of identity matrices, regardless of what the program is).

The Matrix A. We now proceed to incorporate the $\mathcal{X}, \mathcal{Z}$ intervals (including the bookends) and analyze the matrix $\mathbf{A} = [a_{i,j}]_{i,j}$. For any fixed i, j, let us denote the product of the $\mathcal{X}$ interval matrices in the two branches (including the bookend) by $\alpha_x^{(i)} \cdot \mathbf{J}(\tilde{\mathbf{B}}_{\mathcal{X}}^{(i)} + g \cdot \tilde{\mathbf{F}}_{\mathcal{X}}^{(i)})$ and $\alpha_x'^{(i)} \cdot \mathbf{J}'(\tilde{\mathbf{B}}'^{(i)}_{\mathcal{X}} + g \cdot \tilde{\mathbf{F}}'^{(i)}_{\mathcal{X}})$, respectively. Similarly for the $\mathcal{Z}$ interval we denote the products in the two branches by $\alpha_z^{(j)}(\tilde{\mathbf{B}}_{\mathcal{Z}}^{(j)} + g \cdot \tilde{\mathbf{F}}_{\mathcal{Z}}^{(j)})\mathbf{L}$ and $\alpha_z'^{(j)}(\tilde{\mathbf{B}}'^{(j)}_{\mathcal{Z}} + g \cdot \tilde{\mathbf{F}}'^{(j)}_{\mathcal{Z}})\mathbf{L}'$, respectively.

By construction—for the case where the $\mathcal{Y}$ interval includes just the steps u, v, w—we have $\alpha_x^{(i)}\alpha_z^{(j)} = \alpha_x'^{(i)}\alpha_z'^{(j)}$, and we denote this product by $\alpha^{(i,j)}$. (In the more general case we have the same equality, except it includes also the constants α_y, α_y' due to the fixed steps in the $\mathcal{Y}$ interval.) With these notations, we have

$$
\begin{aligned}
\mathsf{Eval}&\left(x^{(i)}\sigma\tau z^{(j)}\right) \\
&= \alpha^{(i,j)}\beta_\sigma\beta'_\tau \cdot \frac{\eta}{g}\Big(\mathbf{J}(\tilde{\mathbf{B}}_{\mathcal{X}}^{(i)} + g \cdot \tilde{\mathbf{F}}_{\mathcal{X}}^{(i)})(\tilde{\mathbf{B}}_{\mathcal{Y}}^{\sigma\tau} + g \cdot \tilde{\mathbf{F}}_{\mathcal{Y}}^{\sigma\tau})(\tilde{\mathbf{B}}_{\mathcal{Z}}^{(j)} + g \cdot \tilde{\mathbf{F}}_{\mathcal{Z}}^{(j)})\mathbf{L} \\
&\quad - \mathbf{J}'(\tilde{\mathbf{B}}'^{(i)}_{\mathcal{X}} + g \cdot \tilde{\mathbf{F}}'^{(i)}_{\mathcal{X}})(\tilde{\mathbf{B}}'^{\sigma\tau}_{\mathcal{Y}} + g \cdot \tilde{\mathbf{F}}'^{\sigma\tau}_{\mathcal{Y}})(\tilde{\mathbf{B}}'^{(j)}_{\mathcal{Z}} + g \cdot \tilde{\mathbf{F}}'^{(j)}_{\mathcal{Z}})\mathbf{L}'\Big) \qquad (25) \\
&= \alpha^{(i,j)}\beta_\sigma\beta'_\tau \cdot \eta\Big(\mathbf{J}\left(\tilde{\mathbf{B}}_{\mathcal{X}}^{(i)}\tilde{\mathbf{B}}_{\mathcal{Y}}^{\sigma\tau}\tilde{\mathbf{F}}_{\mathcal{Z}}^{(j)} + \tilde{\mathbf{B}}_{\mathcal{X}}^{(i)}\tilde{\mathbf{F}}_{\mathcal{Y}}^{\sigma\tau}\tilde{\mathbf{B}}_{\mathcal{Z}}^{(j)} + \tilde{\mathbf{F}}_{\mathcal{X}}^{(i)}\tilde{\mathbf{B}}_{\mathcal{Y}}^{\sigma\tau}\tilde{\mathbf{B}}_{\mathcal{Z}}^{(j)}\right)\mathbf{L} \\
&\quad - \mathbf{J}'\left(\tilde{\mathbf{B}}'^{(i)}_{\mathcal{X}}\tilde{\mathbf{B}}'^{\sigma\tau}_{\mathcal{Y}}\tilde{\mathbf{F}}'^{(j)}_{\mathcal{Z}} + \tilde{\mathbf{B}}'^{(i)}_{\mathcal{X}}\tilde{\mathbf{F}}'^{\sigma\tau}_{\mathcal{Y}}\tilde{\mathbf{B}}'^{(j)}_{\mathcal{Z}} + \tilde{\mathbf{F}}'^{(i)}_{\mathcal{X}}\tilde{\mathbf{B}}'^{\sigma\tau}_{\mathcal{Y}}\tilde{\mathbf{B}}'^{(j)}_{\mathcal{Z}}\right)\mathbf{L}'\Big) \quad (\bmod \mathcal{I})
\end{aligned}
$$

where the last equality follows since $x^{(i)}\sigma\tau z^{(j)}$ is a zero of the function, and hence the "free term" without any factor of g is equal to zero. Using Eq. (26) we can re-write $a_{i,j}$ as

$$
\begin{aligned}
a_{i,j} = {}&\alpha^{(i,j)}\beta_1\beta'_1\eta(\cdots)\gamma_{00}\nu_1\nu_0 - \alpha^{(i,j)}\beta_1\beta'_0\eta(\cdots)\gamma_{00}\nu_1\nu_1 \\
&- \alpha^{(i,j)}\beta_0\beta'_1\eta(\cdots)\gamma_{11}\nu_0\nu_0 + \alpha^{(i,j)}\beta_0\beta'_0\eta(\cdots)\gamma_{11}\nu_0\nu_1 \quad (\bmod \mathcal{I})
\end{aligned}
$$

where the $(\cdots)$'s refer to the parenthesized expression from Eq. (26) relative to the appropriate bits σ, τ. This is where we use the ratios that we recovered in

Step II, by definition we have that

$$\beta_1\beta_1' \cdot \gamma_{00}\nu_1\nu_0 = \beta_1\beta_0' \cdot \gamma_{00}\nu_1\nu_1 = \beta_0\beta_1' \cdot \gamma_{11}\nu_0\nu_0 = \beta_0\beta_0' \cdot \gamma_{11}\nu_0\nu_1 \pmod{\mathcal{I}},$$

so the four terms above (with i, j fixed) all have the same scalar multiple. Moreover that scalar is bilinear in i, j, so we just fold it into the matrices corresponding to $x^{(i)}, z^{(j)}$ and ignore it from now on. Thus we can further re-write the expression for $a_{i,j}$ as

$$
\begin{aligned}
a_{i,j} = \mathbf{J}\bigg(& \tilde{\mathbf{B}}_{\mathcal{X}}^{(i)} \left(\tilde{\mathbf{B}}_{\mathcal{Y}}^{11} - \tilde{\mathbf{B}}_{\mathcal{Y}}^{10} - \tilde{\mathbf{B}}_{\mathcal{Y}}^{01} + \tilde{\mathbf{B}}_{\mathcal{Y}}^{00} \right) \tilde{\mathbf{F}}_{\mathcal{Z}}^{(j)} \\
& \overbrace{\quad\qquad\qquad\qquad\qquad\qquad}^{=\mathbf{H}} \\
& + \tilde{\mathbf{B}}_{\mathcal{X}}^{(i)} \left(\tilde{\mathbf{F}}_{\mathcal{Y}}^{11} - \tilde{\mathbf{F}}_{\mathcal{Y}}^{10} - \tilde{\mathbf{F}}_{\mathcal{Y}}^{01} + \tilde{\mathbf{F}}_{\mathcal{Y}}^{00} \right) \tilde{\mathbf{B}}_{\mathcal{Z}}^{(j)} \\
& + \tilde{\mathbf{F}}_{\mathcal{X}}^{(i)} \left(\tilde{\mathbf{B}}_{\mathcal{Y}}^{11} - \tilde{\mathbf{B}}_{\mathcal{Y}}^{10} - \tilde{\mathbf{B}}_{\mathcal{Y}}^{01} + \tilde{\mathbf{B}}_{\mathcal{Y}}^{00} \right) \tilde{\mathbf{B}}_{\mathcal{Z}}^{(j)} \bigg)\mathbf{L} \\
& - \mathbf{J}'\bigg(\tilde{\mathbf{B}}'^{(i)}_{\mathcal{X}} \left(\tilde{\mathbf{B}}'^{11}_{\mathcal{Y}} - \tilde{\mathbf{B}}'^{10}_{\mathcal{Y}} - \tilde{\mathbf{B}}'^{01}_{\mathcal{Y}} + \tilde{\mathbf{B}}'^{00}_{\mathcal{Y}} \right) \tilde{\mathbf{F}}'^{(j)}_{\mathcal{Z}} \\
& \overbrace{\quad\qquad\qquad\qquad\qquad\qquad}^{=\mathbf{H}'} \\
& + \tilde{\mathbf{B}}'^{(i)}_{\mathcal{X}} \left(\tilde{\mathbf{F}}'^{11}_{\mathcal{Y}} - \tilde{\mathbf{F}}'^{10}_{\mathcal{Y}} - \tilde{\mathbf{F}}'^{01}_{\mathcal{Y}} + \tilde{\mathbf{F}}'^{00}_{\mathcal{Y}} \right) \tilde{\mathbf{B}}'^{(j)}_{\mathcal{Z}} \\
& + \tilde{\mathbf{F}}'^{(i)}_{\mathcal{X}} \left(\tilde{\mathbf{B}}'^{11}_{\mathcal{Y}} - \tilde{\mathbf{B}}'^{10}_{\mathcal{Y}} - \tilde{\mathbf{B}}'^{01}_{\mathcal{Y}} + \tilde{\mathbf{B}}'^{00}_{\mathcal{Y}} \right) \tilde{\mathbf{B}}'^{(j)}_{\mathcal{Z}} \bigg)\mathbf{L}' \pmod{\mathcal{I}}
\end{aligned}
\tag{26}
$$

Next, we denote:

$$
\begin{aligned}
&\tilde{\mathbf{B}}_{\mathcal{Y}}^{\Delta} := \tilde{\mathbf{B}}_{\mathcal{Y}}^{11} - \tilde{\mathbf{B}}_{\mathcal{Y}}^{10} - \tilde{\mathbf{B}}_{\mathcal{Y}}^{01} + \tilde{\mathbf{B}}_{\mathcal{Y}}^{00}, \; \tilde{\mathbf{B}}'^{\Delta}_{\mathcal{Y}} := \tilde{\mathbf{B}}'^{11}_{\mathcal{Y}} - \tilde{\mathbf{B}}'^{10}_{\mathcal{Y}} - \tilde{\mathbf{B}}'^{01}_{\mathcal{Y}} + \tilde{\mathbf{B}}'^{00}_{\mathcal{Y}} \\
&\mathbf{x}_i := \mathbf{J}\tilde{\mathbf{B}}_{\mathcal{X}}^{(i)}, \qquad \mathbf{z}_j := \tilde{\mathbf{B}}_{\mathcal{Z}}^{(j)}\mathbf{L}, \quad \mathbf{x}'_i := \mathbf{J}'\tilde{\mathbf{B}}'^{(i)}_{\mathcal{X}}, \qquad \mathbf{z}'_j := \tilde{\mathbf{B}}'^{(j)}_{\mathcal{Z}}\mathbf{L}', \\
&\mathbf{e}_i := \mathbf{J}\tilde{\mathbf{F}}_{\mathcal{X}}^{(i)}, \qquad \mathbf{f}_j := \tilde{\mathbf{F}}_{\mathcal{Z}}^{(j)}\mathbf{L}, \quad \mathbf{e}'_i := \mathbf{J}'\tilde{\mathbf{F}}'^{(i)}_{\mathcal{X}}, \qquad \mathbf{f}'_j := \tilde{\mathbf{F}}'^{(j)}_{\mathcal{Z}}\mathbf{L}'
\end{aligned}
$$

and so we can write

$$a_{i,j} = \underbrace{\mathbf{x}_i\tilde{\mathbf{B}}_{\mathcal{Y}}^{\Delta}\mathbf{f}_j + \mathbf{x}_i\mathbf{H}\mathbf{z}_j + \mathbf{e}_i\tilde{\mathbf{B}}_{\mathcal{Y}}^{\Delta}\mathbf{z}_j}_{:=d_{i,j}} - \underbrace{\mathbf{x}'_i\tilde{\mathbf{B}}'^{\Delta}_{\mathcal{Y}}\mathbf{f}'_j + \mathbf{x}'_i\mathbf{H}'\mathbf{z}'_j + \mathbf{e}'_i\tilde{\mathbf{B}}'^{\Delta}_{\mathcal{Y}}\mathbf{z}'_j}_{:=d'_{i,j}} \pmod{\mathcal{I}}. \tag{27}$$

Denoting $\mathbf{D} = [d_{i,j}]_{i,j}$ and $\mathbf{D}' = [d'_{i,j}]_{i,j}$, we have $\mathbf{A} = \mathbf{D} - \mathbf{D}'$, and so the rank of $\mathbf{A}$ is at most $\mathsf{rank}(\mathbf{D}) + \mathsf{rank}(\mathbf{D}')$. Recalling the structure of the various components again, we note that they contain many zeros. In particular for the program $\mathcal{B}$ we have $\mathbf{x}_i, \mathbf{x}'_i \in (0^m \; *^m \; *^w)$, $\mathbf{z}_j, \mathbf{z}'_j \in (*^m \; 0^m \; *^w)^t$, and also

$$
\tilde{\mathbf{B}}_{\mathcal{Y}}^{\Delta}, \; \tilde{\mathbf{B}}'^{\Delta}_{\mathcal{Y}} \in \begin{pmatrix} *^{m\times m}, & 0^{m\times m}, & 0^{m\times w} \\ 0^{m\times m}, & *^{m\times m}, & 0^{m\times w} \\ 0^{w\times m}, & 0^{w\times m}, & 0^{w\times w} \end{pmatrix}, \quad \mathbf{H}, \mathbf{H}' \in \begin{pmatrix} *^{m\times m}, & *^{m\times m}, & *^{m\times w} \\ *^{m\times m}, & *^{m\times m}, & *^{m\times w} \\ *^{w\times m}, & *^{w\times m}, & 0^{w\times w} \end{pmatrix},
$$

and for $\mathcal{B}'$ we have almost the same thing except that $\mathbf{H}$ can be arbitrary. Our goal is to detect this difference in the form of $\mathbf{H}$ given sufficiently many $a_{i,j}$'s. Let us analyze first only the term from the functional branch, $\mathbf{D} = [d_{i,j}]_{i,j}$, where we pick $\zeta \geq 2m+1$ different i's and j's.

$$\mathbf{D} = \mathbf{X}\tilde{\mathbf{B}}_{\tilde{y}}^{\Delta}\mathbf{F} + \mathbf{X}\mathbf{H}\mathbf{Z} + \mathbf{E}\tilde{\mathbf{B}}_{\tilde{y}}^{\Delta}\mathbf{Z}$$

$$= \begin{bmatrix} 0 & \mathbf{X}_2 & \mathbf{X}_3 \end{bmatrix} \begin{bmatrix} \tilde{\mathbf{B}}_{1,1} & 0 & 0 \\ 0 & \tilde{\mathbf{B}}_{2,2} & 0 \\ 0 & 0 & 0 \end{bmatrix} \begin{bmatrix} \mathbf{F}_1 \\ \mathbf{F}_2 \\ \mathbf{F}_3 \end{bmatrix} + \begin{bmatrix} 0 & \mathbf{X}_2 & \mathbf{X}_3 \end{bmatrix} \begin{bmatrix} \mathbf{H}_{1,1} & \mathbf{H}_{1,2} & \mathbf{H}_{1,3} \\ \mathbf{H}_{2,1} & \mathbf{H}_{2,2} & \mathbf{H}_{2,3} \\ \mathbf{H}_{3,1} & \mathbf{H}_{3,2} & 0 \end{bmatrix} \begin{bmatrix} \mathbf{Z}_1 \\ 0 \\ \mathbf{Z}_3 \end{bmatrix} \quad (28)$$

$$+ \begin{bmatrix} \mathbf{E}_1 & \mathbf{E}_2 & \mathbf{E}_3 \end{bmatrix} \begin{bmatrix} \tilde{\mathbf{B}}_{1,1} & 0 & 0 \\ 0 & \tilde{\mathbf{B}}_{2,2} & 0 \\ 0 & 0 & 0 \end{bmatrix} \begin{bmatrix} \mathbf{Z}_1 \\ 0 \\ \mathbf{Z}_3 \end{bmatrix}$$

where $\{\tilde{\mathbf{B}}_{k,\ell}, \mathbf{H}_{k,\ell}\}_{k,\ell \in [3]}$ are blocks of $\tilde{\mathbf{B}}_{\tilde{y}}^{\Delta}$, $\mathbf{H}$ with dimensions $[m|m|w] \times [m|m|w]$. $\{\mathbf{X}_k, \mathbf{E}_k\}_{k \in [3]}$ are blocks of $\mathbf{X}$, $\mathbf{E}$ with dimensions $\zeta \times [m|m|w]$. $\{\mathbf{Z}_\ell, \mathbf{F}_\ell\}_{\ell \in [3]}$ are blocks of $\mathbf{Z}$, $\mathbf{F}$ with dimensions $[m|m|w] \times \zeta$.

Observe that many of the blocks in Eq. (28) do not contribute to the result, since they are only multiplied by zeros in the adjacent matrices. For example, E_3 in the last term above does not contribute to the evaluation since the entries in the 3rd blocked rows of $\tilde{\mathbf{B}}_{\tilde{y}}^{\Delta}$ are all zeros. We can therefore treat these blocks as if they were zeros themselves, so we get

$$\mathbf{D} = \begin{bmatrix} 0 & \mathbf{X}_2 & 0 \end{bmatrix} \begin{bmatrix} 0 & 0 & 0 \\ 0 & \tilde{\mathbf{B}}_{2,2} & 0 \\ 0 & 0 & 0 \end{bmatrix} \begin{bmatrix} 0 \\ \mathbf{F}_2 \\ 0 \end{bmatrix} + \begin{bmatrix} 0 & \mathbf{X}_2 & \mathbf{X}_3 \end{bmatrix} \begin{bmatrix} 0 & 0 & 0 \\ \mathbf{H}_{2,1} & 0 & \mathbf{H}_{2,3} \\ \mathbf{H}_{3,1} & 0 & 0 \end{bmatrix} \begin{bmatrix} \mathbf{Z}_1 \\ 0 \\ \mathbf{Z}_3 \end{bmatrix}$$

$$+ \begin{bmatrix} \mathbf{E}_1 & 0 & 0 \end{bmatrix} \begin{bmatrix} \tilde{\mathbf{B}}_{1,1} & 0 & 0 \\ 0 & 0 & 0 \\ 0 & 0 & 0 \end{bmatrix} \begin{bmatrix} \mathbf{Z}_1 \\ 0 \\ 0 \end{bmatrix} \quad (29)$$

From there we get

$$\mathbf{D} = \mathbf{X}_2\tilde{\mathbf{B}}_{2,2}\mathbf{F}_2 + \begin{bmatrix} \mathbf{X}_2\mathbf{H}_{2,1} + \mathbf{X}_3\mathbf{H}_{3,1} & 0 & \mathbf{X}_2\mathbf{H}_{2,3} \end{bmatrix} \begin{bmatrix} \mathbf{Z}_1 \\ 0 \\ \mathbf{Z}_3 \end{bmatrix} + \mathbf{E}_1\tilde{\mathbf{B}}_{1,1}\mathbf{Z}_1$$

$$= \mathbf{X}_2\tilde{\mathbf{B}}_{2,2}\mathbf{F}_2 + (\mathbf{X}_2\mathbf{H}_{2,1} + \mathbf{X}_3\mathbf{H}_{3,1})\mathbf{Z}_1 + \mathbf{X}_2\mathbf{H}_{2,3}\mathbf{Z}_3 + \mathbf{E}_1\tilde{\mathbf{B}}_{1,1}\mathbf{Z}_1 \quad (30)$$

$$= \mathbf{X}_2(\tilde{\mathbf{B}}_{2,2}\mathbf{F}_2 + \mathbf{H}_{2,3}\mathbf{Z}_3) + (\mathbf{X}_2\mathbf{H}_{2,1} + \mathbf{X}_3\mathbf{H}_{3,1} + \mathbf{E}_1\tilde{\mathbf{B}}_{1,1})\mathbf{Z}_1$$

The ranks of block matrices $\mathbf{X}_2$ and $\mathbf{Z}_1$ are upper-bounded by m, which means $\mathbf{D}$ is the sum of two matrices of rank m, hence the maximum rank is $2m$.

For $\mathcal{B}'$, the rank of $\mathbf{D}$ is $2m+1$ whp. To see the difference in the analysis, in Eq. (29) the potential $\mathbf{H}_{3,3}$ block is non-zero, so whp $\mathbf{D}$ is not decomposable to the sum of 2 matrices of rank m like for $\mathcal{B}$.

The analysis of $\mathbf{D}'$ for both $\mathcal{B}$ and $\mathcal{B}'$ is analogous to the analysis of $\mathbf{D}$ in $\mathcal{B}$, i.e. in both cases the rank of $\mathbf{D}'$ is at most $2m$. So we are able to distinguish $\mathcal{B}$ and $\mathcal{B}'$ by obtaining $\mathbf{A} = [a_{i,j}]_{i,j}$ from picking $\zeta \geq 4m+1$ different i's and j's, and computing the rank of $\mathbf{A}$.

3.5   Discussions of Recent Immunizations

Recently Garg et al. [24] (merged from two similar proposals [25,36]) propose immunization mechanisms against the annihilation attack. The common feature of the immunizations is to pad random $2m$-by-$2m$ matrices instead of entries on the diagonal (i.e. change the matrices $\mathbf{V}_{i,b}$ and $\mathbf{V}'_{i,b}$ in Eq. (1) from diagonal to fully random), so as to encode a pseudorandom function in the noises. The difference of the two proposals lies in the ways to instantiate the paradigm.

We observe that the immunizations do not stop the attack if the branching program is input-partitioning. The observation does not contradict the proofs of security in the weakened idealized model from [24], since they require dual-input branching programs (which are not input-partitioning).

Below we briefly describe the two immunizations. In the immunization proposed by Miles, Sahai and Zhandry [36], the bookend vectors are changed to

$$\mathbf{J}, \mathbf{J}' \in \left[0^{2m}, \$^w \right], \quad \mathbf{L}, \mathbf{L}' \in \left[\$^{2m}, \$^w \right]^T \tag{31}$$

These changes do not affect the algorithms and analyses in Steps I and II. In Step III, the analysis of the matrix $\mathbf{H}$ in Eq. (21) remains the same. The analysis of the rank of $\mathbf{D}$ in Eq. (28) changes slightly. For program $\mathcal{B}$ in Example 1,

$$
\begin{aligned}
\mathbf{D} &= \mathbf{X}\tilde{\mathbf{B}}_y^\Delta \mathbf{F} + \mathbf{X}\mathbf{H}\mathbf{Z} + \mathbf{E}\tilde{\mathbf{B}}_y^\Delta \mathbf{Z} \\
&= \left[0 \ \mathbf{X}_2 \right] \begin{bmatrix} \tilde{\mathbf{B}}_{1,1} & 0 \\ 0 & 0 \end{bmatrix} \begin{bmatrix} \mathbf{F}_1 \\ \mathbf{F}_2 \end{bmatrix} + \left[0 \ \mathbf{X}_2 \right] \begin{bmatrix} \mathbf{H}_{1,1} & \mathbf{H}_{1,2} \\ \mathbf{H}_{2,1} & 0 \end{bmatrix} \begin{bmatrix} \mathbf{Z}_1 \\ \mathbf{Z}_2 \end{bmatrix} + \left[\mathbf{E}_1 \ \mathbf{E}_2 \right] \begin{bmatrix} \tilde{\mathbf{B}}_{1,1} & 0 \\ 0 & 0 \end{bmatrix} \begin{bmatrix} \mathbf{Z}_1 \\ \mathbf{Z}_2 \end{bmatrix} \tag{32} \\
&= 0 + \mathbf{X}_2 \mathbf{H}_{2,1} \mathbf{Z}_1 + \mathbf{E}_1 \tilde{\mathbf{B}}_{1,1} \mathbf{Z}_1 = (\mathbf{X}_2 \mathbf{H}_{2,1} + \mathbf{E}_1 \tilde{\mathbf{B}}_{1,1}) \mathbf{Z}_1
\end{aligned}
$$

where $\{\tilde{\mathbf{B}}_{k,\ell}, \mathbf{H}_{k,\ell}\}_{k,\ell \in [2]}$ are blocks of $\tilde{\mathbf{B}}_y^\Delta$, $\mathbf{H}$ with dimensions $[2m|w] \times [2m|w]$. $\{\mathbf{X}_k, \mathbf{E}_k\}_{k \in [2]}$ are blocks of $\mathbf{X}$, $\mathbf{E}$ with dimensions $\zeta \times [2m|w]$. $\{\mathbf{Z}_\ell, \mathbf{F}_\ell\}_{\ell \in [2]}$ are blocks of $\mathbf{Z}$, $\mathbf{F}$ with dimensions $[2m|w] \times \zeta$. The rank of $\mathbf{D}$ is thus upper-bounded by $2m$. For program $\mathcal{B}'$, the potential non-zero $\mathbf{H}_{2,2}$ contributes to an additional term $\mathbf{X}_2 \mathbf{H}_{2,2} \mathbf{Z}_2$. So the same algorithm from Sect. 3.3 distinguishes $\mathcal{B}$ and $\mathcal{B}'$.

More changes are made in the immunization proposed by Garg, Mukherjee and Srinivasan [25]. The plaintext space is set to be $R/\mathcal{J}$ where $\mathcal{J} = \langle g^2 \rangle$. The encoding of $s \in R/\mathcal{J}$ is a short representative of the coset $s + \mathbf{J}$. The zero-test parameter remains the same: $p_{zt} = \eta z^\kappa / g$. The bookend vectors are changed to

$$
\begin{aligned}
\mathbf{J}, \mathbf{J}' &:= \left[g \cdot \mathbf{J}_1, \mathbf{J}_2 \right], \left[g \cdot \mathbf{J}'_1, \mathbf{J}'_2 \right] \in \left[g \cdot \$^{2m}, \$^w \right] \\
\mathbf{L}, \mathbf{L}' &:= \left[\mathbf{L}_1, \mathbf{L}_2 \right]^T, \left[\mathbf{L}'_1, \mathbf{L}'_2 \right]^T \in \left[\$^{2m}, \$^w \right]^T.
\end{aligned} \tag{33}
$$

where $\mathbf{J}_2 \mathbf{L}_2 = \mathbf{J}'_2 \mathbf{L}'_2$. An honest evaluation analogous to Eq. (18) can be expressed as

$$\mathsf{Eval}(u) := \frac{\eta}{g} \left(\mathbf{J}(\prod_{k \in [h]} \mathbf{C}_{k,u_{\iota(k)}}) \mathbf{L} - \mathbf{J}'(\prod_{k \in [h]} \mathbf{C}'_{k,u_{\iota(k)}}) \mathbf{L}' \right) \tag{34}$$

$$= \frac{\eta}{g} \left(\prod_{k \in [h]} \alpha_{k,u_{\iota(k)}} \mathbf{J}_2 \mathbf{L}_2 - \prod_{k \in [h]} \alpha'_{k,u_{\iota(k)}} \mathbf{J}'_2 \mathbf{L}'_2 + g \cdot r_1(u) + g^2 \cdot r_2(u) + \dots \right)$$

where if u is a zero of the function then

$$\prod_{k\in[h]} \alpha_{k,u_{\iota(k)}} \mathbf{J}_2\mathbf{L}_2 - \prod_{k\in[h]} \alpha'_{k,u_{\iota(k)}} \mathbf{J}'_2\mathbf{L}'_2 = 0.$$

The immunization changes the coefficient of g^1 into

$$r_1 = \left(\mathbf{J}_1 \prod_{k\in[h]} \mathbf{V}_{k,u_{\iota(k)}} \mathbf{L}_1 - \mathbf{J}'_1 \prod_{k\in[h]} \mathbf{V}'_{k,u_{\iota(k)}} \mathbf{L}'_1 \right),$$

and pushes all the information about the secrets up to the coefficients of higher order terms. This is the rationale of Garg, Mukherjee and Srinivasan's [25] immunization against annihilation attacks.

Still, for branching programs with input-partitioning, these immunizations do not affect the algorithms and the analyses in Steps I and II, except that we obtain a basis of $\langle g^2 \rangle$ and (possibly big) representatives of scalars α in the coset $\alpha + \langle g^2 \rangle$. In Step III, we analyze $\mathbf{H}$ and $\mathbf{A}$ modulo $\mathcal{J}$ instead of modulo $\mathcal{I}$. The feature of $\mathbf{H}$ remains the same. To $\mathbf{A}$, the expression of each $a_{i,j}$ from Eq. (27) shall be modified to (the following expression still contains coefficients of g^2 that will be removed later)

$$a_{i,j} = \underbrace{\frac{\eta}{g} \left(\mathbf{x}_i \tilde{\mathbf{B}}_{\tilde{y}}^{\Delta} \mathbf{z}_j + g^2 \cdot \left(\mathbf{x}_i \tilde{\mathbf{B}}_{\tilde{y}}^{\Delta} \mathbf{f}_j + \mathbf{x}_i \mathbf{H} \mathbf{z}_j + \mathbf{e}_i \tilde{\mathbf{B}}_{\tilde{y}}^{\Delta} \mathbf{z}_j \right) \right)}_{:=d_{i,j}} \qquad (35)$$

$$\underbrace{- \frac{\eta}{g} \left(\mathbf{x}'_i \tilde{\mathbf{B}}'^{\Delta}_{\tilde{y}} \mathbf{z}'_j + g^2 \left(\mathbf{x}'_i \tilde{\mathbf{B}}'^{\Delta}_{\tilde{y}} \mathbf{f}'_j + \mathbf{x}'_i \mathbf{H}' \mathbf{z}'_j + \mathbf{e}'_i \tilde{\mathbf{B}}'^{\Delta}_{\tilde{y}} \mathbf{z}'_j \right) \right)}_{:=d'_{i,j}} \pmod{\mathcal{J}}.$$

Examining the functional component $\mathbf{D}$ for $\mathcal{B}$, with the same blocked dimensions as Eq. (32):

$$\mathbf{D} = \frac{\eta}{g} \left(\mathbf{X} \tilde{\mathbf{B}}_{\tilde{y}}^{\Delta} \mathbf{Z} + g^2 \cdot \left(\mathbf{X} \tilde{\mathbf{B}}_{\tilde{y}}^{\Delta} \mathbf{F} + \mathbf{X} \mathbf{H} \mathbf{Z} + \mathbf{E} \tilde{\mathbf{B}}_{\tilde{y}}^{\Delta} \mathbf{Z} \right) \right)$$

$$= \frac{\eta}{g} \begin{bmatrix} g\mathbf{X}_1 & \mathbf{X}_2 \end{bmatrix} \begin{bmatrix} \tilde{\mathbf{B}}_{1,1} & 0 \\ 0 & 0 \end{bmatrix} \begin{bmatrix} \mathbf{Z}_1 \\ \mathbf{Z}_2 \end{bmatrix} + \eta g \left(\begin{bmatrix} g\mathbf{X}_1 & \mathbf{X}_2 \end{bmatrix} \begin{bmatrix} \tilde{\mathbf{B}}_{1,1} & 0 \\ 0 & 0 \end{bmatrix} \begin{bmatrix} \mathbf{F}_1 \\ \mathbf{F}_2 \end{bmatrix} \right.$$

$$\left. + \begin{bmatrix} g\mathbf{X}_1 & \mathbf{X}_2 \end{bmatrix} \begin{bmatrix} \mathbf{H}_{1,1} & \mathbf{H}_{1,2} \\ \mathbf{H}_{2,1} & 0 \end{bmatrix} \begin{bmatrix} \mathbf{Z}_1 \\ \mathbf{Z}_2 \end{bmatrix} + \begin{bmatrix} \mathbf{E}_1 & \mathbf{E}_2 \end{bmatrix} \begin{bmatrix} \tilde{\mathbf{B}}_{1,1} & 0 \\ 0 & 0 \end{bmatrix} \begin{bmatrix} \mathbf{Z}_1 \\ \mathbf{Z}_2 \end{bmatrix} \right) \qquad (36)$$

$$= \eta \mathbf{X}_1 \tilde{\mathbf{B}}_{1,1} \mathbf{Z}_1 + \eta g (\mathbf{X}_2 \mathbf{H}_{2,1} + \mathbf{E}_1 \tilde{\mathbf{B}}_{1,1}) \mathbf{Z}_1 + \eta g^2 (\dots)$$

$$= \eta \left(\mathbf{X}_1 \tilde{\mathbf{B}}_{1,1} + g (\mathbf{X}_2 \mathbf{H}_{2,1} + \mathbf{E}_1 \tilde{\mathbf{B}}_{1,1}) \right) \mathbf{Z}_1 \pmod{\mathcal{J}}.$$

The rest of the analysis is analogous. The rank of $\mathbf{A}$ modulo $\mathcal{J}$ distinguishes $\mathcal{B}$ and $\mathcal{B}'$.

4   Cryptanalysis of the GGH15-Based Candidate

4.1   The GGH15 Encoding Scheme

We use here notations similar to [29] for "GGH15 with safeguards". The encoding scheme from [26] is parametrized by a directed graph $G = (V, E)$ (with a single sink) and some integer parameters k, n, r, q (with $r \gg k$). Its plaintext space are matrices $\mathbf{S} \in R^{k \times k}$ (whose entries must be much smaller than q), and the encodings themselves are matrices $\mathbf{D} \in (R_n/qR_n)^{r \times r}$, and both plaintext and encoding matrices are associated with edges (or paths) in the graph.

For each vertex u in the graph we choose a random matrix $\mathbf{A}_u \in R_n^{k \times r}$ together with some trapdoor information τ_u [1,28,33], and another random invertible matrix $\mathbf{P}_u \in (R_n/qR_n)^{r \times r}$. For the source s and sink t we choose random small "bookend vectors" $\mathbf{J}_s$ and $\mathbf{L}_t$ and publish the two transformed vectors $\tilde{\mathbf{J}}_s := \mathbf{J}_s \cdot \mathbf{A}_s \cdot \mathbf{P}_s^{-1} \pmod q$ and $\tilde{\mathbf{L}}_t := \mathbf{P}_t \cdot \mathbf{L}_t$, to be used for zero-testing.

To encode a matrix $\mathbf{S} \in R^{k \times k}$ w.r.t. a path $(u \rightsquigarrow v)$, sample a low-norm error matrix $\mathbf{E} \in R_n^{k \times r}$, use the trapdoor τ_u to sample a small solution $\mathbf{D}$ to $\mathbf{A}_u \mathbf{D} = \mathbf{S}\mathbf{A}_v + \mathbf{E} \pmod q$, and finally output the encoding matrix $\mathbf{C} := \mathbf{P}_u \mathbf{D} \mathbf{P}_v^{-1} \bmod q$.

This scheme supports adding encoded matrices relative to the same path, and multiplying matrices relative to consecutive paths (with the result being defined relative to the concatenation of the two paths). The encoding invariant is that an encoding $\mathbf{C}$ of plaintext matrix $\mathbf{S}$ relative to the path $u \rightsquigarrow v$ satisfies $\mathbf{A}_u \cdot (\mathbf{P}_u^{-1} \mathbf{C} \mathbf{P}_v) = \mathbf{S}\mathbf{A}_v + \mathbf{E} \pmod q$ where $\mathbf{S}, \mathbf{E}$ and $\mathbf{D} := \mathbf{P}_u^{-1} \mathbf{C} \mathbf{P}_v \bmod q$ all have norm much smaller than q. The encoding scheme also supports a zero-test of encoding $\mathbf{C}$ relative to a path $s \rightsquigarrow t$, by checking that $\tilde{\mathbf{J}}_s \mathbf{C} \tilde{\mathbf{L}}_t$ is small, which holds when $\mathbf{S} = 0$ since $\tilde{\mathbf{J}}_s \mathbf{C} \tilde{\mathbf{L}}_t = \mathbf{J}_s \mathbf{A}_s \mathbf{P}_s^{-1} \cdot \mathbf{C} \cdot \mathbf{P}_t \mathbf{L}_t = \mathbf{J}_s (\mathbf{S}\mathbf{A}_s + \mathbf{E})\mathbf{L}_t = \mathbf{J}_s \mathbf{E} \mathbf{L}_t \pmod q$.

Consider two consecutive paths $s \rightsquigarrow u$ and $u \rightsquigarrow t$ and two encoding matrices $\mathbf{C}_1, \mathbf{C}_2$, encoding $\mathbf{S}_1, \mathbf{S}_2$ relative to these two paths, respectively. Then $\mathbf{C}_1 \mathbf{C}_2$ is an encoding of $\mathbf{S}_1, \mathbf{S}_2$ relative to $s \rightsquigarrow t$, which means that $\mathbf{A}_s \cdot (\mathbf{P}_s^{-1} \mathbf{C}_1 \mathbf{C}_2 \mathbf{P}_t) = \mathbf{S}_1 \mathbf{S}_2 \mathbf{A}_t + \mathbf{E}' \pmod q$, but we can say more about the structure of the resulting noise $\mathbf{E}'$. Specifically, it is not hard to verify that (after zero-testing) we have

$$\tilde{\mathbf{J}}_s \mathbf{C}_1 \mathbf{C}_2 \tilde{\mathbf{L}}_t = \mathbf{J}_s (\mathbf{S}_1 \mathbf{S}_2 \mathbf{A}_t + \underbrace{\mathbf{S}_1 \mathbf{E}_2 + \mathbf{E}_1 \mathbf{D}_2}_{\mathbf{E}'}) \mathbf{L}_t$$

$$= \mathbf{J}_s \cdot [\mathbf{S}_1 | \mathbf{E}_1] \begin{bmatrix} \mathbf{S}_2 \mathbf{A}_t + \mathbf{E}_2 \\ \mathbf{D}_2 \end{bmatrix} \cdot \mathbf{L}_t \pmod q, \tag{37}$$

where $\mathbf{J}_s, \mathbf{L}_t$ are the bookend vectors, $\mathbf{E}_1, \mathbf{E}_2$ are the error matrices corresponding to the encoding $\mathbf{C}_1, \mathbf{C}_2$ respectively, and $\mathbf{D}_2 = \mathbf{P}_u^{-1} \mathbf{C}_2 \mathbf{P}_t$ (all of which have low norm). Similarly, if we have three intervals $s \rightsquigarrow u$, $u \rightsquigarrow v$ and $u \rightsquigarrow t$ and three encoding matrices $\mathbf{C}_1, \mathbf{C}_2, \mathbf{C}_3$ for $\mathbf{S}_1, \mathbf{S}_2, \mathbf{S}_3$ relative to these paths, respectively, then

$$\tilde{\mathbf{J}}_s \mathbf{C}_1 \mathbf{C}_2 \mathbf{C}_3 \tilde{\mathbf{L}}_t = \mathbf{J}_s \cdot [\mathbf{S}_1, \mathbf{E}_1] \begin{bmatrix} \mathbf{S}_2, \mathbf{E}_2 \\ 0, \mathbf{D}_2 \end{bmatrix} \begin{bmatrix} \mathbf{S}_3 \mathbf{A}_t + \mathbf{E}_3 \\ \mathbf{D}_3 \end{bmatrix} \cdot \mathbf{L}_t \pmod q. \tag{38}$$

The GGHRSW Obfuscator Over GGH15. When using the GGH15 encoding scheme in the context of the GGHRSW obfuscator, we use a simple graph with two parallel chains leading to a sink. One minor technical issue to reconcile is that the plaintext space of GGH15 consists of only $k \times k$ matrices, while the GGHRSW construction needs to also encode the bookend vectors $\mathbf{J}, \mathbf{L}, \mathbf{J}', \mathbf{L}'$. This is best handled by combining these bookends with the GGH15 bookends $\mathbf{J}_s, \mathbf{L}_t$ from above. Namely we choose the bookends as matrices rather than vectors (but still keep the same structure for the rows/columns of these matrices), and then these matrices will be multiplied by the GGH15 bookends $\mathbf{J}_s, \mathbf{L}_t$ during zero-test, resulting in vectors $\mathbf{J}, \mathbf{L}, \mathbf{J}', \mathbf{L}'$ with the same structure as in Eq. (2).

Another technical issue is that the GGH15 plaintext matrices must be small, whereas the GGHRSW construction requires that we multiply the plaintext matrices by the Kilian randomization matrices $\mathbf{K}, \mathbf{K}^{-1}$. Gentry et al. describe in [26, Sect. 5.2.1] a method for choosing "random matrices" where both $\mathbf{K}, \mathbf{K}^{-1}$ are small, but in fact a closer look at the error terms that we get reveals that the construction will still work even if only $\mathbf{K}^{-1}$ was small but $\mathbf{K}$ was not (as long as as we set $\mathbf{K}_0 = \mathbf{I}$). We stress that the structure of $\mathbf{K}$ plays no role in our attacks, so in the rest of the manuscript we ignore this issue.

4.2   Overview of Our Attacks on the GGH15-Based Obfuscator

The main ingredient in our attack on the GGH15-based branching-program obfuscator is a method to recover some information about the scalars $\alpha_{i,b}$ (and $\alpha'_{i,b}$) that are used in this construction. Specifically, we use a zeroing technique adapted from the work of Coron, Lee, Lepoint and Tibouchi [17], to recover the ratios of (the products of) these $\alpha_{i,b}$'s for some equivalent subbranches, as we describe in Sect. 4.3 below. (Setting up the CLLT-style system of equations relies on the input-partitioning feature of underlying branching program.)

This step is completely algebraic, and hence the ratios that we recover do not give us a small representation of these scalars. Namely, while we learn the ratio β/γ for some small β, γ (each of them is a product of some α's), we do not recover the small β, γ themselves.

One way to mount a full attack to the obfuscator is to directly use factoring and principle-ideal-problem solvers to recover the $\alpha_{i,b}$'s from the known ratio β/γ. Once the bundling scalars $\alpha_{i,b}$ are known, we can mount an input-mixing attack to break the obfuscation. This yields classical sub-exponential time or quantum polynomial time attacks.

4.3   Step I: Recovering Ratios of the Bundling Scalars

Step I.1: Accumulating CLLT-Style Equations. Let $\mathcal{X} || \mathcal{Z} = [h]$ be a 2-partition of the branching program steps. Below we use honest evaluation of the branching program on many inputs of the form $u^{(i,j)} = x^{(i)} z^{(j)}$, where all the bits that only affect steps in $\mathcal{X}$ are in the $x^{(i)}$ part, all the bits that only affect steps in $\mathcal{Z}$ are in the $z^{(j)}$ part, and all the other bits are fixed. This notation *does not mean* that all the bits of $x^{(i)}$ must come before all the bits of $z^{(j)}$ in $u^{(i,j)}$, but it does mean that $u^{(i,j)}$ and $u^{(i,j')}$ can only differ in bits that affect steps in $\mathcal{Z}$, and similarly $u^{(i,j)}$ and $u^{(i',j)}$ can only differ in bits that affect steps in $\mathcal{X}$.

For such an input $u = xz$, we denote by $\mathbf{S}_x$ the plaintext product matrix of functional branch in the the $\mathcal{X}$ interval and by $\mathbf{S}_z$ the plaintext product matrix of the functional branch in the $\mathcal{Z}$ interval (including the bookends), and similarly for encodings $\mathbf{C}_x, \mathbf{C}_z$ and for the dummy branch. That is, we denote

$$\mathbf{S}_x := \mathbf{J} \cdot (\textstyle\prod_{i \in \mathcal{X}} \mathbf{S}_{i,u_\iota(i)}), \ \mathbf{S}_z := (\prod_{i \in \mathcal{Z}} \mathbf{S}_{i,u_\iota(i)}) \cdot \mathbf{L},$$

$$\mathbf{S'}_x := \mathbf{J'} \cdot (\textstyle\prod_{i \in \mathcal{X}} \mathbf{S'}_{i,u_\iota(i)}), \ \mathbf{S'}_z := (\prod_{i \in \mathcal{Z}} \mathbf{S'}_{i,u_\iota(i)}) \cdot \mathbf{L'},$$

$$\mathbf{C}_x := \tilde{\mathbf{J}} \cdot (\textstyle\prod_{i \in \mathcal{X}} \mathbf{C}_{i,u_\iota(i)}), \ \mathbf{C}_z := (\prod_{i \in \mathcal{Z}} \mathbf{C}_{i,u_\iota(i)}) \cdot \tilde{\mathbf{L}},$$

$$\mathbf{C'}_x := \tilde{\mathbf{J}}' \cdot (\textstyle\prod_{i \in \mathcal{X}} \mathbf{C'}_{i,u_\iota(i)}), \ \mathbf{C'}_z := (\prod_{i \in \mathcal{Z}} \mathbf{C'}_{i,u_\iota(i)}) \cdot \tilde{\mathbf{L}}'$$

with the encoding arithmetic modulo q. We also denote by $\mathbf{E}_x, \mathbf{E}_z, \mathbf{E'}_x, \mathbf{E'}_z$ the error matrices in $\mathbf{C}_x, \mathbf{C}_z, \mathbf{C'}_x, \mathbf{C'}_z$ and

$$\mathbf{D}_x := \mathbf{C}_x \mathbf{P}_v, \ \mathbf{D'}_x := \mathbf{C'}_x \mathbf{P}_{v'}, \ \mathbf{D}_z := \mathbf{P}_v^{-1} \mathbf{C}_z \text{ and } \mathbf{D'}_z := \mathbf{P}_{v'}^{-1} \mathbf{C'}_z$$

(where v, v' are the vertices between $\mathcal{X}, \mathcal{Z}$ on the functional and dummy branches).

Following Eq. (37) above, the honest evaluation of branching program on input $u = xz$ yields the element

$$\mathbf{w} := \mathbf{C}_x \mathbf{C}_z - \mathbf{C'}_x \mathbf{C'}_z = [\mathbf{S}_x, \mathbf{E}_x, -\mathbf{S'}_x, -\mathbf{E'}_x] \begin{bmatrix} \mathbf{S}_z \mathbf{A}_t + \mathbf{E}_z \\ \mathbf{D}_z \\ \mathbf{S'}_z \mathbf{A}_t + \mathbf{E'}_z \\ \mathbf{D'}_z \end{bmatrix} \pmod{q}. \quad (39)$$

If $u = xz$ is a zero of the function, then by construction we have $\mathbf{S}_x \mathbf{S}_z = \mathbf{S'}_x \mathbf{S'}_z = \beta \mathbf{J} \mathbf{L}$ for some scalar β, and in this case Eq. (39) holds over the base ring R_n, not just modulo q.

We begin the attack by collecting many instances of Eq. (39) for many $x^{(i)}$'s and $z^{(j)}$'s for which $u^{(i,j)} = x^{(i)} z^{(j)}$ is a zero, and put the corresponding w elements in a matrix. This yields the matrix equation:

$$\mathbf{W} := \mathbf{XZ} := \begin{bmatrix} \mathbf{S}_{x^{(1)}}, \mathbf{E}_{x^{(1)}}, -\mathbf{S'}_{x^{(1)}}, -\mathbf{E'}_{x^{(1)}} \\ \dots, \dots, \dots, \ \dots \\ \mathbf{S}_{x^{(i)}}, \mathbf{E}_{x^{(i)}}, -\mathbf{S'}_{x^{(i)}}, -\mathbf{E'}_{x^{(i)}} \\ \dots, \dots, \dots, \ \dots \\ \mathbf{S}_{x^{(k)}}, \mathbf{E}_{x^{(k)}}, -\mathbf{S'}_{x^{(k)}}, -\mathbf{E'}_{x^{(k)}} \end{bmatrix} \begin{bmatrix} \dots, \mathbf{S}_{z^{(j)}} \mathbf{A}_t + \mathbf{E}_{z^{(j)}}, \dots \\ \dots, \mathbf{D}_{z^{(j)}}, \quad \dots \\ \dots, \mathbf{S'}_{z^{(j)}} \mathbf{A}_t + \mathbf{E'}_{z^{(j)}}, \dots \\ \dots, \mathbf{D'}_{z^{(j)}}, \quad \dots \end{bmatrix}.$$

$$(40)$$

Since all the inputs are zeros of the function, then Eq. (40) holds not only modulo q but also over the base ring R_n. As discussed in [29, Sect. 5.2], the matrix $\mathbf{Z}$ is inherently non-full-rank when considered modulo q, but will have full rank over R_n with high probability (since the $\mathbf{E}_z$'s are random and the $\mathbf{D}_z$'s are chosen at random in some cosets after the $\mathbf{E}_z$'s are fixed). Taking sufficiently many $z^{(j)}$'s, we can therefore ensure that the left kernel of $\mathbf{Z}$ is trivial, i.e., consisting of only the all-zero vector.[4] On the other hand, by using enough $x^{(i)}$'s we can ensure that the left-kernel of $\mathbf{W}$ (and therefore of $\mathbf{X}$) is non-trivial. The thrust of our attack will consist of collecting many vectors in this left-kernel, and using them to recover information about (ratios of) the $\alpha_{i,b}$'s.

Step I.2: Computing the Left-Kernel of $\mathbf{W}$. The CLLT-type attack computes the left-kernel (abbreviated as kernel in the rest of this paper) of $\mathbf{W}$, i.e. vectors $\mathbf{p}$ over R_n s.t. $\mathbf{p}\mathbf{W} = \mathbf{0}$. Since $\mathbf{Z}$ has full rank, then such vector $\mathbf{p}$ must also be in the kernel of $\mathbf{X}$, so it is orthogonal to all its columns. In our attack we only use the fact that these vectors $\mathbf{p}$ are orthogonal to the $\mathbf{S}$'s parts of $\mathbf{X}$, namely we denote

$$\mathbf{Q} := \begin{bmatrix} \mathbf{S}_{x^{(1)}}, & -\mathbf{S}'_{x^{(1)}} \\ \cdots, & \cdots \\ \mathbf{S}_{x^{(i)}}, & -\mathbf{S}'_{x^{(i)}} \\ \cdots, & \cdots \\ \mathbf{S}_{x^{(k)}}, & -\mathbf{S}'_{x^{(k)}} \end{bmatrix} \tag{41}$$

and use the fact that every vector in the kernel of $\mathbf{X}$ must be in particular also in the kernel of $\mathbf{Q}$. We next recall the structure of $\mathbf{S}_{x^{(i)}}, \mathbf{S}'_{x^{(i)}}$ from Eq. (3), namely we have

$$\begin{aligned} \mathbf{S}_{x^{(i)}} &= \alpha_{x^{(i)}}\mathbf{J} \times \mathrm{diag}(u_{x^{(i)}}, v_{x^{(i)}}, \mathbf{B}_{x^{(i)}}) \times \mathbf{K}_z; \\ \mathbf{S}'_{x^{(i)}} &= \alpha'_{x^{(i)}}\mathbf{J}' \times \mathrm{diag}(u'_{x^{(i)}}, v'_{x^{(i)}}, \mathbf{I}) \times \mathbf{K}'_z \end{aligned} \tag{42}$$

where $\mathbf{B}_{x^{(i)}}$ is the product of the branching-program matrices $\mathbf{B}_{i,b}$ over the interval $\mathcal{X}$, $u_{x^{(i)}}, v_{x^{(i)}}, u'_{x^{(i)}}, v'_{x^{(i)}}$ are the random diagonal entries, $\alpha_{x^{(i)}}, \alpha'_{x^{(i)}}$ are the products of the α's on both the functional and dummy branches, and $\mathbf{K}_z, \mathbf{K}'_z$ are the Kilian randomization matrix at the beginning of the $\mathcal{Z}$ interval on the two branches.

Importantly, since all the $x^{(i)}z^{(j)}$'s are zeros of the function, then by Lemma 1 all the $\mathbf{B}_{x^{(i)}}$'s must be equal. We denote that matrix simply by $\mathbf{B}$, namely we have $\mathbf{S}_{x^{(i)}} = \alpha_{x^{(i)}}\mathbf{J} \times \mathrm{diag}(u_{x^{(i)}}, v_{x^{(i)}}, \mathbf{B}) \times \mathbf{K}_z$ for all i. Moreover, all the ratios of $\alpha_{x^{(i)}}/\alpha'_{x^{(i)}}$, $i \in [k]$ must also be equal due to Lemma 2, and below we denote that ratio by δ.

We can therefore re-write Eq. (5) as follows:

$$\forall i \in [k],\ [\mathbf{S}_{x^{(i)}}, -\mathbf{S}'_{x^{(i)}}] \tag{43}$$

$$= \alpha_{x^{(i)}}[\mathbf{J} \times \mathrm{diag}(u_{x^{(i)}}, v_{x^{(i)}}, \mathbf{B}) \times \mathbf{K}_z, -\delta\mathbf{J}' \times \mathrm{diag}(u'_{x^{(i)}}, v'_{x^{(i)}}, \mathbf{I}) \times \mathbf{K}'_z]$$

$$= \alpha_{x^{(i)}} \cdot [\ \underbrace{0, \tilde{v}_{x^{(i)}}, \mathbf{b}}_{\mathbf{J}\times\mathrm{diag}(u_{x^{(i)}}, v_{x^{(i)}}, \mathbf{B})}\ ,\ \underbrace{0, \tilde{v}'_{x^{(i)}}, \mathbf{b}'}_{\mathbf{J}'\times\mathrm{diag}(u'_{x^{(i)}}, v'_{x^{(i)}}, \mathbf{I})}\ ] \times \underbrace{\tilde{K}_z}_{\text{an invertible matrix}}$$

---

[4] With typical parameters it is sufficient to use only four different $z^{(j)}$'s for that purpose.

(recall that by design, the first columns of $\mathbf{J}$ and $\mathbf{J}'$ are zero to erase the u and u' terms on the diagonal). The only $x^{(i)}$-sensitive terms are $\alpha_{x^{(i)}}$, $\tilde{v}_{x^{(i)}}$, and $\tilde{v}'_{x^{(i)}}$, and thus the rank of Q is exactly 3. The Kernel of $\mathbf{Q}$ therefore has dimension $k - 3$, and it is contained in the ($(k-1)$-dimensional) space spanned by the vectors $[\alpha_{x^{(2)}}, -\alpha_{x^{(1)}}, 0, \ldots, 0]$, $[\alpha_{x^{(3)}}, 0, -\alpha_{x^{(1)}}, \ldots, 0]$, $\ldots [\alpha_{x^{(k)}}, 0, 0, \ldots, -\alpha_{x^{(1)}}]$. In other words, every vector $\mathbf{p} = [p_1, p_2, \ldots, p_k]$ in this kernel must in particular satisfy the condition $\sum_i p_i \alpha_{x^{(i)}} = 0$.

Of course, the kernel of $\mathbf{X}$ (which is the linear space that our attack can recover) is only a subspace of the kernel of $\mathbf{Q}$, and hence it has an even lower dimension. However, the difference in dimension between $\mathsf{kernel}(\mathbf{Q})$ and $\mathsf{kernel}(\mathbf{X})$ is bounded by the dimensions of the error matrices $\mathbf{E}_{x^{(i)}}, \mathbf{E}'_{x^{(i)}}$, which is independent of the number of $x^{(i)}$'s. Namely, the number of columns in $\mathbf{E}_{x^{(i)}}, \mathbf{E}'_{x^{(i)}}$ together is only $2r$, hence the dimension of $\mathsf{kernel}(\mathbf{X})$ is at least $\dim(\mathsf{kernel}(\mathbf{Q})) - 2r = k - 3 - 2r$, where k is the number of $x^{(i)}$'s. If we have enough zeros of the branching-program, then we can take k to be much much larger than $2r + 3$.

Step I.3: Extracting the Ratios. The kernel of $\mathbf{W}$ (or equivalently of $\mathbf{X}$) is a subspace of dimension at least $k - (2r + 3)$, all of which is orthogonal to the vector of $\alpha_{x^{(i)}}$'s. However, we do not have enough equations to recover the $\alpha_{x^{(i)}}$'s themselves, since there are k of them and we only have $k - (2r + 3)$ equations. Here we take advantage of the fact that the $\alpha_{x^{(i)}}$'s are not really k independent variables, rather each $\alpha_{x^{(i)}}$'s is obtained as a subset product of the $\alpha_{i,b}$'s that are used in the construction.

Specifically, let $J_x \subset [\ell]$ be set of input bits that only affect steps of the branching program in the interval $\mathcal{X}$, and for any just input bit $j \in J_x$ let us denote $\beta_{j,0} = \prod_{\iota(i')=j} \alpha_{i',0}$ and $\beta_{j,1} = \prod_{\iota(i')=j} \alpha_{i',1}$. Also, recalling that all the input bits outside J_x are fixed, we denote by β_0 the product of the $\alpha_{i',b}$ scalars that are used in all the steps that are not controlled by bits in J_x. Then every $\alpha_{x^{(i)}}$ can be written as a subset product

$$\alpha_{x^{(i)}} = \beta_0 \cdot \prod_{j,b} \beta_{j,b}^{e(i,j,b)}$$

where the exponents $e(i,j,b)$ are all in $\{0,1\}$. This implies in particular that the number of $\alpha_{x^{(i)}}$ is at most $2^{2|J_x|}$.

Consider now what happens if we take all the products of two equations from the kernel. This will give us a set of at least $(k - 2r - 3)^2$ equations in the product variables $\gamma_{i_1,i_2} = \alpha_{x^{(i_1)}} \cdot \alpha_{x^{(i_2)}}$. But the γ_{i_1,i_2} are perhaps not all distinct: each of them can be written as a product $\gamma_{i_1,i_2} = \beta_0^2 \cdot \prod_{j,b} \beta_{j,b}^{e(i_1,i_2,j,b)}$ with the exponents in $\{0,1,2\}$, so the total number of distinct γ_{i_1,i_2} is at most $3^{2|J_x|}$ (which is smaller than $(2^{2|J_x|})^2$).

More generally, we can take products of upto c of our equations, and this will give us at least $(k - 2r - 3)^c$ equations, but the number of variables will still be upper-bounded by $(c + 1)^{2|J_x|}$. If for some constant c we get $(k - 2r - 3)^c > (c + 1)^{2|J_x|}$, then we have more equations than variables (which heuristically should still be linearly independent[5]) and we can solve the system and recover all the products $\gamma_{i_1,i_2,\ldots,i_c}$.

For example, in the extreme case where *every setting* of the input bits in J_x yields a zero of the function, we can collect as many as $k = 2^{|J_x|}$ equations from the kernel. If in addition $|J_x| > 1 + \log(2r+3)$ then $k > 2 \cdot (2r+3)$ and therefore $k - 2r - 3 > k/2 = 2^{|J_x|-1}$. In this case taking $c = 7$ is sufficient to get more equations than variables, since

$$(k - 2r - 3)^c = (k - 2r - 3)^7 > 2^{7(|J_x|-1)} > 2^{6|J_x|} = 8^{2|J_x|} = (c+1)^{2|J_x|}$$

as needed. We note that in this extreme case, we can get more equations than variables already when multiplying pairs of equations (i.e. let $c = 2$) from the kernel if we are careful about which pairs to multiply.

Once we have all the γ's, we can divide them by each other to get ratios of smaller products of the $\beta_{j,b}$'s from above (which are in turn products of the $\alpha_{i',b}$'s from the construction). In particular we can get ratios of individual β's, of the form $\beta_{j,b}/\beta_{j',b'}$, but we cannot get any better granularity. In particular we cannot separate the different $\alpha_{i',b}$ that are multiplied to form the $\beta_{j,b}$'s.

4.4   Step II: Attacking the Obfuscator

If we have a quantum computer, or we are willing to run a classical subexponential-time attack, we can implement a factoring oracle and a principal-ideal-problem solver, using [9,10,20,30,38]. Together, these solvers make it possible to recover (some of) the small scalars $\alpha_{i,b}$, $\alpha_{i',b'}$ from the ratios $\beta_{j,b}/\beta_{j',b'}$. Once we have these $\alpha_{i,b}$'s, we can use them in mixed-input attacks on the obfuscator. Namely, in some steps that are controlled by the j'th input bit we take the 0 matrix, and in some other steps we take the 1 matrix, and this lets us (at least) check if these two matrices are the same.

In GGH15 (following GGHRSW), the small bundling scalars $\alpha_{i,b}$ and $\alpha'_{i,b}$ s.t. $\prod \alpha_{i,b} = \prod \alpha'_{i,b}$ are chosen by first generating a set of random small ζ_k's, let each $\alpha_{i,b}$ be a product of one or two of these ζ_k's, so that each of the product $\prod \alpha_{i,b}, \prod \alpha'_{i,b}$ correspond to the same subset of ζ_k's. A factoring oracle will recover the ideals generated by all the ζ_k's that happen to be prime (which happens with noticeable probability). Then a PIP solver will find the small ζ_k's themselves.[6] As each $\alpha_{i,b}$ is a product of very few of the ζ_k's, we would get some of the $\alpha_{i,b}$'s by trying all singletons and pairs of the ζ_k's.

Acknowledgments. We thank Rex Fernando, Peter Rasmussen, Amit Sahai, and the anonymous reviewers for their comments on a previous version of this report. Y.C. is supported by NSF grants CNS1012798, CNS1012910, CNS1413920 and AF1218461.

---

[5] This heuristic argument is similar to the one used for the multi-variate Coppersmith attack [15].

[6] While the PIP solver will succeed in finding the prime ζ_k's, it will likely fail when applied to factors of the non-prime ζ_k's (since those are unlikely to be principal and/or very small).

Part of the research conducted while visiting Tel Aviv University, IST Austria, and IBM Thomas J. Watson Research Center. C.G. and S.H. are supported by the Defense Advanced Research Projects Agency (DARPA) and Army Research Office(ARO) under Contract No. W911NF-15-C-0236.

References

1. Ajtai, M.: Generating hard instances of the short basis problem. In: Wiedermann, J., Emde Boas, P., Nielsen, M. (eds.) ICALP 1999. LNCS, vol. 1644, pp. 1–9. Springer, Heidelberg (1999). doi:10.1007/3-540-48523-6_1

2. Albrecht, M., Bai, S., Ducas, L.: A subfield lattice attack on overstretched NTRU assumptions. In: Robshaw, M., Katz, J. (eds.) CRYPTO 2016. LNCS, vol. 9814, pp. 153–178. Springer, Heidelberg (2016). doi:10.1007/978-3-662-53018-4_6

3. Ananth, P., Gupta, D., Ishai, Y., Sahai, A.: Optimizing obfuscation: avoiding Barrington's theorem. In: Proceedings of the ACM SIGSAC Conference on Computer and Communications Security, pp. 646–658. ACM (2014)

4. Ananth, P., Jain, A.: Indistinguishability obfuscation from compact functional encryption. In: Gennaro, R., Robshaw, M. (eds.) CRYPTO 2015. LNCS, vol. 9215, pp. 308–326. Springer, Heidelberg (2015). doi:10.1007/978-3-662-47989-6_15

5. Apon, D., Döttling, N., Garg, S., Mukherjee, P.: Cryptanalysis of indistinguishability obfuscations of circuits over GGH13. Cryptology ePrint Archive, report 2016/1003 (2016)

6. Applebaum, B., Brakerski, Z.: Obfuscating circuits via composite-order graded encoding. In: Dodis, Y., Nielsen, J.B. (eds.) TCC 2015. LNCS, vol. 9015, pp. 528–556. Springer, Heidelberg (2015). doi:10.1007/978-3-662-46497-7_21

7. Badrinarayanan, S., Miles, E., Sahai, A., Zhandry, M.: Post-zeroizing obfuscation: new mathematical tools, and the case of evasive circuits. In: Fischlin, M., Coron, J.-S. (eds.) EUROCRYPT 2016. LNCS, vol. 9666, pp. 764–791. Springer, Heidelberg (2016). doi:10.1007/978-3-662-49896-5_27

8. Barak, B., Garg, S., Kalai, Y.T., Paneth, O., Sahai, A.: Protecting obfuscation against algebraic attacks. In: Nguyen, P.Q., Oswald, E. (eds.) EUROCRYPT 2014. LNCS, vol. 8441, pp. 221–238. Springer, Heidelberg (2014). doi:10.1007/978-3-642-55220-5_13

9. Biasse, J.-F., Fieker, C.: Subexponential class group, unit group computation in large degree number fields. LMS J. Comput. Math. 17(A), 385–403 (2014)

10. Biasse, J.F., Song, F.: Efficient quantum algorithms for computing class groups and solving the principal ideal problem in arbitrary degree number fields. In: Proceedings of the Twenty-Seventh Annual ACM-SIAM Symposium on Discrete Algorithms, pp. 893–902. SIAM (2016)

11. Bitansky, N., Vaikuntanathan, V.: Indistinguishability obfuscation from functional encryption. In: FOCS, pp. 171–190. IEEE Computer Society (2015)

12. Brakerski, Z., Rothblum, G.N.: Virtual black-box obfuscation for all circuits via generic graded encoding. In: Lindell, Y. (ed.) TCC 2014. LNCS, vol. 8349, pp. 1–25. Springer, Heidelberg (2014). doi:10.1007/978-3-642-54242-8_1

13. Cheon, J.H., Han, K., Lee, C., Ryu, H., Stehlé, D.: Cryptanalysis of the multilinear map over the integers. In: Oswald, E., Fischlin, M. (eds.) EUROCRYPT 2015. LNCS, vol. 9056, pp. 3–12. Springer, Heidelberg (2015). doi:10.1007/978-3-662-46800-5_1

14. Cheon, J.H., Jeong, J., Lee, C.: An algorithm for CSPR problems and cryptanalysis of the GGH multilinear map without an encoding of zero. To be appear at ANTS (2016)
15. Coppersmith, D.: Small solutions to polynomial equations, and low exponent RSA vulnerabilities. J. Cryptol. **10**(4), 233–260 (1997)
16. Coron, J.-S., et al.: Zeroizing without low-level zeroes: new MMAP attacks and their limitations. In: Gennaro, R., Robshaw, M. (eds.) CRYPTO 2015. LNCS, vol. 9215, pp. 247–266. Springer, Heidelberg (2015). doi:10.1007/978-3-662-47989-6_12
17. Coron, J.-S., Lee, M.S., Lepoint, T., Tibouchi, M.: Cryptanalysis of GGH15 multilinear maps. In: Robshaw, M., Katz, J. (eds.) CRYPTO 2016. LNCS, vol. 9815, pp. 607–628. Springer, Heidelberg (2016). doi:10.1007/978-3-662-53008-5_21
18. Coron, J.-S., Lee, M.S., Lepoint, T., Tibouchi, M.: Zeroizing attacks on indistinguishability obfuscation over CLT13. To be appear at PKC 2017. Cryptology ePrint Archive, report 2016/1011 (2016)
19. Coron, J.-S., Lepoint, T., Tibouchi, M.: Practical multilinear maps over the integers. In: Canetti, R., Garay, J.A. (eds.) CRYPTO 2013. LNCS, vol. 8042, pp. 476–493. Springer, Heidelberg (2013). doi:10.1007/978-3-642-40041-4_26
20. Cramer, R., Ducas, L., Peikert, C., Regev, O.: Recovering short generators of principal ideals in cyclotomic rings. In: Fischlin, M., Coron, J.-S. (eds.) EUROCRYPT 2016. LNCS, vol. 9666, pp. 559–585. Springer, Heidelberg (2016). doi:10.1007/978-3-662-49896-5_20
21. Garg, S., Gentry, C., Halevi, S.: Candidate multilinear maps from ideal lattices. In: Johansson, T., Nguyen, P.Q. (eds.) EUROCRYPT 2013. LNCS, vol. 7881, pp. 1–17. Springer, Heidelberg (2013). doi:10.1007/978-3-642-38348-9_1
22. Garg, S., Gentry, C., Halevi, S., Raykova, M., Sahai, A., Waters, B.: Candidate indistinguishability obfuscation and functional encryption for all circuits. In: FOCS, pp. 40–49 (2013)
23. Garg, S., Gentry, C., Halevi, S., Zhandry, M.: Functional encryption without obfuscation. In: Kushilevitz, E., Malkin, T. (eds.) TCC 2016. LNCS, vol. 9563, pp. 480–511. Springer, Heidelberg (2016). doi:10.1007/978-3-662-49099-0_18
24. Garg, S., Miles, E., Mukherjee, P., Sahai, A., Srinivasan, A., Zhandry, M.: Secure obfuscation in a weak multilinear map model. In: Hirt, M., Smith, A. (eds.) TCC 2016. LNCS, vol. 9986, pp. 241–268. Springer, Heidelberg (2016). doi:10.1007/978-3-662-53644-5_10
25. Garg, S., Mukherjee, P., Srinivasan, A.: Obfuscation without the vulnerabilities of multilinear maps. Technical report (2016)
26. Gentry, C., Gorbunov, S., Halevi, S.: Graph-induced multilinear maps from lattices. In: Dodis, Y., Nielsen, J.B. (eds.) TCC 2015. LNCS, vol. 9015, pp. 498–527. Springer, Heidelberg (2015). doi:10.1007/978-3-662-46497-7_20
27. Garg, S., Gentry, C., Halevi, S., Raykova, M., Sahai, A., Waters, B.: Indistinguishability obfuscation from the multilinear subgroup elimination assumption. In: IEEE 56th Annual Symposium on Foundations of Computer Science (FOCS), pp. 151–170. IEEE (2015)
28. Gentry, C., Peikert, C., Vaikuntanathan, V.: Trapdoors for hard lattices and new cryptographic constructions. In: STOC, pp. 197–206 (2008)
29. Halevi, S.: Graded encoding, variations on a scheme. Cryptology ePrint Archive, report 2015/866 (2015)
30. Lenstra, A.K., Lenstra, H.W., Manasse, M.S., Pollard, J.M.: The number field sieve. In: STOC, pp. 564–572. ACM (1990)

31. Lin, H.: Indistinguishability obfuscation from constant-degree graded encoding schemes. In: Fischlin, M., Coron, J.-S. (eds.) EUROCRYPT 2016. LNCS, vol. 9665, pp. 28–57. Springer, Heidelberg (2016). doi:10.1007/978-3-662-49890-3_2
32. Lin, H., Vaikuntanathan, V.: Indistinguishability obfuscation from DDH-like assumptions on constant-degree graded encodings. In: FOCS, pp. 11–20. IEEE Computer Society (2016)
33. Micciancio, D., Peikert, C.: Trapdoors for lattices: simpler, tighter, faster, smaller. In: Pointcheval, D., Johansson, T. (eds.) EUROCRYPT 2012. LNCS, vol. 7237, pp. 700–718. Springer, Heidelberg (2012). doi:10.1007/978-3-642-29011-4_41
34. Miles, E., Sahai, A., Weiss, M.: Protecting obfuscation against arithmetic attacks. IACR Cryptology ePrint Archive 2014:878 (2014)
35. Miles, E., Sahai, A., Zhandry, M.: Annihilation attacks for multilinear maps: cryptanalysis of indistinguishability obfuscation over GGH13. In: Robshaw, M., Katz, J. (eds.) CRYPTO 2016. LNCS, vol. 9815, pp. 629–658. Springer, Heidelberg (2016). doi:10.1007/978-3-662-53008-5_22
36. Miles, E., Sahai, A., Zhandry, M.: Secure obfuscation in a weak multilinear map model: a simple construction secure against all known attacks. IACR Cryptology ePrint Archive 2016, 588 (2016)
37. Pass, R., Seth, K., Telang, S.: Indistinguishability obfuscation from semantically-secure multilinear encodings. In: Garay, J.A., Gennaro, R. (eds.) CRYPTO 2014. LNCS, vol. 8616, pp. 500–517. Springer, Heidelberg (2014). doi:10.1007/978-3-662-44371-2_28
38. Shor, P.W.: Polynomial-time algorithms for prime factorization and discrete logarithms on a quantum computer. SIAM Rev. 41(2), 303–332 (1999)
39. Zimmerman, J.: How to obfuscate programs directly. In: Oswald, E., Fischlin, M. (eds.) EUROCRYPT 2015. LNCS, vol. 9057, pp. 439–467. Springer, Heidelberg (2015). doi:10.1007/978-3-662-46803-6_15

Quantum Cryptography

Quantum Authentication and Encryption with Key Recycling

Or: How to Re-use a One-Time Pad Even if $P = NP$ — Safely & Feasibly

Serge Fehr[1]([✉]) and Louis Salvail[2]

[1] Centrum Wiskunde & Informatica (CWI), Amsterdam, The Netherlands
serge.fehr@cwi.nl
[2] Université de Montréal (DIRO), Montréal, Canada

Abstract. We propose an information-theoretically secure encryption scheme for classical messages with quantum ciphertexts that offers *detection* of eavesdropping attacks, and *re-usability of the key* in case no eavesdropping took place: the entire key can be securely re-used for encrypting new messages as long as no attack is detected. This is known to be impossible for fully classical schemes, where there is no way to detect plain eavesdropping attacks.

This particular application of quantum techniques to cryptography was originally proposed by Bennett, Brassard and Breidbart in 1982, even before proposing quantum-key-distribution, and a simple candidate scheme was suggested but no rigorous security analysis was given. The idea was picked up again in 2005, when Damgård, Pedersen and Salvail suggested a new scheme for the same task, but now with a rigorous security analysis. However, their scheme is much more demanding in terms of quantum capabilities: it requires the users to have a quantum computer.

In contrast, and like the original scheme by Bennett *et al.*, our new scheme requires from the honest users merely to prepare and measure single BB84 qubits. As such, we not only show the first provably-secure scheme that is within reach of current technology, but we also confirm Bennett *et al.*'s original intuition that a scheme in the spirit of their original construction is indeed secure.

1 Introduction

BACKGROUND. Classical information-theoretic encryption (like the one-time pad) and authentication (like Carter-Wegman authentication) have the serious downside that the key can be re-used only a small number of times, e.g. only *once* in case of the one-time pad for encryption or a strongly universal$_2$ hash function for authentication. This is inherent since by simply *observing* the communication, an eavesdropper Eve inevitably learns a substantial amount of information

---

L. Salvail—Funded by Canada's NSERC discovery grant and NSERC discovery accelerator.

J.-S. Coron and J.B. Nielsen (Eds.): EUROCRYPT 2017, Part III, LNCS 10212, pp. 311–338, 2017.
DOI: 10.1007/978-3-319-56617-7_11

on the key. Furthermore, there is no way for the communicating parties, Alice and Bob, to *know* whether Eve is present and has observed the communication or not, so they have to assume the worst.

This situation changes radically when we move to the quantum setting and let the ciphertext (or authentication tag) be a quantum state: then, by the fundamental properties of quantum mechanics, an Eve that *observes* the communicated state inevitably *changes* it, and so it is potentially possible for the receiver Bob to detect this, and, vice versa, to conclude that the key is still secure and thus can be safely re-used in case everything looks as it is supposed to be.

This idea of key re-usability by means of a quantum ciphertext goes back to a manuscript titled "*Quantum Cryptography II: How to re-use a one-time pad safely even if $P = NP$*" by Bennett, Brassard and Breidbart written in 1982. However, their paper was originally not published, and the idea was put aside after two of the authors discovered what then became known as BB84 quantum-key-distribution [2].[1] Only much later in 2005, this idea was picked up again by Damgård, Pedersen and Salvail in [5] (and its full version in [6]), where they proposed a new such encryption scheme and gave a rigorous security proof — in contrast, Bennett *et al.*'s original reasoning was very informal and hand-wavy.

The original scheme by Bennett *et al.* is simple and natural: you one-time-pad encrypt the message, add some redundancy by encoding the ciphertext using an error correction (or detection) code, and encode the result bit-wise into what we nowadays call BB84 qubits. The scheme by Damgård *et al.* is more involved; in particular, the actual quantum encoding is not done by means of single qubits, but by means of states that form a set of mutually unbiased bases in a Hilbert space of large dimension. This in particular means that their scheme requires a *quantum computer* to produce the quantum ciphertexts and to decrypt them.

OUR RESULTS. We are interested in the question of whether one can combine the simplicity of the originally proposed encryption scheme by Bennett *et al.* with a rigorous security analysis as offered by Damgård *et al.* for their scheme; in particular, whether there is a provably secure scheme that is within reach of being implementable with current technology — and we answer the question in the affirmative.

We start with the somewhat simpler problem of finding an *authentication* scheme that allows to re-use the key in case no attack is detected, and we show a very simple solution. In order to authenticate a (classical) message msg, we encode a random bit string $x \in \{0,1\}^n$ into BB84 qubits $H^\theta |x\rangle$, where $\theta \in \{0,1\}^n$ is part of the shared secret key, and we compute a tag $t = \mathsf{MAC}(k, msg\|x)$ of the message concatenated with x, where MAC is a classical information-theoretic one-time message authentication code, and its key k is the other part of the shared secret key. The qubits $H^\theta |x\rangle$ and the classical tag t are then sent along with msg, and the receiver verifies correctness of the received message in the obvious way by measuring the qubits to obtain x and checking t.

---

[1] A freshly typeset version of the original manuscript was then published more than 30 years later in [3].

One-time security of the scheme is obvious, and the intuition for key-recycling is as follows. Since Eve does not know θ, she has a certain minimal amount of uncertainty in x, so that, if MAC has suitable extractor-like properties, the tag t is (almost) random and *independent* of k and θ, and thus gives away no information on k and θ. Furthermore, if Eve tries to gain information on k and θ by measuring some qubits, she disturbs these qubits and is likely to be detected. A subtle issue is that if Eve measures only *very few* qubits then she has a good chance of not being detected, while still learning a little bit on θ by the fact that she has not been detected. However, as long as her uncertainty in θ is large enough this should not help her (much), and the more information on θ she tries to collect this way the more likely it is that she gets caught.

We show that the above intuition is correct. Formally, we prove that as long as the receiver Bob accepts the authenticated message, the key-pair (k, θ) can be safely re-used, and if Bob rejects, it is good enough to simply refresh θ. Our proof is based on techniques introduced in [19] and extensions thereof.

Extending our authentication scheme to an encryption scheme is intuitively quite easy: we simply extract a one-time-pad key from x, using a strong extractor (with some additional properties) with a seed that is also part of the shared secret key. Similarly to above, we can prove that as long as the receiver Bob accepts, the key can be safely re-used, and if Bob rejects it is good enough to refresh θ.

In our scheme, the description length[2] of θ is $m + 3\lambda$, where m is the length of the encrypted message *msg* and λ is the security parameter (so that the scheme fails with probability at most $2^{-\lambda}$). Thus, with respect to the number of fresh random bits that are needed for the key refreshing, i.e. for updating the key in case Bob rejects, our encryption scheme is comparable to the scheme by Damgård et al.[3] and optimal in terms of the dependency on the message length m.

Our schemes can be made *noise robust* in order to deal with a (slightly) noisy quantum communication; the generic solution proposed in [5,6] of using a quantum error correction code is not an option for us as it would require a quantum computer for en- and decoding. Unfortunately, using straightforward error correction techniques, like sending along the syndrome of x with respect to a suitable error correcting code, renders our proofs invalid beyond an easy fix, though it is unclear whether the scheme actually becomes insecure. However, we can deal with the issue by means of using error correction "without leaking partial information", as introduced by Dodis and Smith [8] and extended to the quantum setting by Fehr and Schaffner [9]. Doing error correction in a more standard way, which would offer more freedom in choosing the error correction code and allow for a larger amount of noise, remains an interesting open problem.

ENCRYPTION with KEY RECYCLING vs QKD. A possible objection against the idea of encryption with key recycling is that one might just as well use QKD

---

[2] In our scheme, θ is not uniformly random in $\{0, 1\}^n$ but is chosen to be a code word, as such, its description length is smaller than its physical bit length, and given by the dimension of the code.

[3] Their scheme needs $m + \ell$ fresh random bits for key refreshing, where ℓ is a parameter in their construction, and their scheme fails with probability approximately $2^{-\ell/2}$.

to produce a new key, rather than re-using the old one. However, there *are* subtle advantages of using encryption with key recycling instead. For instance, encryption with key recycling is (almost) non-interactive and requires only *1 bit* of authenticated feedback: "accept" or "reject", that can be provided *offline*, i.e., after the communication of the private message, as long as it is done before the scheme is re-used. This opens the possibility to provide the feedback by means of a different channel, like by confirming over the phone. In contrast, for QKD, a *large amount* of data needs to be authenticated *online* and in *both directions*. If no physically authenticated channel is available, then the authenticated feedback can actually be done very easily: Alice appends a random token to the message she communicates to Bob in encrypted form, and Bob confirms that no attack is detected by returning the token back to Alice — in plain — and in case he detected an attack, he sends a reject message instead.[4] Furthermore, encryption with key recycling has the potential to be *more efficient* than QKD in terms of communication. Even though this is not the case for our scheme, there is certainly potential, because no sifting takes place and hence there is no need to throw out a fraction of the quantum communication. Altogether, on a stable quantum network for instance, encryption with key-recycling would be the preferred choice over QKD. Last but not least, given that the re-usability of a one-time-pad-like encryption key was one of the very first proposed applications of quantum cryptography — even before QKD — we feel that giving a satisfactory answer should be of intellectual interest.

RELATED WORK. Besides the work of Brassard *et al.* and of Damgård *et al.*, who focus on encrypting *classical* messages, there is a line of work, like [11,13,15], that considers key recycling in the context of authentication and/or encryption of *quantum* messages. However, common to almost all this work is that only *part* of the key can be re-used if no attack is detected, or a new but *shorter* key can be extracted. The only exceptions we know of are the two recent works by Garg *et al.* [10] and by Portmann [16], which consider and analyze authentication schemes for quantum messages that do offer re-usability of the entire key in case no attack is detected. However, these schemes are based on techniques (like unitary designs) that require the honest users to perform quantum computations also when restricting to classical messages. Actually, [16] states it as an explicit open problem to "find a prepare-and-measure scheme to encrypt and authenticate a classical message in a quantum state, so that all of the key may be recycled if it is successfully authenticated". On the other hand, their schemes offer security against *superposition attacks*, where the adversary may trick the sender into authenticating a *superposition* of classical messages; this is something we do not consider here — as a matter of fact, it would be somewhat unnatural for us since such superposition attacks require the sender (wittingly or unwittingly) to hold a quantum computer, which is exactly what we want to avoid.

---

[4]   Of course, when Bob sends back the token to confirm, Eve can easily replace it by the reject message and so prevent Alice and Bob from finding agreement, but this is something that Eve can always achieve by "altering the last message", also in QKD.

2   Preliminaries

2.1   Basic Concepts of Quantum Information Theory

We assume basic familiarity; we merely fix notation and terminology here.

QUANTUM STATES. The state of a quantum system with state space $\mathcal{H}$ is specified by a *state vector* $|\varphi\rangle \in \mathcal{H}$ in case of a pure state, or, more generally in case of a mixed state, by a *density matrix* ρ acting on $\mathcal{H}$. The set of density matrices acting on $\mathcal{H}$ is denoted $\mathcal{D}(\mathcal{H})$. We typically identify different quantum systems by means of labels A, B etc., and we write ρ_A for the state of system A and $\mathcal{H}_A$ for its state space, etc. The joint state of a bipartite system AB is given by a density matrix ρ_{AB} in $\mathcal{D}(\mathcal{H}_A \otimes \mathcal{H}_B)$; it is then understood that ρ_A and ρ_B are the respective *reduced* density matrices $\rho_A = \mathrm{tr}_B(\rho_{AB})$ and $\rho_B = \mathrm{tr}_A(\rho_{AB})$.

We also consider states that consist of a classical and a quantum part. Formally, $\rho_{XE} \in \mathcal{D}(\mathcal{H}_X \otimes \mathcal{H}_E)$ is called a *cq*-state (for classical-quantum), if it is of the form

$$\rho_{XE} = \sum_{x \in \mathcal{X}} P_X(x)|x\rangle\langle x| \otimes \rho_E^x,$$

where $P_X : \mathcal{X} \to [0,1]$ is a probability distribution, $\{|x\rangle\}_{x \in \mathcal{X}}$ is a fixed orthonormal basis of $\mathcal{H}_X$, and $\rho_E^x \in \mathcal{D}(\mathcal{H}_E)$. Throughout, we will slightly abuse notation and express this by writing $\rho_{XE} \in \mathcal{D}(\mathcal{X} \otimes \mathcal{H}_E)$.

In the context of such a cq-state ρ_{XE}, an *event* Λ is specified by means of a decomposition $\rho_{XE} = P[\Lambda] \cdot \rho_{XE|\Lambda} + P[\neg\Lambda] \cdot \rho_{XE|\neg\Lambda}$ with $P[\Lambda], P[\neg\Lambda] \geq 0$ and $\rho_{XE|\Lambda}, \rho_{XE|\neg\Lambda} \in \mathcal{D}(\mathcal{X} \otimes \mathcal{H}_E)$. Associated to such an event Λ is the *indicator random variable* 1_Λ, i.e., the cq-state $\rho_{X1_\Lambda E} \in \mathcal{D}(\mathcal{X} \otimes \{0,1\} \otimes \mathcal{H}_E)$, defined in the obvious way. Note that, for any cq-state ρ_{XE} and any $x \in \mathcal{X}$, the event $X = x$ is naturally defined and $\rho_{XE|X=x} = |x\rangle\langle x| \otimes \rho_E^x$ and $\rho_{E|X=x} = \rho_E^x$.

If a state ρ_X is purely classical, meaning that $\rho_X = \sum_x P_X(x)|x\rangle\langle x|$ and expressed as $\rho_X \in \mathcal{D}(\mathcal{X})$, we may refer to standard probability notation so that probabilities like $P[X = x]$ are well understood. Finally, we write $\mu_\mathcal{X}$ for the *fully mixed state* $\mu_\mathcal{X} = \frac{1}{|\mathcal{X}|} \sum_x |x\rangle\langle x| = \frac{1}{|\mathcal{X}|}\mathbb{I}_\mathcal{X} \in \mathcal{D}(\mathcal{X})$.

GENERAL QUANTUM OPERATIONS. Operations on quantum systems are described by *CPTP maps*. To emphasize that a CPTP map $\mathcal{Q} : \mathcal{D}(\mathcal{H}_A) \to \mathcal{D}(\mathcal{H}_{A'})$ acts on density matrices in $\mathcal{D}(\mathcal{H}_A)$, we sometimes write $\mathcal{Q}_A$, and we say that it "acts on A". Also, we may write $\mathcal{Q}_{A \to A'}$ in order to be explicit about the range too. If $\mathcal{Q}$ is a CPTP map acting on A, we often abuse notation and simply write $\mathcal{Q}_A(\rho_{AB})$ or $\rho_{\mathcal{Q}(A)B}$ for $(\mathcal{Q}_A \otimes id_B)(\rho_{AB})$, where id_B is the identity map on $\mathcal{D}(\mathcal{H}_B)$.

In line with our notation for cq-states, $\mathcal{Q} : \mathcal{D}(\mathcal{X} \otimes \mathcal{H}_E) \to \mathcal{D}(\mathcal{X}' \otimes \mathcal{H}_{E'})$ is used to express that $\mathcal{Q}$ maps any cq-state $\rho_{XE} \in \mathcal{D}(\mathcal{X} \otimes \mathcal{H}_E)$ to a cq-state $\mathcal{Q}(\rho_{X'E})$ in $\mathcal{D}(\mathcal{X}' \otimes \mathcal{H}_{E'})$. We say that a CPTP map $\mathcal{Q} : \mathcal{D}(\mathcal{X} \otimes \mathcal{H}_E) \to \mathcal{D}(\mathcal{X} \otimes \mathcal{H}_{E'})$ is "controlled by X and acts on E" if on a cq-state $\rho_{XE} \in \mathcal{D}(\mathcal{X} \otimes \mathcal{H}_E)$ it acts as

$$\mathcal{Q}(\rho_{XE}) = \sum_x P_X(x)|x\rangle\langle x| \otimes \mathcal{Q}^x(\rho_E^x)$$

with "conditional" CPTP maps $Q^x : D(\mathcal{H}_E) \to D(\mathcal{H}_{E'})$. Note that in this case we write $Q_{XE \to E'}$ rather than $Q_{XE \to XE'}$, as it is understood that Q keeps X alive. For concreteness, we require that such a Q is of the form $Q = \sum_x P_{|x\rangle\langle x|} \otimes Q^x$ where $P_{|x\rangle\langle x|}(\rho) = |x\rangle\langle x| \rho |x\rangle\langle x|$ for any $\rho \in D(\mathcal{H}_X)$.[5] As such, Q is fully specified by means of the conditional CPTP maps Q^x. Finally, for any function $f : \mathcal{X} \to \mathcal{Y}$, we say that $Q : D(\mathcal{X} \otimes \mathcal{H}_E) \to D(\mathcal{X} \otimes \mathcal{H}_{E'})$ is "controlled by $f(X)$" if it is controlled by X, but $Q^x = Q^{x'}$ for any $x, x' \in \mathcal{X}$ with $f(x) = f(x')$.

MARKOV-CHAIN STATES. Let $\rho_{XYE} \in D(\mathcal{X} \otimes \mathcal{Y} \otimes \mathcal{H}_E)$ be a cq-state with two classical subsystems X and Y. Following [7], we define $\rho_{X \leftrightarrow Y \leftrightarrow E}$ to be the "Markov-chain state"

$$\rho_{X \leftrightarrow Y \leftrightarrow E} := \sum_{x,y} P_{XY}(x,y)|x\rangle\langle x| \otimes |y\rangle\langle y| \otimes \rho_E^y$$

with $\rho_E^y = \sum_x P_{X|Y}(x|y) \rho_E^{x,y}$. If the state ρ_{XYE} is clear from the context we write $X \leftrightarrow Y \leftrightarrow E$ to express that $\rho_{XYE} = \rho_{X \leftrightarrow Y \leftrightarrow E}$. It is an easy exercise to verify that the Markov-chain condition $X \leftrightarrow Y \leftrightarrow E$ holds if and only if $\rho_{XYE} = Q_{Y\varnothing \to E}(\rho_{XY})$ for a CPTP map $Q_{Y\varnothing \to E} : D(\mathcal{Y}) \to D(\mathcal{Y} \otimes \mathcal{H}_E)$ that is controlled by Y and acts on the "empty" system $\varnothing$, i.e., the conditional maps act as $Q_{\varnothing \to E}^y : D(\mathbb{C}) \to D(\mathcal{H}_E)$.

QUANTUM MEASUREMENTS. We model a *measurement* of a quantum system A with outcome in $\mathcal{X}$ by means of a CPTP map $M : D(\mathcal{H}_A) \to D(\mathcal{X})$ that acts as

$$M(\rho) = \sum_{x \in \mathcal{X}} \mathrm{tr}(E_x \rho)|x\rangle\langle x|,$$

where $\{|x\rangle\}_{x \in \mathcal{X}}$ is a fixed basis, and $\{E_x\}_{x \in \mathcal{X}}$ forms a *POVM*, i.e., a family of positive-semidefinite operators that add up to the identity matrix $\mathbb{I}_{\mathcal{X}}$. A measurement $M : D(\mathcal{Z} \otimes \mathcal{H}_A) \to D(\mathcal{Z} \otimes \mathcal{X})$ is said to be a "measurement of A controlled by Z" if it is controlled by Z and acts on A as a CPTP map. It is easy to see that in this case the conditional CPTP maps $M^z : D(\mathcal{H}_A) \to D(\mathcal{X})$ are measurements too, referred to as "conditional measurements".

Note that whenever $M : D(\mathcal{H}_Z \otimes \mathcal{H}_A) \to D(\mathcal{X})$ is an *arbitrary* measurement of Z and A that is applied to a cq-state $\rho_{ZA} \in D(\mathcal{Z} \otimes \mathcal{H}_A)$, we may assume that M first "produces a copy of Z", and thus we may assume without loss of generality that $M : D(\mathcal{Z} \otimes \mathcal{H}_A) \to D(\mathcal{Z} \otimes \mathcal{X})$ is *controlled* by Z.

For a given $n \in \mathbb{N}$, $M_{\Theta A \to X}^{\mathrm{BB84}}$ denotes the *BB84 measurement* of an n-qubit system A controlled by Θ. Formally, for every $\theta \in \{0,1\}^n$ the corresponding conditional measurement is specified by the POVM $\{H^\theta |x\rangle\langle x| H^\theta\}$ with x ranging over $\{0,1\}^n$. Here, H is the *Hadamard matrix*, and $H^\theta |x\rangle$ is a short hand for $H^{\theta_1}|x_1\rangle \otimes \cdots \otimes H^{\theta_n}|x_n\rangle \in \mathcal{H}_A = (\mathbb{C}^2)^{\otimes n}$, where $\{|0\rangle, |1\rangle\}$ is the *computational basis* of the qubit system $\mathbb{C}^2$.

TRACE DISTANCE. We capture the distance between two states $\rho, \sigma \in D(\mathcal{H})$ in terms of their *trace distance* $\delta(\rho, \sigma) := \frac{1}{2}\|\rho - \sigma\|_1$, where $\|K\|_1 := \mathrm{tr}(\sqrt{K^\dagger K})$ is the *trace norm* of an arbitrary operator K. If the states ρ_A and $\rho_{A'}$ are clear

---

[5] This means that the system X is actually measured (in the fixed basis $\{|x\rangle\}_{x \in \mathcal{X}}$).

from context, we may write $\delta(A, A')$ instead of $\delta(\rho_A, \rho_{A'})$. Also, for any cq-state ρ_{XE} in $\mathcal{D}(\mathcal{X} \otimes \mathcal{H}_E)$, we write $\delta(X, U_{\mathcal{X}}|E)$ as a short hand for $\delta(\rho_{XE}, \mu_{\mathcal{X}} \otimes \rho_E)$. Obviously, $\delta(X, U_{\mathcal{X}}|E)$ captures how far away X is from uniformly random on $\mathcal{X}$ when given the quantum system E.

It is well known that the trace distance is monotone under CPTP maps, and it is easy to see that if two cq-states $\rho_{XE}, \sigma_{XE} \in \mathcal{D}(\mathcal{X} \otimes \mathcal{H}_E)$ coincide on their classical subsystems, meaning that $\rho_X = \sigma_X$, then $\delta(\rho_{XE}, \sigma_{XE})$ decomposes into $\delta(\rho_{XE}, \sigma_{XE}) = \sum_x P_X(x)\, \delta(\rho_E^x, \sigma_E^x)$.

2.2 The Guessing Probability

An important concept in the technical analysis of our scheme(s) is the following notion of guessing probability, which is strongly related to the (conditional) min-entropy as introduced by Renner [17], but turns out to be more convenient to work with for our purpose. Let $\rho_{XE} \in \mathcal{D}(\mathcal{X} \otimes \mathcal{H}_E)$ be a cq-state.

Definition 1. *The* guessing probability *of X given E is*

$$\mathrm{Guess}(X|E) := \max_{\mathcal{M}} P[\mathcal{M}(E) = X],$$

where the maximum is over all measurements $\mathcal{M} : \mathcal{D}(\mathcal{H}_E) \to \mathcal{D}(\mathcal{X})$ of E with outcome in $\mathcal{X}$.[6]

Note that if Λ is an event, then $\mathrm{Guess}(X|E, \Lambda)$ is naturally defined by means of applying the above to the "conditional state" $\rho_{XE|\Lambda} \in \mathcal{D}(\mathcal{X} \otimes \mathcal{H}_E)$.

We will make use of the following elementary properties of the guessing probability. In all the statements, it is understood that $\rho_{XE} \in \mathcal{D}(\mathcal{X} \otimes \mathcal{H}_E)$, respectively $\rho_{XZE} \in \mathcal{D}(\mathcal{X} \otimes \mathcal{Z} \otimes \mathcal{H}_E)$ in Property 2.

Property 1. $\mathrm{Guess}(X|\mathcal{Q}(E)) \leq \mathrm{Guess}(X|E)$ for any CPTP map $\mathcal{Q}$ acting on E.

Property 2. $\mathrm{Guess}(X|ZE) = \sum_z P_Z(z)\, \mathrm{Guess}(X|E, Z=z)$.

Property 3. $\mathrm{Guess}(X|E, \Lambda) \leq \mathrm{Guess}(X|E)/P[\Lambda]$ for any event Λ.

Note that Property 2 implies that $\mathrm{Guess}(X|E, \Lambda) \leq \mathrm{Guess}(X|1_\Lambda E)/P[\Lambda]$, but the statement of Property 3 is stronger since $\mathrm{Guess}(X|E) \leq \mathrm{Guess}(X|1_\Lambda E)$.

Proof (of Property 3). It holds that[7] $P[\Lambda] \cdot \rho_{XE|\Lambda} \leq \rho_{XE}$, and hence that for any measurement $\mathcal{M}$ on E

$$P[\Lambda] \cdot P[\mathcal{M}(E) = X | \Lambda] \leq P[\mathcal{M}(E) = X] \leq \mathrm{Guess}(X|E),$$

which implies the claim. $\qquad\square$

---

[6] By our conventions, the probability $P[\mathcal{M}(E) = X]$ is to be understood as $P[X' = X]$ for the (purely classical) state $\rho_{XX'} = \rho_{X\mathcal{M}(E)} = (id_X \otimes \mathcal{M})(\rho_{XE}) \in \mathcal{D}(\mathcal{X} \otimes \mathcal{X})$.

[7] Here and throughout, for operators K and L, the inequality $K \leq L$ means that $L - K$ is positive-semidefinite.

Property 4. There exists $\sigma_E \in \mathcal{D}(\mathcal{H}_E)$ so that

$$\rho_{XE} \leq \mathrm{Guess}(X|E) \cdot \mathbb{I}_{\mathcal{X}} \otimes \sigma_E = \mathrm{Guess}(X|E) \cdot |\mathcal{X}| \cdot \mu_{\mathcal{X}} \otimes \sigma_E.$$

Proof. The claim follows from Renner's original definition of the conditional min-entropy as

$$H_\infty(X|E) := \max_{\sigma_E} \max_\lambda \{\lambda \mid \rho_{XE} \leq 2^{-\lambda} \cdot \mathbb{I}_{\mathcal{X}} \otimes \sigma_E\}$$

and the identity $H_\infty(X|E) = -\log \mathrm{Guess}(X|E)$, as shown in [12]. □

3   Enabling Tools

In this section, we introduce and discuss the main technical tools for the constructions and analyses of our key-recycling authentication and encryption schemes.

3.1   On Guessing the Outcome of Quantum Measurements

We consider different "guessing games", where one or two players need to guess the outcome of a quantum measurement. The bounds are derived by means of the techniques of [19].

TWO-PLAYER GUESSING. Here, we consider a game where two parties, Bob and Charlie, need to *simultaneously* and without communication guess the outcome of BB84 measurements performed by Alice (on n qubits prepared by Bob and Charlie), when given the bases that Alice chose. This is very similar to the *monogamy game* introduced and studied in [19], but in the version we consider here, the sequence of bases is not chosen from $\{0,1\}^n$ but from a code $\mathcal{C} \subset \{0,1\}^n$ with minimal distance d. It is useful to think of d to be much larger than $\log|\mathcal{C}|$, i.e., the dimension of the code in case of a linear code. The following shows that in case of a uniformly random choice of the bases in $\mathcal{C}$, Bob and Charlie cannot do much better than to agree on a guess for the bases and to give Alice qubits in those bases.

Proposition 1. *Let $\mathcal{H}_A$ be a n-qubit system, and let $\mathcal{H}_B$ and $\mathcal{H}_C$ be arbitrary quantum systems. Consider a state $\rho_{\Theta ABC} = \mu_C \otimes \rho_{ABC} \in \mathcal{D}(\mathcal{C} \otimes \mathcal{H}_A \otimes \mathcal{H}_B \otimes \mathcal{H}_C)$, and let*

$$\rho_{\Theta XX'X''} = \mathcal{N}_{\Theta C \to X''} \circ \mathcal{N}_{\Theta B \to X'} \circ \mathcal{M}^{\mathrm{BB84}}_{\Theta A \to X}(\rho_{\Theta ABC})$$

where $\mathcal{M}^{\mathrm{BB84}}_{\Theta A \to X}$ is the BB84-measurement of the system A (controlled by Θ), and $\mathcal{N}_{\Theta B \to X'}$ and $\mathcal{N}_{\Theta C \to X''}$ are arbitrary (possibly different) measurements of the respective systems B and C, both controlled by Θ. Then, it holds that

$$P[X' = X \wedge X'' = X] \leq \frac{1}{|\mathcal{C}|} + \frac{1}{2^{d/2}}.$$

Proof. The proof uses the techniques from [19]. By Naimark's theorem, we may assume without loss of generality that the conditional measurements $\mathcal{N}^\theta_{B\to X'}$ and $\mathcal{N}^\theta_{C\to X''}$ are specified by families $\{P^\theta_x\}_x$ and $\{Q^\theta_x\}_x$ of *projections*. Then, defining for every $\theta \in \mathcal{C}$ the projection $\Pi^\theta = \sum_x H^\theta|x\rangle\langle x|H^\theta \otimes P^\theta_x \otimes Q^\theta_x$, we see that

$$P[X'=X \wedge X''=X] \le \frac{1}{|\mathcal{C}|}\left\|\sum_\theta \Pi^\theta\right\| \le \frac{1}{|\mathcal{C}|}\sum_\delta \max_\theta\|\Pi^\theta \Pi^{\theta\oplus\delta}\|,$$

where $\|\cdot\|$ refers to the standard operator norm, and the second inequality is by Lemma 2.2 in [19]. For any $\theta, \theta' \in \mathcal{C}$, bounding Π^θ and $\Pi^{\theta'}$ by

$$\Pi^\theta \le \Gamma^\theta := \sum_x H^\theta|x\rangle\langle x|H^\theta \otimes P^\theta_x \otimes \mathbb{I}$$

and

$$\Pi^{\theta'} \le \Delta^{\theta'} := \sum_x H^{\theta'}|x\rangle\langle x|H^{\theta'} \otimes \mathbb{I} \otimes Q^{\theta'}_x,$$

it is shown in [19] (in the proof of Theorem 3.4) that

$$\|\Pi^\theta \Pi^{\theta'}\| \le \|\Gamma^\theta \Delta^{\theta'}\| \le \frac{1}{2^{d_H(\theta,\theta')/2}} \le \frac{1}{2^{d/2}}$$

where the last inequality holds unless $\theta = \theta'$, from which the claim follows. $\square$

Remark 1. If we restrict $\mathcal{H}_B$ to be a n-qubit system too, and replace the (arbitrary) measurement $\mathcal{N}_{\Theta B}$ by a BB84 measurement $\mathcal{M}^{\text{BB84}}_{\Theta B}$, i.e., "Bob measures correctly", then we get

$$P[X'=X \wedge X''=X] \le \frac{1}{|\mathcal{C}|} + \frac{1}{2^d}.$$

TWO-PLAYER GUESSING WITH QUANTUM SIDE INFORMATION. Now, we consider a version of the game where Alice's choice for the bases is not uniformly random, and, additionally, Bob and Charlie may hold some quantum side information on Alice's choice at the time when they can prepare the initial state (for Alice, Bob and Charlie).

Corollary 1. *Let* $\mathcal{H}_A$ *be a* n-*qubit system, and let* $\mathcal{H}_B, \mathcal{H}_C$ *and* $\mathcal{H}_E$ *be arbitrary quantum systems. Consider a state* $\rho_{\Theta E} \in \mathcal{D}(\mathcal{C} \otimes \mathcal{H}_E)$, *and let*

$$\rho_{\Theta ABC} = \mathcal{Q}_{E\to ABC}(\rho_{\Theta E}) \in \mathcal{D}(\mathcal{C} \otimes \mathcal{H}_A \otimes \mathcal{H}_B \otimes \mathcal{H}_C)$$

where $\mathcal{Q}_{E\to ABC}$ *is a CPTP map acting on* E *(only), and let*

$$\rho_{\Theta XX'X''} = \mathcal{N}_{\Theta C\to X''} \circ \mathcal{N}_{\Theta B\to X'} \circ \mathcal{M}^{\text{BB84}}_{\Theta A\to X}(\rho_{\Theta ABC})$$

as in Proposition 1 above. Then, it holds that

$$P[X'=X \wedge X''=X] \le \text{Guess}(\Theta|E) + \frac{\text{Guess}(\Theta|E) \cdot |\mathcal{C}|}{2^{d/2}}.$$

Proof. By Proposition 1, the claim holds as $P[X'\!=\!X \wedge X''\!=\!X] \leq 1/|\mathcal{C}| + 1/2^{d/2}$ for the special case where $\rho_{\Theta E}$ is of the form $\rho_{\Theta E} = \mu_\mathcal{C} \otimes \sigma_E$. Furthermore, by Property 4 we know that an arbitrary $\rho_{\Theta E} \in \mathcal{D}(\mathcal{C} \otimes \mathcal{H}_E)$ is bounded by

$$\rho_{\Theta E} \leq \mathrm{Guess}(\Theta|E) \cdot |\mathcal{C}| \cdot \mu_\mathcal{C} \otimes \sigma_E.$$

Therefore, since the composed map

$$\mathcal{D}(\mathcal{C} \otimes \mathcal{H}_E) \to \mathcal{D}(\{0,1\}), \quad \rho_{\Theta E} \mapsto \rho_{\Theta ABC} \mapsto \rho_{XX'X''} \mapsto \rho_{1_{X=X' \wedge X=X''}}$$

is still a CPTP map, it holds that for arbitrary $\rho_{\Theta E} \in \mathcal{D}(\mathcal{C} \otimes \mathcal{H}_E)$

$$P[X'\!=\!X \wedge X''\!=\!X] \leq \mathrm{Guess}(\Theta|E) \cdot |\mathcal{C}| \cdot \left(\frac{1}{|\mathcal{C}|} + \frac{1}{2^{d/2}} \right),$$

which proves the claim.   □

Remark 2. Similarly to the remark above, the bound relaxes to

$$P[X'\!=\!X \wedge X''\!=\!X] \leq \mathrm{Guess}(\Theta|E) + \frac{\mathrm{Guess}(\Theta|E) \cdot |\mathcal{C}|}{2^d},$$

when "Bob measure correctly".

SINGLE-PLAYER GUESSING (WITH QUANTUM SIDE INFORMATION). Corollary 1 immediately gives us control also over a slightly different game, where only one party needs to guess Alice's measurement outcome, but here he is *not* given the bases. Indeed, any strategy here gives a strategy for the above simultaneous-guessing game, simply by "pre-measuring" B, and having Bob and Charlie each keep a copy of the measurement outcome.

Corollary 2. *Let $\mathcal{H}_A$ be a n-qubit system, and let $\mathcal{H}_B$ and $\mathcal{H}_E$ be arbitrary quantum systems. Consider a state $\rho_{\Theta E} \in \mathcal{D}(\mathcal{C} \otimes \mathcal{H}_E)$ and let*

$$\rho_{\Theta AB} = \mathcal{Q}_{E \to AB}(\rho_{\Theta E}) \in \mathcal{D}(\mathcal{C} \otimes \mathcal{H}_A \otimes \mathcal{H}_B)$$

where $\mathcal{Q}_{E \to AB}$ is a CPTP map acting on E, and let

$$\rho_{\Theta XX''} = \mathcal{N}_{B \to X''} \circ \mathcal{M}^{\mathrm{BB84}}_{\Theta A \to X}(\rho_{\Theta AB})$$

where $\mathcal{N}_{B \to X''}$ is an arbitrary measurement of B (with no access to Θ). Then, it holds that

$$P[X''\!=\!X] \leq \mathrm{Guess}(\Theta|E) + \frac{\mathrm{Guess}(\Theta|E) \cdot |\mathcal{C}|}{2^{d/2}}.$$

In other words, for the state $\rho_{\Theta XB} = \mathcal{M}^{\mathrm{BB84}}_{\Theta A \to X}(\rho_{\Theta AB})$ we have that

$$\mathrm{Guess}(X|B) \leq \mathrm{Guess}(\Theta|E) + \frac{\mathrm{Guess}(\Theta|E) \cdot |\mathcal{C}|}{2^{d/2}}.$$

Remark 3. If we restrict the side information E to be *classical* then, using slightly different techniques, we can improve the bounds from Corollaries 1 and 2 to

$$\mathrm{Guess}(\Theta|E) + \frac{1}{2^{d/2}}.$$

Whether this improved bound also holds in case of quantum side information is an open question.

3.2   Hash Functions with Message-Independence and Key-Privacy

The goal of key-recycling is to be able to *re-use* a cryptographic key. For this to be possible, it is necessary — actually, not necessary but sufficient — that a key *stays secure*, i.e., that the primitive that uses the key does *not* reveal anything on the key, or only very little. We introduce here a general notion that captures this, i.e., that ensures that the key stays secure *as long as* there is *enough uncertainty* in the message the primitive is applied to — in our construction(s), this uncertainty will then be derived from the quantum part.

Consider a keyed hash function $H : \mathcal{K} \times \mathcal{X} \to \mathcal{Y}$ with key space $\mathcal{K}$, message space $\mathcal{X}$, and range $\mathcal{Y}$. We define the following properties on such a hash function.

Definition 2. *We say that* H *is* message-independent *if for a uniformly random key* K *in* $\mathcal{K}$, *the distribution of the hash value* $Y = H(K, x)$ *is independent of the message* $x \in \mathcal{X}$. *And, we say that* H *is* uniform *if it is message-independent and* $Y = H(K, x)$ *is uniformly random on* $\mathcal{Y}$.

Thus, message-independence simply ensures that if the key is uniformly random and independent of the message, then the hash of the message is independent of the message too. The key-privacy property below on the other hand ensures that for any adversary that has arbitrary but *limited* information on the message and the hash value — but no direct information on the key — has (almost) no information on the key.

Definition 3. *We say that* H *offers* ν-key-privacy *if for any state* ρ_{KXYE} *in* $\mathcal{D}(\mathcal{K} \otimes \mathcal{X} \otimes \mathcal{Y} \otimes \mathcal{H}_E)$ *with the properties that* $\rho_{KX} = \mu_{\mathcal{K}} \otimes \rho_X$, $Y = H(K, X)$ *and* $K \leftrightarrow XY \leftrightarrow E$, *it holds that*

$$\delta(K, U_{\mathcal{K}}|YE) \leq \frac{\nu}{2}\sqrt{\mathrm{Guess}(X|YE) \cdot |\mathcal{Y}|}.$$

We say that H *offers* ideal key-privacy *if it offers* 1-*key-privacy*.

Remark 4. Note that if H is message-independent then for X, Y and E as above in Definition 3, we have that $\mathrm{Guess}(X|YE) = \mathrm{Guess}(X|E)$.

Not so surprisingly, the joint notion of uniformity and key-privacy is closely related to that of a *strong extractor* [14]. Indeed, if H is uniform and offers key-privacy then it is a strong extractor: for $\rho_{KXE} = \mu_{\mathcal{K}} \otimes \rho_{XE}$ and $Y = H(K, X)$, the condition on ρ_{KXYE} in Definition 3 is satisfied, and thus we have the promised bound on $\delta(\rho_{KYE}, \mu_{\mathcal{K}} \otimes \rho_{YE}) = \delta(\rho_{KYE}, \mu_{\mathcal{K}} \otimes \mu_{\mathcal{Y}} \otimes \rho_E)$, where the equality is due to uniformity. As such, [18] shows that the required bound on $\delta(K, U_{\mathcal{K}}|YE)$ is the best one can hope for. On the other hand, the following shows that from every strong extractor we can easily construct a hash function that offers uniformity and key-privacy.

Proposition 2. *Let* $\mathrm{Ext} : \mathcal{K} \times \mathcal{X} \to \mathcal{Y}$ *be a strong extractor, meaning that for* $\rho_{KXE} = \mu_{\mathcal{K}} \otimes \rho_{XE} \in \mathcal{D}(\mathcal{K} \otimes \mathcal{X} \otimes \mathcal{H}_E)$ *and for* Y *computed as* $Y = \mathrm{Ext}(K, X)$ *it* holds that $\delta(\rho_{KYE}, \mu_K \otimes \rho_{YE}) \leq \frac{\nu}{2}\sqrt{\mathrm{Guess}(X|E) \cdot |\mathcal{Y}|}$. *Furthermore, we assume that the range $\mathcal{Y}$ forms a group. Then, the keyed hash function*[8]

$$\mathsf{H} : (\mathcal{K} \times \mathcal{Y}) \times \mathcal{X} \to \mathcal{Y}, \quad (k\|k', x) \mapsto \mathsf{Ext}(k, x) + k'$$

with key space $\mathcal{K} \times \mathcal{Y}$ satisfies uniformity and ν-key-privacy.

Proof. Uniformity is clear. For key-privacy, consider a state ρ_{KXYE} with the properties as in Definition 3. We fix an arbitrary $y \in \mathcal{Y}$ and condition on $Y = y$. Conditioning on $X = x$ as well for an arbitrary $x \in \mathcal{X}$, the key (K, K') is uniformly distributed subject to $\mathsf{H}(K, x) + K' = y$. In other words, K is uniformly random in $\mathcal{K}$, and $K' = y - \mathsf{H}(K, x)$. Therefore, making use of the Markov-chain property, conditioning on $Y = y$ only, K is uniformly random in $\mathcal{K}$ and independent of X and E, and $K' = y - \mathsf{H}(K, X)$. Thus, by the extractor property, $\delta(\rho_{K'KE|Y=y}, \mu_{\mathcal{Y}} \otimes \mu_K \otimes \rho_{E|Y=y}) \leq \frac{\nu}{2}\sqrt{\mathrm{Guess}(X|E, Y=y) \cdot |\mathcal{Y}|}$. The claim follows by averaging over y, and applying Jensen's inequality and Property 2.   □

The following technical result will be useful.

Lemma 1. *Let $\mathsf{H} : \mathcal{K} \times \mathcal{X} \to \mathcal{Y}$ be a keyed hash function that satisfies message-independence. Furthermore, let ρ_{KXYE} be a state with the properties as in Definition 3. Then*

$$\mathrm{Guess}(X|KYE) \leq \mathrm{Guess}(X|YE) \cdot |\mathcal{Y}|.$$

Proof. Note that the Markov-chain property $K \leftrightarrow XY \leftrightarrow E$ can be understood in that E is obtained by acting on XY only: $E = \mathcal{Q}(XY)$. For the purpose of the argument, we extend the state ρ_{XKYE} to a state $\rho_{XKK'YY'EE'}$ as follows. We choose a uniformly random and independent K' in $\mathcal{K}$, and set $Y' = \mathsf{H}(K', X)$ and $E' = \mathcal{Q}(XY')$. Note that ρ_{XKYE} coincides with $\rho_{XK'Y'E'}$. Therefore,

$$\mathrm{Guess}(X|YE) = \mathrm{Guess}(X|Y'E') = \mathrm{Guess}(X|KY'E'),$$

where the second equality is by the independence of K. Furthermore, by Property 3, we have that

$$\mathrm{Guess}(X|KY'E') \geq P[Y = Y'] \, \mathrm{Guess}(X|KY'E', Y = Y')$$
$$= P[Y = Y'] \, \mathrm{Guess}(X|KYE, Y = Y').$$

Finally, by the message-independence of H, it holds that Y' is independent of $KXYE$ (and with the same distribution as Y), and therefore $P[Y = Y'] \geq 1/|\mathcal{Y}|$ and $\mathrm{Guess}(X|KYE, Y = Y') = \mathrm{Guess}(X|KYE)$. Altogether, this gives us the bound $\mathrm{Guess}(X|KYE) \leq \mathrm{Guess}(X|YE) \cdot |\mathcal{Y}|$, which concludes the proof.   □

Equipped with Lemma 1, we can now show the following composition results.

---

[8] Here, and similarly in other occasions, $k\|k'$ is simply a synonym for the element (k, k') in the Cartesian product of, here, $\mathcal{K}$ and $\mathcal{Y}$, and is mainly used to smoothen notation and avoid expressions like $((k, k'), x)$.

Proposition 3 (Parallel Composition). *Consider two keyed hash functions* $H_1 : \mathcal{K}_1 \times \mathcal{X} \to \mathcal{Y}_1$ *and* $H_2 : \mathcal{K}_2 \times \mathcal{X} \to \mathcal{Y}_2$ *with the same message space* $\mathcal{X}$, *and*

$$H : (\mathcal{K}_1 \times \mathcal{K}_2) \times \mathcal{X} \to \mathcal{Y}_1 \times \mathcal{Y}_2, \ (k_1 \| k_2, x) \mapsto \big(H_1(k_1, x), H_2(k_2, x)\big)$$

with key space $\mathcal{K} = \mathcal{K}_1 \times \mathcal{K}_2$ *and range* $\mathcal{Y} = \mathcal{Y}_1 \times \mathcal{Y}_2$. *If* H_1 *and* H_2 *are both message-independent (or uniform) and respectively offer* ν_1- *and* ν_2-*key privacy, then* H *is message-independent (or uniform) and offers* $(\nu_1 + \nu_2)$-*key privacy.*

Proof. That message-independence/uniformity is preserved is clear. To argue key-privacy, assume that we have $\rho_{K_1 K_2 X} = \rho_{K_1} \otimes \rho_{K_2} \otimes \rho_X$, $Y_1 = H_1(K_1, X)$ and $Y_2 = H(K_2, X)$, and $K_1 K_2 \leftrightarrow XY_1Y_2 \leftrightarrow E$. We need to bound the distance of $K_1 K_2$ from uniform when given $Y_1 Y_1 E$, which we can decompose into

$$\delta(K_1 K_2, U_{\mathcal{K}_1} U_{\mathcal{K}_2} | Y_1 Y_1 E) \leq \delta(K_1, U_{\mathcal{K}_1} | Y_1 Y_2 E) + \delta(K_2, U_{\mathcal{K}_2} | K_1 Y_1 Y_2 E).$$

The above conditions on $\rho_{K_1 K_2 X Y_1 Y_2 E}$ imply that $K_1 \leftrightarrow XY_1 \leftrightarrow K_2 Y_2 E$ holds, and thus also $K_1 \leftrightarrow XY_1 \leftrightarrow Y_2 E$. Indeed, $K_1 K_2 \leftrightarrow XY_1 Y_2 \leftrightarrow E$ implies that also $K_1 \leftrightarrow XY_1 K_2 Y_2 \leftrightarrow E$, which together with $K_1 \leftrightarrow XY_1 \leftrightarrow K_2 Y_2$ (which holds by choice of K_2 and Y_2) implies that $K_1 \leftrightarrow XY_1 \leftrightarrow K_2 Y_2 E$. Therefore, by the key-privacy property of H_1, setting $E_1 = Y_2 E$, we see that

$$\delta(K_1, U_{\mathcal{K}_1} | Y_1 Y_2 E) \leq \frac{\nu_1}{2} \sqrt{\mathrm{Guess}(X | Y_1 Y_2 E) \cdot |\mathcal{Y}_1|}.$$

Similarly, $K_2 \leftrightarrow XY_2 \leftrightarrow K_1 Y_1 E$, and so by the key-privacy property of H_2, setting $E_2 = K_1 Y_1 E$, we conclude that

$$\delta(K_2, U_{\mathcal{K}_2} | K_1 Y_1 Y_2 E) \leq \frac{\nu_2}{2} \sqrt{\mathrm{Guess}(X | Y_2 K_1 Y_1 E) \cdot |\mathcal{Y}_2|}$$

$$\leq \frac{\nu_2}{2} \sqrt{\mathrm{Guess}(X | Y_2 Y_1 E) \cdot |\mathcal{Y}_1| \cdot |\mathcal{Y}_2|},$$

which proves the claim. □

Proposition 4 ("Sequarallel" Composition). *Consider two keyed hash functions* $H_1 : \mathcal{K}_1 \times \mathcal{X} \to \mathcal{Y}_1$ *and* $H_2 : \mathcal{K}_2 \times (\mathcal{X} \otimes \mathcal{Y}_1) \to \mathcal{Y}_2$ *with message spaces as specified, and*

$$H : (\mathcal{K}_1 \times \mathcal{K}_2) \times \mathcal{X} \to \mathcal{Y}_1 \times \mathcal{Y}_2, \ (k_1 \| k_2, x) \mapsto \big(H_1(k_1, x), H_2(k_2, x \| H_1(k_1, x))\big)$$

with key space $\mathcal{K} = \mathcal{K}_1 \times \mathcal{K}_2$ *and range* $\mathcal{Y} = \mathcal{Y}_1 \times \mathcal{Y}_2$. *If* H_1 *and* H_2 *are both message-independent (or uniform) and respectively offer* ν_1- *and* ν_2-*key privacy, then* H *is message-independent (or uniform) and offers* $(\nu_1 + \nu_2)$-*key privacy.*

Proof. The proof goes along the same lines as the proof of Proposition 3, except that in the reasoning for the bound on $\delta(K_2, U_{\mathcal{K}_2} | K_1 Y_1 E)$, we append Y_1 to X, with the consequence that we get a bound that is in terms of $\mathrm{Guess}(XY_1 | Y_2 Y_1 E)$, but this obviously coincides with $\mathrm{Guess}(X | Y_2 Y_1 E)$, and thus we end up with the same bound. □

4   Message Authentication with Key-Recycling

We first consider the problem of *message authentication* with key-recycling. It turns out that — at least with our approach — this is the actual challenging problem, and extending to (authenticated) encryption is then quite easy.

4.1   The Semantics

We quickly specify the semantics of a quantum authentication code (or scheme) with key-recycling.[9]

Definition 4. *A quantum authentication code (with key recycling) QMAC with message space $\mathcal{MSG}$ and key space $\mathcal{KEY}$ is made up of the following components: (1) A CPTP map* Auth *that is controlled by a message $msg \in \mathcal{MSG}$ and a key $key \in \mathcal{KEY}$, and that acts on an empty system and outputs a quantum authentication tag (with a fixed state space), (2) a measurement* Verify *that is controlled by $msg \in \mathcal{MSG}$ and $key \in \mathcal{KEY}$, and that acts on a quantum authentication tag and outputs a decision bit $d \in \{0,1\}$, and (3) a randomized function* Refresh $: \mathcal{KEY} \rightarrow \mathcal{KEY}$.

We will often identify an authentication code, formalized as above, with the obvious *authenticated-message-transmission protocol* $\pi_{\mathsf{QMAC}}(msg)$, where Alice and Bob start with a shared key $key \in \mathcal{KEY}$, and Alice sends the message msg along with its quantum authentication tag prepared by means of Auth to Bob over a channel that is controlled by the adversary Eve, and, upon reception of the (possibly modified) message and tag, Bob verifies correctness using Verify and accordingly accepts or rejects. If he rejects, then Alice and Bob replace key by $key' := \mathsf{Refresh}(key)$.[10] Note that, for any message $msg \in \mathcal{MSG}$ and any strategy for Eve on how to interfere with the communication, the protocol $\pi_{\mathsf{QMAC}}(msg)$ induces a CPTP map $\mathcal{Exe}[\pi_{\mathsf{QMAC}}(msg)] : \mathcal{D}(\mathcal{KEY} \otimes \mathcal{H}_E) \rightarrow \mathcal{D}(\mathcal{KEY} \otimes \mathcal{H}_{E'})$ that describes the evolution of the shared key key and Eve's local system as a result of the execution of $\pi_{\mathsf{QMAC}}(msg)$.

Our goal will be to show that, for our construction given below, and for any behavior of Eve, the CPTP map $\mathcal{Exe}[\pi_{\mathsf{QMAC}}(msg)]$ maps a key about which Eve has little information into a (possibly updated) key about which Eve still has little information — what it means here to "have little information" needs to be specified, but it will in particular imply that it still allows Bob to detect a modification of the message. This then ensures re-usability of the quantum authentication code — with the same key as long as Bob accepts the incoming messages, and with the updated key in case he rejects.

---

[9] Our definition is tailored to our goal that the key can be re-used unchanged in case the message is accepted by the recipient, Bob, and only needs to be refreshed in case he rejects. In the literature, key-recycling sometimes comes with *two* refresh procedures, one for the case Bob rejects and one for the case he accepts.

[10] Obviously, this requires Alice and Bob to exchange fresh randomness, i.e., the randomness for executing Refresh, in a *reliable* and *private* way; how this is done is not relevant here.

4.2 The Scheme

Let $\mathcal{MSG}$ be an arbitrary non-empty finite set. We are going to construct a quantum message authentication code QMAC with message space $\mathcal{MSG}$. To this end, let $\mathsf{MAC} : \mathcal{K} \times (\mathcal{MSG} \times \{0,1\}^n) \to \mathcal{T}$ be a classical one-time message authentication code with a message space $\mathcal{MSG} \times \{0,1\}^n$ for some $n \in \mathbb{N}$. We require MAC to be secure in the standard sense, meaning a modified message will be detected except with small probability $\varepsilon_{\mathsf{MAC}}$. Additionally, we require MAC to satisfy message-independence and ideal key-privacy, as discussed in Sect. 3.2. Actually, it is sufficient if $\mathsf{MAC}(\cdot, msg\| \cdot)$, i.e., the hash function $\mathcal{K} \times \{0,1\}^n \to \mathcal{T}$, $(k,x) \mapsto \mathsf{MAC}(k, msg\|x)$ obtained by fixing msg, satisfies message-independence and ideal key-privacy for any $msg \in \mathcal{MSG}$. Assuming that $\mathcal{MSG}$ consists of bit strings of fixed size so that $\mathcal{MSG} \times \{0,1\}^n = \{0,1\}^N$ for some $N \in \mathbb{N}$, the canonical message authentication codes $\mathsf{MAC} : \left(\mathbb{F}_2^{\ell \times N} \times \mathbb{F}_2^\ell\right) \times \mathbb{F}_2^N \to \mathbb{F}_2^\ell$, $(A\|b,x) \mapsto Ax+b$ and $\mathsf{MAC} : \left(\mathbb{F}_{2^N} \times \mathbb{F}_2^\ell\right) \times \mathbb{F}_{2^N} \to \mathbb{F}_2^\ell$, $(a\|b,x) \mapsto trunc(a \cdot x)+b$, where $trunc : \mathbb{F}_{2^N} \to \mathbb{F}_2^\ell$ is an arbitrary surjective $\mathbb{F}_2$-linear map, are suitable choices; this follows directly from Proposition 2. Finally, let $\mathcal{C} \subset \{0,1\}^n$ be a code with large minimal distance d.

Then, our quantum message authentication code QMAC has a key space $\mathcal{KEY} = \mathcal{K} \times \mathcal{C}$, where for a key $k\|\theta \in \mathcal{K} \times \mathcal{C}$ we refer to k as the "MAC key" and to θ as the "basis key", and QMAC works as described in Fig. 1.

QMAC.Auth($k\|\theta, msg$):   Choose a uniformly random $x \in \{0,1\}^n$ and output n qubits $B_\circ$ in state $H^\theta |x\rangle$ and the classical tag $t = \mathsf{MAC}(k, msg\|x)$.

QMAC.Verify($k\|\theta, msg, t$):   Measure the qubits $B_\circ$ in bases θ to obtain x' (supposed to be x), check that $t = \mathsf{MAC}(k, msg\|x')$, and output 0 or 1 accordingly.

QMAC.Refresh($k\|\theta$):   Choose a uniformly random $\theta' \in \mathcal{C}$ and output $k\|\theta'$.

Fig. 1. The quantum message authentication code MAC.

It is clear that as long as the MAC key k is "secure enough", the classical MAC takes care of an Eve that tries to modify the message msg, and it ensures that such an attack is detected by Bob, except with small probability. What is non-trivial to argue is that the MAC key (together with the basis key) indeed stays "secure enough" over multiple executions of $\pi_{\mathsf{QMAC}}(msg)$; this is what we show below.

4.3 Analysis

We consider an execution of the authenticated-message-transmission protocol $\pi_{\mathsf{QMAC}}(msg)$ for a fixed message msg. Let $\rho_{K\Theta E} \in \mathcal{D}(\mathcal{K} \otimes \mathcal{C} \otimes \mathcal{H}_E)$ be the joint state *before* the execution, consisting of the MAC key K, the basis key Θ, and Eve's local quantum system E. The joint state $\mathcal{E}xe[\pi_{\mathsf{QMAC}}(msg)](\rho_{K\Theta E})$ *after* the execution is given by $\rho_{K\Theta'TDC} \in \mathcal{D}(\mathcal{K} \otimes \mathcal{C} \otimes \mathcal{T} \otimes \{0,1\} \otimes \mathcal{H}_C)$, where Θ' is the (possibly) updated basis key, T is the classical tag, D is Bob's decision to accept or reject, and C is Eve's new quantum system. Eve's complete information after the execution of the scheme is thus given by $E' = TDC$.

Recall that TDC is obtained as follows from $K\Theta E$. Alice prepares BB84-qubits B_o for uniformly random bits X and with bases determined by Θ, and she computes the tag $T := \mathsf{MAC}(K, msg\|X)$. Then, Eve acts on $B_o E$ (in a way that may depend on T) and keeps one part, C, of the resulting state, and Bob measures the other part, B, to obtain X' and checks with the (possibly modified) tag T to decide on D.

Note that by a standard reasoning, we can think of the BB84 qubits B_o not as being prepared by first choosing the classical bits X and then "encoding" them into qubits with the prescribed bases Θ, but by first preparing n EPR pairs $\Phi^+_{AB_o}$ and then measuring the qubits in A in the prescribed bases to obtain X, i.e.,

$$\rho_{K\Theta XB_o E} = \mathcal{M}^{\mathsf{BB84}}_{\Theta A \to X}\big(\Phi^+_{AB_o} \otimes \rho_{K\Theta E}\big).$$

The following captures the main security property of the scheme.

Theorem 1. *If the state before the execution of $\pi_{\mathsf{QMAC}}(msg)$ is of the form $\rho_{K\Theta E} = \mu_K \otimes \rho_{\Theta E}$, then for any Eve the state $\rho_{K\Theta'E'} = \mathcal{E}xe[\pi_{\mathsf{QMAC}}(msg)](\rho_{K\Theta E})$ after the execution satisfies*

$$\mathrm{Guess}(\Theta'|E') \le \mathrm{Guess}(\Theta|E) + \frac{1}{|\mathcal{C}|}$$

and

$$\delta(K, U_K|\Theta'E') \le 2\varepsilon_{\mathsf{MAC}} + \frac{\sqrt{2}}{2}\sqrt{\mathrm{Guess}(\Theta|E)\left(1 + \frac{|\mathcal{C}|}{2^{d/2}}\right)|\mathcal{T}|}.$$

This means that if *before* the execution of $\pi_{\mathsf{QMAC}}(msg)$, it holds that Eve's guessing probability on Θ is small and K looks perfectly random to her (even when given Θ), then *after* the execution, Eve's guessing probability on (the possibly refreshed) Θ' is still small and K still looks *almost* perfectly random to her. As such, we may then consider a *hypothetical* refreshing of K that has almost no impact, but which brings us back to the position to apply Theorem 1 again, and hence allows us to re-apply this "preservation of security" for the next execution, and so on. This in particular allows us to conclude that in an arbitrary *sequence* of executions, the MAC key K stays almost perfectly random for Eve, and thus any tampering with an authenticated message will be detected by Bob except with small probability by the security of MAC (see Sect. 4.4 for more details).

Remark 5. For simplicity, in Theorem 1 and in the remainder of this work, we assume the message msg to be arbitrary but *fixed*. However, it is not hard to see that we may also allow msg to be obtained by means of a measurement, applied to Eve's system E before the execution of $\pi_{\mathsf{QMAC}}(msg)$, i.e., Eve can choose it. The bounds of Theorem 1 then hold *on average over the measured msg*. This follows directly from Property 2 for the bound the guessing probability, and from a similar decomposition property for the trace distance, together with Jensen's inequality for the bound on the trace distance. We emphasize however, that we do assume *msg*, even when provided by Eve, to be *classical*, i.e., we do not consider so-called superposition attacks.

The formal proof of Theorem 1 is given below; the intuition is as follows. For the bound on the guessing probability of the (possibly updated) basis key, we have that in case Bob rejects and so the basis key is re-sampled from $\mathcal{C}$, Eve has obviously guessing probability $1/|\mathcal{C}|$. In case Bob accepts, the fact that Bob accepts may increase Eve's guessing probability. For instance, Eve may measure one qubit in, say, the computational basis, and forward the correspondingly collapsed qubit to Bob; if Bob then accepts it is more likely that this qubit had been prepared in the computational basis by Alice, giving Eve some (new) information on the basis key. However, the resulting increase in guessing probability is *inverse proportional* to the probability that Bob actually accepts, so that this advantage is "canceled out" by the possibility that Bob will not accept. For the bound on the "freshness" of K (given the basis key Θ'), by key privacy it is sufficient to argue that Eve has small guessing probability for X. In case Bob rejects, the (refreshed) basis key is useless to her for guessing X, and so the task of guessing X reduces to winning the game considered in Corollary 2. Similarly, the case where Bob accepts fits into the game in Corollary 1. In both cases, we get that the guessing probability of X essentially coincides with $\mathrm{Guess}(\Theta|E)$.

Proof. For the first claim, we simply observe that

$$\mathrm{Guess}(\Theta'|TDC) = \sum_{d=0}^{1} P_D(d)\, \mathrm{Guess}(\Theta'|TC, D=d) \qquad \text{(by Property 2)}$$

$$= P_D(0)\frac{1}{|\mathcal{C}|} + P_D(1)\, \mathrm{Guess}(\Theta|TC, D=1)$$

$$\leq P_D(0)\frac{1}{|\mathcal{C}|} + \mathrm{Guess}(\Theta|TC) \qquad \text{(by Property 3)}$$

$$\leq \frac{1}{|\mathcal{C}|} + \mathrm{Guess}(\Theta|TB_\circ E) \qquad \text{(by Property 1)}$$

$$= \frac{1}{|\mathcal{C}|} + \mathrm{Guess}(\Theta|B_\circ E) \qquad \text{(by Definition 2)}$$

$$= \frac{1}{|\mathcal{C}|} + \mathrm{Guess}(\Theta|E).$$

where the second equality holds because Θ' is freshly chosen in case Bob rejects and $\Theta' = \Theta$ in case he accepts, and the final equality holds because of the fact that $\rho_{B_\circ \Theta E} = \mathrm{tr}_X \circ \mathcal{M}_{\Theta A \to X}^{\mathrm{BB84}}\left(\Phi_{AB_\circ}^+ \otimes \rho_{\Theta E}\right) = \mathrm{tr}_A\left(\Phi_{AB_\circ}^+\right) \otimes \rho_{\Theta E} = \mu_{B_\circ} \otimes \rho_{\Theta E}$.

For the second claim, consider $\tilde{D}$ and $\tilde{\Theta}'$ as follows. $\tilde{D}$ is 1 if $X = X'$ and Eve has not modified the tag T nor the message *msg*, and $\tilde{D}$ is 0 otherwise (i.e., $\tilde{D}$ is an "ideal version" of Bob's decision), and $\tilde{\Theta}'$ is freshly chosen if and only $\tilde{D} = 0$. The states of $K\Theta'TDC$ and $K\tilde{\Theta}'T\tilde{D}C$ are identical except for when $D = 1$ but

$X \neq X'$ or Eve has modified T or msg, which happens with probability at most $\varepsilon_{\mathsf{MAC}}$ by the security of MAC, and thus the two states are $\varepsilon_{\mathsf{MAC}}$-close. Therefore, $\delta(K, U_{\mathcal{K}} | \Theta' T D C) \leq \delta(K, U_{\mathcal{K}} | \tilde{\Theta}' T \tilde{D} C) + 2\varepsilon_{\mathsf{MAC}}$, and so it suffices to analyze the state of $K \tilde{\Theta}' T \tilde{D} C$. Furthermore, we may assume that Eve's state C contains the information of whether she modified T or msg, so that $\tilde{D}$ can be computed from $1_{X=X'}$ when given C, and thus $\delta(K, U_{\mathcal{K}} | \tilde{\Theta}' \tilde{D} T C) \leq \delta(K, U_{\mathcal{K}} | \tilde{\Theta}' 1_{X=X'} \tilde{D} T C) = \delta(K, U_{\mathcal{K}} | \tilde{\Theta}' 1_{X=X'} T C)$.

Now, since K is random and independent of $X \Theta B_{\circ} E$, T is computed as $T = \mathsf{MAC}(K, msg | X)$, and $\tilde{\Theta} 1_{X=X'} C$ is obtained by acting on T and $X \Theta B_{\circ} E$ only (and not on K), we see that the conditions required in Definition 3 are satisfied. Therefore, by the key-privacy of MAC, and recalling Remark 4,

$$\delta(K, U_{\mathcal{K}} | T \tilde{\Theta}' 1_{X=X'} C) \leq \frac{1}{2} \sqrt{\mathrm{Guess}(X | \tilde{\Theta}' 1_{X=X'} C) |\mathcal{T}|}.$$

Furthermore, by Property 2, and noting that $\tilde{\Theta}'$ is freshly chosen when $X \neq X'$ and equal to Θ otherwise,

$$\mathrm{Guess}(X | \tilde{\Theta}' 1_{X=X'} C) = P[X \neq X'] \mathrm{Guess}(X | C, X \neq X') + P[X = X'] \mathrm{Guess}(X | \Theta C, X = X').$$

For the first term, we see that

$$
\begin{aligned}
P[X \neq X'] \mathrm{Guess}(X | C, X \neq X') &\leq \mathrm{Guess}(X | C) &&\text{(by Property 3)} \\
&\leq \mathrm{Guess}(X | T B_{\circ} E) &&\text{(by Property 1)} \\
&\leq \mathrm{Guess}(X | B_{\circ} E) &&\text{(by Definition 2)} \\
&\leq \mathrm{Guess}(\Theta | E)\left(1 + \tfrac{|\mathcal{C}|}{2^{d/2}}\right),
\end{aligned}
$$

where the final inequality follows from Corollary 2 by recalling that $\rho_{\Theta X B_{\circ} E} = \mathcal{M}^{\mathsf{BB84}}_{\Theta A \to X}\left(\Phi^{+}_{AB_{\circ}} \otimes \rho_{\Theta E}\right)$. Similarly, writing X'' for the measurement outcome when measuring C using an optimal measurement $\mathcal{N}_{\Theta C}$ (controlled by Θ), we obtain

$$
\begin{aligned}
P[X = X'] \mathrm{Guess}(X | \Theta C, X = X') &\leq P[X = X'] P[X = X'' | X = X'] \\
&\leq P[X = X' \wedge X = X''] \\
&\leq \mathrm{Guess}(\Theta | E)\left(1 + \tfrac{|\mathcal{C}|}{2^{d/2}}\right),
\end{aligned}
$$

where the final inequality follows from Corollary 1 by observing that, using uniformity of MAC (Definition 2) in the second equality,

$$
\begin{aligned}
\rho_{\Theta X X X''} &= \mathcal{N}_{\Theta C \to X''} \circ \mathcal{M}^{\mathsf{BB84}}_{\Theta B \to X'} \circ \mathcal{Q}_{T B_{\circ} E \to B C}\left(\rho_{\Theta X T B_{\circ} E}\right) \\
&= \mathcal{N}_{\Theta C \to X''} \circ \mathcal{M}^{\mathsf{BB84}}_{\Theta B \to X} \circ \mathcal{Q}_{T B_{\circ} E \to B C}\left(\rho_{\Theta X B_{\circ} E} \otimes \rho_{T}\right) \\
&= \mathcal{N}_{\Theta C \to X''} \circ \mathcal{M}^{\mathsf{BB84}}_{\Theta B \to X'} \circ \mathcal{Q}_{T B_{\circ} E \to B C} \circ \mathcal{M}^{\mathsf{BB84}}_{\Theta A \to X}\left(\Phi^{+}_{AB_{\circ}} \otimes \rho_{\Theta E} \otimes \rho_{T}\right) \\
&= \mathcal{N}_{\Theta C \to X''} \circ \mathcal{M}^{\mathsf{BB84}}_{\Theta B \to X'} \circ \mathcal{M}^{\mathsf{BB84}}_{\Theta A \to X} \circ \mathcal{Q}_{T B_{\circ} E \to B C}\left(\Phi^{+}_{AB_{\circ}} \otimes \rho_{\Theta E} \otimes \rho_{T}\right) \\
&= \mathcal{N}_{\Theta C \to X''} \circ \mathcal{M}^{\mathsf{BB84}}_{\Theta B \to X'} \circ \mathcal{M}^{\mathsf{BB84}}_{\Theta A \to X} \circ \mathcal{Q}'_{E \to ABC}\left(\rho_{\Theta E}\right)
\end{aligned}
$$

where $\mathcal{Q}'_{E \to ABC}$ is the CPTP map $\mathcal{Q}'_{E \to ABC}(\sigma_{E}) = \mathcal{Q}_{T B_{\circ} E \to B C}\left(\Phi^{+}_{AB_{\circ}} \otimes \sigma_{E} \otimes \rho_{T}\right)$. Collecting the terms gives the claimed bound. $\qquad\square$

4.4   Re-usability of QMAC

We formally argue here that Theorem 1, which analyses a *single* usage of QMAC, implies *re-usability*. The reason why this is not completely trivial is that after one execution of π_{QMAC}, the MAC key K is not perfectly secure anymore but "only" almost-perfectly secure, so that Theorem 1 cannot be directly applied anymore for a second execution. However, taking care of this is quite straightforward.

Formally, we have the following result regarding the re-usability of QMAC.

Proposition 5. *If Alice and Bob start off with a uniformly random key, then for a sequence* $\pi_{\text{QMAC}}(msg_1), \pi_{\text{QMAC}}(msg_2), \ldots$ *of sequential executions of protocol* π_{QMAC}, *and for any strategy for Eve and any* $i \in \mathbb{N}$, *the probability* ε_i *that Eve modifies* msg_i *in the execution of* $\pi_{\text{QMAC}}(msg_i)$ *yet Bob accepts is bounded by*

$$\varepsilon_i \leq (2i - 1) \cdot \varepsilon_{\text{MAC}} + \frac{\sqrt{2}}{2} \sum_{j < i} \sqrt{\frac{j}{|\mathcal{C}|} \left(1 + \frac{|\mathcal{C}|}{2^{d/2}} \right) |\mathcal{T}|}.$$

Proof. In case $i = 1$, the statement reduces to $\varepsilon_i \leq \varepsilon_{\text{MAC}}$, which holds by construction of QMAC. To argue the general case, let $\rho_{K_1 \Theta_1 E_1}, \rho_{K_2 \Theta_2 E_2}, \ldots$ describe the evolution of the MAC key and the basis key and Eve's information on them, given that we start with a perfect key $\rho_{K_0 \Theta_0 E_0} = \mu_{\mathcal{K}} \otimes \mu_{\mathcal{C}} \otimes \rho_{E_0}$. Formally, $\rho_{K_i \Theta_i E_i}$ is inductively defined as $\rho_{K_i \Theta_i E_i} = \mathcal{E}xe[\pi_{\text{QMAC}}(msg_i)](\rho_{K_{i-1} \Theta_{i-1} E_{i-1}})$. For the sake of the argument, we also consider $\tilde{\rho}_{K_1 \Theta_1 E_1}, \tilde{\rho}_{K_2 \Theta_2 E_2}, \ldots$ obtained by means of setting $\tilde{\rho}_{K_0 \Theta_0 E_0} = \rho_{K_0 \Theta_0 E_0}$ and $\tilde{\rho}_{K_i \Theta_i E_i} = \mathcal{E}xe[\pi_{\text{QMAC}}(msg_i)](\mu_{\mathcal{K}} \otimes \tilde{\rho}_{\Theta_{i-1} E_{i-1}})$, i.e., the evolution of the keys and Eve's information in a hypothetical setting where the MAC key is refreshed before every new execution. For these latter states $\tilde{\rho}_{K_i \Theta_i E_i}$, we can inductively apply Theorem 1 and conclude that

$$\text{Guess}(\Theta_i | E_i) \leq \frac{i + 1}{|\mathcal{C}|}$$

and

$$\delta(\tilde{\rho}_{K_i \Theta_i E_i}, \mu_{\mathcal{K}} \otimes \tilde{\rho}_{\Theta_i E_i}) \leq \delta_i := 2\varepsilon_{\text{MAC}} + \frac{\sqrt{2}}{2} \sqrt{\frac{i}{|\mathcal{C}|} \left(1 + \frac{|\mathcal{C}|}{2^{d/2}} \right) |\mathcal{T}|}$$

for any $i \in \mathbb{N}$. But now, for the original states $\rho_{K_1 \Theta_1 E_1}, \rho_{K_2 \Theta_2 E_2}, \ldots$, from the triangle inequality we obtain that

$$\delta(\rho_{K_i \Theta_i E_i}, \mu_{\mathcal{K}} \otimes \tilde{\rho}_{\Theta_i E_i}) \leq \delta(\rho_{K_i \Theta_i E_i}, \tilde{\rho}_{K_i \Theta_i E_i}) + \delta(\tilde{\rho}_{K_i \Theta_i E_i}, \mu_{\mathcal{K}} \otimes \tilde{\rho}_{\Theta_i E_i})$$

$$\leq \delta(\rho_{K_{i-1} \Theta_{i-1} E_{i-1}}, \mu_{\mathcal{K}} \otimes \tilde{\rho}_{\Theta_{i-1} E_{i-1}}) + \delta_i$$

$$\leq \sum_{j \leq i} \delta_j,$$

where the last inequality is by induction (where the base case $i = 0$ is trivially satisfied). It now follows by basic properties of the trace distance that we have $\varepsilon_{i+1} \leq \varepsilon_{\text{MAC}} + \sum_{j \leq i} \delta_j$. This proves the claim. $\qquad \square$

4.5   Choosing the Parameters

Let $\lambda \in \mathbb{N}$ be the security parameter. Consider a MAC with $\varepsilon_{\mathsf{MAC}} = 2^{-\lambda}$ and $|\mathcal{T}| = 2^{\lambda}$. This can for instance be achieved with the constructions suggested in Sect. 4.2. Also, consider a code $\mathcal{C}$ with $|\mathcal{C}| = 2^{3\lambda}$ and $d = 6\lambda$, so that $|\mathcal{C}|/2^{d/2} = 1$. The description of the basis key θ thus requires 3λ bits, and, by Singleton bound, it is necessary that $n \geq 9\lambda - 1$. Then, the bound in Proposition 5 becomes

$$\varepsilon_{i+1} \leq (2i+1) \cdot 2^{-\lambda} + \sum_{j \leq i} \frac{\sqrt{2}}{2} \sqrt{\frac{j}{2^{3\lambda}} \left(1 + \frac{2^{3\lambda}}{2^{3\lambda}}\right)} \, 2^{\lambda} = \left(2i + 1 + \sum_{j \leq i} \sqrt{j}\right) 2^{-\lambda}.$$

Hence, the error probability increases at most as $(i\sqrt{i} + 2i + 1)2^{-\lambda}$ with the number i of executions.

5   Extensions and Variations

We show how to modify our scheme QMAC as to offer encryption as well, i.e., to produce an authenticated encryption of *msg*, and how to deal with noise in the quantum communication; we start with the latter since this is more cumbersome. At the end of the section, we show how to tweak our schemes so as to be able to authenticate and/or encrypt *quantum* messages as well, and we discuss some variations.

5.1   Dealing with Noise

In order to deal with noise in the quantum communication, we introduce the following primitive. We consider a keyed hash function $\mathsf{SS} : \mathcal{L} \times \{0,1\}^n \to \mathcal{S}$ that has the property that given the key ℓ, the *secure sketch* $s = \mathsf{SS}(\ell, x)$ of the message x, and a "noisy version" x' of x, i.e., such that $d_H(x, x') \leq \varphi \cdot n$ for some given noise parameter $\varphi < \frac{1}{2}$, the original message x can be recovered except with probability $\varepsilon_{\mathsf{SS}}$. Additionally, we want SS to satisfy the message-independence and ideal key-privacy properties from Definitions 2 and 3. Such constructions exist for small enough $\varphi > 0$, as discussed in Appendix B, based on techniques by Dodis and Smith [8].

Then, the key for our noise-tolerant quantum message authentication code QMAC* consists of a (initially) uniformly random MAC key $k \in \mathcal{K}$ for MAC, an (initially) uniformly random secure-sketch key $\ell \in \mathcal{L}$ for SS, and an (initially) random and independent basis key θ, chosen from the code $\mathcal{C} \subset \{0,1\}^n$, and the scheme works as described in Fig. 2.

Theorem 2. *If the state before the execution of $\pi_{\mathsf{QMAC}^*}(msg)$ is of the form $\rho_{K\Theta E} = \mu_{\mathcal{K}} \otimes \rho_{\Theta E}$, then for any Eve the state $\rho_{K\Theta'E'} = \mathcal{E}xe[\pi_{\mathsf{QMAC}^*}(msg)](\rho_{K\Theta E})$ after the execution satisfies*

$$\mathrm{Guess}(\Theta'|E') \leq \mathrm{Guess}(\Theta|E) + \frac{1}{|\mathcal{C}|}$$

and

$$\delta(KL, U_{\mathcal{K}\times\mathcal{L}}|\Theta'E') \leq 2\varepsilon_{\mathsf{MAC+SS}} + \sqrt{\mathrm{Guess}(\Theta|E)\left(2+\frac{|\mathcal{C}|}{2^{d/2}}+\frac{|\mathcal{C}|\cdot 2^{h(\varphi)n}}{2^d}\right)|\mathcal{T}||\mathcal{S}|}$$

where $\varepsilon_{\mathsf{MAC+SS}} = \varepsilon_{\mathsf{MAC}} + \varepsilon_{\mathsf{SS}}$.

QMAC*.Auth($k\|\ell\|\theta, msg$): Choose a uniformly random $x \in \{0,1\}^n$ and output n qubits B_o in state $H^\theta|x\rangle$ together with the secure sketch $s = \mathsf{SS}(\ell, x)$ and the tag $t = \mathsf{MAC}(k, msg\|x\|s)$.

QMAC*.Verify($k\|\ell\|\theta, msg, t$): Measure the qubits B_o in bases θ to obtain x', recover (what is supposed to be) x using the secure sketch s, and check the tag t. If this check fails or $d_H(x, x') > \varphi \cdot n$ then output 0, else 1.

QMAC*.Refresh($k\|\ell\|\theta$): Choose uniformly random $\theta' \in \mathcal{C}$ and output $k\|\ell\|\theta'$.

Fig. 2. The noise-tolerant quantum message authentication code MAC*.

Proof. The proof of the first statement, i.e., the bound on $\mathrm{Guess}(\Theta'|E')$ is exactly like in the proof of Theorem 1, with the only exception that in the one expression where the tag T appears (i.e. in the expression obtained by using Property 1), now S appears as well (along with T); but like T, it disappears again in the next step due to message-independence.

For the bound on $\delta(KL, U_{\mathcal{K}\times\mathcal{L}}|\Theta'E')$ we follow closely the proof of Theorem 1 but with the following modifications.

1. The key K is replaced by the key pair (K, L), and the tag T by the tag-secure-sketch pair (T, S), and we observe that we can understand (T, S) to be the hash of the input X under key (K, L) with respect to a hash function that satisfies message-independent and (almost) key-privacy. Indeed, this composed hash function can be understood as being obtained by means of Proposition 4. As such, whenever we argue by means of message-independence (Definition 2) or key-privacy (Definition 3) in the proof of Theorem 1, we can still do so, except that we need to adjust the bound on the uniformity of the key to the new — and now composite — hash function.

2. The auxiliary random variable $\tilde{D}$, and correspondingly $\tilde{\Theta}'$, is defined in a slightly different way: $\tilde{D}$ is 1 if $X \approx_\varphi X'$ and Eve has not modified the tag T, the secure-sketch S, nor the message msg. The "real" state with D and Θ' is then ($\varepsilon_{\mathsf{MAC}} + \varepsilon_{\mathsf{SS}}$)-close to the modified one with $\tilde{D}$ and $\tilde{\Theta}'$ instead. Correspondingly, the decomposition of the distance to be bounded is then done with respect to the indicated random variable $1_{X\approx_\varphi X'}$ instead of $1_{X=X'}$.

3. When bounding the probability $P[X \approx_\varphi X' \wedge X = X'']$, we refer to the game analyzed in Corollary 3 in Appendix A, which applies to the situation here where some slack is given for Bob's guess.

The claimed bound is then obtained by adjusting terms according to the above changes: update the bounds obtained by applying Definition 3 to the updated bound $\sqrt{\text{Guess}(X | \cdots)} |\mathcal{T}| |\mathcal{S}|$, obtained by means of Proposition 4, and inserting the $2^{h(\varphi)n}$ blow-up when using Corollary 3 instead of Corollary 1, but making use of the observation in Remark 6. $\square$

In essence, compared to the case with no noise, we have an additional loss due to the $|\mathcal{S}|$ term, whereas we can neglect the term with $2^{h(\varphi)n}$ for small enough φ. To compensate for this additional loss, we need to have $\varsigma = \log|\mathcal{S}|$ additional bits of entropy in Θ, i.e., we need to choose $\mathcal{C}$ with $|\mathcal{C}| = 2^{3\lambda+\varsigma}$ and $d = 6\lambda + 2\varsigma$. By Singleton bound, this requires $n \geq 9\lambda + 3\varsigma - 1$, and thus puts a bound $\varsigma < n/3$ on the size of the secure sketch, and thus limits the noise parameter φ.[11]

5.2 Adding Encryption

Adding encryption now works pretty straightforwardly. Concretely, our quantum encryption scheme with key recycling QENC* is obtained by means of the following modifications to QMAC*. Alice and Bob extract additional randomness from x using an extractor that offers message-independence and key-privacy, and use the extracted randomness as one-time-pad key to en-/decrypt msg. Finally, the resulting ciphertext c is authenticated along with x and s; this is in order to offer authenticity as well and can be omitted if privacy is the only concern.

Security can be proven along the same lines as Theorem 1, respectively Theorem 2 for the noise-tolerant version, and Proposition 5: we simply observe by means of Propositions 3 and 4 that the composition of computing the triple c, s and $t = \text{MAC}(k, x\|c\|s)$ from x constitutes a keyed hash function that offers message-independence and key-privacy, and then we can argue exactly as above to show that the (possibly refreshed) key stays secure over many executions. Also, given that the key is secure before an execution, we can control the min-entropy in X as in the proof of Theorem 2 and argue almost-perfect security of the extracted one-time-pad key, implying privacy of the communicated message.

In order to accommodate for the additional entropy that is necessary to extract this one-time-pad key, which is reflected in the adjusted range of the composed keyed hash function, we now have to choose $\mathcal{C}$ with $|\mathcal{C}| = 2^{3\lambda+\varsigma+m}$ and $d = 6\lambda + 2\varsigma + 2m$, where $m = \log|\mathcal{MSG}|$; this requires $n \geq 9\lambda + 3\varsigma + 3m - 1$ by Singleton bound.

---

[11] We recall that, when using a δ-biased family of codes to construct the secure sketch SS, as discussed in the Appendix B, then ς does not correspond exactly to the size of the syndrome given by the code, but is determined by the parameter δ, and is actually somewhat larger than the size of the syndrome.

5.3 Optimality of the Key Recycling

Our aim was, like in [5,6], to minimize the number of fresh random bits needed for the key refreshing. In our constructions, where the key is refreshed simply by choosing a new basis key θ, this number is obviously given by the number of bits needed to represent θ, i.e., in the above encryption scheme QENC*, it is

$$\log |\mathcal{C}| = 3\lambda + \varsigma + m.$$

This is close to optimal for large messages and assuming almost no noise, so that $m \gg \lambda, \varsigma$. Indeed, assuming that Eve knows the encrypted message, i.e., we consider a known-plaintext attack, it is not hard to see that for any scheme that offers (almost) perfect privacy of the message, by simply keeping everything that is communicated from Alice to Bob, in particular by keeping all qubits that Alice communicates (which will most likely trigger Bob to reject), Eve can always learn (almost) m bits of Shannon information on the key. As such, it is obviously necessary that the key is updated with (almost) m fresh bits of randomness in case Bob rejects, since otherwise Eve will soon have accumulated too much information on the key.

Note that [5,6] offers a rigorously proven bound (of roughly m) on the number of fresh bits necessary for key refreshing. However, their notion of key refreshing is stronger than what we require: they require that the refreshed key is close to random and independent of Eve, whereas we merely require that the refreshed key is "secure enough" as to ensure security of the primitive, i.e., authenticity in QMAC or QMAC*, and privacy (and authenticity) in QENC*. Indeed, in our construction we do not require that the basis key is close to random, only that it is hard to guess. However, the above informal argument shows that the bound still applies.

Similarly, one can argue that in any message authentication scheme with error probability $2^{-\lambda}$, by keeping everything Eve can obtain λ bits of information on the key. Thus, in case of almost no noise, our scheme QMAC* is optimal up to the factor 3.

In our constructions, the number of fresh random bits needed for the key refreshing increases with larger noise. In particular in QMAC*, ς will soon be the dominating term in case we increase the noise level. We point out that it is not clear whether such a dependency is necessary, as we briefly mention in Sect. 6.

5.4 Supporting Quantum Messages

The approach in Sect. 5.2 of extracting a (one-time pad) key also gives us the means to authenticate and/or encrypt *quantum* messages: we simply use the extracted key as quantum-one-time-pad key [1], or as key for a quantum message authentication code [4]. However, when considering arbitrary quantum messages, honest users anyway need a quantum computer, so one might just as well use the scheme by Damgård *et al.* to communicate a secret key and use this key for a quantum-one-time-pad or for quantum message authentication, or resort to [10,16], which additionally offer security against superposition attacks.

5.5   Variations

We briefly mention a few simple variations of our schemes. The first variation is as follows. In QMAC, instead of choosing x uniformly at random and computing the tag t as $t = \mathsf{MAC}(k, msg\|x)$, we can consider a fixed tag $t_\circ \in \mathcal{T}$, and choose x uniformly at random subject to $\mathsf{MAC}(k, msg\|x) = t_\circ$. Since $t_\circ$ is fixed, it does not have to be sent along. In case the classical MAC, as a keyed hash function, is of the form as in Proposition 2, meaning that the tag is one-time-pad encrypted (which in particular holds for the canonical examples suggested in Sect. 4.2), then Theorem 1 and Proposition 5 still hold. Indeed, if MAC is of this form then the concrete choice of $t_\circ$ is irrelevant for security: if Theorem 1 would fail for one particular choice of $t_\circ$ then it would fail for any choice, and thus also for a randomly chosen tag, which would then contradict Theorem 1 for the original QMAC. Similarly, in QMAC* and QENC* we can fix the tag t and the secure sketch s (and ciphertext c), and choose x subject to the corresponding restrictions.

A second variation is to choose the basis key θ not as a code word, but uniformly random from $\{0,1\}^n$. As a consequence, the bounds on the games analyzed in Sect. 3.1 change — indeed, the game analyzed in Proposition 1 then becomes the monogamy-of-entanglement game considered and analyzed in [19] — and therefore we get different bounds in Theorem 1, but conceptually everything should still work out. Our goal was to minimize the number of fresh random bits needed for the key refreshing, which corresponds to the number of bits necessary to describe θ; this allows us to compare our work with [5,6] and show that our encryption scheme performs (almost) as good as theirs in this respect. And with this goal in mind, it makes sense to choose θ as a codeword: it gives the same guessing probability for x but asks for less entropy in θ. Choosing θ uniformly random from $\{0,1\}^n$ seems to be the preferred choice for minimizing the quantum communication instead, which would be a very valid objective too.

As an interesting side remark, we observe that with the above variations, our constructions can be understood as following the design principle of the scheme originally proposed by Bennett *et al.* of encrypting and adding redundancy to the message, and encoding the result into BB84 qubits.

Finally, a last variation we mention is to use the *six-state* encoding instead of the BB84 encoding. Since the three bases of the six-state encoding have the same so-called maximal overlap, the bounds in Sect. 3.1 carry over unchanged, but we get more freedom in choosing the code $\mathcal{C}$ in $\{0,1,2\}^n$ so that fewer qubits need to be communicated for the same amount of entropy in x. Also, when choosing the bases uniformly at random in $\{0,1,2\}^n$, as in the variation above, we get a slightly larger entropy for x when using the six-state encoding.

6   Conclusion, and Open Problems

We reconsider one of the very first problems that was posed in the context of quantum cryptography, even before QKD, and we give the first solution that offers a rigorous security proof *and* does not require any sophisticated quantum computing capabilities from the honest users. However, our solution is not the end of the story yet. An intriguing open problem is whether it is possible to do the error correction in a more straightforward way, by just sending the syndrome of x with respect to a *fixed* suitable code, rather than relying on the techniques from [8]. In return, the scheme would be simpler, it could take care of more noise — Dodis and Smith are not explicit about the amount of noise their codes can correct but it appears to be rather low — and, potentially, the number of fresh random bits needed for key refreshing might not grow with the amount of noise. Annoyingly, it *looks* like our scheme should still be secure when doing the error correction in the straightforward way, but our proof technique does not work anymore, and there seems to be no direct fix.

From a practical perspective, it would be interesting to see to what extent it is possible to optimize the quantum communication rather than the key refreshing, e.g., by using BB84 qubits with fully random and independent bases, and whether is it possible to beat QKD in terms of quantum communication.

Acknowledgments. The authors would like to thank Ivan Damgård and Christian Schaffner for interesting discussions related to this work, and Christopher Portmann for comments on an earlier version of the paper.

Appendix

A   Yet Another (Version of the) Guessing Game

We consider a variant of the guessing game from Sect. 3.1 where Bob and Charlie need to guess Alice's measurement outcome. In the variation considered here, we give some slack to Bob in that it is good enough if his guess is close enough (in Hamming distance) to Alice's measurement outcome, and Charlie is given some (deterministic) classical side information on Alice's measurement outcome before he has to announce his guess.[12] We show that, if the minimal distance d of the code $\mathcal{C}$ is large enough, this does not help Bob and Charlie significantly. This is in line with the intuition that, for large enough d, the optimal strategy for Bob and Charlie is to pre-guess Alice's choice of bases.

Proposition 6. *Let $\mathcal{H}_A$ be a n-qubit system, and let $\mathcal{H}_B$ and $\mathcal{H}_C$ be arbitrary quantum systems. Also, let $0 \leq \varphi \leq \frac{1}{2}$ be a parameter and $f : \{0,1\}^n \to \mathcal{Y}$ a function. Consider a state $\rho_{\Theta ABC} = \mu_C \otimes \rho_{ABC} \in \mathcal{D}(\mathcal{C} \otimes \mathcal{H}_A \otimes \mathcal{H}_B \otimes \mathcal{H}_C)$, and let*

$$\rho_{\Theta XX'X''} = \mathcal{N}_{\Theta f(X)C \to X''} \circ \mathcal{N}_{\Theta B \to X'} \circ \mathcal{M}^{\mathrm{BB84}}_{\Theta A \to X} (\rho_{\Theta ABC})$$

where $\mathcal{N}_{\Theta B \to X'}$ is an arbitrary measurement of system B controlled by Θ, and $\mathcal{N}_{\Theta f(X)C \to X''}$ is an arbitrary measurement of system C controlled by Θ and $f(X)$.

---

[12] Taking care of such side information, given to Charlie, on Alice's measurement outcome is not needed for our application, but we get it almost for free.

Then, *it holds that*

$$P[X' \approx_\varepsilon X \wedge X'' = X] \leq \frac{1}{|\mathcal{C}|} + \frac{2^{h(\varphi)n} \cdot |\mathcal{Y}|}{2^{d/2}},$$

where h is the binary entropy function.

Proof. Here, we can write

$$P[X' \approx_\varepsilon X \wedge X'' = X] = \frac{1}{|\mathcal{C}|} \left\| \sum_\theta \tilde{\Pi}^\theta \right\| \leq \frac{1}{|\mathcal{C}|} \sum_\delta \max_\theta \| \tilde{\Pi}^\theta \tilde{\Pi}^{\theta \oplus \delta} \|$$

for projectors

$$\tilde{\Pi}^\theta = \sum_x H^\theta |x\rangle\langle x| H^\theta \otimes \left(\sum_{e \in B_\varphi^n} P_{x \oplus e}^\theta \right) \otimes Q_x^{\theta, f(x)},$$

where $B_\varphi^n \subset \{0,1\}^n$ denotes the set of stings with Hamming weight at most φn. For any $\theta \neq \theta' \in \mathcal{C}$, we can upper bound $\tilde{\Pi}^\theta$ and $\tilde{\Pi}^{\theta'}$ by

$$\tilde{\Pi}^\theta \leq \tilde{\Gamma}^\theta := \sum_x H^\theta |x\rangle\langle x| H^\theta \otimes \left(\sum_{e \in B_\varphi^n} P_{x \oplus e}^\theta \right) \otimes \mathbb{I} = \sum_{e \in B_\varphi^n} \Gamma_e^\theta$$

and

$$\tilde{\Pi}^{\theta'} \leq \tilde{\Delta}^{\theta'} := \sum_x H^{\theta'} |x\rangle\langle x| H^{\theta'} \otimes \mathbb{I} \otimes \left(\sum_{y \in \mathcal{Y}} Q_x^{\theta', y} \right) = \sum_{y \in \mathcal{Y}} \Delta_y^{\theta'}$$

respectively, where Γ_e^θ and $\Delta_y^{\theta'}$ are like Γ^θ and $\Delta^{\theta'}$, as defined in the proof of Proposition 1, for certain concrete choices of the POVM's $\{P_x^\theta\}_x$ and $\{Q_x^{\theta'}\}_x$ that depend on e and y, respectively. As such, we get that

$$\| \tilde{\Pi}^\theta \tilde{\Pi}^{\theta'} \| \leq \| \tilde{\Gamma}^\theta \tilde{\Delta}^{\theta'} \| \leq \sum_{e,y} \| \Gamma_e^\theta \Delta_y^{\theta'} \| \leq \frac{|B_\varphi^n| \cdot |\mathcal{Y}|}{2^{d/2}} \leq \frac{2^{h(\varphi)n} \cdot |\mathcal{Y}|}{2^{d/2}}.$$

Since we still have that $\| \tilde{\Pi}^\theta \tilde{\Pi}^\theta \| = \| \tilde{\Pi}^\theta \| = 1$, the claim follows. □

By means of the techniques from Sect. 3.1, we can extend the result to the case where Bob and Charlie have a-priori quantum side information on Alice's choice of bases.

Corollary 3. *Let $\mathcal{H}_A$ be a n-qubit system, and let $\mathcal{H}_B, \mathcal{H}_C$ and $\mathcal{H}_E$ be arbitrary quantum systems. Also, let $0 \leq \varphi \leq \frac{1}{2}$ be a parameter and $f : \{0,1\}^n \to \mathcal{Y}$ a function. Consider a state $\rho_{\Theta E} \in \mathcal{D}(\mathcal{C} \otimes \mathcal{H}_E)$, and let*

$$\rho_{\Theta ABC} = \mathcal{Q}_{E \to ABC}(\rho_{\Theta E}) \in \mathcal{D}(\mathcal{C} \otimes \mathcal{H}_A \otimes \mathcal{H}_B \otimes \mathcal{H}_C)$$

where $\mathcal{Q}_{E \to ABC}$ is a CPTP map acting on E, and let

$$\rho_{\Theta X X' X''} = \mathcal{N}_{\Theta f(X) C \to X''} \circ \mathcal{N}_{\Theta B \to X'} \circ \mathcal{M}_{\Theta A \to X}^{\mathrm{BB84}}(\rho_{\Theta ABC})$$

as in Proposition 6 above. Then, it holds that

$$P[X' \approx_\varepsilon X \wedge X'' = X] \leq \mathrm{Guess}(\Theta|E) + \frac{\mathrm{Guess}(\Theta|E) \cdot |\mathcal{C}| \cdot 2^{h(\varphi)n} \cdot |\mathcal{Y}|}{2^{d/2}}.$$

Remark 6. In line with the remarks in Sect. 3.1, if Bob "measures correctly" but is still given some slack, and, say, Charlie is given no side information on Alice's outcome, the bound relaxes to

$$P[X' \approx_\varepsilon X \wedge X'' = X] \leq \mathrm{Guess}(\Theta|E) + \frac{\mathrm{Guess}(\Theta|E) \cdot |\mathcal{C}| \cdot 2^{h(\varphi)n}}{2^d}.$$

B On the Existence of Suitable Secure Sketches

In [8, Lemma 5], Dodis and Smith show that for any constant $0 < \lambda < 1$, there exists an explicitly constructible family of binary linear codes $\{\mathcal{C}_i\}_{i \in \mathcal{I}}$ in $\{0,1\}^n$ with dimension k that efficiently correct a constant fraction of errors and have square *bias* $\delta^2 \leq 2^{-\lambda n}$. Their Lemma 4 then shows that the keyed hash function $\mathsf{Ext} : \mathcal{I} \times \{0,1\}^n \to \mathcal{SYN} = \{0,1\}^{n-k}$, $(i,x) \mapsto syn_i(x)$ is a strong extractor, where $syn_i(x)$ is the syndrome with respect to the code $\mathcal{C}_i$. More precisely, the generalization of their result to quantum side information by Fehr and Schaffner [9] shows that if $\rho_{IXE} = \mu_I \otimes \rho_{XE} \in \mathcal{D}(\mathcal{I} \otimes \mathcal{X} \otimes \mathcal{H}_E)$ then

$$\delta(\rho_{\mathsf{Ext}(I,X)IE}, \mu_{\mathcal{SYN}} \otimes \rho_I \otimes \rho_E) \leq \frac{1}{2}\sqrt{\mathrm{Guess}(X|E)\, \delta^2\, 2^n}.$$

It follows from Proposition 2 that the secure sketch

$$\mathsf{SS} : \mathcal{L} \times \{0,1\}^n \to \mathcal{SYN}, \quad (i\|b,x) \mapsto syn_i(x) + b$$

where $\mathcal{L} := \mathcal{I} \times \mathcal{SYN}$, offers uniformity and ν-key-privacy with parameter $\nu = \delta 2^{n/2}/\sqrt{|\mathcal{SYN}|} = \delta 2^k$.

Dodis and Smith are not explicit about the *size* $n - k$ of the syndrome in their construction, but looking at the details, we see that $n - k \leq \log(\delta^2\, 2^n)$. As such, by artificially extending the range $\mathcal{SYN} = \{0,1\}^{n-k}$ of SS to a set $\mathcal{S} = \{0,1\}^\varsigma$ of bit strings of size $\varsigma := \log(\delta^2\, 2^n)$, and re-defining SS to map $(i\|b,x)$ to $syn_i(x) + b$ *padded with sufficiently many 0's*, we get that the secure sketch $\mathsf{SS} : \mathcal{L} \times \{0,1\}^n \to \mathcal{S}$ is message-independent and offers *ideal* key-privacy.[13]

---

[13] Alternatively, we could simply stick to $\mathsf{SS} : \mathcal{L} \times \{0,1\}^n \to \mathcal{SYN}$ but carry along the non-ideal parameter ν; however, we feel that this additional parameter would make things more cumbersome — but of course would lead to the same end result.

References

1. Ambainis, A., Mosca, M., Tapp, A., De Wolf, R.: Private quantum channels. In: 41st IEEE FOCS, pp. 547–553 (2000)
2. Bennett, C.H., Brassard, G.: Quantum cryptography: public key distribution and coin tossing. In: IEEE International Conference on Computers, Systems and Signal Processing, pp. 175–179 (1984)
3. Bennett, C.H., Brassard, G., Breidbart, S.: Quantum cryptography II: how to re-use a one-time pad safely even if $P = NP$. Nat. Comput. **13**(4), 453–458 (2014)
4. Barnum, H., Crépeau, C., Gottesman, D., Smith, A., Tapp, A.: Authentication of quantum messages. In: 43rd IEEE FOCS, pp. 449–458 (2002)
5. Damgård, I., Pedersen, T.B., Salvail, L.: A quantum cipher with near optimal key-recycling. In: Shoup, V. (ed.) CRYPTO 2005. LNCS, vol. 3621, pp. 494–510. Springer, Heidelberg (2005). doi:10.1007/11535218_30
6. Damgård, I., Brochmann Pedersen, T., Salvail, L.: How to re-use a one-time pad safely and almost optimally even if $P = NP$. Nat. Comput. **13**(4), 469–486 (2014)
7. Damgård, I.B., Fehr, S., Salvail, L., Schaffner, C.: Secure identification and QKD in the bounded-quantum-storage model. In: Menezes, A. (ed.) CRYPTO 2007. LNCS, vol. 4622, pp. 342–359. Springer, Heidelberg (2007). doi:10.1007/978-3-540-74143-5_19
8. Dodis, Y., Smith, A.: Correcting errors without leaking partial information. In: 37th ACM STOC, pp. 654–663 (2005)
9. Fehr, S., Schaffner, C.: Randomness extraction via delta-biased masking in the presence of a quantum attacker. In: Canetti, R. (ed.) TCC 2008. LNCS, vol. 4948, pp. 465–481. Springer, Heidelberg (2008). doi:10.1007/978-3-540-78524-8_26
10. Garg, S., Yuen, H., Zhandry, M.: New security notions and feasibility results for authentication of quantum data. Manuscript (2016). arXiv:1607.07759v1
11. Hayden, P., Leung, D., Mayers, D.: Universal composable security of quantum message authentication with key recycling. Talk at QCRYPT 2011, Zürich (2011)
12. König, R., Renner, R., Schaffner, C.: The operational meaning of min- and max-entropy. IEEE Trans. Inf. Theor. **55**(9), 4337–4347 (2009)
13. Leung, D.: Quantum Vernam cipher. Quantum Inf. Comput. **2**(1), 14–34 (2002)
14. Nisan, N., Zuckerman, D.: Randomness is linear in space. J. Comput. Syst. Sci. **52**(1), 43–52 (1996)
15. Oppenheim, J., Horodecki, M.: How to reuse a one-time pad and other notes on authentication, encryption and protection of quantum information. Phys. Rev. A **72**, 042309 (2005)
16. Portmann, C.: Quantum authentication with key recycling. Manuscript. arXiv.org/abs/1610.03422v1 (2016). Also to appear in these proceedings
17. Renner, R.: Security of quantum key distribution. Ph.D. thesis, ETH Zürich, No. 16242 (2005)
18. Radhakrishnan, J., Ta-Shma, A.: Bounds for dispersers, extractors, and depth-two superconcentrators. SIAM J. Comput. **13**(1), 2–24 (2000)
19. Tomamichel, M., Fehr, S., Kaniewski, J., Wehner, S.: One-sided device-independent QKD and position-based cryptography from monogamy games. In: Johansson, T., Nguyen, P.Q. (eds.) EUROCRYPT 2013. LNCS, vol. 7881, pp. 609–625. Springer, Heidelberg (2013). doi:10.1007/978-3-642-38348-9_36

Quantum Authentication with Key Recycling

Christopher Portmann$^{(\boxtimes)}$

Institute for Theoretical Physics, ETH Zurich, 8093 Zurich, Switzerland
chportma@ethz.ch

Abstract. We show that a family of quantum authentication protocols introduced in [Barnum et al., FOCS 2002] can be used to construct a secure quantum channel and additionally recycle all of the secret key if the message is successfully authenticated, and recycle part of the key if tampering is detected. We give a full security proof that constructs the secure channel given only insecure noisy channels and a shared secret key. We also prove that the number of recycled key bits is optimal for this family of protocols, i.e., there exists an adversarial strategy to obtain all non-recycled bits. Previous works recycled less key and only gave partial security proofs, since they did not consider all possible distinguishers (environments) that may be used to distinguish the real setting from the ideal secure quantum channel and secret key resource.

1 Introduction

1.1 Reusing a One-Time Pad

A one-time pad can famously be used only once [31], i.e., a secret key as long as the message is needed to encrypt it with information-theoretic security. But this does not hold anymore if the honest players can use quantum technologies to communicate. A quantum key distribution (QKD) protocol [5,30] allows players to expand an initial short secret key, and thus encrypt messages that are longer than the length of the original key. Instead of first expanding a key, and then using it for encryption, one can also swap the order if the initial key is long enough: one first encrypts a message, then recycles the key. This is possible due to the same physical principles as QKD: quantum states cannot be cloned, so if the receiver holds the exact cipher that was sent, the adversary cannot have a copy, and thus does not have any information about the key either, so it may be reused. This requires the receiver to verify the authenticity of the message received, and if this process fails, a net key loss occurs—the same happens in QKD: if an adversary tampers with the communication, the players have to abort and also lose some of the initial secret key.

1.2 Quantum Authentication and Key Recycling

Some ideas for recycling encryption keys using quantum ciphers were already proposed in 1982 [6]. Many years later, Damgård et al. [13] (see also [14,18])

© International Association for Cryptologic Research 2017
J.-S. Coron and J.B. Nielsen (Eds.): EUROCRYPT 2017, Part III, LNCS 10212, pp. 339–368, 2017.
DOI: 10.1007/978-3-319-56617-7_12

showed how to encrypt a classical message in a quantum state and recycle the key. At roughly the same time, the first protocol for authenticating quantum messages was proposed by Barnum et al. [3], who also proved that quantum authentication necessarily encrypts the message as well. Gottesman [20] then showed that after the message is successfully authenticated by the receiver, the key can be leaked to the adversary without compromising the confidentiality of the message. And Oppenheim and Horodecki [25] adapted the protocol of [3] to recycle key. But the security definitions in these initial works on quantum authentication have a major flaw: they do not consider the possibility that an adversary may hold a purification of the quantum message that is encrypted. This was corrected by Hayden, Leung and Mayers [21], who give a composable security definition for quantum authentication with key recycling. They then show that the family of protocols from [3] are secure, and prove that one can recycle part of the key if the message is accepted.

The security proof from [21] does however not consider all possible environments. Starting in works by Simmons in the 80's and then Stinson in the 90's (see, for example, [33–36]) the classical literature on authentication studies two types of attacks: *substitution attacks*—where the adversary obtains a valid pair of message and cipher[1] and attempts to substitute the cipher with one that will decode to a different message—and *impersonation attacks*—where the adversary directly sends a forged cipher to the receiver, without knowledge of a valid message-cipher pair. To the best of our knowledge, there is no proof showing that security against impersonation attacks follows from security against substitution attacks, hence the literature analyzes both attacks separately.[2] This is particularly important in the case of composable security, which aims to prove the security of the protocol when used in any arbitrary environment, therefore also in an environment that first sends a forged cipher to the receiver, learns wether it is accepted or rejected, then provides a message to the sender to be authenticated, and finally obtains the cipher for this message. This is all the more crucial when key recycling is involved, since the receiver will already recycle (part of) the key upon receiving the forged cipher, which is immediately given to the environment. The work of Hayden et al. [21] only considers environments that perform substitution attacks—i.e., first provide the sender with a message, then change the cipher, and finally learn the outcome of the authentication as well as receive the recycled key. Hence they do not provide a complete

---

[1] Here we use the term *cipher* to refer to the authenticated message, which is often a pair of the original message and a tag or message authentication code (MAC), but not necessarily.

[2] In fact, one can construct examples where the probability of a successful impersonation attack is higher than the probability of a successful substitution attack. This can occur, because any valid cipher generated by the adversary is considered a successful impersonation attack, whereas only a cipher that decrypts to a different message is considered a successful substitution attack.

composable security proof of quantum authentication, which prevents the protocol from being composed in an arbitrary environment.[3]

More recently, alternative security definitions for quantum authentication have been proposed, both without [9,17] and with [19] key recycling (see also [2]). These still only consider substitution attacks, and furthermore, they are, strictly speaking, not composable. While it is possible to prove that these definitions imply security in a composable framework (if one restricts the environment to substitution attacks), the precise way in which the error ε carries over to the framework has not been worked out in any of these papers. If two protocols with composable errors ε and δ are run jointly (e.g., one is a subroutine of the other), the error of the composed protocol is bounded by the sum of the individual errors, $\varepsilon + \delta$. If a security definition does not provide a bound on the composable error, then one cannot evaluate the new error after composition.[4] For example, quantum authentication with key recycling requires a backwards classical authentic channel, so that the receiver may tell the sender that the message was accepted, and allow her to recycle the key. The error of the complete protocol is thus the sum of errors of the quantum authentication and classical authentication protocols. Definitions such as those of [9,17,19] are not sufficient to directly obtain a bound on the error of such a composed protocol.

In the other direction, it is immediate that if a protocol is ε-secure according to the composable definition used in this work, then it is secure according to [9,17,19] with the same error ε. More precisely, proving that the quantum authentication scheme constructs a secure channel is sufficient to satisfy [9,17]— i.e., the ideal functionality is a secure channel which only allows the adversary to decide if the message is delivered, but does not leak any information about the message to the adversary except its length (confidentiality), nor does it allow the adversary to modify the message (authenticity). And proving that the scheme constructs a secure channel that additionally generates fresh secret key is sufficient to satisfy the definition of *total authentication* from [19]. Garg et al. [19] also propose a definition of *total authentication with key leakage*, which can be captured in a composable framework by a secure channel that generates fresh key and leaks some of it to the adversary. This is however a somewhat unnatural ideal functionality, since it requires a deterministic leakage function, which may be unknown or not exist, e.g., the bits leaked can depend on the adversary's behavior—this is the case for the *trap code* [8,9], which we discuss further in Sect. 4. The next natural step for players in such a situation is to extract a secret key from the partially leaked key, and thus the more natural ideal functionality is what one obtains after this privacy amplification step [7,29]: a secure

---

[3] For example, QKD can be broken if the underlying authentication scheme is vulnerable to impersonation attacks, because Eve could trick Alice into believing that the quantum states have been received by Bob so that she releases the basis information.

[4] In an asymptotic setting, one generally does not care about the exact error, as long as it is negligible. But for any (finite) implementation, the exact value is crucial, since without it, it is impossible to set the parameters accordingly, e.g., how many qubits should one send to get an error $\varepsilon \leq 10^{-18}$.

channel that generates fresh secret key, but where the key generated may be shorter than the key consumed. The ideal functionality used in the current work provides this flexibility: the amount of fresh key generated is a parameter which may be chosen so as to produce less key than consumed, the same amount, or even more.[5] Hence, with one security definition, we encompass all these different cases—no key recycling, partial key recycling, total key recycling, and even a net gain of secret key. Furthermore, having all these notions captured by ideal functionalities makes for a particularly simple comparison between the quite technical definitions appearing in [9,17,19].

1.3   Contributions

In this work we use the Abstract Cryptography (AC) framework [23] to model the composable security of quantum authentication with key recycling. AC views cryptography as a resource theory: a protocol constructs a (strong) resource given some (weak) resources. For example, the quantum authentication protocols that we analyze construct two resources: a secure quantum channel—a channel that provides both *confidentiality* and *authenticity*—and a secret key resource that shares a fresh key between both players. In order to construct these resources, we require shared secret key, an insecure (noiseless) quantum channel and a backwards authentic classical channel. These are all resources, that may in turn be constructed from weaker resources, e.g., the classical authentic channel can be constructed from a shared secret key and an insecure channel, and noiseless channels are constructed from noisy channels. Due to this constructive aspect of the framework, it is also called *constructive cryptography* in the literature [22,24].

Although this approach is quite different from the Universal Composability (UC) framework [10,11], in the setting considered in this work—with one dishonest player and where recipients are denoted by classical strings[6]—the two frameworks are essentially equivalent and the same results could have been derived with a quantum version of UC [37]. In UC, the constructed resource would be called *ideal functionality*, and the resources used in the construction are setup assumptions.

We thus first formally define the ideal resources constructed by the quantum authentication protocol with key recycling—the secure channel and key resource mentioned in this introduction—as well as the resources required by this construction. We then prove that a family of quantum authentication protocols proposed by Barnum et al. [3] satisfy this construction, i.e., no distinguisher (called environment in UC) can distinguish the real system from the ideal resources and simulator except with an advantage ε that is exponentially small in the security parameter. This proof considers all distinguishers allowed by quantum mechanics, including those that perform impersonation attacks.

---

[5] One may obtain more key than consumed by using the constructed secure channel to share secret key between the players. We use this technique to compensate for key lost in a classical authentication subroutine, that cannot be recycled.

[6] In a more general setting, a message may be in a superposition of "sent" and "not sent" or a superposition of "sent to Alice" and "sent to Bob", which cannot be modeled in UC, but is captured in AC [28].

We show that in the case where the message is accepted, every bit of key may be recycled. And if the message is rejected, one may recycle all the key except the bits used to one-time pad the cipher.[7] We prove that this is optimal for the family of protocols considered, i.e., an adversary may obtain all non-recycled bits of key. This improves on previous results, which recycled less key and only considered a subset of possible environments. More specifically, Hayden et al. [21], while also analyzing protocols from [3], only recycle part of the key in case of an accept, and lose all the key in case of a reject. Garg et al. [19] propose a new protocol, which they prove can recycle all of the key in the case of an accept, but do not consider key recycling in the case of a reject either. The protocols we analyze are also more key efficient than that of [19]. We give two instances which need $\Theta(m + \log 1/\varepsilon)$ bits of initial secret key, instead of the $\Theta((m + \log 1/\varepsilon)^2)$ required by [19], where m is the length of the message and ε is the error. Independently from this work, Alagic and Majenz [2] proved that one of the instances analyzed here satisfies the weaker security definition of [19].

Note that the family of protocols for which we provide a security proof is a subset of the (larger) family introduced in [3]. More precisely, Barnum et al. [3] define quantum authentication protocols by composing a quantum one-time pad and what they call a *purity testing code*—which, with high probability, will detect any noise that may modify the encoded message—whereas we require a stricter notion, a *strong purity testing code*—which, with high probability, will detect any noise. This restriction on the family of protocols is necessary to recycle all the key. In fact, there exists a quantum authentication scheme, the *trap code* [8,9], which is a member of the larger class from [3] but not the stricter class analyzed here, and which leaks part of the key to the adversary, even upon a successful authentication of the message—this example is discussed in Sect. 4.

We then give two explicit instantiations of this family of quantum authentication protocols. The first is the construction used in [3], which requires an initial key of length $2m + 2n$, where m is the length of the message and n is the security parameter, and has error $\varepsilon \leq 2^{-n/2+1}\sqrt{2m/n + 2}$. The second is an explicit unitary 2-design [15,16] discovered by Chau [12], which requires $5m + 4n$ bits of initial key[8] and has error $\varepsilon \leq 2^{-n/2+1}$. Both constructions have a net loss of $2m + n$ bits of key if the message fails authentication. Since several other explicit quantum authentication protocols proposed in the literature are instances of this family of schemes, our security proof is a proof for these protocols as well—this is discussed further in Sect. 4.

In the full version of this paper [27], we additionally show how to construct the resources used by the protocol from nothing but insecure noisy channels and shared secret key, and calculate the joint error of the composed protocols. We

---

[7] Key recycling in the case of a rejected message is not related to any quantum advantage. A protocol does not leak more information about the key than (twice) the length of the cipher, so the rest may be reused. The same holds for classical authentication [26].

[8] The complete design would require $5m + 5n$ bits of key, but we show that some of the unitaries are redundant when used for quantum authentication and can be dropped.

also show how to compensate for the bits of key lost in the construction of the backwards authentic channel, so that the composed protocol still has a zero net key consumption if no adversary jumbles the communication. Finally, the full version [27] also contains a security proof of quantum without key recycling, which is valid for weak purity testing codes and achieves an optimal error.

1.4 Structure of This Paper

In Sect. 2 we give a brief introduction to the main concepts of AC, which are necessary to understand the notion of cryptographic construction and corresponding security defintion. In Sect. 3 we then define the resources constructed and used by a quantum authentication scheme with key recycling. We introduce the family of protocols from [3] that we analyze in this work, and then prove that they construct the corresponding ideal resources. We also prove that the number of recycled bits is optimal. Finally, in Sect. 4 we discuss the relation between some quantum authentication schemes that have appeared in the literature and those analyzed here, as well as some open problems.

2 Constructive Cryptography

As already mentioned in Sect. 1.3, the AC framework [23] models cryptography as a resource theory. In this section we give a brief overview of how these constructive statements are formalized. We illustrate this with an example taken from [26], namely authentication of classical messages with message authentication codes (MAC). An expanded version of this introduction to AC is provided in the full version of this paper [27].

In an n player setting, a *resource* is an object with n interfaces, that allows every player to input messages and receive other messages at her interface. The objects depicted in Fig. 1 are examples of resources. The insecure channel in Fig. 1a allows Alice to input a message at her interface on the left and allows Bob to receive a message at his interface on the right. Eve can intercept Alice's message and insert a message of her choosing at her interface. The authentic channel resource depicted in Fig. 1b also allows Alice to send a message and Bob to receive a message, but Eve's interface is more limited than for the insecure channel: she can only decide if Bob receives the message or not, but not tamper with the message being sent. The key resource drawn in Fig. 1c provides each player with a secret key when requested. If two resources $\mathcal{K}$ and $\mathcal{C}$ are both available to the players, we write $\mathcal{K}\|\mathcal{C}$ for the resource resulting from their parallel composition—this is to be understood as the resources being merged into one: the interfaces belonging to player i are simultaneously accessible to her as one new interface, which we depict in Fig. 1d. In the full version of this work [27] we provide a more detailed description of the resources from Fig. 1 along a discussion of how to model them mathematically.

Converters capture operations that a player might perform locally at her interface. For example, if the players share a key resource and an insecure channel, Alice might decide to append a MAC to her message. This is modeled as

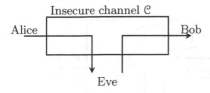

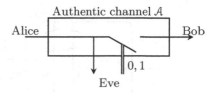

(a) An insecure channel from Alice (on the left) to Bob (on the right) allows Eve (below) to intercept the message and insert a message of her own.

(b) An authentic channel from Alice (on the left) to Bob (on the right) allows Eve (below) to receive a copy of the message and choose whether Bob receives it or an error symbol.

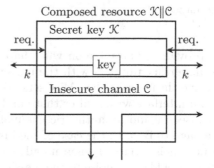

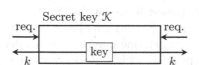

(c) A secret key resource distributes a perfectly uniform key k to the players when they send a request **req**.

(d) If two resources $\mathcal{K}$ and $\mathcal{C}$ are available to the players, we denote the composition of the two as the new resource $\mathcal{K}\|\mathcal{C}$.

Fig. 1. Some examples of resources. The insecure channel on the top left could transmit either classical or quantum messages. The authentic channel on the top right is necessarily classical, since it clones the message.

a converter π_A^{auth} that obtains the message x at the outside interface, obtains a key at the inside interface from a key resource $\mathcal{K}$ and sends $(x, h_k(x))$ on the insecure channel $\mathcal{C}$, where h_k is taken from a family of strongly 2-universal hash functions [36,39]. We illustrate this in Fig. 2. Converters are always drawn with rounded corners. If a converter α_i is connected to the i interface of a resource $\mathcal{R}$, we write $\alpha_i \mathcal{R}$ or $\mathcal{R}\alpha_i$ for the new resource obtained by connecting the two.[9]

A protocol is then defined by a set of converters, one for every honest player. Another type of converter that we need is a *filter*. The resources illustrated in Fig. 1 depict a setting with an adversary that has some control over these resources. For a cryptographic protocol to be useful it is not sufficient to provide guarantees on what happens when an adversary is present, one also has

---

[9] In this work we adopt the convention of writing converters at the A and B interfaces on the left and converters at the E interface on the right, though there is no mathematical difference between $\alpha_i \mathcal{R}$ and $\mathcal{R}\alpha_i$.

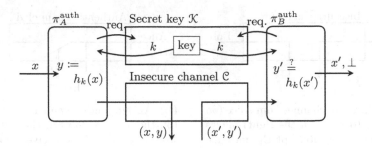

Fig. 2. The real system for a MAC protocol. Alice authenticates her message by appending a MAC to it. Bob checks if the MAC is correct and either accepts or rejects the message.

to provide a guarantee on what happens when no adversary is present, e.g., if no adversary tampers with the message on the insecure channel, then Bob will receive the message that Alice sent. We model this setting by covering the adversarial interface with a filter that emulates an honest behavior. In Fig. 3 we draw an insecure and an authentic channel with filters $\sharp_E$ and $\Diamond_E$ that transmit the message to Bob. In the case of the insecure channel, one may want to model an honest noisy channel when no adversary is present. This is done by having the filter $\sharp_E$ add some noise to the message. A dishonest player removes this and has access to a noiseless channel as in Fig. 1a.

We use the term *filtered resource* to refer to a pair of a resource $\mathcal{R}$ and a filter $\sharp_E$, and often write $\mathcal{R}_\sharp = (\mathcal{R}, \sharp_E)$. Such an object can be thought of as having two modes: it is characterized by the resource $\mathcal{R}\sharp_E$ when no adversary is present and by the resource $\mathcal{R}$ when the adversary is present.

The final object that is required by the AC framework to define the notion of construction and prove that it is composable, is a (pseudo-)metric defined on the space of resources that measures how close two resources are. In the following, we use a distinguisher based metric, i.e., the maximum advantage a distinguisher has in guessing whether it is interacting with resource $\mathcal{R}$ or $\mathcal{S}$, which we write $d(\mathcal{R}, \mathcal{S})$. More specifically, let $\mathcal{D}$ be a distinguisher, and le $\mathcal{D}[\mathcal{R}]$ and $\mathcal{D}[\mathcal{S}]$ be the binary random variables corresponding to $\mathcal{D}$'s output when connected to $\mathcal{R}$ and $\mathcal{S}$, respectively. Then the distinguishing advantage between $\mathcal{R}$ and $\mathcal{S}$ is defined as

$$d(\mathcal{R}, \mathcal{S}) := \sup_{\mathcal{D}} |\Pr[\mathcal{D}[\mathcal{R}] = 0] - \Pr[\mathcal{D}[\mathcal{S}] = 0]|.$$

Since we study information-theoretic security in this work, the supremum is taken over the set of all possible distinguishers allowed by quantum mechanics. This is discussed further in the full version of this work [27].

We are now ready to define the security of a cryptographic protocol. We do so in the three player setting, for honest Alice and Bob, and dishonest Eve. Thus, in the following, all resources have three interfaces, denoted A, B and E, and

Insecure channel $\mathcal{C}$ Authentic channel $\mathcal{A}$

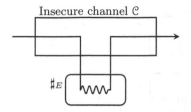
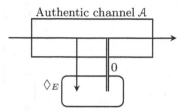

(a) When no adversary is present, Alice's message is delivered to Bob. In the case of a noisy channel, this noise is introduced by the filter $\natural_E$.

(b) When no adversary is present, Bob receives the message sent by Alice.

Fig. 3. Channels with filters. The two channels from Fig. 1a and b are represented with filters on Eve's interface emulating an honest behavior, i.e., when no adversary is present.

a protocol is then given by a pair of converters (π_A, π_B) for the honest players. We refer to [23] for the general case, when arbitrary players can be dishonest.

Definition 1 (Cryptographic security [23]). *Let $\pi_{AB} = (\pi_A, \pi_B)$ be a protocol and $\mathcal{R}_\natural = (\mathcal{R}, \natural)$ and $\mathcal{S}_\Diamond = (\mathcal{S}, \Diamond)$ denote two filtered resources. We say that π_{AB} constructs $\mathcal{S}_\Diamond$ from $\mathcal{R}_\natural$ within ε, which we write $\mathcal{R}_\natural \xrightarrow{\pi, \varepsilon} \mathcal{S}_\Diamond$, if the two following conditions hold:*

(i) We have

$$d(\pi_{AB}\mathcal{R}\natural_E, \mathcal{S}\Diamond_E) \leq \varepsilon.$$

(ii) There exists a converter[10] σ_E—which we call simulator—such that

$$d(\pi_{AB}\mathcal{R}, \mathcal{S}\sigma_E) \leq \varepsilon.$$

If it is clear from the context what filtered resources $\mathcal{R}_\natural$ and $\mathcal{S}_\Diamond$ are meant, we simply say that π_{AB} is ε-secure.

The first of these two conditions measures how close the constructed resource is to the ideal resource in the case where no malicious player is intervening, which is often called *correctness* in the literature. The second condition captures *security* in the presence of an adversary. For example, to prove that the MAC protocol π_{AB}^{auth} constructs an authentic channel $\mathcal{A}_\Diamond$ from a (noiseless) insecure channel $\mathcal{C}_\Box$ and a secret key $\mathcal{K}$ within ε, we need to prove that the real system (with filters) $\pi_{AB}^{auth}(\mathcal{K}\|\mathcal{C}\Box_E)$ cannot be distinguished from the ideal system $\mathcal{A}\Diamond_E$ with advantage greater than ε, and we need to find a converter σ_E^{auth} such that the real system (without filters) $\pi_{AB}^{auth}(\mathcal{K}\|\mathcal{C})$ cannot be distinguished from the

---

[10] For a protocol with information-theoretic security to be composable with a protocol that has computational security, one additionally requires the simulator to be efficient.

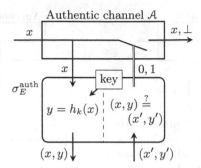

Fig. 4. The ideal system with simulator for a MAC protocol. The simulator σ_E^{auth} picks its own key and generates the MAC. If the value input by Eve is different from the output at her interface (or is input before an output is generated), the simulator prevents Bob from getting Alice's message.

ideal system $\mathcal{A}\sigma_E^{\text{auth}}$ with advantage greater than ε. For the MAC protocol, correctness is satisfied with error 0 and the simulator σ_E^{auth} drawn in Fig. 4 satisfies the second requirement if the family of hash functions $\{h_k\}_k$ is ε-almost strongly 2-universal [26].

Remark 2. The protocols and simulators discussed in this work are all efficient. The protocols we consider are either trivially efficient or taken from other work, in which case we refer to these other works for proofs of efficiency. The efficiency of the simulator used to prove the security of quantum authentication has been analyzed in [9]. All other simulators used in the security proofs run the corresponding honest protocols, and are thus efficient because the protocols are. We therefore do not discuss efficiency any further in this work.

3   Quantum Authentication

We start with some technical preliminaries in Sect. 3.1, where we introduce (strong) purity testing codes, which are a key component of the family of quantum authentication protocols of [3]. In Sect. 3.2 we give a constructive view of quantum authentication with key recycling: we define the resources that such a protocol is expected to construct, as well as the resources that are required to achieve this. In Sect. 3.3 we describe the family of protocols that we analyze in this work, along with a variant in which the order of the encryption and encoding operations has been swapped, which we prove to be equivalent. In Sect. 3.4 we give a security proof for the family of quantum authentication protocols defined earlier. And in Sect. 3.5 we show that the number of recycled key bits is optimal. Finally, in Sect. 3.6 we give two explicit constructions of purity testing codes and get the exact parameters of the quantum authentication protocols with these codes.

3.1   Technical Preliminaries

Pauli Operators. To denote a Pauli operator on n qubits we write either $P_{x,z}$ or P_ℓ, where x and z are n-bit strings indicating in which positions bit and phase flips occur, and $\ell = (x, z)$ is the concatenation of x and z, which is used when we do not need to distinguish between x and z. Two Pauli operators P_j and P_ℓ with $j = (x, z)$ and $\ell = (x', z')$ commute (anti-commute) if the symplectic inner product

$$(j, \ell)_{\mathrm{Sp}} := x \cdot z' - z \cdot x' \tag{1}$$

is 0 (is 1),where $x \cdot z$ is the scalar product of the vectors and the arithmetic is done modulo 2. Hence, for any P_j and P_ℓ

$$P_j P_\ell = (-1)^{(j,\ell)_{\mathrm{Sp}}} P_\ell P_j.$$

We use several times the following equality

$$\sum_{j \in \{0,1\}^n} (-1)^{(j,\ell)_{\mathrm{Sp}}} = \begin{cases} 2^n & \text{if } \ell = 0, \\ 0 & \text{otherwise,} \end{cases} \tag{2}$$

where $\ell = 0$ means that all bits of the string ℓ are 0.

Purity Testing Code. An error correcting code (ECC) that encodes an m qubit message in a $m+n$ qubit code word is generally defined by an isomorphism from $\mathbb{C}^{2^m}$ to $\mathbb{C}^{2^{m+n}}$. In this work we define an ECC by a unitary $U : \mathbb{C}^{2^{m+n}} \to \mathbb{C}^{2^{m+n}}$. The code word for a state $|\psi\rangle$ is obtained by appending a n qubit state $|0\rangle$ to the message, and applying U, i.e., the encoding of $|\psi\rangle$ is $U(|\psi\rangle \otimes |0\rangle)$. We do not need to use the decoding properties of ECCs in this work, we only use the them to detect errors, i.e., given a state $|\varphi\rangle \in \mathbb{C}^{2^{m+n}}$, we apply the inverse unitary $U^\dagger$ and measure the last n qubits to see if they are $|0\rangle$ or not.

The first property we require of our codes, is that they map any Pauli error P_ℓ into another Pauli error $P_{\ell'}$, i.e.,

$$U^\dagger P_\ell U = e^{i\theta_\ell} P_{\ell'}, \tag{3}$$

for some global phase $e^{i\theta_\ell}$. This is always the case for any U that can be implemented with Clifford operators. In particular, all stabilizer codes have this property, which are used in [3] to define purity testing codes. Note that the mapping from ℓ to ℓ' defined by (3) is a permutation on the set of indices $\ell \in \{0,1\}^{2m+2n}$ that depends only on the choice of code.

A code will detect an error P_ℓ if $P_{\ell'} = P_{x,z} \otimes P_{s,z'}$ for $s \neq 0$, where $P_{x,z}$ acts on the first m qubits and $P_{s,z'}$ on the last n. Measuring these last qubits would yield the syndrome s, since $P_{s,z'}$ flips the bits in the positions corresponding to the bits of s. And an error P_ℓ will act trivially on the message if $P_{\ell'} = P_{0,0} \otimes P_{s,z}$. In particular, if $P_{\ell'} = P_{0,0} \otimes P_{0,z}$, then this error will not be detected, but not change the message either.

For a code indexed by a key k, we denote by $\mathcal{P}_k$ the set of Pauli errors that are not detected by this code, and by $\mathcal{Q}_k \subset \mathcal{P}_k$ we denote the undetected errors which act trivially on the message. A purity testing code is a set of codes $\{U_k\}_{k \in \mathcal{K}}$ such that when a code U_k is selected uniformly at random, it will detect with high probability all Pauli errors which act non-trivially on the message.

Definition 3 (Purity testing code [3]). *A purity testing code with error ε is a set of codes $\{U_k\}_{k \in \mathcal{K}}$, such that for all Pauli operators P_ℓ,*

$$\frac{|\{k \in \mathcal{K} : P_\ell \in \mathcal{P}_k \setminus \mathcal{Q}_k\}|}{|\mathcal{K}|} \leq \varepsilon.$$

As mentioned in Sect. 1.3, we use a stricter definition of purity testing code in this work. We require that all non-identity Paulis get detected with high probability, even those that act trivially on the message. Intuitively, the reason for this is that, with the original definition of purity testing, if the adversary introduces some noise P_ℓ, by learning whether the message was accepted or not, she will learn whether that error acts trivially on the message or not, and thus learn something about the ECC used. This means that the adversary learns something about the key used to choose the ECC, and hence it cannot be recycled in its entirety.[11]

Definition 4 (Strong purity testing code). *A strong purity testing code with error ε is a set of codes $\{U_k\}_{k \in \mathcal{K}}$, such that for all non-identity Pauli operators P_ℓ,*

$$\frac{|\{k \in \mathcal{K} : P_\ell \in \mathcal{P}_k\}|}{|\mathcal{K}|} \leq \varepsilon.$$

In Sect. 3.6 we provide explicit constructions of strong purity testing codes.

3.2 Secure Channel and Secret Key Resource

The main result in this paper is a proof that the family of quantum authentication protocols of Barnum et al. [3] restricted to strong purity testing codes can be used to construct a resource that corresponds to the parallel composition of a secure quantum channel $\mathcal{S}^m$ and a secret key resource $\bar{\mathcal{K}}^{\nu_{\mathrm{rej}}, \nu_{\mathrm{acc}}}$, which are illustrated in Fig. 5 and explained in more detail in the following paragraphs.

The secure quantum channel, $\mathcal{S}^m$, drawn in Fig. 5a, allows an m-qubit message ρ to be transmitted from Alice to Bob, which Alice may input at her interface. Since in general the players cannot prevent Eve from learning that a message has been sent, Eve's interface has one output denoted by a dashed arrow, which notifies her that Alice has sent a message. But the players cannot prevent Eve from jumbling the communication lines either, which is captured in the resource $\mathcal{S}^m$ by allowing Eve to input a bit that decides if Bob gets the

---

[11] We conjecture that in this case only 1 bit of the key is leaked, see the discussion in Sect. 4.

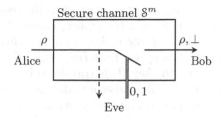

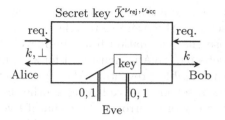

(a) A secure channel $\mathcal{S}^m$ is very similar to the authentic channel from Fig. 1b. It allows Alice (on the left) to send an m-qubit message, and Eve (below) to decide if Bob (on the right) gets it. But this time, Eve only receives a notification that the message has been sent (denoted by the dashed arrow), not a copy.

(b) A slightly weaker secret key resource than that from Fig. 1c, $\bar{\mathcal{K}}^{\nu_{rej},\nu_{acc}}$. It allows Eve (below) to choose the length of the key generated, either $|k| = \nu_{rej}$ or $|k| = \nu_{acc}$. Furthermore, Eve can prevent Alice (on the left) from getting the key at all.

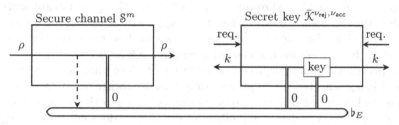

(c) When no adversary is present, the filter $\flat_E$ covers Eve's interface of the resource $\mathcal{S}^m \| \bar{\mathcal{K}}^{\nu_{acc},\nu_{rej}}$. Once $\flat_E$ is notified that a message has been sent, it allows the message through and notifies the secret key resource to prepare a key of length ν_{acc}.

Fig. 5. We depict here the filtered resource $(\mathcal{S}^m \| \bar{\mathcal{K}}^{\nu_{acc},\nu_{rej}}, \flat_E)$ constructed by the quantum authentication protocols analyzed in this work. It can be seen as the composition of a secure channel $\mathcal{S}^m$ (Fig. 5a) and a secret key resource $\bar{\mathcal{K}}^{\nu_{acc},\nu_{rej}}$ (Fig. 5b). The filter $\flat_E$ that emulates an honest behavior is drawn in Fig. 5c.

message or an error symbol $\perp$ —Eve may also decide not to provide this input (Eve cuts the communication lines), in which case the system is left waiting and Bob obtains neither the message nor an error. Note that the order in which messages are input to the resource $\mathcal{S}^m$ is not fixed, Eve may well provide her bit before Alice inputs a message. In this case, Bob immediately receives an error $\perp$ regardless of the value of Eve's bit.

The secret key resource, $\bar{\mathcal{K}}^{\nu_{rej},\nu_{acc}}$, depicted in Fig. 5b distributes a uniformly random key to Alice and Bob. Unlike the simplified key resource from Fig. 1c, here the adversary has some control over the length of the key produced. This is because in the real setting Eve can prevent the full key from being recycled by jumbling the message. This is reflected at Eve's interface of $\bar{\mathcal{K}}^{\nu_{rej},\nu_{acc}}$ allowing her to decide if the key generated is of length ν_{rej} or ν_{acc}. Furthermore, if in the real setting Alice were to recycle her key before Bob receives the cipher, Eve could use the information from the recycled key to modify the cipher without being detected. So Alice must wait for a confirmation of reception from Bob, which Eve can jumble, preventing Alice from ever recycling the key. This translates in the ideal setting to Eve having another control bit, deciding whether Alice receives the key or an error $\perp$. Note that if Eve provides her two bits in the wrong order, Alice always gets an error $\perp$. This key resource is modeled so that the honest players must request the key to obtain its value. If Bob does this before Eve has provided the bit deciding the key length, he gets an error instead of a key. If Alice makes the request before Eve has provided both her bits, she also gets an error. Otherwise they get the key k.

If no adversary is present, a filter $\flat_E$ covers Eve's interface of the resources $\mathcal{S}^m$ and $\bar{\mathcal{K}}^{\nu_{\text{rej}},\nu_{\text{acc}}}$, which is drawn in Fig. 5c. This filter provides the inputs to the resources that allow Bob to get Alice's message and generate a key of length ν_{acc} that is made available to both players.

To construct the filtered resource $(\mathcal{S}^m \| \bar{\mathcal{K}}^{\nu_{\text{rej}},\nu_{\text{acc}}})_\flat$, the quantum authentication protocol will use a shared secret key to encrypt and authenticate the message. This means that the players must share a secret key resource. For simplicity we assume the players have access to a resource $\mathcal{K}^\mu$ as depicted in Fig. 1c, that always provides them with a key of length μ.[12] Note that the security of the protocol is not affected if the players only have a weaker resource which might shorten the key or not deliver it to both players—such as the one constructed by the protocol, $\bar{\mathcal{K}}^{\nu_{\text{rej}},\nu_{\text{acc}}}$—because if either of the players does not have enough key, they simply abort, which is an outcome Eve could already achieve by cutting or jumbling the communication.

They also need to share an insecure quantum channel, which is used to send the message, and is illustrated in Fig. 1a without a filter and in Fig. 3a with a filter. The authentication protocol we consider is designed to catch any error, so if it is used over a noisy channel, it will always abort, even though no adversary is tampering with the message. We thus assume that the players share a noiseless channel, which we denote $\mathcal{C}_\square$, i.e., $\mathcal{C}$ is controlled by the adversary as in Fig. 1a. But if no adversary is present, the filter $\square_E$ is noiseless. In the full version of this work [27] we explain how to compose the protocol with an error correcting code so as to run it over a noisy channel.

Finally, the players need a backwards authentic channel, that can send one bit of information from Bob to Alice. This is required so that Alice may learn whether the message was accepted and recycle the corresponding amount of key. The authentic channel and its filter $\mathcal{A}_\lozenge$ are drawn in Figs. 1b and 3b. Putting all this together in the case of an active adversary, we get Fig. 6, where the converters for Alice's and Bob's parts of the quantum authentication protocol are labeled $\pi_A^{\text{q-auth}}$ and $\pi_B^{\text{q-auth}}$, respectively.

---

[12] Since Eve's interface of $\mathcal{K}^\mu$ is empty, this resource has a trivial empty filter, which we do not write down.

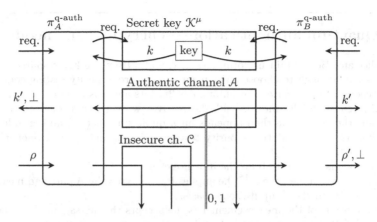

Fig. 6. The real system for quantum authentication with key recycling. Upon receiving a message ρ, $\pi_A^{\text{q-auth}}$ encrypts it with a key that it obtains from K^μ and sends it on the insecure channel. Upon receiving a quantum state on the insecure channel, $\pi_B^{\text{q-auth}}$ checks whether it is valid, and outputs the corresponding message ρ' or an error message $\perp$. It may then recycle (part of) the key, k', and uses the authentic channel to notify $\pi_A^{\text{q-auth}}$ whether the message was accepted or not. $\pi_A^{\text{q-auth}}$ then recycles the key as well. Concrete protocols for this are given in Sect. 3.3.

According to Definition 1, a protocol $\pi_{AB}^{\text{q-auth}} = (\pi_A^{\text{q-auth}}, \pi_B^{\text{q-auth}})$ is then a quantum authentication protocol (with key recycling) with error $\varepsilon^{\text{q-auth}}$ if it constructs $(\mathcal{S}^m \| \bar{\mathcal{K}}^{\nu_{\text{rej}}, \nu_{\text{acc}}})_\flat$ from $\mathcal{C}_\square \| \mathcal{A}_\Diamond \| \mathcal{K}^\mu$ within $\varepsilon^{\text{q-auth}}$, i.e.,

$$\mathcal{C}_\square \| \mathcal{A}_\Diamond \| \mathcal{K}^\mu \xrightarrow{\pi_{AB}^{\text{q-auth}}, \varepsilon^{\text{q-auth}}} (\mathcal{S}^m \| \bar{\mathcal{K}}^{\nu_{\text{rej}}, \nu_{\text{acc}}})_\flat. \tag{4}$$

In Sect. 3.3 we describe the protocol, and in Sect. 3.4 we prove that (4) is satisfied and provide the parameters $\mu, \nu_{\text{rej}}, \nu_{\text{acc}}, \varepsilon^{\text{q-auth}}$.

3.3 Generic Protocol

The family of quantum authentication protocols from [3] consists in first encrypting the message to be sent with a quantum one-time pad, then encoding it with a purity testing code and a random syndrome. We do the same, but with a strong purity testing code. We also extend the protocol so that the players recycle all the key if the message is accepted, and the key used to select the strong purity testing code if the message is rejected. So that Alice may also recycle the key, Bob uses the backwards authentic classical channel to notify her of the outcome. We refer to this as the "encrypt-then-encode" protocol, the details of which are provided in Fig. 7.

Alternatively, one may perform the encoding and encryption in the opposite order: Alice first encodes her message with the strong purity testing code with syndrome 0, then does a quantum one-time pad on the resulting $m + n$ qubit state. This "encode-then-encrypt" protocol is described in Fig. 8.

Quantum authentication — encrypt-then-encode

1. Alice and Bob obtain uniform keys k, ℓ, and s from the key resource, where k is long enough to choose an element from a strong purity testing code that encodes m qubits in $m + n$ qubits, ℓ is $2m$ bits and s is n bits.
2. Alice encrypts the message ρ^A she receives with a quantum one-time pad using the key ℓ. She then appends an n qubit state $|s\rangle\langle s|^S$, and encodes the whole thing with a strong purity testing code, obtaining the cipher $\sigma^{AS} = U_k(P_\ell \rho^A P_\ell \otimes |s\rangle\langle s|^S)U_k^\dagger$.
3. Alice sends σ^{AS} to Bob on the insecure channel.
4. Bob receives a message $\tilde{\sigma}^{AS}$, he applies $U_k^\dagger$, decrypts the A part and measures the S part in the computational basis.
5. If the result of the measurement is s, he accepts the message and recycles k, ℓ and s. If the result is not s, he rejects the message, and recycles k.
6. Bob sends Alice a bit on the backwards authentic channel to tell her if he accepted or rejected the message.
7. When Alice receives Bob's bit, she either recycles all the keys or only k.

Fig. 7. This protocol is identical to the scheme from [3], except that the players use a strong purity testing code, recycle key, and have a backwards authentic channel so that Alice may learn the outcome.

The pseudo-code described in Figs. 7 and 8 can easily be translated into converters as used in the AC formalism, i.e., the objects $\pi_A^{\text{q-auth}}$ and $\pi_B^{\text{q-auth}}$ from Fig. 6. More precisely, if $\pi_A^{\text{q-auth}}$ receives a message at its outer interface, it requests a key from the key resource, encrypts the message as described and sends the cipher on the insecure channel. It may receive three symbols from the backwards authentic channel: an error $\perp$, in which case it does not recycle any key, a message 0 saying that $\pi_B^{\text{q-auth}}$ did not receive the correct state, in which case it recycles the part of the key used to choose the code, or a message 1 saying that $\pi_B^{\text{q-auth}}$ did receive the correct state, in which case it recycles all the key. If $\pi_A^{\text{q-auth}}$ first receives a message on the backwards authentic channel before receiving a message to send, it will not recycle any key. Similarly, when $\pi_B^{\text{q-auth}}$ receives a cipher on the insecure channel, it requests a key from the key resource, performs the decryption, outputs either the message or an error depending on the result of the decryption, and sends this result back to $\pi_A^{\text{q-auth}}$ on the authentic channel.

The encode-then-encrypt protocol uses n bits more key, and since these bits are not recycled in case of a reject, it is preferable to use the encrypt-then-encode protocol. These protocols are however identical: no external observer can detect which of the two is being run. This holds, because the encode-then-encrypt protocol performs phase flips on a syndrome that is known to be in a computational basis state $|s\rangle$. Thus, they have no effect and can be skipped. Likewise, Bob performs phase flips on S before measuring in the computational basis—he might as well skip these phase flips, since they have no effect either. We formalize this

---

Quantum authentication — encode-then-encrypt

1. Alice and Bob obtain uniform keys k and ℓ from the key resource, where k is long enough to choose an element from a strong purity testing code that encodes m qubits in $m + n$ qubits and ℓ is $2m + 2n$ bits long.
2. Alice appends a n qubit state $|0\rangle\langle0|$ to the message ρ^A she receives, encodes it with a strong purity testing code chosen according to the key k, and encrypts the whole thing with a quantum one-time pad using the key ℓ. She thus obtains the cipher $\sigma^{AS} = P_\ell U_k(\rho^A \otimes |0\rangle\langle0|^S)U_k^\dagger P_\ell$.
3. Alice sends σ^{AS} to Bob on the insecure channel.
4. Bob receives a message $\tilde{\sigma}^{AS}$, he applies P_ℓ, then $U_k^\dagger$, and measures the S part in the computational basis.
5. If the result of the measurement is 0, he accepts the message and recycles k and ℓ. Otherwise, he rejects the message, and recycles k.
6. Bob sends Alice a bit on the backwards authentic channel to tell her if he accepted or rejected the message.
7. When Alice receives Bob's bit, she either recycles all the keys or only k.

Fig. 8. This protocol is similar to the protocol from Fig. 7, except that the order of the encryption and encoding have been reversed. To do this, the players need an extra n bits of key.

statement by proving (in Lemma 5) that the converters corresponding to the two different protocols are indistinguishable. This result is similar in spirit to proofs that some prepare-and-measure quantum key distribution (QKD) protocols are indistinguishable from entanglement-based QKD protocols, and thus security proofs for one are security proofs for the other [32].

Since these two protocols are indistinguishable, we provide a security proof in Sect. 3.4 for the encode-then-encrypt protocol. However, in Sect. 3.6, when we count the number of bits of key consumed, we count those of the encrypt-then-encode protocol.

Lemma 5. *Let* $(\bar{\pi}_A^{q\text{-}auth}, \bar{\pi}_B^{q\text{-}auth})$ *and* $(\pi_A^{q\text{-}auth}, \pi_B^{q\text{-}auth})$ *denote the pairs of converters modeling Alice's and Bob's behavior in the encrypt-then-encode and encode-then-encrypt protocols, respectively. Then*

$$d(\bar{\pi}_A^{q\text{-}auth}, \pi_A^{q\text{-}auth}) = d(\bar{\pi}_B^{q\text{-}auth}, \pi_B^{q\text{-}auth}) = 0.$$

Proof. We start with Alice's part of the protocol. Let $\bar{\pi}_A^{q\text{-}auth}$ and $\pi_A^{q\text{-}auth}$ receive keys k, ℓ and s as in the protocol from Fig. 7, as well as an extra key z of length n that is needed by $\pi_A^{q\text{-}auth}$, since it requires more key. The distinguisher prepares a state ρ^{RA}, and sends the A part to the system. $\bar{\pi}_A^{q\text{-}auth}$ outputs

$$U_k^{AS} P_\ell^A \left(\rho^{RA} \otimes |s\rangle\langle s|^S \right) P_\ell^A \left(U_k^{AS} \right)^\dagger$$

$$= U_k^{AS} \left(P_\ell^A \otimes P_{s,0}^S \right) \left(\rho^{RA} \otimes |0\rangle\langle 0|^S \right) \left(P_\ell^A \otimes P_{s,0}^S \right) \left(U_k^{AS} \right)^\dagger$$

$$= U_k^{AS} \left(P_\ell^A \otimes P_{s,z}^S \right) \left(\rho^{RA} \otimes |0\rangle\langle 0|^S \right) \left(P_\ell^A \otimes P_{s,z}^S \right) \left(U_k^{AS} \right)^\dagger$$

$$= P_{\ell'}^{AS} U_k^{AS} \left(\rho^{RA} \otimes |0\rangle\langle 0|^S \right) \left(U_k^{AS} \right)^\dagger P_{\ell'}^{AS},$$

where in the last line we used (3). This is exactly the state output by $\pi_A^{\text{q-auth}}$ if when receiving the key k, ℓ, s, z, the protocol uses the Pauli $P_{\ell'}$ for the quantum one-time pad.

For Bob's part of the protocol, let the distinguisher prepare a state σ^{RAS} and send the AS part to the system. The subnormalized state held jointly by $\pi_B^{\text{q-auth}}$ and the distinguisher after decoding and performing the measurement is given by

$$\langle s | P_\ell^A \left(U_k^{AS} \right)^\dagger \sigma^{RAS} U_k^{AS} P_\ell^A | s \rangle$$

$$= \langle 0 | \left(P_\ell^A \otimes P_{s,0}^S \right) \left(U_k^{AS} \right)^\dagger \sigma^{RAS} U_k^{AS} \left(P_\ell^A \otimes P_{s,0}^S \right) | 0 \rangle$$

$$= \langle 0 | \left(P_\ell^A \otimes P_{s,z}^S \right) \left(U_k^{AS} \right)^\dagger \sigma^{RAS} U_k^{AS} \left(P_\ell^A \otimes P_{s,z}^S \right) | 0 \rangle$$

$$= \langle 0 | \left(U_k^{AS} \right)^\dagger P_{\ell'}^{AS} \sigma^{RAS} P_{\ell'}^{AS} U_k^{AS} | 0 \rangle.$$

We again obtain the state that is jointly held by $\pi_B^{\text{q-auth}}$ and the distinguisher if when receiving the key k, ℓ, s, z, the protocol uses the Pauli $P_{\ell'}$ for the quantum one-time pad.   □

Remark 6. If part of the message is classical—i.e., it is diagonal in the computational basis and known not to have a purification held be the distinguisher—then running the same proof as Lemma 5, one can show that it is sufficient to perform bit flips on that part of the message, the phase flips are unnecessary. This is used in the full version of this work [27] to save some key in a construction that involves a message that is part classical.

3.4   Security Proof

Suppose that there exists a strong purity testing code $\{U_k\}_{k \in \mathcal{K}}$ of size $\log |\mathcal{K}| = \nu$ and with error ε that encodes an m qubit message in an $m + n$ qubit cipher. And let $\pi_{AB}^{\text{q-auth}} = (\pi_A^{\text{q-auth}}, \pi_A^{\text{q-auth}})$ denote Alice and Bob's converters when running the encode-then-encrypt protocol from Fig. 8. We are now ready to state the main theorem, namely that $\pi_{AB}^{\text{q-auth}}$ is a secure authentication scheme with key recycling.

Theorem 7. *Let $\pi_{AB}^{q\text{-}auth}$ denote converteres corresponding to the protocol from Fig. 8. Then $\pi_{AB}^{q\text{-}auth}$ constructs the secure channel and secret key filtered resource*

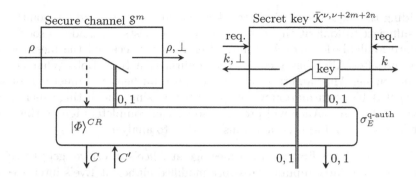

Fig. 9. The ideal quantum authentication system consisting of the constructed resources S^m and $\bar{\mathcal{K}}^{\nu,\nu+2m+2n}$, and the simulator $\sigma_E^{\text{q-auth}}$.

$(S^m\|\bar{\mathcal{K}}^{\nu,\nu+2m+2n})_\flat$, *given an insecure quantum channel* $\mathcal{C}_\square$, *a backwards authentic channel* $\mathcal{A}_\lozenge$ *and a secret key* $\mathcal{K}^{\nu+2m+2n}$, *i.e.,*

$$\mathcal{C}_\square\|\mathcal{A}_\lozenge\|\mathcal{K}^{\nu+2m+2n} \xrightarrow{\;\pi_{AB}^{q\text{-}auth},\varepsilon^{q\text{-}auth}\;} (S^m\|\bar{\mathcal{K}}^{\nu,\nu+2m+2n})_\flat,$$

with $\varepsilon^{q\text{-}auth} = \sqrt{\varepsilon} + \varepsilon/2$, *where* ε *is the error of the strong purity testing code.*

In order to prove this theorem, we need to find a simulator such that the real and ideal systems are indistinguishable except with advantage $\sqrt{\varepsilon} + \varepsilon/2$. The simulator that we use is illustrated in Fig. 9, and works as follows. When it receives a notification from the ideal resource that a message is sent, it generates EPR pairs $|\Phi\rangle^{CR}$ and outputs half of each pair (the C register) at its outer interface. Once it receives a modified cipher (denoted C' in the picture), it measures this state and the half of the EPR pairs it kept in the Bell basis to decide if they were modified. It accordingly activates the switches on the two resources controlling whether Bob gets the message and the length of the key generated, and outputs the bit of backward communication from Bob to Alice—which is always leaked to Eve. If it first receives the register C' before generating the EPR pairs, it always notifies the ideal resource to output an error and outputs 0 as the leak on the backwards authentic channel.

Proof. It is trivial to show that correctness holds with error 0, namely that

$$d\left(\pi_{AB}^{\text{q-auth}}\left(\mathcal{C}\square_E\|\mathcal{A}\lozenge_E\|\mathcal{K}^{\nu+2m+2n}\right),(S^m\|\bar{\mathcal{K}}^{\nu,\nu+2m+2n})_{\flat E}\right) = 0. \tag{5}$$

We now prove the case of security, i.e.,

$$d\left(\pi_{AB}^{\text{q-auth}}\left(\mathcal{C}\|\mathcal{A}\|\mathcal{K}^{\nu+2m+2n}\right),(S^m\|\bar{\mathcal{K}}^{\nu,\nu+2m+2n})\sigma_E^{\text{q-auth}}\right) \leq \sqrt{\varepsilon}+\varepsilon/2. \tag{6}$$

The real and ideal systems, drawn in Figs. 6 and 9 have 5 inputs. The distinguisher thus has the choice between 5! possible orders for providing inputs. However, most of these orders are redundant and do not need to be analyzed.

Providing the requests for the secret keys before they are ready is pointless. So it is sufficient to look at the case where these requests are made as soon as the keys are available for recycling, i.e., after Bob has received the message from Alice and after Alice has received the confirmation from Bob. What is more, neither sending Alice an error on the backwards authentic channel nor allowing her to get Bob's confirmation will help either way, since the distinguisher already knows what output Alice will produce, so we can completely ignore this input. That leaves only 2 in-ports, and thus 2 orders to analyze:

1. The distinguisher first inputs a message at Alice's interface, gets the cipher at Eve's interface, inputs a possibly modified cipher at Eve's interface, gets the output at Bob's interface, and requests the recycled key.
2. The distinguisher first inputs a fake cipher at Eve's interface, gets the output at Bob's interface, makes a request for his recycled key, then inputs a message at Alice's interface and receives the cipher for that message.

We start with the first case, the initial message is sent to Alice. The distinguisher prepares a message $|\psi\rangle^{ME}$ and inputs the M part at Alice's interface. The ideal channel then notifies the simulator that a message has been input. The simulator prepares a maximally entangled state $|\Phi\rangle^{CR}$ of dimension 2^{2m+2n} and outputs the C register at Eve's interface. The distinguisher now holds a bipartite state in CE, to which it applies a unitary U^{CE}. Without loss of generality, one may write the unitary as $U^{CE} = \sum_j P_j^C \otimes E_j^E$, where P_j^C are Paulis acting on the cipher register C and E_j^E act on the distinguisher's internal memory E. The resulting state in the C register is input back in the E interface. The simulator now measures CR in the Bell basis defined by the projectors $\{P_j \otimes I|\Phi\rangle\langle\Phi|^{CR}P_j \otimes I\}_j$. If the outcome is $j = 0$—where $P_0 = I$—it tells the two resources that the cipher was not modified. In which case the contents of the register M is output at Bob's interface with an acc flag. Furthermore, it generates a fresh uniform key (k, ℓ), where $|k| = \nu$ and $|\ell| = 2m + 2n$. If the outcome is $j \neq 0$, then the simulator notifies the channel to delete the message and output a rej flag, and tells the key resource to prepare only the shorter key k. The distinguisher then sends a request to obtain the fresh key. So the final state held by the distinguisher interacting with the ideal system is

$$\zeta = |\text{acc}\rangle\langle\text{acc}| \otimes \tau^K \otimes \tau^L \otimes \left[\left(I^M \otimes E_0^E \right) |\psi\rangle\langle\psi|^{ME} \left(I^M \otimes \left(E_0^E \right)^\dagger \right) \right]$$
$$+ \sum_{j \neq 0} |\text{rej}\rangle\langle\text{rej}| \otimes \tau^K \otimes E_j^E \rho^E \left(E_j^E \right)^\dagger, \qquad (7)$$

where τ^K and τ^L are fully mixed states and $\rho^E = \text{tr}_M(|\psi\rangle\langle\psi|^{ME})$. One could append states $\perp^L$ and $\perp^M$ in the rej branch of (7) so that both terms have the same number of registers; we omit them for simplicity.

In the real system, for the secret key (k, ℓ), the state before Bob's measurement of the syndrome is given by

$$|\varphi_{k,\ell}\rangle^{SME} = \sum_j \left((U_k^{SM})^\dagger P_\ell^{SM} P_j^{SM} P_\ell^{SM} U_k^{SM} \otimes E_j^E \right) |0\rangle^S |\psi\rangle^{ME}$$

$$= \sum_j (-1)^{(j,\ell)_{\mathrm{Sp}}} \left((U_k^{SM})^\dagger P_j^{SM} U_k^{SM} \otimes E_j^E \right) |0\rangle^S |\psi\rangle^{ME},$$

where $(\cdot, \cdot)_{\mathrm{Sp}}$ denotes the symplectic product defined in (1). Let $\mathcal{J}_s^k$ be the set of indices j such that the error P_j^{SM} produces a syndrome s when code k is used, i.e., $(U_k^{SM})^\dagger P_j^{SM} U_k^{SM} = e^{i\theta_{k,j}} P_{s,z}^S \otimes P_{j'}^M$ for some $\theta_{k,j}$ (see (3) and discussion thereafter). For $j \in \mathcal{J}_s^k$, let

$$|s\rangle^S |\psi_{j,k}\rangle^{ME} := \left((U_k^{SM})^\dagger P_j^{SM} U_k^{SM} \otimes E_j^E \right) |0\rangle^S |\psi\rangle^{ME}$$

$$= e^{i\theta_{k,j}} \left(P_{s,z}^S \otimes P_{j'}^M \otimes E_j^E \right) |0\rangle^S |\psi\rangle^{ME}.$$

Then

$$|\varphi_{k,\ell}\rangle = \sum_s \sum_{j \in \mathcal{J}_s^k} (-1)^{(j,\ell)_{\mathrm{Sp}}} \left((U_k^{SM})^\dagger P_j^{SM} U_k^{SM} \otimes E_j^E \right) |0\rangle^S |\psi\rangle^{ME}$$

$$= \sum_s \sum_{j \in \mathcal{J}_s^k} (-1)^{(j,\ell)_{\mathrm{Sp}}} |s\rangle^S |\psi_{j,k}\rangle^{ME}.$$

The next step in Bob's protocol consists in measuring the syndrome. If $s = 0$ is obtained, he outputs the message as well as the key (k, ℓ) and a flag acc. Otherwise he deletes the message, outputs k with the flag rej. The final state held be the distinguisher in this case is

$$\xi = |\mathrm{acc}\rangle\langle\mathrm{acc}| \otimes \frac{1}{2^{\nu+2m+2n}} \sum_{k,\ell} |k,\ell\rangle\langle k,\ell|$$

$$\otimes \sum_{j_1,j_2 \in \mathcal{J}_0^k} (-1)^{(j_1 \oplus j_2, \ell)_{\mathrm{Sp}}} |\psi_{j_1,k}\rangle\langle\psi_{j_2,k}|^{ME}$$

$$+ |\mathrm{rej}\rangle\langle\mathrm{rej}| \otimes \frac{1}{2^{\nu+2m+2n}} \sum_{k,\ell} |k\rangle\langle k|$$

$$\otimes \sum_{s \neq 0} \sum_{j_1,j_2 \in \mathcal{J}_s^k} (-1)^{(j_1 \oplus j_2, \ell)_{\mathrm{Sp}}} E_{j_1}^E \rho^E (E_{j_2}^E)^\dagger,$$

where we have used $|\psi_{j,k}\rangle^{ME} = \left(V_{k,j}^M \otimes E_j^E \right) |\psi\rangle^{ME}$ for some unitary $V_{k,j}^M$.

Setting

$$\zeta^{\text{acc}} := \left(I^M \otimes E_0^E\right) |\psi\rangle\langle\psi|^{ME} \left(I^M \otimes \left(E_0^E\right)^\dagger\right),$$

$$\zeta^{\text{rej}} := \sum_{j\neq 0} E_j^E \rho^E \left(E_j^E\right)^\dagger,$$

$$\xi_{k,\ell}^{\text{acc}} := \sum_{j_1,j_2\in\mathcal{J}_0^k} (-1)^{(j_1\oplus j_2,\ell)_{\text{SP}}} |\psi_{j_1,k}\rangle\langle\psi_{j_2,k}|^{ME},$$

$$\xi_k^{\text{rej}} := \frac{1}{2^{2m+2n}} \sum_{\ell,s\neq 0} \sum_{j_1,j_2\in\mathcal{J}_s^k} (-1)^{(j_1\oplus j_2,\ell)_{\text{SP}}} E_{j_1}^E \rho^E \left(E_{j_2}^E\right)^\dagger,$$

the distance between real and ideal systems may be written as

$$\frac{1}{2} \|\zeta - \xi\|_{\text{tr}} = \frac{1}{2\cdot 2^{\nu+2m+2n}} \sum_{k,\ell} \left\|\zeta^{\text{acc}} - \xi_{k,\ell}^{\text{acc}}\right\|_{\text{tr}} + \frac{1}{2\cdot 2^\nu} \sum_k \left\|\zeta^{\text{rej}} - \xi_k^{\text{rej}}\right\|_{\text{tr}}.$$

ζ^{acc} and $\xi_{k,\ell}^{\text{acc}}$ are both pure states, so using the fact that[13]

$$\frac{1}{2} \left\||\psi\rangle\langle\psi| - |\varphi\rangle\langle\varphi|\right\|_{\text{tr}} \leq \||\psi\rangle - |\varphi\rangle\|, \tag{8}$$

we bound their distance as

$$\frac{1}{2}\left\|\zeta^{\text{acc}} - \xi_{k,\ell}^{\text{acc}}\right\|_{\text{tr}} \leq \left\|\left(I^M \otimes E_0^E\right) |\psi\rangle^{ME} - \sum_{j\in\mathcal{J}_0^k} (-1)^{(j,\ell)_{\text{SP}}} |\psi_{j,k}\rangle^{ME}\right\|$$

$$= \left\|\sum_{j\in\mathcal{J}_0^k\setminus\{0\}} (-1)^{(j,\ell)_{\text{SP}}} |\psi_{j,k}\rangle^{ME}\right\|$$

$$= \sqrt{\sum_{j_1,j_2\in\mathcal{J}_0^k\setminus\{0\}} (-1)^{(j_1\oplus j_2,\ell)_{\text{SP}}} \langle\psi_{j_1,k}|\psi_{j_2,k}\rangle},$$

where $\||a\rangle\| = \sqrt{\langle a|a\rangle}$ is the vector 2-norm and we used the fact that $|\psi_{0,k}\rangle^{ME} = \left(I^M \otimes E_0^E\right) |\psi\rangle^{ME}$. From Jensen's inequality and using (2) we obtain

$$\frac{1}{2\cdot 2^{\nu+2m+2n}} \sum_{k,\ell} \left\|\zeta^{\text{acc}} - \xi_{k,\ell}^{\text{acc}}\right\|_{\text{tr}}$$

$$\leq \sqrt{\frac{1}{2^{\nu+2m+2n}} \sum_{k,\ell} \sum_{j_1,j_2\in\mathcal{J}_0^k\setminus\{0\}} (-1)^{(j_1\oplus j_2,\ell)_{\text{SP}}} \langle\psi_{j_1,k}|\psi_{j_2,k}\rangle}$$

$$= \sqrt{\frac{1}{2^\nu} \sum_k \sum_{j\in\mathcal{J}_0^k\setminus\{0\}} \langle\psi_{j,k}|\psi_{j,k}\rangle}.$$

---

[13] See the full version of this work [27] for a proof that (8) holds.

Finally, because the code is a strong purity testing code with error ε and that $\langle \psi_{j,k} | \psi_{j,k} \rangle = \mathrm{tr}(E_j^E \rho^E (E_j^E)^\dagger) =: p_j$ with $\sum_j p_j = 1$, we get

$$\frac{1}{2|\mathcal{K}||\mathcal{L}|} \sum_{k,\ell} \left\| \zeta^{\mathrm{acc}} - \xi_{k,\ell}^{\mathrm{acc}} \right\|_{\mathrm{tr}} \leq \sqrt{\frac{1}{|\mathcal{K}|} \sum_{j \neq 0} \sum_{k : j \in \partial_0^k} \langle \psi_{j,k} | \psi_{j,k} \rangle}$$

$$= \sqrt{\frac{1}{|\mathcal{K}|} \sum_{j \neq 0} \sum_{k : j \in \partial_0^k} p_j}$$

$$\leq \sqrt{\sum_{j \neq 0} \varepsilon p_j} \leq \sqrt{\varepsilon}.$$

In the reject branch of the real system we have

$$\xi_k^{\mathrm{rej}} = \frac{1}{2^{2m+2n}} \sum_{\ell, s \neq 0} \sum_{j_1, j_2 \in \partial_s^k} (-1)^{(j_1 \oplus j_2, \ell)_{\mathrm{Sp}}} E_{j_1}^E \rho^E (E_{j_2}^E)^\dagger$$

$$= \sum_{s \neq 0} \sum_{j \in \partial_s^k} E_j^E \rho^E (E_j^E)^\dagger$$

$$= \sum_{j \notin \partial_0^k} E_j^E \rho^E (E_j^E)^\dagger,$$

where we used again (2). Thus

$$\frac{1}{2 \cdot 2^\nu} \sum_k \left\| \zeta^{\mathrm{rej}} - \xi_k^{\mathrm{rej}} \right\|_{\mathrm{tr}} = \frac{1}{2 \cdot 2^\nu} \sum_k \left\| \sum_{j \in \partial_0^k \setminus \{0\}} E_j^E \rho^E (E_j^E)^\dagger \right\|_{\mathrm{tr}}$$

$$\leq \frac{1}{2 \cdot 2^\nu} \sum_k \sum_{j \in \partial_0^k \setminus \{0\}} p_j \leq \varepsilon/2.$$

Putting all this together we get

$$\frac{1}{2} \left\| \zeta - \xi \right\|_{\mathrm{tr}} \leq \sqrt{\varepsilon} + \varepsilon/2.$$

We now consider the second case: the distinguisher first prepares a state $|\psi\rangle^{CE}$ and inputs the C part at Eve's interface, then obtains the output at Bob's interface. Note that in the ideal case the channel always outputs a rej message at Bob's interface. Thus, if the cipher is accepted by Bob—who outputs a state ζ^{acc}—the distinguisher must be interacting with the real system and can already output this guess. In the case of a rejection, it now holds a bipartite system KE—the recycled key K and its purifying system E. It then applies an isometry $U : \mathcal{H}_{KE} \to \mathcal{H}_{KME}$ to this system and inputs the M part of the resulting state at Alice's interface. After which it obtains a cipher at Eve's interface and holds the tripartite system KCE—the recycled key K, the cipher

C and its internal memory E. We denote this state ζ in the ideal case and ξ^{rej} in the real case, and we need to bound

$$\frac{1}{2} \left\| \zeta - \xi^{\text{rej}} \right\|_{\text{tr}} + \frac{1}{2} \left\| \xi^{\text{acc}} \right\|_{\text{tr}}.$$

In a first step, we assume that the state $|\psi\rangle^{CE}$ prepared by the distinguisher is an antisymmetric fully entangled state, which we denote $|\Psi^-\rangle^{CE} = \sum_x (-1)^{w(x)} |x, \bar{x}\rangle^{CE}$, where $w(x)$ is the Hamming weight of $x \in \{0, 1\}^{m+n}$ and $\bar{x}$ is the string x with all bits flipped. In the ideal case the simulator notifies the channel to reject the cipher, and the state $|\text{rej}\rangle\langle\text{rej}| \otimes \tau^K$ is output at Bob's interface. The distinguisher then holds $\zeta = \tau^K \otimes \tau^E$. In the real case, Bob applies the decoding algorithm, i.e., first a Pauli P_ℓ^C, then a unitary $(U_k^C)^\dagger$ and finally measures n bits of the syndrome in the computational basis. Since the antisymmetric state is invariant under $U \otimes U$, one could equivalently apply the inverse operation, $P_\ell U_k$, to the E system, i.e., the state after Bob's measurement is given by

$$\frac{1}{2^{\nu+3m+3n}} \sum_{k,\ell,s,x_1,x_2} (-1)^{w(x_1) \oplus w(x_2)} |k, \ell\rangle\langle k, \ell|$$
$$\otimes \left(I^C \otimes P_\ell^E U_k^E \right) |s, x_1, \bar{s}, \bar{x}_1\rangle\langle s, x_2, \bar{s}, \bar{x}_1|^{CE} \left(I^C \otimes (U_k^E)^\dagger P_\ell^E \right).$$

If $s = 0$ Bob accepts the cipher as being valid, which happens with probability 2^{-n}, i.e., $\|\xi^{\text{acc}}\|_{\text{tr}} = 2^{-n}$. In the case where $s \neq 0$, he deletes the cipher, so the remaining state is given by

$$\frac{1}{2^{\nu+3m+3n}} \sum_{k,\ell,s\neq 0,x} |k, \ell\rangle\langle k, \ell| \otimes \left(I^C \otimes P_\ell^E U_k^E \right) |\bar{s}, \bar{x}\rangle\langle\bar{s}, \bar{x}|^{CE} \left(I^C \otimes (U_k^E)^\dagger P_\ell^E \right)$$
$$= \tau^K \otimes \tau^L \otimes \tau^E - \rho^{KLE},$$

where

$$\rho^{KLE} = \frac{1}{2^{\nu+3m+3n}} \sum_{k,\ell,x} |k, \ell\rangle\langle k, \ell| \otimes P_\ell^E U_k^E |\bar{0}, \bar{x}\rangle\langle\bar{0}, \bar{x}|^E (U_k^E)^\dagger P_\ell^E,$$

K is made public and the L system is the part of the key kept secret by the players.

Let $\mathcal{E}$ denote the completely positive, trace-preserving (CPTP) map consisting of the distinguisher's next step—the isometry $U : \mathcal{H}_{KE} \to \mathcal{H}_{KME}$—and the final operation of the ideal system—deleting the message system M that is input at Alice's interface and outputting a fully mixed state τ^C. Let $\mathcal{F}$ denote the CPTP map consisting of the distinguisher's next step and the final operation of the real system—encoding the message system M according to the protocol and outputting the resulting cipher. We have $\zeta = \mathcal{E} \left(\tau^K \otimes \tau^E \right)$ and $\xi^{\text{rej}} = \mathcal{F} \left(\tau^K \otimes \tau^L \otimes \tau^E \right) - \mathcal{F} \left(\rho^{KLE} \right)$. Thus,

$$\frac{1}{2} \left\| \zeta - \xi^{\text{rej}} \right\|_{\text{tr}} \leq \frac{1}{2} \left\| \mathcal{E} \left(\tau^K \otimes \tau^E \right) - \mathcal{F} \left(\tau^K \otimes \tau^L \otimes \tau^E \right) \right\|_{\text{tr}} + \frac{1}{2} 2^{-n},$$

since $\left\| \rho^{KLE} \right\|_{\mathrm{tr}} = 2^{-n}$. Finally, note that we have

$$\mathcal{E}\left(\tau^K \otimes \tau^E\right) = \mathcal{F}\left(\tau^K \otimes \tau^L \otimes \tau^E\right) = \tau^C \otimes \sigma^{KE}$$

for $\sigma^{KE} = \mathrm{tr}_M\left[U\left(\tau^K \otimes \tau^E\right)U^\dagger\right]$, since the random Pauli P_ℓ applied by the encryption algorithm completely decouples the cipher from KE. Putting this together, we get

$$\frac{1}{2}\left\|\zeta - \xi\right\|_{\mathrm{tr}} \leq 2^{-n} \leq \sqrt{\varepsilon} \ ,$$

since a strong purity testing code will always have an error $\varepsilon \geq \frac{2^{2m+n}-1}{2^{2m+2n}-1} \geq 2^{-2n}$.

The final case that remains to consider is when the distinguisher prepares a state $|\psi\rangle^{CE}$ that is not the antisymmetric state. We will reduce this case to that of the entangled antisymmetric by using the entangled state $|\Psi^-\rangle^{CE}$ to teleport the C' part of any state $|\psi\rangle^{C'E'}$. Due to space restrictions, the proof of this case is provided in the full version of this work [27]. □

3.5 Optimality of the Recycled Key Length

It follows from Lemma 5 that Theorem 7 is also a proof of security for the encrypt-then-encode protocol from Fig. 7, i.e.,

$$\mathcal{C}_\square\|\mathcal{A}_\lozenge\|\mathcal{K}^{\nu+2m+n} \xrightarrow{\ \pi_{AB}^{\mathrm{q\text{-}auth}},\varepsilon^{\mathrm{q\text{-}auth}}\ } \left(\mathcal{S}^m\|\bar{\mathcal{K}}^{\nu,\nu+2m+n}\right)_\flat,$$

with $\varepsilon^{\mathrm{q\text{-}auth}} = \sqrt{\varepsilon} + \varepsilon/2$. Thus, in the case where the message is not accepted by Bob, $2m+n$ bits of key are lost. We prove here that this is optimal: one cannot recycle any extra bit of key.

Lemma 8. *There exists an adversarial strategy to obtain all the secret bits that are not recycled in the encrypt-then-encode protocol.*

Proof. The distinguisher prepares EPR pairs $|\Phi\rangle^{ME}$ and provides the M part to Alice. It then receives the cipher and thus holds the state

$$U_k^{SM} P_\ell^M \left(|s\rangle^S \otimes |\Phi\rangle^{ME}\right),$$

which it keeps. It then sends a bogus cipher to Bob, and obtains the key k after Bob recycles it. It applies the decoding unitary $\left(U_k^{SM}\right)^\dagger$, measures the S register to get the secret key s and measures the joint ME register in the Bell basis to get the secret key ℓ. □

3.6 Explicit Constructions

The protocols we have given in Sect. 3.3 use strong purity testing codes, and the parameters of the key used, key recycled and error depend on the parameters of these codes. In this section we give two constructions of purity testing codes.

The first requires less initial secret key, the second has a better error parameter. Both have the same net consumption of secret key bits.

The first construction is from Barnum et al. [3]. They give an explicit strong purity testing code with $\nu = n$ and $\varepsilon = \frac{2m/n+2}{2^n}$.[14] Plugging this in the parameters from Theorem 7 with the encrypt-then-encode protocol, we get the following.

Corollary 9. *The encrypt-then-encode protocol with the purity testing code of [3] requires an initial key of length $2m + 2n$. It recycles all bits if the message is accepted, and n bits if the message is rejected. The error is*

$$\varepsilon^{q\text{-}auth} = \sqrt{\frac{2m/n+2}{2^n} + \frac{m/n+1}{2^n}}.$$

The second construction we give is based on an explicit purity testing code by Chau [12]—though he does not name it this way. Chau [12] finds a set of unitaries $\mathcal{U} = \{U_k\}$ in dimension d such that, if k is chosen uniformly at random, any non-identity Pauli is mapped to every non-identity Pauli with equal frequency, i.e., $\forall P_j, P_\ell$ with $P_j \neq I$ and $P_\ell \neq I$,

$$\left|\left\{U_k \in \mathcal{U} : U_k P_j U_k^\dagger = e^{i\theta_{j,k,\ell}} P_\ell\right\}\right| = \frac{|\mathcal{U}|}{d^2 - 1},$$

where $e^{i\theta_{j,k,\ell}}$ is some global phase.

We prove in the full version of this work [27] that this is a strong purity testing code with $\varepsilon = 2^{-n}$ for $d = 2^{m+n}$. It also has $|\mathcal{U}| = 2^{m+n}\left(2^{2m+2n} - 1\right)$, hence $\nu = m+n+\log\left(2^{2m+2n} - 1\right) \leq 3m + 3n$. Note that when composed with Paulis as in the encode-then-encrypt protocol, $\{P_\ell U_k\}_{k,\ell}$ is a unitary 2-design [15,16]. It follows that any (approximate) unitary t-design is a good quantum authentication scheme (see the full version of this work [27] for a formal proof).

Corollary 10. *The encrypt-then-encode protocol with the purity testing code of [12] requires an initial key of length $5m + 4n$. It recycles all bits if the message is accepted, and $3m + 3n$ bits if the message is rejected. The error is $\varepsilon^{q\text{-}auth} = 2^{-n/2} + 2^{-n-1}$.*

4   Discussion and Open Questions

The family of quantum authentication protocols of Barnum et al. [3] as well as the subset analyzed in this work are large classes, which include many protocols appearing independently in the literature. The signed polynomial code [1,4], the Clifford code [1,9,17] (which is a unitary 3-design [38,40]) and the unitary 8-design scheme from [19] and all instances which use a strong purity testing code.

---

[14] In fact, [3] only prove that their construction is a purity testing code, not a strong one. But one can easily verify that it is strong with the same parameters. What is more, their construction has $\nu = \log(2^n + 1)$ and $\varepsilon = \frac{2m/n+2}{2^n+1}$. We remove one of the keys (and thus increase the error), so as to get simpler final expressions.

Our results apply directly to the Clifford and unitary 8-design schemes—which have in the same error as the unitary 2-design scheme from Corollary 10. But the signed polynomial code uses an ECC on qudits, not qubits, so our proof does not cover this case, and would have to be adapted to do so.

The trap code [8,9] is an example of a quantum authentication scheme that uses a purity testing code that is not a strong purity testing code, i.e., errors which do not modify the message do not necessarily provoke an abort. For example, if the adversary performs a simple bit flip in one position, this will provoke an abort with probability 2/3 in the variant from [8] and with probability 1/3 in the variant from [9], but leaves the message unmodified if no abort occurs. If the adversary learns whether Bob accepted the message or not, she will learn whether the ECC used detects that specific bit flip or not, and thus learn something about the key used to select the ECC. Hence, the players cannot recycle the entire key, even in the case where the message is accepted. The restriction to strong purity testing codes is thus necessary to recycle every bit. It remains open how many bits of key can be recycled with the trap code, but we conjecture that this bit leaked due to the decision to abort or not is the only part of the key leaked, and the rest can be recycled.

Another quantum authentication scheme, Auth-QFT-Auth, has been proposed in [19], where the authors prove that some of the key can be recycled as well. We do not know if this scheme fits in the family from [3] or not.

In the classical case, almost strongly 2-universal hash functions [36,39] are used for authentication, and any new family of such functions immediately yields a new MAC. Likewise, any new purity testing code provides a new quantum authentication scheme. However, it is unknown whether all quantum authentication schemes can be modeled as a combination of a one-time pad and a purity testing code, or whether there exist interesting schemes following a different pattern.

We have proven that a loss of $2m + n$ bits of key is inevitable with these schemes if the adversary tampers with the channel. In the case of the unitary 2-design scheme, which has the smallest error, this is $2m + 2\log 1/\varepsilon + 2$ bits of key which are consumed. A loss of $2m$ bits will always occur, since these are required to one-time pad the message. It remains open whether there exist other schemes—which do not fit the one-time pad + purity testing code model—which recycle more key.

The initial preprint of this work suggested that one should also investigate whether it is possible to find a prepare-and-measure scheme to encrypt and authenticate a classical message in a quantum state, so that all of the key may be recycled if it is successfully authenticated. At the time of writing, a possible solution had already been found by Fehr and Salvail [18]. Their protocol is however not known to be composable, and it remains open to prove that it achieves the desired result in such a setting.

Acknowledgments. CP would like to thank Anne Broadbent, Frédéric Dupuis and Debbie Leung for useful discussions.

CP is supported by the European Commission FP7 Project RAQUEL(grant No. 323970), US Air Force Office of Scientific Research (AFOSR) via grant FA9550-16-1-0245, the Swiss National Science Foundation (via the National Centre of Competence in Research 'Quantum Science and Technology') and the European Research Council – ERC (grant No. 258932).

References

1. Aharonov, D., Ben-Or, M., Eban, E.: Interactive proofs for quantum computations. In: Proceedings of Innovations in Computer Science, ICS 2010, pp. 453–469. Tsinghua University Press (2010)
2. Alagic, G., Majenz, C.: Quantum non-malleability and authentication (2016). http://www.arxiv.org/abs/1610.04214, eprint
3. Barnum, H., Crepeau, C., Gottesman, D., Smith, A., Tapp, A.: Authentication of quantum messages. In: Proceedings of the 43rd Symposium on Foundations of Computer Science, FOCS 2002, pp. 449–458. IEEE (2002)
4. Ben-Or, M., Crepeau, C., Gottesman, D., Hassidim, A., Smith, A.: Secure multiparty quantum computation with (only) a strict honest majority. In: Proceedings of the 47th Symposium on Foundations of Computer Science, FOCS 2006, pp. 249–260 (2006)
5. Bennett, C.H., Brassard, G.: Quantum cryptography: public key distribution and coin tossing. In: Proceedings of IEEE International Conference on Computers, Systems, and Signal Processing, pp. 175–179 (1984)
6. Bennett, C.H., Brassard, G., Breidbart, S.: Quantum cryptography II: how to reuse a one-time pad safely even if P = NP (1982). http://www.arxiv.org/abs/1407.0451, original unpublished manuscript uploaded to arXiv in 2014
7. Bennett, C.H., Brassard, G., Crépeau, C., Maurer, U.: Generalized privacy amplification. IEEE Trans. Inf. Theor. 41(6), 1915–1923 (1995)
8. Broadbent, A., Gutoski, G., Stebila, D.: Quantum one-time programs. In: Canetti, R., Garay, J.A. (eds.) CRYPTO 2013. LNCS, vol. 8043, pp. 344–360. Springer, Heidelberg (2013). doi:10.1007/978-3-642-40084-1_20
9. Broadbent, A., Wainewright, E.: Efficient simulation for quantum message authentication. In: Nascimento, A.C.A., Barreto, P. (eds.) ICITS 2016. LNCS, vol. 10015, pp. 72–91. Springer, Heidelberg (2016). doi:10.1007/978-3-319-49175-2_4
10. Canetti, R.: Universally composable security: a new paradigm for cryptographic protocols. In: Proceedings of the 42nd Symposium on Foundations of Computer-Science, FOCS 2001, pp. 136–145. IEEE (2001)
11. Canetti, R.: Universally composable security: a new paradigm for cryptographic protocols. Cryptology ePrint Archive, Report 2000/067 (2013). http://eprint.iacr.org/2000/067, updated version of [10]
12. Chau, H.F.: Unconditionally secure key distribution in higher dimensions by depolarization. IEEE Trans. Inf. Theor. 51(4), 1451–1468 (2005)
13. Damgård, I., Pedersen, T.B., Salvail, L.: A quantum cipher with near optimal key-recycling. In: Shoup, V. (ed.) CRYPTO 2005. LNCS, vol. 3621, pp. 494–510. Springer, Heidelberg (2005). doi:10.1007/11535218_30
14. Damgård, I., Pedersen, T.B., Salvail, L.: How to re-use a one-time pad safely and almost optimally even if P = NP. Nat. Comput. 13(4), 469–486 (2014)
15. Dankert, C.: Efficient simulation of random quantum states and operators. Master's thesis, University of Waterloo (2005)

16. Dankert, C., Cleve, R., Emerson, J., Livine, E.: Exact and approximate unitary 2-designs and their application to fidelity estimation. Phys. Rev. A **80**, 012304 (2009)
17. Dupuis, F., Nielsen, J.B., Salvail, L.: Actively secure two-party evaluation of any quantum operation. In: Safavi-Naini, R., Canetti, R. (eds.) CRYPTO 2012. LNCS, vol. 7417, pp. 794–811. Springer, Heidelberg (2012). doi:10.1007/978-3-642-32009-5_46
18. Fehr, S., Salvail, L.: Quantum authentication and encryption with key recycling (2016). http://www.arxiv.org/abs/1610.05614, eprint
19. Garg, S., Yuen, H., Zhandry, M.: New security notions and feasibility results for authentication of quantum data (2016). http://www.arxiv.org/abs/1607.07759, eprint
20. Gottesman, D.: Uncloneable encryption. Quantum Inf. Comput. **3**, 581 (2003)
21. Hayden, P., Leung, D., Mayers, D.: The universal composable security of quantum message authentication with key recycling (2011). http://www.arxiv.org/abs/1610.09434, eprint, presented at QCrypt 2011
22. Maurer, U.: Constructive cryptography – a new paradigm for security definitions and proofs. In: Mödersheim, S., Palamidessi, C. (eds.) TOSCA 2011. LNCS, vol. 6993, pp. 33–56. Springer, Heidelberg (2012). doi:10.1007/978-3-642-27375-9_3
23. Maurer, U., Renner, R.: Abstract cryptography. In: Proceedings of Innovations in Computer Science, ICS 2011, pp. 1–21. Tsinghua University Press (2011)
24. Maurer, U., Renner, R.: From indifferentiability to constructive cryptography (and back). In: Hirt, M., Smith, A. (eds.) TCC 2016. LNCS, vol. 9985, pp. 3–24. Springer, Heidelberg (2016). doi:10.1007/978-3-662-53641-4_1
25. Oppenheim, J., Horodecki, M.: How to reuse a one-time pad and other notes on authentication, encryption, and protection of quantum information. Phys. Rev. A **72**, 042309 (2005)
26. Portmann, C.: Key recycling in authentication. IEEE Trans. Inf. Theor. **60**(7), 4383–4396 (2014)
27. Portmann, C.: Quantum authentication with key recycling (2016). http://www.arxiv.org/abs/1610.03422, eprint, full version of the current paper
28. Portmann, C., Matt, C., Maurer, U., Renner, R., Tackmann, B.: Causal boxes: quantum information-processing systems closed under composition (2017). http://www.arxiv.org/abs/1512.02240, to appear in IEEE Trans. Inf. Theory
29. Renner, R., König, R.: Universally composable privacy amplification against quantum adversaries. In: Kilian, J. (ed.) TCC 2005. LNCS, vol. 3378, pp. 407–425. Springer, Heidelberg (2005). doi:10.1007/978-3-540-30576-7_22
30. Scarani, V., Bechmann-Pasquinucci, H., Cerf, N.J., Dusek, M., Lutkenhaus, N., Peev, M.: The security of practical quantum key distribution. Rev. Modern Phys. **81**, 1301–1350 (2009)
31. Shannon, C.: Communication theory of secrecy systems. Bell Syst. Tech. J. **28**(4), 656–715 (1949)
32. Shor, P.W., Preskill, J.: Simple proof of security of the BB84 quantum key distribution protocol. Phys. Rev. Lett. **85**, 441–444 (2000)
33. Simmons, G.J.: Authentication theory/coding theory. In: Blakley, G.R., Chaum, D. (eds.) CRYPTO 1984. LNCS, vol. 196, pp. 411–431. Springer, Heidelberg (1985). doi:10.1007/3-540-39568-7_32
34. Simmons, G.J.: A survey of information authentication. Proc. IEEE **76**(5), 603–620 (1988)
35. Stinson, D.R.: The combinatorics of authentication and secrecy codes. J. Cryptol. **2**(1), 23–49 (1990)

36. Stinson, D.R.: Universal hashing and authentication codes. Des. Codes Crypt. **4**(3), 369–380 (1994)
37. Unruh, D.: Universally composable quantum multi-party computation. In: Gilbert, H. (ed.) EUROCRYPT 2010. LNCS, vol. 6110, pp. 486–505. Springer, Heidelberg (2010). doi:10.1007/978-3-642-13190-5_25
38. Webb, Z.: The Clifford group forms a unitary 3-design. Quantum Inf. Comput. **16**(15&16), 1379–1400 (2015)
39. Wegman, M.N., Carter, L.: New hash functions and their use in authentication and set equality. J. Comput. Syst. Sci. **22**(3), 265–279 (1981)
40. Zhu, H.: Multiqubit Clifford groups are unitary 3-designs (2015). http://www.arxiv.org/abs/1510.02619, eprint

Relativistic (or 2-Prover 1-Round) Zero-Knowledge Protocol for **NP** Secure Against Quantum Adversaries

André Chailloux[✉] and Anthony Leverrier

Inria, Paris, France
{andre.chailloux,anthony.leverrier}@inria.fr

Abstract. In this paper, we show that the zero-knowledge construction for HAMILTONIAN CYCLE remains secure against quantum adversaries in the relativistic setting. Our main technical contribution is a tool for studying the action of consecutive measurements on a quantum state which in turn gives upper bounds on the value of some entangled games. This allows us to prove the security of our protocol against quantum adversaries. We also prove security bounds for the (single-round) relativistic string commitment and bit commitment in parallel against quantum adversaries. As an additional consequence of our result, we answer an open question from [Unr12] and show tight bounds on the quantum knowledge error of some Σ-protocols.

Keywords: Relativistic cryptography · Zero-knowledge protocols · Quantum security

1 Introduction

1.1 Context

The goal of relativistic cryptography is to exploit the no superluminal signaling (NSS) principle in order to perform various cryptographic tasks. NSS states that no information carrier can travel faster than the speed of light. Note that this principle is closely related to the non-signaling principle that says that a local action performed in a laboratory cannot have an *immediate* influence outside of the lab. NSS is more precise since it gives an upper bound on the speed at which such an influence can propagate. Apart from this physical principle, we want to ensure information-theoretic security meaning that the schemes proposed cannot be attacked by any classical (or quantum) computers, even with unlimited computing power.

The idea of using the NSS principle for cryptographic protocols originated in a pioneering work by Kent in 1999 [Ken99] as a way to physically enforce a no-communication constraint between the different agents of one party (the idea of splitting up a party into several agents dates back to [BOGKW88], but without any explicit implementation proposal). The original goal of Kent

© International Association for Cryptologic Research 2017
J.-S. Coron and J.B. Nielsen (Eds.): EUROCRYPT 2017, Part III, LNCS 10212, pp. 369–396, 2017.
DOI: 10.1007/978-3-319-56617-7_13

was to bypass the no-go theorems for quantum bit-commitment [May97, LC97]. More recently, quantum relativistic bit commitment protocols were developed where the parties exchange quantum systems, with the hope that combining the NSS principle together with quantum theory would lead to more secure (but less practical) protocols [Ken11, Ken12, KTHW13]. In particular, the protocol [Ken12] was implemented in [LKB+13]. We note that the scope of relativistic cryptography is not limited to bit commitment. For instance, there was recently some interest (sparked again by Kent) for position-verification protocols [KMS11, LL11, Unr14] but contrary to the case of bit commitment, it was shown that secure position-verification is impossible both in the classical and the quantum settings [CGMO09, BCF+14].

The original idea of [BOGKW88] was recently revisited by Crépeau *et al.* in [CSST11] (see also [Sim07]). Based on this work, Lunghi *et al.* devised a bit commitment protocol involving only four agents, two for Alice and two for Bob [LKB+15]. Their protocol is secure against quantum adversaries and a multi-round variant, with longer duration time, was shown to be secure against classical adversaries [LKB+15, CCL15, FF15]. While those protocols only seemed of theoretical interest at first, recent implementations have convincingly demonstrated that the required timing and location constraints can be efficiently enforced. In [VMH+16], the authors performed a 24-hour-long bit commitment with the pairs of agents standing 8 km apart.

The security analysis against quantum adversaries of [LKB+15] and against classical adversaries of [CCL15] relies on the study of variants of **CHSH** games where the inputs and outputs belong to the field $\mathbb{F}_Q$ for some large prime power Q, instead of $\{0,1\}$ for the usual **CHSH** game. In many cases, the (quantum) security of a relativistic protocol can be derived from the value of an (entangled) 2-player game. Because the relativistic constraint essentially boils down to 2 non-communicating provers, a relativistic protocol can also be seen as a 2-prover interactive protocol.

—

The above results are promising for relativistic cryptography but very limited in scope. Indeed, bit commitment schemes are used as parts of larger cryptosystems. The only study of the composability of the $\mathbb{F}_Q$ bit commitment scheme was done in [FF15] but mainly with itself, in order to increase the commit time. There has not been any proposition to use this scheme for a more general purpose.

One natural application of bit commitment are zero-knowledge protocols. With such a protocol, a prover wishes to convince a verifier that a given statement is true without revealing any extra information. A zero-knowledge protocol is already a more advanced cryptographic primitive and has more direct applications such as identification schemes [GMR89] for instance. Here, we will consider the zero-knowledge construction for HAMILTONIAN CYCLE, which is an NP complete problem. The prover will convince the verifier that a given graph $G = (V, E)$ has a Hamiltonian cycle, *i.e.* a cycle going through each vertex exactly once, without revealing any information, in particular no information about this cycle. Since HAMILTONIAN CYCLEis NP complete, a zero-knowledge protocol for this problem can be used to obtain a zero-knowledge protocol for arbitrary NP problems.

There is a known zero-knowledge protocol for HAMILTONIAN CYCLE using bit commitment first presented by Blum [Blu86] which we recall now.

Zero-knowledge protocol for HAMILTONIAN CYCLE using bit commitment

1. The prover picks a random permutation $\Pi : V \to V$. He commits to each of the bits of the adjacency matrix $M_{\Pi(G)}$ of $\Pi(G)$.
2. The verifier sends a random bit (called the challenge) $chall \in \{0, 1\}$ to the prover.
3. – If $chall = 0$, the prover decommits to all the elements of $M_{\Pi(G)}$, and reveals Π.
   – If $chall = 1$, he reveals only the bits (of value 1) of the adjacency matrix that correspond to a Hamiltonian cycle $\mathcal{C}'$ of $\Pi(G)$.
4. The verifier checks that these decommitments are valid and correspond, for $chall = 0$ to $M_{\Pi(G)}$ and, for $chall = 1$, to a Hamiltonian cycle.

It is natural to combine this zero-knowledge protocol with the $\mathbb{F}_Q$ relativistic bit commitment protocol mentioned above. The (single-round) $\mathbb{F}_Q$ relativistic bit commitment protocol is secure against quantum adversaries but it doesn't directly imply that the zero-knowledge protocol remains secure. Indeed, the security definition considered for the bit commitment is fairly weak and composes poorly with other protocols. The soundness of the protocol against entangled provers will be reduced to a 2-player entangled game. Proving zero-knowledge against a quantum verifier can sometimes be complicated because of the presence of a quantum auxiliary input. In this case, however, due to properties of the relativistic $\mathbb{F}_Q$ bit commitment, we will not need any rewinding from our simulator and the simulation will actually be rather simple.

———

The goal of this paper is to show that it is indeed possible to plug in the $\mathbb{F}_Q$ relativistic bit commitment protocol into Blum's zero-knowledge protocol for HAMILTONIAN CYCLE. This widens the possible applications for relativistic cryptography and will encourage further implementations.

The main contribution of this paper is a technical analysis involving successive measurements on a quantum system. Indeed, to prove that the above scheme is secure against quantum adversaries, we use the fact that an adversary who can answer both challenges at the same time can guess the value of a string on which he has no information, due to non-signaling. This naturally involves consecutive measurements on a quantum system, and leads us to analyze how the first measurement disturbs the system before the second measurement.

1.2   Relativistic Zero-Knowledge Protocol for Hamiltonian Cycle

Here, we show how the final protocol will look like and where exactly we rely on the physical NSS principle. The final protocol is the following:

Relativistic zero knowledge protocol for HAMILTONIAN CYCLE

Input — The provers and the verifiers are given a graph $G = (V, E)$.
Auxiliary Input — The provers P_1 and P_2 know a Hamiltonian cycle $\mathcal{C}$ of G.
Preprocessing — P_1 and P_2 agree beforehand on a random permutation $\Pi : V \to V$ and on an $n \times n$ matrix $A \in \mathcal{M}_n^{\mathbb{F}_Q}$ where each element of A is chosen uniformly at random in $\mathbb{F}_Q$.
Protocol —

1. Commitment to each bit of $M_{\Pi(G)}$: V_1 sends a matrix $B \in \mathcal{M}_n^{\mathbb{F}_Q}$ where each element of B is chosen uniformly at random in $\mathbb{F}_Q$. P_1 outputs the matrix $Y \in \mathcal{M}_n^{\mathbb{F}_Q}$ such that $\forall i, j \in [n]$, $Y_{i,j} = A_{i,j} + (B_{i,j} * (M_{\Pi(G)})_{i,j})$.
2. The verifier sends a random bit (called the challenge) $chall \in \{0, 1\}$ to the prover.
3.   – If $chall = 0$, P_2 decommits to all the elements of $M_{\Pi(G)}$, i.e. he sends all the elements of A to V_2 and reveals Π.
    – If $chall = 1$, P_2 reveals only the bits (of value 1) of the adjacency matrix that correspond to a Hamiltonian cycle $\mathcal{C}'$ of $\Pi(G)$, i.e. for all edges (u, v) of $\mathcal{C}'$, he sends $A_{u,v}$ as well as $\mathcal{C}'$.
4. The verifier checks that those decommitments are valid and correspond to what the provers have declared. He also checks that the timing constraint of the bit commitment is satisfied. This means that
    – if $chall = 0$, the prover's opening A must satisfy $\forall i, j \in [n]$, $Y_{i,j} = A_{i,j} + (B_{i,j} * (M_{\Pi(G)})_{i,j})$.
    – if $chall = 1$, the prover's opening A must satisfy $\forall (u, v) \in \mathcal{C}'$, $Y_{u,v} = A_{u,v} + B_{u,v}$.

The above protocol is obtained by plugging in the $\mathbb{F}_Q$ relativistic bit commitment protocol into Blum's zero-knowledge protocol for HAMILTONIAN CYCLE. We discuss more the setting in Sects. 4 and 5. We just want here to briefly present in which way we use the no superluminal signaling condition in this protocol.

In order for the protocol to be secure, we require the following:

1. Both the prover and the verifier are split into 2 agents, respectively P_1, P_2 and V_1, V_2.
2. V_1 and V_2 are far apart (we discuss this later).
3. The opening phase (steps 2 and 3) must be performed as soon as the commit phase (step 1) is completed.

Constraints (2) and (3) are here to enforce that during step 3 of the protocol, *the message P_2 sends to V_2 does not depend on the matrix B sent by V_1.*

Because information travels at most at light speed, by synchronizing the steps well enough, the verifiers can enforce this condition. For instance, it is sufficient to check that V_2 receives the message from P_2 before the information on B sent by V_1 had time to reach V_2. If this is the case, then it guarantees that P_2's answer to the challenge cannot depend on the value of the matrix B, since otherwise it would violate the NSS. An important consequence is that we do not require the verifiers to know anything about the spatial locations of the provers: it is sufficient for the verifiers to know their own relative position.

As said before, there were already several experiments made that showed how to achieve the above constraints. The most notable one [VMH+16] succeeded in performing the above bit commitment protocol by having V_1 and V_2 being 8 km apart, which shows it can be achievable in real life conditions.

In summary, the main contribution of the paper is to prove the security of the above protocol for HAMILTONIAN CYCLE against quantum adversaries. The main challenge is to prove the soundness property, *i.e.* security against a cheating prover on an input which *does not* contain a Hamiltonian cycle. Here, we have two dishonest provers P_1 and P_2 that want to pass the protocol even though the input graph does not contain a Hamiltonian cycle. A cheating prover that would be able to answer simultaneously to both challenges could break the underlying string commitment scheme, which is a consequence of the special soundness property of the scheme.

To prove the security of the above protocol against quantum adversaries, we will, from a cheating strategy, construct a strategy that will successfully answer both challenges by consecutively applying the cheating strategy for each challenge, which is expressed by our consecutive measurement theorem (see Theorem 1 below). We can also view this cheating scenario as a 2-player entangled game and we will show how in general our theorem regarding quantum consecutive measurements can be translated into a bound on the entangled value of 2-player games.

1.3   Consecutive Measurements

Our main technical contribution is expressed by the following theorem

Theorem 1. *Consider n projectors $P_1, \ldots, P_n$ such that for each i, we can write $P_i := \sum_{s=1}^{S} P_i^s$ where the $\{P_i^s\}_s$ are orthogonal projectors for each i, i.e. for each i and s, s', we have $P_i^s P_i^{s'} = \delta_{s,s'} P_i^s$. Let σ be any quantum state, let $V := \frac{1}{n} \sum_{i=1}^{n} tr(P_i \sigma)$, and let $E := \frac{1}{n(n-1)} \sum_{i,j \neq i} \sum_{s,s'=1}^{S} tr(P_j^{s'} P_i^s \sigma P_i^s P_j^{s'})$. Then it holds that $E \geq \frac{1}{64S} \left(V - \frac{1}{n} \right)^3$.*

Such a statement can be fairly easily transposed to the context of games: see Proposition 1 below. This theorem can be seen as a generalization of the *gentle measurement* lemma [Win99], which is similar to the above with $n = 2$ and $S = 1$. The case of $n = 2$ can be seen as a worst-case consecutive measurement theorem: how much can the first measurement disturb the measured state before the second measurement? However, for larger values of n, this shows that when we pick 2 measurements out of n, the disturbance is much smaller, as shown by the dependence of the lower bound in n. Our theorem also improves on known results since it deals with larger values of S.

Interestingly, this kind of statement has already appeared previously in a paper by Unruh [Unr12], who studied quantum sigma protocols and in particular quantum proofs of knowledge. He showed the following

Theorem 2 [Unr12]. *Consider n projectors $P_1, \ldots, P_n$ and an arbitrary quantum state σ. Let $V := \frac{1}{n} \sum_{i=1}^{n} tr(P_i \sigma)$, and let $E := \frac{1}{n(n-1)} \sum_{i,j \neq i} tr(P_j P_i \sigma P_i P_j)$. If $V \geq \frac{1}{\sqrt{n}}$ then $E \geq V(V^2 - \frac{1}{n})$.*

Let us now compare our main theorem to Unruh's one, for the case of $S = 1$ where they are comparable. If $V \gg \frac{1}{\sqrt{n}}$ then both bounds give essentially the same bound $E \geq \Omega(V^3)$ which will translate into the relation $\omega^*(G_{coup}) \geq \Omega(\omega^*(G)^3)$ for the entangled values of a game G and its coupled version G_{coup} (see below). However, Theorem 2 is only valid when $V \geq \frac{1}{\sqrt{n}}$ while Theorem 1 works for any $V \geq \frac{1}{n}$. Moreover, Theorem 1 is tight in its extremal point in the sense that there exist a quantum state and n projectors such that $V = \frac{1}{n}$ and $E = 0$, as can be seen by considering for example $\sigma = |\phi\rangle\langle\phi|$ with $|\phi\rangle := \frac{1}{\sqrt{n}} \sum_i |i\rangle$ and $P_i = |i\rangle\langle i|$.

—

A natural application of our consecutive measurement theorem is to bound the value of some entangled games. The phrasing in terms of nonlocal games is sometimes more comfortable to use. In this paper, our security proofs will usually reduce to bounding the entangled value of such game, that is the maximum winning probability for a pair of players allowed to share arbitrary entangled states as a resource. For any game G on the uniform distribution (meaning that the inputs of the game are drawn independently from the uniform distribution), we define the game G_{coup} (consisting of a certain *couple* of instances of G) as follows:

– In G, Alice and Bob respectively receive x and y taken from the uniform distribution on the sets I_A and I_B, respectively, and output a and b such that $V(a, b|x, y) = 1$ for some valuation function V specified by G.
– In G_{coup}, Alice receives a random x as in G and Bob receives a pair of distinct random inputs (y, y'). Alice outputs a and Bob outputs a pair (b, b'). They win the game if $V(a, b|x, y) = 1$ and $V(a, b'|x, y') = 1$, that is, if they win both instances of the game G, but for the same input/output pair of Alice.

In many cases, upper bounding the value of G_{coup} will follow directly from a non-signaling argument of the form: "If the players are able to win G_{coup} with probability p then Bob can learn some (or all) bits of x with probability p and no-signaling implies that $p \leq 1/|I_A|$". What is left to do is to relate the entangled values of both games, $\omega^*(G)$ and $\omega^*(G_{coup})$. To do this, we construct the following strategy for G_{coup}: Alice follows the same strategy as for G; on inputs (y, y'), Bob performs the same strategy (measurement) as for G on input y to get output b and then on input y' to get b'. Note here that the non trivial part is that Bob's second measurement is applied on the post-measurement state resulting from his first measurement. Because we are in the quantum setting, this first measurement will generally perturb the state shared by Alice and Bob, which makes it non trivial to relate the success probability of this strategy for G_{coup} with the entangled value $\omega^*(G)$ of the original game G.

A similar construction of *squared games* was introduced in [DS14, DSV15] to study projective classical and entangled games. There, the input x is not revealed to the players but they receive respectively y and y' and output b and b'. They win if there exists a such that $V(a, b|x, y) = V(a, b'|x, y') = 1$. It would be interesting to see the similarities and differences between those two approaches.

We show the following.

Proposition 1. *For any game G on the uniform distribution which is S-projective, we have $\omega^*(G_{coup}) \geq \frac{1}{S \cdot 64} \cdot (\omega^*(G) - \frac{1}{n})^3$ where n is dimension of Bob's input.*

A game G is said to be S-projective if for all x, y, a, there are at most S possible outputs for Bob that allow them to win the game, i.e. $\max_{x,y,a} |\{b : V(a, b|x, y) = 1\}| \leq S$.

In order to prove this statement, we need to analyze the strategy that we presented above. As already mentioned, the main difficulty is that the first measurement from Bob will modify the common shared state and therefore we cannot directly bound the probabilities related to the second measurement. One way of analyzing these consecutive measurements would be to use a kind of gentle measurement lemma but unfortunately, this would only work when the winning probability $\omega^*(G)$ is close to 1, which isn't the case for the games we consider.

Fortunately, Theorem 1 is tailored for this kind of applications and can be used directly to prove the above proposition. We can notice the exact transposition of the parameters of Theorem 1 to Proposition 1.

1.4 Applications of the Bound

1. First, we prove that the extensions of the $\mathbb{F}_Q$ bit commitment to string commitment and its parallel repetition remain secure against quantum adversaries with using the sum-binding definition. This is a direct consequence on upper bounds on the entangled value of **CHSH** variants, like the $\mathbf{CHSH_Q}(P)$ game introduced in [CCL15].
2. We show that the presented relativistic zero-knowledge protocol for HAMILTONIAN CYCLE is secure against quantum adversaries. This also implies a 2-prover 1-round zero-knowledge protocol for HAMILTONIAN CYCLE also secure against quantum adversaries.
3. Finally, as a direct corollary of our consecutive measurement claim, we answer an open question from Unruh regarding quantum proofs of knowledge [Unr12]. We show tight bounds on the quantum knowledge error of a Σ-protocol with

strict and special soundness as function of the challenge size, matching the classical bound. We will not discuss in detail this result as it just requires to plug our bound in the proof of [Unr12] and is a bit beyond the scope of this paper. However, this shows that our results are useful beyond just the study of relativistic protocols or entangled games.

The last point shows that our bound could find even more applications when considering security against quantum adversaries. Indeed, when studying cryptographic protocols, for instance Σ-protocols, a notion that often appears is *special soundness* which roughly states that an attacker shouldn't be able to simultaneously answer successfully to 2 verifier's challenges. The relativistic zero-knowledge protocol we study is one example of this and Unruh's quantum proofs of knowledge setting is another one but there are more where our theorem could be useful.

Organisation of the Paper. In Sect. 2, we prove our main consecutive measurement theorem. In Sect. 3, we show how to use this bound for proving upper bounds on the entangled value of nonlocal games. In Sect. 4, we present in more detail the relativistic model and the $\mathbb{F}_Q$ relativistic bit commitment protocol. Finally, in Sect. 5, we describe the protocol obtained by plugging this bit commitment into Blum's zero-knowledge protocol for HAMILTONIAN CYCLE and we prove that it remains secure, even against quantum adversaries.

2   Consecutive Measurement Theorems

We first present some useful lemmata in the preliminaries. Then, we dive in directly in the proof of our consecutive measurements theorems.

2.1   Preliminaries

Lemma 1. *Let* $|\phi\rangle$ *a quantum pure state,* $P \le \mathbb{I}$ *a projector acting on* $|\phi\rangle$ *and* $|\psi\rangle := \frac{P(|\phi\rangle)}{||P(|\phi\rangle)||}$. *We have* $|\langle\phi|\psi\rangle|^2 = ||P(|\phi\rangle)||^2 = tr(P|\phi\rangle\langle\phi|)$.

Proof. We write $|\phi\rangle = P(|\phi\rangle) + (\mathbb{I} - P)(|\phi\rangle) = ||P(|\phi\rangle)|| \; |\psi\rangle + (\mathbb{I} - P)(|\phi\rangle)$. By noticing that $\langle\psi|\mathbb{I} - P|\phi\rangle = 0$, we get $|\langle\phi|\psi\rangle|^2 = ||P(|\phi\rangle)||^2 = tr(P|\phi\rangle\langle\phi|)$.

Lemma 2. *Let* $|\phi\rangle$ *a quantum pure state,* $P \le \mathbb{I}$ *a projector acting on* $|\phi\rangle$ *and* $|\psi\rangle$ *such that* $P|\psi\rangle = |\psi\rangle$. *We have* $|\langle\phi|\psi\rangle|^2 \le tr(P|\phi\rangle\langle\phi|)$.

Proof. We decompose $|\phi\rangle$ in order to make $|\psi\rangle$ appear. We write $|\phi\rangle = \alpha|\psi\rangle + \beta|\psi^\perp\rangle$ with $|\alpha|^2 + |\beta|^2 = 1$ and $\langle\psi|\psi^\perp\rangle = 0$. This gives us $P|\phi\rangle = \alpha|\psi\rangle + \beta P|\psi^\perp\rangle$. Notice that we also have $\langle\psi|P|\psi^\perp\rangle = 0$. From there, we conclude

$$tr(P|\phi\rangle\langle\phi|) = ||P|\phi\rangle||^2 = |\alpha|^2 + |\beta|^2||P|\psi^\perp\rangle||^2 \ge |\alpha|^2 = |\langle\phi|\psi\rangle|^2.$$

2.2   Single Outcome Case : S = 1

We first prove the theorem for the case where $S = 1$.

Theorem 3. *Consider n projectors $P_1, \ldots, P_n$ and a quantum mixed state σ in some Hilbert space $\mathcal{B}$. Let $V := \frac{1}{n} \sum_{i=1}^{n} tr(P_i \sigma)$, and let*

$$E := \frac{1}{n(n-1)} \sum_{i,j \neq i} tr(P_j P_i \sigma P_i P_j).$$

Then it holds that $E \geq \frac{1}{64} \left(V - \frac{1}{n} \right)^3$.

Proof. We fix a quantum mixed state σ in some Hilbert space $\mathcal{B}$ and n projectors $P_1, \ldots, P_n$ acting on $\mathcal{B}$. We first move to the realm of pure states which will be easier to analyze by adding an extra Hilbert space $\mathcal{E}$. We consider a purification $|\phi\rangle$ of σ in some space $\mathcal{BE} = \mathcal{B} \otimes \mathcal{E}$. We define

$$|\phi_i\rangle := \frac{(P_i \otimes \mathbb{1}_\mathcal{E})|\phi\rangle}{||(P_i \otimes \mathbb{1}_\mathcal{E})|\phi\rangle||}.$$

The state $|\phi_i\rangle$ corresponds to the normalized projection of $|\phi\rangle$ using P_i. We first express E and V as inner products of the quantum pure states we defined:

Lemma 3. $E \geq \frac{1}{n(n-1)} \sum_{i,j \neq i} |\langle \phi | \phi_i \rangle|^2 |\langle \phi_i | \phi_j \rangle|^2$ *and* $V = \frac{1}{n} \sum_{i=1}^{n} |\langle \phi | \phi_i \rangle|^2.$

Proof. We write

$$E = \frac{1}{n(n-1)} \sum_{i,j \neq i} tr(P_j P_i \sigma P_i P_j)$$

$$= \frac{1}{n(n-1)} \sum_{i,j \neq i} tr \left((P_j \otimes \mathbb{1}_\mathcal{E})(P_i \otimes \mathbb{1}_\mathcal{E})|\phi\rangle\langle\phi|(P_i \otimes \mathbb{1}_\mathcal{E})(P_j \otimes \mathbb{1}_\mathcal{E}) \right)$$

Here, by using Lemma 1, notice that

$$(P_i \otimes \mathbb{1}_\mathcal{E})|\phi\rangle\langle\phi|(P_i \otimes \mathbb{1}_\mathcal{E}) = ||(P_i \otimes \mathbb{1}_\mathcal{E})|\phi\rangle||^2 |\phi_i\rangle\langle\phi_i| = |\langle\phi|\phi_i\rangle|^2 |\phi_i\rangle\langle\phi_i|.$$

From there, we can continue have

$$E = \frac{1}{n(n-1)} \sum_{i,j \neq i} |\langle\phi|\phi_i\rangle|^2 tr \left((P_j \otimes \mathbb{1}_\mathcal{E})|\phi_i\rangle\langle\phi_i| \right)$$

$$\geq \frac{1}{n(n-1)} \sum_{i,j \neq i} |\langle\phi|\phi_i\rangle|^2 |\langle\phi_i|\phi_j\rangle|^2$$

where the last inequality comes from Lemma 2. Notice also that we immediately have $V = \frac{1}{n} \sum_{i=1}^{n} tr(P_i \sigma) = \sum_i |\langle\phi|\phi_i\rangle|^2$.

Our goal is to relate E and V. We will deal with the terms $|\langle\phi_i|\phi_j\rangle|^2$ using the following proposition on almost orthogonal states.

Proposition 2. *Consider n quantum pure states $|\phi_1\rangle, \ldots, |\phi_n\rangle$. Let*

$$S := \max_{|\Omega\rangle} \sum_{i=1}^{n} |\langle \Omega|\phi_i\rangle|^2 \quad and \quad C := \sum_{i,j \neq i}^{n} |\langle \phi_i|\phi_j\rangle|^2.$$

We have $S \leq 1 + \sqrt{\frac{(n-1)C}{n}} \leq 1 + \sqrt{C}$.

Proof. Let $M = \sum_{i=1}^{n} |\phi_i\rangle\langle\phi_i|$. M is a positive semi-definite matrix of dimension at most n. Let $\lambda_1 \geq \lambda_2 \geq \cdots \geq \lambda_n$ the n eigenvalues of M in decreasing order. We have $\sum_i \lambda_i = tr(M) = \sum_j tr(|\phi_j\rangle\langle\phi_j|) = n$. Moreover, notice that $S = \max_{|\Omega\rangle} \sum_i |\langle\Omega|\phi_i\rangle|^2 = \lambda_1$.

We write $M^2 = \sum_{i,j} \langle\phi_i|\phi_j\rangle|\phi_i\rangle\langle\phi_j|$ and $tr(M^2) = \sum_{i,j} |\langle\phi_i|\phi_j\rangle|^2 = n + C$. Moreover, we have $tr(M^2) = \sum_i \lambda_i^2$. This gives us

$$n + C = tr(M^2) = \sum_{i=1}^{n} \lambda_i^2 = \lambda_1^2 + \sum_{i=2}^{n} \lambda_i^2 \geq \lambda_1^2 + (n-1)\left(\frac{n-\lambda_1}{n-1}\right)^2$$

$$= \lambda_1^2 + \frac{(n-\lambda_1)^2}{n-1} = S^2 + \frac{(n-S)^2}{n-1}$$

where the inequality comes from the convexity of the square function. From there, we have

$$(n-1)S^2 + (n-S)^2 - n(n-1) \leq (n-1)C$$

Using $(n-1)S^2 + (n-S)^2 - n(n-1) = n(S-1)^2$, we conclude that $n(S-1)^2 \leq (n-1)C$ or equivalently $S \leq 1 + \sqrt{\frac{(n-1)C}{n}}$.

In particular, the above proposition implies that

$$V \leq \frac{1}{n} + \frac{n-1}{n}\sqrt{\frac{1}{n(n-1)}\sum_{i,j\neq i}|\langle\phi_i|\phi_j\rangle|^2}.$$

The term in the squared root is very similar to E. Unfortunately, the expression for E contains an extra factor $|\langle\phi|\phi_i\rangle|^2$ in the sum under the square-root. If the quantity $|\langle\phi|\phi_i\rangle|^2$ was independent of i, it would be equal to V and we would be able to conclude. However, this is not always the case and this adds a difficulty in the proof. In order to overcome it, we will use Proposition 2 only with the states for which $|\langle\phi|\phi_i\rangle|^2$ is not too small. We will choose a threshold κ (that will be fixed later) and consider only the indices i for which $|\langle\phi|\phi_i\rangle|^2 \geq V/\kappa$. This is the goal of the next proposition.

Proposition 3. $\forall \kappa > 1$, $V \leq \left(1 + \frac{1}{\kappa-1}\right)\left(\frac{1}{n} + \sqrt{\frac{\kappa E}{V}}\right)$.

Proof. For all i, let $p_i := |\langle\phi|\phi_i\rangle|^2$. We have by definition $V = \sum_i p_i$. We fix $\kappa > 1$ and define the set $Z := \{i \in [n] : p_i \geq \frac{V}{\kappa}\}$. We have

$$\frac{1}{n}\sum_{i\notin Z} p_i \leq \frac{1}{n}\sum_{i\notin Z}\frac{V}{\kappa} \leq \frac{V}{\kappa},$$

which implies

$$\frac{1}{n}\sum_{i\in Z} p_i \geq (1 - \frac{1}{\kappa})V. \tag{1}$$

We write

$$E \geq \frac{1}{n(n-1)}\sum_{i,j\neq i} p_i|\langle\phi_i|\phi_j\rangle|^2 \geq \frac{1}{n(n-1)}\sum_{\substack{i,j\in Z \\ i\neq j}} p_i|\langle\phi_i|\phi_j\rangle|^2 \tag{2}$$

$$\geq \frac{V}{\kappa}\cdot\frac{1}{n(n-1)}\sum_{\substack{i,j\in Z \\ i\neq j}}|\langle\phi_i|\phi_j\rangle|^2. \tag{3}$$

Now, starting from Eq. 1, we have

$$V \leq \frac{1}{1-\frac{1}{\kappa}}\frac{1}{n}\sum_{i\in Z} p_i = \frac{1}{1-\frac{1}{\kappa}}\frac{1}{n}\sum_{i\in Z}|\langle\phi|\phi_i\rangle|^2 \leq \left(1+\frac{1}{\kappa-1}\right)\max_{|\Omega\rangle}\frac{1}{n}\sum_{i\in Z}|\langle\Omega|\phi_i\rangle|^2$$

$$\leq \left(1+\frac{1}{\kappa-1}\right)\left(\frac{1}{n}+\sqrt{\frac{1}{n(n-1)}\sum_{\substack{i,j\in Z \\ i\neq j}}|\langle\phi_i|\phi_j\rangle|^2}\right) \tag{4}$$

$$\leq \left(1+\frac{1}{\kappa-1}\right)\left(\frac{1}{n}+\sqrt{\frac{\kappa E}{V}}\right) \tag{5}$$

where we used Lemma 2 in Eqs. 4 and 2 for the last inequality. This proves the proposition.

We can now use Proposition 3 to prove our theorem. We distinguish two cases:

1. If $(\frac{V}{n^2 E})^{1/3} > 2$. We take $\kappa = (\frac{V}{n^2 E})^{1/3} > 2$ which implies $\kappa(\frac{n^2 E}{V})^{1/3} = 1$ and $(\kappa\frac{n^2 E}{V})^{\frac{1}{2}} = \frac{1}{\kappa}$. We get

$$V \leq \frac{1}{n}\left(1+\frac{1}{\kappa-1}\right)\left(1+\left[\kappa\frac{n^2 E}{V}\right]^{1/2}\right) = \frac{1}{n}\left(1+\frac{1}{\kappa-1}\right)\left(1+\frac{1}{\kappa}\right)$$

$$\leq \frac{1}{n}(1+\frac{4}{\kappa}) = \frac{1}{n}+4\left(\frac{E}{nV}\right)^{1/3}.$$

This gives $E \geq \frac{nV}{64}(V - \frac{1}{n})^3$ which implies $E \geq \frac{1}{64}(V - \frac{1}{n})^3$. To see this last implication, consider the following two cases: if $V \geq \frac{1}{n}$ then the equality comes immediately from the previous inequality. If $V \leq \frac{1}{n}$, we immediately have $E \geq 0 \geq \frac{1}{64}(V - \frac{1}{n})^3$.

2. If $(\frac{V}{n^2 E})^{1/3} \leq 2$. This implies $(\frac{V}{E})^{1/2} \leq n \cdot 2^{3/2}$. We take $\kappa = 2$ and obtain

$$
V \leq \left(1 + \frac{1}{\kappa - 1}\right)\left(\frac{1}{n} + \sqrt{\frac{\kappa E}{V}}\right) = 2(\frac{1}{n} + \sqrt{\frac{2E}{V}})
$$

$$
\leq 2(2^{2/3}\sqrt{\frac{E}{V}} + \sqrt{\frac{2E}{V}}) \leq 6\sqrt{\frac{E}{V}}
$$

which implies $E \geq \frac{V^3}{36} \geq \frac{1}{64}(V - \frac{1}{n})^3$.

2.3   General Case

We can now show our theorem for any S. The general case will be a direct corollary of the following.

Proposition 4. *Let a projector* $P := \sum_{i=1}^{m} P_i$ *where* $\{P_i\}_{i \in [m]}$ *are orthogonal projectors. For any pure state* $|\psi\rangle$, *we have*

$$
\sum_{i=1}^{m} P_i|\psi\rangle\langle\psi|P_i \geq \frac{1}{m}P|\psi\rangle\langle\psi|P.
$$

We note that this result can be obtained as an application of the pinching inequality [Hay02, SBT16], but we provide a proof here for completeness.

Proof. We define the following *unnormalized states* $|\psi^P\rangle = P(|\psi\rangle)$ and $|\psi_i^P\rangle = P_i(|\psi\rangle)$. Because $P = \sum_i P_i$, we have $|\psi^P\rangle = \sum_i |\psi_i^P\rangle$. This gives

$$
\sum_{i=1}^{m} P_i|\psi\rangle\langle\psi|P_i = \sum_{i=1}^{m} |\psi_i^P\rangle\langle\psi_i^P|
$$

$$
P|\psi\rangle\langle\psi|P^\dagger = |\psi^P\rangle\langle\psi^P|
$$

Consider now any state $|\phi\rangle = \sum_i \alpha_i|\psi_i^P\rangle + |\xi\rangle$ where $|\xi\rangle$ is orthogonal to all the $|\psi_i^P\rangle$. We have

$$
\langle\phi|\sum_{i=1}^{m} P_i|\psi\rangle\langle\psi|P_i|\phi\rangle = \sum_i |\langle\psi_i^P|\phi\rangle|^2 = |\alpha_i|^2 \left|\langle\psi_i^P|\psi_i^P\rangle\right|^2
$$

and

$$
\langle\phi|P|\psi\rangle\langle\psi|P|\phi\rangle = |\langle\psi^P|\phi\rangle|^2 = \left|\sum_i \alpha_i\langle\psi_i^P|\psi_i^P\rangle\right|^2
$$

From there, we can conclude. We have:

$$\langle\phi|\sum_{i=1}^{m} P_i|\psi\rangle\langle\psi|P_i|\phi\rangle = \sum_{i}|\alpha_i|^2|\langle\psi_i^P|\psi_i^P\rangle|^2$$

$$\geq \frac{1}{m}\left|\sum_{i}|\alpha_i||\langle\psi_i^P|\psi_i^P\rangle|\right|^2 \quad \text{(from Cauchy-Schwarz)}$$

$$\geq \frac{1}{m}\langle\phi|P|\psi\rangle\langle\psi|P|\phi\rangle$$

Since this holds for any state $|\phi\rangle$, we can conclude that

$$\sum_{i=1}^{m} P_i|\psi\rangle\langle\psi|P_i \geq \frac{1}{m}P|\psi\rangle\langle\psi|P^\dagger.$$

From there, and using the previous theorem, we can show our main technical result.

Theorem 1. *Consider n projectors $P_1,\ldots,P_n$ such that for each i, we can write $P_i := \sum_{s=1}^{S} P_i^s$ where the $\{P_i^s\}_s$ are orthogonal projectors for each i, i.e. for each i and s, s', we have $P_i^s P_i^{s'} = \delta_{s,s'} P_i^s$. Let σ be any quantum state, let $V := \frac{1}{n}\sum_{i=1}^{n} tr(P_i\sigma)$, and let $E := \frac{1}{n(n-1)}\sum_{i,j\neq i}\sum_{s,s'=1}^{S} tr(P_j^{s'} P_i^s \sigma P_i^s P_j^{s'})$. Then it holds that $E \geq \frac{1}{64S}\left(V - \frac{1}{n}\right)^3$.*

Proof. We fix n projectors $P_1,\ldots,P_n$ such that for each i, we can write $P_i := \sum_{s=1}^{S} P_i^s$ where the $\{P_i^s\}_s$ are orthogonal projectors for each i. We fix a quantum state σ. We have

$$E = \frac{1}{n(n-1)}\sum_{i,j\neq i}\sum_{s,s'=1}^{S} tr(P_j^{s'} P_i^s \sigma(P_i^{s'})(P_j^s))$$

$$= \frac{1}{n(n-1)}\sum_{i,j\neq i}\sum_{s=1}^{S} tr(P_j P_i^s \sigma(P_i^{s'})P_j)$$

$$\geq \frac{1}{Sn(n-1)}\sum_{i,j\neq i} tr(P_j P_i \sigma(P_i)P_j) \quad \text{(from Proposition 4)}$$

$$\geq \frac{1}{64S}\left(V - \frac{1}{n}\right)^3 \quad \text{(from Theorem 3)}$$

3   Entangled Games

The goal of this section is to use the consecutive measurement theorems of the previous section to establish upper bounds on the value of entangled games. For a game G on the uniform distribution, we will define a game G_{coup} which corresponds to a couple of instances of G where Alice plays twice with the same input and Bob receives two distinct inputs and they need to win both instances in order to win the game G_{coup}. In the cases we consider, upper bounding G_{coup} will be easily done from non-signaling. Our learning lemmata will allow us to relate the winning probabilities of G and G_{coup}. These two steps together will give us bounds on the value of G.

3.1   First Definitions

Definition 1. A game $G = (I_A, I_B, O_A, O_B, V, p)$ is defined by

- 2 input sets I_A, I_B which are respectively Alice's and Bob's input sets.
- 2 output sets sets O_A, O_B which are respectively Alice's and Bob's output sets.
- A valuation function $V : I_A \times I_B \times O_A \times O_B \to \{0, 1\}$ which indicates whether the game is won for some fixed input and outputs. The game is won if the value of V is 1.
- A probability function $p : I_A \times I_B \to [0, 1]$ which corresponds to the input distribution. We have $\sum_{(x,y) \in I_A \times I_B} p_{xy} = 1$.

Definition 2. A game $G = (I_A, I_B, O_A, O_B, V, p)$ is said to be on the uniform distribution if $\forall (x, y) \in I_A \times I_B$, $p_{xy} = \frac{1}{|I_A||I_B|}$.

Definition 3. A game $G = (I_A, I_B, O_A, O_B, V, p)$ is projective if

$$\forall (x, y) \in I_A \times I_B \text{ st. } p_{xy} \neq 0, \ \forall a \in O_A, \ \exists! \ b \in O_B, \ st. \ V(x, y, a, b) = 1.$$

A game G is S-projective if

$$\forall (x, y) \in I_A \times I_B \text{ st. } p_{xy} \neq 0, \forall a \in O_A, |\{b \in O_B : V(x, y, a, b) = 1\}| \leq S.$$

In particular, a projective game is 1-projective.

In the case where Alice and Bob are classical and want to win a game G, it is known that their optimal strategy to win is to perform a deterministic strategy. Notice that a projective game is asymmetric in Alice and Bob.

Definition 4. For a game $G = (I_A, I_B, O_A, O_B, V, p)$, we denote by $\omega^*(G)$ its entangled value, i.e. the maximum winning probability for the game when Alice and Bob are quantum and share an entangled state.

In order to study this maximal winning probability, it is enough to consider the case where Alice and Bob perform projective measurements.

In order to prove upper bounds on $\omega^*(G)$ for a game G on the uniform distribution, we introduce the notion of coupled game G_{coup}.

Definition 5. For any game $G = (I_A, I_B, O_A, O_B, V, p)$ on the uniform distribution we define G_{coup} as follows:

- Alice receives a random $x \in_R I_A$. Bob receives a random pair of different inputs (y, y') from I_B.
- Alice outputs $a \in O_A$. Bob outputs $b, b' \in O_B$.
- They win the game if $V(x, y, a, b) = V(x, y', a, b') = 1$.

3.2   Relating G and G_{coup}

In this section, we use our results from the previous section to relate the values of G and G_{coup}.

Proposition 1. *For any game G on the uniform distribution which is S-projective, we have $\omega^*(G_{coup}) \geq \frac{1}{S \cdot 64}(\omega^*(G) - \frac{1}{n})^3$ where $n = |I_B|$.*

Proof. Consider an optimal strategy for Alice and Bob for the game G. In particular, for each y, let $Q^y = \{Q_b^y\}$ the projective measurement that corresponds to his strategy for input y. Fix an input/output pair (x, a) for Alice and let σ^{xa} be the state held by Bob, conditioned on this pair. For each y, let $W_y = \{b : V(a, b|x, y) = 1\}$ be the set of winning outputs for Bob. Since G is S-projective, we have $|W_y| \leq S$. We define $Q_W^y = \sum_{b \in W_y} Q_b^y$.

We denote by V^{xa} the probability that Alice and Bob win the game for a fixed x, a. Notice that $\omega^*(G) = \mathbb{E}_{xa}[V^{xa}]$. We have

$$V^{xa} = \frac{1}{n} \sum_y tr(Q_W^y \sigma^{xa}(Q_W^y)),$$

since y is uniformly distributed over the set I_B of size n.

We now consider the following quantum strategy for G_{coup}: Alice and Bob share the same initial state as in the optimal strategy for G; Alice performs the same measurement strategy as for G; on inputs y, y, Bob applies the first measurement Q^y and obtains outcome b, then applies the measurement $P^{y'}$ on his resulting state and gets outcome b'. Bob outputs (b, b'). Let E^{xa} be the probability that Alice and Bob win G_{coup} using this strategy for a fixed x, a. Notice that $\omega^*(G_{coup}) \geq \mathbb{E}_{xa}[E^{xa}]$ since the value $\mathbb{E}_{xa}[E^{xa}]$ is achievable. We have

$$E^{xa} = \frac{1}{n(n-1)} \sum_{\substack{y, y' \neq y}} \sum_{\substack{b:V(ab|xy)=1 \\ b':V(ab'|xy')=1}} tr(Q_{b'}^{y'} Q_b^y \sigma^{xa} Q_b^y Q_{b'}^{y'})$$

$$\geq Pos(\frac{1}{64S}(V^{xa} - \frac{1}{n})^3) \qquad\qquad \text{from Theorem 1}$$

where $Pos(x) := \max(x, 0)$ is the positive part of x. By taking the expectation on each side, we obtain

$$\omega^*(G_{coup}) = \mathbb{E}_{xa}[E^{xa}] \geq \mathbb{E}_{xa}[Pos(\frac{1}{64S}(V^{xa} - \frac{1}{n})^3)] \geq Pos(\frac{1}{64S}(\omega^*(G) - \frac{1}{n})^3)$$

$$\geq \frac{1}{64S}(\omega^*(G) - \frac{1}{n})^3$$

where we used the convexity of the function $x \mapsto Pos(x^3)$.

3.3   Retrieving the Value of Certain Entangled Games

We now use the technique developed above in order to obtain upper bounds on games based on the $\mathbb{F}_Q$ variant of $CHSH$.

CHSH^Q(P) — We consider the nonlocal game called $CHSH^Q(P)$ with $P \leq Q$. Here, Alice and Bob receive inputs x and y, where x is a uniformly random element in $\mathbb{F}_Q$ and y is an element of $\mathbb{F}_Q$ taken uniformly at random from $\{0, \ldots, P-1\}$. They output values $a, b \in \mathbb{F}_Q$ and win if $a + b = x * y$, where the addition and multiplication are with respect to $\mathbb{F}_Q$. Notice that $CHSH^Q(P)$ is a projective game on the uniform distribution.

Let's analyze $CHSH^Q(P)_{coup}$. Fix an input/output pair (x, a) and a pair (y, y') of inputs for Bob with $y \neq y'$. Let b, b' Bob's output. If Alice and Bob win the game then we have $a + b = x * y$ and $a + b' = x * y'$ which implies that $(b - b') * (y - y')^{-1} = x$. This means that Bob can use any strategy for $CHSH^Q(P)_{coup}$ as a strategy to guess x with the same winning probability. Because of non-signaling, this happens with probability at most $\frac{1}{Q}$. We therefore have $\omega^*(CHSH^Q(P)_{coup}) \leq \frac{1}{Q}$. Using Proposition 1 (we have $S = 1$ in this setting), we obtain $\omega^*(CHSH^Q(P)) \leq \frac{1}{P} + \frac{4}{Q^{1/3}}$.

CHSH^Q(2)$^{\otimes n}$ — This is the parallel repetition of **CHSH^Q** where Alice and Bob receive n uniform strings $x_1, \cdots, x_n$ and $y_1, \cdots, y_n \in \{0, 1\}$ and output strings $a_1, \cdots, a_n$ and $b_1, \cdots, b_n$, respectively. They win the **CHSH^Q**$(2)^{\otimes n}$ game if they win all n instances of the **CHSH^Q** games, *i.e.* if $a_i + b_i = x_i * y_i$ for all $i \in \{1, \cdots, n\}$. Consider now the coupled version of this game. For any two inputs $y = y_1, \ldots, y_n, y' = y'_1, \ldots, y'_n$ given to Bob, if Alice and Bob win the game then similarly as in **CHSH**, Bob can recover Alice's input bits x_i for each i where $y_i \neq y'_i$. From non signaling, this happens with probability at most $Q^{-|y-y'|_H}$, where $|y - y'|_H$ is the Hamming distance between strings y and y', counting in how many indices both strings differ. Therefore, we have $\omega^*(CHSH^Q(2)_{coup}^{\otimes n}) = \mathbb{E}_{y,y' \neq y}[Q^{-|y-y'|_H}] = \frac{1}{2^n}\left((1 + \frac{1}{Q})^n - 1\right)$. If $Q > n$, we have

$$\omega^*(CHSH^Q(2)_{coup}^{\otimes n}) \leq \frac{2n}{Q2^n} \quad \text{which gives} \quad \omega(CHSH^Q(2)^{\otimes n}) \leq \frac{1}{2^n} + 4(\frac{2n}{Q2^n})^{1/3}.$$

In particular, if we take $Q = \frac{64 \cdot 2^{2n}}{2n\varepsilon^3}$, we obtain $\omega(CHSH^Q(2)^{\otimes n}) \leq \frac{1}{2^n}(1 + \varepsilon)$.

4   Relativistic Bit and String Commitment

In this section, we will review the relativistic $\mathbb{F}_Q$ bit commitment scheme and its natural extension to string commitment. We will show how the sum-binding property (with worst parameters) is preserved when considering string commitment or the parallel repetition of bit commitment. This is showed by Propositions 5 and 6.

4.1 Bit Commitment

Bit commitment is a cryptographic primitive between two distrustful parties Alice and Bob which consists of 2 phases: a *Commit phase* and a *Reveal phase*. Alice has a bit d at the beginning of the protocol. In the commit phase, Alice will commit to this value d by performing some communication protocol such that at end of the commit phase, Bob has no information about d (hiding property). In the second phase, the reveal phase, Alice and Bob also perform some communication which results in Alice revealing d. A desired property here is that Alice is unable to reveal a bit different from the one chosen during the commit phase (binding property).

In some sense, a bit commitment protocol simulates a digital safe. In the commit phase, Alice writes her input d on a piece of paper, puts that paper into the safe and sends the safe to Bob. If Bob doesn't hold the key of the safe then he cannot open it and therefore has no information about d. In the reveal phase, Alice would send to Bob the key to open the safe. But she cannot change the value of the bit in the safe because Bob has control of the safe. This primitive has been widely studied. However, bit commitment can only be performed with computational security in the most usual models.

We now define more formally a bit commitment scheme.

Definition 6. *A quantum commitment scheme is an interactive protocol between Alice and Bob with two phases, a Commit phase and a Reveal phase.*

- Commit phase. *Alice chooses a uniformly random input d that she wants to commit to. To do so, Alice and Bob perform a communication protocol that corresponds to this commit phase.*
- Reveal phase. *Alice interacts with Bob in order to reveal d. To do so, they perform a second communication protocol where at the end, Bob should know the value revealed by Alice. Bob, depending on this revealed value and the interaction with Alice, outputs either "Accept" or "Reject".*

A commitment scheme $\Pi = (COMM, OPEN)$ is the description of the protocol followed by the honest parties during both the commit and the open phases. All protocols that we will consider will be perfectly hiding and we will only be interested in the binding property. Therefore, we only consider the case of a cheating Alice, which will be described through her cheating strategy $Str^* = (Comm^*, Open^*)$ in both phases of the protocol. The binding property we consider is the standard sum-property, that was also used in previous work regarding relativistic bit commitment [LKB+15,FF15,CCL15].

Definition 7 (Sum-binding). *We say that a bit commitment protocol Π is ε-sum-binding if*

$$\forall \, Comm^*, \sum_{d=0}^{1} \max_{Open^*} \left(\Pr[Alice \; successfully \; reveals \; d \mid (Comm^*, Open^*)] \right) \leq 1+\varepsilon.$$

In the case of string commitment, meaning Alice wants to commit/reveal to a string of dimension P (*i.e.* $\lceil \log(P) \rceil bits$), we can extend the sum-binding property as follows.

Definition 8 (String sum-binding). *We say that a P-string commitment protocol Π is ε-sum-binding if*

$$\forall \, \text{Comm}^*, \sum_{d=0}^{P-1} \max_{\text{Open}^*} \left(\Pr[Alice \; successfully \; reveals \; d \, | \, (\text{Comm}^*, \text{Open}^*)] \right) \leq 1 + \varepsilon.$$

The sum-binding property for bit commitment is a relatively weak one. Indeed, it is very hard to use this definition when combining it with other primitives. For example, when committing to n bits in parallel, it is not always the case that this overall commitment, seen as a 2^n-string commitment, satisfies a good string sum-binding property. On the other hand, the string sum-binding for strings seems more exploitable.

4.2   Relativistic Bit Commitment

A relativistic bit commitment scheme is a commitment scheme where we use physical property that no information carrier can travel faster than the speed of light. In order to take advantage of this principle, we split Alice (resp. Bob) into 2 agents $\mathcal{A}_1$ and $\mathcal{A}_2$ (respectively $\mathcal{B}_1$ and $\mathcal{B}_2$). For each $i \in \{1, 2\}$, $\mathcal{A}_i$ interacts only with $\mathcal{B}_i$. If we put the two pairs $(\mathcal{A}_1, \mathcal{B}_1)$ and $(\mathcal{A}_2, \mathcal{B}_2)$ far apart, and use some timing constraints, we can enforce some non-signaling type scenarios. Here, we will only use the property that the two honest Bob's know their respective location. In particular, there is no trust needed regarding the location of the cheating parties.

The security definitions for relativistic bit commitment are the ones we presented above: Definitions 7 and 8. We will now describe the $\mathbb{F}_Q$ relativistic bit commitment scheme. This scheme will consist of 4 phases, the preparation phase, the commit phase, the sustain phase and the reveal phase. The preparation phase is some preprocessing phase that can be done anytime before the protocol. The sustain phase can be seen as a part of the reveal phase, and corresponds to the time where the committed bit is safe. We assume here that the two Alices learn at the beginning of the sustain phase the bit d they should try to reveal (which doesn't necessarily correspond to the bit, if any, they committed to).

The Single-Round $\mathbb{F}_Q$ Protocol. The single-round version corresponds **CHSH$_Q$** to the protocol introduced by Crépeau *et al.* [CSST11] (see also [Sim07]). Both players, Alice and Bob, have agents $\mathcal{A}_1, \mathcal{A}_2$ and $\mathcal{B}_1, \mathcal{B}_2$ present at two spatial locations, 1 and 2, separated by a distance D. We consider the case where Alice makes the commitment. The protocol (followed by honest players) consists of 4 phases: preparation, commit, sustain and reveal. The sustain phase in the single-round protocol is trivial and simply consists in waiting for a time less than D/c, which is the time needed for light to travel between the two locations. The bit commitment protocol goes as follows.

1. *Preparation phase:* $\mathcal{A}_1, \mathcal{A}_2$ (resp. $\mathcal{B}_1, \mathcal{B}_2$) share a random number $a \in \mathbb{F}_Q$ (resp. $x \in \mathbb{F}_Q$).
2. *Commit phase:* $\mathcal{B}_1$ sends b to $\mathcal{A}_1$, who immediately returns $y = a + d * x$ where $d \in \{0, 1\}$ is the committed bit.
3. *Sustain phase:* $\mathcal{A}_1$ and $\mathcal{A}_2$ wait for some time $\tau < D/c$, where c is the speed of light. Crucially, for any time less than D/c, the NSS principle guarantees that $\mathcal{A}_2$ has no information about the value of b.
4. *Reveal phase:* $\mathcal{A}_2$ reveals the values of d and a to $\mathcal{B}_2$ who checks that $y = a + d * tx$.

This relativistic bit commitment protocol is known to be $O(\frac{1}{\sqrt{Q}})$-sum-binding [LKB+15]. It can be easily extended to a P-string commitment where d is an element of $\mathbb{F}_P$ instead of an element of $\{0, 1\}$. The above construction is well defined as long as $Q \geq P$ (all the operations are still the modular operations in $\mathbb{F}_Q$).

Proposition 5. *The above relativistic P-string commitment protocol is ε-sum-binding with $\varepsilon = \frac{4P}{Q^{1/3}}$.*

Proof. Consider a P-string commitment Π and a cheating strategy $Str^* = (Comm^*, Open^*)$. In this strategy, $\mathcal{A}_1$ and $\mathcal{A}_2$ share an entangled state $|\psi\rangle$. After receiving b, $\mathcal{A}_1$ performs a measurement on her part of the state to produce an output y which she sends to $\mathcal{B}_1$. For a random d that $\mathcal{A}_2$ wants to reveal, she performs a measurement on her part of the state to produce an output a. We have

$$\frac{1}{P} \sum_{d=0}^{P-1} (\Pr[\text{Alice successfully reveals } d \mid (Comm^*, Open^*)]) = \Pr[a + y = b * d].$$

One can directly use the above strategy to construct a strategy for a **CHSH$_Q$**(P) game (defined in Sect. 3), with respective inputs $b \in F_Q$, $d \in F_P$ and with respective outputs y and a. We have immediately

$$\Pr[a + y = b * d] \leq \omega^*(\mathbf{CHSH_Q}(P)) \leq \frac{1}{P} + \frac{4}{Q^{1/3}},$$

where the bound on the entangled is the one from Sect. 3. This gives us

$$\sum_{d=0}^{P-1} \max_{Open^*} (\Pr[\text{Alice successfully reveals } d \mid (Comm^*, Open^*)]) \leq 1 + \frac{4P}{Q^{1/3}}$$

which proves the desired proposition.

If we want to perform an ε-sum-binding P-string commitment protocol then we need to send $\log(Q) = \log(\frac{64P^3}{\varepsilon^3}) = 3(\log(P) + |\log(\varepsilon)|) + 8$ bits for each round of the protocol.

4.3   Parallel Repetiton of *RBC*

The problem with string commitment is that it is not possible to reveal only some bits of the string: by construction, one has to reveal the whole string. In order to circumvent this issue, we need to consider performing a bit commitment n times in parallel. This then allows one to reveal only a fraction of the bits. The scheme will still feature sum-binding property but the scaling in parameters – although still polynomial – will not be as good as for string commitment.

1. *Preparation phase*: $\mathcal{A}_1, \mathcal{A}_2$ (resp. $\mathcal{B}_1, \mathcal{B}_2$) share n random bits $a_1, \ldots, a_n \in \mathbb{F}_Q$ (resp. $b_1, \ldots, b_n \in \mathbb{F}_Q$).
2. *Commit phase*: $\mathcal{B}_1$ sends each b_i to $\mathcal{A}_1$, who returns for each i $y_i = a_i + d_i * b_i$ where $d_1, \ldots, d_n \in \{0,1\}$ is the sequence of committed bits.
3. *Sustain phase*: $\mathcal{A}_1$ and $\mathcal{A}_2$ wait for some time $\tau \leq D/c$.
4. *Reveal phase*: Let S be the subset of indices Alice wants to reveal. $\mathcal{A}_2$ indicates S to $\mathcal{B}_2$ and reveals the values $\{a_i\}_{i \in S}$ and $\{d_i\}_{i \in S}$ to $\mathcal{B}_2$ who checks that for each $i \in S$, the relation $y_i = a_i + d_i * b_i$ holds.

Proposition 6. *Fix a subset S of indices Alice will reveal to. Relative to S, the above protocol is ε-sum binding with $\varepsilon = 4(\frac{2|S|2^{2|S|}}{Q})^{1/3} \leq 4(\frac{2n2^{2n}}{Q})^{1/3}$.*

Proof. Fix a subset S. As before, we can use a strategy for the relativistic bit commitment to solve an instance of $\mathbf{CHSH_Q}(2)^{\otimes|S|}$ which implies

$$\sum_{d \in \{0,1\}^{|S|}} \max_{Open^*} (\Pr[\text{Alice successfully reveals } d \mid (Comm^*, Open^*)])$$

$$\leq 2^{|S|}\omega^*(\mathbf{CHSH_Q}(2)^{\otimes|S|}).$$

Since we know that $\omega^*(\mathbf{CHSH_Q}(2)^{\otimes|S|}) \leq \frac{1}{2^{|S|}} + 4(\frac{2|S|}{Q2^{|S|}})^{1/3}$, we can immediately conclude that

$$\sum_{d \in \{0,1\}^{|S|}} \max_{Open^*} (\Pr[\text{Alice successfully reveals } d \mid (Comm^*, Open^*)]) \leq 2^{|S|}$$

$$\leq 1 + 4(\frac{2|S|2^{2|S|}}{Q})^{1/3}.$$

If we want the above protocol to be ε-sum-binding, we need to send $n \log(Q) = O(n^2 \log(n) + n|\log(\varepsilon)|)$ bits at each round.

5   Relativistic Zero-Knowledge

In this section, we present our relativistic zero-knowledge protocol for NP. Our protocol will be based on the well known protocol for the NP-complete problem HAMILTONIAN CYCLE, which uses bit commitment.

5.1   The Zero-Knowledge Hamiltonian Cycle Protocol

Here, we present the zero-knowledge Hamiltonian cycle protocol and its adaptation to the relativistic setting. Let S_n the set of permutation on $\{1, \ldots, n\}$.

Definition 9. *A cycle of* $\{1, \ldots, n\}$ *is a set of couples*

$$\{(\Pi(1), \Pi(2)), (\Pi(2), \Pi(3)), \ldots, (\Pi(n-1), \Pi(n)), (\Pi(n), \Pi(1))\}$$

for a permutation $\Pi \in S_n$. *We denote by* Γ_n *the set of cycles of* $\{1, \ldots, n\}$. *We have* $|\Gamma_n| = (n-1)!$. *For a cycle* $\mathcal{C} = \{(u, v)\}$ *and a permutation* Π, *we also define* $\Pi(\mathcal{C}) := \{(\Pi(u), \Pi(v))\}$.

Definition 10. *A Hamiltonian cycle of a graph* $G = (V, E)$ *is a cycle* $\mathcal{C}$ *of* $\{1, \ldots, |V|\}$ *such that* $\mathcal{C} \in E$ *i.e.* $\forall (i, j) \in \mathcal{C}, (i, j) \in E$.

Determining whether a graph G has a Hamiltonian cycle or not is an NP-complete problem. The corresponding decision problem is HAMILTONIAN CYCLE and $G \in$ HAMILTONIAN CYCLE means that the graph contains a Hamiltonian cycle.

5.2   The Protocol

We recall the zero-knowledge protocol for HAMILTONIAN CYCLE first presented by Blum [Blu86].

Zero knowledge protocol for HAMILTONIAN CYCLE

Input — The prover and the verifier are given a graph $G = (V, E)$.
Auxiliary Input — The prover knows a Hamiltonian cycle $\mathcal{C}$ of G.
Protocol —

1. The prover picks a random permutation $\Pi : V \to V$. He commits to each of bit of the adjacency matrix $M_{\Pi(G)}$ of $\Pi(G)$.
2. The verifier sends a random bit (called the challenge) $chall \in \{0, 1\}$ to the prover.
3. – If $chall = 0$, the prover decommits to all the elements of $M_{\Pi(G)}$, and reveals Π.
   – If $chall = 1$, he reveals only the bits (of value 1) of the adjacency matrix that correspond to a Hamiltonian cycle $\mathcal{C}' = \Pi(\mathcal{C})$ of $\Pi(G)$.
4. The verifier checks that these decommitments are valid and correspond, for $chall = 0$ to $M_{\Pi(G)}$ and, for $chall = 1$, to a Hamiltonian cycle.

We now present the relativistic zero-knowledge protocol, that uses the $\mathbb{F}_Q$ bit commitment.

Relativistic zero knowledge protocol for HAMILTONIAN CYCLE

Input — The provers and the verifiers are given a graph $G = (V, E)$.

Auxiliary Input — The provers P_1 and P_2 know a Hamiltonian cycle C of G.

Preprocessing — P_1 and P_2 agree beforehand on a random permutation $\Pi : V \to V$ and on an $n \times n$ matrix $A \in \mathcal{M}_n^{\mathbb{F}_Q}$ where each element of A is chosen uniformly at random in $\mathbb{F}_Q$.

Protocol —

1. Commitment to each bit of $M_{\Pi(G)}$: V_1 sends a matrix $B \in \mathcal{M}_n^{\mathbb{F}_Q}$ where each element of B is chosen uniformly at random in $\mathbb{F}_Q$. P_1 outputs the matrix $Y \in \mathcal{M}_n^{\mathbb{F}_Q}$ such that $\forall i, j \in [n]$, $Y_{i,j} = A_{i,j} + (B_{i,j} * (M_{\Pi(G)})_{i,j})$.
2. The verifier V_2 sends a random bit (called the challenge) $chall \in \{0,1\}$ to the prover P_2.
3.   – If $chall = 0$, P_2 decommits to all the elements of $M_{\Pi(G)}$, *i.e.* he sends all the elements of A to V_2 and reveals Π.
     – If $chall = 1$, P_2 reveals only the bits (of value 1) of the adjacency matrix that correspond to a Hamiltonian cycle C' of $\Pi(G)$, *i.e.* for all edges (u, v) of C', he sends $A_{u,v}$ as well as C'.
4. The verifier checks that those decommitments are valid and correspond to what the provers have declared. He also checks that the timing constraint of the bit commitment is satisfied. This means that
     – if $chall = 0$, the prover's opening A must satisfy $\forall i, j \in [n]$, $Y_{i,j} = A_{i,j} + (B_{i,j} * (M_{\Pi(G)})_{i,j})$.
     – if $chall = 1$, the prover's opening A must satisfy $\forall (u, v) \in C'$, $Y_{u,v} = A_{u,v} + B_{u,v}$.

5.3   Proof of Security

Our goal is to show that the above protocol is a relativistic zero-knowledge protocol for HAMILTONIAN CYCLE. In order to do this, we show the following

- Completeness: If the prover and the verifier are honest then for any graph G that has a Hamiltonian cycle, the verifier accepts with certainty.
- Soundness: If we take $Q = 64n!2^{3k}$, we have that for any cheating prover, $\forall G \notin$ HAMILTONIAN CYCLE, the verifier accepts with probability at most $\frac{1}{2} + 2^{-k}$. With this parameter Q, the amount of bits sent during the protocol is $\log(Q)$ for each committed bit and is therefore $n^2 \log(Q) = O(kn^3 \log(n))$ at each round.
- Perfect zero-knowledge: for any cheating verifier V^*, there exists a quantum poly-time simulator Σ that can reproduce the cheating verifier's view of the protocol for any input $G \in$ HAMILTONIAN CYCLE and any auxiliary input ρ. More details about this zero-knowledge property can be found in the corresponding subsection.

Completeness. If both players are honest and G contains a Hamiltonian cycle then the protocol always succeeds. Indeed, the original protocol from Blum has perfect completeness. Moreover, the $\mathbb{F}_Q$ bit commitment always succeeds when done honestly.

Soundness. The soundness can be reduced to the following 2-player game $G^{RZK-HAM}$.

- P_1 receives a matrix $B \in \mathcal{M}_n^{\mathbb{F}_Q}$ where each element of B is chosen uniformly at random in $\mathbb{F}_Q$. P_2 receives a random input bit $chall$.
- P_1 outputs a matrix $Y \in \mathcal{M}_n^{\mathbb{F}_Q}$. If $chall = 0$ then P_2 outputs a permutation Π and a matrix $A \in \mathcal{M}_n^{\mathbb{F}_Q}$. If $chall = 1$ then P_2 outputs a cycle $\mathcal{C}'$ and n strings $\{A'_{(u,v)}\}_{(u,v)\in\mathcal{C}'}$ in $\mathbb{F}_Q$.
- If $chall = 0$, the two players win if $\forall i, j \in [n]$, $Y_{i,j} = A_{i,j} + (B_{i,j} * (M_{\Pi(G)})_{i,j})$. If $chall = 1$, the two players win if for all edges (u, v) of $\mathcal{C}'$, $Y_{u,v} = A_{u,v} + B_{u,v}$, which corresponds to revealing 1 for each edge of the cycle $\mathcal{C}'$.

This game is $n!$-projective: once the permutation (or the cycle) is chosen, the winning output is fixed. In order to study this game, we study the game $G^{RZK-HAM}_{coup}$. We fix an input/output pair (B, Y) for P_1 and we consider winning outputs for P_2 for both inputs. For $chall = 0$, we have a permutation Π and a matrix $A \in \mathcal{M}_n^{\mathbb{F}_Q}$ which is a valid opening of $M_{\Pi(G)}$ meaning that

$$\forall(i,j), \ A_{i,j} = Y_{i,j} - B_{i,j} * (M_{\Pi(G)})_{i,j}. \tag{6}$$

For $chall = 1$, we have a cycle $\mathcal{C}'$ of $\{1, \ldots, |V|\}$ as well as openings $A'_{u,v}$ for each $(u, v) \in \mathcal{C}'$. Because it is a winning output, the openings must satisfy

$$\forall(u,v) \in \mathcal{C}', \ A'_{u,v} = Y_{u,v} - B_{u,v}. \tag{7}$$

If the graph G (hence also $\Pi(G)$) does not contain a Hamiltonian cycle then there has to be an edge (u, v) of $\mathcal{C}'$ such that $(M_{\Pi(G)})_{u,v} = 0$. For this specific (u, v), we combine Eqs. 6 and 7 and get:

$$A_{u,v} = Y_{u,v}; \quad A'_{u,v} = Y_{u,v} - B_{u,v}.$$

This implies that $A_{u,v} - A'_{u,v} = B_{u,v}$ which happens with probability at most $\frac{1}{Q}$ from non-signaling. We therefore conclude that $\omega^*(G^{RZK-HAM}_{coup}) \leq \frac{1}{Q}$. From there, we can apply Proposition 1 and obtain

$$\omega^*(G^{RZK-HAM}) \leq \frac{1}{2} + \left(\frac{64n!}{Q}\right)^{1/3}.$$

If we take $Q = 64n!2^{3k}$ then the protocol has soundness $\frac{1}{2} + 2^{-k}$. The amount of bits sent during the protocol is $\log(Q)$ for each committed bit and is therefore $n^2 \log(Q) = O(kn^3 \log(n))$, which shows that the protocol is efficient.

5.4   Zero-Knowledge Property

In this section, we show that the above protocol is zero-knowledge. One of the main difficulties in proving zero-knowledge in the quantum setting arises when requiring the simulator to perform rewinding while preserving an auxiliary state. Here, there is no need for rewinding and the simulation can be done perfectly and quite simply. The simulator will simply simulate each round of the protocol from the first one to the last one. The reason of this simplicity is that in our bit commitment scheme, the verifier and the simulator are able, for any commitment, to reveal an arbitrary value of their choice. This is a rare feature because the prover shouldn't be able to do this to preserve the binding property. In our case, this asymmetry comes from the relativistic constraints imposed on the provers.

Zero-Knowledge in the Relativistic Setting. From the provers' point of view, each of them receives a message and replies. We assume that a cheating verifier can totally bypass the timing constraints. We therefore consider one cheating verifier that interacts with both provers. Moreover, we allow the verifier to send a query to the second prover after receiving the answer from the first prover or vice-versa. All of this is meant to have a cheating verifier as strong as possible. Proving the zero-knowledge property in this setting will therefore be stronger in this model. Also, this will show the zero-knowledge property both for relativistic zero-knowledge and for the (very related) 2-prover 1-round multi-prover interactive proof model.

A cheating verifier V^* is modeled by a polynomial-time uniform family of pairs of circuits $\{(V_1^*(n), V_2^*(n)\}$ where each $V_i^*(n)$ represents the verifier action towards prover P_i on input size n. The verifier sends a query to each prover in respective classical registers Q_1 and Q_2 and gets responses in respective classical registers R_1 and R_2. The verifier also has access to private quantum register $\mathcal{V}$, which initially contains a quantum auxiliary state ρ.

Fix a cheating verifier V^*. For any message $B \in \mathcal{M}_n^{\mathbb{F}_Q}$ sent from the verifier to P_1, the message from P_1 is a uniformly random matrix $Y = \mathcal{M}_n^{\mathbb{F}_Q}$ while the message from P_2 consists of:

- if $chall = 0$, a random permutation Π and a matrix A satisfying $Y = A + B *$ $M_{\Pi(G)}$ where the multiplication is the entry-wise matrix multiplication.
- if $chall = 1$, a random cycle $\mathcal{C}'$ and a family of strings $\{A'_{u,v}\}_{(u,v)\in\mathcal{C}'}$ satisfying

$$\forall (u,v) \in \mathcal{C}', \ Y_{u,v} = A'_{u,v} + B_{u,v}.$$

The verifier receives as a first message a random matrix Y and as second message a random permutation (for chall = 0) or a cycle (for chall = 1) with a uniquely determined message A or A' that he can perfectly infer from the information available to him. Notice that in the soundness analysis, the prover doesn't know what message he has to send because of relativistic constraints which do not apply for the verifier (as we said, this only increases our claim on zero-knowledge).

All of the above remains true for any strategy for the cheating verifier and with any auxiliary input, and even if the verifier queries a prover depending on the answer of the other prover. Moreover, simulating the interaction between V^* and the provers can be done step by step following V^*'s actions, without any need for rewinding. This therefore shows the perfect zero-knowledge property of our scheme. In order to illustrate this, we present below a step by step simulation of the verifier's view in a more formal way than what we did above.

Step by Step Simulation of the Verifier's View of the Protocol. For a cheating verifier V^*, we construct a quantum poly-time simulator such that on any input $G \in$ HAMILTONIAN CYCLE and auxiliary input ρ, the simulator can recreate the verifier's view of the protocol perfectly. The simulator will use V^* as a black box and will mimic the verifier's view of the protocol after each round. When considering the interaction between the verifier and the provers, we will always distinguish 2 cases

1. The action of V_2^* depends on the interaction with P_1.
2. The action of V_1^* depends on the interaction with P_2.

Note that both of these events cannot happen simultaneously. In the analysis below, we will consider case 1 but the other one can be treated in the exact same way.

We first describe the different view for a cheating verifier V^* and then show how to perform the simulation. Let σ_i be the verifier's view at step i of the protocol.

– At the beginning of the protocol, the verifier's view consists of $\sigma_0 := \rho_V$.
– After the verifier's first message to P_1, the verifier's view is

$$\sigma_1 := V_1^*(\rho) = \sum_{B \in \mathcal{M}_n^{\mathbb{F}_Q}} p_B |B\rangle\langle B|_{Q_1} \otimes \rho(B)_V.$$

– After the first prover's answer, the shared state between the provers and the verifier is

$$\sigma_2 := \frac{1}{n!}\frac{1}{Q^{n^2}} \sum_{\Pi \in S_n} \sum_{A \in \mathcal{M}_n^{\mathbb{F}_Q}} \sum_{B \in \mathcal{M}_n^{\mathbb{F}_Q}} p_B \, |Y(\Pi, A)\rangle\langle Y(\Pi, A)|_{R_1} \otimes |B\rangle\langle B|_{Q_1} \otimes \rho(B)_V.$$

where $Y(\Pi, A) := A + B * \Pi(G)$ with $*$ being the entry wise matrix multiplication.
– Now, the verifier sends his challenge bit, which can depend on everything that happened before. His view becomes

$$\sigma_3 := \frac{1}{n!}\frac{1}{Q^{n^2}} \sum_{\Pi \in S_n} \sum_{A \in \mathcal{M}_n^{\mathbb{F}_Q}} \sum_{B \in \mathcal{M}_n^{\mathbb{F}_Q}} \sum_{c \in \{0,1\}} p_{B,c} |Y(\Pi, A)\rangle\langle Y(\Pi, A)|_{R_1} \otimes |c\rangle\langle c|_{Q_2}$$

$$\otimes |B\rangle\langle B|_{Q_1} \otimes \rho(B, c, Y(\Pi, A))_V.$$

– After the final message from the prover, the verifier's view becomes

$$\sigma_4 := \frac{1}{n!}\frac{1}{Q^{n^2}} \sum_{\Pi \in S_n} \sum_{A \in \mathcal{M}_n^{\mathbb{F}_Q}} \sum_{B \in \mathcal{M}_n^{\mathbb{F}_Q}} |Y(\Pi, A)\rangle\langle Y(\Pi, A)|_{R_1} \otimes |B\rangle\langle B|_{Q_1} \otimes$$

$$\Big(p_{B,0}|0\rangle\langle 0|_{Q_2} \otimes |\Pi, A\rangle\langle \Pi, A|_{R_2} \otimes \rho(B, 0, Y(\Pi, A))$$

$$+ p_{B,1}|1\rangle\langle 1|_{Q_2} \otimes |\Pi(\mathcal{C}), A_{\Pi(\mathcal{C})}\rangle\langle \Pi(\mathcal{C}), A_{\Pi(\mathcal{C})}|_{R_2} \otimes \rho(B, 1, Y(\Pi, A)) \Big).$$

Notice that we are interested here in the verifier's view on a 'Yes' instance, meaning that on challenge $'1'$ in register Q_2, the answer $|\Pi(\mathcal{C}), A_{\Pi(\mathcal{C})}\rangle\langle \Pi(\mathcal{C}), A_{\Pi(\mathcal{C})}|$ satisfies

$$\forall (i,j) \in \Pi(\mathcal{C}), \; Y_{i,j} = A_{i,j} + B_{i,j}.$$

meaning that the prover revealed the output bit '1' for entry $\Pi(G)_{i,j}$. Notice also that for a fixed cycle $\mathcal{C}$, the mapping $\Pi \to \Pi(\mathcal{C})$ is a bijection between the set of permutation and the set of cycles.

We show now how to simulate the view of the verifier. The simulator can easily simulate σ_0 and σ_1 since he has a copy of ρ and knows V_1^*. Notice that in σ_2, the message from the prover is a uniform random matrix because of the randomness A. Therefore, we have

$$\sigma_2 = \frac{1}{Q^{n^2}} \sum_{Y \in \mathcal{M}_n^{\mathbb{F}_Q}} \sum_{B \in \mathcal{M}_n^{\mathbb{F}_Q}} p_B \, |Y\rangle\langle Y|_{R_1} \otimes |B\rangle\langle B|_{Q_1} \otimes \rho(B)_V.$$

This can be easily created by the simulator by just tensoring the totally mixed state in register R_1 to σ_1. In order to construct σ_3, the simulator just applies V_2^* to transform σ_2 into σ_3 as the cheating verifier would and gets exactly

$$\sigma_3 = \frac{1}{Q^{n^2}} \sum_{Y \in \mathcal{M}_n^{\mathbb{F}_Q}} \sum_{B \in \mathcal{M}_n^{\mathbb{F}_Q}} \sum_{c \in \{0,1\}} p_{B,c} \, |Y\rangle\langle Y|_{R_1} \otimes |c\rangle\langle c|_{Q_2} \otimes |B\rangle\langle B|_{Q_1} \otimes \rho(B, c, Y)_V.$$

Finally, in order to construct σ_4, the simulator does the following

– conditioned on $c = 0$ in register Q_2, the simulator picks a random permutation Π and puts $|\Pi, A(\Pi, B, Y)\rangle\langle \Pi, A(\Pi, B, Y)|$ in register R_2 where $A(\Pi, B, Y) := Y - B * \Pi(G)$, with $*$ being the entry wise matrix multiplication.
– conditioned on $c = 1$ in register Q_2, the simulator picks a random cycle $\mathcal{C}'$ and outputs $|\mathcal{C}', A'(\mathcal{C}', B, Y)\rangle\langle \mathcal{C}', A'(\mathcal{C}', B, Y)|$ such that for all $(i, j) \in \mathcal{C}'$, it holds that $A'(\mathcal{C}', B, Y)_{i,j} := Y_{i,j} - B_{i,j}$.

The state constructed by the simulator is therefore

$$\frac{1}{Q^{n^2}} \sum_{A \in \mathcal{M}_n^{\mathbb{F}_Q}} \sum_{B \in \mathcal{M}_n^{\mathbb{F}_Q}} |Y\rangle\langle Y|_{R_1} \otimes |B\rangle\langle B|_{Q_1} \otimes$$

$$(p_{B,0}|0\rangle\langle 0|_{Q_2} \otimes \frac{1}{n!} \sum_{\Pi \in S_n} |\Pi, A(\Pi, B, Y)\rangle\langle \Pi, A(\Pi, B, Y)|_{R_2} \otimes \rho_{B,0,Y}$$

$$+ p_{b,1}|1\rangle\langle 1|_{Q_2} \otimes \frac{1}{(n-1)!} \sum_{\mathcal{C}' \in \Gamma_n} |\mathcal{C}', A'(\mathcal{C}', B, Y)\rangle\langle \mathcal{C}', A'(\mathcal{C}', B, Y)|_{R_2} \otimes \rho_{B,1,Y}).$$

By simple changes of variables, we can see that the above state is actually exactly equal to σ_4. Therefore, we succeeded in the simulation and we can conclude that our protocol is perfectly zero-knowledge against quantum adversaries.

Acknowledgements. The authors were partially supported by ANR DEREC <ANR-16-CE39-0001-01>.

References

[BCF+14] Buhrman, H., Chandran, N., Fehr, S., Gelles, R., Goyal, V., Ostrovsky, R., Schaffner, C.: Position-based quantum cryptography: impossibility and constructions. SIAM J. Comput. **43**(1), 150–178 (2014)

[Blu86] Blum, M.: How to prove a theorem so no one else can claim it. In: Proceedings of the International Congress of Mathematicians, vol. 1, p. 2. Citeseer (1986)

[BOGKW88] Ben-Or, M., Goldwasser, S., Kilian, J., Wigderson, A.: Multi-prover interactive proofs: how to remove intractability assumptions. In: Proceedings of the twentieth annual ACM symposium on Theory of computing, pp. 113–131. ACM (1988)

[CCL15] Chakraborty, K., Chailloux, A., Leverrier, A.: Arbitrarily long relativistic bit commitment. arXiv preprint arXiv:1507.00239 (2015)

[CGMO09] Chandran, N., Goyal, V., Moriarty, R., Ostrovsky, R.: Position based cryptography. In: Halevi, S. (ed.) CRYPTO 2009. LNCS, vol. 5677, pp. 391–407. Springer, Heidelberg (2009). doi:10.1007/978-3-642-03356-8_23

[CSST11] Crépeau, C., Salvail, L., Simard, J.-R., Tapp, A.: Two provers in isolation. In: Lee, D.H., Wang, X. (eds.) ASIACRYPT 2011. LNCS, vol. 7073, pp. 407–430. Springer, Heidelberg (2011). doi:10.1007/978-3-642-25385-0_22

[DS14] Dinur, I., Steurer, D.: Analytical approach to parallel repetition. In: Proceedings of the 46th Annual ACM Symposium on Theory of Computing, STOC 2014, New York, NY, USA, pp. 624–633. ACM (2014)

[DSV15] Dinur, I., Steurer, D., Vidick, T.: A parallel repetition theorem for entangled projection games. Comput. Complex. **24**(2), 201–254 (2015)

[FF15] Fehr, S., Fillinger, M. On the composition of two-prover commitments, applications to multi-round relativistic commitments. arXiv preprint arXiv:1507.00240v1 (2015)

[GMR89] Goldwasser, S., Micali, S., Rackoff, C.: The knowledge complexity of interactive proof systems. SIAM J. Comput. **18**, 186–208 (1989)

[Hay02] Hayashi, M.: Optimal sequence of quantum measurements in the sense of Stein's lemma in quantum hypothesis testing. J. Phys. A: Math. Gen. **35**(50), 10759 (2002)

[Ken99] Kent, A.: Unconditionally secure bit commitment. Phys. Rev. Lett. **83**, 1447–1450 (1999)

[Ken11] Kent, A.: Unconditionally secure bit commitment with flying qudits. New J. Phys. **13**(11), 113015 (2011)

[Ken12] Kent, A.: Unconditionally secure bit commitment by transmitting measurement outcomes. Phys. Rev. Lett. **109**, 130501 (2012)

[KMS11] Kent, A., Munro, W.J., Spiller, T.P.: Quantum tagging: authenticating location via quantum information and relativistic signaling constraints. Phys. Rev. A **84**, 012326 (2011)

[KTHW13] Kaniewski, J., Tomamichel, M., Hanggi, E., Wehner, S.: Secure bit commitment from relativistic constraints. IEEE Trans. Inf. Theory **59**(7), 4687–4699 (2013)

[LC97] Lo, H.-K., Chau, H.F.: Is quantum bit commitment really possible? Phys. Rev. Lett. **78**(17), 3410–3413 (1997)

[LKB+13] Lunghi, T., Kaniewski, J., Bussières, F., Houlmann, R., Tomamichel, M., Kent, A., Gisin, N., Wehner, S., Zbinden, H.: Experimental bit commitment based on quantum communication and special relativity. Phys. Rev. Lett. **111**, 180504 (2013)

[LKB+15] Lunghi, T., Kaniewski, J., Bussières, F., Houlmann, R., Tomamichel, M., Wehner, S., Zbinden, H.: Practical relativistic bit commitment. Phys. Rev. Lett. **115**, 030502 (2015)

[LL11] Lau, H.-K., Lo, H.-K.: Insecurity of position-based quantum-cryptography protocols against entanglement attacks. Phys. Rev. A **83**(1), 012322 (2011)

[May97] Mayers, D.: Unconditionally secure quantum bit commitment is impossible. Phys. Rev. Lett. **78**(17), 3414–3417 (1997)

[SBT16] Sutter, D., Berta, M., Tomamichel, M.: Multivariate trace inequalities. arXiv preprint arXiv:1604.03023 (2016)

[Sim07] Simard, J.R.: Classical and quantum strategies for bit commitment schemes in the two-prover model. Master's thesis, McGill University (2007)

[Unr12] Unruh, D.: Quantum proofs of knowledge. In: Pointcheval, D., Johansson, T. (eds.) EUROCRYPT 2012. LNCS, vol. 7237, pp. 135–152. Springer, Heidelberg (2012). doi:10.1007/978-3-642-29011-4_10

[Unr14] Unruh, D.: Quantum position verification in the random oracle model. In: Garay, J.A., Gennaro, R. (eds.) CRYPTO 2014. LNCS, vol. 8617, pp. 1–18. Springer, Heidelberg (2014). doi:10.1007/978-3-662-44381-1_1

[VMH+16] Verbanis, E., Martin, A., Houlmann, R., Boso, G., Bussières, F., Zbinden, H.: 24-hour relativistic bit commitment. Phys. Rev. Lett. arXiv:1605.07442 (2016, to appear)

[Win99] Winter, A.: Coding theorem and strong converse for quantum channels. IEEE Trans. Inf. Theory **45**(7), 2481–2485 (1999)

Multiparty Computation III

Faster Secure Two-Party Computation in the Single-Execution Setting

Xiao Wang[1]([⊠]), Alex J. Malozemoff[2], and Jonathan Katz[1]

[1] University of Maryland, College Park, USA
{wangxiao,jkatz}@cs.umd.edu
[2] Galois, Portland, USA
amaloz@galois.com

Abstract. We propose a new protocol for two-party computation, secure against malicious adversaries, that is significantly faster than prior work in the single-execution setting (i.e., non-amortized and with no pre-processing). In particular, for computational security parameter κ and statistical security parameter ρ, our protocol uses only ρ garbled circuits and $O(\rho + \kappa)$ public-key operations, whereas previous work with the same number of garbled circuits required either $O(\rho \cdot n + \kappa)$ public-key operations (where n is the input/output length) or a second execution of a secure-computation sub-protocol. Our protocol can be based on the decisional Diffie-Hellman assumption in the standard model.

We implement our protocol to evaluate its performance. With $\rho = 40$, our implementation securely computes an AES evaluation in 65 ms over a local-area network using a single thread without any pre-computation, 22× faster than the best prior work in the non-amortized setting. The relative performance of our protocol is even better for functions with larger input/output lengths.

1 Introduction

Secure multi-party computation (MPC) allows multiple parties with private inputs to compute some agreed-upon function such that all parties learn the output while keeping their inputs private. Introduced in the 1980s [38], MPC has become more practical in recent years, with several companies now using the technology. A particularly important case is secure *two-party computation* (2PC), which is the focus of this work.

Many existing applications and implementations of 2PC assume that all participants are *semi-honest*, that is, they follow the protocol but can try to learn sensitive information from the protocol transcript. However, in real-world applications this assumption may not be justified. Although protocols with stronger

X. Wang and J. Katz—Research supported by NSF awards #1111599 and #1563722.
A.J. Malozemoff—Conducted in part with Government support through the National Defense Science and Engineering Graduate (NDSEG) Fellowship, 32 CFG 168a, awarded by DoD, Air Force Office of Scientific Research. Work done while at the University of Maryland.

© International Association for Cryptologic Research 2017
J.-S. Coron and J.B. Nielsen (Eds.): EUROCRYPT 2017, Part III, LNCS 10212, pp. 399–424, 2017.
DOI: 10.1007/978-3-319-56617-7_14

security guarantees exist, 2PC protocols secure against malicious adversaries are relatively slow, especially when compared to protocols in the semi-honest setting. To address this, researchers have considered variants of the classical, "single-execution" setting for secure two-party computation, including both the *batch* setting [18,28,29,33] (in which the computational cost is amortized over multiple evaluations of the same function) and the *offline/online* setting [28,29,33] (in which parties perform pre-processing when the circuit—but not the parties' inputs—is known). The best prior result [33] (done concurrently and independently of our own work) relies on both amortization and pre-processing (as well as extensive parallelization) to achieve an overall amortized time of 6.4 ms for evaluating AES with 40-bit statistical security. Due to the pre-processing, however, it introduces a latency of 5222 ms until the first execution can be done.

In addition, existing 2PC schemes with security against malicious adversaries perform poorly on even moderate-size inputs or very large circuits. For example, the schemes of Lindell [24] and Afshar et al. [1] require a number of public-key operations at least proportional to the statistical security parameter *times* the sum of one party's input length and the output length. The schemes tailored to the batch, offline/online setting [29,33] do not scale well for large circuits due to memory constraints: the garbled circuits created during the offline phase need either to be stored in memory, in which case evaluating very large circuits is almost impossible, or else must be written/read from disk, in which case the online time incurs a huge penalty[1] due to disk I/O (see Sect. 5).

Motivated by these issues, we design a new 2PC protocol with security against malicious adversaries that is tailored for the *single-execution* setting (i.e., no amortization) without any pre-processing. Our protocol uses the cut-and-choose paradigm [25] and the input-recovery approach introduced by Lindell [24], but the number of public-key operations required is independent of the input/output length. Overall, we make the following contributions.

– Our protocol is more efficient, and often much more efficient, than the previous best protocol with malicious security in the single-execution setting (see Table 1). Concretely, our protocol takes only 65 ms to evaluate an AES circuit over a local-area network, better than the most efficient prior work in the same setting.
– We identify and fix bottlenecks in various building blocks for secure computation; these fixes may prove useful in subsequent work. As an example, we use Streaming SIMD Extensions (SSE) to improve the performance of oblivious-transfer extension, and improve the efficiency of the XOR-tree technique to avoid high (non-cryptographic) complexity when applied to large inputs. Our optimizations reduce the cost of processing the circuit evaluator's input by $1000\times$ for 2^{16}-bit inputs, and even more for larger inputs.
– We release an open-source implementation, EMP-toolkit [36], with the aim of providing a benchmark for secure computation and allowing other researchers to experiment with, use, and extend our code.

---

[1] The performance numbers reported in [29,33] do not take this into account.

Table 1. Times for two-party computation of AES, with security against malicious adversaries, in the single-execution setting. The statistical security parameter is ρ. All numbers except for [29] are taken directly from the cited paper, and thus are based on different hardware/network configurations. The numbers for [29] are from our own experiments, using the same hardware/network configuration as for our own implementation. We do not include [33] here because it is not in the single-execution setting. See Sect. 5 for more details.

Protocol	ρ	Time	Notes
PSSW09 [32]	40	1,114 s	
SS11 [34]	40	192 s	
NNOB12 [31]	55	4,000 ms	
KSS12 [23]	80	1,400 ms	256 CPUs/party
FN13 [13]	39	1,082 ms	GPU
AMPR14 [1]	40	5,860 ms	
FJN14 [11]	40	455 ms	GPU
LR15 [29]	40	1,442 ms	
Here	40	65 ms	

1.1 High-Level Approach

Our protocol is based on the cut-and-choose paradigm. Let f be the circuit the parties want to compute. At a high level, party P_1, also called the *circuit garbler*, begins by generating s garbled circuits for f and sending those to P_2, the *circuit evaluator*. Some portion of those circuits (the *check circuits*) are randomly selected and checked for correctness by the evaluator, and the remaining circuits (the *evaluation circuits*) are evaluated. The outputs of the evaluation circuits are then processed in some way to determine the output.

The Input-Recovery Technique. To achieve statistical security $2^{-\rho}$, early cut-and-choose protocols [26,34,35] required $s \approx 3\rho$. Lindell [24] introduced the *input-recovery technique* and demonstrated a protocol requiring only $s = \rho$ garbled circuits of f (plus additional, smaller garbled circuits computing another function). At a high level, the input-recovery technique allows P_2 to obtain P_1's input x if P_1 cheats; having done so, P_2 can then compute the function itself to learn the output. For example, in one way of instantiating this approach [24], every garbled circuit uses the same output-wire labels for a given output wire i, and moreover the labels on every output wire share the same XOR difference Δ. That is, for every wire i, the output-wire label $Z_{i,0}$ corresponding to '0' is random whereas the output-wire label $Z_{i,1}$ corresponding to '1' is set to $Z_{i,1} := Z_{i,0} \oplus \Delta$. (The protocol is set up so that Δ is not revealed by the check circuits.) If P_2 learns different outputs for some output wire i in two different gabled circuits—which means that P_1 cheated—then P_2 recovers Δ. The parties then run a *second* 2PC protocol in which P_2 learns x if it knows Δ; here, *input-consistency checks* are used to enforce that P_1 uses the same input x as before.

Afshar et al. [1] designed an input-recovery mechanism that does not require a secondary 2PC protocol. In their scheme, P_1 first commits to its input bit-by-bit using ElGamal encryption; that is, for each bit $x[i]$ of its input, P_1 sends $(g^r, h^r g^{x[i]})$ to P_2, where $h := g^\omega$ for some ω known only to P_1. As part of the protocol, P_1 sends $\{Z_{i,b} + \omega_b\}_{b \in \{0,1\}}$ to P_2 (where, as before, $Z_{i,b}$ is the label corresponding to bit b for output wire i), with $\omega = \omega_0 + \omega_1$. Now, if P_2 learns two different output-wire labels for some output wire, P_2 can recover ω and hence recover x. Afshar et al. use homomorphic properties of ElGamal encryption to enable P_2 to efficiently check that the $\{Z_{i,b} + \omega_b\}_{b \in \{0,1\}}$ are computed correctly, and for this bit-by-bit encryption of the input is required. Overall, $O(\rho \cdot n)$ public-key operations (where $|x| = n$) are needed.

Our construction relies on the same general idea introduced by Afshar et al., but our key innovation is that we are able to replace most of the public-key operations with symmetric-key operations, overall using only $O(\rho)$ public-key operations rather than $O(\rho \cdot n)$; see Sect. 3.1 for details.

Input Consistency. One challenge in the cut-and-choose approach with the input-recovery technique is that P_2 needs to enforce that P_1 uses the same input x in all the evaluation circuits, as well as in the input-recovery phase. Afshar et al. address this using zero-knowledge proofs to demonstrate (in part) that the ElGamal ciphertexts sent by P_1 all commit to the same bit across all evaluation circuits. We observe that it is not actually necessary to ensure that P_1 uses the same input x across all evaluation circuits and in the input-recovery step; rather, it is sufficient to enforce that the input x used in the input-recovery step is used in *at least one* of the evaluation circuits. This results in a dramatic efficiency improvement; see Sect. 3.1 for details.

Preventing a Selective-Failure Attack. 2PC protocols must also prevent a selective-failure attack whereby a malicious P_1 uses one valid input-wire label and one invalid input-wire label (for one of P_2's input wires) in the oblivious-transfer step. If care is not taken, P_1 could potentially use this to learn a bit of P_2's input by observing whether or not P_2 aborts. Lindell and Pinkas [25] proposed to deal with this using the *XOR-tree approach* in which P_2 replaces each bit y_i of its input by ρ random bits that XOR to y_i. By doing so, it can be shown that the probability with which P_2 aborts is (almost) *independent* of its actual input. This approach increases the number of oblivious transfers by a factor of ρ, but this can be improved by using a ρ-probe matrix [25,35], which only increases the length of the effective input by a constant factor.

Nevertheless, this constant-factor blow-up in the number of (effective) input bits corresponds to a *quadratic* blow-up in the number of XOR operations required. Somewhat surprisingly (since these XORs are non-cryptographic operations), this blow-up can become quite prohibitive. For example, for inputs as small as 4096 bits, we find that the time to compute all the XORs required for a ρ-probe matrix is over *3 s*! We resolve this bottleneck by breaking P_2's input into small chunks and constructing smaller ρ-probe matrices for each chunk, thereby reducing the overall processing required. See Sect. 4 for details.

Results. Combining the above ideas, as well as other optimizations identified in Sect. 4, we obtain a new 2PC protocol with provable security against malicious adversaries; see Sect. 3.2 for a full description. Implementing this protocol, we find that it outperforms prior work by up to several orders of magnitude in the single-execution setting; see Table 1 and Sect. 5.

Subsequent Work. In our extended version [37] we adopt ideas by David et al. [10] to further improve the efficiency of our protocol—especially when communication is the bottleneck—by reducing the communication required for the check circuits (as in [14]).

1.2 Related Work

Since the first implementation of a 2PC protocol with malicious security [27], many implementations with better performance (including those already discussed in the introduction) have been developed [11,13,23,31,32,35]. Although other approaches have been proposed for two-party computation with malicious security (e.g., [9,12,31]), here we focus on protocols using the *cut-and-choose* paradigm that is currently the most efficient approach in the single-execution setting when pre-processing is not used. Lindell and Pinkas [25] first showed how to use the cut-and-choose technique to achieve malicious security. Their construction required 680 garbled circuits for statistical security 2^{-40}, but this has been improved in a sequence of works [1,7,11,17,24,26,34] to the point where currently only 40 circuits are required.

2 Preliminaries

Let κ be the computational security parameter and let ρ be the statistical security parameter. For a bit-string x, let $x[i]$ denote the ith bit of x. We use the notation $a := f(\cdots)$ to denote the output of a deterministic function, $a \leftarrow f(\cdots)$ to denote the output of a randomized function, and $a \in_R S$ to denote choosing a uniform value from set S. Let $[n] = \{1, \ldots, n\}$. We use the notation $(\mathsf{c}, \mathsf{d}) \leftarrow \mathsf{Com}(x)$ for a commitment scheme, where c and d are the commitment and decommitment of x, respectively.

In Figs. 1 and 2, we show functionalities $\mathcal{F}_{\mathsf{OT}}$ and $\mathcal{F}_{\mathsf{cOT}}$ for parallel oblivious transfer (OT) and a weak flavor of committing OT used also by Jawurek et al. [19]. $\mathcal{F}_{\mathsf{cOT}}$ can be made compatible with OT extension as in [19].

Throughout this paper, we use P_1 and P_2 to denote the circuit garbler and circuit evaluator, respectively. We let n_1, n_2, and n_3 denote P_1's input length, P_2's input length, and the output length, respectively.

Two-Party Computation. We use a (standard) ideal functionality for two-party computation in which the output is only given to P_2; this can be extended to deliver (possibly different) outputs to both parties using known techniques [25,34].

Functionality $\mathcal{F}_{OT}$

Private inputs: P_1 has input $x \in \{0,1\}^n$ and P_2 has input $\{X_{i,b}\}_{i\in[n],b\in\{0,1\}}$.

1. Upon receiving x from P_1 and $\{X_{i,b}\}_{i\in[n],b\in\{0,1\}}$ from P_2, send $\{X_{i,x[i]}\}_{i\in[n]}$ to P_1.

Fig. 1. Functionality $\mathcal{F}_{OT}$ for oblivious transfer.

Functionality $\mathcal{F}_{cOT}$

Private inputs: P_1 has input $x \in \{0,1\}^n$ and P_2 has input $\{X_{i,b}\}_{i\in[n],b\in\{0,1\}}$.

1. Upon receiving x from P_1 and $\{X_{i,b}\}_{i\in[n],b\in\{0,1\}}$ from P_2, send $\{X_{i,x[i]}\}_{i\in[n]}$ to P_1.
2. Upon receiving open from P_2, send $\{X_{i,b}\}_{i\in[n],b\in\{0,1\}}$ to P_1.

Fig. 2. Reactive functionality $\mathcal{F}_{cOT}$ for committing oblivious transfer.

Building Blocks. Our implementation of garbled circuits uses all recent optimizations [5,21,22,32,39]. Our implementation uses the base OT protocol of Chou and Orlandi [8], and the OT extension protocol of Asharov et al. [4].

ρ-**probe Matrix.** A ρ-probe matrix, used to prevent selective-failure attacks, is a binary matrix $M \in \{0,1\}^{n_2 \times m}$ such that for any $L \subseteq [n_2]$, the Hamming weight of $\bigoplus_{i\in L} M_i$ (where M_i is the ith row of M) is at least ρ. If P_2's actual input is $y \in \{0,1\}^{n_2}$, then P_2 computes its effective input by sampling a random $y' \in \{0,1\}^m$ such that $y = My'$.

The original construction by Lindell and Pinkas [25] has $m = \max\{4n_2, 8\rho\}$. shelat and Shen [35] improved this to $m = n_2 + O(\rho + \log(n_2))$. Lindell and Riva [29] proposed to append an identity matrix to M to ensure that M is full rank, and to make it easier to find y' such that $y = My'$.

3 Our Protocol

3.1 Protocol Overview

We describe in more detail the intuition behind the changes we introduce. This description is not complete, but only illustrates the main differences from prior work. Full details of our protocol are given in Sect. 3.2.

In our protocol, the two parties first run ρ instances of OT, where in the jth instance P_1 sends a random key key_j and a random seed seed_j, while P_2 chooses whether to learn key_j (thereby choosing to make the jth garbled circuit an evaluation circuit) or seed_j (thereby choosing to make the jth garbled circuit a check circuit). The protocol is designed such that key_j can be used to recover the input-wire labels associated with P_1's input in the jth garbled circuit, whereas seed_j can be used to recover all the randomness used to generate the jth garbled circuit. Thus far, the structure of our protocol is similar to that of Afshar et al. [1]. However, we differ in how we recover P_1's input if P_1 is caught cheating, and in how we ensure input consistency for P_1's input.

Input Recovery. Recall that we want to ensure that if P_2 detects cheating by P_1, then P_2 can recover P_1's input. This is done by encoding some trapdoor in the output-wire labels of the garbled circuits such that if P_2 learns *both* labels for some output wire (in different garbled circuits) then P_2 can recover the trapdoor and thus learn P_1's input. In slightly more detail, input recovery consists of the following high-level steps:

1. P_1 "commits to" its input x using some trapdoor.
2. P_1 sends garbled circuits and the input-wire labels associated with x, using an input-consistency protocol (discussed below) to enforce that consistent input-wire labels are used.
3. P_1 and P_2 run some protocol such that if P_2 detects cheating by P_1, then P_2 gets the trapdoor without P_1 learning this fact.
4. P_2 either (1) detects cheating, recovers x using the trapdoor, and locally computes (and outputs) $f(x, y)$, or (2) outputs the (unique) output of the evaluated garbled circuits, which is $f(x, y)$.

In Afshar et al. [1], the above is done using ElGamal encryption and efficient zero-knowledge checks to enforce input consistency. However, this approach requires $O(\rho \cdot (n_1 + n_3))$ public-key operations. In contrast, our protocol achieves the same functionality with only $O(\rho)$ public-key operations.

Our scheme works as follows. Assume for ease of presentation that P_1's input x is a single bit and the output of the function is also a single bit. The parties run an OT protocol in which P_1 inputs x and P_2 inputs two random labels M_0, M_1, with P_1 receiving M_x. Then, for the jth garbled circuit, P_1 "commits" to x by computing $R_{j,x} := \text{PRF}_{\text{seed}_j}(\text{"}R\text{"}) \oplus M_x$ and sending to P_2 an encryption of $R_{j,x}$ under key_j. Note that P_1 cannot "commit" to $1 - x$ unless P_1 can guess M_{1-x}. Also, if P_1 is honest then x remains hidden from P_2 because P_2 knows either key_j or seed_j for each j, but not both. The value seed_j for any evaluation circuit j serves as a trapdoor since, in conjunction with the value key_j that P_2 already has, it allows P_2 to learn M_x (and hence determine x).

The next step is to devise a way for P_2 to recover seed_j if it learns *inconsistent* output-wire labels in two different evaluation circuits. We do this as follows. First, P_1 chooses random $\Delta, \Delta_0, \Delta_1$ such that $\Delta = \Delta_0 \oplus \Delta_1$. Then, for all j it encrypts Δ_0 using $Z_{j,0}$ and encrypts Δ_1 using $Z_{j,1}$, where $Z_{j,0}, Z_{j,1}$ are the two output-wire labels of the jth garbled circuit. It sends all these encryptions to P_2. Thus, if P_2 learns $Z_{j_1,0}$ for some j_1 it can recover Δ_0, and if it learns $Z_{j_2,1}$ for some j_2 it can recover Δ_1. If it learns *both* output-wire labels, it can then of course recover Δ.

P_1 and P_2 then run a protocol that guarantees that if P_2 knows Δ it recovers seed_j, and otherwise it learns nothing. This is done as follows. P_2 sets $\Omega := \Delta$

Table 2. Notation used in our protocol.

Notation	Meaning
$\mathcal{E}$	evaluation set
E	ρ-probe matrix
GC_j	jth garbled circuit
$\{A_{j,i,b}\}_b$	ith input-wire labels for P_1 in GC_j
$\{B_{j,i,b}\}_b$	ith input-wire labels for P_2 in GC_j
$\{Z_{j,i,b}\}_b$	ith output-wire labels in GC_j
$\{T_{j,i,b}\}_b$	ith output-mapping table for GC_j
$\{R_{j,i,b}\}_{j,b}$	commitments for the ith bit of P_1's input
C_j, D_j	input-recovery elements

if it learned Δ, and sets $\Omega := 1$ otherwise. P_2 then computes $(h, g_1, h_1) := (g^\omega, g^r, h^r \Omega)$, for random ω and r, and sends (h, g_1, h_1) to P_1. Then, for each index j, party P_1 computes $C_j := g^{s_j} h^{t_j}$ and $D_j := g_1^{s_j}(h_1/\Delta)^{t_j}$ for random s_j, t_j, and sends C_j along with an encryption of seed_j under D_j. Note that if $\Omega = \Delta$, then $C_j^r = D_j$ and thus P_2 can recover seed_j, whereas if $\Omega \neq \Delta$ then P_2 learns nothing (in an information-theoretic sense).

Of course, the protocol as described does not account for the fact that P_1 can send invalid messages or otherwise try to cheat. However, by carefully integrating appropriate correctness checks as part of the cut-and-choose process, we can guarantee that if P_1 tries to cheat then P_2 either aborts (due to detected cheating) or learns P_1's input with high probability without leaking any information.

Input Consistency. As discussed in Sect. 1.1, prior schemes enforce that P_1 uses the same input x for all garbled circuits and also for the input-recovery sub-protocol. However, we observe that this is not necessary. Instead, it suffices to ensure that P_1 uses the same input in the input-recovery sub-protocol and *at least one* of the evaluated garbled circuit. Even if P_1 cheats by using different inputs in two different evaluated garbled circuits, P_2 still obtains the correct output: if P_2 learns only one output then this is the correct output; if P_2 learns multiple outputs, then the input-recovery procedure ensures that P_2 learns x and so can compute the correct output.

We ensure the above weaker notion of consistency by integrating the consistency check with the cut-and-choose process as follows. Recall that in our input-recovery scheme, P_1 sends to P_2 a "commitment" $R_{j,x} := \mathsf{PRF}_{\mathsf{seed}_j}("R") \oplus M_x$ for each index j. After these commitments are sent, we now have P_2 reveal $M_0 \oplus M_1$ to P_1 (we use committing OT for this purpose), so P_1 learns both M_0 and M_1. P_1 then computes and sends (in a randomly permuted order) $\mathsf{Com}(R_{j,0}, A_{j,0})$ and $\mathsf{Com}(R_{j,1}, A_{j,1})$, where $A_{j,0}, A_{j,1}$ are P_1's input-wire labels in the jth garbled circuit and the commitments are generated using randomness derived from seed_j. P_1 also sends $\mathsf{Enc}_{\mathsf{key}_j}(\mathsf{Decom}(\mathsf{Com}(R_{j,x}, A_{j,x})))$. Note that (1) if P_2 chose j as a check circuit then it can check correctness of the commitment pair, since

---

Protocol Π_{2pc}

Private inputs: P_1 has input $x \in \{0,1\}^{n_1}$ and P_2 has input $y \in \{0,1\}^{n_2}$.

Common inputs:

ρ-probe matrix $E \in \{0,1\}^{n_2 \times m}$, where $m = O(n_2)$;

Circuit $f : \{0,1\}^{n_1} \times \{0,1\}^{n_2} \to \{0,1\}^{n_3}$;

Circuit $f' : \{0,1\}^{n_1} \times \{0,1\}^{m} \to \{0,1\}^{n_3}$ such that $f'(x, y') = f(x, Ey')$;

Prime q with $|q| = \mathsf{poly}(\kappa)$.

Protocol:

1. P_1 picks random κ-bit strings $\{\mathsf{key}_j, \mathsf{seed}_j\}_{j \in [\rho]}$, and sends them to $\mathcal{F}_{\mathsf{OT}}$. P_2 picks $\mathcal{E} \in_R \{0,1\}^{\rho}$, sends $\mathcal{E}$ to $\mathcal{F}_{\mathsf{OT}}$, and receives $\{\mathsf{seed}_j\}_{j \notin \mathcal{E}}$ and $\{\mathsf{key}_j\}_{j \in \mathcal{E}}$.

2. P_1 computes $\{B_{j,i,b} := \mathsf{PRF}_{\mathsf{seed}_j}(i, b, \text{``}B\text{''})\}_{j \in [\rho], i \in [m], b \in \{0,1\}}$ and sends $\{B_{1,i,b} \| \cdots \| B_{\rho,i,b}\}_{i \in [m], b \in \{0,1\}}$ to $\mathcal{F}_{\mathsf{OT}}$. P_2 chooses random $y' \in_R \{0,1\}^{m}$ such that $y = Ey'$, sends y' to $\mathcal{F}_{\mathsf{OT}}$, and receives $\{B_{1,i,y'[i]} \| \cdots \| B_{\rho,i,y'[i]}\}_{i \in [m]}$.

3. P_2 sends random labels $\{M_{i,b}\}_{i \in [n_1], b \in \{0,1\}}$ to $\mathcal{F}_{\mathsf{cOT}}$. P_1 sends x to $\mathcal{F}_{\mathsf{cOT}}$ and receives $\{M_{i,x[i]}\}_{i \in [n_1]}$. For $j \in [\rho], i \in [n_1]$, P_1 computes $R_{j,i,x[i]} := \mathsf{PRF}_{\mathsf{seed}_j}(i, \text{``}R\text{''}) \oplus M_{i,x[i]}$, and sends $\mathsf{Enc}_{\mathsf{key}_j}(\{R_{j,i,x[i]}\}_{i \in [n_1]})$ to P_2. P_2 sends **open** to $\mathcal{F}_{\mathsf{cOT}}$ (which sends $\{M_{i,0}, M_{i,1}\}_{i \in [n_1]}$ to P_1), and for $j \in \mathcal{E}$ uses key_j to decrypt and learn $R_{j,i,x[i]}$.

4. For $j \in [\rho], i \in [n_1]$, P_1 computes $R_{j,i,1-x[i]} := R_{j,i,x[i]} \oplus M_{i,0} \oplus M_{i,1}$, $\{A_{j,i,b} := \mathsf{PRF}_{\mathsf{seed}_j}(i, b, \text{``}A\text{''})\}_{b \in \{0,1\}}$, and $\{(\mathsf{c}^{R}_{j,i,b}, \mathsf{d}^{R}_{j,i,b}) \leftarrow \mathsf{Com}(R_{j,i,b}, A_{j,i,b})\}_{b \in \{0,1\}}$ using randomness derived from seed_j, and sends $\{(\mathsf{c}^{R}_{j,i,0}, \mathsf{c}^{R}_{j,i,1})\}$ (in random permuted order) and $\mathsf{Enc}_{\mathsf{key}_j}(\{\mathsf{d}^{R}_{j,i,x[i]}\}_{i \in [n_1]})$ to P_2. For $j \in \mathcal{E}, i \in [n_1]$, P_2 opens $\mathsf{c}^{R}_{j,i,x[i]}$ to obtain $R_{j,i,x[i]}$ and $A_{j,i,x[i]}$, and checks that $R_{j,i,x[i]}$ equals the value from Step 3. If any decommitment is invalid or any check fails, P_2 aborts.

5. P_1 picks random κ-bit labels Δ, $\{\Delta_{i,0}\}_{i \in [n_3]}$, sets $\{\Delta_{i,1} := \Delta_{i,0} \oplus \Delta\}_{i \in [n_3]}$, and sends $\{H(\Delta_{i,b})\}_{i \in [n_3], b \in \{0,1\}}$ to P_2. For $j \in [\rho]$, P_1 computes garbled circuit GC_j for function f' using $A_{j,i,b}, B_{j,i,b}$ as the input-wire labels and randomness derived from seed_j for internal wire labels. Let $Z_{j,i,b}$ denote the output-wire labels. P_1 computes $\{T_{j,i,b} := \mathsf{Enc}_{Z_{j,i,b}}(\Delta_{i,b})\}_{i \in [n_3], b \in \{0,1\}}$ and $(\mathsf{c}^{T}_j, \mathsf{d}^{T}_j) \leftarrow \mathsf{Com}(\{T_{j,i,b}\}_{i \in [n_3], b \in \{0,1\}})$ using randomness derived from seed_j, and sends GC_j, c^{T}_j, and $\mathsf{Enc}_{\mathsf{key}_j}(\mathsf{d}^{T}_j)$ to P_2.

6. For $j \in \mathcal{E}$, P_2 decrypts to learn d^{T}_j and opens c^{T}_j to learn $\{T_{j,i,b}\}_{i \in [n_3], b \in \{0,1\}}$; if any decommitment is invalid, P_2 aborts. P_2 evaluates GC_j using labels $\{A_{j,i,x[i]}\}_{i \in [n_1]}$ and $\{B_{j,i,y'[i]}\}_{i \in [m]}$, and obtains output-wire labels $\{Z_{j,i}\}_i$. P_2 checks validity of these labels by checking if $H(\mathsf{Dec}_{Z_{j,i}}(T_{j,i,b}))$ matches $H(\Delta_{i,b})$ for some $b \in \{0,1\}$, and if so sets $z'_j[i] := b$; else it sets $z'_j[i] := \bot$.
   - *Invalid circuits.* If, for every $j \in \mathcal{E}$, there is some i with $z'_j[i] = \bot$, then P_2 sets $\Omega := 1, z := \bot$.
   - *Inconsistent output labels.* Else if, for some $i \in [n_3], j_1, j_2 \in \mathcal{E}$, P_2 obtains $z'_{j_1}[i] = 0$ and $z'_{j_2}[i] = 1$, then P_2 sets $\Omega := \mathsf{Dec}_{Z_{j_1,i}}(T_{j_1,i,0}) \oplus \mathsf{Dec}_{Z_{j_2,i}}(T_{j_2,i,1})$. If different Ωs are obtained, P_2 sets $z := \bot$.
   - *Consistent output labels.* Else, for all i, set $z[i] := z'_j[i]$ for the first index j such that $z'_j[i] \neq \bot$, and set $\Omega := 1$.

Fig. 3. The full description of our malicious 2PC protocol, part 1.

Protocol Π_{2pc} continued

Protocol:

7. P_2 picks $\omega, r \in_R \mathbb{F}_q$, and sends $(h, g_1, h_1) := (g^\omega, g^r, h^r \Omega)$ to P_1. P_1 sends Δ and $\{\Delta_{i,b}\}_{i \in [n_3], b \in \{0,1\}}$ to P_2, who checks that $\{\Delta = \Delta_{i,0} \oplus \Delta_{i,1}\}_{i \in [n_3]}$ and that $H(\Delta_{i,b})$ matches the values P_1 sent in Step 5; if any check fails, P_2 aborts. For $j \in [\rho]$, P_1 picks $s_j, t_j \in_R \mathbb{F}_q$ using randomness derived from seed_j, computes $C_j := g^{s_j} h^{t_j}, D_j := g_1^{s_j} \left(\frac{h_1}{\Delta}\right)^{t_j}$, and sends C_j and $\mathsf{Enc}_{D_j}(\mathsf{seed}_j)$ to P_2. For $j \in \mathcal{E}$, P_2 uses C_j^r to decrypt and obtains some seed_j'.

8. If $\Omega \neq 1$, P_2 recovers x as follows: For $j \in \mathcal{E}, i \in [n_1]$, if $R_{j,i,x[i]} = \mathrm{PRF}_{\mathsf{seed}_j'}(i, \text{"}R\text{"}) \oplus M_{i,0}$, P_2 sets $x_j[i] := 0$; if $R_{j,i,x[i]} = \mathrm{PRF}_{\mathsf{seed}_j'}(i, \text{"}R\text{"}) \oplus M_{i,1}$, P_2 sets $x_j[i] := 1$; and otherwise, P_2 sets $x_j[i] := \perp$. If no valid x_j is obtained, or more than two different x_j are obtained, P_2 sets $z := \perp$; otherwise P_2 sets $z := f(x_j, y)$.

9. If all the following checks hold for all $j \notin \mathcal{E}$, then P_2 outputs z; otherwise, P_2 aborts.

   (a) For $i \in [m]$, the $B_{j,i,y'[i]}$ value received in Step 2 equals $\mathrm{PRF}_{\mathsf{seed}_j}(i, y'[i], \text{"}B\text{"})$.

   (b) GC_j is computed correctly using $A_{j,i,b} := \mathrm{PRF}_{\mathsf{seed}_j}(i, b, \text{"}A\text{"})$ and $B_{j,i,b} := \mathrm{PRF}_{\mathsf{seed}_j}(i, b, \text{"}B\text{"})$ as input-wire labels and randomness derived from seed_j.

   (c) Compute $T_{j,i,b}$ using $Z_{j,i,b}$ from GC_j and $\Delta_{i,b}$ sent by P_1, and check that c_j^T is computed correctly with randomness derived from seed_j.

   (d) The $C_j, \mathsf{Enc}_{D_j}(\mathsf{seed}_j)$ values in Step 7 are correctly computed, using Δ and seed_j.

   (e) For $i \in [n_1], b \in \{0,1\}$, $\mathsf{c}_{j,i,b}^R$ is correctly computed using seed_j, $A_{j,i,b}$, and $R_{j,i,b}$ (which are themselves computed from seed_j).

Fig. 4. The full description of our malicious 2PC protocol, part 2.

everything is computed from seed_j, and (2) if P_2 chose j as an evaluation circuit then it can open the appropriate commitment to recover $R_{j,x}$, and check that this matches the value sent before.

3.2 Protocol Details and Proof of Security

We present the full details of our protocol in Figs. 3 and 4. To aid in understanding the protocol, we also present a graphical depiction in Fig. 5. We summarize some important notations in Table 2 for reference.

Our protocol, including the optimizations detailed in Sect. 4, requires a total of $O(\rho \cdot (n_1 + n_2 + n_3 + |C|))$ symmetric-key operations and $O(\rho + \kappa)$ group operations. Most of the symmetric-key operations, including circuit garbling and computing the PRFs, can be accelerated using hardware AES instructions.

Theorem 1. *Let* Com *be a computationally hiding/binding commitment scheme, let the garbling scheme satisfy authenticity, privacy, and obliviousness*

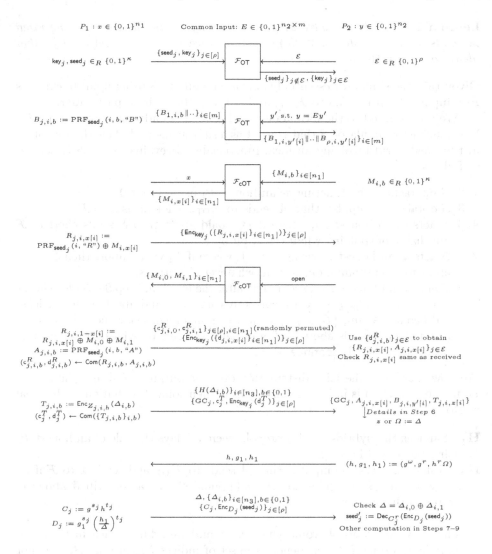

Fig. 5. Graphical depiction of our protocol.

(cf. [6]), let H be collision-resistant, and assume the decisional Diffie-Hellman assumption holds. Then the protocol in Figs. 3 and 4 securely computes f in the $(\mathcal{F}_{\mathsf{OT}}, \mathcal{F}_{\mathsf{cOT}})$-hybrid model with security $2^{-\rho} + \mathsf{negl}(\kappa)$.

Proof. We consider separately the case where P_1 or P_2 is malicious.

Malicious P_1. Our proof is based on the fact that with all but negligible probability, P_2 either aborts or learns the output $f(x, y)$, where x is the input P_1 sent to $\mathcal{F}_{\mathsf{cOT}}$ in Step 3 and y is P_2's input. We rely on the following lemma, which we prove in Sect. 3.3.

Lemma 1. *Consider an adversary $\mathcal{A}$ corrupting P_1 and denote x as the input $\mathcal{A}$ sends to $\mathcal{F}_{cOT}$ in Step 3. With probability at least $1 - 2^{-\rho} - \mathsf{negl}(\kappa)$, P_2 either aborts or learns $f(x, y)$.*

Given this, the simulator essentially acts as an honest P_2 using input 0, extracts P_1's input x from the call to $\mathcal{F}_{cOT}$, and outputs $f(x, y)$ if no party aborts.

We now proceed to the formal details. Let $\mathcal{A}$ be an adversary corrupting P_1. We construct a simulator $\mathcal{S}$ that runs $\mathcal{A}$ as a subroutine and plays the role of P_1 in the ideal world involving an ideal functionality $\mathcal{F}$ evaluating f. $\mathcal{S}$ is defined as follows.

1–2 $\mathcal{S}$ interacts with $\mathcal{A}$, acting as an honest P_2 using input 0.

3 $\mathcal{S}$ obtains the input x that $\mathcal{A}$ sends to $\mathcal{F}_{cOT}$. It forwards x to $\mathcal{F}$.

4–6 $\mathcal{S}$ acts as an honest P_2, where if P_2 would abort then $\mathcal{S}$ sends abort to $\mathcal{F}$ and halts, outputting whatever $\mathcal{A}$ outputs.

7–8 $\mathcal{S}$ acts as an honest P_2 using $\Omega := 1$, where if P_2 would abort then $\mathcal{S}$ sends abort to $\mathcal{F}$ and halts, outputting whatever $\mathcal{A}$ outputs.

9 $\mathcal{S}$ acts as an honest P_2, except that after the check in Step 9a, $\mathcal{S}$ also checks if $\{B_{j,i,b}\}_{j \notin \mathcal{E}, i \in [m], b \in \{0,1\}}$ are correctly computed and aborts if, for at least ρ different $i \in [m]$, $\{B_{j,i,b}\}_{j \notin \mathcal{E}, b \in \{0,1\}}$ contains incorrect values. If P_2 would abort then $\mathcal{S}$ sends abort to $\mathcal{F}$ and halts, outputting whatever $\mathcal{A}$ outputs; otherwise, $\mathcal{S}$ sends continue to $\mathcal{F}$.

We now show that the joint distribution over the outputs of $\mathcal{A}$ and the honest P_2 in the real world is indistinguishable from their joint distribution in the ideal world.

H$_1$. Same as the hybrid-world protocol, where $\mathcal{S}$ plays the role of an honest P_2 using the actual input y.

H$_2$. $\mathcal{S}$ now extracts the input x that $\mathcal{A}$ sends to $\mathcal{F}_{cOT}$ and sends x to $\mathcal{F}$ if no party aborts. $\mathcal{S}$ also performs the additional checks as described above in Step 9 of the simulator.

There are two ways $\mathcal{A}$ would cheat here, and we address each in turn. For simplicity, we let $I \subset [m]$ denote the set of indices i such that $B_{j,i,b}$ is not correctly computed.

1. $\mathcal{A}$ launches a selective-failure attack with $|I| < \rho$. Lemma 1 ensures (in **H$_1$**) that P_2 either aborts or learns $f(x, y)$ with probability at least $1 - 2^{-\rho}$. In **H$_2$**, note that P_2 either aborts or learns $f(x, y)$ with probability 1. Further, since fewer than ρ wires are corrupted, the probability of an abort due to the selective-failure attack is exactly the same in both hybrids. Therefore the statistical distance between **H$_1$** and **H$_2$** is at most $2^{-\rho}$.

2. $\mathcal{A}$ launches a selective-failure attack with $|I| \geq \rho$. By the security of the ρ-probe matrix [29], $\mathcal{S}$ aborts in **H$_1$** with probability at least $1 - 2^{-\rho}$. (If $\mathcal{A}$ cheats elsewhere, the probability of abort can only increase.) But in **H$_2$** in this case P_2 aborts with probability 1, and so there is at most a $2^{-\rho}$ difference between **H$_1$** and **H$_2$**.

H₃. Same as **H₂**, except $\mathcal{S}$ uses $y := 0$ throughout of protocol and sets $\Omega := 1$ in Step 7.

In [**H₃**], P_2 sends $(h, g_1, h_1) := (g^\omega, g^r, g^{\omega r})$, which is indistinguishable from $(g^\omega, g^r, g^{\omega r} \Omega)$ by the decisional Diffie-Hellman problem. Computationally indistinguishability of **H₂** and **H₃** follows.

As **H₃** is the ideal-world execution, the proof is complete. □

Malicious P_2. Here, we need to simulate the correct output $f(x, y)$ that P_2 learns. Rather than simulate the garbled circuit, as is done in most prior work, we modify the output-mapping tables $\{T_{j,i,b}\}$ to encode the correct output. At a high level, the simulator acts as an honest P_1 with input 0, which lets P_2 learn the output-wire labels for $f(0, y)$ when evaluating the garbled circuits. The simulator then "tweaks" the output mapping tables $\{T_{j,i,b}\}$ to ensure that P_2 reconstructs the "correct" output $f(x, y)$.

We now proceed to the formal details. Let $\mathcal{A}$ be an adversary corrupting P_2; we construct a simulator $\mathcal{S}$ as follows.

1. $\mathcal{S}$ acts as an honest P_1 and obtains the set $\mathcal{E}$ that $\mathcal{A}$ sends to $\mathcal{F}_{\mathsf{OT}}$.
2. $\mathcal{S}$ acts as an honest P_1, and obtains the input y' that $\mathcal{A}$ sends to $\mathcal{F}_{\mathsf{OT}}$. $\mathcal{S}$ computes y from y', sends (input, y) to $\mathcal{F}$, which sends back $z := f(x, y)$ to $\mathcal{S}$.
3. $\mathcal{S}$ acts as an honest P_1 with input $x := 0$. That is, $\mathcal{S}$ receives $\{M_{i,b}\}$ labels and sends $\{\mathsf{Enc}_{\mathsf{key}_j}(\{R_{j,i,0}\}_{i \in [n_1]})\}_{j \in [\rho]}$ to $\mathcal{A}$. If $\mathcal{A}$ send abort to $\mathcal{F}_{\mathsf{cOT}}$, $\mathcal{S}$ aborts, outputting whatever $\mathcal{A}$ outputs.
4. $\mathcal{S}$ acts as an honest P_1 with input $x := 0$. That is, $\mathcal{S}$ sends $\{(c^R_{j,i,0}, c^R_{j,i,1})\}$ (in random permuted order) and $\mathsf{Enc}_{\mathsf{key}_j}(\{d^R_{j,i,0}\}_{i \in [n_1]})$ to $\mathcal{A}$.
5. $\mathcal{S}$ acts as an honest P_1, except as follows. $\mathcal{S}$ computes $z' := f(0, y)$ and for $j \in \mathcal{E}$, $i \in [n_3]$, and $b \in \{0, 1\}$, sets $T_{j,i,b} := \mathsf{Enc}_{Z_{j,i,b}}(\Delta_{i,1-b})$ if $z[i] \neq z'[i]$.

6–7. $\mathcal{S}$ acts as an honest P_1.

We now show that the joint distribution over the outputs of the honest P_1 and $\mathcal{A}$ in the real world is indistinguishable from their joint distribution in the ideal world.

H₁. Same as the hybrid-world protocol, where $\mathcal{S}$ plays the role of an honest P_1.

H₂. $\mathcal{S}$ extracts P_2's input y from $\mathcal{F}_{\mathsf{OT}}$ and sends (input, y) to $\mathcal{F}$, receiving back z. $\mathcal{S}$ uses $x := 0$ throughout the simulation and "tweaks" $\{T_{j,i,b}\}$ as is done by the simulator using knowledge of z.

H₁ and **H₂** are the same except:

1. In **H₁**, P_2 learns $\{R_{j,i,x[i]}\}_{j \in \mathcal{E}}$, while P_2 learns $\{R_{j,i,0}\}_{j \in \mathcal{E}}$ in **H₂**. Note that these values are computed such that $R_{j,i,b} := \mathsf{PRF}_{\mathsf{seed}_j}(i, "R") \oplus M_{i,b}$. Since P_2 does not know any $\{\mathsf{seed}_j\}_{j \in \mathcal{E}}$, $\mathsf{PRF}_{\mathsf{seed}_j}(i, "R")$ looks random to P_2. Because only one of $\{R_{j,i,b}\}_{b \in \{0,1\}}$ is given in both **H**s, $R_{j,i,x[i]}$ in **H₁** and $R_{j,i,0}$ in **H₂** are uniformly random to P_2.
2. In **H₁**, party P_2 gets $Z_{j,i,z[i]}$ and $T_{j,i,z[i]} := \mathsf{Enc}_{Z_{j,i,z[i]}}(\Delta_{i,z[i]})$, while in **H₂**, if $z[i] \neq z'[i]$ then P_2 gets $Z_{j,i,1-z[i]}$ and $T_{j,i,1-z[i]} := \mathsf{Enc}_{Z_{j,i,1-z[i]}}(\Delta_{i,z[i]})$ instead. In both hybrids, P_2 cannot learn any information about the other

output-wire label due to the authenticity property of the garbled circuit. By the obliviousness property of the garbling scheme, $Z_{j,i,0}$ and $Z_{j,i,1}$ are indistinguishable. Likewise, by the security of the encryption scheme the values $T_{j,i,0}$ and $T_{j,i,1}$ are indistinguishable.

As $\mathbf{H_2}$ is the same as the ideal-world execution, the proof is complete.

3.3   Proving Lemma 1

We now prove a series of lemmas toward proving Lemma 1. We begin with a definition of what it means for an index $j \in [\rho]$ to be "good."

Definition 1. *Consider an adversary $\mathcal{A}$ corrupting P_1, and denote $\{\mathsf{seed}_j\}$ as the labels $\mathcal{A}$ sent to $\mathcal{F}_{\mathsf{OT}}$. An index $j \in [\rho]$ is good if and only if all the following hold.*

1. *The $B_{j,i,y'[i]}$ values $\mathcal{A}$ sent to $\mathcal{F}_{\mathsf{OT}}$ in Step 2 are computed honestly using seed_j.*
2. *The commitments $\{c_{j,i,b}^R\}_{i \in [n_1], b \in \{0,1\}}$ that $\mathcal{A}$ sent to P_2 in Step 4 are computed honestly using seed_j.*
3. *GC_j is computed honestly using $\{A_{j,i,b}\}$ and $\{B_{j,i,b}\}$ as the input-wire labels and seed_j.*
4. *The values C_j and $\mathsf{Enc}_{D_j}(\mathsf{seed}_j)$ are computed honestly using seed_j and the Δ value sent by $\mathcal{A}$ in Step 7.*
5. *The commitment c_j^T is computed honestly using $\Delta_{i,b}$ and seed_j.*

It is easy to see the following.

Fact. *If an index $j \in [\rho]$ is not good then it cannot pass all the checks in Step 9.* We first show that P_2 is able to recover the correct output-wire labels for a good index.

Lemma 2. *Consider an adversary $\mathcal{A}$ corrupting P_1, and denote x as the input $\mathcal{A}$ sent to $\mathcal{F}_{\mathsf{cOT}}$. If an index $j \in \mathcal{E}$ is good and P_2 does not abort, then with all but negligible probability P_2 learns output labels $Z_{j,i,z[i]}$ with $z = f(x, y)$.*

Proof. Since j is good, we know that P_2 receives an honestly computed GC_j and $T_{j,i,b}$ from $\mathcal{A}$ and honest $B_{j,i,y'[i]}$ from $\mathcal{F}_{\mathsf{OT}}$. However, it is still possible that P_2 does not receive correct input labels for P_1's input that corresponds to the input x that $\mathcal{A}$ sent to $\mathcal{F}_{\mathsf{cOT}}$. We will show that this can only happen with negligible probability.

Note that if j is good, then the commitments $\{c_{j,i,b}^R\}$ are computed correctly. Since P_2 obtains the $A_{j,i,x[i]}$ labels by decommitting one of these commitments, the labels P_2 gets are valid input-wire labels, although they may not be consistent with the input x that $\mathcal{A}$ sent to $\mathcal{F}_{\mathsf{cOT}}$.

Assume that for some $i \in [n_1]$, P_2 receives $A_{j,i,1-x[i]}$. This means P_2 also receives $R_{j,i,1-x[i]}$ from the same decommitment, since $c_{j,i,b}$ is computed honestly. However, if P_2 does not abort, then we know that P_2 receives the same label $R_{j,i,1-x[i]}$ in Step 3 since the checks pass. We also know that

$$R_{j,i,1-x[i]} = \mathsf{PRF}_{\mathsf{seed}_j}(i, \text{``}R\text{''}) \oplus M_{i,1-x[i]}.$$

Therefore $\mathcal{A}$ needs to guess $M_{i,1-x[i]}$ correctly before P_2 sends both labels, which happens with probability at most $2^{-\kappa}$.

We next show that P_2 can recover x if P_1 tries to cheat on a good index.

Lemma 3. *Consider an adversary $\mathcal{A}$ corrupting P_1, and denote x as the input $\mathcal{A}$ sent to $\mathcal{F}_{\mathsf{cOT}}$. If an index $j \in \mathcal{E}$ is good and P_2 learns $\Omega = \Delta$, then P_2 can recover $x_j = x$ in Step 8 if no party aborts.*

Proof. Since j is good, we know that C_j and $\mathsf{Enc}_{D_j}(\mathsf{seed}_j)$ are constructed correctly, where seed_j is the one P_1 sent to $\mathcal{F}_{\mathsf{OT}}$ in Step 1. Therefore, P_2 can recompute seed_j from them. We just need to show that P_2 is able to recover x from a good index using seed_j.

Using a similar argument as the previous proof, we can show that the label $R_{j,i,x[i]}$ that P_2 learns in Step 4 is a correctly computed label using x that P_1 sent to $\mathcal{F}_{\mathsf{cOT}}$ in Step 3: Since j is good, the $c^R_{j,i,b}$ values are all good, which means that the $R_{j,i,x[i]}$ labels P_2 learns are valid. However, P_1 cannot "flip" the wire label unless P_1 guesses a random label correctly, which happens with negligible probability.

In conclusion, P_2 has the correct $R_{j,i,x[i]} = \mathsf{PRF}_{\mathsf{seed}_j}(i, \text{``}R\text{''}) \oplus M_{i,x[i]}$ and the seed_j used in the computation. Further P_2 has $M_{i,0}, M_{i,1}$. Therefore P_2 can recover x that P_1 sent to $\mathcal{F}_{\mathsf{cOT}}$ if P_2 has $\Omega = \Delta$.

Note that given the above lemma, it may still be possible that a malicious P_1 acts in such a way that P_2 recovers different x's from different indices. In the following we show this only happens with negligible probability.

Lemma 4. *Consider an adversary $\mathcal{A}$ corrupting P_1 and denote x as the input P_1 sends to $\mathcal{F}_{\mathsf{cOT}}$ in Step 3. If P_2 does not abort, then P_2 recovers some $x' \neq x$ with negligible probability.*

Proof. Our proof is by contradiction. Assume that P_2 does not abort and recovers some $x' \neq x$ for some $j \in \mathcal{E}$. Let i be an index at which $x'[i] \neq x[i]$; we will show in the following that $\mathcal{A}$ will have to complete some task that is information-theoretically infeasible, and thus a contradiction.

Since P_2 does not abort at Step 4, we can denote $R_{j,i,x[i]}$ as the label P_1 learns in Step 3, which also equals the one decommitted to in Step 4. P_2 recovering some x' means that

$$R_{j,i,x[i]} = \mathsf{PRF}_{\mathsf{seed}'_j}(i, \text{``}R\text{''}) \oplus M_{i,x'[i]},$$

where seed'_j is the seed P_2 recovers in Step 7. Therefore we conclude that

$$\mathsf{PRF}_{\mathsf{seed}'_j}(i, \text{``}R\text{''}) = R_{j,i,x[i]} \oplus M_{i,x'[i]}$$
$$= R_{j,i,x[i]} \oplus M_{i,1-x[i]}.$$

Although $\mathcal{A}$ receives $M_{i,x[i]}$ in Step 3, $M_{i,1-x[i]}$ remains completely random before $\mathcal{A}$ sends $R_{j,i,x[i]}$. Further, $\mathcal{A}$ receives $M_{i,b}$ only after sending $R_{j,i,x[i]}$.

Therefore, the value of $R_{j,i,x[i]} \oplus M_{i,1-x[i]}$ is completely random to $\mathcal{A}$. If $\mathcal{A}$ wants to "flip" a bit in x, $\mathcal{A}$ needs to find some seed'_j such that $\{\mathrm{PRF}_{\mathsf{seed}'_j}(i,$ "R")$\}_{i \in [n_1]}$ equals a randomly chosen string, which is information theoretically infeasible if $n_1 > 1$.

Finally, we are ready to prove Lemma 1, namely, that P_2 either aborts or learns $f(x, y)$, regardless of P_1's behavior.

Lemma 1. *Consider an adversary $\mathcal{A}$ corrupting P_1 and denote x as the input P_1 sends to $\mathcal{F}_{\mathsf{cOT}}$ in Step 3. With probability at least $1 - (2^{-\rho} + \mathsf{negl}(\kappa))$, P_2 either aborts or learns $f(x, y)$.*

Proof. Denote the set of P_1's good circuits as $\mathcal{E}'$ and consider the following three cases:

- $\bar{\mathcal{E}} \cap \bar{\mathcal{E}}' \neq \emptyset$. In this case P_2 aborts because P_2 checks some $j \notin \mathcal{E}'$ which is not a good index.
- $\mathcal{E} \cap \mathcal{E}' \neq \emptyset$. In this case, there is some $j \in \mathcal{E} \cap \mathcal{E}'$, which means P_2 learns $z := f(x, y)$ and $Z_{j,i,z[i]}$ from the jth garbled circuit (by Lemma 2). However, it is still possible that P_2 learns more than one valid z. If this is the case, P_2 learns Δ. Lemma 3 ensures that P_2 obtains x; Lemma 4 ensures that P_2 cannot recover any other valid x' even from bad indices.
- $\mathcal{E} = \mathcal{E}'$. This only happens when $\mathcal{A}$ guesses $\mathcal{E}$ correctly, which happens with at most $2^{-\rho}$ probability.

This completes the proof.

3.4   Universal Composability

Note that in our proof of security, the simulators do not rewind in any of the steps. Similarly, none of the simulators in the hybrid arguments need any rewinding. Therefore, if we instantiate all the functionalities using UC-secure variants then the resulting 2PC protocol is UC-secure.

4   Optimizations

We now discuss several optimizations we discovered in the course of implementing our protocol, some of which may be applicable to other malicious 2PC implementations.

4.1   Optimizing the XOR-tree

We noticed that when using a ρ-probe matrix to reduce the number of OTs needed for the XOR-tree, we incurred a large performance hit when P_2's input was large. In particular, processing the XOR gates introduced by the XOR-tree, which are always assumed to be free due to the free-XOR technique [22], takes a significant amount of time. The naive XOR-tree [25] requires ρn OTs and ρn XOR gates; on the other hand, using a ρ-probe matrix of dimension $n \times cn$, with $c \ll \rho$, requires cn OTs but cn^2 XOR gates. We observe that this quadratic blowup becomes prohibitive as P_2's input size increases: for a 4096-bit input, it takes more than 3 s to compute *just* the XORs in the ρ-probe matrix of Lindell and Riva [29] across all circuits. Further, it also introduces a large memory overhead: it takes gigabytes of memory just to store the matrix for 65,536-bit inputs.

In the following we introduce two new techniques to both asymptotically reduce the number of XOR gates required and concretely reduce the hidden constant factor in the ρ-probe matrix.

A General Transformation to a Sparse Matrix. We first asymptotically reduce the number of XORs needed. Assuming a ρ-probe matrix with dimensions $n \times cn$, we need $c\rho n^2$ XOR gates to process the ρ-probe matrices across all ρ circuits. Our idea to avoid this quadratic growth in n is to break P_2's input into small chunks, each of size k. When computing the random input y', or recovering y in the garbled circuits, we process each chunk individually. By doing so, we reduce the complexity to $\rho \cdot \frac{n}{k} c(k)^2 = ck\rho n$. By choosing $k = 2\rho$, this equates to a $51\times$ decrease in computation even for just 4096-bit inputs. This also eliminates the memory issue, since we only need a very small matrix for any input size.

A Better ρ-probe Matrix. After applying the above technique, our problem is reduced to finding an efficient ρ-probe matrix for k-bit inputs for some small k, while maintaining a small blowup c. We show that a combination of the previous solutions [25, 29] with a new tighter analysis results in a better solution, especially for small k. Our solution can be written as $A = [M \| I_k]$, where $M \in \{0, 1\}^{k \times (c-1)k}$ is a random matrix and I_k is an identity matrix of dimension k. The use of I_k makes it easy to find a random y' such that $y = Ay'$ for any y, and ensures that A is full rank [29]. However, we show that it also helps to reduce c. The key idea is that the XOR of any i rows of A has Hamming weight at least i, contributed by I_k, so we do not need as much Hamming weight from the random matrix as in prior work [25].

In more detail, for each $S \subseteq [k]$, denote $M_S := \bigoplus_{i \in S} M_i$ and use random variable X_S to denote the number of ones in M_S. In order to make A a ρ-probe matrix, we need to ensure that $X_S + |S| \geq \rho$ for any $S \subseteq [k]$, because XORing any $|S|$ rows from I_k gives us a Hamming weight of $|S|$.

X_S is a random variable with distribution $\mathrm{Bin}(ck - k, \frac{1}{2})$. Therefore, we can compute the probability that A is not a ρ-probe matrix as follows:

$$
\begin{aligned}
\Pr[A \text{ is bad}] &= \Pr\left[\bigcup_{S \subseteq [k]} X_S < \rho - |S| \right] \\
&\leq \sum_{S \subseteq [k]} \Pr[X_S < \rho - |S|] \\
&= \sum_{S \subseteq [k]} cdf(\rho - |S| - 1) = \sum_{i=1}^{k} \binom{k}{i} cdf(\rho - i - 1),
\end{aligned}
$$

Table 3. Values of c as a function of chunk size k for an ρ-probe matrix with $\rho = 40$.

Scheme	k						
	40	65	80	103	143	229	520
LP07 [25]	6.66	4.1	4	4	4	4	4
sS13 [35]	7.95	5.2	4.5	4.1	3.2	2.4	1.6
This work	5.68	4	3.5	3	2.5	2	1.5

where $cdf()$ is the cumulative distribution function for $\mathrm{Bin}(ck - k, \frac{1}{2})$. Now, for each k we can find the smallest c such that $\Pr[A \text{ is bad}] \leq 2^{-\rho}$; we include some results in Table 3. We see that our new ρ-probe matrix achieves smaller c than prior work [25,35]. Note that the number of XORs is $c\rho kn$ and the number of OTs is cn. In our implementation we use $k = 232$ and $c = 2$ to achieve the maximum overall efficiency.

Performance Results. See Fig. 6 for a comparison between our approach and the best previous scheme [35]. When the input is large, the cost of computing the ρ-probe matrix over all circuits dominates the overall cost. As we can see, our design is about $10\times$ better for 1,024-bit inputs and can be $1000\times$ better for 65,536-bit inputs. We are not able to compare beyond this point, because just storing the ρ-probe matrix for 262,144 bits for the prior work takes at least 8.59 GB of memory.

4.2 Other Optimizations

Oblivious Transfer with Hardware Acceleration. As observed by Asharov et al. [3], matrix transposition takes a significant amount of the time during the execution of OT extension. Rather than adopting their solution using cache-friendly matrix transposition, we found that a better speedup can be obtained by using matrix transposition routines based on Streaming SIMD Extensions (SSE) instructions [30]. The use of SSE-based matrix transposition in the OT extension protocol is also independently studied in a concurrent work by Keller et al. [20] in a multi-party setting.

Given a 128-bit vector of the form $a[0], \ldots, a[15]$ where each $a[i]$ is an 8-bit number, the instruction _mm_movemask_epi8 returns the concatenation of the highest bits from all $a[i]$s. This makes it possible to transpose a matrix of dimension 8×16 very efficiently in 15 instructions (8 instructions to "assemble the matrix" and 7 instructions to shift the vector left by one bit). By composing such an approach, we achieve very efficient matrix transposition, which leads to highly efficient OT extension protocols; see Sect. 5.1 for performance results.

Reducing OT Cost. Although our protocol requires three instantiations of OT, we only need to construct the base OTs once. The OTs in Steps 1 and 2 can be done together, and further, by applying the observation by Asharov et al. [3] that the "extension phase" can be iterated, we can perform more random OTs along with the OTs for Steps 1 and 2 to be used in the OTs of Step 3.

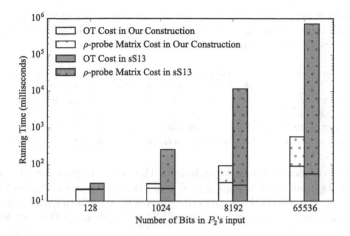

Fig. 6. Comparing the cost of our ρ-probe matrix design with the prior best scheme [35]. When used in a malicious 2PC protocol, computing the ρ-probe matrix needs to be done ρ times, and OT extension needs to process a cn-bit input because of the blowup of the input caused by the ρ-probe matrix.

Pipelining. Pipelining garbled circuits was first introduced by Huang et al. [16] to reduce memory usage and hence improve efficiency. We adopt a similar idea for our protocol. While as written we have P_2 conduct most of the correctness checks at the end of the protocol, we note that P_2 can do most of the checks much earlier. In our implementation, we "synchronize" P_1 and P_2's computation such that P_2's checking is pipelined with P_1's computation. Pipelining also enables us to evaluate virtually any sized circuit (as long as the width of the circuit is not too large). As shown in Sect. 5.4, we are able to evaluate a 4.3 billion-gate circuit without any memory issue, something that offline/online protocols [29] cannot do without using lots of memory or disk I/O.

Pushing Computation Offline. Although the focus of our work is better efficiency in the absence of pre-processing, it is still worth noting that several steps of our protocol can be pushed offline (i.e., before the parties' inputs are known) when that is an option. Specifically:

1. In addition to the base OTs, most of the remaining public-key operations can also be done offline. P_2 can send $(h, g_1) := (g^\omega, g^r)$ before knowing the input to P_1, who can compute the C_j values and half the D_j values. During the online phase, P_1 and P_2 only need to perform ρ exponentiations.
2. Garbled circuits can be computed, sent, and checked offline. P_2 can also decommit c_j^T to learn the output translation tables for the evaluation circuits.

Table 4. Performance of common functions over various networks. SE stands for "single execution." All numbers are in milliseconds. Offline time includes disk I/O. For online time, disk I/O is shown separately in the parentheses.

	n_1	n_2	n_3	$|C|$	Localhost			LAN			WAN		
					SE	Offline	Online	SE	Offline	Online	SE	Offline	Online
ADD	32	32	33	127	29	60	6 (0.2)	39	27	12 (0.2)	1060	474	697 (0.2)
AES	128	128	128	6,800	50	82	14 (2)	65	62	21 (3)	1513	867	736 (2)
SHA1	256	256	160	37,300	136	156	48 (32)	200	206	52 (27)	3439	2705	820 (20)
SHA256	256	256	256	90,825	277	356	85 (144)	438	497	92 (128)	6716	5990	856 (99)

Table 5. Performance of our building blocks. The first row gives the running time of P_2 recovering its input when using a ρ-probe matrix. The second row gives the running time of garbling, and the third row gives the running time for both garbling and sending. The remaining rows give the performance of OT and malicious OT extension.

Building block	Localhost	LAN	WAN
ρ-probe matrix for 2^{15}-bit input	5.8 ms	—	—
Garble 10^4 AES circuits	3.42 s	—	—
Garble and send 10^4 AES circuits	4.83 s	7.53 s	87.4 s
2^{10} malicious base OTs	113 ms	133 ms	249 ms
8×10^6 malicious OT extension	4.99 s	5.64 s	25.6 s

5   Implementation and Evaluation

We implemented our protocol in C++ using RELIC [2] for group operations, OpenSSL libssl for instantiating the hash function, and libgarble for garbling [15]. We adopted most of the recent advances in the field [4,5,8,29,39] as well as the optimizations introduced in Sect. 4. We instantiate the commitment scheme as $(\text{SHA-1}(x, r), r) \leftarrow \text{Com}(x)$, though when x has sufficient entropy we use SHA-1(x) alone as the commitment.

Evaluation Setup. All evaluations were performed with a single-threaded program with computational security parameter $\kappa = 128$ and statistical security parameter $\rho = 40$. We evaluated our system in three different network settings:

1. **localhost.** Experiments were run on the same machine using the loopback network interface.
2. **LAN.** Experiments were run on two c4.2xlarge Amazon EC2 instances with 2.32 Gbps bandwidth as measured by iperf and less than 1 ms latency as measured by ping.
3. **WAN.** Experiments were run on two c4.2xlarge Amazon EC2 instances with total bandwidth throttled to 200 Mbps and 75 ms latency.

All numbers are average results of 10 runs. We observed very small variance between multiple executions.

Table 6. Single execution performance. All numbers are in milliseconds. Numbers for [29] were obtained by adapting their implementation to the single-execution setting, using the same hardware as our results. Numbers for [1] are taken from their paper and are for a single execution, not including any I/O time.

	Our Protocol	[29]	[1]
ADD	39	1034	—
AES	65	1442	5860
SHA1	200	2007	—
SHA256	438	2621	7580

5.1 Subprotocol Performance

Because of the various optimizations mentioned in Sect. 4, as well as a carefully engineered implementation, many parts of our system perform better than previously reported implementations. We summarize these results in Table 5.

The garbling speed is about 20 million AND gates per second. When both garbling and sending through localhost, this reduces to 14 million AND gates per second due to the overhead of sending all the data through the loopback interface. Over LAN the speed is roughly 9.03 million gates per second, reaching the theoretical upper bound of $2.32 \cdot 10^9/256 = 9.06 \cdot 10^6$ gates per second.

For oblivious transfer, our malicious OT extension reports 5.64 s for 8 million OTs. Our implementation takes 0.133 s for 1024 base OTs. We observe that when two machines are involved, bandwidth is the main bottleneck.

5.2 Overall Performance

We now discuss the overall performance of our protocol. Table 4 presents the running time of our protocol on several standard 2PC benchmark circuits for various network settings. For each network condition, we report a *single execution* running time, which includes all computation for one 2PC invocation, and an *offline/online* running time. In order to be comparable with Lindell and Riva [29], the offline time includes disk I/O and the online time does not; the time to preload all garbled circuits before the online stage starts is reported separately in parentheses. We tried to compare with the implementation by Rindal and Rosulek [33]; however, their implementation is inherently parallelized, making comparisons difficult. Estimations suggest that their implementation is faster than Lindell and Riva but still less efficient than ours in the single-execution setting.

In Table 6, we compare the performance of our protocol with the existing state-of-the-art implementations. The most efficient implementation for single execution of malicious 2PC without massive parallelization or GPUs we are aware of is by Afshar et al. [1]. They reported 5860 ms of computation time for AES and 7580 ms for SHA256, with disk and network I/O excluded, whereas we achieve 65 ms and 438 ms, respectively, with all I/O included. Thus our result is 17× to 90× better than their result, although ours includes network cost while theirs does not.

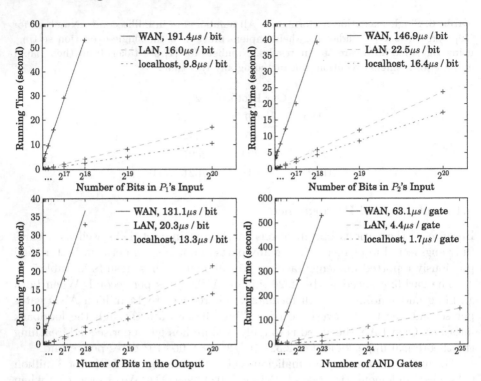

Fig. 7. The performance of our protocol while modifying the input lengths, output length and the circuit size. Input and output lengths are set to 128 bits initially and circuit size is set as 16,384 AND gates. Numbers in the figure show the slope of the lines, namely the cost to process an additional bit or gate.

We also evaluated the performance of the implementation by Lindell and Riva [29] using the same hardware with one thread and parameters tuned for single execution, i.e., 40 main circuits and 132 circuits for input recovery. Their implementation is about 3× to 4× better than Afshar et al., but still 6× to 26× slower than our LAN results.

Table 7. Scalability of our protocol. All numbers are in microseconds per bit or microseconds per gate.

	Localhost	LAN	WAN
Time per P_1's input bit	9.8	16	191.4
Time per P_2's input bit	16.4	22.5	146.9
Time per output bit	13.3	20.3	131.1
Time per AND gate	1.7	4.4	63.1

Table 8. Performance of our implementation on additional examples. *Running Time* reports the performance of our single execution over LAN; *Projected Time* is calculated using the formula in Sect. 5.3; *Total Comm.* is the total communication as measured by our implementation; and *Non-GC Comm.* is the percentage of communication not used for garbled circuits.

Example	n_1	n_2	n_3	$|C|$	Running time	Projected time	Total comm.	Non-GC comm.
16384-bit cmp	16,384	16,384	1	16,383	0.67 s	0.72 s	128 MB	84%
128-bit sum	128	128	128	127	0.04 s	0.03 s	1.8 MB	91%
256-bit sum	256	256	256	255	0.05 s	0.04 s	3.4 MB	90%
1024-bit sum	1024	1,024	1,024	1,023	0.08 s	0.09 s	11.2 MB	88%
128-bit mult	128	128	128	16,257	0.13 s	0.1 s	22.4 MB	7%
256-bit mult	256	256	256	65,281	0.4 s	0.37 s	86.6 MB	3%
Sort 1024 32-bit ints	32,768	32,768	32,768	1,802,240	9.43 s	9.8 s	2.6 GB	11.5%
Sort 4096 32-bit ints	131,072	131,072	131,072	10,223,616	53.7 s	52.7 s	14.2 GB	7.7%
1024-bit modular exp	1,024	1,024	1,024	4,305,443,839	5.3 h	5.26 h	5.5 TB	0.0002%

5.3   Scalability

In order to understand the cost of each component of our construction, we investigated the scalability as one modifies the input lengths, output length, and circuit size. We set input and output lengths to 128 bits and circuit size as 16,384 AND gates and increase each the variables separately. In Fig. 7, we show how the performance is related to these parameters.

Not surprisingly, the cost increases linearly for each parameter. We can thus provide a realistic estimate of the running time (in μs) of a given circuit of size $|C|$ with input lengths n_1 and n_2 and output length n_3 through the following formula (which is specific to the LAN setting):

$$T = 16n_1 + 22.5n_2 + 20.3n_3 + 4.4|C| + 23,000.$$

The coefficients for other network settings can be found in Table 7, with the same constant cost of the base OTs.

5.4   Additional Examples

Finally, in Table 8 we report the performance of our implementation in the LAN setting on several additional examples. We also show the projected time calculated based on the formula in the previous section. We observe that over different combinations of input, output and circuit sizes, the projected time matches closely to the real results we get.

We further report the total communication and the percentage of the communication not spent on garbled circuits. We can see the percentage stays low except when the circuit is linear to the input lengths.

Acknowledgments. The authors thank Samuel Ranellucci and Yan Huang for helpful discussions and comments.

References

1. Afshar, A., Mohassel, P., Pinkas, B., Riva, B.: Non-interactive secure computation based on cut-and-choose. In: Nguyen, P.Q., Oswald, E. (eds.) EUROCRYPT 2014. LNCS, vol. 8441, pp. 387–404. Springer, Heidelberg (2014). doi:10.1007/978-3-642-55220-5_22

2. Aranha, D.F., Gouvêa, C.P.L.: RELIC is an Efficient Library for Cryptography. https://github.com/relic-toolkit/relic

3. Asharov, G., Lindell, Y., Schneider, T., Zohner, M.: More ecient oblivious transfer and extensions for faster secure computation. In: 20th ACM Conference on Computer and Communications Security (CCS), pp. 535–548. ACM Press (2013)

4. Asharov, G., Lindell, Y., Schneider, T., Zohner, M.: More efficient oblivious transfer extensions with security for malicious adversaries. In: Oswald, E., Fischlin, M. (eds.) EUROCRYPT 2015. LNCS, vol. 9056, pp. 673–701. Springer, Heidelberg (2015). doi:10.1007/978-3-662-46800-5_26

5. Bellare, M., Hoang, V.T., Keelveedhi, S., Rogaway, P.: Efficient garbling from a fixed-key blockcipher. In: 2013 IEEE Symposium on Security and Privacy, pp. 478–492. IEEE (2013)

6. Bellare, M., Hoang, V.T., Rogaway, P.: Foundations of garbled circuits. In: ACM Conference on Computer and Communications Security (CCS), pp. 784–796. ACM Press (2012)

7. Brandão, L.T.A.N.: Secure two-party computation with reusable bit-commitments, via a cut-and-choose with forge-and-lose technique. In: Sako, K., Sarkar, P. (eds.) ASIACRYPT 2013. LNCS, vol. 8270, pp. 441–463. Springer, Heidelberg (2013). doi:10.1007/978-3-642-42045-0_23

8. Chou, T., Orlandi, C.: The simplest protocol for oblivious transfer. In: Lauter, K., Rodríguez-Henríquez, F. (eds.) LATINCRYPT 2015. LNCS, vol. 9230, pp. 40–58. Springer, Cham (2015). doi:10.1007/978-3-319-22174-8_3

9. Damgård, I., Lauritsen, R., Toft, T.: An empirical study and some improvements of the minimac protocol for secure computation. In: Abdalla, M., Prisco, R. (eds.) SCN 2014. LNCS, vol. 8642, pp. 398–415. Springer, Heidelberg (2014). doi:10.1007/978-3-319-10879-7_23

10. David, B.M., Nishimaki, R., Ranellucci, S., Tapp, A.: Generalizing efficient multi-party computation. In: Lehmann, A., Wolf, S. (eds.) ICITS 2015. LNCS, vol. 9063, pp. 15–32. Springer, Heidelberg (2015). doi:10.1007/978-3-319-17470-9_2

11. Frederiksen, T.K., Jakobsen, T.P., Nielsen, J.B.: Faster maliciously secure two-party computation using the GPU. In: Abdalla, M., Prisco, R. (eds.) SCN 2014. LNCS, vol. 8642, pp. 358–379. Springer, Heidelberg (2014). doi:10.1007/978-3-319-10879-7_21

12. Frederiksen, T.K., Jakobsen, T.P., Nielsen, J.B., Nordholt, P.S., Orlandi, C.: MiniLEGO: efficient secure two-party computation from general assumptions. In: Johansson, T., Nguyen, P.Q. (eds.) EUROCRYPT 2013. LNCS, vol. 7881, pp. 537–556. Springer, Heidelberg (2013). doi:10.1007/978-3-642-38348-9_32

13. Frederiksen, T.K., Nielsen, J.B.: Fast and maliciously secure two-party computation using the GPU. In: Jacobson, M., Locasto, M., Mohassel, P., Safavi-Naini, R. (eds.) ACNS 2013. LNCS, vol. 7954, pp. 339–356. Springer, Heidelberg (2013). doi:10.1007/978-3-642-38980-1_21

14. Goyal, V., Mohassel, P., Smith, A.: Efficient two party and multi party computation against covert adversaries. In: Smart, N. (ed.) EUROCRYPT 2008. LNCS, vol. 4965, pp. 289–306. Springer, Heidelberg (2008). doi:10.1007/978-3-540-78967-3_17

15. Groce, A., Ledger, A., Malozemoff, A.J., Yerukhimovich, A.: CompGC: efficient offline/online semi-honest two-party computation. Cryptology ePrint Archive, Report 2016/458 (2016). http://eprint.iacr.org/2016/458

16. Huang, Y., Evans, D., Katz, J., Malka, L.: Faster secure two-party computation using garbled circuits. In: 20th USENIX Security Symposium. USENIX Association (2011)

17. Huang, Y., Katz, J., Evans, D.: Efficient secure two-party computation using symmetric cut-and-choose. In: Canetti, R., Garay, J.A. (eds.) CRYPTO 2013. LNCS, vol. 8043, pp. 18–35. Springer, Heidelberg (2013). doi:10.1007/978-3-642-40084-1_2

18. Huang, Y., Katz, J., Kolesnikov, V., Kumaresan, R., Malozemoff, A.J.: Amortizing garbled circuits. In: Garay, J.A., Gennaro, R. (eds.) CRYPTO 2014. LNCS, vol. 8617, pp. 458–475. Springer, Heidelberg (2014). doi:10.1007/978-3-662-44381-1_26

19. Jawurek, M., Kerschbaum, F., Orlandi, C.: Zero-knowledge using garbled circuits: how to prove non-algebraic statements efficiently. In: ACM Conference on Computer and Communications Security (CCS), pp. 955–966. ACM Press (2013)

20. Keller, M., Orsini, E., Scholl, P.: MASCOT: faster malicious arithmetic secure computation with oblivious transfer, pp. 830–842. ACM Press (2016)

21. Kolesnikov, V., Mohassel, P., Rosulek, M.: FleXOR: flexible garbling for XOR gates that beats free-XOR. In: Garay, J.A., Gennaro, R. (eds.) CRYPTO 2014. LNCS, vol. 8617, pp. 440–457. Springer, Heidelberg (2014). doi:10.1007/978-3-662-44381-1_25

22. Kolesnikov, V., Schneider, T.: Improved garbled circuit: free XOR gates and applications. In: Aceto, L., Damgård, I., Goldberg, L.A., Halldórsson, M.M., Ingólfsdóttir, A., Walukiewicz, I. (eds.) ICALP 2008. LNCS, vol. 5126, pp. 486–498. Springer, Heidelberg (2008). doi:10.1007/978-3-540-70583-3_40

23. Kreuter, B., Shelat, A., Shen, C.H.: Billion-gate secure computation with malicious adversaries. In: USENIX Security Symposium, pp. 285–300. USENIX Association (2012)

24. Lindell, Y.: Fast cut-and-choose based protocols for malicious and covert adversaries. In: Canetti, R., Garay, J.A. (eds.) CRYPTO 2013. LNCS, vol. 8043, pp. 1–17. Springer, Heidelberg (2013). doi:10.1007/978-3-642-40084-1_1

25. Lindell, Y., Pinkas, B.: An efficient protocol for secure two-party computation in the presence of malicious adversaries. In: Naor, M. (ed.) EUROCRYPT 2007. LNCS, vol. 4515, pp. 52–78. Springer, Heidelberg (2007). doi:10.1007/978-3-540-72540-4_4

26. Lindell, Y., Pinkas, B.: Secure two-party computation via cut-and-choose oblivious transfer. In: Ishai, Y. (ed.) TCC 2011. LNCS, vol. 6597, pp. 329–346. Springer, Heidelberg (2011). doi:10.1007/978-3-642-19571-6_20

27. Lindell, Y., Pinkas, B., Smart, N.P.: Implementing two-party computation efficiently with security against malicious adversaries. In: Ostrovsky, R., Prisco, R., Visconti, I. (eds.) SCN 2008. LNCS, vol. 5229, pp. 2–20. Springer, Heidelberg (2008). doi:10.1007/978-3-540-85855-3_2

28. Lindell, Y., Riva, B.: Cut-and-choose yao-based secure computation in the online/offline and batch settings. In: Garay, J.A., Gennaro, R. (eds.) CRYPTO 2014. LNCS, vol. 8617, pp. 476–494. Springer, Heidelberg (2014). doi:10.1007/978-3-662-44381-1_27

29. Lindell, Y., Riva, B.: Blazing fast 2PC in the offline/online setting with security for malicious adversaries. In: 22nd ACM Conference on Computer and Communications Security (CCS), pp. 579–590. ACM Press (2015)

30. Mischasan: What is SSE good for? Transposing a bit matrix. https://mischasan.
    wordpress.com/2011/07/24/what-is-sse-good-for-transposing-a-bit-matrix/. Acce
    ssed 10 Dec 2015
31. Nielsen, J.B., Nordholt, P.S., Orlandi, C., Burra, S.S.: A new approach to prac-
    tical active-secure two-party computation. In: Safavi-Naini, R., Canetti, R. (eds.)
    CRYPTO 2012. LNCS, vol. 7417, pp. 681–700. Springer, Heidelberg (2012). doi:10.
    1007/978-3-642-32009-5_40
32. Pinkas, B., Schneider, T., Smart, N.P., Williams, S.C.: Secure two-party compu-
    tation is practical. In: Matsui, M. (ed.) ASIACRYPT 2009. LNCS, vol. 5912, pp.
    250–267. Springer, Heidelberg (2009). doi:10.1007/978-3-642-10366-7_15
33. Rindal, P., Rosulek, M.: Faster malicious 2-party secure computation with
    online/offline dual execution. In: USENIX Security Symposium, pp. 297–314.
    USENIX Association (2016)
34. Shelat, A., Shen, C.: Two-output secure computation with malicious adversaries.
    In: Paterson, K.G. (ed.) EUROCRYPT 2011. LNCS, vol. 6632, pp. 386–405.
    Springer, Heidelberg (2011). doi:10.1007/978-3-642-20465-4_22
35. Shelat, A., Shen, C.H.: Fast two-party secure computation with minimal assump-
    tions. In: ACM Conference on Computer and Communications Security (CCS),
    pp. 523–534. ACM Press (2013)
36. Wang, X., Malozemoff, A.J., Katz, J.: EMP-toolkit: efficient multiparty computa-
    tion toolkit. https://github.com/emp-toolkit
37. Wang, X., Ranellucci, S., Malozemoff, A.J., Katz, J.: Faster secure two-party
    computation in the single-execution setting. Cryptology ePrint Archive, Report
    2016/762 (2016). http://eprint.iacr.org/2016/762
38. Yao, A.C.C.: Protocols for secure computations. In: 23rd Annual Symposium on
    Foundations of Computer Science (FOCS), pp. 160–164. IEEE (1982)
39. Zahur, S., Rosulek, M., Evans, D.: Two halves make a whole—reducing data trans-
    fer in garbled circuits using half gates. In: Oswald, E., Fischlin, M. (eds.) EURO-
    CRYPT 2015. LNCS, vol. 9057, pp. 220–250. Springer, Heidelberg (2015). doi:10.
    1007/978-3-662-46803-6_8

Non-interactive Secure 2PC in the Offline/Online and Batch Settings

Payman Mohassel[1](✉) and Mike Rosulek[2]

[1] Visa Research, Palo Alto, USA
pmohasse@visa.com
[2] Oregon State University, Corvallis, USA
rosulekm@eecs.oregonstate.edu

Abstract. In cut-and-choose protocols for two-party secure computation (2PC) the main overhead is the number of garbled circuits that must be sent. Recent work (Lindell and Riva; Huang et al. Crypto 2014) has shown that in a batched setting, when the parties plan to evaluate the same function N times, the number of garbled circuits per execution can be reduced by a $O(\log N)$ factor compared to the single-execution setting. This improvement is significant in practice: an order of magnitude for N as low as one thousand. Besides the number of garbled circuits, communication round trips are another significant performance bottleneck. Afshar et al. (Eurocrypt 2014) proposed an efficient cut-and-choose 2PC that is round-optimal (one message from each party), but in the single-execution setting.

In this work we present new malicious-secure 2PC protocols that are round-optimal and also take advantage of batching to reduce cost. Our contributions include:

- A 2-message protocol for batch secure computation (N instances of the same function). The number of garbled circuits is reduced by a $O(\log N)$ factor over the single execution case. However, other aspects of the protocol that depend on the input/output size of the function do not benefit from the same $O(\log N)$-factor savings.
- A 2-message protocol for batch secure computation, in the random oracle model. All aspects of this protocol benefit from the $O(\log N)$-factor improvement, except for small terms that do not depend on the function being evaluated.
- A protocol in the offline/online setting. After an offline preprocessing phase that depends only on the function f and N, the parties can securely evaluate f, N times (not necessarily all at once). Our protocol's online phase is only 2 messages, and the total online communication is only $\ell + O(\kappa)$ bits, where ℓ is the input length of f and κ is a computational security parameter. This is only $O(\kappa)$ bits more than the information-theoretic lower bound for malicious 2PC.

M. Rosulek—Partially supported by NSF awards 1149647 and 1617197.

J.-S. Coron and J.B. Nielsen (Eds.): EUROCRYPT 2017, Part III, LNCS 10212, pp. 425–455, 2017.
DOI: 10.1007/978-3-319-56617-7_15

1   Introduction

Secure two-party computation (2PC) allows two parties to compute a function of their inputs without revealing any other information. Yao's garbled circuit protocol [39] provides an efficient general-purpose 2PC in presence of semi-honest adversaries and has been the subject of various optimization [22, 23, 34, 41]. The most common approach for obtaining security against malicious adversaries is the cut-and-choose paradigm wherein multiple circuits are garbled and a subset of them are opened to check for correctness, while the remaining circuits are evaluated to obtain the final output. A large body of work has focused on making cut-and-choose 2PC more efficient by (i) reducing the number of garbled circuits [15, 24–26, 36], (ii) minimizing rounds of interaction [1, 9], and (iii) optimizing techniques for checking consistency of inputs to the computation [25, 27–29, 36, 37].

Until recently, all protocols for cut-and-choose 2PC required at least 3λ garbled circuits in order to ensure the majority output is correct with probability $1 - 2^{-\lambda}$. Lindell [24] proposed a new technique for recovering from cheating that only relied on evaluation of one correct garbled circuit, hence reducing the number of garbled circuits to λ. The recent independent work of Lindell and Riva [26], and Huang et al. [15], building on ideas from earlier work of [11, 31], showed how to further reduce the number of circuits to $\lambda/O(\log N)$ per execution, when performing N instances of 2PC for the same function. This leads to significant reduction in amortized communication and computation. For example for $N = 1024$, only 4 garbled circuits per execution are sufficient to achieve cheating probability of less than 2^{-40}. However, the proposed constructions require at least 4 rounds of interaction between the parties, rendering round complexity the main bottleneck when communicating over the internet as demonstrated in the recent implementation of [27].

Previous Two-Round 2PC and Shortcomings. A **non-interactive secure computation (NISC)** protocol for general computation can be constructed from Yao's garbled circuit, non-interactive zero-knowledge proofs (NIZK), and fully-secure one-round oblivious transfer (OT): P_1, who is the evaluator of the circuit, sends the first message of the OT protocol. P_2, who is the circuit constructor, returns a garbled circuit, the second message of the OT protocol, and a NIZK proof that its message is correct. (See, for example, [7, 14] for such protocols.) Unfortunately, the NIZK proof in this case requires a *non black-box* use of cryptographic primitives (namely, it must prove the correctness of each encryption in each gate of the circuit).

Efficient NISC protocols that do not require such non black-box constructions are presented in [17] based on the MPC-in-the-head technique of [18]. The complexity of the NISC protocol of [17] is $|C| \cdot poly(\log(|C|), \log(\lambda)) + depth(C) \cdot poly(\log(|C|), \lambda)$ invocations of a Pseudo-Random Generator (PRG), where C is a boolean circuit that computes the function of interest. (Another protocol presented in that work uses only $O(|C|)$ PRG invocations, but is based on a relaxed security notion.) Although the protocols in [17] are very efficient asymptotically, their practicality is unclear and left as an open question in [17]. For instance, the protocols combine several techniques that are very efficient asymptotically, such as scalable MPC and using expanders in a non black-box way, each of which contributes large constant factors to the concrete complexity.

Afshar et al. [1], proposed a cut-and-choose 2PC with only two rounds of interaction, with concrete efficiency comparable to the state-of-the-art single-execution cut-and-choose 2PC. It is not clear how to adapt their solution to the batched execution setting to achieve better amortized efficiency. In particular, in batched cut-and-choose protocols, the sender generates and sends many garbled circuits. The receiver chooses a random subset of these circuits to check, and randomly arranges the remaining circuits into *buckets*. The kth bucket contains the circuits that will be evaluated in the kth execution. A main step for turning such a protocol into a NISC is a non-interactive mechanism for the "cut-and-choose" step and the bucket assignment. While in the single-execution setting this can be easily done using one OT per circuit [1], the task is more challenging when assigning many circuits to N buckets.

However, a bigger challenge is that the sender has no way of knowing *a priori* to which execution (i.e., which bucket) the ith circuit will be assigned. We must design a mechanism whereby the receiver can learn garbled inputs of the ith circuit that encode the input to kth execution, *if and only if* circuit i is assigned to the kth execution. Furthermore, in a typical cut-and-choose protocol, different mechanisms must be designed for checking consistency of the sender's and the receiver's inputs. For example, the sender must convince the receiver that all circuits in a particular bucket are evaluated with the same input, even though the sender does not know in advance the association between circuits and inputs (and other sibling circuits). Similarly, cheating-recovery enables the receiver to learn the sender's input if two valid circuits return different outputs in the same bucket. However, existing techniques implicitly assume the sender knows all circuits assigned to the same bucket, for example, by using the same wire labels on output wires of those circuits.

To further highlight the difficulty, consider a simple solution where for each garbled circuit GC_i, the sender prepares its garbled inputs and the input-consistency gadgets for all N possible bucket assignments and all inputs x_k, $k \in [N]$. Then, for each circuit parties perform a 1-out-of-N OT where the receiver's input is the index k such that GC_i is assigned to bucket k, and the sender's inputs are the N input garblings/gadgets for GC_i. First, note that this is prohibitively expensive as it needs to be repeated for each circuit and incurs a multiplicative factor of $N^2\lambda/\log N$ on input-related gadgets/commitments (compared to the expected $N\lambda/logN$ or $N\lambda$). Second, this still does not address how to route receiver's garbled input, and more importantly, how to incorporate cheating-recovery techniques since the existing solutions also depend on the choice of sibling circuits that are assigned to the same bucket.

Our Results. As discussed above, with current techniques, one either obtains a two-round cut-and-choose 2PC that requires λ circuits per execution or a multiple-round 2PC that requires $O(\lambda/\log N)$ circuits per execution. *The main question motivating this work is whether we can obtain the best of both worlds while maintaining concrete efficiency.* Our results are several protocols that achieve different combinations of features (summarized in Table 1):

- We propose the first cut-and-choose 2PC with two rounds of interaction that only requires $O(N\lambda/\log N)$ garbled circuits to evaluate a function N times in a single batch. The protocol is both asymptotically and concretely efficient and can be instantiated in the standard model and using only symmetric-key operations in the OT-hybrid model.
- In the above protocol, the number of garbled circuits is reduced by a factor $O(\log N)$ compared to the single-execution setting. This is the only part of the protocol whose cost depends on the size of the circuit for f. However, several mechanisms in the protocol depend on the input/output length of f, and these mechanisms scale as $O(N\lambda)$ instead of $O(N\lambda/\log N)$.

  We therefore describe a two-round protocol for batched 2PC *in the random oracle model*, in which all aspects of the protocol benefit from batching. That is, apart from protocol features that do not depend on f at all, the entire protocol scales with $O(\kappa N/\log N)$ rather than $O(\kappa N)$. Unfortunately, the number of garbled circuits now depends on the (larger) computational security parameter κ rather than the statistical security parameter λ as before. This is due to technical reasons (see Sect. 6.2).
- In the offline-online setting, parties perform dedicated offline preprocessing that depends only on the function f and number of times N they would like to evaluate it. Then, when inputs are known, the parties can engage in an online phase to securely obtain the output. The online phases need not be performed in a single batch—they can happen asynchronously.

  We describe a 2PC protocol in this offline-online setting. As in other offline-online protocols [15,26,35], the total costs are reduced by a $O(\log N)$ factor (and the number of circuits is dependent on the statistical security parameter λ). Unlike previous protocols, our online phase consist of only 2 rounds. The total online communication can be reduced to only $|x| + |y| + O(\kappa)$ bits, where x is the sender's input, y is the receiver's input, and κ is a computational security parameter. We note that $|x| + |y|$ bits of communication are required for malicious-secure 2PC,[1] so our protocol has nearly optimal online communication complexity.

Our Techniques. Our main NISC construction takes advantage of a two-round protocol for obliviously mapping garbled circuits and their associated input/output gadgets to many buckets while hiding from the garbler the bucket assignment and consequently what inputs a circuits would be evaluated on. As a result, we need to extend and adapt all existing techniques for obtaining garbled inputs, performing input consistency checks and cheating-recovery to this new setting.

---

[1] Each party must send a message at least as long as his/her input, otherwise it is information-theoretically impossible for the simulator to extract a corrupt party's input.

Table 1. Asymptotic efficiency of our protocols. n_{in}, n_{out} are number of input/output wires. $n_{both} = n_{in} + n_{out}$. Rounds are listed as offline + online. κ is the computational security parameter, and λ is the statistical security parameter.

	NISC	RO-NISC	Online-offline
Rounds	$0 + 2$	$0 + 2$	$2 + 2$
# GC	$O(N\lambda/\log N)$	$O(N\kappa/\log N)$	$O(N\lambda/\log N)$
# plain commit	$O(n_{in}N\lambda/\log N)$	$O(n_{in}N\kappa/\log N)$	$O(n_{in}N\lambda/\log N)$
# hom commit	$O(n_{out}N\lambda/\log N)$	$O(n_{out}N\kappa/\log N)$	$O(n_{out}N\lambda/\log N)$
OSN OTs	$O(n_{both}N\lambda)$	-	-
Other OTs	$O(n_{in}N)$	$O(n_{in}N)$	$O(n_{in}N)$

Another main ingredient of our constructions is a homomorphic commitment scheme with homomorphic properties on the decommitment strings. Such a primitive can be efficiently instantiated using both symmetric-key and public-key primitives, trading-off communication for computation. We show how such a commitment scheme combined with an oblivious switching network protocol [30] allows a sender to obliviously open linear relations between various committed values without *a priori* knowledge of the choice of committed values. See Sect. 4.1 for a detailed overview of the techniques used in our main protocol.

2 Preliminaries

2.1 Garbled Circuits

Garbled Circuits were first introduced by Yao [40]. A garbling scheme consists of a garbling algorithm that takes a random seed σ and a function f and generates a garbled circuit F and a decoding table dec; the encoding algorithm takes input x and the seed σ and generates garbled input $\widehat{x}$; the evaluation algorithm takes $\widehat{x}$ and F as input and returns the garbled output $\widehat{z}$; and finally, a decoding algorithm that takes the decoding table dec and $\widehat{z}$ and returns $f(x)$. We require the garbling scheme to satisfy the standard security properties formalized in [6]. Our construction uses the garbling scheme in a black-box way and hence can incorporate all recent optimizations proposed in the literature. In the offline-online setting, the scheme needs to adaptively secure in the sense of [5].

2.2 Commitments

A standard commitment scheme Com allow a party to commit to a message m, by computing $C = \mathsf{Com}(m; d)$ using a decommitment d. To open a commitment $C = \mathsf{Com}(m; d)$, the committer reveals (m, d). The verifier recomputes the commitment and accepts if it obtains the same C, and rejects otherwise. We require standard standalone security properties of a commitment scheme:

- *Hiding:* For any a, b, the distributions $\mathsf{Com}(a; d_a)$ and $\mathsf{Com}(a; d_b)$, induced by random choice of d_a, d_b, are indistinguishable.
- *Binding:* It is computationally infeasible to compute $m \neq m', d, d'$ such that $\mathsf{Com}(m; d) = \mathsf{Com}(m'; d')$.

Homomorphic Commitments. In a homomorphic commitment scheme HCom, we further require the scheme to be homomorphic with respect to an operation on the message space denoted by $\oplus$. In particular given two commitments $C_a = \mathsf{HCom}(a, d_a)$ and $C_b = \mathsf{HCom}(b, d_b)$, the committer can open $a \oplus b$ (revealing nothing beyond $a \oplus b$) by giving $d_a \oplus d_b$.

Note that here we have assumed that the homomorphic operation also operates on the decommitment values. This is indeed the case for most instantiations of homomorphic commitments, as we discuss in Sect. 5.2. The security properties are extended for homomorphic commitments as follows:

- *Hiding:* For a set of values $v_1, \ldots, v_n$ and a set $S \subseteq [n]$, define $v(S) = \oplus_{i \in S} v_i$. Then, informally, the hiding property is that commitments to $v_1, \ldots, v_n$ and openings of $v(S_1), \ldots, v(S_k)$ reveal no more than the $v(S_1), \ldots, v(S_k)$ values. More formally, for all $\boldsymbol{v} = (v_1, \ldots, v_n), \boldsymbol{v'} = (v'_1, \ldots, v'_n)$, and sets $S_1, \ldots, S_k$ where $v(S_j) = v'(S_j)$ for each j, the following distributions are indistinguishable:

$$(\mathsf{Com}(v_1; d_1), \ldots, \mathsf{Com}(v_n; d_n); d(S_1), \ldots, d(S_k)),$$
$$\text{and } (\mathsf{Com}(v'_1; d_1), \ldots, \mathsf{Com}(v'_n; d_n); d(S_1), \ldots, d(S_k))$$

- *Binding:* Intuitively, it should be hard to decommit to inconsistent values. More formally, it should be hard to generate commitments $C_1, \ldots, C_n$ and values $\{(S_j, d_j, m_j)\}_j$ such that d_j is a valid decommitment of $\bigoplus_{i \in S_j} C_i$ to the value m_j, and yet there is no solution (in the x_i's) to the system of equations defined by equations: $\left\{ \bigoplus_{i \in S_j} x_i = m_j \right\}_j$.

2.3 Probe-Resistant Input Encoding

In garbled-circuit-based 2PC, the receiver uses oblivious transfers to pick up his garbled inputs. A standard problem is that a malicious sender can give incorrect wire labels in these OTs. Furthermore, if the sender gives an incorrect value for only one of the pair of wire labels, then the receiver picks up incorrect values (and presumably aborts), *based on his private input.* Hence, a malicious sender causes the receiver to abort, depending on the receiver's private input. This cannot be simulated in the ideal world, so it is indeed an attack.

A standard way to deal with this is the idea of a probe-resistant matrix:

Definition 1 [25,37]. *A boolean matrix $M \in \{0, 1\}^{n \times n'}$ is λ-**probe resistant** if for all $R \subseteq [n]$, the Hamming weight of $\bigoplus_{i \in R} M_i$ is at least λ, where M_i denotes the ith row of M.*

The idea is for Bob, with input y to choose a random encoding $\tilde{y}$ such that $M\tilde{y} = y$. Then the parties will evaluate the function $\tilde{f}(x, \tilde{y}) = f(x, M\tilde{y}) = f(x, y)$. The matrix M can be public, so the computation $M\tilde{y}$ uses only XOR operations (free in a typical garbling scheme [23]).

Suppose the parties perform n' OTs. In each OT the sender provides two items, and the receiver uses the bits of $\tilde{y}$ to select one. The items can be either *good* or *bad*, and the receiver will abort if it receives any *bad* item. If for any single OT, both inputs are *bad*, then the receiver will always abort. However, if every OT has at least one *good* item, then the receiver will abort based on $\tilde{y}$.

Lemma 2 [25,37]. *Suppose M is λ-probe-resistant, and fix a set of sender's inputs to the OTs as described above. Let $P(y)$ denote the probability that the receiver aborts (i.e., sees a bad item) when it chooses a random $\tilde{y}$ such that $M\tilde{y} = y$, and uses $\tilde{y}$ as the choice bits in the OTs. Then for all y, y', we have $|P(y) - P(y')| = O(2^{-\lambda})$.*

Hence, the abort probability is *nearly independent* of the receiver's input, when using this probe-resistant technique.

2.4 Secure Computation and the NISC Model

We consider security in the *universal composability* framework of Canetti [8]. We refer the reader to that work for detailed security definitions. Roughly speaking, the definition considers a *real interaction* and an *ideal one*.

In the *real* interaction, parties interact in the protocol. Their inputs are chosen by an *environment*, and their outputs are given to the environment. An adversary who attacks the protocol takes control of one of the parties and causes it to arbitrarily deviate from the protocol. The adversary may also communicate arbitrarily with the environment before/during/after the protocol interaction.

In the *ideal* interaction, parties simply forward their inputs to a trusted party called a *functionality*. They receive output from the functionality which they forward to the environment.

A protocol **UC-securely realizes** an ideal functionality if, for all adversaries attacking the real world, there exists an adversary in the ideal world (called a simulator) such that for all environments, the view of the environment is indistinguishable between the real and ideal interactions.

NISC. Ishai et al. [17] defined a special model of secure computation called *non-interactive secure computation (NISC)*. A protocol is NISC if it consists of a single message from one party to the other, possibly with some (static, parallel) calls to some ideal functionality (typically an oblivious transfer functionality).

One can think of replacing the calls to an ideal oblivious transfer functionality with a two-round secure OT protocol (like that of [33]). Then the NISC protocol becomes a two-message protocol: in the first message the OT receiver sends the first OT protocol message. In the second message, the OT sender sends the OT response along with the single NISC protocol message.

2.5   Correlation Robust

One of our techniques requires a correlation-robust hash function. This property was defined in Ishai et al. [16].

Definition 3 [16]. *A function $H : \{0,1\}^\kappa \to \{0,1\}^n$ is **correlation robust** if $F(s, x) = H(x \oplus s)$ is a weak PRF (with s as the seed). In other words, the distribution of: $\left(x_1, \ldots, x_m; H(x_1 \oplus s), \ldots, H(x_m \oplus s) \right)$ is pseudorandom, for random choice of x_i's and s.*

2.6   Compressed Garbled Inputs

Applebaum et al. [2] described a technique for randomized encodings with low online complexity. In the language of garbled circuits, this corresponds to a way to compress garbled inputs in the online phase of a protocol, at the expense of more data in an offline phase. We abstract their primitive as a **garbled input compression** scheme, as follows.

Let $e = (e_{1,0}, e_{1,1}, \ldots, e_{n,0}, e_{n,1})$ be a set of wire labels (i.e., $e_{j,b}$ is the wire label encoding value b on wire j). In a traditional protocol, the garbled encoding of a string x is $(e_{1,x_1}, \ldots, e_{n,x_n})$, which is sent in the online phase of the protocol. Using the approach of [2], we can do the following to reduce the online cost:

- In an offline phase, the garbler runs $\mathsf{Compress}(e) \to (sk, \widehat{e})$, and sends $\widehat{e}$ to the evaluator.
- In the online phase, when garbled encoding of x is needed, the garbler runs $\mathsf{Online}(sk, x) \to \widehat{x}$ and sends $\widehat{x}$ to the evaluator.
- The evaluator runs $\mathsf{Decompress}(\widehat{e}, x, \widehat{x})$, which returns the garbled encoding $(e_{1,x_1}, \ldots, e_{n,x_n})$.

The security of the compression scheme is that $(\widehat{e}, \widehat{x}, x)$ can be simulated given only the garbled encoding $(e_{1,x_1}, \ldots, e_{n,x_n})$. In other words, the compressed encoding reveals no more than the expected garbled encoding.

In a traditional garbling scheme, the size of the garbled encoding is $n\kappa$. Applebaum et al. [2] give constructions where the online communication $\widehat{x}$ has size only $n + O(\kappa)$. These constructions are proven secure under a variety of assumptions (DDH, LWE, RSA). We refer the reader to their paper for details.

3   Switching Networks

3.1   Definitions

A **switching network** is a circuit of gates that we call **switches**, whose behavior is described below. The network as a whole has n *primary* inputs (strings, or more generally, elements from some group) and p *programming* inputs (bits). All wires in the network have no branching. Each **switch** has two inputs and two outputs. A switch is parameterized by an index $j \in [p]$. The behavior of an individual switch is that when its primary input wires have values (X, Y) and the jth programming input to the circuit is 0, then the outputs are (X, Y); otherwise (the jth programming input is 1) the outputs are (Y, X).

Note that many switches can be tied to the same programming input. When $\mathcal{S}$ is a switching network and π is a programming string, we let $\mathcal{S}^\pi(X_1, \ldots, X_n)$ denote the output of the switching network when the primary inputs are $X_1, \ldots, X_n$ and its programming input is π.

3.2   Oblivious Switching Network Protocol

In the full version, we describe the **oblivious switching network (OSN) protocol** of [30]. The idea is that the parties agree on a switching network $\mathcal{S}$. The sender has inputs $(X_1, \ldots, X_n)$ and $(Z_1, \ldots, Z_m)$. The receiver has input π, and learns $\mathcal{S}^\pi(X_1, \ldots, X_n) \oplus (Z_1, \ldots, Z_m)$. The sender learns nothing.

The cost of the protocol is essentially a 1-out-of-2 OT (for values on the switching network's wires) for each switch in the network. All of the OTs can be performed in parallel, and hence the protocol can be realized as a NISC protocol in the OT-hybrid model.

This protocol will be used as a subroutine in our main NISC functionality. Yet we do *not* abstract the OSN protocol in terms of an ideal functionality. This is because the protocol does not ensure that a malicious sender acts consistently with the switching network. However, this turns out to be non-problematic in our larger NISC protocol. We simply abstract out the properties of this subprotocol as follows:

Observation 4. *When the sender is honest and the receiver is corrupt, the simulator can extract the corrupt receiver's programming string π. When the underlying OTs are performed in parallel, the simulator extracts π before simulating any outputs from these OTs.*

Observation 5. *When the sender is honest, the receiver's view can be simulated given only π and the output $\mathcal{S}^\pi(X_1, \ldots, X_n) \oplus (Z_1, \ldots, Z_m)$.*

While we described the OSN protocol for the $\oplus$ operation, we note that it is easy to replace $\oplus$ for any group operations. In particular, we also use the protocol in scenarios where $\oplus$ represent homomorphic operations on message domain and/or decommitment domain of a homomorphic commitment.

4   Batched NISC

In this section we describe a protocol for securely evaluating many instances of the same function f in a single batch. The ideal functionality we achieve is described in Fig. 1.

We let N denote the number of instances of 2PC being executed, $\widehat{N}$ the number of garbled circuits computed and B the number of garbled circuits assigned to each execution/bucket. For a full treatment of these parameters, we refer the

---

Parameters: A function f and number N of instances.

Behavior: On input $(y_1, \ldots, y_N)$ from the receiver, internally record these values and send (input) to the sender. Later, on input $(x_1, \ldots, x_N)$ from the sender, do the following. If $x_i = \bot$ for any i, then give output $\bot$ to the receiver. Otherwise compute $z_i = f(x_i, y_i)$ for $i \in [N]$ and give $(z_1, \ldots, z_N)$ to the receiver.

Fig. 1. Ideal functionality for batch 2PC

reader to [26]. For our purposes, we will assume that the parameters satisfy the following combinatorial property: The adversary generates $\widehat{N}$ items, some *good*, some *bad*. The items are randomly assigned into N buckets of B items each. The remaining $\widehat{N} - NB$ items are *opened*. Then the probability that all opened items are *good* while there exists a bucket with *all bad* items is at most $2^{-\lambda}$. Here λ is a statistical security parameter (often $\lambda = 40$). Asymptotically, $\widehat{N} = O(\lambda N / \log N)$ and $B = O(\lambda / \log N)$.

Regarding our conventions for notation: we use i to index a garbled circuit, j to index a wire in the circuit computing f, k to index a bucket (an evaluation of f, or the special "check bucket" defined below), and l to index a position within a bucket. We let SendInpWires, RecvInpWires, OutWires denote the set of wire indices corresponding to inputs of Alice, inputs of Bob, and outputs of f, respectively.

4.1   Overview of Techniques

Bucket-Coupling via Switching Networks. Recall that the receiver must choose randomly which circuits are checked, and which circuits are mapped to each bucket. For simplicity, let us say that checked circuits are assigned to "bucket #0." Recall that the cut-and-choose statistical bounds require the receiver to choose a random assignment of circuits into buckets. Suppose the cut-and-choose parameters call for N buckets, B circuits per bucket, and $\widehat{N} > NB$ total circuits (with $\widehat{N} - NB$ circuits being checked). Think of this process as first randomly permuting the $\widehat{N}$ circuits, assigning the first $\widehat{N} - NB$ circuits to bucket #0, assigning the next B circuits to bucket #1, and so on. More formally, we can define public functions bkt and pos so that, *after randomly permuting* the circuits, the ith circuit will be the pos(i)'th circuit placed in bucket bkt(i).

A main building block in our NISC protocol is one we call **bucket coupling**, which is a non-interactive way to bind information related to garbled circuits to information related to a particular bucket, under a bucketing-assignment chosen by the receiver. Suppose the parties use the OSN subprotocol of Sect. 3, on a universal switching network $\mathcal{S}$, where the sender's input is $(A_1, \ldots, A_{\widehat{N}}), (B_1, \ldots, B_{\widehat{N}})$, and the receiver's input is the programming string for a random permutation π. Then the receiver will learn $A_{\pi(i)} \oplus B_i$.

Interpret π as the receiver's random permutation of circuits when assigning circuits to buckets as described above. Then we can interchangeably use B_v and

$B_{\mathsf{bkt}(v),\mathsf{pos}(v)}$, since there is a one-to-one correspondence between these ways of indexing. We have the following generic functionality:

Bucket coupling: The sender has an item A_i for each circuit i, and an item $B_{k,l}$ for each position l in the kth bucket. The receiver holds a bucketing assignment π. The receiver learns $A_i \oplus B_{k,l}$ if and only if π assigns circuit i to position l of bucket k.

We can perform many such couplings, *all with respect to the same permutation* π. Simply imagine a switching network that is a disjoint union of many universal switching networks, but where corresponding switches are programmed by the same programming bit (this is enforced in the OSN protocol).

Of course, our OSN protocol does not guarantee consistent behavior by the sender. Furthermore, the sender might not even use the expected inputs to the OSN protocol. However, we argue that these shortcomings do not lead to problems in our larger NISC protocol. Intuitively, the worst the sender can do is to cause inconsistent outputs for the receiver in a way that depends on the receiver's choice of bucket-assignments π. But π is chosen independently of his input to the *NISC protocol!* Hence the simulator can exactly simulate the abort probability of the honest receiver, by sampling a uniform π just as the honest receiver does.

Basic Cut-and-Choose. The sender Alice generates $\widehat{N}$ garblings $\{F_i\}_i$ of f (along with some other associated data, described below). Let σ_i denote the seed used to generate all the randomness for the ith circuit. The parties can perform a coupling whereby **Bob learns σ_i if and only if circuit i is assigned to bucket 0** (in the notation above, $A_i = \sigma_i$ and $B_{0,l} = 0^\kappa$ and $B_{k,l}$ random for $k \neq 0$). Then every circuit mapped to bucket 0 (i.e., every check circuit) can be verified by Bob.

Delivering the Receiver's Garbled Input. Let RecvInpWires denote the set of input wires corresponding to Bob's input to f. Let $\mathsf{in}_{i,j,b}$ denote the input wire label on the jth wire of the ith circuit, encoding logical bit b. When circuit i is mapped to bucket k, we must let Bob obtain his garbled input value $\mathsf{in}_{i,j,b}$, where b is the jth bit of Bob's input for the kth execution. Recall that the association between circuits (i) and executions (k) is not known to Alice.

Alice commits to each input wire label as follows, and sends the commitments to Bob:

$$C_{i,j,b}^{\mathsf{in}} \leftarrow \mathsf{Com}(\mathsf{in}_{i,j,b}; d_{i,j,b}^{\mathsf{in}})$$

The randomness for these commitments is derived from σ_i, so that the commitments can be checked by Bob if circuit i is assigned to be a check-circuit.

Then, for each execution $k \in [N]$ and each $j \in \mathsf{RecvInpWires}$, Alice chooses random *input tokens* $\mathsf{tok}_{k,j,0}$ and $\mathsf{tok}_{k,j,1}$. The parties use an instance of OT so that Bob picks up the correct $\mathsf{tok}_{k,j,b}$, where b is Bob's input value on wire j in the kth evaluation of f.

Let PRF be a PRF. Then for each $b \in \{0,1\}, j \in$ RecvInpWires the parties perform a coupling in which **Bob learns** $d^{in}_{i,j,b} \oplus$ **PRF**(tok$_{k,j,b}$; l) **if and only if circuit** i **is assigned to position** l **of bucket** k. If Bob has input bit b on the jth wire in the kth evaluation of f, then he holds tok$_{k,j,b}$ and can decrypt the corresponding $d^{in}_{i,j,b}$ and use it to decommit to the appropriate input wire label for the ith garbled circuit. If he does not have input bit b, then these outputs of the coupling subprocess look independently pseudorandom by the guarantee of the PRF.

If Alice sends inconsistent values into the coupling, then Bob may not receive the decommitment values $d^{in}_{i,j,b}$ he expects. If this happens, then Bob aborts. Because this abort event would then depend on Bob's private input, we have Bob encode his input in a λ-probe-resistant encoding, following the discussion in Sect. 2.3. This standard technique makes Bob's abort probability independent of his private input.

Enforcing Consistency of Sender's Inputs. We must ensure that Alice uses the same input for all of the circuits mapped to a particular bucket $k \neq 0$, despite Alice not knowing which circuits will be assigned to that bucket. This must furthermore be done without leaking Alice's input to Bob in the process.

We use an approach similar to [27] based on a XOR-homomorphic commitment scheme. But here the sender does not know *a priori* which committed values' XOR it needs to open. Hence, we need a mechanism for letting the receiver obliviously learn the decommitment strings for XOR of the appropriate committed values.

For each circuit i, we have Alice choose a random string s_i and commit individually to all of her input wire labels, permuted according to s_i. More precisely, she computes commitments:

$$C^{in}_{i,j,0} \leftarrow \text{Com}(\text{in}_{i,j,s_{i,j}}; d^{in}_{i,j,0})$$
$$C^{in}_{i,j,1} \leftarrow \text{Com}(\text{in}_{i,j,\overline{s_{i,j}}}; d^{in}_{i,j,1})$$

Here $s_{i,j}$ denotes the jth bit of s_i. Hence $C^{in}_{i,j,b}$ is a commitment to the input wire label representing *truth value* $b \oplus s_{i,j}$.

Alice also commits to s_i under a homomorphic commitment scheme $C^s_i \leftarrow$ HCom$(s_i; d^s_i)$. As before, the randomness used in all of these commitments is derived from σ_i so the commitments can be checked in the cut-and-choose.

For each bucket k, Alice gives a homomorphic commitment to x_k, her input in that execution—$C^x_k \leftarrow$ HCom$(x_k; d^x_k)$. The parties perform a coupling so that **Bob learns** $d^s_i \oplus d^x_k$ **iff circuit** i **is assigned to bucket** k. The result is a decommitment value that Bob can use to learn $s_i \oplus x_k$. The soundness of the commitment scheme ensures that Bob knows values $o_i = s_i \oplus x_k$ for a *consistent* x_k. Given that the commitments to Alice's input wires ($C^{in}_{i,j,b}$) are arranged/permuted using s_i (a property enforced with high probability by the cut-and-choose), the commitments indexed by o_i correspond to the garbled inputs that encode the logical value x_k. Hence, to ensure that Alice uses consistent inputs within each bucket, Bob expects Alice to open the commitments indexed by o_i.

Routing the Sender's Inputs. We must let Bob obtain garbled inputs encoding Alice's inputs to the ith garbled circuit. As above, when circuit i is mapped to bucket k, it suffices to let Bob learn the decommitment to $C^{in}_{i,j,o_{i,j}}$ where $o_i = s_i \oplus x_k$. The challenge is to accomplish this without Alice knowing *a priori* which circuit i will be assigned to which bucket k, and hence which input x_k needs to be garbled. We propose a novel and efficient technique for this step that, for each input wire, only requires one symmetric-key operation and the routing of one string of length κ through the switching network.

For each wire $j \in \mathsf{SendInpWires}$, Alice chooses random Δ_j. As a matter of notation, when b is a bit, we let $b\Delta_j$ denote the value [if $b = 0$ then 0^κ else Δ_j].

For each circuit i and wire $j \in \mathsf{SendInpWires}$, Alice chooses random $r_{i,j}$ and sends an encryption $e_{i,j,b} = H(r_{i,j} \oplus b\Delta_j) \oplus d^{in}_{i,j,b}$ to Bob. Here H is a correlation-robust hash function (Sect. 2.5).

For each wire $j \in \mathsf{SendInpWires}$ the parties perform a coupling in which **Bob learns $(r_{i,j} \oplus s_{i,j}\Delta_j) \oplus x_{k,j}\Delta_j$ if and only if circuit i is assigned to bucket k.** Simplifying, we see that Bob learns:

$$K_{i,j} = (r_{i,j} \oplus s_{i,j}\Delta_j) \oplus x_{k,j}\Delta_j = r_{i,j} \oplus (s_{i,j} \oplus x_{k,j})\Delta_j = r_{i,j} \oplus o_{i,j}\Delta_j$$

Indeed, this is the key that Bob can use to decrypt $e_{i,j,o_{i,j}}$ to obtain $d^{in}_{i,j,o_{i,j}}$. He can then use this value to decommit to the wire label encoding truth value $x_{k,j}$, as desired. Bob will abort if he is unable to decommit to the expected wire labels in this way. Here, the abort probability depends only on Alice's behavior, and is not influenced by Bob's input in any way.

Note that the decommitment values for the "other" wire labels are masked by a term of the form $H(K_{i,j} \oplus \Delta_j)$, where Δ_j is unknown to Bob. Even though the same Δ_j is used for many such ciphertexts, the correlation-robustness of H ensures that these masks look random to Bob.

Cheating Recovery. Lindell [24] introduced a *cheating recovery* technique, where if the receiver detects the sender cheating, the receiver is able to learn the sender's input (and hence evaluate the function in the clear). This technique is crucial in reducing the number of garbled circuits, since now only a *single* circuit in a bucket needs to be correctly generated. Our protocol also adapts this technique, but in a non-interactive setting. The approach here is similar to that used in [1], but it is describe more generally in terms of any homomorphic commitment scheme and of course adapted to the batch setting.

For each output bit j and each bucket k, Alice generates $\mathsf{w}_{k,j,0}$ at random and sets $\mathsf{w}_{k,j,1} = x_k - \mathsf{w}_{k,j,0}$. The main idea is two-fold:

- We will arrange so that if Bob evaluates *any* circuit in bucket k and obtains output b on wire j, then Bob will learn $\mathsf{w}_{k,j,b}$.

– Then, if Bob evaluates two circuits in the same bucket that disagree on their output—say, they disagree on output bit j—then Bob can recover Alice's input $x_k = \mathsf{w}_{k,j,0} + \mathsf{w}_{k,j,1}$.

For technical reasons, we must introduce *pre-output* and *post-output* wire labels for each garbled circuit. When evaluating a garbled circuit, the evaluator obtains *pre-output* wire labels. We denote by $d^{\mathsf{out}}_{i,j,b}$ the pre-output wire label for wire j of circuit i encoding truth value b. We use this notation since the pre-output wire labels are used as decommitment values.

Alice chooses random *post-output* wire labels, $\{\mathsf{out}_{i,j,b}\}$ and generates a homomorphic commitment to them using the pre-output labels as the randomness:

$$C^{\mathsf{out}}_{i,j,b} \leftarrow \mathsf{HCom}(\mathsf{out}_{i,j,b}; d^{\mathsf{out}}_{i,j,b})$$

The technical reason for having both pre- and post-output labels is so that there is a homomorphic commitment that is bound to each output wire of each circuit, that *can be checked* in the cut-and-choose. Indeed, these commitments can be checked in the cut-and-choose, since they use the circuit's [pre-]output wire labels as their randomness.

Separately, for each bucket $k \neq 0$, Alice generates and sends homomorphic commitments:

$$C^{\mathsf{w}}_{k,j,b} \leftarrow \mathsf{HCom}(\mathsf{w}_{k,j,b}; d^{\mathsf{w}}_{k,j,b})$$

She sends a homomorphic opening to the linear expression $\mathsf{w}_{k,j,0} + \mathsf{w}_{k,j,1} - x_k$, to prove that this expression is all-zeroes (*i.e.*, to prove that $\mathsf{w}_{k,j,0} + \mathsf{w}_{k,j,1} = x_k$).

Then, for each $j \in outpwires$ and $b \in \{0,1\}$ the parties do a coupling in which **Bob learns $d^{\mathsf{out}}_{i,j,b} \oplus d^{\mathsf{w}}_{k,j,b}$ when circuit i is assigned to bucket k**. Bob can use the result to decommit to the value of $\mathsf{out}_{i,j,b} \oplus \mathsf{w}_{k,j,b}$.

Putting things together, Bob evaluates a circuit i assigned to bucket k. He learns the corresponding pre-output wire labels $d^{\mathsf{out}}_{i,j,b}$, which he uses to decommit to the post-output wire labels $\mathsf{out}_{i,j,b}$. Since he has learned $\mathsf{out}_{i,j,b} \oplus \mathsf{w}_{k,j,b}$ from the coupling, he can therefore compute $\mathsf{w}_{k,j,b}$ (a *bucket-specific* value, whereas $\mathsf{out}_{i,j,b}$ was a *circuit-specific* value). If any two circuits disagree in their output, he can recover the sender's input x_k as described above and compute the correct output. Otherwise, since at least one circuit in the bucket is guaranteed (by the cut-and-choose bounds) to be generated honestly, Bob can uniquely identify the correct output.

4.2 Detailed Protocol Description

We present our complete protocol in Fig. 2. We refer the reader to the full version for the proof of the following Theorem.

Theorem 6. *The protocol in Fig. 2 is a UC-secure realization of the functionality in Fig. 1.*

Parameters: A function f and number N of instances. $\widehat{N}$ denotes the number of garbled circuits, chosen according to the discussion in the text. λ is the statistical security parameter.

Inputs: Alice has inputs $(x_1, \ldots, x_N)$ and Bob has inputs $(y_1, \ldots, y_N)$.

1. Bob chooses a random permutation π, and uses it as input to all coupling sub-protocols below (i.e., all couplings are performed in parallel and bound to the same π). The parties agree on a λ-probe resistant matrix M, and Bob encodes each y_k as $\tilde{y}_k$ where $M\tilde{y}_k = y_k$.
2. For each circuit $i \in [\widehat{N}]$: Alice chooses a PRF seed σ_i and uses it to derive *all randomness used in this step* of the protocol:
   Alice generates a garbling of the function $\tilde{f}(x, \tilde{y}) = f(x, M\tilde{y})$; let F_i denote the garbled circuit, and let $\mathsf{in}_{i,j,b}$ (resp. $d^{\mathsf{out}}_{i,j,b}$) denote the input (resp. output) wire label encoding truth value b on wire j of circuit i. She sends each F_i to Bob. Alice chooses random "post-output" keys $\{\mathsf{out}_{i,j,b}\}_{j \in \mathsf{OutWires}, b \in \{0,1\}}$. She generates and sends the following commitments (where d^{in} and d^s values are derived randomly from σ_i):

$$C^{\mathsf{in}}_{i,j,b} \leftarrow \mathsf{Com}(\mathsf{in}_{i,j,b \oplus s_{i,j}}; d^{\mathsf{in}}_{i,j,b \oplus s_{i,j}}) \qquad \text{for } j \in \mathsf{SendInpWires}, b \in \{0,1\}$$

$$C^{\mathsf{in}}_{i,j,b} \leftarrow \mathsf{Com}(\mathsf{in}_{i,j,b}; d^{\mathsf{in}}_{i,j,b}) \qquad \text{for } b \in \{0,1\}, j \in \mathsf{RecvInpWires}$$

$$C^{\mathsf{out}}_{i,j,b} \leftarrow \mathsf{HCom}(\mathsf{out}_{i,j,b}; d^{\mathsf{out}}_{i,j,b}) \qquad \text{for } b \in \{0,1\}, j \in \mathsf{OutWires}$$

$$C^s_i \leftarrow \mathsf{HCom}(s_i; d^s_i)$$

3. The parties perform a coupling with input for Alice $\{\sigma_i\}_i$, all-zeroes masks for bucket #0, and random masks for other buckets. Bob learns σ_i if circuit i is mapped to bucket 0. For such i, Bob checks that F_i and corresponding commitments from the previous step are generated using randomness derived from σ_i, and aborts if this is not the case.
4. For $j \in \mathsf{SendInpWires}$, Alice chooses a random Δ_j. For $j \in \mathsf{SendInpWires}, i \in [\widehat{N}]$, Alice chooses a random $r_{i,j}$. Alice generates and sends input-encryptions:

$$e_{i,j,b} = H(r_{i,j} \oplus b\Delta_j) \oplus d^{\mathsf{in}}_{i,j,b}$$

5. For $k \in [N], j \in \mathsf{OutWires}$, Alice chooses random $\mathsf{w}_{k,j,0}$ and sets $\mathsf{w}_{k,j,1} = x_k \oplus \mathsf{w}_{k,j,0}$ (recall x_k is her input to the kth execution). Alice generates and sends commitments:

$$C^{\mathsf{w}}_{k,j,b} \leftarrow \mathsf{HCom}(\mathsf{w}_{k,j,b}; d^{\mathsf{w}}_{k,j,b}) \qquad \text{for } k \in [N], j \in \mathsf{OutWires}, b \in \{0,1\}$$

$$C^x_k \leftarrow \mathsf{HCom}(x_k; d^x_k) \qquad \text{for } k \in [N]$$

Alice also gives homomorphic decommitments:

$$d^{\mathsf{w}}_{k,j,0} \oplus d^{\mathsf{w}}_{k,j,1} \oplus d^x_k \qquad \text{for } k \in [N], j \in \mathsf{OutWires}$$

Bob aborts if these values do not decommit $C^{\mathsf{w}}_{k,j,0} \oplus C^{\mathsf{w}}_{k,j,1} \oplus C^x_k$ to the all-zeroes string.

(protocol description continues...)

Fig. 2. Batch NISC protocol

6. For $k \in [N], j \in \mathsf{RecvInpWires}$, Alice chooses random $\mathsf{tok}_{k,j,0}, \mathsf{tok}_{k,j,1}$. Parties engage in an instance of OT with inputs $(\mathsf{tok}_{k,j,0}, \mathsf{tok}_{k,j,1})$ for Alice and $\tilde{y}_{k,j}$ (i.e., jth bit of $\tilde{y}_k$) for Bob. Bob gets input $\mathsf{tok}_{k,j,\tilde{y}_{k,j}}$.

7. For $k \in [N], j \in \mathsf{RecvInpWires}, b \in \{0,1\}$, the parties perform a coupling with inputs $\{d_{i,j,b}^{\mathsf{in}}\}_i, \{\mathsf{PRF}(\mathsf{tok}_{k,j,b}; l)\}_{k,l}$ for Alice. Bob learns $\beta_{i,j,b} = d_{i,j,b}^{\mathsf{in}} \oplus \mathsf{PRF}(\mathsf{tok}_{k,j,b}; l)$ when circuit i is assigned to position l of bucket k. Bob aborts if $\beta_{i,j,\tilde{y}_{i,j}} \oplus \mathsf{PRF}(\mathsf{tok}_{k,j,\tilde{y}_{i,j}}; l)$ is not a valid decommitment of $C_{i,j,\tilde{y}_{i,j}}^{\mathsf{in}}$. Otherwise, Bob sets $\mathsf{in}_{i,j}^*$ to be the result of the decommitment.

8. The parties perform a coupling with input $\{d_i^s\}_i, \{d_k^x\}_k$ for Alice. Bob learns $d_i^s \oplus d_k^x$ when circuit i is assigned to bucket k, and aborts if this is not a valid opening of $C_i^s \oplus C_k^x$. Otherwise, Bob sets o_i to be the result of this decommitment.

9. For $k \in [N], j \in \mathsf{SendInpWires}$ the parties perform a coupling with input $\{r_{i,j} \oplus s_{i,j}\Delta_j\}_i, \{x_{k,j}\Delta_j\}_k$ for Alice. Bob learns $K_{i,j} = (r_{i,j} \oplus s_{i,j}\Delta_j) \oplus x_{k,j}\Delta_j$ when circuit i is assigned to bucket k.

   For $i \in [\widehat{N}], j \in \mathsf{SendInpWires}$, Bob aborts if $e_{i,j,o_{i,j}} \oplus H(K_{i,j})$ is not a valid decommitment to $C_{i,j,o_{i,j}}^{\mathsf{in}}$. Otherwise, Bob sets $\mathsf{in}_{i,j}^*$ to be the result of this decommitment.

10. For $j \in \mathsf{OutWires}, b \in \{0,1\}$ the parties perform a coupling with input $\{d_{i,j,b}^{\mathsf{out}}\}_i, \{d_{k,j,b}^{\mathsf{w}}\}_k$ for Alice. Bob gets $d_{i,j,b}^{\mathsf{out}} \oplus d_{k,j,b}^{\mathsf{w}}$ if circuit i is assigned to bucket k. Bob aborts if this value is not a valid decommitment to $C_{i,j,b}^{\mathsf{out}} \oplus C_{k,j,b}^{\mathsf{w}}$. Otherwise, Bob sets $\delta_{i,j,b}$ to be the result of the decommitment.

11. For $i \in [\widehat{N}]$, where circuit i has not been mapped to bucket #0: Bob evaluates garbled circuit F_i with input wire labels $\{\mathsf{in}_{i,j}^*\}_{j \in \mathsf{SendInpWires} \cup \mathsf{RecvInpWires}}$. The result is plain output z_i and corresponding pre-output wire labels $\{d_{i,j,z_{i,j}}^{\mathsf{out}}\}$. If for some j, $d_{i,j,z_{i,j}}^{\mathsf{out}}$ is not a valid decommitment of $C_{i,j,z_{i,j}}^{\mathsf{out}}$ then Bob changes $z_i = \perp$. Otherwise, Bob opens the commitments to obtain $\mathsf{out}_{i,j,z_{i,j}}$ values.

12. For each bucket $k \neq 0$: If $z_i = \perp$ for all i assigned to this bucket, then abort. If there are $z_i \neq z_{i'}$, neither of them $\perp$, in this bucket, then let j be some position for which $z_{i,j} \neq z_{i',j}$. Bob computes

$$\tilde{x}_k = (\mathsf{out}_{i,j,z_{i,j}} \oplus \delta_{i,j,z_{i,j}}) \oplus (\mathsf{out}_{i',j,z_{i',j}} \oplus \delta_{i',j,z_{i',j}})$$

   and sets $z_k^* = f(\tilde{x}_k, \tilde{y}_k)$. Otherwise, let z_k^* be the unique value such that $z_i \in \{\perp, z_k^*\}$ for all i in this bucket.

13. Bob outputs $z_1^*, \ldots, z_N^*$.

Fig. 2. (*Continued*)

5   Protocol Efficiency and Choice of Commitments

We review the efficiency of our construction. First, we note that besides the calls to an ideal OT (in the main protocol and also in the OSN subprotocol), the protocol consists of a monolothic message from Alice to Bob (containing garbled circuits, commitments, etc.). All instances of OT are performed in parallel. Hence, ours is a NISC protocol in the sense of [17]. Concretely, the OT can be instantiated with a two-round protocol such as that of [33], making our protocol also a two-round protocol (Bob sends the first OT message, Alice sends the second OT message along with her monolothic NISC protocol message.)

5.1   Effect of Oblivious Switching Network

From Table 1 we see that the parts of the protocol that involve the oblivious switching network (OSN) scale with $N\lambda$, whereas everything else scales with $N\lambda/\log N$ (or independent of λ altogether). The $\log N$ term in the denominator is a result of savings by batching the cut-and-choose step. In particular, the number of garbled circuits (which is the main communication overhead in general), as well as their associated commitments, benefits from batching.

However, information related to the various commitments is sent as input into the OSN. The OSN incurs a $\log \widehat{N}$ overhead which "cancels out" the benefits of batching, for these values. We elaborate on this fact:

We instantiate the OSN with a Waksman network [38], which is a universal switching network (i.e., it can be programmed to realize any permutation). Each "bucket coupling" step requires a permutation on $\widehat{N}$ items, leading to a Waksman network with $O(\widehat{N} \log \widehat{N}) = O(N\lambda)$ switches.

Note that only decommitment and similar values are processed via the OSN subprotocol (bucket coupling steps). The garbled circuits and their associated commitments are not.

5.2   Instantiating Homomorphic Commitments

Pedersen Commitment. Let g be the generator for a prime order group G where the discrete-log problem is hard, and let $h = g^x$ for a random secret x. In our setting g, h can either be chosen by the receiver and sent along with its first OT message, or it can be part of a CRS.

In Pedersen commitments [32], to commit to a message m, we let $\mathsf{Com}(m; r) = g^m h^r$ for a random r. The decommitment string is (m, r). The scheme is statistically hiding and computationally binding. It is also homomorphic (with respect to addition over $\mathbb{Z}_p$) on the message space and the decommitment. In particular, given $\mathsf{Com}(m; r)$ and $\mathsf{Com}(m'; r')$, we can decommit to $m + m'$ by sending $(m+m', r+r')$ to the receiver who can check whether $\mathsf{Com}(m; r) \cdot \mathsf{Com}(m'; r') = g^{m+m'} h^{r+r'}$.

Regarding their suitability for our scheme: Clearly Pedersen commitments have optimal communication overhead (commitment length is equal to the message length). However, they require exponentiations in a DH group. In practice these operations are much slower than symmetric-key primitives like hash functions or block ciphers, which would be preferred. Pedersen commitments are homomorphic over the group $(\mathbb{Z}_p, +)$. For many of the commitments in our scheme (in particular, the $\mathsf{out}_{i,j,b}$ and $\mathsf{w}_{k,j,b}$ values) the choice of group is not crucial, but we actually require the commitments to x_k and s_i to be combined with respect to bitwise XOR. Later in this section we discuss techniques for combining Pedersen commitments with other kinds of homomorphic commitments.

OT-Based Homomorphic Commitments. We discuss a paradigm for homomorphic commitments based on simple OTs.

Starting Point. Our starting point is an XOR-homomorphic commitment of Lindell and Riva [27], that is further based on a technique of Kilian [20] for proving equality of committed values (*i.e.*, proving that the XOR of two commitments is zero). The Lindell-Riva commitment has an interactive opening phase, but we will show how to make it non-interactive.

Let Com be a regular commitment. To generate a homomorphic commitment to message m, the sender secret shares $m_0 \oplus m_1 = m$ and generates plain commitments $\mathsf{Com}(m_0)$ and $\mathsf{Com}(m_1)$.

Suppose commitments to m and m' exist (*i.e.*, there are plain commitments to m_0, m_1, m'_0, m'_1). To open $m \oplus m'$ the parties do the following:

- Preamble: the sender gives $\Delta = m \oplus m'$ (the claimed xor of the two commitments) and $\delta = m_0 \oplus m'_0$
- Challenge: receiver chooses random $b \leftarrow \{0,1\}$
- Response: sender opens $\mathsf{Com}(m_b)$ and $\mathsf{Com}(m'_b)$. Receiver checks: $m_b \oplus m'_b \stackrel{?}{=} \delta \oplus b\Delta$

This scheme has soundness $1/2$, but can be repeated in parallel λ times to achieve soundness $2^{-\lambda}$.

If we settle for the Fiat-Shamir technique to generate the challenge bits, the above scheme can easily become non-interactive. Similarly, in the offline-online variant of our construction where the commitments and preambles can all be sent in the offline phase, the online phase will be non-interactive (challenge and response). But for our main construction in the standard model, we need to make the above scheme non-interactive.

Making It Non-interactive. In our NISC application, we already assume access to an ideal oblivious transfer functionality. Then the above approach can be modified to both do away with the standalone commitments and to make a non-interactive decommitment phase.

The idea is to replace commitments and a public challenge with an instance of OT. To commit to m, the commitment phase proceeds as follows:

- The receiver chooses a random string $b = b_1 \cdots b_\lambda$ and uses the bits of b as choice bits to λ instances of OT.
- The sender chooses λ pairs $(m_{1,0}, m_{1,1}), \ldots, (m_{\lambda,0}, m_{\lambda,1})$ so that $m_{i,0} \oplus m_{i,1} = m$. The sender uses these pairs as inputs to the instances of OT. Hence, the receiver picks up m_{i,b_i}.

We note that when committing to many values as is the case in our constructions, the *same OTs are used for all commitments*. That is, the same challenge bits b are used for all commitments.

Suppose two such commitments have been made in this way, to m and to m'. Then to decommit to $\Delta = m \oplus m'$ the sender can simply send Δ and $\delta = (\delta_1, \ldots, \delta_\lambda) = (m_{1,0} \oplus m'_{1,0}, \ldots, m_{\lambda,0} \oplus m'_{\lambda,0})$. The receiver can check the soundness equations:

$$m_{i,b_i} \oplus m'_{i,b_i} \stackrel{?}{=} \delta_i \oplus b_i\Delta$$

Note that the same b_i challenges are shared for all commitments, so the receiver will indeed have m_{i,b_i} and m'_{i,b_i} for a consistent b_i. Since the sender's view is independent of the receiver's challenge b, soundness follows from the same reasoning as above.

In this way, the decommitment string for a commitment to m is $(m, m_{1,0}, \ldots, m_{\lambda,0})$. Furthermore, to decommit to $m \oplus m'$, the decommitment value is the XOR of the individual decommitment values. In other words, the scheme satisfies the homomorphic-opening property described in Sect. 2.2. Finally, note that since we use the same challenge bits for all commitments, it easy to prove multiple XOR relations involving the same committed value.

Code-Based Homomorphic Commitments. A recent series of works [10,12] construct homomorphic commitments from an oblivious-transfer-based setup.

Looking abstractly at our presentation of the Lindell-Riva commitment above, their construction takes the payload m and generates $(m_{1,0}, m_{1,1}, \ldots, m_{\lambda,0}, m_{\lambda,1})$, where $(m_{1,0} \oplus m_{1,1}, \ldots, m_{\lambda,0} \oplus m_{\lambda,1})$ is an encoding of m. In this case, the encoding is a λ-repetition encoding.

The idea behind [12] is to choose an encoding with better rate. Namely, the sender generates $(m_{1,0}, m_{1,1}, \ldots, m_{n,0}, m_{n,1})$, where $(m_{1,0} \oplus m_{1,1}, \ldots, m_{n,0} \oplus m_{n,1})$ encodes m in some error-correcting code. Here the total length of the encoding may be much smaller than $2\lambda|m|$ as in the Lindell-Riva scheme. The binding property of the construction is related to the minimum distance of this code. We refer the reader to [12] for details about the construction and how to choose an appropriate error-correcting code. Instead, we point out some facts that are relevant to our use of homomorphic commitments:

- When the error-correcting code is *linear*, then the commitments are additively homomorphic. Following our pattern, the decommitment value for a commitment is the vector $(m, m_{1,0}, \ldots, m_{n,0})$. These decommitment values are indeed homomorphic in the sense we require.
- The *rate* of a commitment scheme is the length of the commitment's payload divided by the communication cost of the commitment. For example, the Lindell-Riva scheme has rate $O(1/\lambda)$. By a suitable choice of error-correcting codes, the *rate* of the scheme in [12] can be made constant, or even $1 + o(1)$. Concretely, to commit to 128 bits requires the committer to send only 262 bits when using an appropriate BCH code, leading to a rate 0.49.

Unfortunately, unlike the Lindell-Riva construction, the scheme of [12] requires some additional interaction in the setup phase. In particular, there must be some mechanism to ensure that the sender is indeed using valid codewords. The sender can violate binding, for instance, by choosing a non-codeword that is "halfway between" two valid codewords. In [12], after the parties have performed the OTs of the setup phase, the receiver challenges the sender to open some random combination of values to ensure that they are consistent with valid codewords.

Removing this interaction turns out to be problematic. In our offline/online application the extra interaction is in the offline phase, and so not a problem.

In our offline/online application we therefore use this highly efficient commitments. However, we cannot afford the extra round of interaction in our batch NISC application. Hence, our options are: (1) use less efficient homomorphic commitment schemes like the Lindell-Riva one; (2) remove the round of interaction using the Fiat-Shamir heuristic, since the receiver's challenge is random.

5.3   Reducing Cost of Homomorphic Commitments

We described our main protocol without specifying exactly which homomorphic commitment to use. Based on the previous discussion, we have several options, none of them ideal:

- Pedersen commitments, which are rate 1, but require public-key operations and are homomorphic only with respect to addition in $\mathbb{Z}_p$.
- Lindell-Riva-style commitments based on OTs, which have rate $O(1/\lambda)$ and are homomorphic with respect to XOR.
- FJNT [12] commitments, which have constant rate and are homomorphic with respect to XOR, but require some interaction in the initialization step (unless one is satisfied with the Fiat-Shamir heuristic).

The only "off-the-shelf" choice that is compatible with our construction is the Lindell-Riva-style commitments, which are the least efficient in terms of communication.

We therefore describe two methods to significantly improve the efficiency related to homomorphic commitments in our construction.

Linking Short-to-Long Commitments. The protocol performs homomorphic decommitments that combine x_k and $w_{k,j,b}$ values—hence, these values must have the same length ($|\mathsf{SendInpWires}|$). There is an $w_{k,j,b}$ value for each bucket $k \in [N]$ and each circuit output wire $j \in \mathsf{OutWires}$. Accounting for the total communication cost for these commitments in the Lindell-Riva-style scheme, we get $O(\lambda N |\mathsf{OutWires}| \cdot |\mathsf{SendInpWires}|)$. For circuits with relatively long inputs/outputs, the cost $|\mathsf{OutWires}| \cdot |\mathsf{SendInpWires}|$ is undesirable.

We propose a technique for reducing this cost when $|\mathsf{SendInpWires}|$ is long (longer than a computational security parameter κ). Recall that the purpose of the $w_{k,j,b}$ values is that if the receiver learns both $w_{k,j,0}$ and $w_{k,j,1}$, then he can combine them to learn x_k. We modify the construction so that the sender gives a homomorphic commitment to a random ("short") w_k for each bucket k, where

$$w_{k,j,0} \oplus w_{k,j,1} = w_k \qquad (\forall j \in \mathsf{OutWires})$$

(i.e., we have replaced x_k with w_k in the above expression). The $w_k, w_{k,j,b}$ values have length κ, so the total cost of these commitments to $w_{k,j,b}$ is $O(\kappa \lambda N |\mathsf{OutWires}|)$.

Now we must modify the protocol so that if the receiver ever learns w_k, then he (non-interactively) can recover x_k, where w_k is "short" and x_k is "long." Recall that to commit to x_k and w_k in the Lindell-Riva scheme, the sender needs to generate λ independent additive sharings: $\{x_{k,0,i}, x_{k,1,i}\}_{i \in [\lambda]}$ and $\{w_{k,0,i}, w_{k,1,i}\}_{i \in [\lambda]}$ and using them as sender's inputs to the challenge OTs. To "link" w_k to x_k, we simply have the sender also send ciphertexts $\{\mathsf{Enc}(w_{k,b,i}; x_{k,b,i})\}_{i \in [\lambda], b \in \{0,1\}}$, i.e. encrypting each additive share of x_k using the corresponding share of w_k as the key. Note that Enc can be a symmetric-key encryption scheme and therefore relatively fast.

In the Lindell-Riva scheme, the receiver learns one share from each pair of shares. He learns either $(w_{k,0,i}, x_{k,0,i})$ or $(w_{k,b,i}, x_{k,b,i})$. Hence, the receiver can check half of the ciphertexts sent by the sender, and abort if any are not correct/consistent. If the receiver doesn't abort, this guarantees that with high probability the majority of these linking encryptions are correct. To bound the probability of error to $2^{-\lambda}$, we must increase the number of parallel repetitions to $\sim 3\lambda$.

Now if the receiver learns w_k at some later time, it can solve for both shares $w_{k,0,i}, w_{k,1,i}$ for every i, and use them to decrypt the shares $x_{k,0,i}, x_{k,1,i}$. The receiver thus recovers x_k as the *majority value* among all $x_{k,0,i} \oplus x_{k,1,i}$.

If the receiver never cheats, then the value of x_k remains hidden by the semantic security of the Enc-encryptions.

Replacing with Pedersen Commitments. We can reduce the $\kappa\lambda$ term in the communication complexity to κ by using Pedersen commitments for the output wires. In particular, $O(|\mathsf{OutWires}|\widehat{N})$ Pedersen commitments are sufficient for committing to the w_k, $w_{k,j,b}$, and $\mathsf{out}_{i,j,h}$ values. However, we also need to "link" w_k to x_k. To do so, we can use the input consistency check technique used in [1] that uses an El Gamal encryption of x_k, and algebraically links the Pedersen commitments to the output wires with the Elgamal encryption of the input. We refer the reader to [1] for details of this approach.

We note that the use of Pedersen Commitments provides a trade-off between communication and computation as the computation cost will likely increase due the public-key operations required by the scheme, but we save on the communication requires for cheating-recovery.

Reducing Communication Using the Seed Technique. In the full version of the paper, we show how to further reduce communication of our protocol by incorporating the seed technique of [1,13] wherein only the garbled circuits that are evaluated are communicated in full.

6 Optimizations for the Offline-Online Setting and Random Oracle Model

Using the random oracle model, we can remove or improve several sources of inefficiency in our construction. To introduce these improvements, we first describe a 2PC protocol in the offline-online setting, which may be of independent interest.

6.1   Offline-Online Protocol

The Setting. In this setting, the parties know that they will securely evaluate some function f, N times (perhaps not altogether in a single batch). In an *offline phase* they perform some pre-processing that depends only on f and N. Then, when it comes time to securely evaluate an instance of f, they perform an *online phase* that is as inexpensive as possible, and depends on their inputs to this evaluation of f.

We will describe how to modify our NISC protocol to obtain an offline-online 2PC protocol where:

- The offline phase is constant-round.
- Each online phase is two rounds, consisting of a length-$|y|$ message from the receiver Bob followed by a message of length $(|x| + |y|)\kappa$ (or $|x| + |y| + O(\kappa)$, after further optimization) from Alice.
- The total cost of N secure evaluations of f is $O(N/\log N)$ times that of a single secure evaluation. In particular, batching improves *all aspects* of the protocol by a $\log N$ factor (unlike in the NISC protocol where the cost associated with circuit inputs/outputs did not have a $\log N$-factor saving).

Removing the Switching Network. Recall that in our NISC protocol the costs associated with *garbled circuits* scale as $O(N/\log N)$, while the costs associated with circuit inputs/outputs scales as $O(N)$. The reason is that decommitment information related to inputs/outputs is sent through the oblivious switching network (OSN). The switching network has $\log \widehat{N}$ depth, and incurs a $\log \widehat{N}$ factor overhead that cancels out the $\log N$ savings incurred by the batch cut-and-choose.

The main reason for the oblivious switching network protocol was to non-interactively choose an assignment of circuits to buckets. We showed how to perform this task using a two-round OSN protocol in the standard model. However, the assignment of circuits to buckets can be done in the offline phase, as it does not depend on the parties' inputs to f.

Let π denote the receiver's assignment of circuits to buckets. In the non-interactive setting, it was necessary to hide π from the sender—the sender cannot know in advance which circuits will be checked in the cut-and-choose. However, in principle π does not need to be completely secret; it merely suffices for it to be chosen *after* the sender commits to the garbled circuits.

When we allow more interaction in the offline phase, we can do away with the oblivious switching network (and its $\log \widehat{N}$ overhead on garbled inputs/outputs) altogether. The main changes to remove the OSN subprotocols are as follows:

- The receiver chooses a random assignment π and commits to it.
- For the coupling subprotocols involving σ_i, we instead have the sender commit individually to each σ_i. The sender also sends all of the garbled circuits and various commitments, just as in the NISC protocol.
- After the σ_i's are committed, the receiver opens the commitment of π.

– The sender opens the commitments to σ_i for i assigned to be checked. This allows the receiver to learn the σ_i's while avoiding the bucket-coupling subprotocol involving these values. For all other couplings, the sender simply sends whatever the receiver's output would have been in the NISC protocol. This is possible since the sender knows π.

In this way, we remove all invocations of the switching network, and their associated $O(\log \widehat{N})$ overhead.

To argue that the protocol is still secure, we need to modify the simulator for the NISC protocol. When the sender is corrupt, the simulator extracts the σ_i values, but does not use any special capabilities for the other couplings—it merely runs these couplings honestly and uses only their output. In this offline/online modification, the simulator can still extract the σ_i's from the commitments. Then it can receive the other values (formerly obtained via the couplings) directly from the sender. To simulate a corrupt receiver, the simulator need only extract the commitment to π in the first step, similar to how the NISC simulator extracts π as its first operation. However, in this setting the inputs to the function are chosen *after* the receiver has seen the garbled circuits. Hence, we require a garbling scheme that has **adaptive security** [5].

Note that in this setting we can apply the optimization of Goyal et al. [13]: the sender can initially send only a *hash* of each garbled circuit. For circuits that are assigned to be checked, it is not necessary to send the entire garbled circuit – the receiver can simply recompute the circuit from the seed and compare to the hash. Only circuits that are actually evaluated must be sent. This optimization reduces concrete cost by a significant constant factor.

Optimizing Sender's Garbled Input. First, we can do away with the encryptions $e_{i,j,b}$ (step 4) and the associated coupling (step 9). These were needed only to route the sender's inputs to the correct buckets without a priori knowledge of the bucketing assignment. Instead, the sender (after learning the bucketing assignment) can simply directly send the decommitments to the correct commitments to her garbled input.

Besides this optimization, we observe that the NISC protocol uses the sender's input x_k in several places. We briefly describe ways to move the bulk of these operations to the offline phase.

Offline commitments to sender's input: In the NISC protocol the sender gives a homomorphic commitment to x_k. For each circuit i assigned to bucket k, the receiver learns the decommitment to $s_i \oplus x_k$, and in the online phase will expect the sender to open commitments indexed by $s_i \oplus x_k$, since these will be the commitments to wire labels holding truth value x_k. Furthermore, the sender chooses bucket-wide values $\mathsf{w}_{k,j,b}$ so that $x_k = \mathsf{w}_{k,j,0} \oplus \mathsf{w}_{k,j,1}$. The idea is that, if the receiver obtains conflicting outputs within a bucket, he can learn x_k.

To reduce the online dependence on x_k, we make the following change. Instead of giving a homomorphic commitment to x_k, the sender uses a random value μ_k. Since μ_k is unrelated to her input x_k, all of the commitments and homomorphic openings can be done in the offline phase. In other words, the homomorphic commitments are arranged so that the receiver learns $\mu_k \oplus s_i$, and so that the receiver learns μ_k if he obtains conflicting outputs in the bucket. Then in the online phase, the sender simply gives $x_k \oplus \mu_k$ *in the clear.* The receiver will expect the sender to open commitments indexed by $(x_k \oplus \mu_k) \oplus (\mu_k \oplus s_i) = x_k \oplus s_i$. If cheating is detected, the receiver learns μ_k and thus obtains $x_k = (x_k \oplus \mu_k) \oplus \mu_k$.

Packaging together sender's garbled inputs: Suppose there are B circuits assigned to each bucket. The receiver will be expecting the sender to decommit to B values for each input bit ($j \in$ SendInpWires). This leads to $O(B|x|\kappa)$ communication from the sender in each execution.

But since the sender knows the bucket-assignment in the offline phase, she can "package" the corresponding decommitment values together in the following way. For each of her input wires $j \in$ SendInpWires and value $b \in \{0, 1\}$ the sender can choose a bucket-specific token $\text{tok}_{k,j,b}$. Then, in she can encrypt all of the B different openings that will be necessary in the event that she has truth value b on wire j in the online phase. She can generate these ciphertexts in the offline phase and send them (in a random order with respect to the b-values). Then in the online phase, she need only send a single $\text{tok}_{k,j,b}$ value for each bit of her input, at a cost of only $|x|\kappa$ per execution.

Optimizing Receiver's Garbled Input. In the online phase, the parties must perform the OTs for the receiver's inputs. As in the NISC protocol these OTs are already on bucket-wide "tokens" and not B sets of wire labels per input wire.

Note that the number of OTs per execution is $|\tilde{y}_k|$, where $\tilde{y}_k$ is the λ-probe-resistant encoding of the receiver's true input y_k. Indeed, $\tilde{y}_k$ is longer than y_k by a significant constant factor in practice. However, we can reduce the online cost to $|y_k|$ by using an optimization proposed by Lindell and Riva [27] in their offline/online protocol, which we describe below:

Recall that M is the λ-probe-resistant matrix, and the parties are evaluating the function $\tilde{f}(x, \tilde{y}) = f(x, M\tilde{y})$. We instead ask the parties to evaluate the function $g(x, \tilde{r}, m) = f(x, m \oplus M\tilde{r})$. Note that $\tilde{r}$ is the length of a λ-probe-resistant-encoded input, while m has the same length as y. The idea is for Bob to choose an encoding $\tilde{r}$ of a *random* r, in the offline phase. The parties can perform OTs for $\tilde{r}$ in the offline phase. Then in the online phase, Bob announces $m = r \oplus y$ in the clear. Alice must then decommit to the input wire labels corresponding to m (in the protocol description we refer to these input wires of g as PubInpWires). As above, the decommitments for all B circuits in this bucket can be "packaged" together with encryptions sent in the offline phase. Therefore, the online cost attributed to the receiver's input is the receiver sending m and the sender sending $|m|$ encryption keys (where $|m| = |y|$).

Futher Compressing the Online Phase. Using the optimizations listed above, each online phase consists only of a length-$|y|$ message from Bob and a reply from Alice of length $(|x| + |y|)\kappa$. However, we point out that the message from Alice can be shortened *even further* using a technique of Applebaum et al. [2] that we summarize in Sect. 2.6. As a result, the *total communication* in the online phase is $|x| + |y| + O(\kappa)$ bits—only $O(\kappa)$ bits less than the information-theoretic minimum for secure computation.

Protocol Description. The detailed protocol description is given in Fig. 4. For simplicity this description does not include the technique of Applebaum et al. [2] for compressing garbled inputs in the online phase. This optimization can be applied in a black-box manner to our protocol.

Theorem 7. *The protocol in Fig. 4 is a UC-secure realization of the functionality in Fig. 3. The online phase is 2 rounds, and requires a length-$|y|$ message from the receiver and length-$(|x| + O(\kappa))$ message from the sender.*

Parameters: A function f and number N of instances.

Behavior: On input **setup** from the sender, give output **setup** to the receiver. Then do the following N times: wait for input x from the sender and y from the receiver. Then give output $f(x, y)$ to the receiver.

Fig. 3. Ideal functionality for offline/online 2PC

6.2 NISC, Optimized for Random Oracle Model

In the offline/online protocol we just described, the receiver first commits to π, receives garbled circuits and commitments, then opens π. Suppose we remove the commitment to π from the protocol. In other words, suppose the offline phase begins with the sender giving the garbled circuits and associated commitments, and then the receiver sends a random π in the clear.

This modified offline phase is then *public-coin* for the verifier. The only messages sent by the verifier are the random π and a random challenge for the FJNT homomorphic commitment scheme setup (not explicitly shown in the protocol description). We can therefore apply the **Fiat-Shamir** technique to make the protocol non-interactive again, in the programmable random oracle model.[2] In doing so we obtain a batch-NISC protocol that is considerably more efficient than our standard-model protocol. In particular:

- The RO protocol makes no use of the switching network, so avoids the associated overhead on garbled inputs/outputs.
- The RO protocol can be instantiated with the lightweight homomorphic commitments of [12].
- The RO protocol avoids communication for garbled circuits that are assigned to be checked.
- Unlike in the NISC setting, the offline/online protocol can take advantage of efficient OT extension techniques [3,4,16,19,21] which greatly reduce the cost of the (many) OTs in the protocol, but require interaction. This property is of course shared by all 2PC protocols that allow for more than 2 rounds.

---

[2] When considering a corrupt receiver, instead of extracting π from the commitment, the simulator can simply choose π upfront and then program the random oracle to output π on the appropriate query.

Parameters: A function f and number N of instances. $\widehat{N}$ denotes the number of garbled circuits, chosen according to the discussion in the text. λ is the statistical security parameter.

Offline phase:

1. Bob chooses a random permutation π, and commits to it.
2. For each circuit $i \in [\widehat{N}]$: Alice chooses a PRF seed σ_i and uses it to derive *all randomness used in this step* of the protocol:
   Alice generates a garbling of the function $\tilde{f}(x, \tilde{r}, m) = f(x, m \oplus M\tilde{r})$; let F_i denote the garbled circuit, and let $\mathsf{in}_{i,j,b}$ (resp. $d^{\mathsf{out}}_{i,j,b}$) denote the input (resp. output) wire label encoding truth value b on wire j of circuit i. She computes $h_i = H(F_i)$ where H is a CRHF, and sends h_i to Bob.
   Alice chooses random "post-output" keys $\{\mathsf{out}_{i,j,b}\}_{j \in \mathsf{OutWires}, b \in \{0,1\}}$. She generates and sends the following commitments (where d^{in} and d^{s} values are derived randomly from σ_i):

$$C^{\mathsf{in}}_{i,j,b} \leftarrow \mathsf{Com}(\mathsf{in}_{i,j,b \oplus s_{i,j}}; d^{\mathsf{in}}_{i,j,b \oplus s_{i,j}}) \qquad \text{for } j \in \mathsf{SendInpWires}, b \in \{0,1\}$$

$$C^{\mathsf{in}}_{i,j,b} \leftarrow \mathsf{Com}(\mathsf{in}_{i,j,b}; d^{\mathsf{in}}_{i,j,b}) \qquad \text{for } b \in \{0,1\}, j \in \mathsf{RecvInpWires}$$

$$C^{\mathsf{out}}_{i,j,b} \leftarrow \mathsf{HCom}(\mathsf{out}_{i,j,b}; d^{\mathsf{out}}_{i,j,b}) \qquad \text{for } b \in \{0,1\}, j \in \mathsf{OutWires}$$

$$C^{s}_i \leftarrow \mathsf{HCom}(s_i; d^{s}_i)$$

3. For each $i \in [\widehat{N}]$, Alice commits to each σ_i.
4. Bob opens the commitment to π.
5. For all i assigned to be checked by π, Alice opens the commitment to σ_i. Bob checks that h_i and corresponding commitments from the previous step are generated using randomness derived from σ_i, and aborts if this is not the case.
   For all i *not* assigned to be checked, Alice sends F_i; Bob aborts if $h_i \neq H(F_i)$.
6. For $k \in [N]$, Alice chooses a random μ_k. For $k \in [N], j \in \mathsf{OutWires}$, Alice chooses random $\mathsf{w}_{k,j,0}$ and sets $\mathsf{w}_{k,j,1} = \mu_k \oplus \mathsf{w}_{k,j,0}$. Alice generates and sends commitments:

$$C^{\mathsf{w}}_{k,j,b} \leftarrow \mathsf{HCom}(\mathsf{w}_{k,j,b}; d^{\mathsf{w}}_{k,j,b}) \qquad \text{for } k \in [N], j \in \mathsf{OutWires}, b \in \{0,1\}$$

$$C^{\mu}_k \leftarrow \mathsf{HCom}(\mu_k; d^{x}_k) \qquad \text{for } k \in [N]$$

   Alice also gives homomorphic decommitments:

$$d^{\mathsf{w}}_{k,j,0} \oplus d^{\mathsf{w}}_{k,j,1} \oplus d^{\mu}_k \qquad \text{for } k \in [N], j \in \mathsf{OutWires}$$

   Bob aborts if these values do not decommit $C^{\mathsf{w}}_{k,j,0} \oplus C^{\mathsf{w}}_{k,j,1} \oplus C^{\mu}_k$ to the all-zeroes string.
7. For $j \in \mathsf{OutWires}, b \in \{0,1\}$ and all circuits i assigned to bucket k, Alice sends $d^{\mathsf{out}}_{i,j,b} \oplus d^{\mathsf{w}}_{k,j,b}$. Bob aborts if this is not a valid decommitment to $C^{\mathsf{out}}_{i,j,b} \oplus C^{\mathsf{w}}_{k,j,b}$; otherwise he sets $\delta_{i,j,b}$ to be the result of this decommitment.
8. For all circuits i assigned to bucket k, Alice sends $d^{\mu}_k \oplus d^{s}_i$. Bob aborts if this is not a valid decommitment to $C^{\mu}_k \oplus C^{s}_i$; otherwise he sets o_i to be the result of this decommitment.

(protocol description continues. . .)

Fig. 4. Online-offline protocol

9. For all $k \in [N]$, Bob chooses a random λ-probe-resistant encoding $\tilde{r}_k$. For $j \in \mathsf{RecvInpWires}$, the parties engage in an instance of OT with inputs $(\{d^{\mathsf{in}}_{i,j,0}\}_i, \{d^{\mathsf{in}}_{i,j,b}\}_i)$ for Alice and input $\tilde{r}_{k,j}$ for Bob. Here the index i ranges over circuits assigned to bucket k.

Hence Bob learns input wire labels $\{d^{\mathsf{in}}_{i,j,\tilde{r}_{k,j}}\}_i$. He aborts if these are not valid decommitments to $\{C^{\mathsf{in}}_{i,j,\tilde{r}_{k,j}}\}_i$. Otherwise he sets $\mathsf{in}^*_{i,j}$ to be the corresponding decommitted values.

10. For $k \in [N], j \in \mathsf{SendInpWires}, b \in \{0,1\}$, Alice chooses a random token $\mathsf{tok}_{k,j,b}$, generates and sends an encryption:

$$e_{k,j,b} = \mathsf{Enc}\Big(\mathsf{tok}_{k,j,b}; \{d^{\mathsf{in}}_{i,j,\mu_{k,j}}\}_i\Big)$$

Here the index i ranges over circuits assigned to bucket k. These are decommitments to wire labels indexed by μ_k, hence wire labels having truth value $\mu_k \oplus s_i$. Similarly, for $k \in [N], j \in \mathsf{PubInpWires}, b \in \{0,1\}$, Alice chooses a random token $\mathsf{tok}_{k,j,b}$, generates and sends an encryption:

$$e_{k,j,b} = \mathsf{Enc}\Big(\mathsf{tok}_{k,j,b}; \{d^{\mathsf{in}}_{i,j,b}\}_i\Big)$$

11. For $k \in [N]$, Alice generates compressed garbled encodings of the tokens for her input wires and public input wires:

$$(sk_k, \widehat{e}_k) \leftarrow \mathsf{Compress}(\{\mathsf{tok}_{k,j,b} \mid j \in \mathsf{SendInpWires} \cup \mathsf{PubInpWires}; b \in \{0,1\}\})$$

She sends $\widehat{e}_k$ to Bob.

(protocol description continues…)

Fig. 4. (*Continued*)

Unfortunately, in this protocol we must use the *computational* security parameter κ (e.g., 128), and not the statistical security parameter λ (e.g., 40) to determine the bucket sizes. In the other protocols, the sender is committed to her choice of garbled circuits before the cut-and-choose challenge and bucketing assignment are chosen. Hence, cheating in the cut-and-choose phase is a one-time opportunity. In this Fiat-Shamir protocol, the sender can generate many candidate first protocol messages, until it finds one whose hash is favorable (i.e., it allows her to cheat undetected). Since this step involves no interaction, she has as many opportunities to try to find an advantageous first protocol message as her computation allows. Hence the probability of undetected cheating in the cut-and-choose step must be bound by the computational security parameter.

We note that the garbled-input-compressing technique of Applebaum et al. [2] is not useful in NISC since it increases total cost to improve online cost. In the NISC setting, there is no distinction between offline and online, so their technique simply increases the cost.

Online phase: For the kth time the online phase is invoked, Alice has input x_k and Bob has input y_k.

1. Bob computes $m_k = y_k \oplus M\tilde{r}_k$ and sends it to Alice.
2. Alice computes $\gamma_k = x_k \oplus \mu_k$. She computes online compressed garbled encoding $\hat{v}_k \leftarrow \mathsf{Online}(sk_k, m_k \| \gamma_k)$, and sends both γ_k and $\hat{v}_k$ to Bob.
3. Bob decompresses the garbled encodings:

$$\{\mathsf{tok}_{k,j,m_{k,j}} \mid j \in \mathsf{PubInpWires}\} \cup \{\mathsf{tok}_{k,j,\gamma_{k,j}} \mid j \in \mathsf{SendInpWires}\}$$
$$\leftarrow \mathsf{Decompress}(\hat{e}_k, m_k \| \gamma_k, \hat{v}_k)$$

4. Bob decrypts the corresponding ciphertexts as follows:

$$\{d^{\mathsf{in}}_{i,j,\gamma_{k,j}}\}_i = \mathsf{Dec}\Big(\mathsf{tok}_{k,j,\gamma_{k,j}}; e_{k,j,\gamma_{k,j}}\Big) \qquad \text{for } j \in \mathsf{SendInpWires}$$
$$\{d^{\mathsf{in}}_{i,j,m_{k,j}}\}_i = \mathsf{Dec}\Big(\mathsf{tok}_{k,j,m_{k,j}}; e_{k,j,m_{k,j}}\Big) \qquad \text{for } j \in \mathsf{PubInpWires}$$

   Bob aborts if the $d^{\mathsf{in}}_{i,j,b}$ values are not valid decommitments of the corresponding $C^{\mathsf{in}}_{i,j,b}$ commitments. Otherwise, Bob sets $\mathsf{in}^*_{i,j}$ to be the result of decommitment. Now, for all circuits i in this bucket, Bob has a complete garbled input (with wire labels for $\mathsf{RecvInpWires}$ obtained in step 9 of the offline phase).
5. For each circuit i assigned to bucket k, Bob evaluates garbled circuit F_i with input wire labels $\{\mathsf{in}^*_{i,j}\}_j$. The result is plain output z_i and corresponding pre-output wire labels $\{d^{\mathsf{out}}_{i,j,z_{i,j}}\}$. If for some j, $d^{\mathsf{out}}_{i,j,z_{i,j}}$ is not a valid decommitment of $C^{\mathsf{out}}_{i,j,z_{i,j}}$ then Bob changes $z_i = \bot$. Otherwise, Bob opens the commitments to obtain $\mathsf{out}_{i,j,z_{i,j}}$ values.
6. If $z_i = \bot$ for all i assigned to this bucket, then abort. If there are $z_i \neq z_{i'}$, neither of them $\bot$, in this bucket, then let j be some position for which $z_{i,j} \neq z_{i',j}$. Bob computes

$$\tilde{x}_k = (\mathsf{out}_{i,j,z_{i,j}} \oplus \delta_{i,j,z_{i,j}}) \oplus (\mathsf{out}_{i',j,z_{i',j}} \oplus \delta_{i',j,z_{i',j}}) \oplus \gamma_k$$

   and outputs $z^*_k = f(\tilde{x}_k, y_k)$. Otherwise, Bob outputs the unique value z^*_k such that $z_i \in \{\bot, z^*_k\}$ for all i in this bucket.

Fig. 4. (*Continued*)

Theorem 8. *There is a UC-secure batch NISC protocol in the programmable random oracle model, that evaluates N instances of f with total cost $N/O(\log N)$ times more than a single evaluation of f (plus some small additive terms that do not depend on f).*

References

1. Afshar, A., Mohassel, P., Pinkas, B., Riva, B.: Non-interactive secure computation based on cut-and-choose. In: Nguyen, P.Q., Oswald, E. (eds.) EUROCRYPT 2014. LNCS, vol. 8441, pp. 387–404. Springer, Heidelberg (2014). doi:10.1007/978-3-642-55220-5_22

2. Applebaum, B., Ishai, Y., Kushilevitz, E., Waters, B.: Encoding functions with constant online rate or how to compress garbled circuits keys. In: Canetti, R., Garay, J.A. (eds.) CRYPTO 2013. LNCS, vol. 8043, pp. 166–184. Springer, Heidelberg (2013). doi:10.1007/978-3-642-40084-1_10
3. Asharov, G., Lindell, Y., Schneider, T., Zohner, M.: More efficient oblivious transfer extensions with security for malicious adversaries. In: Oswald, E., Fischlin, M. (eds.) EUROCRYPT 2015. LNCS, vol. 9056, pp. 673–701. Springer, Heidelberg (2015). doi:10.1007/978-3-662-46800-5_26
4. Asharov, G., Lindell, Y., Schneider, T., Zohner, M.: More efficient oblivious transfer and extensions for faster secure computation. In: ACM CCS 2013, pp. 535–548. ACM Press, November 2013
5. Bellare, M., Hoang, V.T., Rogaway, P.: Adaptively secure garbling with applications to one-time programs and secure outsourcing. In: Wang, X., Sako, K. (eds.) ASIACRYPT 2012. LNCS, vol. 7658, pp. 134–153. Springer, Heidelberg (2012). doi:10.1007/978-3-642-34961-4_10
6. Bellare, M., Hoang, V.T., Rogaway, P.: Foundations of garbled circuits. In: ACM CCS 2012, pp. 784–796. ACM Press, October 2012
7. Cachin, C., Camenisch, J., Kilian, J., Müller, J.: One-round secure computation and secure autonomous mobile agents. In: Montanari, U., Rolim, J.D.P., Welzl, E. (eds.) ICALP 2000. LNCS, vol. 1853, pp. 512–523. Springer, Heidelberg (2000). doi:10.1007/3-540-45022-X_43
8. Canetti, R.: Universally composable security: a new paradigm for cryptographic protocols. In: 42nd FOCS, pp. 136–145. IEEE Computer Society Press, October 2001
9. Canetti, R., Jain, A., Scafuro, A.: Practical UC security with a global random oracle. In: Proceedings of ACM CCS 2014, pp. 597–608. ACM (2014)
10. Cascudo, I., Damgård, I., David, B., Giacomelli, I., Nielsen, J.B., Trifiletti, R.: Additively homomorphic UC commitments with optimal amortized overhead. In: Katz, J. (ed.) PKC 2015. LNCS, vol. 9020, pp. 495–515. Springer, Heidelberg (2015). doi:10.1007/978-3-662-46447-2_22
11. Frederiksen, T.K., Jakobsen, T.P., Nielsen, J.B., Nordholt, P.S., Orlandi, C.: MiniLEGO: efficient secure two-party computation from general assumptions. In: Johansson, T., Nguyen, P.Q. (eds.) EUROCRYPT 2013. LNCS, vol. 7881, pp. 537–556. Springer, Heidelberg (2013). doi:10.1007/978-3-642-38348-9_32
12. Frederiksen, T.K., Jakobsen, T.P., Nielsen, J.B., Trifiletti, R.: On the complexity of additively homomorphic UC commitments. In: Kushilevitz, E., Malkin, T. (eds.) TCC 2016. LNCS, vol. 9562, pp. 542–565. Springer, Heidelberg (2016). doi:10.1007/978-3-662-49096-9_23
13. Goyal, V., Mohassel, P., Smith, A.: Efficient two party and multi party computation against covert adversaries. In: Smart, N. (ed.) EUROCRYPT 2008. LNCS, vol. 4965, pp. 289–306. Springer, Heidelberg (2008). doi:10.1007/978-3-540-78967-3_17
14. Horvitz, O., Katz, J.: Universally-composable two-party computation in two rounds. In: Menezes, A. (ed.) CRYPTO 2007. LNCS, vol. 4622, pp. 111–129. Springer, Heidelberg (2007). doi:10.1007/978-3-540-74143-5_7
15. Huang, Y., Katz, J., Kolesnikov, V., Kumaresan, R., Malozemoff, A.J.: Amortizing garbled circuits. In: Garay, J.A., Gennaro, R. (eds.) CRYPTO 2014. LNCS, vol. 8617, pp. 458–475. Springer, Heidelberg (2014). doi:10.1007/978-3-662-44381-1_26
16. Ishai, Y., Kilian, J., Nissim, K., Petrank, E.: Extending oblivious transfers efficiently. In: Boneh, D. (ed.) CRYPTO 2003. LNCS, vol. 2729, pp. 145–161. Springer, Heidelberg (2003). doi:10.1007/978-3-540-45146-4_9

17. Ishai, Y., Kushilevitz, E., Ostrovsky, R., Prabhakaran, M., Sahai, A.: Efficient non-interactive secure computation. In: Paterson, K.G. (ed.) EUROCRYPT 2011. LNCS, vol. 6632, pp. 406–425. Springer, Heidelberg (2011). doi:10.1007/978-3-642-20465-4_23

18. Ishai, Y., Prabhakaran, M., Sahai, A.: Founding cryptography on oblivious transfer – efficiently. In: Wagner, D. (ed.) CRYPTO 2008. LNCS, vol. 5157, pp. 572–591. Springer, Heidelberg (2008). doi:10.1007/978-3-540-85174-5_32

19. Keller, M., Orsini, E., Scholl, P.: Actively secure OT extension with optimal overhead. In: Gennaro, R., Robshaw, M. (eds.) CRYPTO 2015. LNCS, vol. 9215, pp. 724–741. Springer, Heidelberg (2015). doi:10.1007/978-3-662-47989-6_35

20. Kilian, J.: Founding cryptography on oblivious transfer. In: 20th ACM STOC, pp. 20–31. ACM Press, May 1988

21. Kolesnikov, V., Kumaresan, R.: Improved OT extension for transferring short secrets. In: Canetti, R., Garay, J.A. (eds.) CRYPTO 2013. LNCS, vol. 8043, pp. 54–70. Springer, Heidelberg (2013). doi:10.1007/978-3-642-40084-1_4

22. Kolesnikov, V., Mohassel, P., Rosulek, M.: FleXOR: flexible garbling for XOR gates that beats free-XOR. In: Garay, J.A., Gennaro, R. (eds.) CRYPTO 2014. LNCS, vol. 8617, pp. 440–457. Springer, Heidelberg (2014). doi:10.1007/978-3-662-44381-1_25

23. Kolesnikov, V., Schneider, T.: Improved garbled circuit: Free XOR gates and applications. In: Aceto, L., Damgård, I., Goldberg, L.A., Halldórsson, M.M., Ingólfsdóttir, A., Walukiewicz, I. (eds.) ICALP 2008. LNCS, vol. 5126, pp. 486–498. Springer, Heidelberg (2008). doi:10.1007/978-3-540-70583-3_40

24. Lindell, Y.: Fast cut-and-choose based protocols for malicious and covert adversaries. In: Canetti, R., Garay, J.A. (eds.) CRYPTO 2013. LNCS, vol. 8043, pp. 1–17. Springer, Heidelberg (2013). doi:10.1007/978-3-642-40084-1_1

25. Lindell, Y., Pinkas, B.: An efficient protocol for secure two-party computation in the presence of malicious adversaries. In: Naor, M. (ed.) EUROCRYPT 2007. LNCS, vol. 4515, pp. 52–78. Springer, Heidelberg (2007). doi:10.1007/978-3-540-72540-4_4

26. Lindell, Y., Riva, B.: Cut-and-choose Yao-based secure computation in the online/offline and batch settings. In: Garay, J.A., Gennaro, R. (eds.) CRYPTO 2014. LNCS, vol. 8617, pp. 476–494. Springer, Heidelberg (2014). doi:10.1007/978-3-662-44381-1_27

27. Lindell, Y., Riva, B.: Blazing fast 2PC in the offline, online setting with security for malicious adversaries. In: Ray, I., Li, N., Kruegel, C. (eds.) ACM CCS 2015, pp. 579–590. ACM Press, October 2015

28. Mohassel, P., Franklin, M.: Efficiency tradeoffs for malicious two-party computation. In: Yung, M., Dodis, Y., Kiayias, A., Malkin, T. (eds.) PKC 2006. LNCS, vol. 3958, pp. 458–473. Springer, Heidelberg (2006). doi:10.1007/11745853_30

29. Mohassel, P., Riva, B.: Garbled circuits checking garbled circuits: more efficient and secure two-party computation. In: Canetti, R., Garay, J.A. (eds.) CRYPTO 2013. LNCS, vol. 8043, pp. 36–53. Springer, Heidelberg (2013). doi:10.1007/978-3-642-40084-1_3

30. Mohassel, P., Sadeghian, S.: How to hide circuits in MPC an efficient framework for private function evaluation. In: Johansson, T., Nguyen, P.Q. (eds.) EUROCRYPT 2013. LNCS, vol. 7881, pp. 557–574. Springer, Heidelberg (2013). doi:10.1007/978-3-642-38348-9_33

31. Nielsen, J.B., Orlandi, C.: LEGO for two-party secure computation. In: Reingold, O. (ed.) TCC 2009. LNCS, vol. 5444, pp. 368–386. Springer, Heidelberg (2009). doi:10.1007/978-3-642-00457-5_22

32. Pedersen, T.P.: Non-interactive and information-theoretic secure verifiable secret sharing. In: Feigenbaum, J. (ed.) CRYPTO 1991. LNCS, vol. 576, pp. 129–140. Springer, Heidelberg (1992). doi:10.1007/3-540-46766-1_9
33. Peikert, C., Vaikuntanathan, V., Waters, B.: A framework for efficient and composable oblivious transfer. In: Wagner, D. (ed.) CRYPTO 2008. LNCS, vol. 5157, pp. 554–571. Springer, Heidelberg (2008). doi:10.1007/978-3-540-85174-5_31
34. Pinkas, B., Schneider, T., Smart, N.P., Williams, S.C.: Secure two-party computation is practical. In: Matsui, M. (ed.) ASIACRYPT 2009. LNCS, vol. 5912, pp. 250–267. Springer, Heidelberg (2009). doi:10.1007/978-3-642-10366-7_15
35. Rindal, P., Rosulek, M.: Faster malicious 2-party secure computation with online/offline dual execution. In: Holz, T., Savage, S. (eds.) 25th USENIX Security Symposium, pp. 297–314. USENIX Association (2016)
36. Shelat, A., Shen, C.-H.: Two-output secure computation with malicious adversaries. In: Paterson, K.G. (ed.) EUROCRYPT 2011. LNCS, vol. 6632, pp. 386–405. Springer, Heidelberg (2011). doi:10.1007/978-3-642-20465-4_22
37. Shelat, A., Shen, C.-H.: Fast two-party secure computation with minimal assumptions. In: ACM CCS 2013, pp. 523–534. ACM Press, November 2013
38. Waksman, A.: A permutation network. J. ACM (JACM) 15(1), 159–163 (1968)
39. Yao, A.C.-C.: Protocols for secure computations (extended abstract). In: 23rd FOCS, pp. 160–164. IEEE Computer Society Press, November 1982
40. Yao, A.C.-C.: How to generate and exchange secrets (extended abstract). In: 27th FOCS, pp. 162–167. IEEE Computer Society Press, October 1986
41. Zahur, S., Rosulek, M., Evans, D.: Two halves make a whole. In: Oswald, E., Fischlin, M. (eds.) EUROCRYPT 2015. LNCS, vol. 9057, pp. 220–250. Springer, Heidelberg (2015). doi:10.1007/978-3-662-46803-6_8

Hashing Garbled Circuits for Free

Xiong Fan[1], Chaya Ganesh[2], and Vladimir Kolesnikov[3(✉)]

[1] Cornell University, Ithaca, NY, USA
xfan@cs.cornell.edu
[2] New York University, New York, NY, USA
ganesh@cs.nyu.edu
[3] Bell Labs, Murray Hill, NJ, USA
kolesnikov@research.bell-labs.com

Abstract. We introduce *Free Hash*, a new approach to generating Garbled Circuit (GC) hash at no extra cost during GC generation. This is in contrast with state-of-the-art approaches, which hash GCs at computational cost of up to 6× of GC generation. GC hashing is at the core of the cut-and-choose technique of GC-based secure function evaluation (SFE).

Our main idea is to intertwine hash generation/verification with GC generation and evaluation. While we *allow* an adversary to generate a GC $\widehat{GC}$ whose hash collides with an honestly generated GC, such a $\widehat{GC}$ w.h.p. will fail evaluation and cheating will be discovered. Our GC hash is simply a (slightly modified) XOR of all the gate table rows of GC. It is compatible with Free XOR and half-gates garbling, and can be made to work with many cut-and-choose SFE protocols.

With today's network speeds being not far behind hardware-assisted fixed-key garbling throughput, eliminating the GC hashing cost will significantly improve SFE performance. Our estimates show substantial cost reduction in typical settings, and up to factor 6 in specialized applications relying on GC hashes.

We implemented GC hashing algorithm and report on its performance.

1 Introduction

Today Garbled Circuit (GC) is the main technique for secure computation. It has advantages of high performance, low round complexity/low latency, and, importantly, relative engineering simplicity. Both core GC (garbling), as well as the meta-protocols, such as Cut-and-Choose (C&C), have been thoroughly investigated and are today highly optimized. Particularly in the semi-honest model there have been few asymptotic/qualitative improvements since the original protocols of Yao [Yao86] and Goldreich et al. [GMW87]. Possibly the most important development in the area of practical SFE since the 1980 s was the very efficient oblivious transfer (OT) extension technique of Ishai et al. [IKNP03]. This allowed the running of an arbitrarily large number of OTs by executing a small (security parameter) number of (possibly inefficient) "bootstrapping" OT instances and a number of symmetric key primitives. The cheap OTs made a

© International Association for Cryptologic Research 2017
J.-S. Coron and J.B. Nielsen (Eds.): EUROCRYPT 2017, Part III, LNCS 10212, pp. 456–485, 2017.
DOI: 10.1007/978-3-319-56617-7_16

dramatic difference for securely computing functions with large inputs relative to the size of the function, as well as for GMW-like approaches, where OTs are performed in each level of the circuit. Another important GC core improvement is the Free-XOR algorithm [KS08a], which allowed for the evaluation of all XOR gates of a circuit without any computational or communication costs.

As SFE moves from theory to practice, even "small" factor improvements can have a significant effect.

1.1   Motivation of Efficient GC Hashing: Cut-and-Choose (C&C) and Other Uses

In this work we improve (actually show how to achieve it for free) a core garbling feature of GC, circuit hashing. We discuss how this improves standard GC-based SFE protocols. We also discuss evaluation of certified functions, and motivate this use case.

GC hashing is an essential tool for C&C, and is employed in many uses of C&C. We start with describing C&C at the high level.

C&C. According to the "Cut-and-Choose Protocol" entry of the Encyclopedia of Cryptography and Security [TJ11], a (non-zero-knowledge) C&C protocol was first mentioned in the protocol of Rabin [Rab77] where this concept was used to convince a party that the other party sent it a specially formed integer n. The expression "cut and choose" was introduced later by Brassard et al. in [BCC88] in analogy to a popular cake-sharing problem: given a cake to be divided among two distrustful players, one of them cuts the cake in two shares, and lets the other one choose.

Recall, the basic GC protocol is not secure against cheating GC generator, who can submit a maliciously garbled circuit. Today, C&C is the standard tool in achieving malicious security in secure computation. At the high level, it proceeds as follows. CC generator generates a number of garbled circuits $GC_1, ..., GC_n$ and sends them to GC evaluator, who chooses a subset of them (say, half) at random to be opened (with the help of the generator) and verifies the correctness of circuit construction. If all circuits were constructed correctly, the players proceed to securely evaluate the unopened circuits, and take the majority output. It is easy to see that the probability of GC generator succeeding in submitting a maliciously garbled circuit is exponentially small in n. We note that significant improvement in the concrete values of n required for a specific probability guarantee was achieved by relatively recent C&C techniques [LP11, Lin13, HKE13, Bra13, LR14, HKK+14, AO12, KM15].

Using GC Hashing for C&C. What motivates our work is the following natural idea, which was first formalized in Goyal et al. [GMS08]. To save on communication (usually a more scarce resource than computation), GC generator, firstly, generates all the circuits $GC_1, ..., GC_n$ from PRG seeds $s_1, ..., s_n$. Then, instead of sending the circuits $GC_1, ..., GC_n$, it sends their hashes $H(GC_1), ..., H(GC_n)$. Finally, while the evaluation circuits will need to be sent in full over the network, only the seeds $s_1, ..., s_n$ need to be sent to verify that GC generator did not cheat in the generation of the opened circuits, saving a significant amount of communication at the cost of computing and checking $H(\mathsf{GC}_i)$ for all n circuits.

On many of today's computing architectures (e.g. Intel PC CPUs, with or without hardware AES), the cost of hashing the GC can be up to 6× greater than the cost of fixed-key garbling. At the same time, today's network speeds are comparable in throughput with hardware-assisted fixed-key garbling (see our calculations in Sect. 5.3). Hence, eliminating the GC hashing cost will improve SFE performance by eliminating the (smaller of the) cost of hashing or sending the open circuits. We stress that the use of our Free Hash requires syntactic changes in C&C protocols and it provides a security guarantee somewhat distinct from collision-resistant hash. Hence its use in C&C protocols should be evaluated for security. See Sect. 5.1 for more details.

Additionally, we show that a new computation/communication cost ratio offered by our free GC hash will allow for reduced communication, computation, and execution time, while achieving the same cheating probability.

SFE of Private Certified Functions. One advantage offered by GC is the hiding of the evaluated function from the evaluator. To be more precise, the circuit topology of the function is revealed, but this information leakage can be removed or mitigated by using techniques such as universal circuit [Val76, KS08b, LMS16, KS16] or circuit branch overlay [KKW16].

In practical scenarios, evaluated functions are to be selected as allowed by a mutually agreed policy, e.g., to prevent evaluation of identity function outputting player's private input. Then evaluating a hidden function presumes either a semi-honest GC generator, or employing a method for preventing/deterring out-of-policy GC generation. An efficient C&C approach does not seem to help prevent cheating here, since check circuits will reveal the evaluated function and will not be acceptable to the GC generator. Further, depending on policy/application, the zero-knowledge proofs of correctly constructing the circuits may be very expensive.

In many scenarios, Certificate Authorities (CA) may be used to certify the correct generation of GCs. Indeed, this is quite feasible at small to medium scale. Our motivating application here is the private attribute-based credential (ABC) checking. Very recent concurrent works [CGM16, KKL+16] showed for the first time that ABCs can be based on GCs. While both [CGM16, KKL+16] discuss public policy only, their GC-based constructions will not preclude achieving private policy. We note that this is a novel property in the ABC literature, where all previous work (in addition to supporting very small policies only) relied in an essential manner on the policy being known to both prover and verifier.

At the high level, the architecture/steps for evaluation of private CA-certified functions is as follows.

1. CA generates seeds $s_1, ... s_n$ and, for $i = 1, ... n$, CA generates GCs GC_i, GC hashes $H(\mathsf{GC}_i)$ and signatures $\sigma_i = \mathsf{Sign}_{CA}(H(\mathsf{GC}_i))$. It sends all $s_i, H(\mathsf{GC}_i), \sigma_i$ to ABC verifier V.

2. Prover P and V proceed with execution of the ABC protocols [CGM16, KKL+16], with the following modification:
   (a) Whenever GC GC_i needs to be sent by V, instead V generates GC_i from s_i and sends to P the pair (GC_i, σ_i).
   (b) P computes $H(GC)$ and verifies the signature σ_i prior to continuing. If the verification or GC evaluation fails, P outputs abort.

Free Hash will allow to significantly (up to factor 6) reduce the computational effort required by the CA to support such an application. Indeed the cost of the signature generation can be small and ignored in cases where the signed circuits are large, or a single signature can certify a number of circuits. The latter would be the case where two players may be expected to evaluate a number of circuits.

Importantly, evaluation of certified functions may be essential in scenarios where legislative and/or operational demands require high degree of accountability and auditability (recall, digital signatures are a recognized legal instrument in many countries [Wik]). These scenarios may frequently arise in government, intelligence or military applications.

Technical results of this work will have direct impact, up to factor 6 improvement, in the bottleneck (CA load) in many scenarios discussed above.

1.2   Our Contributions and Outline of the Work

We start the presentation with a brief discussion of related work and then providing a high-level technical overview of our approach. Then, in Sect. 2, we introduce existing definitions and constructions required for this work. In Sect. 3 we discuss definitional aspects, assumptions and parameter choices of our work.

We start technical Sect. 4 with introducing our proposed definition of GC hash security. Our definition is weaker than the standard hash collision guarantees, yet it is possible to make free hashing work with several standard GC constructions (cf. Sect. 5.1 for discussion about its C&C use). We then present hashed garbling algorithms for standard garbling (based on Just Garble of [BHKR13]) as well as for half-gates garbling of [ZRE15]. Our main contribution is the improvement of the state-of-the-art half-gates; we consider hashed Just Garble a valuable generalization and an instructional example.

In Sect. 5, we discuss the impact of Free Hash garbling and C&C. We report on our implementation and its performance evaluation. We discuss the application to certified circuits. We propose a unified cost metric (time) and show higher speeds/smaller computation and communication for the same error probability. We estimate total execution time reduction of about 43% for the C&C components of [LP11], and of about 64% for [AO12, KM15] in settings we consider (1Gbps channel and hardware AES).

1.3   Technical Overview of Free GC Hash

In this section we present the main intuition behind our technical approach.

We take advantage of the observation that the input to the hash is a garbled circuit GC, which must be evaluatable using the garbled circuit Eval function. We will not require standard hash collision resilience of GC strings, achieving which is very costly relative to the cost of GC generation. Instead, we guarantee that if an adversary can find another string $\widehat{\mathsf{GC}}$ that matches the hash of a correctly garbled GC, then with high probability, the garbled circuit property of $\widehat{\mathsf{GC}}$ is broken and its evaluation will fail.

We present our intuition iteratively; we start with a naive efficient approach, which we then refine and arrive at a secure hashed garbling. Recall, we start with a correctly generated GC GC with the set of output decoding labels d. Adversary's goal is to generate a circuit $\widehat{\mathsf{GC}}$ with the same hash as GC, and which will not fail evaluation/decoding given *the same* output labels d. This hash guarantee is sufficient for certain GC-based SFE protocols. A syntactic difference with [GMS08] C&C hashing is that verification of Free Hash involves GC evaluation, and is only possible once input labels are received (e.g., after OT of input labels). More importantly, Free Hash, as applied to C&C, provides a security guarantee subtly distinct from collision-resistant hash. Hence, drop-in replacement of [GMS08] C&C hashing with Free Hash may not be always possible, and in general should be done by hand and original proofs re-checked. See Sect. 5.1 for additional discussion.

We present the intuition for the classical four-row GC; we use similar ideas to achieve half-gates GC hashing as well. We present and prove secure both Free Hash constructions.

The first Free Hash idea is to simply set the hash of the garbled circuit to be the XOR of all garbled table (GT) rows of GC. This is clearly problematic, since a cheating garbler $\mathcal{A}$ can mount, for example, the following attack. $\mathcal{A}$ will set one GT entry to be the encryption of the wrong wire label. This affects the XOR hash as follows $H(\widehat{\mathsf{GC}}) = H(\mathsf{GC}) \oplus \Delta$. Now suppose the garbler knows (or guesses) which GT entry anywhere in GC will not be used in evaluation (inactive GT row). Now $\mathcal{A}$ simply replaces the inactive GT row X with value $X \oplus \Delta$. This will restore the hash to the desired value, and since this entry will not be used in the evaluation, the garbler will not be caught.

The following refinement of this approach counters the above attack: we make the gate's output wire key depend (in an efficient manner) on *all* GT rows of that gate. The idea is that XOR hash correction, such as above, will necessarily involve modification to an active GT row, which will affect the computed wire key on that gate. Importantly, because wire keys and GT rows are related via a random (albeit known) function, a GT row offset by Δ (needed to "fix" the hash) will result in effectively randomizing the output wire label of the gate. Because a non-failing evaluation requires output wire labels to be consistent with the fixed decoding information d, $\mathcal{A}$ will now be stuck.

We attempt this by starting with a secure garbling scheme $\mathcal{G}$, and modifying the way the wire labels are defined, as follows. The two wire labels w_i^0, w_i^1 associated with gate G_i's output wire will now be treated as temporary labels. A label W_i^j of the new scheme will be obtained from the w_i^j simply by XORing it with all the GT rows of G_i.

This is not quite sufficient, as it still allows the attacker to modify a GT row and then correct it within *the same* gate table. This is possible since a "fix" for the hash does not disrupt the validity of the wire label, as both the hash and the new wire label are defined in the same manner (as XOR of all the GT rows of G_i). Our final idea, is to use the GT rows as XOR pads in a different manner for computing the GC hash and for offsetting the wire values. This way, the fix for the hash w.h.p. will not simultaneously keep the wire label valid. We achieve this by malleating GT rows prior to using them as XOR pads in wire value computation.

It is not hard to show that the above changes preserve the privacy and authenticity properties of the garbling scheme.

We summarize the intuition for the hash security of the above construction. Consider a $\widehat{GC} \neq GC$ that collides under the above hash. Then, the evaluation of $\widehat{GC}$ will deviate from that of GC w.r.t. some wire label. Importantly, $\widehat{GC}$ evaluation can subsequently either return to a valid wire label or to a correct running hash, but not both. Thus, evaluation of $\widehat{GC}$ using encoding information $\widehat{e}$ cannot go back to both the wire label and the hash being correct.

1.4   Related Work

To our knowledge, there is no prior work specifically addressing hashing of GCs. At the same time, significant research effort has been expended on optimizing core GC performance. Work includes algorithmic GC improvements, such as Free XOR [KS08a], FleXOR [KMR14], half-gates [ZRE15], as well as optimizing underlying primitives, such as JustGarble [BHKR13]. Our work complements the existing GC improvement work.

Of course, the natural GC hashing approach works: just hash the generated GC. The problem with this is, of course, its cost. Relative cost of fixed-key cipher garbling and hashing are strongly architecture-dependent. They can be almost the same (e.g., when both AES and SHA are implemented in hardware). In another extreme, Intel's white paper [GGO+] reports that AES-NI evaluation of 16-byte blocks is 23× faster that of SHA1 (35, 965.9 vs 793, 718.7 KB/sec). In our experiments reported in Sect. 5.2, we observed about 6× performance difference between AES-NI and SHA1.

Improving on this, and motivated in part by the availability of fast hardware AES implementations, there was a short series of works [BRS02, RS08b, RS08a, BÖS11], implementing a hash function with three fixed-key AES function calls. A recent work of Rogaway and Steinberger [RS08a] constructs a class of linearly-determined, permutation-based compression functions $\{0, 1\}^{mn} \rightarrow \{0, 1\}^{rn}$ making k calls to the different permutations π_i for $i \in [k]$, where they named their construction as LPmkr. The fastest construction LP362 (12.09 cycles per byte) [BÖS11], with 6 calls to fixed-key AES would cost about 6× of that of fast garbling. Davies-Meyer-based hash construction [Win84] in the ideal cipher model considered in literature is reported to have similar speeds [BÖS11].

In comparison, our work eliminates the cost of hash whatsoever, while adding no cost to garbling or GC evaluation.

C&C and Uses of Hashed GC. There is a long sequence of GC-based SFE work, e.g. [Lin13, HKE13, Bra13, LR14, HKK+14, KM15], most of which uses some form of C&C or challenging the GC generator. Based on [GMS08], these works will benefit from our result, to varying degree. The exact performance benefit will depend on where the Free Hash is used, the ratio of evaluated/test circuits, as well as the computational/communication resources available to the players. In Sect. 5, we calculate performance improvement in several C&C protocols due to our GC hash.

2   Preliminaries

Notation. Let PPT denote probabilistic polynomial time. We let λ be the security parameter, $[n]$ denote the set $\{1, ..., n\}$, and $|t|$ denote the number of bits in a string. We denote the i-th bit value of a string s by $s[i]$, use $\|$ to denote concatenation of bit strings. We write $x \xleftarrow{R} \mathcal{X}$ to mean sampling a value x uniformly from the set $\mathcal{X}$. For a bit string s, we let $s^{\ll i}$ denote the bit string obtained by shifting s by i bits to the left. Throughout, by shift we mean a *circular* shift, where the vacant bit positions are filled not by zeros but by the shifted bits. $\mathsf{lsb}(s)$ denotes the least significant bit of string s. We say a function $f(\cdot)$ is negligible if $\forall c \in \mathbb{N}$, there exists $n_0 \in \mathbb{N}$ such that $\forall n \geq n_0$, it holds that $f(n) < n^{-c}$.

Let S be an infinite set and $X = \{X_s\}_{s \in S}, Y = \{Y_s\}_{s \in S}$ be distribution ensembles. We say X and Y are computationally indistinguishable, if for any PPT distinguisher $\mathcal{D}$ and all sufficiently large $s \in S$, we have $|\Pr[\mathcal{D}(X_s) = 1] - \Pr[\mathcal{D}(Y_s) = 1]| < 1/p(|s|)$ for every polynomial $p(\cdot)$.

Ideal Cipher Model. The Ideal Cipher Model (ICM) is an idealized model of computation, similar to the random oracle model (ROM) [BR93]. In ICM, one has a publicly accessible random block cipher (or ideal cipher). This is a block cipher with a k-bit key and a n-bit input/output, that is chosen uniformly at random among all block ciphers of this form; this is equivalent to having a family of 2^k independent random permutations. All parties including the adversary can make both encryption and decryption queries to the ideal block cipher, for any given key. ICM is shown to be equivalent to ROM [CPS08].

Collision-Resistant Hash Function. A hash function family $\mathcal{H}$ is a collection of functions, where each $H \in \mathcal{H}$ is a mapping from $\{0,1\}^m$ to $\{0,1\}^n$, such that $m > n$ and m, n are polynomials in security parameter λ. An instance $H \in \mathcal{H}$ can be described by a key which is public known. We say a hash function family $\mathcal{H}$ is *collision-resistant* if for any PPT adversary $\mathcal{A}$

$$\mathbf{Pr}[H \xleftarrow{R} \mathcal{H}, (x, x') \leftarrow \mathcal{A}(H) : x \neq x' \wedge H(x) = H(x')] = \mathsf{negl}(\lambda)$$

2.1   Yao's Construction

A comprehensive treatment of Yao's construction of garbled circuits, was given in [LP09]. At a high-level, in Yao's construction, each wire of the boolean circuit is associated with two random strings called wire labels or wire keys that encode logical 0 and 1 wire values. A garbled truth table is constructed for every gate in the circuit, where each combination of input wire labels is used to encrypt the appropriate output wire label as per the gate functionality. This results in four ciphertexts per gate, one for each input combination of the gate. The evaluator knows only one label for each input wire, and can therefore, open only one of the four ciphertexts.

2.2 Garbled Circuits

We make use of the abstraction of garbling schemes [BHR12] introduced by Bellare et al. At a high-level, a garbling scheme consists of the following algorithms: Gb takes a circuit as input and outputs a garbled circuit, encoding information, and decoding information. En takes an input x and encoding information and outputs a garbled input X. Eval takes a garbled circuit and garbled input X and outputs a garbled output Y. Finally, De takes a garbled output Y and decoding information and outputs a plain circuit-output (or an error $\perp$).

We note that this deviates from the definition of [BHR12], in that, we include $\perp$ in the range of the decoding algorithm De, so it now outputs a plain output value corresponding to a garbled output value or $\perp$ if the garbled output value is invalid. [JKO13] add an additional verification algorithm Ve to the garbling scheme. Formally, we define a *verifiable garbling scheme* by a tuple of functions $\mathcal{G} = (\mathsf{Gb}, \mathsf{En}, \mathsf{Eval}, \mathsf{De}, \mathsf{Ve})$ with each function defined as follows.

- *Garbling* algorithm $\mathsf{Gb}(1^\lambda, \mathcal{C})$: A randomized algorithm which takes as input the security parameter and a circuit $\mathcal{C} : \{0,1\}^n \rightarrow \{0,1\}^m$ and outputs a tuple of strings $(\mathsf{GC}, \{X_j^0, X_j^1\}_{j \in [n]}, \{Z_j^0, Z_j^1\}_{j \in [m]})$, where GC is the garbled circuit, the values $\{X_j^0, X_j^1\}_{j \in [n]}$ denote the input-wire labels, and the values $\{Z_j^0, Z_j^1\}_{j \in [m]}$ denote the output-wire labels.
- *Encode* algorithm $\mathsf{En}(x, \{X_j^0, X_j^1\}_{j \in [n]})$: a deterministic algorithm that outputs the input wire labels $\mathbf{X} = \{X_i^{x[i]}\}_{i \in [n]}$ corresponding to input x.
- *Evaluation* algorithm $\mathsf{Eval}(\mathsf{GC}, \{X_j\}_{j \in [n]})$: A deterministic algorithm which evaluates garbled circuit GC on input-wire labels $\{X_j\}_{j \in [n]}$, and outputs a garbled output $\mathbf{Y}$.
- *Decode* algorithm $\mathsf{De}(\mathbf{Y}, \{Z_j^0, Z_j^1\}_{j \in [m]})$: A deterministic algorithm that outputs the plaintext output corresponding to $\mathbf{Y}$ or $\perp$ signifying an error if the garbled output $\mathbf{Y}$ is invalid.
- *Verification* algorithm $\mathsf{Ve}(\mathcal{C}, \mathsf{GC}, \{Z_j^0, Z_j^1\}_{j \in [m]}, \{X_j^0, X_j^1\}_{j \in [n]})$: A deterministic algorithm which takes as input a circuit $\mathcal{C}$, garbled circuit GC, input-wire labels $\{X_j^0, X_j^1\}_{j \in [n]}$, and output-wire labels $\{Z_j^0, Z_j^1\}_{j \in [m]}$ and outputs accept if GC is a valid garbling of $\mathcal{C}$ and reject otherwise.

A verifiable garbling scheme may satisfy several properties such as *correctness, privacy, obliviousness, authenticity and verifiability*. We now review some of these notions: (1) *correctness*, (2) *privacy* (3) *authenticity*, and (4) *verifiability*. The definitions for correctness and authenticity are standard: correctness enforces that a correctly garbled circuit, when evaluated, outputs the correct output of the underlying circuit; authenticity enforces that the evaluator can only learn the output label that corresponds to the value of the function. *Verifiability* [JKO13] allows one to check that the garbled circuit indeed implements the specified plaintext circuit $\mathcal{C}$.

We include the definitions of these properties for completeness.

Definition 2.1 *(Correctness).* *A garbling scheme $\mathcal{G}$ is **correct** if for all input lengths $n \leq \mathsf{poly}(\lambda)$, circuits $\mathcal{C} : \{0,1\}^n \to \{0,1\}^m$ and inputs $x \in \{0,1\}^n$, the following probability is negligible in λ:*

$$\Pr(\mathsf{De}(\mathsf{Eval}(\mathsf{GC}, \{X_j^{x_j}\}_{j \in [n]}), \{Z_j^0, Z_j^1\}_{j \in [m]}) \neq \mathcal{C}(x) :$$

$$(\mathsf{GC}, \{X_j^0, X_j^1\}_{j \in [n]}, \{Z_j^0, Z_j^1\}_{j \in [m]}) \leftarrow \mathsf{Gb}(1^\lambda, \mathcal{C}))$$

Definition 2.2 *(Privacy).* *A garbling scheme $\mathcal{G}$ has **privacy** if for all input lengths $n \leq \mathsf{poly}(\lambda)$, circuits $\mathcal{C} : \{0,1\}^n \to \{0,1\}^m$, there exists a PPT simulator Sim such that for all inputs $x \in \{0,1\}^n$, for all probabilistic polynomial-time adversaries $\mathcal{A}$, the following two distributions are computationally indistinguishable:*

- REAL(f, x) : *run* $(\mathsf{GC}, e, d) \leftarrow \mathsf{Gb}(1^\lambda, \mathcal{C})$, *and output* $(\mathsf{GC}, \mathsf{En}(x, e), d)$.
- IDEAL$_{\mathsf{Sim}}(\mathcal{C}, f(x))$: *output* $\mathsf{Sim}(1^\lambda, \mathcal{C}, \mathcal{C}(x))$

Definition 2.3 *(Authenticity).* *A garbling scheme $\mathcal{G}$ is **authentic** if for all input lengths $n \leq \mathsf{poly}(\lambda)$, circuits $\mathcal{C} : \{0,1\}^n \to \{0,1\}^m$, inputs $x \in \{0,1\}^n$, and all probabilistic polynomial-time adversaries $\mathcal{A}$, the following probability is negligible in λ:*

$$\Pr \begin{pmatrix} \widehat{Y} \neq \mathsf{Eval}(\mathsf{GC}, \{X_j^{x_j}\}_{j \in [n]}) & (\mathsf{GC}, \{X_j^0, X_j^1\}_{j \in [n]}, \{Z^0, Z^1\}_{j \in [m]}) \leftarrow \mathsf{Gb}(1^\lambda, \mathcal{C}) \\ \wedge \mathsf{De}(\widehat{Y}, \{Z_j^0, Z_j^1\}_{j \in [m]}) \neq \perp & \widehat{Y} \leftarrow \mathcal{A}(\mathcal{C}, x, \mathsf{GC}, \{X_j^{x_j}\}_{j \in [n]}) \end{pmatrix}$$

Definition 2.4 *(Verifiability).* *A garbling scheme $\mathcal{G}$ is **verifiable** if for all input lengths $n \leq \mathsf{poly}(\lambda)$, circuits $\mathcal{C} : \{0,1\}^n \to \{0,1\}^m$, inputs $x \in \{0,1\}^n$, and all probabilistic polynomial-time adversaries $\mathcal{A}$, the following probability is negligible in λ:*

$$\Pr \begin{pmatrix} \mathsf{De}(\mathsf{Eval}(\mathsf{GC}, \mathsf{En}(x, e)), d) \neq \mathcal{C}(x) : & (\mathsf{GC}, e, d) \leftarrow \mathcal{A}(1^\lambda, \mathcal{C}) \\ & \mathsf{Ve}(\mathcal{C}, \mathsf{GC}, d, e) = \mathsf{accept} \end{pmatrix}$$

In the definition of verifiability above, we give the decoding information explicitly to the verification algorithm since in our construction the garbled circuit includes only the garbled tables and not the decoding information. We note that a natural and efficient way to obtain a verifiable garbling scheme is to generate GC by using the output of a pseudorandom generator on a seed as the random tape for Gb, and then provide the seed to the verification procedure Ve. Ve will regenerate the GC and the encoding and decoding tables, and will output accept for a garbled circuit if and only if it is equal to the generated one.

2.3   Free-XOR and Other Optimizations

Several works have studied optimizations to reduce the size of a garbled gate down from four ciphertexts. Garbled row-reduction was introduced by Naor et al. [NPS99]. There, instead of choosing the wire labels at random for each wire, they are chosen such that the first ciphertext will be the all-zero string, and hence need not be sent. In [PSSW09], the authors describe a way to further reduce the number of ciphertexts per gate to 2, by applying polynomial interpolation at each gate. Kolesnikov and Schneider [KS08a] introduced the Free XOR approach, allowing evaluation of XOR gates without any cost. Here, the idea is to choose wire labels such that the two labels on the same wire have the same (secret) offset across the entire circuit. The two labels for a given wire are of the form $(A, A \oplus \Delta)$, where Δ is secret and common to all wires. Now, as first proposed in [Kol05], an evaluator who has one of $(A, A \oplus \Delta)$ and one of $(B, B \oplus \Delta)$ can compute the XOR by simply XORing the wire labels. The result is either C or $C \oplus \Delta$ where $C = A \oplus B$ and correctly represents the result of XOR. Thus, no ciphertexts are needed for the XOR gate. Kolesnikov, Mohassel and Rosulek proposed a generalization of Free XOR called FleXOR [KMR14]. In FleXOR, each XOR gate can be garbled using 0, 1, or 2 ciphertexts, depending on certain structural properties of the circuit. In [ZRE15], the authors present a method built on Free XOR that can garble an AND gate using only two ciphertexts. This technique is also compatible with Free XOR. The idea is to write an AND gate as a combination of XOR and two *half-gates*, where a half-gate is an AND gate for which one party knows one of the inputs. The half-gates can be garbled with one ciphertext each, and the resulting AND gate, in combination with free-XOR, uses two ciphertexts.

3   Preliminary Discussion

3.1   Our Treatment of GC Topology and Formalization of the GC Representation

A formalization of what precisely the GC description string GC includes is often natural and hence is usually omitted from discussion. In our setting this an important aspect, as we focus on the collision resilience-related properties of GC strings, as well as on minimizing the size of GC and its computation time.

Firstly, we remind the reader that in the BHR [BHR12] notation the function Gb outputs the garbling *function* F. Since it is problematic to operate on functions, BHR regards Gb as operating on strings representing and defining the corresponding functions. In our notation, Gb outputs GC, which we treat as a string defining the evaluation process as well.

Clearly, GC will contain a set of garbled tables; the question is how to treat the circuit topology, i.e. exactly how to describe/define how Eval should process GC. One choice is to treat the plaintext circuit/topology as a part of GC. Because we focus on size/computation, this approach would cause some waste. Indeed, in most scenarios, the circuit and topology is known to both players, and hence could be implicit in GC.

Instead, we opt to consider the circuit description, including the locations of the free XOR gates as an externally generated string. It is certainly the case in SFE where the evaluated function is known to both players, and players can *a priori* adopt a convention on how to map the GC garbled gates to the circuit gates, hence defining the evaluation process. In PFE, which is the case in our certified function evaluation scenario (see Sect. 1.1), the evaluated function is *not* known to the evaluator. In this case, we still treat the topology/evaluation instructions as external to GC and assume that they are correctly delivered to the evaluator.

We note that in the certified function case, this can be naturally achieved by the CA signing the topology with a unique identifier, and including this identifier with GC and the hash of GC.

3.2 Our Assumptions

Our work optimizes high-performance primitives, and it is important to be clear on the assumptions we require of them so as to properly compare to related work.

We use the same primitives, and nearly identical constructions as JustGarble [BHKR13] and half-gates [ZRE15]. As a result, privacy and authenticity properties of our schemes hold under the same assumptions as [BHKR13, ZRE15], namely that the Davies-Meyer (DM) construction is a primitive meeting the guarantee of the random-permutation model (RPM). While [BHKR13] proves the security of their construction in RPM directly, [ZRE15] abstracts the DM security property as a variant of correlation-robust function. Our first (auxiliary) construction, namely, the privacy property, is proven under assumption that DM is correlation-robust.

To achieve hash security, we need to assume collision resistance of DM. We note that collision resistance of DM can be achieved e.g., by assuming that DM meets the requirement of the ideal-cipher model (ICM) [BRS02].

3.3 Cipher Instantiation

As noted above, we instantiate the key derivation function (KDF) calls as do [BHKR13, ZRE15], with the Davies-Meyer construction. Namely, the input X to KDF $H(X, i)$ are the 128-bit long wire keys, and i is an internal integer that simply increments per hash function call. We set $H_\pi(X, i) = \pi(K) \oplus K$, where $K = 2x \oplus i$ (π is assumed to be an ideal cipher, instantiated with 128-bit AES with randomly chosen key).

3.4 Hash Security Parameters

We use $\lambda = 128$-bit security parameter, which is standard for encryption and GCs. However, 128-bit hash domain is often seen as insufficient. This is because of the birthday attack, which provides time-space tradeoff for an attacker. Specifically, a collision-finding attacker can precompute and store a square-root number of hash images. Then by birthday paradox, a random collision will be found among these images with significant probability. This attack requires 2^{64} hash computations and efficiently accessible storage for 2^{64} hash values.

We argue that 128-bit hash security is nevertheless acceptable in SFE, if used carefully.

Firstly, we note that computing 2^{64} hashes is an extremely expensive task. Indeed, recent Bitcoin reports [Bra] suggest that world's hashing power recently peaked at 1 PetaHash per second (i.e. $1000^5 < 2^{50}$ hashes/sec). That is, global Blockchain hashing power can compute 2^{64} hashes in the order of 2^{14} s (or 4.5 h). Much more importantly, storage systems operate many orders of magnitude slower than CPUs and hashing ASICs, implying that storing and searching these hashes will take 10^3–10^6 times more time than generating them. Thus, extremely conservatively, we estimate that today a random hash collision may be found by engaging the *entire* Bitcoin mining system fitted with global-scale storage system in 4500 h (about 6 months).

In the majority of applications, the time and financial expense to achieve such a task will not be feasible.

Importantly, SFE hash checks have an *online* property, meaning that we can set up the system such that preprocessing or post-processing will not aid the attacker. Indeed, consider the SFE scenario and the following solution. In the existing fixed-key cipher-based protocols it is specified that the fixed key is chosen at random prior to GC generation. We can simply explicitly require that *both* players contribute to key generation, and that the selected key will be the one defining the fixed-key permutation used in GC. This will render any precomputation useless. Post-computation, while a threat to the privacy and, perhaps, authenticity of GC, is not helping the attacker, since the GC evaluator decision to accept or reject reached during the execution, is irrevocable. GC evaluator can set a generous time limit (e.g. several seconds or even minutes) after which it will abort the execution. The probability of $\mathcal{A}$ cheating via finding a 128-bit hash collision in this period of time sufficiently small, even given entire world's resources available to $\mathcal{A}$.

In sum, we have argued that using 128-bit hash security is appropriate for SFE and the applications we discuss in this work. Further, as eventually we move from 128-bit AES to next-generation of ciphers, our hash security guarantee will benefit from the transition.

4   GC Hashing Scheme

In this section, we define our hashed garbled circuit scheme. We capture the security guarantees we require from this new notion, and then present our construction that outputs a garbled circuit and its hash. Our garbled circuit construction satisfies the properties of correctness, authenticity and privacy. We then show that our construction is secure according to our hash security definition.

4.1   Hashed Garbled Circuit Security

Recall, we want to define hash security of garbled circuits with the same topology (cf. Sect. 3.1). We require that if the hash of such two garbled circuits collide, and one of them verifies correctly, then with high probability the other garbled circuit will fail evaluation. We now formalize this intuition in the definition below.

Definition 4.1 *(Hash security). A garbling scheme $\mathcal{G}$ is hash-secure with respect to a hash function $\mathcal{H}$ if for every boolean circuit $\mathcal{C}$, input x and PPT adversary $\mathcal{A}$,*

$$\Pr\left(\mathsf{De}(\mathsf{Eval}(\widehat{\mathsf{GC}},\mathsf{En}(x,\widehat{e})),d)\neq\bot:\begin{array}{c}\left(\mathsf{GC},\widehat{\mathsf{GC}},e=\{X_j^0,X_j^1\}_{j\in[m]},\right.\\ \left.\widehat{e}=\{\widehat{X}_j^0,\widehat{X}_j^1\}_{j\in[m]},d,h\right)\leftarrow\mathcal{A}(\mathcal{C},1^\lambda),\\ \mathsf{GC}\neq\widehat{\mathsf{GC}},\\ \mathsf{Topology}(\mathsf{GC})=\mathsf{Topology}(\widehat{\mathsf{GC}}),\\ \mathsf{Ve}(\mathcal{C},\mathsf{GC},d,e)=\mathsf{accept},\\ \mathcal{H}(\mathsf{GC})=\mathcal{H}(\widehat{\mathsf{GC}})=h)\end{array}\right)$$

is negligible in λ.

We point out that the decoding information d that results in failed decoding of $\widehat{\mathsf{GC}}$ is the same decoding information with respect to which GC successfully verifies, and this is essential to hash security. If we did not place this requirement, then an adversary can change d to $\widehat{d}$ which decodes any string that Eval on $\widehat{\mathsf{GC}}$ returns. We note that in full generality it is not necessary to require $\mathcal{A}$ to generate a GC passing the verification Ve of a specific circuit $\mathcal{C}$. We can achieve that if an $\mathcal{A}$ generates two unequal GCs with the same hash, at least one of them will always output $\bot$. However, the above Definition 4.1 reflects the typical use of GCs, and is sufficient for our construction.

In this work we consider verifiable garbling schemes with hash security. That is, $\mathcal{G} = (\mathsf{Gb}, \mathsf{En}, \mathsf{Eval}, \mathsf{De}, \mathsf{Ve}, \mathcal{H})$. Because we apply our constructions to secure computation, we will need schemes additionally satisfying the properties of correctness (cf. Definition 2.1) and privacy (cf. Definition 2.2). If needed, the authenticity property of GC (cf. Definition 2.3) can be achieved as well.

4.2   Our Construction

We now formalize the intuition of Sect. 1.3 on how to generate a GC hash for free when garbling. The full construction is presented in Fig. 1; here we provide additional intuition. Recall, in Sect. 1.3, we explained that after we generated (temporary) GC tables, we need to XOR their GT entries into the GC hash in one manner, and into the GC wire labels in another manner. In our construction, we do so by bitwise shifting the GT entries C_i prior to XORing them into the wire labels.

We note that we use bit shifting because it is fast and easy to implement, but a more general condition is sufficient for security of our scheme[1].

In presenting our construction, we adopt the approach used by [BHKR13] and others, where the gates are garbled as $H(w_i||w_j||r) \oplus w_k$, where w_i and w_j are wire labels on input wires, r is a nonce and w_k is a wire label on the output wire. H is a key-derivation function modeled as a random oracle.

The scheme we present below follows the standard point-and-permute optimization. This was introduced by Beaver et al. in [BMR90], where a select bit is appended to each wire label, such that the two labels on each wire have opposite select bits. This association between select bits and the logical truth values is random and kept secret. Now the garbled truth table can be arranged by these public select bits. The evaluator can select the correct ciphertext to decrypt based on the select bit instead of trying all four. For each wire label w, its least significant bit $\mathsf{lsb}(w)$ is reserved as a select bit that is used as in the point-and-permute technique, and complementary wire labels have opposite select bits. For the ith wire, define $p_i = \mathsf{lsb}(w_i^0)$. When using Free XOR, the global randomly chosen offset R is such that $\mathsf{lsb}(R) = 1$. Since $w_i^0 \oplus w_i^1 = R$ holds for each i in the circuit, we have that $\mathsf{lsb}(w_i^0) \neq \mathsf{lsb}(w_i^1)$.

To simplify presentation, in our constructions and notation we set the decoding information simply to be the output wire labels. We note, this does not preserve the authenticity property of GC. Authenticity can be easily achieved in our scheme, e.g. by instead setting the decoding information to be the collision-resistant hashes of the output labels. In more detail, let H be a collision-resistant hash function. The output translation table for a wire will now be $\{H(w_i^0), H(w_i^1)\}$. Given a garbled value w_i^b on an output wire, it is possible to determine whether it corresponds to the 0 or 1 key by computing $H(w_i^b)$ and checking whether it is equal to the first or second value in the pair. However, given this output translation table, it is not feasible to find the actual garbled values.

Let $H : \{0,1\}^* \to \{0,1\}^\lambda$ be a function, satisfying properties discussed in Sect. 3.2. For a function represented by a circuit $\mathcal{C} : \{0,1\}^n \to \{0,1\}^m$, we use $W_{\mathsf{in}}, W_{\mathsf{out}}$ to denote the input and output wires of f respectively, and G_{inter} for

---

[1] This condition is as follows. We set the wire labels of a gate output wire as a function of its temporary wire labels and the entries of the garbled gate table. Consider functions f_i such that, if

$$\bigoplus_{i=1}^{4} C_i = \bigoplus_{i=1}^{4} \widehat{C}_i$$

for $C_i \neq \widehat{C}_i$. Then,

$$\Pr[\bigoplus_{i=1}^{4} f_i(C_i) = \bigoplus_{i=1}^{4} f_i(\widehat{C}_i)]$$

is negligible. As we will later see in the proof, this is the property that we use in proving the hash security of our construction in proof of Theorem 4.6.

intermediate gates. The Free Hash garbling scheme $h\mathcal{G} = (\mathsf{Gb}, \mathsf{En}, \mathsf{De}, \mathsf{Eval}, \mathsf{Ve}, \mathcal{H})$ is described in Fig. 1.

The construction in Fig. 1 satisfies the properties of authenticity (cf. Definition 2.3), privacy (cf. Definition 2.2) and hash security (cf. Definition 4.1).

Theorem 4.2. *The Free Hash garbling scheme* $h\mathcal{G}$ *described in Fig. 1 satisfies privacy as in Definition 2.2 assuming the correlation robustness of* H.

Theorem 4.3. *The Free Hash garbling scheme* $h\mathcal{G}$ *described in Fig. 1 satisfies authenticity as in Definition 2.3 assuming the correlation robustness of* H.

We omit the proofs of privacy and authenticity in the main body, since our changes to the standard construction do not affect them, and closely follow the arguments of [BHKR13, ZRE15]. We include the proofs in the full version.

Hash Security. We now state and prove a technical lemma on which we rely for proving hash security (Theorem 4.6). The lemma below captures the following useful fact about GC and $\widehat{\mathsf{GC}}$: a gate in $\widehat{\mathsf{GC}}$ whose $\mathsf{pad}_{i,2}$ (XOR hash of the gate table) collides with that of the gate in GC will not be evaluated correctly (i.e. will not produce a valid label on the output wire) if the gate table is different, or if the input wire keys of the gate are different, or both. We say that a wire label, obtained during evaluation on input x encoded using $\widehat{e}$, is valid if it is one of the two possible wire labels for the same wire in GC. For presentation, we slightly abuse notation, by writing g_i to mean both the gate and the garbled table corresponding to the gate. It will be clear from context, which of the two is meant.

Definition 4.4 *(Valid key).* *Let* $(\mathsf{GC}, e, \widehat{\mathsf{GC}}, \widehat{e}, d, h)$ *be such that* $\mathsf{GC} \neq \widehat{\mathsf{GC}}$, $\mathsf{Topology}(\mathsf{GC}) = \mathsf{Topology}(\widehat{\mathsf{GC}}), \mathcal{H}(\widehat{\mathsf{GC}}) = \mathcal{H}(\mathsf{GC}) = h$ *and* $\mathsf{Ve}(\mathsf{GC}, d, e) =$ accept. *An internal wire key* $\widehat{K}_i^b$ *obtained on wire* w_i *during* Eval *of* $\widehat{\mathsf{GC}}$ *is called valid if* $\widehat{K}_i^b \in \{K_i^0, K_i^1\}$ *where* (K_i^0, K_i^1) *are the wire keys corresponding to* 0 *and* 1 *on wire* w_i *in* GC.

Lemma 4.5. *Let* $(\mathsf{GC}, e, \widehat{\mathsf{GC}}, \widehat{e}, d, h) \leftarrow \mathcal{A}(1^\lambda)$ *be such that* $\mathsf{GC} \neq \widehat{\mathsf{GC}}, \mathsf{Topology}(\mathsf{GC}) = \mathsf{Topology}(\widehat{\mathsf{GC}}), \mathcal{H}(\widehat{\mathsf{GC}}) = \mathcal{H}(\mathsf{GC}) = h$ *and* $\mathsf{Ve}(\mathsf{GC}, d, e) =$ accept. *Assuming* $\mathsf{pad}_{i,2} = \widehat{\mathsf{pad}}_{i,2}$, *evaluation of the garbled gate* $\widehat{g}_i$ *during* Eval *results in a valid wire label for the output wire of the gate with probability* $\mathsf{negl}(\lambda)$ *in the following cases:*

1. *Input wire keys to gate* $\widehat{g}_i$ *are valid, and* $\widehat{g}_i \neq g_i$.
2. *At least one input wire key to gate* $\widehat{g}_i$ *is invalid and* $\widehat{g}_i = g_i$.
3. *At least one input wire key to gate* $\widehat{g}_i$ *is invalid, and* $\widehat{g}_i \neq g_i$.

Proof. Let $g_i = \{C_1, C_2, C_3, C_4\}$ be the ith garbled table in GC and $\widehat{g}_i = \{\widehat{C_1}, \widehat{C_2}, \widehat{C_3}, \widehat{C_4}\}$ the ith garbled table in $\widehat{\mathsf{GC}}$.

- Gb(1^λ, $\mathcal{C}$): On input the security parameter λ and a circuit $\mathcal{C}$, choose $R \leftarrow \{0,1\}^{\lambda-1}\|1$ and set $h = 0$.
  1. For each input wire $W_i \in W_{\text{in}}$ of the circuit $\mathcal{C}$, set garbled labels in the following way: Randomly choose $K_i^0 \in \{0,1\}^\lambda$. Set $K_i^1 = K_i^0 \oplus R$. Set the garbled labels for input wire W_i as $w_i = (K_i^0, K_i^1)$.
  2. For each intermediate gate $G_i : W_c = g_i(W_a, W_b)$ of $\mathcal{C}$ in topological order:
     (a) Parse the garbled input labels as $w_a = (K_a^0, K_a^1)$ and $w_b = (K_b^0, K_b^1)$.
     (b) If G_i is an XOR gate, set garbled labels for the gate output wire W_c as $K_c^0 = K_a^0 \oplus K_b^0$, and $K_c^1 = K_c^0 \oplus R$.
     (c) If G_i is an AND gate
        • Choose temporary garbled labels for the gate output wire W_c as $T_c^0 \in \{0,1\}^\lambda$, and set $T_c^1 = T_c^0 \oplus R$.
        • Create G_i's garbled table: For each possible combination of G_i's input values $v_a, v_b \in \{0,1\}$, set $\tau_{v_a,v_b}^i = H(K_a^{v_a}|K_b^{v_b}|i) \oplus T_c^{g_i(v_a,v_b)}$. Sort entries τ^i in the table by input pointers, and let the entries be $C_{i,1}, C_{i,2}, C_{i,3}, C_{i,4}$.
        • For $d \in \{0,1\}$, compute:

        $$\mathsf{pad}_{i,1} = C_{i,1}^{\lll 1} \oplus C_{i,2}^{\lll 2} \oplus C_{i,3}^{\lll 3} \oplus C_{i,4}^{\lll 4}$$

        $$K_c^0 = T_c^0 \oplus \mathsf{pad}_{i,1}$$

        Set the garbled labels for wire W_c as

        $$w_c = (K_c^0, K_c^1), \text{ where } K_c^1 = K_c^0 \oplus R$$

        • Define

        $$\mathsf{pad}_{i,2} = C_{i,1} \oplus C_{i,2} \oplus C_{i,3} \oplus C_{i,4}$$

        $$h = h \oplus \mathsf{pad}_{i,2}$$

  3. For each output wire $W_i \in W_{\text{out}}$ of $\mathcal{C}$, set $d_i^0 = (0, K_i^0)$ and $d_i^1 = (1, K_i^1)$
  4. Output encoding information e, decoding information d, garbled circuit GC and hash $\mathcal{H}(\text{GC})$ as

     $$e = \{(K_i^0, K_i^1)\}_{W_i \in W_{\text{in}}}, d = \{(d_i^0, d_i^1)\}_{W_i \in W_{\text{out}}}, \text{GC} = \{\tau_{a,b}^i\}_{\substack{a,b \in \{0,1\} \\ G_i \in G_{\text{inter}}}}, \mathcal{H}(\text{GC}) = h$$

- En($\boldsymbol{x}$, e): On input encoding information e and input $\boldsymbol{x}$, output encoding $\mathbf{X} = \{X_i^{m[i]}\}_{i \in [n]}$.
- De($\mathbf{Y}$, d): On input the decoding information d and the garbled output of the circuit $\mathbf{Y} = (Y_1, ..., Y_m)$, for each output wire i of the circuit $\mathcal{C}$, parse d as $d = \{(0, K_i^0), (1, K_i^1)\}_{i \in [m]}$. Then, set $y_i = b$ if $Y_i = K_i^b$ and $y_i = \bot$ if $Y_i \notin \{K_i^0, K_i^1\}$. Output the result $\boldsymbol{y} = (y_1, ..., y_m)$ if $\forall i, y_i \neq \bot$. Else, output $\bot$.
- Eval(GC, $\mathbf{X}$): On input the garbled circuit GC and garbled input $\mathbf{X}$, for each gate $G_i : W_c = g_i(W_a, W_b)$ with garbled inputs $w_a = K_a^{v_a}, w_b = K_b^{v_b}$. If G_i is an XOR gate, compute $w_c^{g_i(v_a,v_b)} = K_a^{v_a} \oplus K_b^{v_b}$. If G_i is an AND gate:

  1. Let C_1, C_2, C_3, C_4 be the table entries. Compute $\mathsf{pad} = \bigoplus_{i=1}^{4} C_i^{\lll i}$.
  2. Decode the temporary output value from garbled table entry τ^i in position (v_a, v_b) as $T_c^{g_i(v_a,v_b)} = H(K_a^{v_a}|K_b^{v_b}|i) \oplus \tau^i$.
  3. Compute the garbled value as $w_c^{g_i(v_a,v_b)} = T_c^{g_i(v_a,v_b)} \oplus \mathsf{pad}$.

- Ve($\mathcal{C}$, GC, d, e): Check that each gate in GC correctly encrypts the gate in $\mathcal{C}$ given the encoding information e. If yes, then output accept, else output reject.
- $\mathcal{H}(GC)$: On input the garbled circuit GC, output h as the XOR of all ciphertexts,

  $$h = \bigoplus_{g_i} (C_{i,1} \oplus C_{i,2} \oplus C_{i,3} \oplus C_{i,4})$$

Fig. 1. The free hash garbling scheme h$\mathcal{G}$

Case 1 Since $\widehat{g}_i \neq g_i$, w.l.o.g., let $C_1 \neq \widehat{C}_1$. Since $\mathsf{pad}_{i,2} = \widehat{\mathsf{pad}}_{i,2}$, there must be (at least) one $j \neq 1$ such that $\widehat{C}_j \neq C_j$. Now, $\mathsf{pad}_{i,2} = \widehat{\mathsf{pad}}_{i,2}$ gives,

$$\widehat{C}_j \oplus \widehat{C}_1 = C_j \oplus C_1 \tag{1}$$

Let $\widehat{K} = (\widehat{K}_a, \widehat{K}_b)$ be the input wire key to gate g_i in $\widehat{\mathsf{GC}}$ during Eval, which by assumption is valid.

For the sake of contradiction, say, one of the ciphertexts, say, $\widehat{C}_1$, in $\widehat{g}_i$ gives a valid output wire key. Let T be the intermediate key obtained by decrypting $\widehat{C}_1$. Now validity of output wire key implies $T \oplus \widehat{\mathsf{pad}}_{i,1} = K \in \{K^0, K^1\}$.

$$T \oplus \widehat{\mathsf{pad}}_{i,1} = K$$
$$\widehat{C}_j^{\lll j} \oplus \widehat{C}_1^{\lll 1} = C_j^{\lll j} \oplus C_1^{\lll 1} \oplus R \tag{2}$$

where $R = T \oplus K \oplus \mathsf{pad}_{i,1}$ is a fixed value, and $T = H(\widehat{K}\|i) \oplus \widehat{C}_1$. Therefore, K is valid only when both (1) and (2) hold. We now argue that this happens with probability $\leq 1/2^\lambda$. By the assumption that $\mathsf{Ve}(\mathsf{GC}, d, e) = \mathsf{accept}$, C_1 and C_j are random keys masked by the outputs of the function H. If, therefore, a $\widehat{C}_1$ and $\widehat{C}_j$ that satisfies (1), also satisfies (2), then we can find r_1 and r_2 such that $r_1 \oplus r_2$ is δ for some fixed δ and $r_1^{\lll} \oplus r_2^{\lll}$ collides with the output of the function H on a fixed value. By collision resistance of the function H, this happens only with probability $\leq 1/2^\lambda$.

Case 2 $g_i = \widehat{g}_i$. Either $\widehat{K}_a \notin \{K_a^0, K_a^1\}$ or $\widehat{K}_b \notin \{K_b^0, K_b^1\}$ or both, where (K_a^0, K_a^1) and (K_b^0, K_b^1) are the wire keys corresponding to the input wires of $\widehat{g}_i$ in GC. Let (K^0, K^1) be the wire keys of the output wire of g_i.

For the sake of contradiction, say, one of the ciphertexts, say, C_1, gives a valid output wire key with $\widehat{K}$ as the input wire keys. Let T be the intermediate key obtained by decrypting C_1. Now validity of output wire key implies $T \oplus \mathsf{pad}_{i,1} = K \in \{K^0, K^1\}$. That is,

$$H(\widehat{K}\|i) \oplus C_1 \oplus \mathsf{pad}_{i,1} = K \tag{3}$$

K is valid when (3) holds, and that happens with negligible probability since we can find a r such that the output of H on r collides with a given value only with probability $\leq 1/2^\lambda$.

Case 3 W.l.o.g., let $C_1 \neq \widehat{C}_1$. Since $\mathsf{pad}_{i,2} = \widehat{\mathsf{pad}}_{i,2}$, there must be (at least) one $j \neq 1$ such that $\widehat{C}_j \neq C_j$.

Either $\widehat{K}_a \notin \{K_a^0, K_a^1\}$ or $\widehat{K}_b \notin \{K_b^0, K_b^1\}$ or both, where (K_a^0, K_a^1) and $(K_b^0 K_b^1)$ are the wire keys corresponding to the input wires of $\widehat{g}_i$ in GC. (K^0, K^1) be the wire keys of the output wire of g_i.

Now, $\mathsf{pad}_{i,2} = \widehat{\mathsf{pad}}_{i,2}$ gives,

$$\widehat{C}_j \oplus \widehat{C}_1 = C_j \oplus C_1 \tag{4}$$

Let $\widehat{K} = (\widehat{K}_a, \widehat{K}_b)$ be the input wire key to gate g_i in $\widehat{\mathsf{GC}}$ during Eval. Since $\widehat{K}$ is invalid by assumption, either $\widehat{K}_a \notin \{K_a^0, K_a^1\}$ or $\widehat{K}_b \notin \{K_b^0, K_b^1\}$ or both, where (K_a^0, K_a^1) and (K_b^0, K_b^1) are the wire keys corresponding to the input wires of $\widehat{g}_i$ in GC. (K^0, K^1) be the wire keys of the output wire of g_i.
For the sake of contradiction, say, one of the ciphertexts, say, $\widehat{C}_1$, in $\widehat{g}_i$ gives a valid output wire key. Let T be the intermediate key obtained by decrypting $\widehat{C}_1$. Now validity of output wire key implies $T \oplus \widehat{\mathsf{pad}}_{i,1} = K \in \{K^0, K^1\}$.

$$T \oplus \widehat{\mathsf{pad}}_{i,1} = K$$
$$\widehat{C_j}^{\ll j} \oplus \widehat{C_1}^{\ll 1} = C_j^{\ll j} \oplus C_1^{\ll 1} \oplus R \tag{5}$$

where $R = T \oplus K \oplus \mathsf{pad}_1$, and $T = H(\widehat{K}\|i) \oplus \widehat{C}_1$. Therefore, K is valid only when both (4) and (5) hold. We now argue that this happens with probability $\leq 1/2^\lambda$. By the assumption that $\mathsf{Ve}(\mathsf{GC}, d, e) = \mathsf{accept}$, C_1 and C_j are random keys masked by the outputs of the function H. If, therefore, $\widehat{K}$, $\widehat{C}_1$ and $\widehat{C}_j$ satisfy (4) and (5), then we can find r, r_1 and r_2 such that the output of the function H on r collides with $r_1^{\ll} \oplus r_2^{\ll}$ and $r_1 \oplus r_2$ is δ for some fixed δ. By collision resistance of the function H, this happens with probability at most $1/2^\lambda$.

When there is more than one $j \neq 1$ such that $\widehat{C}_j \neq C_j$ in cases (1) and (3) above, we will have,

$$\bigoplus_{j \neq 1} \widehat{C_j} \oplus \widehat{C_1} = \bigoplus_{j \neq 1} C_j \oplus C_1$$
$$\bigoplus_{j \neq 1} \widehat{C_j}^{\ll j} \oplus \widehat{C_1}^{\ll 1} = \bigoplus_{j \neq 1} C_j^{\ll j} \oplus C_1^{\ll 1} \oplus R$$

and the same arguments extend.   □

Theorem 4.6. *The Free Hash garbling scheme* h$\mathcal{G}$ *described in Fig. 1 satisfies hash security as defined in Definition 4.1 assuming the collision-resistance of H.*

Proof. Given an adversary $\mathcal{A}$ who outputs $(\mathsf{GC}, e, \widehat{\mathsf{GC}}, \widehat{e}, d, h)$ such that $\mathsf{GC} \neq \widehat{\mathsf{GC}}, \mathcal{H}(\widehat{\mathsf{GC}}) = \mathcal{H}(\mathsf{GC}) = h$, $\mathsf{Ve}(\mathsf{GC}, d, e) = \mathsf{accept}$, we show that $\forall x, \Pr[\mathsf{Eval}(\widehat{\mathsf{GC}}, \mathsf{En}(x, \widehat{e})) \neq \bot] = \mathsf{negl}(\lambda)$. Since $\mathsf{GC} \neq \widehat{\mathsf{GC}}$, they differ in at least one garbled gate. Let g_i be the first gate in topological order that differs in GC and $\widehat{\mathsf{GC}}$. When $\mathsf{pad}_{i,2} = \widehat{\mathsf{pad}}_{i,2}$ for all $\widehat{g}_i \neq g_i$, by case (1) of Lemma 4.5, we have that the output wire key for $\widehat{g}_i$ is invalid. Now, by inductively applying cases (2) and (3) of Lemma 4.5, all wire keys from then on, in topological order of evaluation remain invalid.

Now, when $\mathsf{pad}_{i,2} \neq \widehat{\mathsf{pad}}_{i,2}$, Eval on $\widehat{\mathsf{GC}}$ can return to a valid wire key for the output wire of $\widehat{g}_i \neq g_i$. Let us denote by $\widehat{\mathcal{H}}_i$ the running hash up until gate $\widehat{g}_i$ in $\widehat{\mathsf{GC}}$. Since $\mathsf{pad}_{i,2} \neq \widehat{\mathsf{pad}}_{i,2}$, we have $\widehat{\mathcal{H}}_i \neq \mathcal{H}_i$. By the assumption that $\mathcal{H}(\widehat{\mathsf{GC}}) = \mathcal{H}(\mathsf{GC})$, there must be a gate $\widehat{g}_j \neq g_j$ such that

$$\Delta = \bigoplus_{i:\text{pad}_{i,2} \neq \widehat{\text{pad}}_{i,2}} (\text{pad}_{i,2} \oplus \widehat{\text{pad}}_{i,2}) = \widehat{\mathcal{H}}_i \oplus \mathcal{H}_i \tag{6}$$

$$\widehat{\text{pad}}_{j,2} = \text{pad}_{j,2} \oplus \Delta \tag{7}$$

We now argue that the output wire of $\widehat{g}_j$ is invalid. From an argument similar to case (1) of Lemma 4.5 (since the input wire keys to $\widehat{g}_j$ are valid), (7) imposes a constraint on the ciphertexts of $\widehat{g}_j$. Thus the probability that output wire key is valid is bounded by the probability of finding r_1 and r_2 such that $r_1 \oplus r_2$ is δ for some fixed δ and $r_1^\ll \oplus r_2^\ll$ collides with the output of the function H on a fixed value. By collision resistance of the function H, this happens only with probability $\leq 1/2^\lambda$.

By Lemma 4.5 and the union bound, we have that, $\Pr[\text{De}(\text{Eval}(\widehat{\text{GC}}, \text{En}(x, \widehat{e})), d) \neq \perp] \leq |C|q^2/2^\lambda$, where $|C|$ is the number of gates in the circuit, and q is the number of queries to the function H that $\mathcal{A}$ is allowed to make.

Since the input x that lead to the above wire labels was arbitrary, we have that, given $\mathcal{H}(\widehat{\text{GC}}) = \mathcal{H}(\text{GC}), \text{GC} \neq \widehat{\text{GC}}, \text{Ve}(\text{GC}, d, e) = \text{accept}$,

$$\forall x, \Pr[\text{De}(\text{Eval}(\widehat{\text{GC}}, \text{En}(x, \widehat{e})), d) \neq \perp] = \text{negl}(\lambda)$$

$\square$

As calculated in the proof, the probability of hash collision is bounded by $|C|q^2/2^\lambda$. See Sect. 3.4 for discussion on parameter choices.

4.3 Hashing in Half-Gates Garbling Scheme

The current state of the art for garbled circuit construction is the half-gates scheme of Zahur et al. In the half-gates construction, two ciphertexts are used for each AND gate and the construction is compatible with the free-XOR technique [KS08a]. A half-gate is a garbled AND gate where one of the inputs to the gate is known in clear to one of the parties. Consider an AND gate $c = a \wedge b$. Now suppose the generator chooses a uniformly random bit r, and imagine we can have the evaluator learn the value of $r \oplus b$. We can write c as

$$c = a \wedge b = (a \wedge r) \oplus (a \wedge (r \oplus b))$$

[ZRE15] show how to garble the first AND gate with a generator-half-gate where the generator knows one of the values r, and the second AND gate with evaluator-half-gate since the evaluator know $r \oplus b$. The full AND gate is garbled by taking XOR of the two half-gates. Each garbled half-gate is one ciphertext, and with free-XOR, the full AND gate is two ciphertexts.

Let $\text{GC}' = (\text{Gb}', \text{En}', \text{De}', \text{Eval}')$ be the algorithms of the half-gate garbling procedure in [ZRE15]. The algorithms for encoding and evaluation in our scheme are the same; we only include the garbling and decoding algorithms, Gb and De. We assume that the half-gate garbling scheme outputs wire labels corresponding to both 0 and 1 on the output wires as the decoding information. Gb outputs

- Gb($1^\lambda, \mathcal{C}$): On input security parameter λ and a circuit $\mathcal{C}$, run the half-gate garbling algorithm $(e', d', \mathsf{GC}') \leftarrow \mathsf{Gb}'(1^\lambda, \mathcal{C})$, where $\mathsf{GC}' = \{\tau_{G_i}, \tau_{E_i}\}_{g_i \in G_{\mathsf{inter}}}$, and $d' = \{(d_i^0, d_i^1)\}_{W_i \in W_{\mathsf{out}}}$. Set the encoding information e, decoding information d, garbled circuit GC and hash $\mathcal{H}(\mathsf{GC})$ as

$$ e = e', \quad d = d', \quad \mathsf{GC} = \mathsf{GC}', \quad \mathcal{H}(\mathsf{GC}) = \bigoplus_i (\tau_{G_i} \oplus \tau_{E_i}) $$

- En is defined to be En$'$.
- Eval is defined to be Eval$'$.
- De($\mathbf{Y}, d$): On input the decoding information d and the garbled output of the circuit $\mathbf{Y} = (Y_1, ..., Y_m)$, for each output wire i of the circuit $\mathcal{C}$, parse d as $d = \{d_i^0, d_i^1\}_{i \in [m]}$. Then, set $y_i = b$ if $Y_i = d_i^b$ and $y_i = \bot$ if $Y_i \notin \{d_i^0, d_i^1\}$. Output the result $\boldsymbol{y} = (y_1, ..., y_m)$ if $\forall i, y_i \neq \bot$. Else, output $\bot$.

Fig. 2. The half-gate free hash garbling scheme half$\mathcal{G}$

a garbled circuit, the encoding and decoding information and the hash of the garbled circuit. De returns a decoded output or $\bot$ if the garbled output is invalid.

Note that in the construction of hashed garbling scheme for half-gates above, the hash is the XOR of all the ciphertexts. Unlike our construction for general garbled circuits (cf. Fig. 1), we do not modify the wire keys. Since the garbled circuit is the same as the original half-gates construction, we retain the privacy and authenticity properties. To argue hash security, first observe that in the half-gates scheme both ciphertexts in a garbled gate (one per half-gate) are decrypted and used for output wire computation. Consider an attacker $\mathcal{A}$ which modifies a gate table and changes one entry to decrypt to a wrong label. Then there must be another modified entry to correct the hash, and *both* modified entries need to decrypt correctly during evaluation to produce a valid label. Thus, in the half-gate garbling, the intuition for hash security is similar to that of our original 4-row construction. Namely, any modified gate will break the XOR hash. Further, any gate table that brings back the hash to the correct value will result in an invalid output wire label. We provide a proof sketch below.

Theorem 4.7. *The Half-Gate Free Hash garbling scheme* half$\mathcal{G}$ *described in Fig. 2 satisfies hash security as defined in Definition 4.1 assuming the collision-resistance of H.*

Proof Sketch. Given an adversary $\mathcal{A}$ who outputs $(\mathsf{GC}, e, \widehat{\mathsf{GC}}, \widehat{e}, d, h)$ such that $\mathsf{GC} \neq \widehat{\mathsf{GC}}, \mathcal{H}(\widehat{\mathsf{GC}}) = \mathcal{H}(\mathsf{GC}) = h$, $\mathsf{Ve}(\mathsf{GC}, d, e) = $ accept, we show that $\forall x, \Pr[\mathsf{Eval}(\widehat{\mathsf{GC}}, \mathsf{En}(\widehat{e}, x)) \neq \bot] = \mathsf{negl}(\lambda)$. Since $\mathsf{GC} \neq \widehat{\mathsf{GC}}$, they must differ in at least one garbled gate, and let $g_i \neq \widehat{g}_i$ be the first gate in topological order that differs: $g_i = \{\tau_{G_i}, \tau_{E_i}\}$ and $\widehat{g}_i = \{\widehat{\tau}_{G_i}, \widehat{\tau}_{E_i}\}$. Let $\widehat{\mathcal{H}}_i$ be the running hash up until gate $\widehat{g}_i$ in $\widehat{\mathsf{GC}}$. We consider the following cases:

1. $\widehat{\mathcal{H}}_i = \mathcal{H}_i$ where $\mathcal{H}_i$ is the running hash until gate g_i in GC. Now $g_i \neq \widehat{g}_i$ and $\widehat{\mathcal{H}}_i = \mathcal{H}_i$ implies that both half-gates are modified since $\widehat{g}_i$ is the first gate that differs from GC. That is,

$$\tau_{G_i} \neq \widehat{\tau}_{G_i} \text{ and } \tau_{E_i} \neq \widehat{\tau}_{E_i}$$

Let $(\widehat{K}_a, \widehat{K}_b)$ be the input wire keys of $\widehat{g}_i$. The output wire key of $\widehat{g}_i$ during Eval is given by

$$\widehat{K} = H(\widehat{K}_a) \oplus s_a \widehat{\tau}_{G_i} \oplus H(\widehat{K}_b) \oplus s_b(\widehat{\tau}_{E_i} \oplus \widehat{K}_a)$$

where s_a and s_b are select bits. The probability that $\widehat{K}$ is valid is at most $1/2^\lambda$ by the collision resistance of function H. Now, by inductively using argument similar to cases (2) and (3) of Lemma 4.5, the wire keys of $\widehat{\mathsf{GC}}$ remain invalid.

2. $g_i \neq \widehat{g}_i$, $\widehat{\mathcal{H}}_i \neq \mathcal{H}_i$ and $\mathcal{H}(\mathsf{GC}) = \mathcal{H}(\widehat{\mathsf{GC}})$ implies there must be a gate $\widehat{g}_j \neq g_j$ such that

$$\widehat{\tau}_{G_j} \oplus \widehat{\tau}_{E_j} = \widehat{\mathcal{H}}_i \oplus \mathcal{H}_i \oplus (\tau_{G_j} \oplus \tau_{E_j}) \tag{8}$$

We now argue that the output wire of $\widehat{g}_j$ is invalid: (8) imposes a constraint on the ciphertexts of $\widehat{g}_j$. Thus the probability that output wire key is valid is bounded by the probability of finding r_1 and r_2 such that $r_1 \oplus r_2$ is δ for some fixed δ and r_1 and r_2 collide with the outputs of function H. By collision resistance of H, this happens with probability at most $1/2^\lambda$. Again, inductively all further wire keys of $\widehat{\mathsf{GC}}$ remain invalid.

By the union bound, we have that, $\Pr[\mathsf{De}(\mathsf{Eval}(\widehat{\mathsf{GC}}, \mathsf{En}(\widehat{e}, x)), d) \neq \bot] \leq |C| q^2 / 2^\lambda$, where $|C|$ is the number of gates in the circuit, and q is the number of queries to the function H that $\mathcal{A}$ is allowed to make. □

As calculated in the proof, the probability of hash collision is bounded by $|C| q^2 / 2^\lambda$. See Sect. 3.4 for discussion on parameter choices.

5   Performance and Impact

5.1   Cut-and-Choose Protocols Using h$\mathcal{G}$

As pointed out in [GMS08], an improvement in communication complexity can be achieved by taking the following approach. To compute a garbled circuit, the garbler P_1 generates a random PRG seed. Then the output of the pseudorandom generator is used as the random tape for the garbling algorithm. In C&C, P_1 sends to P_2 only a collision-resistant (CR) hash of each GC. In a later stage of the protocol, if a GC GC is chosen as a check circuit and needs to be opened, P_1 simply sends the seed corresponding to that circuit to P_2.

h$\mathcal{G}$ hash can be used in C&C similarly to standard CR hash of GC. In [GMS08], P_1 commits via a collision resistant hash function to garbled circuits. These GCs can be either *good* or *cheating*. Importantly, due to the CR property of the hash, a malicious P_1 cannot change this designation at a later time. In using h$\mathcal{G}$, P_1 has the same choice: he can compute h$\mathcal{G}$ of either a good or a cheating GC. If he computed and sent the hash h of a good garbled circuit GC, then h cannot be claimed to match a cheating evaluation circuit $\widehat{\mathsf{GC}}$, even if the XOR hash $H(\mathsf{GC}) = H(\widehat{\mathsf{GC}})$. Indeed, w.h.p., evaluation of such a $\widehat{\mathsf{GC}}$ will fail and P_2 will abort, *independently* of P_2's input. Similarly, if P_1 computed and sent the hash of a cheating circuit $\widehat{\mathsf{GC}}$, it cannot be later opened as a good check circuit GC.

We stress that we must be careful when P_2 is allowed to abort, so as to not allow a selective failure attack. Specifically, a malicious P_1 could cause evaluation failure by sending an invalid label on a specific input wire/value pair or by generating a GC which produces an invalid label based on a value of an internal wire. Thus, while it is OK for P_2 to abort if it sees a GC which does not match the $\mathsf{h}\mathcal{G}$-hash, it should not (necessarily) abort simply based on seeing a decoding failure. Instead, this failure should be treated by the C&C procedure. We stress that it is protocol dependent, and protocol security should be evaluated. At the high level, our hashing guarantees that the garbler cannot open/equivocate an "honest" hashed circuit as a valid "malicious" circuit (or vice versa). However, he can open any (i.e. honest or malicious) hashed circuit as a "broken" one (i.e. one which will fail evaluation).

Covert C&C protocols [AL07,KM15], as well as C&C based on majority output, such as [LP11], can be made to work with $\mathsf{h}\mathcal{G}$. Indeed, exercising the extra power the adversary has (turning a good or bad evaluation circuit into a broken evaluation circuit) will simply cause covert evaluator to abort independently of its input. Similarly, in [LP11], the evaluation circuits which were made broken cannot be used to contribute to majority output. Using $\mathsf{h}\mathcal{G}$ with [KM15] requires a bit of care. [KM15] actually already explicitly support using [GMS08]. Using $\mathsf{h}\mathcal{G}$ differs from [GMS08] only in that a cheating P_1 can open an honest evaluation circuit as a broken one, resulting in an abort. However, the same effect could be achieved by P_1 sending an invalid signature on the garbled circuit.

We note that [Lin13] uses [LP11] as a basic step in cheating punishment and our $\mathsf{h}\mathcal{G}$ can be used within the [LP11] subprotocol of [Lin13]. However, it is not immediately clear $\mathsf{h}\mathcal{G}$ can be used elsewhere in [Lin13]. This is because the cheating punishment relies on evaluator having received a good evaluation circuit to recover the cheating garbler's input. However, in our case, malicious garbler can present a broken circuit, preventing input recovery.

Similarly, it is not immediately clear that the dual-execution C&C protocols of [HKE13,KMRR15] can take advantage of $\mathsf{h}\mathcal{G}$. Intuitively, this is because a malicious generator P_1 might produce a single cheating circuit, which is likely to be chosen for evaluation among a number of honest circuits. Then, P_1 will open all honest evaluation circuits as broken ones. Avoiding selective failure attack, P_2 will not abort, and the resulting output will depend on the output of the cheating circuit.

5.2   Implementation

We implemented our scheme using libgarble [Mal] for garbling and report on the performance below. In Table 1, we compare the cost of our GC hashing construction with garbling and then hashing the GC using SHA. We use the AES circuit to garble in the comparisons. The numbers in Table 1 are in cycles per gate. The configuration of the machine we used to run our implementation is: 2.3 GHz Core i5-2410M processor with 4 GB RAM. The processor has AES-NI integrated.

Table 1. Evaluation times of the AES circuits, in cycles per gate.

	Our h$\mathcal{G}$ construction	Garble + SHA	justGarble
Standard garbling	31.1	226.7	29
Half-gates	26.8	157.7	25.3

We believe that free hashing will simplify and speed up GC use particularly in larger systems using GC, such as the Blind Seer encrypted database [PKV+14, FVK+15], where GC processing will be competing for the CPU resource with a number of other tasks.

SFE of Private Certified Functions. We now consider the use case described in Sect. 1.1, where a Certificate Authority (CA) generates and certifies a number of GCs for use by the subscribers of the CA. In this case, clearly, CA is the bottleneck; Table 1 demonstrates over 6× performance improvement for the state-of-the-art half-gates GC, as compared with using standard hashing available with the OpenSSL library. Again, we stress that with half-gates hashing, simple XOR of all rows of all the gate tables provides a secure hash. This allows simple implementation in addition to the performance improvement.

5.3   Impact on Cut-and-Choose

We discuss the SFE performance improvement brought by our work on the example of the state-of-the-art approach of [LP11, KM15]. (Subsequent improvements to [LP11], as well as C&C, covert and other GC protocols will benefit from free GC hashing correspondingly). We review the C&C choices and parameters of [LP11, AO12, KM15] in light of [GMS08] and free hashing allowed by our work. We will show that:

1. Computing and sending additional GC hashes does not increase communication cost (computation cost is minimal due to our work), but significantly reduces cheating probability (see Table 2).
2. Keeping the cheating probability constant, we improve total C&C time by 43–64% by sending circuit hashes instead of circuits as suggested by [GMS08] (See Table 3).

For concreteness, to achieve a cheating probability of, say, 2^{-40}, the number of garbled circuits that need to be sent is n. This incurs a communication cost, in bits, of k, where $k = n\mathsf{C}$, and C is the cost of a garbled circuit.

Sending only the hashes of the garbled circuits in the beginning of the cut-and-choose, let the total number of garbled circuits be $\tilde{n}$. Let h be the size of the hash of a GC, which is the communication cost of a check circuit. Now, we have that the communication bits incurred,

$$\tilde{k} = \tilde{n}\mathsf{h} + \frac{1}{2}\tilde{n}\mathsf{C}$$

Setting the communication complexity to be the same, $\tilde{k} = k = n\mathsf{C}$, we have,

$$n = \tilde{n}\mathsf{q} + \frac{\tilde{n}}{2}$$

where $\mathsf{q} = \frac{\mathsf{h}}{\mathsf{C}}$ is the ratio of the cost of a check circuit and the cost of a garbled circuit. For $q < 1/2$, we have $\tilde{n} = \frac{n}{q+\frac{1}{2}} > n$, thus giving a cheating probability $2^{-\tilde{n}} < 2^{-n}$ for the same communication complexity. For large circuits, we expect $\mathsf{C} \gg \mathsf{h}$, giving concrete improvements in the security at no additional communication cost.

Table 2. Reducing cheating probabilities in [LP11] and [KM15] using h$\mathcal{G}$.

	Communication k	Number of circuits	Cheating probability/deterrence
[LP11]	$k = 125\|GC\|$	$n = 125$	2^{-40}
[LP11] with h$\mathcal{G}$, $q = 1/4$	$k = 125\|GC\|$	$\tilde{n} = 166$	2^{-51}
[LP11] with h$\mathcal{G}$, $q = 1/8$	$k = 125\|GC\|$	$\tilde{n} = 200$	2^{-62}
[KM15] without [GMS08][a]	$k = 10\|GC\|$	$n = 10$	0.9
[KM15] with h$\mathcal{G}$, $q = 1/4$	$k = 10\|GC\|$	$\tilde{n} = 36$	0.972
[KM15] with h$\mathcal{G}$, $q = 1/8$	$k = 10\|GC\|$	$\tilde{n} = 72$	0.986

[a] We note that [KM15] incorporates the [GMS08] hashing in the protocol. As we discussed, sending circuits over a fast channel may only be about 3× slower than hardware-assisted garbling, while computing SHA1 may be up to 6× slower than such garbling. Hence, sending circuits over a fast channel may actually be faster than generating SHA1 hash. Therefore, in our calculations for the fast channel setting as above, we consider [KM15] without [GMS08] hash.

Performance Improvement for Constant Cheating Probability. Consider the task of evaluating a billion-gate circuit (cf. [KSS12]). We show estimated improvement due to our technique as applied to [LP11,KM15]. We do this in terms of expended time by unifying the computation and communication costs of generating and sending garbled circuits. These calculations are not based on specific implementations or protocol definitions. Instead they are based on simple estimates of time needed to generate, hash and send GCs, and adding them together.

We first calculate and explain the computation and communication costs in seconds of our basic tasks.

According to [BHKR13], using JustGarble to garble the AES circuit (6660 non-XOR gates) takes 637 μs. Adjusting for size, we calculate that the time taken for GC generation for a circuit with 1 billion gates to be 95 s. For communication, assuming ideal scenario in 1 Gbps channel, assume we can send 1 billion bits/sec. Thus the time to send a circuit of 1 billion gates is 256 s at (assuming half gates and 2 × 128 bits per gate).

The total number of seconds needed in the cut-and-choose phase to maliciously evaluate a 1 billion-gate circuit with 2^{-40} cheating probability using previous technique and our construction using the optimal parameters. In our calculation we include the costs of generating, hashing (in our scheme) and sending the GCs. We do not include the cost of *regenerating the check circuits* at the evaluator's end that is incurred by our technique. This is because this cost is also incurred by other techniques. Indeed, checking correctness of a circuit that the evaluator already has (directly, or when using [GMS08] hash) is simplest and fastest by receiving its generating seed, reconstructing and comparing. We are concerned only with the cut-and-choose phase, and ignore the time taken for OT and GC evaluation in the protocol and show how our construction allow for reduced execution time in the cut-and-choose phase.

The cost in seconds calculated in Table 3 is obtained by adding the time to generate, hash (if needed) and send all the required garbled circuits. As explained above, we assume that it takes 95 s to generate a 1-Billion gate GC, and 256 s to send it (Table 4).

Finally, we note that even though we don't know whether the dual-execution C&C of Huang et al. [HKE13] could be modified to take advantage of our Free Hash, we point out that an improved balance between the check and evaluation circuits is possible when [HKE13] is used with the [GMS08] hash. We include the calculations of optimal parameters for [HKE13] in Appendix A.

Table 3. A billion-gate circuit. Execution time estimates of cut-and-choose with our improvements to achieve cheating probability of 2^{-40}

	Total number of circuits	Number of check circuits	Circuits sent	Time (in secs)
[LP11]	125	75	125	43875
[LP11] + h$\mathcal{G}$	125	75	50	24675

Table 4. A billion-gate circuit. Execution time estimates of cut-and-choose with our improvements to achieve deterrence of $\epsilon = 0.9$.

	Total number of circuits	Number of check circuits	Circuits sent	Time (in secs)
[AL07]	10	9	10	3510
[AL07] + h$\mathcal{G}$	10	9	1	1260
[KM15] without [GMS08][a]	10	9	10	3510
[KM15] + h$\mathcal{G}$	10	9	1	1260

[a]We note that [KM15] incorporates the [GMS08] hashing in the protocol. As we discussed, sending circuits over a fast channel may only be about 3× slower than hardware-assisted garbling, while computing SHA1 may be up to 6× slower than such garbling. Hence, sending circuits over a fast channel may actually be faster than generating SHA1 hash. Therefore, in our calculations for the fast channel setting as above, we consider [KM15] without [GMS08] hash.

Acknowledgments. We thank the anonymous reviewers of Eurocrypt 2017 for valuable comments. We also thank Mike Rosulek for a discussion on the applicability of the Free Hash. This work was supported by the Office of Naval Research (ONR) contract number N00014-14-C-0113.

A   Performance Calculation for the Protocol of [HKE13]

The idea in the protocol of [HKE13] is to have both parties play the role of the circuit constructor and circuit evaluator respectively in two simultaneous executions of a cut-and-choose of the protocol. The protocol is thus symmetric, and symmetric protocols might be desirable in certain situations since they have less idle time. The number of garbled circuits required in the cut-and-choose to achieve statistical security $2^{-\kappa}$ is $\kappa + O(\log \kappa)$. In the cut-and-choose of [HKE13], a party successfully cheats if it generates exactly $n - c$ incorrect circuits and none of them is checked by the other party. The probability that a cheating garbler succeeds,

$$\Pr[\mathcal{A}\ \text{wins}] = \frac{1}{\binom{n}{c}}$$

where n is the number of garbled circuits and c is the number of check circuits. It is easy to see that the above probability is minimized by setting $c = n/2$. This gives $\Pr[\mathcal{A}\ \text{wins}] = 2^{-n+\log n}$. We now apply the hash optimization in the cut-and-choose phase of the protocol. And now, we want compute the optimal value of c in the case where the communication cost of a circuit is a check circuit is cheaper than the cost of a garbled circuit that is evaluated. Let h be the cost of the hash of a garbled circuit. The cost of a check circuit is just the hash and the cost of an evaluation circuit is the cost of the hash plus the cost of a garbled circuit. Let $q = \frac{h}{C}$ be the ratio of the cost of a check circuit and the cost for an evaluation circuit, where $C = h + |GC|$. We have, the total communication complexity,

$$k = ch + eC$$

where e is the number of evaluation circuits, c the number of check circuits, $n = c + e$ the total number of circuits. Now, for a fixed k, given q we find the c and n that minimizes

$$\Pr[\mathcal{A}\ \text{wins}] = P = \frac{1}{\binom{n}{c}}$$

Using Stirling's approximation, we get,

$$P \approx \frac{(n-c)^{n-c+\frac{1}{2}} c^{c+\frac{1}{2}}}{n^{n+\frac{1}{2}}}$$

Let $r = \frac{c}{n}$ be the optimal fraction. Using $k = ch + eC$ and $q = \frac{h}{C}$, differentiating P with respect to c, and setting the first derivative to 0, we get,

$$r = (1 - r)^q \tag{9}$$

When q = 1, this gives $r = \frac{c}{n} = \frac{1}{2}$ which is indeed the optimal value when no hashes are used and a check circuit costs the same as an evaluation circuit. Now we compare the standard cut-and-choose with the cut-and-choose using hash and using optimal parameters as computed above. For security $2^{-\kappa}$, $\kappa + O(\log \kappa)$ circuits need to be sent in the standard cut-and-choose, which gives a communication $k = |GC|(\kappa + O(\log \kappa))$, (for each party) where $|GC|$ is the cost of a garbled circuit. Now given the cost of a hashed GC to be h, we get cost of a check circuit = h, $C = |GC| + h, q = \frac{h}{C}$. We now solve (9) for r and set

$$n = \frac{k}{rh + (1 - r)C} \text{ and } c = rn$$

This achieves a better cheating probability for the same communication k. In the table below, we compare the cheating probability for values of k and q. Recall, in the protocol, both parties act as sender and send $\kappa + O(\log \kappa)$ number of circuits each. The first column in Table 5 denoting the communication is the *total* communication of the cut-and-choose.

Table 5. Cheating probability in [HKE13] using [GMS08] hash and optimal parameters.

	Communication k	Optimal number of circuits	Optimal number check circuits	Cheating probability
[HKE13]	$k \approx 90|GC|$	$n = 45$	$c = n/2$	2^{-40}
[HKE13] + [GMS08] hash, $q = 1/4$	$k \approx 90|GC|$	$\tilde{n} = 71$	$c = 0.7\tilde{n} \approx 49$	2^{-60}
[HKE13] + [GMS08] hash, $q = 1/8$	$k \approx 90|GC|$	$\tilde{n} = 98$	$c = 0.8\tilde{n} = 78$	2^{-68}

References

[AL07] Aumann, Y., Lindell, Y.: Security against covert adversaries: efficient protocols for realistic adversaries. In: Vadhan, S.P. (ed.) TCC 2007. LNCS, vol. 4392, pp. 137–156. Springer, Heidelberg (2007). doi:10.1007/978-3-540-70936-7_8

[AO12] Asharov, G., Orlandi, C.: Calling out cheaters: covert security with public verifiability. In: Wang, X., Sako, K. (eds.) ASIACRYPT 2012. LNCS, vol. 7658, pp. 681–698. Springer, Heidelberg (2012). doi:10.1007/978-3-642-34961-4_41

[BCC88] Brassard, G., Chaum, D., Crépeau, C.: Minimum disclosure proofs of knowledge. J. Comput. Syst. Sci. **37**(2), 156–189 (1988)

[BHKR13] Bellare, M., Hoang, V.T., Keelveedhi, S., Rogaway, P.: Efficient garbling from a fixed-key blockcipher. In: 2013 IEEE Symposium on Security and Privacy, pp. 478–492. IEEE Computer Society Press, May 2013

[BHR12] Bellare, M., Hoang, V.T., Rogaway, P.: Foundations of garbled circuits. In: Proceedings of the ACM Conference on Computer and Communications Security, pp. 784–796. ACM (2012)

[BMR90] Beaver, D., Micali, S., Rogaway, P.: The round complexity of secure protocols. In: Proceedings of the Twenty-Second Annual ACM Symposium on Theory of Computing, pp. 503–513. ACM (1990)

[BÖS11] Bos, J.W., Özen, O., Stam, M.: Efficient hashing using the AES instruction set. In: Preneel, B., Takagi, T. (eds.) CHES 2011. LNCS, vol. 6917, pp. 507–522. Springer, Heidelberg (2011). doi:10.1007/978-3-642-23951-9_33

[BR93] Bellare, M., Rogaway, P.: Random oracles are practical: a paradigm for designing efficient protocols. In: Ashby, V. (ed.) ACM CCS 1993, pp. 62–73. ACM Press, November 1993

[Bra] Bradbury, D.: Bitcoin mining difficulty soars as hashing power nudges 1 petahash. http://www.coindesk.com/bitcoin-mining-difficulty-soars-hashing-power-nudges-1-petahash/. Accessed 3 Feb 2017

[Bra13] Brandão, L.T.A.N.: Secure two-party computation with reusable bit-commitments, via a cut-and-choose with forge-and-lose technique. In: Sako, K., Sarkar, P. (eds.) ASIACRYPT 2013. LNCS, vol. 8270, pp. 441–463. Springer, Heidelberg (2013). doi:10.1007/978-3-642-42045-0_23

[BRS02] Black, J., Rogaway, P., Shrimpton, T.: Black-box analysis of the block-cipher-based hash-function constructions from PGV. In: Yung, M. (ed.) CRYPTO 2002. LNCS, vol. 2442, pp. 320–335. Springer, Heidelberg (2002). doi:10.1007/3-540-45708-9_21

[CGM16] Chase, M., Ganesh, C., Mohassel, P.: Efficient zero-knowledge proof of algebraic and non-algebraic statements with applications to privacy preserving credentials. In: Robshaw, M., Katz, J. (eds.) CRYPTO 2016. LNCS, vol. 9816, pp. 499–530. Springer, Heidelberg (2016). doi:10.1007/978-3-662-53015-3_18

[CPS08] Coron, J.-S., Patarin, J., Seurin, Y.: The random oracle model and the ideal cipher model are equivalent. In: Wagner, D. (ed.) CRYPTO 2008. LNCS, vol. 5157, pp. 1–20. Springer, Heidelberg (2008). doi:10.1007/978-3-540-85174-5_1

[FVK+15] Fisch, B.A., Vo, B., Krell, F., Kumarasubramanian, A., Kolesnikov, V., Malkin, T., Bellovin, S.M.: Malicious-client security in blind seer: a scalable private DBMS. In: 2015 IEEE Symposium on Security and Privacy, pp. 395–410. IEEE Computer Society Press, May 2015

[GGO+] Gopal, V., Guilford, J., Ozturk, E., Gulley, S., Feghali, W.: Improving OpenSSL performance. https://software.intel.com/sites/default/files/open-ssl-performance-paper.pdf. Accessed 3 Feb 2017

[GMS08] Goyal, V., Mohassel, P., Smith, A.: Efficient two party and multi party computation against covert adversaries. In: Smart, N. (ed.) EUROCRYPT 2008. LNCS, vol. 4965, pp. 289–306. Springer, Heidelberg (2008). doi:10.1007/978-3-540-78967-3_17

[GMW87] Goldreich, O., Micali, S., Wigderson, A.: How to play any mental game or a completeness theorem for protocols with honest majority. In: Aho, A. (ed.) 19th ACM STOC, pp. 218–229. ACM Press, May 1987

[HKE13] Huang, Y., Katz, J., Evans, D.: Efficient secure two-party computation using symmetric cut-and-choose. In: Canetti, R., Garay, J.A. (eds.) CRYPTO 2013. LNCS, vol. 8043, pp. 18–35. Springer, Heidelberg (2013). doi:10.1007/978-3-642-40084-1_2

[HKK+14] Huang, Y., Katz, J., Kolesnikov, V., Kumaresan, R., Malozemoff, A.J.: Amortizing garbled circuits. In: Garay, J.A., Gennaro, R. (eds.) CRYPTO

2014. LNCS, vol. 8617, pp. 458–475. Springer, Heidelberg (2014). doi:10. 1007/978-3-662-44381-1_26

[IKNP03] Ishai, Y., Kilian, J., Nissim, K., Petrank, E.: Extending oblivious transfers efficiently. In: Boneh, D. (ed.) CRYPTO 2003. LNCS, vol. 2729, pp. 145–161. Springer, Heidelberg (2003). doi:10.1007/978-3-540-45146-4_9

[JKO13] Jawurek, M., Kerschbaum, F., Orlandi, C.: Zero-knowledge using garbled circuits: how to prove non-algebraic statements efficiently. In: Proceedings of the 2013 ACM SIGSAC Conference on Computer and Communications Security, pp. 955–966. ACM (2013)

[KKL+16] Kolesnikov, V., Krawczyk, H., Lindell, Y., Malozemoff, A.J., Rabin, T.: Attribute-based key exchange with general policies. Cryptology ePrint Archive, Report 2016/518 (2016). http://eprint.iacr.org/2016/518

[KKW16] Kennedy, W.S., Kolesnikov, V., Wilfong, G.: Overlaying circuit clauses for secure computation. Cryptology ePrint Archive, Report 2016/685 (2016). http://eprint.iacr.org/2016/685

[KM15] Kolesnikov, V., Malozemoff, A.J.: Public verifiability in the covert model (almost) for free. In: Iwata, T., Cheon, J.H. (eds.) ASIACRYPT 2015. LNCS, vol. 9453, pp. 210–235. Springer, Heidelberg (2015). doi:10.1007/ 978-3-662-48800-3_9

[KMR14] Kolesnikov, V., Mohassel, P., Rosulek, M.: FleXOR: flexible garbling for XOR gates that beats free-XOR. In: Garay, J.A., Gennaro, R. (eds.) CRYPTO 2014. LNCS, vol. 8617, pp. 440–457. Springer, Heidelberg (2014). doi:10.1007/978-3-662-44381-1_25

[KMRR15] Kolesnikov, V., Mohassel, P., Riva, B., Rosulek, M.: Richer efficiency/security trade-offs in 2PC. In: Dodis, Y., Nielsen, J.B. (eds.) TCC 2015. LNCS, vol. 9014, pp. 229–259. Springer, Heidelberg (2015). doi:10. 1007/978-3-662-46494-6_11

[Kol05] Kolesnikov, V.: Gate evaluation secret sharing and secure one-round two-party computation. In: Roy, B. (ed.) ASIACRYPT 2005. LNCS, vol. 3788, pp. 136–155. Springer, Heidelberg (2005). doi:10.1007/11593447_8

[KS08a] Kolesnikov, V., Schneider, T.: Improved garbled circuit: free XOR gates and applications. In: Aceto, L., Damgård, I., Goldberg, L.A., Halldórsson, M.M., Ingólfsdóttir, A., Walukiewicz, I. (eds.) ICALP 2008. LNCS, vol. 5126, pp. 486–498. Springer, Heidelberg (2008). doi:10.1007/ 978-3-540-70583-3_40

[KS08b] Kolesnikov, V., Schneider, T.: A practical universal circuit construction and secure evaluation of private functions. In: Tsudik, G. (ed.) FC 2008. LNCS, vol. 5143, pp. 83–97. Springer, Heidelberg (2008). doi:10.1007/ 978-3-540-85230-8_7

[KS16] Kiss, Á., Schneider, T.: Valiant's universal circuit is practical. In: Fischlin, M., Coron, J.-S. (eds.) EUROCRYPT 2016. LNCS, vol. 9665, pp. 699–728. Springer, Heidelberg (2016). doi:10.1007/978-3-662-49890-3_27

[KSS12] Kreuter, B., Shelat, A., Shen, C.-H.: Billion-gate secure computation with malicious adversaries. In: Proceedings of the 21st USENIX Conference on Security Symposium, Security 2012, p. 14. USENIX Association, Berkeley (2012)

[Lin13] Lindell, Y.: Fast cut-and-choose based protocols for malicious and covert adversaries. In: Canetti, R., Garay, J.A. (eds.) CRYPTO 2013. LNCS, vol. 8043, pp. 1–17. Springer, Heidelberg (2013). doi:10.1007/ 978-3-642-40084-1_1

[LMS16] Lipmaa, H., Mohassel, P., Sadeghian, S.: Valiant's universal circuit: improvements, implementation, and applications. Cryptology ePrint Archive, Report 2016/017 (2016). http://eprint.iacr.org/2016/017

[LP09] Lindell, Y., Pinkas, B.: A proof of security of Yao's protocol for two-party computation. J. Cryptol. **22**(2), 161–188 (2009)

[LP11] Lindell, Y., Pinkas, B.: Secure two-party computation via cut-and-choose oblivious transfer. In: Ishai, Y. (ed.) TCC 2011. LNCS, vol. 6597, pp. 329–346. Springer, Heidelberg (2011). doi:10.1007/978-3-642-19571-6_20

[LR14] Lindell, Y., Riva, B.: Cut-and-choose yao-based secure computation in the online/offline and batch settings. In: Garay, J.A., Gennaro, R. (eds.) CRYPTO 2014. LNCS, vol. 8617, pp. 476–494. Springer, Heidelberg (2014). doi:10.1007/978-3-662-44381-1_27

[Mal] Malozemoff, A.J.: libgarble: garbling library based on justgarble. https://github.com/amaloz/libgarble

[NPS99] Naor, M., Pinkas, B., Sumner, R.: Privacy preserving auctions and mechanism design. In: Proceedings of the 1st ACM Conference on Electronic Commerce, pp. 129–139. ACM (1999)

[PKV+14] Pappas, V., Krell, F., Vo, B., Kolesnikov, V., Malkin, T., Choi, S.G., George, W., Keromytis, A.D., Bellovin, S.: Blind seer: a scalable private DBMS. In: 2014 IEEE Symposium on Security and Privacy, pp. 359–374. IEEE Computer Society Press, May 2014

[PSSW09] Pinkas, B., Schneider, T., Smart, N.P., Williams, S.C.: Secure two-party computation is practical. In: Matsui, M. (ed.) ASIACRYPT 2009. LNCS, vol. 5912, pp. 250–267. Springer, Heidelberg (2009). doi:10.1007/978-3-642-10366-7_15

[Rab77] Rabin, M.O.: Digitalized signatures. Foundations of secure computation. In: Richard, A.D., et al. (eds.) Papers Presented at a 3 Day Workshop Held at Georgia Institute of Technology, Atlanta, pp. 155–166. Academic, New York (1977)

[RS08a] Rogaway, P., Steinberger, J.: Constructing cryptographic hash functions from fixed-key blockciphers. In: Wagner, D. (ed.) CRYPTO 2008. LNCS, vol. 5157, pp. 433–450. Springer, Heidelberg (2008). doi:10.1007/978-3-540-85174-5_24

[RS08b] Rogaway, P., Steinberger, J.: Security/efficiency tradeoffs for permutation-based hashing. In: Smart, N. (ed.) EUROCRYPT 2008. LNCS, vol. 4965, pp. 220–236. Springer, Heidelberg (2008). doi:10.1007/978-3-540-78967-3_13

[TJ11] Tilborg, H.C.A., Jajodia, S.: Encyclopedia of Cryptography and Security, 2nd edn. Springer, Heidelberg (2011). doi:10.1007/978-1-4419-5906-5

[Val76] Valiant, L.G.: Universal circuits (preliminary report). In: STOC, pp. 196–203. ACM Press, New York (1976)

[Wik] Wikipedia: Electronic signatures and law. https://en.wikipedia.org/wiki/Electronic_signatures_and_law. Accessed 21 Sept 2016

[Win84] Winternitz, R.S.: A secure one-way hash function built from DES. In: IEEE Symposium on Security and Privacy, p. 88. IEEE (1984)

[Yao86] Yao, A.C.-C.: How to generate and exchange secrets (extended abstract). In: 27th FOCS, pp. 162–167. IEEE Computer Society Press, October 1986

[ZRE15] Zahur, S., Rosulek, M., Evans, D.: Two halves make a whole. In: Oswald, E., Fischlin, M. (eds.) EUROCRYPT 2015. LNCS, vol. 9057, pp. 220–250. Springer, Heidelberg (2015). doi:10.1007/978-3-662-46803-6_8

Public-Key Encryption
and Key-Exchange

Adaptive Partitioning

Dennis Hofheinz[✉]

Karlsruhe Institute of Technology, Karlsruhe, Germany
Dennis.Hofheinz@kit.edu

Abstract. We present a new strategy for partitioning proofs, and use it to obtain new tightly secure encryption schemes. Specifically, we provide the following two conceptual contributions:

- A new strategy for tight security reductions that leads to compact public keys and ciphertexts.
- A relaxed definition of non-interactive proof systems for non-linear ("OR-type") languages. Our definition is strong enough to act as a central tool in our new strategy to obtain tight security, and is achievable both in pairing-friendly and DCR groups.

We apply these concepts in a generic construction of a tightly secure public-key encryption scheme. When instantiated in different concrete settings, we obtain the following:

- A public-key encryption scheme whose chosen-ciphertext security can be tightly reduced to the DLIN assumption in a pairing-friendly group. Ciphertexts, public keys, and system parameters contain 6, 24, and 2 group elements, respectively. This improves heavily upon a recent scheme of Gay et al. (Eurocrypt 2016) in terms of public key size, at the cost of using a symmetric pairing.
- The first public-key encryption scheme that is tightly chosen-ciphertext secure under the DCR assumption. While the scheme is not very practical (ciphertexts carry 28 group elements), it enjoys constant-size parameters, public keys, and ciphertexts.

Keywords: Public-key encryption · Tight security proofs

1 Introduction

Tight Security. Ideally, the only way to attack a cryptographic scheme S should be to solve a well-investigated, presumably hard computational problem P (such as factoring large integers). In fact, most existing constructions of cryptographic schemes provide such security guarantees, by exhibiting a security reduction. A reduction shows that any attack that breaks the scheme with some probability ε_S implies a problem solver that succeeds with probability ε_P. Of course, we would like ε_P to be as large as possible, depending on ε_S.

D. Hofheinz—Supported by DFG grants HO 4534/4-1 and HO 4534/2-2.

J.-S. Coron and J.B. Nielsen (Eds.): EUROCRYPT 2017, Part III, LNCS 10212, pp. 489–518, 2017.
DOI: 10.1007/978-3-319-56617-7_17

Specifically, we could call the quotient $\ell := \varepsilon_S/\varepsilon_P$ the *security loss* of a reduction.[1] A small value of ℓ is desirable, since it indicates a tight coupling of the security of the scheme to the hardness of the computational problem. It is also desirable that ℓ does not depend, e.g., on the number of considered instances of the scheme. Namely, when ℓ is linear in the number of instances, the scheme's security guarantees might vanish quickly in large settings. This can be a problem when being forced to choose concrete key sizes for schemes in settings whose size is not even known at setup time.

Hence, let us call a security reduction *tight* if its security loss ℓ only depends on a global security parameter (but not, e.g., on the number of considered instances, or the number of usages). Most existing cryptographic reductions are not tight. Specifically, it appears to be a nontrivial problem to construct tightly secure public-key primitives, such as public-key encryption, or digital signature schemes. (A high-level explanation of the arising difficulties can be found in [18].)

Existing Work on Tight Security. The importance of a tight security reduction was already pointed out in 2000 by Bellare, Boldyreva, and Micali [4]. However, the first chosen-ciphertext secure (CCA secure) public-key encryption (PKE) scheme with a tight security reduction from a standard assumption was only proposed in 2012, by Hofheinz and Jager [18]. Their scheme is rather inefficient, however, with several hundred group elements in the ciphertext. A number of more efficient schemes were then proposed in [2,3,5,7,12,14,17,21,26,27]. In particular, Chen and Wee [7] introduced a very useful partitioning strategy to conduct tight security reductions. Their strategy leads to very compact ciphertexts (of as few as 3 group elements [12], plus the message size), but also to large public keys. We will describe their strategy in more detail later, when explaining our techniques. Conversely, Hofheinz [17] presented a different partitioning strategy that leads to compact public keys, but larger ciphertexts (of 60 group elements). We give an overview over existing tightly secure PKE schemes (and some state-of-the-art schemes that are not known to be tightly secure for reference) in Fig. 1.

Our Contribution. In this work, we propose a new strategy to obtain tightly secure encryption schemes. This strategy leads to new tightly secure PKE schemes with simultaneously compact public keys and compact ciphertexts (cf. Fig. 1). In particular, our technique yields a practical pairing-based PKE scheme that compares well even with the recent tightly secure PKE scheme of Gay, Hofheinz, Kiltz, and Wee [12]. However, we should also note that our scheme relies on a symmetric pairing (unlike the scheme of [12], which can be instantiated even in DDH groups). Hence, the price we pay for a significantly smaller public key is that the scheme of [12] is clearly superior to ours in terms of

---

[1] Technically, we also need to take into account the complexity of the attacks on S and P. However, for this exposition, let us simply assume that the complexity of these attacks is comparable.

Scheme	$|pk|$	$|C| - |M|$	sec. loss	assumption	pairing
CS98 [8]	3	3	$O(q)$	DDH	no
KD04, HK07 [25, 19]	$k+1$	$k+1$	$O(q)$	k-LIN $(k \geq 1)$	no
HJ12 [18]	$O(1)$	$O(\lambda)$	$O(1)$	DLIN	yes
ADKNO13 [2]	$O(1)$	$O(\lambda)$	$O(1)$	DLIN	yes
HKS15 [21]	$O(\lambda)$	2	$O(\lambda)$	subgroup	yes
LPJY15 [26, 27]	$O(\lambda)$	47	$O(\lambda)$	DLIN	yes
AHY15 [3]	$O(\lambda)$	12	$O(\lambda)$	DLIN	yes
GCDCT16 [14]	$O(\lambda)$	$6k+4$	$O(\lambda)$	k-LIN $(k \geq 1)$	yes
H16 [17]	2	60	$O(\lambda)$	SXDH	yes
GHKW16 [12]	$2k\lambda$	$3k$	$O(\lambda)$	k-LIN $(k \geq 1)$	no
This work	$2k(k+5)$	$k+4$	$O(\lambda)$	k-LIN $(k \geq 2)$	yes
CS02 [9]	9	2	$O(q)$	DCR	—
CS03 [6]	3	2	$O(q)$	DCR	—
This work	20	28	$O(\lambda)$	DCR	—

Fig. 1. Comparison of CCA-secure public-key encryption schemes. λ is the security parameter, and q is the number of challenge ciphertexts. The sizes $|pk|$ and $|C| - |M|$ of public key (excluding public parameters) and ciphertext overhead are counted in group elements. For the ciphertext overhead $|C| - |M|$, we do not count smaller components (like MACs) inherited from the used symmetric encryption scheme.

computational efficiency. Besides, the use of a symmetric pairing might entail larger group sizes for comparable security.

Our technique also yields the first PKE scheme whose security can be tightly reduced to the Decisional Composite Residuosity (DCR [29]) assumption in groups of the form $\mathbb{Z}_{N^2}^*$ for RSA numbers $N = PQ$. To obtain the DCR instance of our scheme, we also introduce a new type of "OR-proofs" (i.e., a proof system to show disjunctions of simpler statements) in the DCR setting. We give more details on these proofs below.

We remark that our main scheme is completely generic, and can be instantiated both with prime-order groups, and in the DCR setting. Only some of our building blocks (such as the "OR-proofs" mentioned above) require setting-dependent instantiations, which we give both in a prime-order, and in the DCR setting.

Hence, we view our main contribution as conceptual. Indeed, in terms of computational efficiency, our encryption schemes do not outperform existing (non-tightly secure) schemes, even when taking into account our tight security reduction in the choice of key sizes. Still, we believe that specializations of our technique can lead to schemes whose efficiency is at least on par with that of existing non-tightly secure schemes.

1.1   Technical Overview

Technical Goal. To explain our approach, consider the following security game with an adversary $\mathcal{A}$. First, $\mathcal{A}$ obtains a public key, and then may ask for many encryptions of arbitrary messages. Depending on a single bit b chosen by the security game, $\mathcal{A}$ then either always gets an encryption of the desired message, or an encryption of a random message. Also, $\mathcal{A}$ has access to a decryption oracle, and is finally supposed to guess b (i.e., whether the encrypted ciphertexts contain the desired, or random messages). If no efficient $\mathcal{A}$ can predict b non-negligibly better than guessing, the used PKE scheme is considered CCA secure in the multi-challenge setting. Note that regular (i.e., single-challenge) CCA security implies CCA security in the multi-challenge setting using a hybrid argument (over the challenge encryptions $\mathcal{A}$ gets), but this hybrid argument incurs a large security loss. Hence, the difficulty in proving multi-challenge security is to randomize many challenge ciphertexts in as few steps as possible.

General Paradigm. All of the mentioned works on tightly secure PKE follow a general paradigm. Namely, in these schemes, each ciphertext $C = (c, \pi)$ carries some kind of "consistency proof" π that the plaintext message encrypted in c is intact. What this concretely means varies in different schemes. For instance, in some works [2,17,18,26,27], π is explicit and proves knowledge of the plaintext *or* of a valid signature on c. In other works [3,5,7,12,14,21], π is implicit, and proves knowledge of the plaintext *or* of a special authentication tag for that ciphertext. All of these works, however, use π to enable the security reduction to get leverage over the adversary $\mathcal{A}$, as follows. For instance, in the signature-based works above, the security reduction will be able to produce proofs π for ciphertexts with unknown plaintexts (by proving knowledge of a signature), while an adversary can only construct proofs from which the plaintext can be extracted. This enables the security reduction to implement a decryption oracle, while being able to randomize plaintexts encrypted for $\mathcal{A}$.

Chen and Wee's Approach. Chen and Wee [7] implement the above approach with an economic partitioning strategy (that in turn draws from an argument of Naor and Reingold [28]). Specifically, in their scheme, π implicitly proves knowledge of the plaintext *or* of a special tag T. Initially, T is constant, and committed to in the public key. In their security analysis, Chen and Wee introduce dependencies of T on the corresponding c. Specifically, in the i-th step of their analysis, they set $T = \mathbf{F}(\tau_{..i})$, where $\mathbf{F}$ is a random function, and $\tau_{..i}$ is the i-bit prefix of the hash τ of c. After a small number of such steps, T is a random value that is individual to each ciphertext. At this point, T is unpredictable for $\mathcal{A}$ on fresh ciphertexts, and hence $\mathcal{A}$'s decryption queries must prove knowledge of the respective plaintext. At the same time, the security game (which defines $\mathbf{F}$) can also prepare valid ciphertexts with unknown messages, and thus randomize all challenge ciphertexts at once.

We call the approach of Chen and Wee a partitioning strategy, since each hybrid step above proceeds as follows:

1. Partition the ciphertext space into two halves (in this case, according to the i-th bit of τ).
2. Change the definition of the "authentication tag" T for all ciphertexts from one half. (Keep the authentication tag for ciphertexts from the other half unchanged.)

In particular, the second step introduces an additional dependency of T on the bit τ_i. Most existing works use a partitioning strategy based on the individual bits of (the hash of) the ciphertext. An exception is the recent work [17], which implements a similar strategy based on an algebraic predicate of the ciphertext. This latter approach leads to shorter public keys, but requires relatively complex proofs π, and thus not only entails larger ciphertexts, but also requires a pairing.

Our Approach. Here, we also follow the generic paradigm sketched above, but refine the partitioning strategy of Chen and Wee. Namely, instead of partitioning the ciphertext space statically (e.g., through the hash of c), we add a special (encrypted) bit to π that determines the half in which the corresponding ciphertext is supposed to be. In contrast to previous works, that bit is not always known, not even to the security reduction itself. This change has several consequences:

- The bit that determines the partitioning in each ciphertext is easily accessible with a suitable decryption key, and so leads to a simple consistency proof π (and thus small ciphertexts). (This is in contrast to the scheme from [17], which proves complex statements in π.)
- The partitioning bit can by changed dynamically in challenge ciphertexts in different steps of the proof. Hence, a single "bit slot" can be used to partition the ciphertext space in many different ways during the proof. Eventually, this leads to compact public keys, since only few statements (about this single bit slot) need to be proven. (This is in contrast to partitioning schemes in which one proof for each bit position is generated.)
- However, since also the adversary can dynamically determine the partitioning of his ciphertexts from decryption queries, the security analysis becomes more complicated. Specifically, the reduction must cope with a situation in which an adversary submits a ciphertext for which the partitioning bit is not known.

In particular the last consequence will require additional measures in our security analysis. Namely, we will in some cases need to accept several authentication tags T in $\mathcal{A}$'s decryption queries, simply because we do not know in which half of the partitioning the corresponding ciphertext is. In fact, we will not be able to force $\mathcal{A}$ to use "the right" authentication tag in his decryption queries. We will only be able to force $\mathcal{A}$ to use an authentication tag T from a previous challenge ciphertext (since all other tags are unpredictable to $\mathcal{A}$). Hence, in order to eventually exclude that $\mathcal{A}$ produces ciphertexts without a proof of knowledge of the corresponding plaintext, we will need to work a bit more.

At this point, our main conceptual idea will be to introduce a dependency of T on a suitable value τ that is individual to each ciphertext. (While the construction in our scheme is slightly more complicated, one can think of τ as being simply the hash of the ciphertext.) Hence, in the first part of our analysis, we force $\mathcal{A}$ to reuse a tag T from a previous challenge ciphertext, while we tie this T to a ciphertext-unique value τ in the second part. When this is done, $\mathcal{A}$'s proofs π from decryption queries must prove knowledge of the encrypted plaintext message, *or* break the collision-resistance of the used hash function. Since the hash function will be assumed to be collision-resistant, $\mathcal{A}$ must prove knowledge of the respective plaintext in each decryption query. Hence, we can proceed with a proof of CCA security as in previous schemes.

Building Blocks. To implement our strategy, we require a variety of building blocks. Specifically, like previous works, we require re-randomizable (chosen-plaintext-secure) encryption, and universal hash proof systems for linear languages. We also require tightly secure one-time signatures, for which we give the first construction in the DCR setting. However, apart from our new partitioning strategy, the main technical innovation from our work is the construction of a non-interactive proof system for disjunctions (of simpler statements) in the DCR setting.

Namely, our proof system allows to prove that, given two ciphertexts c_1, c_2, at least one of them decrypts to zero. (In fact, the syntactics are a little more complicated, and in particular, honest proofs can only be formulated when the *first* ciphertext decrypts to zero. However, proofs that *one* of the two ciphertexts decrypts to zero can always be simulated using a special trapdoor, and we have soundness even in the presence of such simulated proofs.)

Such a proof system for disjunctions already exists in pairing-friendly groups (see [1]). A construction without pairings is far from obvious, though. Intuitively, the reason is that the language of pairs (c_1, c_2) as above (with at least one c_i that encrypts zero) is not closed under addition (of the respective plaintexts). Hence, disjunctions as above do not correspond to linear languages, and most common constructions (e.g., for universal hash proof systems [9,23]) do not apply. Our DCR-based construction thus is not linear, and relies on new techniques.

Concretely, our proof system can be viewed as a randomized variant of a universal hash proof system. Namely, depending on how many of the c_i do not encrypt zero, a valid proof reveals zero, one, or two linear equations about the secret verification key of our system. However, proofs in our system are randomized, and the revealed equations are also blinded with precisely one random value. Hence, up to one equation about the secret key is completely blinded. But as soon as both c_i encrypt nonzero values, a valid proof contains nontrivial information about the secret key. Thus, such proofs cannot be produced by an adversary who only sees proofs for valid statements (with at least one c_i that encrypts zero). Soundness follows as with regular universal hash proof systems.

Roadmap and Additional Content in Full Version

In Sect. 2, we recall some basic notation and definitions. In Sect. 3, we formulate an algebraic setting that allows to express both our DLIN-based and DCR-based schemes in a generic way. In Sect. 4, we recall some existing and construct some new necessary tightly secure building blocks. In Sect. 5, we introduce our notion of "benign" proof systems, and our DCR-based benign proof system for "OR"-like languages. Finally, in Sect. 6, we describe our new generic key encapsulation scheme.

Unfortunately, our work requires several rather technical concepts, and we need to outsource several proofs and additional discussion into the full version [16] of this paper. In particular, in [16], we discuss the security of our scheme in the multi-user setting, analyze its performance, and suggest optimizations. In the full version, we also present a new DCR-based tightly secure one-time signature scheme (which constitutes a technical building block for our main encryption scheme). Moreover, we present details for "more conventional" benign proof systems, and full details of the proof for our encryption scheme.

2   Preliminaries

Notation and Conventions. For a group $\mathbb{G}$ of order $|\mathbb{G}|$, a group element $g \in \mathbb{G}$, and a vector $\mathbf{u} = (u_1, \ldots, u_n)^\top \in \mathbb{Z}_{|\mathbb{G}|}^n$, we write $g^{\mathbf{u}} := (g^{u_1}, \ldots, g^{u_n})^\top \in \mathbb{G}^n$. Similarly, we define $g^{\mathbf{M}} \in \mathbb{G}^{n \times m}$ for matrices $\mathbf{M} \in \mathbb{Z}_{|\mathbb{G}|}^{n \times m}$. For integers $x, N \in \mathbb{Z}$ with $N > 0$, we define $[x]_N := x \bmod N$, and $[x]^N$ to be the unique integer with $x = [x]_N + N \cdot [x]^N$. Furthermore, we define the "absolute modular value" $|x|_N$ through

$$|x|_N := \begin{cases} [x]_N & \text{if } [x]_N < N/2 \\ [-x]_N & \text{if } [x]_N \geq N/2, \end{cases}$$

such that $0 \leq |x|_N \leq N/2$ in any case. Finally, we let $\left(\frac{x}{N}\right)$ denote the Jacobi symbol of x modulo N. For a bit $b \in \{0, 1\}$, we denote with $\bar{b} = 1 - b$ the complement of b. For a bitstring $x = (x_1, \ldots, x_n) \in \{0, 1\}^n$, we denote with $x_{..i} = (x_1, \ldots, x_i)$ the i-bit prefix of x, and with $x_{i..} = (x_i, \ldots, x_n)$ the $(n-i+1)$-bit postfix of x. For random variables $X, Y \in \{0, 1\}^*$, we let $\mathbf{SD}(X; Y)$ denote their statistical distance, and $H_\infty(X)$ the min-entropy of X.

Global Public Parameters. To simplify notation, we assume that all algorithms in this work (including adversaries) implicitly receive public parameters pp as input. In our case, these public parameters will contain the description of algebraic groups and related algorithms, and a collision-resistant and a universal hash function. We give more details on these parameters when we discuss the algebraic setting, collision-resistant hashing, and our key extractor (which uses the universal hash function).

Collision-Resistant Hashing. We require collision-resistant hashing:

Definition 1 (Collision-resistant hashing). *A hash function generator is a PPT algorithm* **CRHF** *that, on input 1^λ, outputs (the description of) an efficiently computable function $H : \{0, 1\}^* \to \{0, 1\}^{\ell_H}$. We say that* **CRHF** *outputs* *collision-resistant hash functions H (or, slightly abusing notation, that* **CRHF** *is collision-resistant), if*

$$\text{Adv}^{\text{crhf}}_{\text{CRHF},\mathcal{A}}(\lambda) = \Pr\left[x \neq x' \land H(x) = H(x') \mid (x, x') \leftarrow \mathcal{A}(1^\lambda, H)\right]$$

(where $H \leftarrow \text{CRHF}(1^\lambda)$) is negligible for every PPT adversary $\mathcal{A}$.

We assume that the public parameters pp contain a function H sampled with a hash function generator **CRHF**.

Universal Hashing, and Randomness Extraction. We also assume a family **UHF** = **UHF**$_\lambda$ of universal hash functions $h : \{0,1\}^* \to \{0,1\}^\lambda$. Since universal hash functions are good randomness extractors, we in particular have that for any random variable X with min-entropy $H_\infty(X) \geq 3\lambda$,

$$\text{SD}\big((h, h(X)); (h, R)\big) \leq 1/2^\lambda,$$

where $h \in \text{UHF}_\lambda$ and $R \in \{0,1\}^\lambda$ are uniformly chosen.

Key Encapsulation Mechanisms, and Multi-challenge Security. A key encapsulation mechanism (KEM) **KEM** consists of PPT algorithms (**Gen, Enc, Dec**). Key generation **Gen**(1^λ) outputs a public key pk and a secret key sk. Encapsulation **Enc**(pk) takes a public key pk, and outputs a ciphertext c, and a session key K. Decapsulation **Dec**(sk, c) takes a secret key sk, and a ciphertext c, and outputs a session key K. For correctness, we require that for all (pk, sk) in the range of **Gen**(1^λ), and all (c, K) in the range of **Enc**(pk), we always have **Dec**$(sk, c) = K$. Security is defined as follows:

Definition 2 (Multi-challenge ciphertext indistinguishability). *Given a key encapsulation scheme* **KEM**, *consider the following game between a challenger $\mathcal{C}$ and an adversary $\mathcal{A}$:*

1. *$\mathcal{C}$ samples a keypair through $(pk, sk) \leftarrow$ **Gen**(1^λ), and chooses a uniform bit $b \leftarrow \{0,1\}$.*
2. *$\mathcal{A}$ is invoked on input $(1^\lambda, pk)$, and with (many-time) access to the following oracles:*
   - *$\mathcal{O}_{\text{enc}}()$ runs $(c, K) \leftarrow$ **Enc**(pk), sets $K_0 = K$, samples a fresh $K_1 \leftarrow \{0,1\}^\lambda$, and returns (c, K_b).*
   - *$\mathcal{O}_{\text{dec}}(c)$ returns $\bot$ if c is a previous output of $\mathcal{O}_{\text{enc}}$. Otherwise, $\mathcal{O}_{\text{dec}}$ returns $K \leftarrow$ **Dec**(sk, c).*
3. *Finally, $\mathcal{A}$ outputs a bit b', and $\mathcal{C}$ outputs 1 iff $b = b'$.*

Let

$$\text{Adv}^{\text{mcca}}_{\text{KEM},\mathcal{A}}(\lambda) = \Pr\left[\mathcal{C} \text{ outputs } 1\right] - 1/2.$$

We say that **KEM** *has indistinguishable ciphertexts under chosen-ciphertext attacks in the multi-challenge setting (short: is IND-MCCA secure) if and only if $\text{Adv}^{\text{mcca}}_{\text{KEM},\mathcal{A}}(\lambda)$ is negligible for all PPT $\mathcal{A}$.*

Secure KEM schemes imply secure PKE schemes [8], and the corresponding security reduction is tight also in the multi-challenge setting. Hence, like [12], we will focus on obtaining an IND-MCCA secure KEM scheme in the following.

3   The Generic Algebraic Setting

3.1   The Generic Setting

Groups and Public Parameters. In the following, let $\mathbb{G}$ be a group of order $|\mathbb{G}|$. We require that $|\mathbb{G}|$ is square-free, and only has prime factors larger than 2^λ. Furthermore, we assume two subgroups $\mathbb{G}_1, \mathbb{G}_2 \subseteq \mathbb{G}$ of order $|\mathbb{G}_1|$ and $|\mathbb{G}_2|$, respectively, and such that $\mathbb{G}_1 \cdot \mathbb{G}_2 = \{h_1 \cdot h_2 \mid h_1 \in \mathbb{G}_1, h_2 \in \mathbb{G}_2\} = \mathbb{G}$. Note that we neither require nor exclude that $|\mathbb{G}|$ (or $|\mathbb{G}_1|$ or $|\mathbb{G}_2|$) is prime, or that $\mathbb{G}_1 \cap \mathbb{G}_2$ is trivial.

We assume that the global public parameters pp include

- (descriptions of) $\mathbb{G}$, $\mathbb{G}_1$, and $\mathbb{G}_2$,
- fixed generators g of $\mathbb{G}$, g_1 of $\mathbb{G}_1$ and g_2 of $\mathbb{G}_2$,
- the group order $|\mathbb{G}_2|$ of $\mathbb{G}_2$,
- a positive integer $\ell_\mathbf{B}$, and a matrix $g_1^\mathbf{B}$, for $\mathbf{B} \in \mathbb{Z}_{|\mathbb{G}_1|}^{\ell_\mathbf{B} \times \ell_\mathbf{B}}$.[2]

We stress that these parameters may depend on λ, and note that $|\mathbb{G}|$, $|\mathbb{G}_1|$, and $\mathbf{B}$ do not need to be public. However, we do require that there are efficient algorithms for the following tasks:

- performing the group operation in $\mathbb{G}$,
- sampling uniformly distributed $\mathbb{Z}_{|\mathbb{G}_1|}$-elements,
- recognizing $\mathbb{G}$ (i.e., deciding group membership in $\mathbb{G}$).

Since we assume $|\mathbb{G}_2|$ to be public, we also have algorithms for deciding membership in $\mathbb{G}_2$, and for uniformly sampling from $\mathbb{Z}_{|\mathbb{G}_2|}$ and $\mathbb{G}_2$, and thus also from $\mathbb{Z}_{|\mathbb{G}|}$ and $\mathbb{G}$.

Computational Assumptions. In our generic setting, we will use an assumption that can be seen as a combination of the Extended Decisional Diffie-Hellman assumption from [15], and the Matrix Decisional Diffie-Hellman assumption from [10].

Definition 3 (Generalized DDH, combining [10,15]). *We say that the Generalized Decisional Diffie-Hellman (GDDH) assumption holds in our setting if the following advantage is negligible for every PPT adversary $\mathcal{A}$, and for uniformly chosen $\boldsymbol{\omega}, \mathbf{r} \in \mathbb{Z}_{|\mathbb{G}_1|}^{\ell_\mathbf{B}}$:*

$$\mathrm{Adv}_{\mathbb{G},\mathcal{A}}^{\mathrm{gddh}}(\lambda) = \frac{1}{2}\Big(\Pr\Big[\mathcal{A}(1^\lambda, g_1^{\boldsymbol{\omega}^\top \mathbf{B}}, g_1^{\mathbf{Br}}, g_1^{\boldsymbol{\omega}^\top \mathbf{Br}}) = 1\Big]$$
$$- \Pr\Big[\mathcal{A}(1^\lambda, g_1^{\boldsymbol{\omega}^\top \mathbf{B}}, g_1^{\mathbf{Br}}, g_1^{\boldsymbol{\omega}^\top \mathbf{Br}} g_2) = 1\Big]\Big).$$

---

[2] How $\ell_\mathbf{B}$ and $\mathbf{B}$ are chosen depends in the concrete instance. In the prime-order setting, $\ell_\mathbf{B}$ and $\mathbf{B}$ determine what concrete computational problem is reduced to. Conversely, in the DCR setting, $\ell_\mathbf{B} = 1$, and $\mathbf{B} = 1$ is trivial.

Besides GDDH, we will also assume that it is infeasible to find a nontrivial element $g_2^u \in \mathbb{G}_2$ that does not already generate $\mathbb{G}_2$:

Definition 4 ($\mathbb{G}_2$-factoring assumption). *We say that the factoring $\mathbb{G}_2$ is hard in our setting if the following advantage is negligible for every PPT adversary $\mathcal{A}$ whose output $(g_2^{u_1}, \dots, g_2^{u_q}) \in \mathbb{G}_2^q$ is always a vector of $\mathbb{G}_2$-elements:*

$$\mathrm{Adv}_{\mathbb{G},\mathcal{A}}^{\mathrm{gddh}}(\lambda) = \frac{1}{2}\left(\Pr\left[\mathcal{A}(1^\lambda, g_1^{\boldsymbol{\omega}^\top \mathbf{B}}, g_1^{\mathbf{Br}}, g_1^{\boldsymbol{\omega}^\top \mathbf{Br}}) = 1 \right] - \Pr\left[\mathcal{A}(1^\lambda, g_1^{\boldsymbol{\omega}^\top \mathbf{B}}, g_1^{\mathbf{Br}}, g_1^{\boldsymbol{\omega}^\top \mathbf{Br}} g_2) = 1 \right] \right).$$

Generalized ElGamal Encryption. To simplify our notation, and to structure our presentation, we consider the following generalized variant of ElGamal:

Keypairs. Keypairs (epk, esk) are of the form $(epk, esk) = (g_1^{\boldsymbol{\omega}^\top \mathbf{B}}, \boldsymbol{\omega})$ for $\boldsymbol{\omega} \in \mathbb{Z}_{|\mathbb{G}_1|}^{\ell_\mathbf{B}}$.

Encryption. To encrypt $u \in \mathbb{Z}_{|\mathbb{G}_2|}$ with random coins $\mathbf{r} \in \mathbb{Z}_{|\mathbb{G}_1|}^{\ell_\mathbf{B}}$, compute

$$\mathbf{E}_{epk}(u; \mathbf{r}) = \mathbf{c} = (\mathbf{c}_0, c_1) = (g_1^{\mathbf{Br}}, g_1^{\boldsymbol{\omega}^\top \mathbf{Br}} g_2^u) \in \mathbb{G}^{\ell_\mathbf{B}} \times \mathbb{G}.$$

If we omit $\mathbf{r}$ and only write $\mathbf{E}_{epk}(u)$, then $\mathbf{r}$ is implicitly chosen uniformly from $\mathbb{Z}_{|\mathbb{G}_1|}^{\ell_\mathbf{B}}$.

Decryption. A ciphertext $\mathbf{c} = (\mathbf{c}_0, c_1) = (g^\gamma, g^\delta)$ is decrypted to

$$\mathbf{D}_{esk}(\mathbf{c}) = g^{\delta - \boldsymbol{\omega}^\top \gamma} \in \mathbb{G}.$$

Note that we encrypt exponents, while decryption only retrieves the respective group element.

It will also be useful to generalize this encryption to vectors of plaintexts with reused random coins: for $\mathbf{pk} = (epk_1, \dots, epk_n)$ and $\mathbf{sk} = (esk_1, \dots, esk_n)$ with $(epk_i, esk_i) = (g_1^{\boldsymbol{\omega}_i^\top \mathbf{B}}, \boldsymbol{\omega}_i)$, and $\mathbf{u} = (u_1, \dots, u_n) \in \mathbb{Z}_{|\mathbb{G}_2|}^n$, let

$$\begin{aligned}
\mathbf{E}_{\mathbf{pk}}(\mathbf{u}; \mathbf{r}) &= (\mathbf{c}_0, (c_1, \dots, c_n)) \\
&= (g_1^{\mathbf{Br}}, (g_1^{\boldsymbol{\omega}_1^\top \mathbf{Br}} g_2^{u_1}, \dots, g_1^{\boldsymbol{\omega}_n^\top \mathbf{Br}} g_2^{u_n})) \in \mathbb{G}^{\ell_\mathbf{B}} \times \mathbb{G}^n \\
\mathbf{D}_{\mathbf{sk}}(\mathbf{c}) &= (g^{\delta_1 - \boldsymbol{\omega}_1^\top \gamma}, \dots, g^{\delta_n - \boldsymbol{\omega}_n^\top \gamma}) \in \mathbb{G}^n \quad \text{for} \quad \mathbf{c} = (g^\gamma, (g^{\delta_1}, \dots, g^{\delta_n})).
\end{aligned}$$

When no confusion is possible, we may write $(\mathbf{c}_0, c_1, \dots, c_n)$ instead of the more cumbersome $(\mathbf{c}_0, (c_1, \dots, c_n))$. Sometimes, it will also be convenient to write

$$\boldsymbol{\Omega} = (\boldsymbol{\omega}_1 || \dots || \boldsymbol{\omega}_n) \in \mathbb{Z}_{|\mathbb{G}_1|}^{\ell_\mathbf{B} \times n}, \text{ such that } \mathbf{pk} = g_1^{\boldsymbol{\Omega}^\top \mathbf{B}} \text{ and}$$

$$\begin{aligned}
\mathbf{E}_{\mathbf{pk}}(\mathbf{u}; \mathbf{r}) &= (g_1^{\mathbf{Br}}, g_1^{\boldsymbol{\Omega}^\top \mathbf{Br}} g_2^{\mathbf{u}}) \\
\mathbf{D}_{\mathbf{sk}}(\mathbf{c}) &= g^{\gamma - \boldsymbol{\Omega}^\top \delta} \quad \text{for} \quad \mathbf{c} = (g^\gamma, g^\delta) \in \mathbb{G}^{\ell_\mathbf{B}} \times \mathbb{G}^n.
\end{aligned}$$

While this variant of ElGamal encryption will mainly be a notational tool, it is also a very simple tightly (chosen-plaintext) secure encryption scheme:

Definition 5 (IND-MCCPA security game for $(\mathbf{E}, \mathbf{D})$). *Consider the following game (which we call the IND-MCCPA security game, for "indistinguishability against multiple (partial) corruptions and chosen-plaintext attacks") between a challenger $\mathcal{C}$ and an adversary $\mathcal{A}$:*

1. $\mathcal{A}(1^\lambda)$ *picks* $n \in \mathbb{N}$, *and an index* $i^* \in \{1, \dots, n\}$.
2. $\mathcal{C}$ *samples* $b \in \{0, 1\}$, *and* $\boldsymbol{\omega}_1, \dots, \boldsymbol{\omega}_n \in \mathbb{Z}_{|\mathbb{G}_1|}^{\ell_\mathbf{B}}$, *and sets* $(epk_i, esk_i) = (g_1^{\boldsymbol{\omega}_i^\top \mathbf{B}}, \boldsymbol{\omega}_i)$, *and* $\mathbf{pk} = (epk_1, \dots, epk_n)$ *and* $\mathbf{sk} = (esk_1, \dots, esk_n)$.
3. *Next,* $\mathcal{A}$ *is run on input* $(epk_i)_{i=1}^{\ell_\mathbf{B}}$, $(esk_i)_{i \neq i^*}$, *and with (many-time) access to the following oracle:*
   - $\mathcal{O}_{\text{enc}}(\mathbf{u}^{(0)}, \mathbf{u}^{(1)})$, *for* $\mathbf{u}^{(j)} = (u_1^{(j)}, \dots, u_n^{(j)}) \in \mathbb{Z}_{|\mathbb{G}_2|}^n$ ($j \in \{0, 1\}$), *first checks that* $u_i^{(0)} = u_i^{(1)}$ *for all* $i \neq i^*$, *and returns* $\perp$ *if not. Then,* $\mathcal{O}_{\text{enc}}$ *computes and returns* $\mathbf{c} = \mathbf{E}_{\mathbf{pk}}(\mathbf{u}^{(b)})$.
4. *If* $\mathcal{A}$ *terminates with output* b', *then* $\mathcal{C}$ *outputs 1 iff* $b = b'$.

Let
$$\text{Adv}_{G,\mathcal{A}}^{\text{mccpa}}(\lambda) = \Pr[\mathcal{C} \text{ outputs } 1] - 1/2.$$

Lemma 1 (Tight security of $(\mathbf{E}, \mathbf{D})$). *For every* $\mathcal{A}$, *there exists an adversary* $\mathcal{B}$ *(of essentially the same complexity as the IND-MCCPA game with* $\mathcal{A}$*) for which*
$$\text{Adv}_{G,\mathcal{B}}^{\text{gddh}}(\lambda) = \text{Adv}_{G,\mathcal{A}}^{\text{mccpa}}(\lambda). \tag{1}$$

Proof. $\mathcal{B}$ gets $epk^* = g_1^{\boldsymbol{\omega}^{*\top}\mathbf{B}}$ and $\mathbf{c}^* = (\mathbf{c}_0^*, c_1^*) = (g_1^{\mathbf{Br}^*}, g_1^{\boldsymbol{\omega}^{*\top}\mathbf{Br}^*} g_2^b)$ (for unknown $b \in \{0, 1\}$) as input. Now $\mathcal{B}$ first runs $\mathcal{A}$ to obtain n and i^*. Then, $\mathcal{B}$ generates public and secret keys as follows:

- For $i \neq i^*$, $\mathcal{B}$ samples $\boldsymbol{\omega}_i \in \mathbb{Z}_{|\mathbb{G}_1|}^{\ell_\mathbf{B}}$, and sets $(epk_i, esk_i) = (g^{\boldsymbol{\omega}_i^\top \mathbf{B}}, \boldsymbol{\omega}_i)$.
- $\mathcal{B}$ sets $epk_{i^*} = g_1^{\boldsymbol{\omega}^{*\top}\mathbf{B}}$, and thus implicitly defines $esk_{i^*} = \boldsymbol{\omega}_{i^*} = \boldsymbol{\omega}^*$.

Then, $\mathcal{B}$ runs $\mathcal{A}$, on input $\mathbf{pk} = (epk_i)_i$ and $(esk_i)_{i \neq i^*}$, and implements oracle $\mathcal{O}_{\text{enc}}$ as follows:

- Upon an $\mathcal{O}_{\text{enc}}(\mathbf{u}^{(0)}, \mathbf{u}^{(1)})$ query with $u_i^{(0)} = u_i^{(1)}$ for $i \neq i^*$, $\mathcal{B}$ first samples a fresh $\mathbf{r}' \in \mathbb{Z}_{|\mathbb{G}_1|}^{\ell_\mathbf{B}}$, implicitly defines $\mathbf{r} = (u_{i^*}^{(1)} - u_{i^*}^{(0)})\mathbf{r}^* + \mathbf{r}'$, and sets up

$$c_0 = g_1^{(u_{i^*}^{(1)} - u_{i^*}^{(0)})\mathbf{Br}^* + \mathbf{Br}'} = g_1^{\mathbf{B}((u_{i^*}^{(1)} - u_{i^*}^{(0)})\mathbf{r}^* + \mathbf{r}')} = g_1^{\mathbf{Br}}$$

$$c_i = g_1^{(u_{i^*}^{(1)} - u_{i^*}^{(0)})\boldsymbol{\omega}_i^\top \mathbf{Br}^*} g_1^{\boldsymbol{\omega}_i^\top \mathbf{Br}'} g_2^{u_i^{(0)}} = g_1^{\boldsymbol{\omega}_i^\top \mathbf{Br}} g_2^{u_i^{(0)}} \quad \text{for } i \neq i^*$$

$$c_i = g_1^{(u_{i^*}^{(1)} - u_{i^*}^{(0)})\boldsymbol{\omega}^{*\top}\mathbf{Br}^* + \boldsymbol{\omega}^{*\top}\mathbf{Br}'} g_2^{(u_{i^*}^{(1)} - u_{i^*}^{(0)}) \cdot b + u_{i^*}^{(0)}} = g_1^{\boldsymbol{\omega}_{i^*}^\top \mathbf{Br}} g_2^{u_{i^*}^{(b)}}$$

For the resulting $\mathbf{c} = (c_0, c_1, \dots, c_n)$, we have that $\mathbf{c} = \mathbf{E}_{\mathbf{pk}}(\mathbf{u}^{(b)}; \mathbf{r})$ for (independently and uniformly distributed) random coins $\mathbf{r} = (u_{i^*}^{(1)} - u_{i^*}^{(0)})\mathbf{r}^* + \mathbf{r}'$. Hence, $\mathcal{O}_{\text{enc}}$ returns $\mathbf{c}$.

Finally, $\mathcal{B}$ relays any guess b' from $\mathcal{A}$ as its own output.

   Observe that $\mathcal{B}$ perfectly simulates the game from Lemma 1 (with the same challenge bit b). We obtain (1).

3.2   The Prime-Order Setting

The Groups. We consider two concrete instantiations of our generic setting. The first is a prime-order setting, in which $\mathbb{G} = \mathbb{G}_1 = \mathbb{G}_2$ has prime order $|\mathbb{G}| = |\mathbb{G}_1| = |\mathbb{G}_2|$. In these cases, we assume that $|\mathbb{G}| > 2^\lambda$ is public, and hence most syntactic requirements from Sect. 3.1 are trivially met. However, we will additionally need to assume that membership in $\mathbb{G}$ is efficiently decidable. We have numerous candidates for such groups (including, e.g., subgroups of $\mathbb{Z}_p^*$, or elliptic curves). In such groups, plausible assumptions include the Decisional Diffie-Hellman (DDH) assumption, the k-Linear (k-LIN) assumption [19,30], or a whole class of assumptions called Matrix-DDH assumptions [10].

Hardness of the GDDH and Factoring Problems. All of the mentioned assumptions imply our GDDH assumption for suitable $\ell_\mathbf{B}$ and $\mathbf{B}$. For instance, GDDH with $\ell_\mathbf{B} = 1$ and uniform $\mathbf{B}$ is nothing but a reformulation of the DDH assumption. More generally, GDDH with uniform $\mathbf{B}$ is actually the so-called $\mathcal{U}_{\ell_\mathbf{B}}$-MDDH assumption. In particular, this means that the k-LIN assumption implies GDDH with $\ell_\mathbf{B} = k$ and uniform $\mathbf{B}$ (see [10]). Additionally, we note that the $\mathbb{G}_2$-factoring assumption we make is trivially satisfied in prime-order settings (since $\mathrm{Adv}_{\mathbb{G}_2,\mathcal{A}}^{\mathrm{fact}}(\lambda) = 0$ for all $\mathcal{A}$ if $|\mathbb{G}_2| = |\mathbb{G}|$ is prime).

Pairing-Friendly Groups. In Sect. 5.4, we also exhibit a building block in the prime-order setting that uses a symmetric pairing $\mathbb{G} \times \mathbb{G} \to \mathbb{G}_T$ (for a suitable target group $\mathbb{G}_T$). Also for such pairing-friendly groups, we have a variety of candidates in case $\ell_\mathbf{B} \geq 2$. (Unfortunately, for $\ell_\mathbf{B} = 1$, a symmetric pairing can be used to trivially break the GDDH assumption.)

3.3   The DCR Setting

The Public Parameters. The second setting we consider is compatible with the Decisional Composite Residuosity (DCR) assumption [29]. In this case, the global public parameters include an integer $N = PQ$, for distinct safe primes P, Q (i.e., such that $P = 2P'+1$ and $Q = 2Q'+1$ for prime $P', Q' > 2^\lambda$).[3] We also assume that P, Q, P', Q' are pairwise different, and that $\gcd(P + Q - 1, N) = 1$ (the latter of which ensures that N is invertible modulo $\varphi(N) = (P-1)(Q-1) = 4P'Q'$).

We implicitly set $\ell_\mathbf{B} = 1$, and the matrix $\mathbf{B} \in \mathbb{Z}_{|\mathbb{G}_1| \times |\mathbb{G}_1|}$ from Sect. 3.1 to be trivial (i.e., the identity matrix). Hence, neither $\ell_\mathbf{B}$ nor $g_1^\mathbf{B}$ will have to be included in the parameters. However, we also include a generator g_1 of $\mathbb{G}_1$ in the public parameters, chosen as described below.

---

[3] We note that our DCR-based OR-proofs from Sect. 5.4 require P, Q to be somewhat larger, although still compatible with practical parameter choices.

The Groups. We now define the groups $\mathbb{G}$, $\mathbb{G}_1$, and $\mathbb{G}_2$. Since $\mathbb{G}$ should only have large prime factors, we should avoid setting $\mathbb{G} = \mathbb{Z}_{N^2}^*$. Instead, we could set $\mathbb{G}_1$ and $\mathbb{G}_2$ to be the subgroups of order $\varphi(N)/4$ and N, respectively, and then $\mathbb{G} = \mathbb{G}_1 \cdot \mathbb{G}_2$. However, in this case, membership in $\mathbb{G}$ would not be efficiently decidable in an obvious way. So here, we define our groups in a slightly more complex way, following the approach of signed quadratic residues [11,13,20].

Equipped with the notation $|x|_N$ and $\left(\frac{x}{N}\right)$ from Sect. 2, we set

$$\mathbb{G}_1 = \left\{ \, |x^N|_{N^2} \mid x \in \mathbb{Z}_{N^2}^*, \, \left(\frac{x^N}{N}\right) = 1 \, \right\} \subseteq \mathbb{Z}_{N^2}^*$$

$$\mathbb{G}_2 = \left\{ \, |(1+N)^e|_{N^2} \mid e \in \mathbb{Z}_N \, \right\} \subseteq \mathbb{Z}_{N^2}^*$$

$$\mathbb{G} = \left\{ \, |y|_{N^2} \mid y \in \mathbb{Z}_{N^2}^*, \, \left(\frac{y}{N}\right) = 1 \, \right\}.$$

These sets are groups, when equipped with the group operation $a \cdot b = |a \cdot b|_{N^2}$. Indeed, since $P, Q = 3 \bmod 4$, we have $\left(\frac{-1}{N}\right) = 1$, and thus $\left(\frac{|y_1 y_2|_{N^2}}{N}\right) = \left(\frac{y_1 y_2}{N}\right) = 1$ for $\left(\frac{y_1}{N}\right) = \left(\frac{y_2}{N}\right) = 1$. Hence, $\mathbb{G}_1$ and $\mathbb{G}$ are closed under group operation. It is then straightforward to check that $\mathbb{G}_1$, $\mathbb{G}_2$ and $\mathbb{G}$ are groups.

A canonical generator g_2 of $\mathbb{G}_2$ is $|1+N|_{N^2}$, and a generator g_1 of $\mathbb{G}_1$ (to be included in the public parameters) can be randomly chosen as $|x^N|_{N^2}$ for a uniform $x \in \mathbb{Z}_{N^2}$.

Properties of the Groups. We claim that $|\mathbb{G}_1| = \varphi(N)/4$. Indeed, we have that

$$\left| \{ \, |x^N|_N \mid x \in \mathbb{Z}_{N^2}^* \, \} \right| = \left| \{ \, |x^N|_{N^2} \mid x \in \mathbb{Z}_{N^2}^* \, \} \right| = \varphi(N)/2.$$

In other words, $|x^N|_N$ uniquely determines $|x^N|_{N^2}$. Furthermore, since N is invertible modulo $\varphi(N)$, the map $f : \mathbb{Z}_{N^2}^* \to \mathbb{Z}_N^*$ with $f(x) = x^N \bmod N$ is surjective. Hence, the set of all $|x^N|_N$ with $\left(\frac{x^N}{N}\right) = 1$ has cardinality $\varphi(N)/4$ (cf. [20, Lemma 1]). Using that $|x^N|_N$ fixes $|x^N|_{N^2}$, we obtain $|\mathbb{G}_1| = \varphi(N)/4$. Moreover, for $e \in \mathbb{Z}_N$, we can write $|(1+N)^e|_{N^2} = |1+eN|_{N^2} = e/|e| + |e|_N N$, and thus $|\mathbb{G}_2| = N$. Finally, we have $\mathbb{G} = \mathbb{G}_1 \cdot \mathbb{G}_2$, since every $|y|_{N^2} \in \mathbb{G}$ can be written as $|y|_{N^2} = |x^N (1+N)^e|_{N^2}$ with $\left(\frac{x^N}{N}\right) = 1$. Hence, since $|\mathbb{G}_1| = \varphi(N)/4 = P'Q'$ and $|\mathbb{G}_2| = N = PQ$ are coprime, $|\mathbb{G}| = |\mathbb{G}_1| \cdot |\mathbb{G}_2| = N \cdot \varphi(N)/4$ is square-free.

We also note that the discrete logarithm problem is easy in $\mathbb{G}_2$. Indeed, for $g_2^u \in \mathbb{G}_2$, we have

$$g_2^u = |(1+N)^e|_{N^2} = |1+eN|_{N^2} = \begin{cases} [e]_N N + 1 & \text{if } [e]_N < N/2 \\ [-e]_N N - 1 & \text{if } [e]_N > N/2. \end{cases}$$

A simple case distinction thus allows to compute $[e]_N$.

Membership Testing and Sampling Exponents. It is left to note that membership in $\mathbb{G}$ can be efficiently decided (by checking that $y \in \mathbb{Z}_{N^2}$ is invertible, lies between $-N^2/2$ and $N^2/2$, and satisfies $\left(\frac{y}{N}\right) = 1$). However, since $|\mathbb{G}_1|$ will not be public, exponents $s \in \mathbb{Z}_{|\mathbb{G}_1|}$ can only be sampled approximatively, e.g., by uniformly sampling $s \in \mathbb{Z}_{\lfloor N/4 \rfloor}$. This incurs a statistical defect of $\mathbf{O}(1/2^\lambda)$ upon each such sampling. In the following, we will silently ignore these statistical defects (and assume that there is an algorithm that uniformly samples $s \in \mathbb{Z}_{\varphi(N)}$) in our generic constructions for simplicity and ease of presentation. However, we note that the concrete bound (8) also holds for such an approximative sampling in the DCR setting.

Hardness of the GDDH and Factoring Problems. We claim that in the setting described above, the Decisional Composite Residuosity (DCR) assumption [29] implies the GDDH assumption. This connection has already been established in [15, Theorem 2] for a slight variant of the groups $\mathbb{G}, \mathbb{G}_1, \mathbb{G}_2$ above. (In their setting, $\mathbb{G}_1$ consists of elements $x^N \in \mathbb{Z}_{N^2}$ with $\left(\frac{x^N}{N}\right) = 1$, instead of elements $|x^N|_{N^2}$ with $\left(\frac{x^N}{N}\right) = 1$.) In fact, their proof applies also to our setting, and we obtain that the DCR assumption implies the GDDH assumption with $\ell = 1$ and trivial $\mathbf{B} = 1$ in $\mathbb{G}$ (as in Definition 3).

Furthermore, we note that the DCR assumption also implies the $\mathbb{G}_2$-factoring assumption (Definition 4). We sketch how any $\mathbb{G}_2$-factoring adversary $\mathcal{A}$ can be transformed into a DCR adversary $\mathcal{B}$. First, $\mathcal{B}$ runs $\mathcal{A}$, and obtains elements $g_2^{u_1}, \ldots, g_2^{u_q}$. Then, $\mathcal{B}$ uses that the discrete logarithm problem is easy in $\mathbb{G}_2$, and retrieves the corresponding $u_1, \ldots, u_q \in \mathbb{Z}_{|\mathbb{G}_2|}$. Now if $\gcd(|\mathbb{G}_2|, u_i) \notin \{1, |\mathbb{G}_2|\}$ for some u_i, then $\gcd(N, u_i) \in \{P, Q\}$ directly allows to factor N. Hence, if $\mathcal{A}$ succeeds, then $\mathcal{B}$ can factor N, and solve its own DCR challenge (e.g., by computing the order of its input).

4   Tightly Secure Building Blocks

In this section, we describe two building blocks for our main KEM construction. The first, tightly secure one-time signature schemes, is fairly standard, but requires a new instantiation in the DCR setting to achieve tight security. The second is, key extractors, is new, but similar building blocks have been used at least in the prime-order setting implicitly in previous works on tight security (e.g., [12]).

4.1   One-Time Signature Schemes

Definition 6 (Digital signature scheme). *A digital signature scheme* $\mathbf{OTS} = (\mathbf{SGen}, \mathbf{SSig}, \mathbf{SVer})$ *consists of the following PPT algorithms:*

- $\mathbf{SGen}(1^\lambda)$ *outputs a keypair* (ovk, osk). *We call* ovk *and* osk *the verification, resp. signing key.*

- **SSig**(osk, M), for a message $M \in \{0,1\}^*$, outputs a signature σ.
- **SVer**(ovk, M, σ), outputs either 0 or 1.

We require correctness in the sense that for all (ovk, osk) that lie in the range of **SGen**(1^λ), all $M \in \{0,1\}^*$, and all σ in the range of **SSig**(osk, M), we always have **SVer**(ovk, M, σ) = 1.

We only require one-time security (and call a signature scheme secure in this sense also a one-time signature scheme):

Definition 7 (EUF-MOTCMA security). *Let **OTS** be a digital signature scheme as in Definition 6, and consider the following game between a challenger $\mathcal{C}$ and an adversary $\mathcal{A}$:*

1. *$\mathcal{C}$ runs $\mathcal{A}$ on input 1^λ, and with (many-time) oracle access to the following oracles:*
   - *$\mathcal{O}_{\mathbf{gen}}()$ samples a fresh keypair (ovk, osk) $\leftarrow$ **SGen**(), and returns ovk.*
   - *$\mathcal{O}_{\mathbf{sig}}(ovk, M)$ first checks if ovk has been generated by $\mathcal{O}_{\mathbf{gen}}$, and returns $\perp$ if not. Next, $\mathcal{O}_{\mathbf{sig}}$ checks if there has been a previous $\mathcal{O}_{\mathbf{sig}}(ovk, \cdot)$ query (i.e., an $\mathcal{O}_{\mathbf{sig}}$ query with the same ovk), and returns $\perp$ if so. Let osk be the corresponding secret key generated alongside ovk. (If ovk has been generated multiple times by $\mathcal{O}_{\mathbf{gen}}$, take the first such osk.) $\mathcal{O}_{\mathbf{sig}}$ returns $\sigma \leftarrow$ **SSig**(osk, M).*
2. *If $\mathcal{A}$ returns (ovk^*, M^*, σ^*), such that **SVer**(ovk^*, M^*, σ^*) = 1, and ovk^* has been returned by $\mathcal{O}_{\mathbf{gen}}$, but σ^* has not been returned by $\mathcal{O}_{\mathbf{sig}}(ovk^*, M^*)$, then $\mathcal{C}$ returns 1. Otherwise, $\mathcal{C}$ returns 0.*

*Let $\mathrm{Adv}^{\mathrm{ots}}_{\mathbf{OTS},\mathcal{A}}(\lambda)$ be the probability that $\mathcal{C}$ finally outputs 1 in the above game. We say that **OTS** is strongly existentially unforgeable under many one-time chosen-message attacks (EUF-MOTCMA secure) iff for every PPT $\mathcal{A}$, the function $\mathrm{Adv}^{\mathrm{ots}}_{\mathbf{OTS},\mathcal{A}}(\lambda)$ is negligible.*

We remark, however, that our security notion is "strong", in the sense that a forger is already successful when he manages to generate a new signature for an already signed message.

A Construction in the Prime-Order Setting. In case $\mathbb{G} = \mathbb{G}_1 = \mathbb{G}_2$ with $|\mathbb{G}|$ prime and public, [18] already give a simple construction of a digital signature scheme that achieves EUF-MOTCMA security under the discrete logarithm assumption. Most importantly for our case, their security reduction is tight (i.e., only loses a constant factor). We refer to their paper for details.

A Construction in the DCR Setting. In the DCR setting (as in Sect. 3.3), there exist simple and efficient EUF-MOTCMA secure signature schemes from the factoring [24] or RSA assumptions [22]. However, these schemes are not known to be tightly secure.

Hence, in the full version [16], we construct a new digital signature scheme whose EUF-MOTCMA security can be tightly reduced to the GDDH assumption in the DCR setting.

4.2   Key Extractors

Intuition. Intuitively, a key extractor derives a pseudorandom key K from a given encryption $\mathbf{c} = \mathbf{E}(0; \mathbf{r})$ of 0. This K can be derived either publicly, using a public extraction key xpk and the witness $\mathbf{r}$, or secretly, using a secret extraction key xsk and only the ciphertext $\mathbf{c}$. We desire security in the sense keys derived secretly (i.e., using xsk) from random ciphertexts $\mathbf{c} = \mathbf{E}(R; \mathbf{r})$ for random R cannot be distinguished from truly random bitstrings K. This should hold even for many such challenges, and in the face of oracle access to xsk on "consistent" ciphertexts $\mathbf{c} = \mathbf{E}(0; \mathbf{r})$.

In this sense, key extractors give a computational form of the soundness guarantee provided by universal hash proof systems. We also note that a similar tool has been implicitly used in [12] for a similar purpose in the prime-order setting. Hence, we abstract and generalize their construction in a straightforward way.

Definition. In the following, fix a function $\ell_{\text{ext}} = \ell_{\text{ext}}(\lambda)$. In the following definition, we will choose the value R encrypted in random ciphertexts uniformly from $\mathbb{Z}_{2^{\ell_{\text{ext}}}}$. Our generic construction of key extractors works for any $\ell_{\text{ext}} \geq 3\lambda$ (and $|\mathbb{G}_2| \geq 2^{3\lambda}$).

Definition 8 (Key extractor). *A key extractor* $\mathbf{EXT} = (\mathbf{ExtGen}, \mathbf{Ext}_{\text{pub}}, \mathbf{Ext}_{\text{priv}})$ *for* $\mathbb{G}$ *consists of the following PPT algorithms*

- $\mathbf{ExtGen}(1^\lambda, epk)$, *on input a public encryption key* $epk = g_1^{\boldsymbol{\omega}^\top \mathbf{B}} \in \mathbb{G}_1^{\ell_{\mathbf{B}}}$ *for* $(\mathbf{E}, \mathbf{D})$ *(as in Sect. 3.1), outputs a keypair* (xpk, xsk). *We call* xpk *the public and* xsk *the private extraction key.*
- $\mathbf{Ext}_{\text{pub}}(xpk, \mathbf{c}, \mathbf{r})$, *for* $\mathbf{c} = \mathbf{E}_{epk}(0; \mathbf{r})$, *outputs a key* $K \in \{0, 1\}^\lambda$.
- $\mathbf{Ext}_{\text{priv}}(xsk, \mathbf{c})$ *also outputs a session key* $K \in \{0, 1\}^\lambda$.

We require the following:

Correctness. *For all* $epk = g_1^{\boldsymbol{\omega}^\top \mathbf{B}}$, *all keypairs* (xpk, xsk) *that lie in the range of* $\mathbf{ExtGen}(1^\lambda, epk)$, *all* $\mathbf{r} \in \mathbb{Z}_{|\mathbb{G}_1|}^{\ell_{\mathbf{B}}}$, *and all* $\mathbf{c} = \mathbf{E}_{epk}(0; \mathbf{r})$, *we always have* $\mathbf{Ext}_{\text{pub}}(xpk, \mathbf{c}, \mathbf{r}) = \mathbf{Ext}_{\text{priv}}(xsk, \mathbf{c})$.

Indistinguishability. *Consider the following game between a challenger* $\mathcal{C}$ *and an adversary* $\mathcal{A}$:

1. $\mathcal{C}$ *uniformly samples* $\boldsymbol{\omega} \in \mathbb{Z}_{|\mathbb{G}_1|}^{\ell_{\mathbf{B}}}$ *and sets* $(epk, esk) = (g_1^{\boldsymbol{\omega}^\top \mathbf{B}}, \boldsymbol{\omega})$. *Then,* $\mathcal{C}$ *generates an* $\mathbf{EXT}$ *keypair* $(xpk, xsk) \leftarrow \mathbf{ExtGen}(1^\lambda, epk)$, *and finally samples* $b \in \{0, 1\}$.
2. $\mathcal{A}$ *is run on input* $(1^\lambda, epk, xpk)$, *with (many-time) access to oracles* $\mathcal{O}_{\text{cha}}$ *and* $\mathcal{O}_{\text{ext}}$ *that operate as follows:*
   - $\mathcal{O}_{\text{cha}}()$ *uniformly chooses a fresh* $R \in \mathbb{Z}_{2^{\ell_{\text{ext}}}}$, *computes* $\mathbf{c} \leftarrow \mathbf{E}_{epk}(R)$ *and* $K_0 = \mathbf{Ext}_{\text{priv}}(xsk, \mathbf{c})$, *and uniformly chooses* $K_1 \in \{0, 1\}^\lambda$. *Finally,* $\mathcal{O}_{\text{cha}}$ *returns* $(\mathbf{c}, K_b)$.

- $\mathcal{O}_{\text{ext}}(\mathbf{c})$ *first checks if* $\mathbf{D}_{esk}(\mathbf{c}) = g_2^0$. *If not, then we say that* $\mathcal{A}$ *fails, and* $\mathcal{C}$ *terminates with output* 0 *immediately. Otherwise,* $\mathcal{O}_{\text{ext}}$ *computes and returns* $K = \mathbf{Ext}_{\text{priv}}(xsk, \mathbf{c})$.
- *Finally,* $\mathcal{A}$ *outputs a bit* b', *and* $\mathcal{C}$ *outputs* 1 *iff* $b = b'$ *(and* 0 *otherwise).*

Let $\mathrm{Adv}^{\text{ext}}_{\mathbf{EXT},\mathcal{A}}(\lambda) = \Pr[\mathcal{C} \text{ outputs } 1] - 1/2$. *We require that for all PPT* $\mathcal{A}$, $\mathrm{Adv}^{\text{snd}}_{\mathbf{PS},\mathcal{A}}(\lambda) \leq \varepsilon$ *for a negligible function* $\varepsilon = \varepsilon(\lambda)$.

A Generic Construction. For our GDDH-based key extractor, we assume that a function h chosen from a family of universal hash functions $\mathbf{UHF}_\lambda$ is made public in the global public parameters pp. Then, our extractor $\mathbf{EXT}^{\text{gddh}} = (\mathbf{ExtGen}^{\text{gddh}}, \mathbf{Ext}^{\text{gddh}}_{\text{pub}}, \mathbf{Ext}^{\text{gddh}}_{\text{priv}})$ is defined as follows:

- $\mathbf{ExtGen}^{\text{gddh}}(1^\lambda, epk)$, for $epk = g_1^{\boldsymbol{\omega}^\top \mathbf{B}}$, uniformly samples $\mathbf{s} \in \mathbb{Z}^{\ell_{\mathbf{B}}}_{|\mathbb{G}|}$ and $t \in \mathbb{Z}_{|\mathbb{G}|}$, and computes $g_1^{\mathbf{w}^\top} := g_1^{\mathbf{s}^\top \mathbf{B} + t \cdot \boldsymbol{\omega}^\top \mathbf{B}} \in \mathbb{G}_1^{\ell_{\mathbf{B}}}$. The output of $\mathbf{ExtGen}^{\text{gddh}}$ is $xpk = g_1^{\mathbf{w}^\top}$ and $xsk = (\mathbf{s}, t)$.
- $\mathbf{Ext}^{\text{gddh}}_{\text{pub}}(xpk, \mathbf{c}, \mathbf{r})$, for xpk as above and $\mathbf{c} = \mathbf{E}_{epk}(0; \mathbf{r})$, outputs $K = h(g_1^{\mathbf{w}^\top \cdot \mathbf{r}})$.
- $\mathbf{Ext}^{\text{gddh}}_{\text{priv}}(xsk, \mathbf{c})$, for $\mathbf{c} = (g^{\gamma}, g^{\delta}) \in \mathbb{G}^{\ell_{\mathbf{B}}} \times \mathbb{G}$, outputs $K = h(g^{\mathbf{s}^\top \gamma + t \cdot \delta})$.

Given $g_1^{\mathbf{w}^\top} = g_1^{\mathbf{s}^\top \mathbf{B} + t \cdot \boldsymbol{\omega}^\top \mathbf{B}}$ and a ciphertext $\mathbf{c} = \mathbf{E}(0; \mathbf{r}) = (g^{\gamma}, g^{\delta}) = (g_1^{\mathbf{Br}}, g^{\boldsymbol{\omega}^\top}_{1} g_1^{\mathbf{Br}})$, we have

$$g_1^{\mathbf{w}^\top \mathbf{r}} = g_1^{\mathbf{s}^\top \mathbf{Br} + t \cdot \boldsymbol{\omega}^\top \mathbf{Br}} = g^{\mathbf{s}^\top \gamma + t \cdot \delta},$$

and correctness follows. Indistinguishability follows from the following lemma:

Lemma 2. *For* $\ell_{\text{ext}} \geq 3\lambda$ *and* $|\mathbb{G}_2| \geq 2^{3\lambda}$, $\mathbf{EXT}^{\text{gddh}}$ *above satisfies the indistinguishability property of Definition 8, assuming GDDH in* $\mathbb{G}$. *Specifically, for every adversary* $\mathcal{A}$ *that makes at most* q *oracle queries, there is an adversary* $\mathcal{B}$ *(with roughly the same complexity as the indistinguishability experiment with* $\mathbf{EXT}^{\text{gddh}}$ *and* $\mathcal{A}$), *such that*

$$\mathrm{Adv}^{\text{ext}}_{\mathbf{EXT}^{\text{gddh}},\mathcal{A}}(\lambda) \leq \mathrm{Adv}^{\text{gddh}}_{\mathbb{G},\mathcal{B}}(\lambda) + q/2^\lambda. \tag{2}$$

Due to lack of space, we outsource a proof of Lemma 2 to the full version [16]. Summing up, we obtain

Theorem 1. *Under the GDDH assumption, and for* $\ell_{\text{ext}} \geq 3\lambda$ *and* $|\mathbb{G}_2| \geq 2^{3\lambda}$, $\mathbf{EXT}^{\text{gddh}}$ *is a key extractor in the sense of Definition 8.*

5   Benign Proof Systems

Intuition. Benign proof systems are the central technical tool in our KEM construction. Intuitively, a benign proof system for some language $\mathcal{L}$ is a non-interactive designated-verifier zero-knowledge proof system with strong soundness guarantees. Concretely, the system guarantees soundness even if simulated proofs for potentially false statements $x \notin \mathcal{L}$ are known. However, we do not quite require "simulation-soundness", in the sense that this should hold for simulated proofs for arbitrary false statements. (We note that simulation-sound proof systems are extremely useful in the context of tight security proofs, but they are also very hard to construct.)

Instead, we only require that no adversary can forge proofs for statements $x \notin \mathcal{L}$ that are "more false" than any statement for which a simulated proof is known. A little more specifically, we require that even if simulated proofs for statements $x \in \mathcal{L}' \supseteq \mathcal{L}$ are known, an adversary cannot forge a proof for some $x \notin \mathcal{L}'$. The main benefit over existing soundness notions is that $\mathcal{L}'$ does not even have to be known during the construction of the scheme. (For instance, our first proof system provides a "graceful soundness degradation", in the sense that it is sound in this sense for arbitrary linear languages $\mathcal{L}' \supseteq \mathcal{L}$.)

Overview Over Our Constructions. Apart from the abstraction, we also provide generic and setting-specific constructions of benign proof systems. Our generic constructions (for a linear, and a "dynamically parameterized" linear language) can be viewed as abstractions and generalizations of universal hash proof systems. For $\mathcal{L}' = \mathcal{L}$, soundness in the above sense follows immediately from the correctness property of hash proof systems. (Indeed, hash proofs for valid instances $x \in \mathcal{L}$ are unique and completely determined by public information.) For $\mathcal{L}' \supsetneq \mathcal{L}$, we will use additional properties of specific (existing) hash proof systems. In fact, the mentioned "graceful degradation" guarantees have already been used implicitly in the work of [12].

However, we also consider a somewhat nonstandard (and in our application crucial) "OR-language". Here, we give a prime-order instantiation in pairing-friendly groups (which is directly implied by the universal hash proof systems for disjunctions from [1]), and a new instance in the DCR setting. This DCR instance will be the key to the DCR-based instantiation of our KEM.

5.1   Definition

Definition 9 (Proof system). *Let* $\mathcal{L} = \{\mathcal{L}_{pars}\}$ *be a family of languages*[4] *with* $\mathcal{L}_{pars} \subseteq \mathcal{X}_{pars}$, *and with efficiently computable witness relation* $\mathcal{R}$. *A non-interactive designated-verifier proof system (NIDVPS)* **PS** = **(PGen, PPrv, PVer, PSim)** *for* $\mathcal{L}$ *consists of the following PPT algorithms:*

- **PGen**$(1^\lambda, pars)$ *outputs a keypair* (ppk, psk). *We call* ppk *the public and* psk *the private key.*
- **PPrv**(ppk, x, w), *for* $x \in \mathcal{L}$ *and* $\mathcal{R}(x, w) = 1$, *outputs a proof* π.
- **PVer**(psk, x, π), *for* $x \in \mathcal{X}$ *and a proof* π, *outputs a verdict* $b \in \{0, 1\}$.
- **PSim**(psk, x), *for* $x \in \mathcal{L}$, *outputs a proof* π.

We require correctness in the following sense:

---

[4] These languages may also implicitly depend on the global public parameters pp.

Completeness. *For all pars, all (ppk, psk) in the range of* $\mathbf{PGen}(1^\lambda, pars)$, *all* $x \in \mathcal{L}$, *and all* w *with* $\mathcal{R}(pars, x, w) = 1$, *we always have* $\mathbf{PVer}(psk, x, \mathbf{PPrv}(ppk, x, w)) = 1$.

All relevant security properties of a NIDVPS are condensed in the following definition.

Definition 10 (Benign proof system). *Let* **PS** *be an NIDVPS for* $\mathcal{L}$ *as in Difinition 9, and let* $\mathcal{L}^{\mathrm{sim}} = \{\mathcal{L}^{\mathrm{sim}}_{pars}\}$, $\mathcal{L}^{\mathrm{ver}} = \{\mathcal{L}^{\mathrm{ver}}_{pars}\}$, *and* $\mathcal{L}^{\mathrm{snd}} = \{\mathcal{L}^{\mathrm{snd}}_{pars}\}$ *be families of languages. We say that* **PS** *is* $(\mathcal{L}^{\mathrm{sim}}, \mathcal{L}^{\mathrm{ver}}, \mathcal{L}^{\mathrm{snd}})$-*benign if the following properties hold:*

(Perfect) zero-knowledge. *For all pars, all (ppk, psk) that lie in the range of* $\mathbf{PGen}(1^\lambda, pars)$, *and all* $x \in \mathcal{L}$ *and* w *with* $\mathcal{R}(pars, x, w) = 1$, *we have the following equivalence of distributions:*

$$\mathbf{PPrv}(ppk, x, w) \quad \equiv \quad \mathbf{PSim}(psk, x).$$

(Statistical) $(\mathcal{L}^{\mathrm{sim}}, \mathcal{L}^{\mathrm{ver}}, \mathcal{L}^{\mathrm{snd}})$-**soundness.** *Consider the following game played between a challenger* $\mathcal{C}$ *and an adversary* $\mathcal{A}$:

1. $\mathcal{A}$ *is run on input* 1^λ, *and chooses pars.*
2. $\mathcal{C}$ *generates* $(ppk, psk) \leftarrow \mathbf{PGen}(1^\lambda, pars)$.
3. $\mathcal{A}$ *is run again on input* $(1^\lambda, ppk)$, *and with (many-time) access to oracles* $\mathcal{O}_{\mathsf{sim}}$ *and* $\mathcal{O}_{\mathsf{ver}}$ *that operate as follows:*
   - $\mathcal{O}_{\mathsf{sim}}(x)$ *checks if* $x \in \mathcal{L}^{\mathrm{sim}}_{pars}$, *and if yes, returns* $\mathbf{PSim}(psk, x)$. *Otherwise,* $\mathcal{O}_{\mathsf{sim}}$ *returns* $\bot$.
   - $\mathcal{O}_{\mathsf{ver}}(x, \pi)$ *checks if* $x \in \mathcal{L}^{\mathrm{ver}}_{pars}$, *and, if so, returns* $\mathbf{PVer}(psk, x, \pi)$. *Otherwise,* $\mathcal{O}_{\mathsf{ver}}$ *returns* $\bot$.

Finally, $\mathcal{A}$ *wins iff it has queried* $\mathcal{O}_{\mathsf{ver}}$ *with* (x, π) *such that* $x \in \mathcal{X}_{pars} \setminus \mathcal{L}^{\mathrm{snd}}_{pars}$ *and* $\mathbf{PVer}(psk, x, \pi) = 1$. *Let* $\mathrm{Adv}^{\mathrm{snd}}_{\mathbf{PS}, \mathcal{A}}(\lambda)$ *the probability that* $\mathcal{A}$ *wins. We require that for all (not necessarily computationally bounded)* $\mathcal{A}$ *that only make a polynomial number of oracle queries,* $\mathrm{Adv}^{\mathrm{snd}}_{\mathbf{PS}, \mathcal{A}}(\lambda)$ *is negligible.*

Intuitively, the soundness condition of Definition 10 thus states that no proofs for $\mathcal{X} \setminus \mathcal{L}^{\mathrm{snd}}_{pars}$-statements can be forged, even when (simulated) proofs for $\mathcal{L}^{\mathrm{sim}}_{pars}$-statements are available, and proofs for $\mathcal{L}^{\mathrm{ver}}_{pars}$-statements can be verified.

5.2   The Generic Linear Language

We will be interested in proof systems for "linear languages", in the sense that instances are vectors of group elements, and the language is closed under vector addition (i.e., componentwise group operation).

In the following, let $D \in \mathbb{N}$ and $\mathbf{pk} = (epk_1, \ldots, epk_D) = (g_1^{\boldsymbol{\omega}_1^\top \mathbf{B}}, \ldots, g_1^{\boldsymbol{\omega}_D^\top \mathbf{B}}) \in (\mathbb{G}_1^{\ell_\mathbf{B}})^D$. For a concise notation, write $\boldsymbol{\Omega} = (\boldsymbol{\omega}_1 || \ldots || \boldsymbol{\omega}_D) \in \mathbb{Z}_{|\mathbb{G}_1|}^{\ell_\mathbf{B} \times D}$. Also, fix a $\mathbb{Z}_{|\mathbb{G}_2|}$-module

$$\mathfrak{U} = \{\mathbf{Mx} \mid \mathbf{x} \in \mathbb{Z}_{|\mathbb{G}_2|}^d\} \subseteq \mathbb{Z}_{|\mathbb{G}_2|}^D \tag{3}$$

defined by a matrix $\mathbf{M} \in \mathbb{Z}_{|\mathbb{G}_2|}^{D \times d}$. Our languages are parameterized over $pars_{\mathrm{lin}} = (\mathbf{pk}, \mathbf{M})$, although $\mathcal{L}_{\mathbf{pk}}^{\mathrm{lin}}$ only depends on $\mathbf{pk}$, and not on $\mathbf{M}$. Namely, consider

$$
\begin{aligned}
\mathcal{L}_{\mathbf{pk}}^{\mathrm{lin}} &= \left\{ \mathbf{E}_{\mathbf{pk}}(\mathbf{u}; \mathbf{r}) \mid \mathbf{r} \in \mathbb{Z}_{|\mathbb{G}_1|}^{\ell_{\mathbf{B}}}, \ \mathbf{u} = \mathbf{0} \in \mathbb{Z}_{|\mathbb{G}_2|}^{D} \right\} \\
\mathcal{L}_{\mathrm{sim},(\mathbf{pk},\mathbf{M})}^{\mathrm{lin}} &= \mathcal{L}_{\mathrm{ver},(\mathbf{pk},\mathbf{M})}^{\mathrm{lin}} = \mathcal{L}_{\mathrm{snd},(\mathbf{pk},\mathbf{M})}^{\mathrm{lin}} \\
&= \left\{ \mathbf{E}_{\mathbf{pk}}(\mathbf{u}; \mathbf{r}) \mid \mathbf{r} \in \mathbb{Z}_{|\mathbb{G}_1|}^{\ell_{\mathbf{B}}}, \ \mathbf{u} \in \mathfrak{U} \right\} \\
\mathcal{X}^{\mathrm{lin}} &= \mathbb{G}^{\ell_{\mathbf{B}}+D},
\end{aligned}
\tag{4}
$$

and set $\mathcal{L}^{\mathrm{lin}} = \{\mathcal{L}_{\mathbf{pk}}^{\mathrm{lin}}\}$ and $\mathcal{L}_{\mathrm{sim}}^{\mathrm{lin}} = \mathcal{L}_{\mathrm{ver}}^{\mathrm{lin}} = \mathcal{L}_{\mathrm{snd}}^{\mathrm{lin}} = \{\mathcal{L}_{\mathrm{sim},(\mathbf{pk},\mathbf{M})}^{\mathrm{lin}}\}$. A witness for $x \in \mathcal{L}_{\mathbf{pk}}^{\mathrm{lin}}$ is $\mathbf{r}$.

In the full version [16], we present a simple GDDH-based construction (based upon hash proof systems) of an $(\mathcal{L}_{\mathrm{sim}}^{\mathrm{lin}}, \mathcal{L}_{\mathrm{ver}}^{\mathrm{lin}}, \mathcal{L}_{\mathrm{snd}}^{\mathrm{lin}})$-benign proof system for $\mathcal{L}^{\mathrm{lin}}$.

5.3   A Dynamically Parameterized Linear Language

In our scheme, we will also use a slight variant of the generic linear language above. Specifically, we will consider a simple "dynamically parameterized" linear language, where one parameter (i.e., coefficient) is determined by the language instance. For a formal description, let $pars_{\mathrm{hash}} = \mathbf{pk} = (epk_1, epk_2) \in (\mathbb{G}_1^{\ell_{\mathbf{B}}})^2$, and

$$
\begin{aligned}
\mathcal{L}_{\mathbf{pk}}^{\mathrm{hash}} &= \left\{ (\mathbf{E}_{\mathbf{pk}}(\mathbf{u}; \mathbf{r}), \tau) \mid \mathbf{u} = \mathbf{0} \in \mathbb{Z}_{|\mathbb{G}_2|}^{2} \right\} \\
\mathcal{L}_{\mathrm{sim},\mathbf{pk}}^{\mathrm{hash}} &= \mathcal{L}_{\mathrm{ver},\mathbf{pk}}^{\mathrm{hash}} = \mathcal{L}_{\mathrm{snd},\mathbf{pk}}^{\mathrm{hash}} \\
&= \left\{ (\mathbf{E}_{\mathbf{pk}}(\mathbf{u}; \mathbf{r}), \tau) \mid \mathbf{u} = (u_1, u_2)^{\top} \in \mathbb{Z}_{|\mathbb{G}_2|}^{2}, \ u_2 = \tau u_1 \right\} \\
\mathcal{X}_{\mathbf{pk}}^{\mathrm{hash}} &= \left\{ (\mathbf{E}_{\mathbf{pk}}(\mathbf{u}; \mathbf{r}), \tau) \mid \mathbf{u} \in \mathbb{Z}_{|\mathbb{G}_2|}^{2} \right\},
\end{aligned}
\tag{5}
$$

where $\mathbf{r}$ and τ always range over $\mathbb{Z}_{|\mathbb{G}_1|}^{\ell_{\mathbf{B}}}$ and $\mathbb{Z}_{|\mathbb{G}_2|}$, respectively. A witness for $x \in \mathcal{L}^{\mathrm{hash}}$ is $\mathbf{r}$. The families $\mathcal{L}^{\mathrm{hash}}$, $\mathcal{L}_{\mathrm{sim}}^{\mathrm{hash}}$, $\mathcal{L}_{\mathrm{ver}}^{\mathrm{hash}}$, and $\mathcal{X}^{\mathrm{hash}}$ are defined in the obvious way.

In the full version [16], we present a simple GDDH-based construction (based upon hash proof systems) of an $(\mathcal{L}_{\mathrm{sim}}^{\mathrm{hash}}, \mathcal{L}_{\mathrm{ver}}^{\mathrm{hash}}, \mathcal{L}_{\mathrm{snd}}^{\mathrm{hash}})$-benign proof system for $\mathcal{L}^{\mathrm{hash}}$.

5.4   The Generic OR-Language

We will also be interested in the following family $\mathcal{L}^{\vee}$, together with its "simulation", "verification" and "soundness" counterparts $\mathcal{L}_{\mathrm{sim}}^{\vee}$, $\mathcal{L}_{\mathrm{ver}}^{\vee}$ and $\mathcal{L}_{\mathrm{snd}}^{\vee}$. Here, the actual languages in $\mathcal{L}^{\vee}$ are linear like those in $\mathcal{L}^{\mathrm{lin}}$. However, soundness also holds when $\mathcal{L}_{\mathrm{sim}}^{\vee}$-instances are simulated, and those instances have an "OR flavor".

The language parameters are $pars_{\vee} = (\mathbf{pk}, \ell_{\vee})$ for $\mathbf{pk} = (epk_1, epk_2) \in (\mathbb{G}_1^{\ell_{\mathbf{B}}})^2$, and a function $\ell_{\vee} = \ell_{\vee}(\lambda)$. The families $\mathcal{L}^{\vee}$, $\mathcal{L}_{\mathrm{sim}}^{\vee}$, $\mathcal{L}_{\mathrm{ver}}^{\vee}$, $\mathcal{L}_{\mathrm{snd}}^{\vee}$, and $\mathcal{X}^{\vee}$

are comprised of the following languages, where we consider all $\mathbf{r} \in \mathbb{Z}_{|\mathbb{G}_1|}^{\ell_{\mathsf{B}}}$, and $\mathbf{u} = (u_1, u_2) \in (\mathbb{Z}_{|\mathbb{G}_2|}^* \cup \{0\})^2$:

$$
\begin{aligned}
\mathcal{L}_{\mathbf{pk}}^{\vee} = \mathcal{L}_{\mathrm{ver},\mathbf{pk}}^{\vee} &= \left\{ \mathbf{E}_{\mathbf{pk}}(\mathbf{u};\mathbf{r}) \mid u_1 = 0 \right\} \\
\mathcal{L}_{\mathrm{sim},(\mathbf{pk},\ell_\vee)}^{\vee} &= \left\{ \mathbf{E}_{\mathbf{pk}}(\mathbf{u};\mathbf{r}) \mid u_1 = 0 \vee (|u_1| < 2^{\ell_\vee} \wedge u_2 = 0) \right\} \\
\mathcal{L}_{\mathrm{snd},\mathbf{pk}}^{\vee} &= \left\{ \mathbf{E}_{\mathbf{pk}}(\mathbf{u};\mathbf{r}) \mid u_1 = 0 \vee u_2 = 0 \right\} \\
\mathcal{X}_{\mathbf{pk}}^{\vee} &= \left\{ \mathbf{E}_{\mathbf{pk}}(\mathbf{u};\mathbf{r}) \right\}.
\end{aligned}
$$

Here, the value $|u_1|$ (in the definition of $\mathcal{L}_{\mathrm{sim},(\mathbf{pk},\ell_\vee)}^{\vee}$ is to be understood simply as the absolute value for signed $\mathbb{Z}_{|\mathbb{G}_2|}$-values in the prime-order setting, and as $|u_1| = |u_1|_N$ in the DCR setting. Observe that $\mathcal{L}_{\mathbf{pk}}^{\vee} \subseteq \mathcal{L}_{\mathrm{sim},(\mathbf{pk},\ell_\vee)}^{\vee} \subseteq \mathcal{L}_{\mathrm{snd},\mathbf{pk}}^{\vee} \subseteq \mathcal{X}_{\mathbf{pk}}^{\vee}$. A valid witness for $x \in \mathcal{L}^{\vee}$ is $\mathbf{r}$.

A Construction in Pairing-Friendly Groups. Now assume that $\mathbb{G} = \mathbb{G}_1 = \mathbb{G}_2$ is a prime-order group equipped with a symmetric pairing. Then, a benign proof system for $\mathcal{L}^{\vee}$ can be constructed from the universal hash proof systems for disjunctions from [1]. Specifically, [1] construct universal hash proof systems for languages of the form $\mathcal{L} = \{(x_1, x_2) \mid x_1 \in \mathcal{L}_1 \vee x_2 \in \mathcal{L}_2\}$, where $\mathcal{L}_i \subseteq \mathbb{G}^{\ell}$ are linear languages (i.e., vector spaces over $\mathbb{Z}_{|\mathbb{G}|}$). In our case, given $\mathbf{pk} = (epk_1, epk_2)$, we can thus set

$$
\begin{aligned}
\mathcal{L}_1 &= \left\{ \mathbf{E}_{epk_1}(0;\mathbf{r}) \right\} \\
\mathcal{L}_2 &= \left\{ \mathbf{E}_{epk_2}(0;\mathbf{r}) \right\} \\
\mathcal{L} &= \left\{ x = (\mathbf{c}_0, \mathbf{c}_1, \mathbf{c}_2) \mid (\mathbf{c}_0, \mathbf{c}_1) \in \mathcal{L}_1 \vee (\mathbf{c}_0, \mathbf{c}_2) \in \mathcal{L}_2 \right\}.
\end{aligned}
\tag{6}
$$

Invoking [1] with these languages yields a NIDVPS $\mathbf{PS}_{\mathrm{pair}}^{\vee}$ that achieves:

Theorem 2. $\mathbf{PS}_{\mathrm{pair}}^{\vee}$ *is an* $(\mathcal{L}_{\mathrm{sim}}^{\vee}, \mathcal{L}_{\mathrm{ver}}^{\vee}, \mathcal{L}_{\mathrm{snd}}^{\vee})$-*benign NIDVPS for* $\mathcal{L}^{\vee}$.

A Construction in the DCR Setting. In the following, we assume an $N = PQ$, and groups $\mathbb{G}, \mathbb{G}_1, \mathbb{G}_2$ as in Sect. 3.3. In particular, we have $\ell_{\mathsf{B}} = 1$, and $\mathbf{B}$ is the trivial (identity) matrix. Furthermore, fix an $\ell_\vee = \ell_\vee(\lambda)$. We additionally assume that $P, Q > 2^{\ell_\vee + 4\lambda}$. Recall that $g_1, epk_1, epk_2 \in \mathbb{G}_1$ are of order $|\mathbb{G}_1| = \varphi(N)/4$, and that $g_2 \in \mathbb{G}_2$ is of order $|\mathbb{G}_2| = N$.

Our $(\mathcal{L}_{\mathrm{sim}}^{\vee}, \mathcal{L}_{\mathrm{ver}}^{\vee}, \mathcal{L}_{\mathrm{snd}}^{\vee})$-benign proof system $\mathbf{PS}_{\mathrm{DCR}}^{\vee}$ for $\mathcal{L}^{\vee}$ is given by the following algorithms:

- $\mathbf{PGen}^{\vee}(1^\lambda)$ uniformly chooses $s_1, s_2 \in \mathbb{Z}_{\lfloor N^2/4 \rfloor}$ and then outputs $ppk_\vee = (epk_1^{s_1}, epk_1^{s_2})$ and $psk_\vee = (s_1, s_2)$.
- $\mathbf{PPrv}^{\vee}(ppk_\vee, x, r)$ (with $ppk_\vee = (epk_1^{s_1}, epk_1^{s_2})$, and $x = (c_0, c_1, c_2) = (g_1^r, epk_1^r, epk_2^r g_2^{u_2}))$ uniformly chooses $t_1, t_2 \in \mathbb{Z}_N$, and outputs

$$
\pi_\vee = (\pi_0, \pi_1, \pi_2) = \left(c_2^{t_1 + N \cdot t_2},\ (epk_1^{s_1})^r \cdot g_2^{t_1},\ (epk_1^{s_2})^r \cdot g_2^{t_2} \right).
$$

- $\mathbf{PVer}^\vee(psk_\vee, x, \pi_\vee)$ (for $psk = (s_1, s_2)$, $x = (c_0, c_1, c_2)$, and $\pi_\vee = (\pi_0, \pi_1, \pi_2)$) first checks that $\pi_1/c_1^{s_1} = g_2^{t_1}$ and $\pi_2/c_1^{s_2} = g_2^{t_2}$ for some $t_1, t_2 \in \mathbb{Z}_N$ (and outputs 0 if not). $\mathbf{PVer}$ then computes[5] these t_1, t_2, and outputs 1 iff $\pi_0 = c_2^{t_1 + N \cdot t_2}$.
- $\mathbf{PSim}^\vee(psk_\vee, x)$ (for $psk = (s_1, s_2)$ and $x = (c_0, c_1, c_2)$) uniformly picks $t_1, t_2 \in \mathbb{Z}_{N^2}$ and outputs

$$\pi_\vee = (\pi_0, \pi_1, \pi_2) = \left(c_2^{t_1 + N \cdot t_2}, \; c_1^{s_1} \cdot g_2^{t_1}, \; c_1^{s_2} \cdot g_2^{t_2}\right).$$

The completeness and zero-knowledge properties of $\mathbf{PS}_{\mathrm{DCR}}^\vee$ follow directly from the fact that $c_1^{s_i} = (epk_1^r)^{s_i} = (epk_1^{s_i})^r$. To show the soundness of $\mathbf{PS}_{\mathrm{DCR}}^\vee$, we prove a helpful technical lemma:

Lemma 3. *Let s_1, s_2, t_1, t_2 be distributed as in $\mathbf{PS}_{\mathrm{DCR}}^\vee$, and fix any $u \in \mathbb{Z}$ with $|u| < 2^{\ell_\vee}$. Let[6]*

$$aux := \left([s_1]_{\varphi(N)/4}, \; [s_2]_{\varphi(N)/4}, \; [t_1 + N \cdot t_2]_{\varphi(N)/4}, \; [us_1 + t_1]_N, \; [us_2 + t_2]_N\right),$$

and write[7] $w_1 := [s_1/\alpha]_N$ (with the division performed in $\mathbb{Z}_N$) for $\alpha := [N]_{\varphi(N)/4}$. Then, for an independently random $R \in \mathbb{Z}_{2^\lambda}$, we have

$$\varepsilon := \mathbf{SD}\big(([w_1]_{2^\lambda}, aux); (R, aux)\big) \leq 3/2^\lambda.$$

In other words, w_1 (and thus s_1) is unpredictable, even given aux.

Proof. Without loss of generality, assume $u \geq 0$. (For $u < 0$, we can invoke the lemma with $-u$, $-t_1$, and $-t_2$ in place of u, t_1 and t_2.) We proceed in steps, in each step modifying aux, and bounding the impact on ε. Specifically, in the following, we will define a number of random variables aux_i, and abbreviate $\varepsilon_i := \mathbf{SD}\big(([w_1]_{2^\lambda}, aux_i); (R, aux_i)\big)$. As a starting point, consider

$$aux_1 := \left([t_1 + N \cdot t_2]_{\varphi(N)/4}, \; [us_1 + t_1]_N, \; [us_2 + t_2]_N\right).$$

Now note that $w_1 = [s_1/\alpha]_N$ and the $[us_i + t_i]_N$ (for $i \in \{1, 2\}$) only depend on $[s_i]_N$. However, our uniform choice of $s_i \in \mathbb{Z}_{\lfloor N^2/4 \rfloor}$ is statistically $2/2^{\ell_\vee + 4\lambda}$-close to a uniform choice of $s_i \in \mathbb{Z}_{N \cdot \varphi(N)/4}$ (in which case $[s_i]_N$ and $[s_i]_{\varphi(N)/4}$ are independently and uniformly random). Hence, the $[s_i]_{\varphi(N)/4}$ are essentially independent of w_1 and aux_1, and we obtain $\varepsilon \leq \varepsilon_1 + 4/2^{\ell_\vee + 4\lambda}$. Next, consider

$$aux_2 := \left([t_1]_\alpha, \; [t_2]^\alpha, \; [t_1]^\alpha + [t_2]_\alpha, \; [us_1 + t_1]_N, \; [us_2 + t_2]_N\right).$$

Since $t_1 + \alpha \cdot t_2 = [t_1]_\alpha + \alpha \cdot ([t_1]^\alpha + [t_2]_\alpha) + \alpha^2 \cdot [t_2]^\alpha$, we have that aux_1 is a function of aux_2, and so $\varepsilon_1 \leq \varepsilon_2$. Similarly, we can refine the last two components of aux_2 to obtain

$$aux_3 := \left([t_1]_\alpha, \; [t_2]^\alpha, \; [t_1]^\alpha + [t_2]_\alpha, \; [us_1 + \alpha \cdot [t_1]^\alpha]_N, \; [us_2 + [t_2]_\alpha]_N\right).$$

---

[5] Here, we implicitly use that computing discrete logarithms in $\mathbb{G}_2$ is easy, see Sect. 3.3.

[6] In this lemma and its proof, we heavily rely on the notation of $[s]_N$ and $[s]^N$ from Sect. 2.

[7] Here, we use our assumption that $[N]_{\varphi(N)/4} = P + Q - 1$ and N are coprime.

Again, $\varepsilon_2 \le \varepsilon_3$ since aux_3 fully defines aux_2. Similar to our first step, now $[t_1]_\alpha$ and $[t_2]^\alpha$ are essentially independent of the remaining parts of aux (up to a statistical defect of at most $2/2^{\ell_\vee + 4\lambda}$ for each). Hence, for

$$aux_4 := ([t_1]^\alpha + [t_2]_\alpha, \ [us_1 + \alpha \cdot [t_1]^\alpha]_N, \ [us_2 + [t_2]_\alpha]_N),$$

we get that $\varepsilon_3 \le \varepsilon_4 + 4/2^{\ell_\vee + 4\lambda}$. Now let $w_2 := [s_2]_N$, and consider

$$aux_5 := ([t_1]^\alpha + [t_2]_\alpha, \ uw_1 + [t_1]^\alpha, \ uw_2 + [t_2]_\alpha).$$

Since aux_4 can be computed from aux_5, we have $\varepsilon_4 \le \varepsilon_5$. Next, we release $w_1 + w_2$ (over $\mathbb{Z}$):

$$aux_6 := ([t_1]^\alpha + [t_2]_\alpha, \ w_1 + w_2, \ uw_1 + [t_1]^\alpha).$$

Again, aux_5 can be computed from aux_6, and hence $\varepsilon_5 \le \varepsilon_6$. Since we consider the statistical distance between $[w_1]_{2^\lambda}$ and R, we can release (and then drop) $[w_1]^{2^\lambda}$. Concretely, consider

$$aux_7 := ([t_1]^\alpha + [t_2]_\alpha, \ [w_1]_{2^\lambda} + w_2, \ u \cdot [w_1]_{2^\lambda} + [t_1]^\alpha, \ [w_1]^{2^\lambda}),$$
$$aux_8 := ([t_1]^\alpha + [t_2]_\alpha, \ [w_1]_{2^\lambda} + w_2, \ u \cdot [w_1]_{2^\lambda} + [t_1]^\alpha)$$
$$aux_9 := ([t_1]^\alpha + [t_2]_\alpha, \ u \cdot [w_1]_{2^\lambda} + [t_1]^\alpha).$$

Here, $\varepsilon_6 \le \varepsilon_7$ since aux_6 can be computed from aux_7. Moreover, recall that $N > 2^{2\ell_\vee + 8\lambda}$ by our choice of $P, Q > 2^{\ell_\vee + 4\lambda}$. Hence, $\varepsilon_7 \le \varepsilon_8 + 1/2^{2\ell_\vee + 7\lambda}$, since $[w_1]_{2^\lambda}$ and $[w_1]^{2^\lambda}$ are independent up to a statistical defect of at most $1/2^{2\ell_\vee + 7\lambda}$. Finally, $\varepsilon_8 \le \varepsilon_9 + 1/2^{2\ell_\vee + 7\lambda}$, since w_2 is uniformly and independently random chosen from $\mathbb{Z}_N$.

Similarly, we can show that $[t_2]_\alpha$ blinds $[[t_1]^\alpha]_{2^{\ell_\vee + 2\lambda}}$:

$$aux_{10} := ([[t_1]^\alpha]_{2^{\ell_\vee + 2\lambda}} + [t_2]_\alpha, \ u \cdot [w_1]_{2^\lambda} + [[t_1]^\alpha]_{2^{\ell_\vee + 2\lambda}}, \ [[t_1]^\alpha]^{2^{\ell_\vee + 2\lambda}}),$$
$$aux_{11} := ([[t_1]^\alpha]_{2^{\ell_\vee + 2\lambda}} + [t_2]_\alpha, \ u \cdot [w_1]_{2^\lambda} + [[t_1]^\alpha]_{2^{\ell_\vee + 2\lambda}}),$$
$$aux_{12} := (u \cdot [w_1]_{2^\lambda} + [[t_1]^\alpha]_{2^{\ell_\vee + 2\lambda}}).$$

With the same reasoning as in aux_7-aux_9 (and using that $\alpha, N/\alpha > 2^{\ell_\vee + 4\lambda}/2$ by $P, Q > 2^{\ell_\vee + 4\lambda}$), we get $\varepsilon_9 \le \varepsilon_{10}$, as well as $\varepsilon_{10} \le \varepsilon_{11} + 1/2^{2\lambda}$, and $\varepsilon_{11} \le \varepsilon_{12} + 1/2^{2\lambda}$. Finally, if we set $aux_{13} := ()$ to be the empty sequence, we get $\varepsilon_{12} \le \varepsilon_{13} + 1/2^\lambda + 2/2^{\ell_\vee + 4\lambda}$, since $[t_1]^\alpha$ is $2/2^{\ell_\vee + 4\lambda}$-close to uniform over $\mathbb{Z}_{\lceil N/\alpha \rceil}$ (which implies that $[[t_1]^\alpha]_{2^{\ell_\vee + 2\lambda}}$ blinds $u \cdot [w_1]_{2^\lambda}$). It is left to observe that $\varepsilon_{13} = \mathbf{SD}([w_1]_{2^\lambda} ; R) \le 1/2^{2\ell_\vee + 7\lambda}$, since $w_1 \in \mathbb{Z}_N$ is uniformly random. Summing up, we get $\varepsilon \le 1/2^\lambda + 2/2^{2\lambda} + 10/2^{\ell_\vee + 4\lambda} + 3/2^{2\ell_\vee + 7\lambda} \le 3/2^\lambda$, as desired.

We can now proceed to show the soundness of $\mathbf{PS}_{\mathrm{DCR}}^\vee$:

Lemma 4. $\mathbf{PS}_{\mathrm{DCR}}^\vee$ *is statistically* $(\mathcal{L}_{\mathrm{sim}}^\vee, \mathcal{L}_{\mathrm{ver}}^\vee, \mathcal{L}_{\mathrm{snd}}^\vee)$*-sound in the sense of Definition 10. Concretely, for an adversary $\mathcal{A}$ that makes at most $q = q(\lambda)$ oracle queries in the soundness game from Definition 10,*

$$\mathrm{Adv}_{\mathbf{PS}_{\mathrm{DCR}}^\vee, \mathcal{A}}^{\mathrm{snd}}(\lambda) \le 4q/2^\lambda. \tag{7}$$

Proof. Fix $\ell_\vee$ and $\mathbf{pk}$, and let $\mathbf{view}_\mathcal{A}$ be $\mathcal{A}$'s view in a run of the computational soundness game from Definition 10. Specifically, $\mathbf{view}_\mathcal{A}$ consists of $\mathcal{A}$'s input $ppk_\vee = (epk_1^{s_1}, epk_1^{s_2})$, as well as all oracle queries (and the corresponding answers). We first consider to what extent $\mathbf{view}_\mathcal{A}$ determines the secret key $psk_\vee = (s_1, s_2)$.

- $\mathcal{A}$'s input $ppk_\vee = (epk_1^{s_1}, epk_1^{s_2})$ only depends on $[s_1]_{\varphi(N)/4}$ and $[s_2]_{\varphi(N)/4}$ (since epk_1 has order $\varphi(N)/4$).
- Each $\mathcal{O}_{\mathbf{sim}}$ oracle query of $\mathcal{A}$ reveals a value $\pi_\vee = (\pi_0, \pi_1, \pi_2) = (c_2^{t_1 + N \cdot t_2}, c_1^{s_1} \cdot g_2^{t_1}, c_1^{s_2} \cdot g_2^{t_2})$ for $\mathcal{A}$-supplied c_1, c_2 and fresh t_1, t_2. We may assume that $c_1 = epk_1^r \cdot g_2^{u_1}$ and $c_2 = epk_2^r \cdot g_2^{u_2}$ with $u_1 = 0$ or $|u_1|_N < 2^{\ell_\vee} \wedge u_2 = 0$ (since otherwise, $\mathcal{O}_{\mathbf{sim}}$ rejects the query). Hence, such a query reveals

$$(\pi_0, \pi_1, \pi_2) = (epk_2^{r(t_1 + N \cdot t_2)} g_2^{u_2 t_1}, \ epk_1^{rs_1} \cdot g_2^{u_1 s_1 + t_1}, \ epk_1^{rs_2} \cdot g_2^{u_1 s_2 + t_2}),$$

  which only depends on $[s_1]_{\varphi(N)/4}$, $[s_2]_{\varphi(N)/4}$, $[t_1 + N \cdot t_2]_{\varphi(N)/4}$, $[u_2 t_1]_N$, as well as $[u_1 s_1 + t_1]_N$ and $[u_1 s_2 + t_2]_N$. Thus, if $u_1 = 0$, the query reveals only $[s_1]_{\varphi(N)/4}$ and $[s_2]_{\varphi(N)/4}$ about (s_1, s_2). But if $u_1 \neq 0$ (and thus $u_2 = 0$), we can apply Lemma 3 with $u := u_1$, where we represent $u_1 \in \mathbb{Z}_N$ as an integer between $-N/2$ and $N/2$. This yields that the query leaves $[w_1]_{2^\lambda}$ undetermined, up to a small statistical defect. A hybrid argument over all of $\mathcal{A}$'s $\mathcal{O}_{\mathbf{sim}}$ queries shows that the overall statistical defect is bounded by $3q/2^\lambda$.
- An $\mathcal{O}_{\mathbf{ver}}$ query on input $(x, \pi_\vee)$ yields $\bot$ unless $x \in \mathcal{L}_{\mathbf{ver},(\mathbf{pk},\ell_\vee)}^\vee = \mathcal{L}_{(\mathbf{pk},\ell_\vee)}^\vee$. But for $x = (c_0, c_1, c_2) = (g_1^r, epk_1^r, epk_2^r g_2^{u_2}) \in \mathcal{L}_{(\mathbf{pk},\ell_\vee)}^\vee$, we get that $\mathcal{O}_{\mathbf{ver}}$'s output only depends on $c_1^{s_i} = epk_1^{rs_i}$, and hence only on $[s_i]_{\varphi(N)/4}$ (for $i = 1, 2$).

To summarize, $\mathbf{view}_\mathcal{A}$ is essentially independent of $[w_1]_{2^\lambda}$, up to a statistical defect of $3q/2^\lambda$.

It remains to prove that any $\mathcal{O}_{\mathbf{ver}}$ query on some $(x, \pi_\vee)$ with $x \in \mathcal{X}^\vee \setminus \mathcal{L}_{\mathbf{snd},\mathbf{pk}}^\vee$ (i.e., an x with $x = (c_0, c_1, c_2) = (g_1^r, epk_1^r \cdot g_2^{u_1}, epk_2^r \cdot g_2^{u_2})$ for $u_1, u_2 \in \mathbb{Z}_N^*$) is invalid in the sense that $\mathbf{PVer}(psk_\vee, x, \pi_\vee) = 0$ with high probability. To this end, write

$$\pi_\vee = (\pi_0, \pi_1, \pi_2) = (epk_2^{\rho_0} \cdot g_2^{\alpha_0}, \ epk_1^{\rho_1} \cdot g_2^{\alpha_1}, \ epk_1^{\rho_2} \cdot g_2^{\alpha_2})$$

for suitable $\rho_0, \rho_1, \rho_2, \alpha_0, \alpha_1, \alpha_2$. Recall that $(x, \pi_\vee)$ is valid only if for $i = 1, 2$, we have $\pi_i / c_1^{s_i} = g_2^{t_i}$ for some $t_i \in \mathbb{Z}_N$, and if $\pi_0 = c_2^{t_1 + N \cdot t_2}$ for those t_i. Hence, if $(x, \pi_\vee)$ is valid, then the following holds for some t_1, t_2:

$$\rho_0 = [r(t_1 + N \cdot t_2)]_{\varphi(N)/4} \qquad \alpha_0 = [u_2 t_1]_N$$
$$\rho_1 = [rs_1]_{\varphi(N)/4} \qquad\qquad \alpha_1 = [u_1 s_1 + t_1]_N$$
$$\rho_2 = [rs_2]_{\varphi(N)/4} \qquad\qquad \alpha_2 = [u_1 s_2 + t_2]_N.$$

By assumption, $u_2 \in \mathbb{Z}_N^*$, and thus α_0 determines t_1. Using also $u_1 \in \mathbb{Z}_N^*$, hence α_0 and α_1 determine $[s_1]_N$, and thus also $w_1 = [s_1/\alpha]_N$. However, as we have argued above, $\mathbf{view}_\mathcal{A}$ is essentially independent of $[w_1]_{2^\lambda}$. The probability that $\mathcal{A}$ correctly guesses an independently and uniformly random $[w_1]_{2^\lambda}$ with a single query is exactly $1/2^\lambda$. Since $\mathcal{A}$ makes at most q guesses, the probability for a correct guess is bounded by $q/2^\lambda$. Taking into account the mentioned statistical defect in $\mathbf{view}_\mathcal{A}$, we obtain (7).

Taking things together, we obtain

Theorem 3. $\mathbf{PS}_{\mathrm{DCR}}^\vee$ *is an* $(\mathcal{L}_{\mathrm{sim}}^\vee, \mathcal{L}_{\mathrm{ver}}^\vee, \mathcal{L}_{\mathrm{snd}}^\vee)$*-benign NIDVPS for* $\mathcal{L}^\vee$.

6   The Key Encapsulation Scheme

In the following, we present our main construction of an IND-MCCA secure key encapsulation (KEM) scheme. (This directly implies a PKE scheme with the same security properties [8].)

6.1   The Construction

Ingredients and Public Parameters. In our construction, we use the following ingredients:

- groups $\mathbb{G}, \mathbb{G}_1, \mathbb{G}_2$ with $|\mathbb{G}_2| > 2^{3\lambda}$ (see Sect. 3.1 for a description of the generic setting),
- the generalized ElGamal scheme $(\mathbf{E}, \mathbf{D})$ implicitly defined through $\mathbb{G}, \mathbb{G}_1, \mathbb{G}_2$ (Sect. 3.1),
- an EUF-MOTCMA secure one-time signature scheme $\mathbf{OTS} = (\mathbf{SGen}, \mathbf{SSig}, \mathbf{SVer})$ (Sect. 4.1),
- a key extractor $\mathbf{EXT} = (\mathbf{ExtGen}, \mathbf{Ext}_{\mathrm{pub}}, \mathbf{Ext}_{\mathrm{priv}})$ for $\mathbb{G}$ (see Sect. 4.2) with $\ell_{\mathrm{ext}} = 3\lambda$,
- an $(\mathcal{L}_{\mathrm{sim}}^{\mathrm{lin}}, \mathcal{L}_{\mathrm{ver}}^{\mathrm{lin}}, \mathcal{L}_{\mathrm{snd}}^{\mathrm{lin}})$-benign proof system denoted with $\mathbf{PS}^{\mathrm{lin}} = (\mathbf{PGen}^{\mathrm{lin}}, \mathbf{PPrv}^{\mathrm{lin}}, \mathbf{PVer}^{\mathrm{lin}}, \mathbf{PSim}^{\mathrm{lin}})$ for $\mathcal{L}^{\mathrm{lin}}$ (Sect. 5.2),
- an $(\mathcal{L}_{\mathrm{sim}}^{\mathrm{hash}}, \mathcal{L}_{\mathrm{ver}}^{\mathrm{hash}}, \mathcal{L}_{\mathrm{snd}}^{\mathrm{hash}})$-benign proof system denoted $\mathbf{PS}^{\mathrm{hash}} = (\mathbf{PGen}^{\mathrm{hash}}, \mathbf{PPrv}^{\mathrm{hash}}, \mathbf{PVer}^{\mathrm{hash}}, \mathbf{PSim}^{\mathrm{hash}})$ for $\mathcal{L}^{\mathrm{hash}}$ (Sect. 5.3),
- an $(\mathcal{L}_{\mathrm{sim}}^\vee, \mathcal{L}_{\mathrm{ver}}^\vee, \mathcal{L}_{\mathrm{snd}}^\vee)$-benign proof system denoted $\mathbf{PS}^\vee = (\mathbf{PGen}^\vee, \mathbf{PPrv}^\vee, \mathbf{PVer}^\vee, \mathbf{PSim}^\vee)$ for $\mathcal{L}^\vee$ (Sect. 5.4) with $\ell_\vee = 3\lambda$, and
- a collision-resistant hash function generator $\mathbf{CRHF}$ (Sect. 2) with $\ell_H = 2\lambda$.[8]

We can use the presented generic constructions for $\mathbf{EXT}$, $\mathbf{PS}^{\mathrm{lin}}$, and $\mathbf{PS}^{\mathrm{hash}}$, and, in the prime-order and DCR settings, the presented concrete constructions for $\mathbf{OTS}$ and $\mathbf{PS}^\vee$. (We note, however, that the DCR-based proof system $\mathbf{PS}_{\mathrm{DCR}}^\vee$ additionally requires that $|\mathbb{G}|$ has no prime factors smaller than $2^{7\lambda}$.) Specifically, we obtain instantiations both in the prime-order (with symmetric pairing) and DCR settings.

We also assume public parameters pp that contain whatever public parameters our building blocks require. Specifically, pp defines groups $\mathbb{G}, \mathbb{G}_1$, and $\mathbb{G}_2$ (as described in Sect. 3.1), and contains a hash function H output by $\mathbf{CRHF}$.

---

[8] Since we assume collision-resistance (and not only target collision-resistance), we will have to take into account, e.g., birthday attacks on the hash function. This unfortunately entails $\ell_H \geq 2\lambda$.

The Algorithms. Now our KEM **KEM** is defined through the following algorithms:

- **Gen**(1^λ) first uniformly picks $\omega_1, \ldots, \omega_4 \in \mathbb{Z}_{|\mathbb{G}_1|}^{\ell_B}$, and sets $(\mathbf{pk}, \mathbf{sk}) = (epk_i, esk_i)_{i=1}^4 = (g_1^{\omega_i^\top \mathbf{B}}, \omega_i)_{i=1}^4$. Next, **Gen** samples

$$(ppk_{\mathrm{lin}}, psk_{\mathrm{lin}}) \leftarrow \mathbf{PGen}^{\mathrm{lin}}(1^\lambda, \mathbf{pk})$$
$$(ppk_{\mathrm{hash}}, psk_{\mathrm{hash}}) \leftarrow \mathbf{PGen}^{\mathrm{hash}}(1^\lambda, (epk_1, epk_2/epk_3))$$
$$(ppk_{\mathrm{V},1}, psk_{\mathrm{V},1}) \leftarrow \mathbf{PGen}^{\vee}(1^\lambda, (epk_1, epk_1))$$
$$(ppk_{\mathrm{V},2}, psk_{\mathrm{V},2}) \leftarrow \mathbf{PGen}^{\vee}(1^\lambda, (epk_4, epk_4))$$
$$(ppk_{\mathrm{V},3}, psk_{\mathrm{V},3}) \leftarrow \mathbf{PGen}^{\vee}(1^\lambda, (epk_4, epk_1))$$
$$(ppk_{\mathrm{V},4}, psk_{\mathrm{V},4}) \leftarrow \mathbf{PGen}^{\vee}(1^\lambda, (epk_2/epk_3, epk_4))$$
$$(ppk_{\mathrm{V},5}, psk_{\mathrm{V},5}) \leftarrow \mathbf{PGen}^{\vee}(1^\lambda, (epk_2/epk_3, epk_4))$$
$$(ppk_{\mathrm{V},6}, psk_{\mathrm{V},6}) \leftarrow \mathbf{PGen}^{\vee}(1^\lambda, (epk_2/epk_3, epk_1))$$
$$(xpk, xsk) \leftarrow \mathbf{ExtGen}(1^\lambda, epk_2),$$

sets $\mathbf{ppk} = (ppk_{\mathrm{lin}}, ppk_{\mathrm{hash}}, ppk_{\mathrm{V},1}, \ldots, ppk_{\mathrm{V},6})$ and $\mathbf{psk} = (psk_{\mathrm{lin}}, psk_{\mathrm{hash}}, psk_{\mathrm{V},1}, \ldots, psk_{\mathrm{V},6})$, and finally outputs

$$pk = (\mathbf{pk}, \mathbf{ppk}, xpk) \qquad\qquad sk = (\mathbf{sk}, \mathbf{psk}, xsk).$$

- **Enc**(pk) (for pk as above) selects a random $\mathbf{r}$, and computes

$$\mathbf{c} = (\mathbf{c}_0, c_1, \ldots, c_4) = \mathbf{E}(\mathbf{pk}, 0; \mathbf{r})$$
$$(ovk, osk) \leftarrow \mathbf{SGen}()$$
$$\tau = H(ovk)$$
$$\pi_{\mathrm{lin}} \leftarrow \mathbf{PPrv}^{\mathrm{lin}}(ppk_{\mathrm{lin}}, \mathbf{c}, \mathbf{r})$$
$$\pi_{\mathrm{hash}} \leftarrow \mathbf{PPrv}^{\mathrm{hash}}(ppk_{\mathrm{hash}}, ((\mathbf{c}_0, c_1, c_2/c_3), \tau), \mathbf{r})$$
$$\pi_{\mathrm{V},1} \leftarrow \mathbf{PPrv}^{\vee}(ppk_{\mathrm{V},1}, (\mathbf{c}_0, c_1, c_1/g_2), \mathbf{r})$$
$$\pi_{\mathrm{V},2} \leftarrow \mathbf{PPrv}^{\vee}(ppk_{\mathrm{V},2}, (\mathbf{c}_0, c_4, c_4/g_2), \mathbf{r})$$
$$\pi_{\mathrm{V},3} \leftarrow \mathbf{PPrv}^{\vee}(ppk_{\mathrm{V},3}, (\mathbf{c}_0, c_4, c_1/g_2), \mathbf{r})$$
$$\pi_{\mathrm{V},4} \leftarrow \mathbf{PPrv}^{\vee}(ppk_{\mathrm{V},4}, (\mathbf{c}_0, c_2/c_3, c_4), \mathbf{r})$$
$$\pi_{\mathrm{V},5} \leftarrow \mathbf{PPrv}^{\vee}(ppk_{\mathrm{V},5}, (\mathbf{c}_0, c_2/c_3, c_4/g_2), \mathbf{r})$$
$$\pi_{\mathrm{V},6} \leftarrow \mathbf{PPrv}^{\vee}(ppk_{\mathrm{V},6}, (\mathbf{c}_0, c_2/c_3, c_1/g_2), \mathbf{r})$$
$$\boldsymbol{\pi} = (\pi_{\mathrm{lin}}, \pi_{\mathrm{hash}}, \pi_{\mathrm{V},1}, \ldots, \pi_{\mathrm{V},6})$$
$$\sigma \leftarrow \mathbf{SSig}(osk, (\mathbf{c}, \boldsymbol{\pi}))$$
$$K = \mathbf{Ext}_{\mathrm{pub}}(xpk, (\mathbf{c}_0, c_2), \mathbf{r}).$$

Here, we interpret $\tau = (\tau_1, \ldots, \tau_{2\lambda}) \in \{0, 1\}^{2\lambda}$ as an integer $\tau = \sum_{i=1}^{2\lambda} 2^{i-1} \tau_i \in \{0, \ldots, 2^{2\lambda} - 1\}$, with τ_1 being interpreted as the least significant bit. The final output of **Enc** is $C = (\mathbf{c}, \boldsymbol{\pi}, ovk, \sigma)$ and K.

- **Dec**(sk, C) (for sk and C as above), first verifies σ and all proofs in $\boldsymbol{\pi}$ using ovk and $\mathbf{sk}$, and, if all are valid, returns

$$K = \mathbf{Ext}_{\mathrm{priv}}(xsk, (\mathbf{c}_0, c_2)).$$

Explanation. The proofs in $\boldsymbol{\pi}$ require some explanation. They prove various (seemingly highly redundant) properties of the vector $\mathbf{u} = (u_i)_{i=1}^4 \in \mathbb{Z}_{|\mathbb{G}_2|}^4$ encrypted in $\mathbf{c}$. Some of these properties will be violated in different steps of our security analysis already by the security game, and we will then rely on the remaining properties. For instance, π_{lin} always guarantees that the vectors $\mathbf{u}$ encrypted in decryption queries lie in the subspace spanned by the vectors $\mathbf{u}$ encrypted in challenge ciphertexts. (That subspace is initially trivial, since honest encryptions contain $\mathbf{u} = \mathbf{0}$, but will be larger in later parts of the analysis.) π_{hash} guarantees that $\tau u_1 = u_2 - u_3$ in $\mathcal{A}$'s decryption queries (unless generated challenge ciphertexts already violate that relation).

The $\mathbf{PS}^{\vee}$-proofs $\pi_{\vee,i}$ are a bit more delicate. First, $\pi_{\vee,1}$ and $\pi_{\vee,2}$ guarantee that $u_1, u_4 \in \{0, 1\}$. The condition $u_1 \in \{0, 1\}$ only simplifies the analysis, but $u_4 \in \{0, 1\}$ is instrumental to enforce our partitioning strategy. In particular, u_4 will be the bit that determines the partitioning of ciphertexts in our partitioning argument. Depending on the value of u_4, $\pi_{\vee,4}$ and $\pi_{\vee,5}$ give further guarantees: $\pi_{\vee,4+b}$ guarantees $u_2 = u_3 \vee u_4 = b$. At each point in our analysis, at least one of these conditions (for one value of b) is never violated. Hence, $u_2 = u_3$ is guaranteed in decryption queries whenever $u_4 \neq b$. Finally, the proofs $\pi_{\vee,3}$ and $\pi_{\vee,6}$ ensure technical conditions ($u_4 = 0 \vee u_1 = 1$ and $u_2 = u_3 \vee u_1 = 1$) that will help to deal with the somewhat limited soundness guarantees of $\mathbf{PS}^{\vee}$. (In particular, these proofs help to cope with the fact that the soundness game of $\mathbf{PS}^{\vee}$ only allows a limited type of verification queries.)

Correctness. The correctness of **KEM** follows directly from the correctness of the underlying primitives.

6.2   Security Analysis

Theorem 4 (Security of KEM). *If the ingredients from Sect. 6.1 are secure, then* **KEM** *is IND-MCCA secure. Specifically, for every IND-MCCA adversary $\mathcal{A}$ that makes at most q oracle queries, there are adversaries $\mathcal{B}^{\mathrm{crhf}}$, $\mathcal{B}^{\mathrm{ots}}$, $\mathcal{B}^{\mathrm{fact}}$, $\mathcal{B}^{\mathrm{mccpa}}$, $\mathcal{B}^{\mathrm{lin}}$, $\mathcal{B}^{\mathrm{hash}}$, and $\mathcal{B}^{\vee}$ with*

$$|\mathrm{Adv}_{\mathbf{KEM},\mathcal{A}}^{\mathrm{mcca}}(\lambda)| \leq \mathrm{Adv}_{\mathbf{CRHF},\mathcal{B}^{\mathrm{crhf}}}^{\mathrm{crhf}}(\lambda) + \mathrm{Adv}_{\mathbf{OTS},\mathcal{B}^{\mathrm{ots}}}^{\mathrm{ots}}(\lambda)$$

$$+\mathbf{O}(\lambda)\mathrm{Adv}_{\mathbb{G}_2,\mathcal{B}^{\mathrm{fact}}}^{\mathrm{fact}}(\lambda) + \mathbf{O}(\lambda)\mathrm{Adv}_{\mathbb{G},\mathcal{B}^{\mathrm{mccpa}}}^{\mathrm{mccpa}}(\lambda) + \mathbf{O}(\lambda)\mathrm{Adv}_{\mathbf{PS}^{\mathrm{lin}},\mathcal{B}^{\mathrm{lin}}}^{\mathrm{snd}}(\lambda)$$

$$+\mathbf{O}(\lambda)\mathrm{Adv}_{\mathbf{PS}^{\mathrm{hash}},\mathcal{B}^{\mathrm{hash}}}^{\mathrm{snd}}(\lambda) + \mathbf{O}(\lambda)\mathrm{Adv}_{\mathbf{PS}^{\vee},\mathcal{B}^{\vee}}^{\mathrm{snd}}(\lambda) + \mathbf{O}(\lambda q)/2^{\lambda}.$$

$$(8)$$

Outline. The goal of our proof will be to randomize all keys handed out by $\mathcal{O}_{\mathbf{enc}}$ along with challenge ciphertexts. In order to do so, we rely on the indistinguishability of the key extractor **EXT**. However, to apply **EXT**'s indistinguishability (Definition 8), we first need to establish a certain kind of "unfairness". Specifically, we will randomize the u_2 component of all challenge ciphertexts, while rejecting all decryption queries with $u_2 \neq 0$. (Note that this in particular means that the experiment does not need to be able to decrypt challenge ciphertexts.)

Establishing this unfairness thus is the key to proving chosen-ciphertext security. But it will also form the main difficulty of the proof, and we will outsource this process into several helper lemmas.

A little more specifically, we will proceed as hinted in the introduction. Namely, to show that all adversarial decryption queries with $u_2 \neq 0$ are rejected, we will gradually add more and more restrictive additional checks on decryption queries with $u_2 \neq 0$. In particular, if $u_2 \neq 0$, then we will require that u_2 "authenticates" the full ciphertext, in the sense that $u_2 = T$ for a ciphertext-dependent "authentication token" T (cf. also the description in the introduction). We will change the definition of T in a series of transformations such that eventually $T = X + \tau$, where X is a fixed secret value, and τ is the (up to **OTS**-forgeries and **CRHF** -collisions ciphertext-unique) value of $\tau = H(ovk)$ from above. However, we will only be able to prove that the adversary must *reuse* a previously used authentication in his decryption queries. This means that the following rules apply:

- Any challenge ciphertext handed to the adversary satisfies $u_2 = X + \tau$.
- Any decryption query with $u_2 \neq 0$ must satisfy $u_2 = X + \tau^{(j)}$ for *some* $\tau^{(j)}$ from a challenge ciphertext. (Hence, the adversary must "reuse" an authentication tag.)

Additionally, all challenge ciphertexts will satisfy $u_1 = 1$ and $u_3 = X$. Hence, using the soundness of our benign proof systems $\mathbf{PS}^{\mathrm{lin}}$ and $\mathbf{PS}^{\vee}$, also any decryption query with $u_1 \neq 0$ will have to satisfy $u_1 = 1$ and $u_3 = X$ (or it is rejected). Finally invoking the soundness of $\mathbf{PS}^{\mathrm{hash}}$ (on the equation $u_2 = u_3 + \tau \cdot u_1$, which is fulfilled in all challenge ciphertexts), we obtain that also decryption queries will have to satisfy $u_2 = X + \tau$ for the respective value τ from that decryption query.

Hence, the requirements on adversarial decryption queries with $u_2 \neq 0$ are now that $u_2 = X + \tau$ and $u_2 = X + \tau^{(j)}$, and thus that $\tau = \tau^{(j)}$ for some $\tau^{(j)}$ from a previous challenge. Since the value τ is ciphertext-unique, we obtain a contradiction. (Thus, any decryption query with $u_2 \neq 0$ is rejected.)

Due to lack of space, we have to postpone our proof (and in particular the more complex argument for establishing the requirement $u_2 = X + \tau^{(j)}$ on adversarial decryption queries) to the full version [16].

Acknowledgements. I would like to thank Antonio Faonio for pointing out a problem in the formulation of Definition 8, and Dingding Jia and Ryo Nishimaki for a careful proofreading. In particular, Dingding spotted a mistake in the description of honest key derivation. I am also indebted to Lin Lyu, who found a flaw in an earlier version of the DCR-based one-time signature scheme $\mathbf{OTS_{DCR}}$, a gap in the proof of a technical lemma from the main proof, and many smaller mistakes in an earlier version in a very thorough proofreading. Finally, I would like to thank the reviewers for helpful comments concerning the presentation.

References

1. Abdalla, M., Benhamouda, F., Pointcheval, D.: Disjunctions for hash proof systems: new constructions and applications. In: Oswald, E., Fischlin, M. (eds.) EUROCRYPT 2015. LNCS, vol. 9057, pp. 69–100. Springer, Heidelberg (2015). doi:10.1007/978-3-662-46803-6_3

2. Abe, M., David, B., Kohlweiss, M., Nishimaki, R., Ohkubo, M.: Tagged one-time signatures: tight security and optimal tag size. In: Kurosawa, K., Hanaoka, G. (eds.) PKC 2013. LNCS, vol. 7778, pp. 312–331. Springer, Heidelberg (2013). doi:10.1007/978-3-642-36362-7_20

3. Attrapadung, N., Hanaoka, G., Yamada, S.: A framework for identity-based encryption with almost tight security. In: Iwata, T., Cheon, J.H. (eds.) ASIACRYPT 2015. LNCS, vol. 9452, pp. 521–549. Springer, Heidelberg (2015). doi:10.1007/978-3-662-48797-6_22

4. Bellare, M., Boldyreva, A., Micali, S.: Public-key encryption in a multi-user setting: security proofs and improvements. In: Preneel, B. (ed.) EUROCRYPT 2000. LNCS, vol. 1807, pp. 259–274. Springer, Heidelberg (2000). doi:10.1007/3-540-45539-6_18

5. Blazy, O., Kiltz, E., Pan, J.: (Hierarchical) identity-based encryption from affine message authentication. In: Garay, J.A., Gennaro, R. (eds.) CRYPTO 2014. LNCS, vol. 8616, pp. 408–425. Springer, Heidelberg (2014)

6. Camenisch, J., Shoup, V.: Practical verifiable encryption and decryption of discrete logarithms. In: Boneh, D. (ed.) CRYPTO 2003. LNCS, vol. 2729, pp. 126–144. Springer, Heidelberg (2003). doi:10.1007/978-3-540-45146-4_8

7. Chen, J., Wee, H.: Fully, (almost) tightly secure IBE and dual system groups. In: Canetti, R., Garay, J.A. (eds.) CRYPTO 2013. LNCS, vol. 8043, pp. 435–460. Springer, Heidelberg (2013). doi:10.1007/978-3-042-40084-1_25

8. Cramer, R., Shoup, V.: Design and analysis of practical public-key encryption schemes secure against adaptive chosen ciphertext attack. SIAM J. Comput. **33**(1), 167–226 (2003)

9. Cramer, R., Shoup, V.: Universal hash proofs and a paradigm for adaptive chosen ciphertext secure public-key encryption. In: Knudsen, L.R. (ed.) EUROCRYPT 2002. LNCS, vol. 2332, pp. 45–64. Springer, Heidelberg (2002). doi:10.1007/3-540-46035-7_4

10. Escala, A., Herold, G., Kiltz, E., Ràfols, C., Villar, J.: An algebraic framework for Diffie-Hellman assumptions. In: Canetti, R., Garay, J.A. (eds.) CRYPTO 2013. LNCS, vol. 8043, pp. 129–147. Springer, Heidelberg (2013)

11. Fischlin, R., Schnorr, C.P.: Stronger security proofs for RSA and rabin bits. In: Fumy, W. (ed.) EUROCRYPT 1997. LNCS, vol. 1233, pp. 267–279. Springer, Heidelberg (1997). doi:10.1007/3-540-69053-0_19

12. Gay, R., Hofheinz, D., Kiltz, E., Wee, H.: Tightly CCA-secure encryption without pairings. In: Fischlin, M., Coron, J.-S. (eds.) EUROCRYPT 2016. LNCS, vol. 9665, pp. 1–27. Springer, Heidelberg (2016). doi:10.1007/978-3-662-49890-3_1

13. Goldwasser, S., Micali, S., Rivest, R.L.: A digital signature scheme secure against adaptive chosen-message attacks. SIAM J. Comput. **17**(2), 281–308 (1988)

14. Gong, J., Chen, J., Dong, X., Cao, Z., Tang, S.: Extended nested dual system groups, revisited. In: Cheng, C.-M., Chung, K.-M., Persiano, G., Yang, B.-Y. (eds.) PKC 2016. LNCS, vol. 9614, pp. 133–163. Springer, Heidelberg (2016)

15. Hemenway, B., Ostrovsky, R.: Extended-DDH and lossy trapdoor functions. In: Fischlin, M., Buchmann, J., Manulis, M. (eds.) PKC 2012. LNCS, vol. 7293, pp. 627–643. Springer, Heidelberg (2012). doi:10.1007/978-3-642-30057-8_37

16. Hofheinz, D.: Adaptive partitioning. IACR ePrint Archive, report 2016/373. http://eprint.iacr.org/2016/373 (2016)

17. Hofheinz, D.: Algebraic partitioning: fully compact and (almost) tightly secure cryptography. In: Kushilevitz, E., Malkin, T. (eds.) TCC 2016. LNCS, vol. 9562, pp. 251–281. Springer, Heidelberg (2016). doi:10.1007/978-3-662-49096-9_11

18. Hofheinz, D., Jager, T.: Tightly secure signatures and public-key encryption. In: Safavi-Naini, R., Canetti, R. (eds.) CRYPTO 2012. LNCS, vol. 7417, pp. 590–607. Springer, Heidelberg (2012)

19. Hofheinz, D., Kiltz, E.: Secure hybrid encryption from weakened key encapsulation. In: Menezes, A. (ed.) CRYPTO 2007. LNCS, vol. 4622, pp. 553–571. Springer, Heidelberg (2007). doi:10.1007/978-3-540-74143-5_31

20. Hofheinz, D., Kiltz, E.: The group of signed quadratic residues and applications. In: Halevi, S. (ed.) CRYPTO 2009. LNCS, vol. 5677, pp. 637–653. Springer, Heidelberg (2009). doi:10.1007/978-3-642-03356-8_37

21. Hofheinz, D., Koch, J., Striecks, C.: Identity-based encryption with (almost) tight security in the multi-instance, multi-ciphertext setting. In: Katz, J. (ed.) PKC 2015. LNCS, vol. 9020, pp. 799–822. Springer, Heidelberg (2015). doi:10.1007/978-3-662-46447-2_36

22. Hohenberger, S., Waters, B.: Realizing hash-and-sign signatures under standard assumptions. In: Joux, A. (ed.) EUROCRYPT 2009. LNCS, vol. 5479, pp. 333–350. Springer, Heidelberg (2009). doi:10.1007/978-3-642-01001-9_19

23. Kiltz, E., Wee, H.: Quasi-adaptive NIZK for linear subspaces revisited. In: Oswald, E., Fischlin, M. (eds.) EUROCRYPT 2015. LNCS, vol. 9057, pp. 101–128. Springer, Heidelberg (2015). doi:10.1007/978-3-662-46803-6_4

24. Krawczyk, H., Rabin, T.: Chameleon signatures. In: Proceedings of NDSS 2000. The Internet Society (2000)

25. Kurosawa, K., Desmedt, Y.: A new paradigm of hybrid encryption scheme. In: Franklin, M. (ed.) CRYPTO 2004. LNCS, vol. 3152, pp. 426–442. Springer, Heidelberg (2004). doi:10.1007/978-3-540-28628-8_26

26. Libert, B., Joye, M., Yung, M., Peters, T.: Concise multi-challenge CCA-secure encryption and signatures with almost tight security. In: Sarkar, P., Iwata, T. (eds.) ASIACRYPT 2014. LNCS, vol. 8874, pp. 1–21. Springer, Heidelberg (2014). doi:10.1007/978-3-662-45608-8_1

27. Libert, B., Peters, T., Joye, M., Yung, M.: Compactly hiding linear spans. In: Iwata, T., Cheon, J.H. (eds.) ASIACRYPT 2015. LNCS, vol. 9452, pp. 681–707. Springer, Heidelberg (2015)

28. Naor, M., Reingold, O.: Number-theoretic constructions of efficient pseudo-random functions. J. ACM **51**(2), 231–262 (2004)

29. Paillier, P.: Public-key cryptosystems based on composite degree residuosity classes. In: Stern, J. (ed.) EUROCRYPT 1999. LNCS, vol. 1592, pp. 223–238. Springer, Heidelberg (1999). doi:10.1007/3-540-48910-X_16

30. Shacham, H.: A cramer-shoup encryption scheme from the linear assumption and from progressively weaker linear variants. IACR ePrint Archive, report 2007/74. http://eprint.iacr.org/2007/74 (2007)

0-RTT Key Exchange with Full Forward Secrecy

Felix Günther[1]($\boxtimes$), Britta Hale[2]($\boxtimes$), Tibor Jager[3]($\boxtimes$), and Sebastian Lauer[4]($\boxtimes$)

[1] Technische Universität Darmstadt, Darmstadt, Germany
guenther@cs.tu-darmstadt.de
[2] NTNU, Norwegian University of Science and Technology, Trondheim, Norway
britta.hale@item.ntnu.no
[3] Paderborn University, Paderborn, Germany
tibor.jager@upb.de
[4] Ruhr-University Bochum, Bochum, Germany
sebastian.lauer@rub.de

Abstract. Reducing latency overhead while maintaining critical security guarantees like forward secrecy has become a major design goal for key exchange (KE) protocols, both in academia and industry. Of particular interest in this regard are 0-RTT protocols, a class of KE protocols which allow a client to send cryptographically protected payload in zero round-trip time (0-RTT) along with the very first KE protocol message, thereby minimizing latency. Prominent examples are Google's QUIC protocol and the upcoming TLS protocol version 1.3. Intrinsically, the main challenge in a 0-RTT key exchange is to achieve forward secrecy and security against replay attacks for the very first payload message sent in the protocol. According to cryptographic folklore, it is impossible to achieve forward secrecy for this message, because the session key used to protect it must depend on a non-ephemeral secret of the receiver. If this secret is later leaked to an attacker, it should intuitively be possible for the attacker to compute the session key by performing the same computations as the receiver in the actual session.

In this paper we show that this belief is actually false. We construct the first 0-RTT key exchange protocol which provides full forward secrecy for all transmitted payload messages and is automatically resilient to replay attacks. In our construction we leverage a puncturable key encapsulation scheme which permits each ciphertext to only be decrypted once. Fundamentally, this is achieved by evolving the secret key after each decryption operation, but without modifying the corresponding public key or relying on shared state.

Our construction can be seen as an application of the puncturable encryption idea of Green and Miers (S&P 2015). We provide a new generic and standard-model construction of this tool that can be instantiated with any selectively secure hierarchical identity-based key encapsulation scheme.

© International Association for Cryptologic Research 2017
J.-S. Coron and J.B. Nielsen (Eds.): EUROCRYPT 2017, Part III, LNCS 10212, pp. 519–548, 2017.
DOI: 10.1007/978-3-319-56617-7_18

1   Introduction

AUTHENTICATED KEY EXCHANGE AND TLS. The Transport Layer Security (TLS) protocol is the most important cryptographic security mechanism on the Internet today, with TLS 1.2 being the most recent standardized version [16] and TLS 1.3 under development [40]. As one core functionality TLS provides an (authenticated) key exchange (AKE) which allows two remote parties to establish a shared cryptographic key over an insecure channel like the Internet. The study of provable security guarantees for AKE protocols was initiated by the seminal work of Bellare and Rogaway [4]; the huge body of work on cryptographic analyses of the TLS key exchange(s) includes [5,17,26,28].

THE DEMAND FOR LOW-LATENCY KEY EXCHANGE. Classical AKE protocols like TLS incur a considerable latency overhead due to exchanging a relatively large number of protocol messages before the first actual (application) data messages can be transmitted under cryptographic protection. Latency is commonly measured in *round-trip time* (RTT), indicating the number of rounds/round trips messaging has to take before the first application data can be sent. Even very efficient examples of high-performance AKE protocols like HMQV [27] need at least two messages (i.e., 1-RTT) before either party can compute the session key.

0-RTT KEY EXCHANGE. Reducing the latency overhead of key exchange protocols to zero round-trip time (0-RTT) while maintaining reasonable security guarantees has become a major design goal both in academia [23,29,36,44] and industry [38,40].[1] In terms of practical designs, the two principal protocols are Google's QUIC protocol [38] and the 0-RTT mode drafted for the upcoming TLS version 1.3 [40]. While the latter is still in development, QUIC is already implemented in recent versions of the Google Chrome and Opera web browsers, is currently used on Google's web servers, and has been proposed as an IETF standard (July 2015).

As authentication and establishment of cryptographic keys in 0-RTT without prior knowledge is impossible, 0-RTT key-exchange protocols must leverage keying material obtained in some prior communication to establish 0-RTT keys. Consequently, one very common approach, employed in particular in QUIC, is based on the Diffie–Hellman key exchange and is essentially comprised of the following steps (see also Fig. 1):

1. From prior communication (which may be a key exchange or some out-of-band communication), the client obtains a "medium-lived" (usually a couple of days) *server configuration*. This server configuration contains a Diffie–Hellman share g^s (with g being a generator of an algebraic group) for which

---

[1] Beyond the pure cryptographic protocol, round trips may also be induced by lower-layer transport protocols. For example, the TCP protocol requires 1-RTT for its own handshake before a higher-level cryptographic key exchange can start. Here we focus on the overhead round-trip time caused by the cryptographic components of the key-exchange protocol.

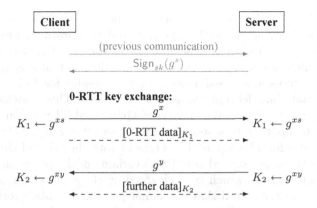

Fig. 1. The typical outline of a 0-RTT key exchange. Key K_1 can be used immediately to send 0-RTT data, key K_2 is used for all further communication.

the server knows s, and is signed under a public signing key certified for belonging to the server.

2. In the 0-RTT key exchange, the client knowing g^s now picks a secret exponent x at random and sends the share g^x to the server. It also directly computes a preliminary, 0-RTT key K_1 from the Diffie–Hellman value g^{xs}. In immediate application, this key can be used to send cryptographically protected (0-RTT) application data along with the client's key-exchange message.

3. The server responds with a freshly chosen, ephemeral Diffie–Hellman share g^y which is used by both the server and the client to compute the actual session key K_2 from g^{xy}. All further communication throughout the session is subsequently protected under K_2.

An alternative approach, pursued in the latest TLS 1.3 drafts, is to derive the 0-RTT key from a pre-shared symmetric key. Note that this requires storing *secret* key information on the client between sessions. In contrast, we consider 0-RTT key establishment protocols, which do not require secret information to be stored between sessions.

ISSUES WITH 0-RTT KEY EXCHANGE. As outlined, the 0-RTT key-exchange design elegantly allows clients to initiate a secure connection with zero latency overhead, addressing an important practical problem. Unfortunately, all protocols that follow this format—including QUIC and TLS 1.3 as well as academic approaches [23,44]—face at least one of the following two very undesirable drawbacks.

Forward Secrecy. Recall that forward secrecy essentially demands that transmitted data remains secret even if an attacker learns the secret key of one communication partner. From contemporary insight, this is considered a standard and crucial security goal of modern key exchange protocols, as it addresses data protection in the presence of passive key compromises or mass surveillance. Observe that a 0-RTT key exchange of the form outlined above, however, cannot provide forward secrecy for the 0-RTT application data transmitted from the client to the server. As such data is protected under the key K_1, derived from g^{xs}, an attacker which eavesdrops on the communication and later compromises the server's secret exponent s (possibly long after the session has finished) can easily compute K_1 and thus decrypt the 0-RTT data sent. This drawback is clearly acknowledged in the design documents of QUIC and TLS 1.3 and one of the main reasons to upgrade to a second, forward-secret key K_2. Notably, the lack of forward secrecy for TLS 1.3 0-RTT is true of both the original Diffie–Hellman-based and the latest pre-shared key (PSK) variants of the protocol, albeit under different assumptions on which key is learned by the attacker [29,39,40,42].

In 2005, Krawczyk stated that it was not possible to obtain forward secrecy for implicitly-authenticated 2-message protocols in a public-key authentication context, if there was no pre-existing shared state [27]. Subsequent works referenced this idea prominently, but often dropped one or more of the original conditions [8,11,30]. Despite modeling changes and arguments to the contrary in relation to 1-round protocols [13,15], and work on forward secrecy for non-interactive key-exchange (NIKE) protocols [37], the assumption that forward secrecy is fundamentally impossible under limited rounds has perpetuated. In particular, the QUIC crypto specification accepts an "upper bound on the forward security of the connection" for 0-RTT handshakes [31]. Likewise, this limitation is accepted as seemingly inherent in academic 0-RTT designs [23,44], and early discussions around the development of TLS 1.3 go so far as to claim that forward secrecy "can't be done in 0-RTT" [43].

Replay Attacks. In a replay attack, an attacker aims at making the receiver accept the same payload twice. Specifically, replay attacks in the example 0-RTT protocol given can take the form of replaying the client's Diffie–Hellman share g^x or the 0-RTT data sent. Observe that, without further countermeasures, an adversary can simply replay (potentially multiple times) a recorded client message g^x, making the server derive the same key K_1 as in the original connection, and then replay the client's 0-RTT data which the server can correctly decrypt and would therefore process. Depending on the application logic, such replays can lead to severe security issues. For example, an authenticated request (e.g., via login credentials or cookie tokens) might allow an adversary to replay client actions like online orders or financial transactions.

One potential countermeasure, implemented in QUIC, is essentially to store all seen client values g^x (in a certain time frame encoded in an additional nonce value) in order to detect and reject repeated requests with the same value and nonce.[2] Notably, this solution induces a substantial management overhead and arguably is acceptable only for certain server configurations. As such, the solution is not elegant, but effectively prevents the same key from being accepted twice by a server. We remark, though, that on a higher level applications may resend

---

[2] In case of Google this approach amounts to a few gigabytes of data to be held in shared state between multiple server instances.

data under a later-derived key in common web scenarios, essentially rendering replay attacks on the application layer unavoidable in such cases [19, 41].

Low-latency key-exchange designs proposed thus far widely accepted the aforementioned drawbacks on forward secrecy and replay protection as inherent to the 0-RTT environment. This assumption paves the way for the following research question for the design of modern, low-latency authenticated key-exchange protocols: *Can a key-exchange protocol establish a cryptographic key in 0-RTT while upholding strong forward-secrecy and replay protection guarantees?*

CONTRIBUTIONS. In this work we introduce the notion of *forward-secret one-pass key exchange* and a *generic construction* of such a protocol, resolving the aforementioned open problem. Notable features of this protocol are summarized as follows.

- The protocol provides *full forward secrecy*, even for the first message transmitted from the client to the server, and is automatically resilient to replay attacks. We provide a rigorous security analysis for which we develop a novel key-exchange model (in the style of Bellare and Rogaway [4]) that captures the peculiarities of forward secrecy and replay protection in 0-RTT key exchange.
- The protocol has the *simplest message* flow imaginable: the client encrypts a session key and sends it to the server. We do not need to distinguish between preliminary and final keys but only derive a single session key. The forward secrecy and replay security of the protocol stem from the fact that the long-term secret key of this scheme is evolved.
- The construction and security proof are completely *generic*, based on any one-time signature scheme and any hierarchical identity-based key encapsulation scheme (HIBKEM) that needs to provide only a weak form of selective-ID security. This allows for flexible instantiation of the protocol with arbitrary cryptographic constructions of these primitives, adjusted with suitable deployment and efficiency parameters for a given application, and based on various hardness assumptions.
- The construction and its security analysis are completely independent of a particular instantiation of building blocks, immediately yielding the first *post-quantum* secure 0-RTT key exchange protocol, via instantiation of the protocol with suitable lattice-based building blocks, such as the HIBE from [1] and the one-time signature from [34].
- More generally, by instantiating the protocol with different HIBKEM schemes, one can easily obtain different "cipher suites", with different security and performance characteristics. Replacement of a cipher suite is easy, as it does not require a new security analysis of the protocol. In contrast, several consecutive research papers were required to establish the security of only the most important basic cipher suites of TLS [26, 28, 32].

Our work is inspired by earlier work of Canetti, Halevi, and Katz [9] on forward-secure public-key encryption and Green and Miers [21] on forward-secret puncturable public-key encryption. The main novelties in this work are:

- We make the conceptual observation that the tool of forward-secret puncturable public-key encryption can be leveraged to enable forward-secret 0-RTT AKE.
- We carve out puncturable forward-secret key encapsulation as a versatile building block and build it in a generic fashion from any HIBKEM scheme, in the standard model, and from a wide range of assumptions. In contrast, the cunning, but involved construction by Green and Miers [21] blends the attribute-based encryption scheme of Ostrovsky, Sahai, and Waters [35] with forward-secret encryption [9]. It therefore relies on specific assumptions and, using the Fujisaki-Okamoto transform [20] to achieve CCA-security, relies on the random-oracle model.
- We formalize 0-RTT key exchange security with forward secrecy. This is a non-trivial extension of previous models (particularly [24]) in that it needs to take evolving state, (semi-)synchronized time, and accordingly conditioned forward secrecy into account in the security experiment.

We consider the established concepts as valuable towards the understanding of forward-secret 0-RTT key exchange, its foundations, and its connection to, in particular, asynchronous messaging.

HIGH-LEVEL PROTOCOL DESCRIPTION. The basic outline of our protocol is the simplest one can imagine. We use a public-key *key encapsulation mechanism* (KEM)[3] to transport a random session key from the client to the server. That is, the server is in possession of a long-term key pair (pk, sk) for the KEM, and the client uses pk to encapsulate a key. This immediately yields a 0-RTT protocol, because we can send encrypted payload data along with the encapsulated key. However, of course, it does not yet provide forward secrecy or security against replay attacks.

The key idea to achieve these additional properties is not to modify the protocol, but to modify the way the server stores and processes its secret key. More precisely, we construct and use a special *puncturable forward-secure* KEM (PFS-KEM). Consider a server with long-term secret key sk. When receiving an encapsulated session key in ciphertext c_1, the server can use this scheme to proceed as follows.

1. It decrypts c_1 using sk.
2. The server then derives a new secret key $sk_{\backslash c_1}$ from sk, which is "punctured at position c_1". This means that $sk_{\backslash c_1}$ can be used to decrypt all ciphertexts *except* for c_1.
3. Finally, the server deletes sk.

This process is executed repeatedly for all ciphertexts received by the server. That is, when the server receives a second ciphertext c_2 from the same or a different client, it again "punctures" $sk_{\backslash c_1}$ to obtain a new secret key $sk_{\backslash c_1, c_2}$,

---

[3] This is essentially a public-key encryption scheme which can only be used to transport random keys, but not to transport payload messages.

which can be used to decrypt all ciphertexts except for c_1 and c_2. Note that this yields forward secrecy, because an attacker that obtains $sk_{\backslash c_1, c_2}$ will not be able to use this key to decrypt c_1 or c_2, and thus will not be able to learn the session key of previous sessions.

The drawback of using this approach naïvely is that the size of secret keys grows linearly with the number of sessions, which is of course impractical. For efficiency reasons, we therefore add an additional *time* component to the protocol, which requires only loosely synchronized clocks between client and server. Within each time slot, the size of the secret key grows linearly with the number of sessions. However, at the end of the time slot, the server is able to "purge" the key, which reduces its size back to a factor logarithmic in the number of time intervals. We stress that the loose time synchronization is included in our protocol's design only for efficiency reasons, but is not needed to achieve the desired security goals.

A particularly beneficial aspect of this approach is that the server's public key pk remains *static* over its entire lifetime (which would typically be 1–2 years in practice, but longer lifetimes are easily possible), because there is no QUIC-like server configuration that needs to be frequently updated at client-side. Thus, this yields a protocol without the need to frequently replace the server configuration g^s at the client.

The maximal size of punctured secret keys, and thus the storage requirement of the protocol, depends on the size of time slots. Longer time slots (several hours or possibly even a few days, depending on the number of connections during this time) require more storage, but only loosely synchronized clocks. Short time slots (a few minutes) require less storage, but more precisely synchronized clocks. These parameters can be chosen depending on the individual characteristics of a server and the services that it provides.

RELATED WORK. The idea of forward-secret encryption based on hierarchical identity-based encryption is due to Canetti, Halevi, and Katz [9]. Pointcheval and Sanders [37] studied forward secrecy for non-interactive key-exchange protocols based on multilinear maps. Both approaches however only provide coarse-grained forward secrecy with respect to time periods, whereas we aim at a fine-grained, immediate notion of forward secrecy in the setting of key exchange.

With a similar goal in mind, the previously mentioned work of Green and Miers [21] achieves forward secrecy in the context of asynchronous messaging.[4] Their construction blends the attribute-based encryption scheme of Ostrovsky, Sahai, and Waters [35] with the scheme of Canetti, Halevi, and Katz [9] or, alternatively, with the scheme of Boneh, Boyen, and Goh [7]. This makes their scheme relatively complex and bound to specific algebraic settings and complexity assumptions. Moreover, their scheme achieves only CPA security, and requires

---

[4] Observe that asynchronous messaging and 0-RTT key exchange are conceptually relatively close. In both settings, a data protection key is to be established while only unilateral communication is possible. While different, e.g., in constraints for latency and storage overhead, this in particular implies that our construction can also be employed in the setting of asynchronous messaging.

the random oracle model [3] and the Fujisaki-Okamoto transform [20] to achieve CCA security. In contrast, we describe a simple, natural and directly CCA-secure construction based on any hierarchical identity-based KEM (HIBKEM), which can be instantiated from any HIBKEM that only needs to provide weak selective-ID security.

The security of the QUIC protocol was formally analyzed by Fischlin and Günther [18] as well as Lychev *et al.* [33]. Krawczyk and Wee [29] described the OPTLS protocol as a foundation for TLS 1.3, including a 0-RTT mode. For TLS 1.3, Cremers *et al.* [14] conducted a tool-supported analysis of TLS 1.3 including a draft 0-RTT handshake mode, and Fischlin and Günther [19] analyzed the provable security of both Diffie–Hellman- and PSK-based 0-RTT handshake drafts. Foundational definitions and generic constructions of 0-RTT key exchange from other cryptographic primitives were given by Hale *et al.* [23]. All these works consider security models and constructions *without* forward secrecy of the first message. In a related, but different direction, Cohn-Gordon *et al.* [12] consider *post-compromise security* for key-exchange protocols that use key ratcheting, where the session key is frequently updated during the lifetime of a single session.

OUTLINE OF THE PAPER. Section 2 introduces the necessary building blocks for our construction as well as puncturable forward-secret key encapsulation (PFSKEM), before we provide a generic PFSKEM construction from HIBE. We formalize forward-secret one-pass key exchange protocols (FSOPKE) in Sect. 3, together with a corresponding security model. In Sect. 4 we provide a generic construction of FSOPKE with server authentication from PFSKEM and prove its security in the FSOPKE model. In Sect. 5 we analyze the size of keys and messages for different deployment parameters.

2   Generic Construction of Puncturable Encryption

2.1   Building Blocks

Let us begin with recapping the definition and security of one-time signature schemes, as well as hierarchical identity-based key encapsulation schemes.

Definition 1 (One-Time Signatures). *A one-time signature scheme* OTSIG *consists of three probabilistic polynomial-time algorithms (*OTSIG.KGen, OTSIG.Sign, OTSIG.Vfy*).*

- OTSIG.KGen(1^λ) *takes as input a security parameter λ and outputs a public key pk_{OT} and a secret key sk_{OT}.*
- OTSIG.Sign(sk_{OT}, m) *takes as input a secret key and a message $m \in \{0,1\}^n$. Output is a signature σ.*
- OTSIG.Vfy(pk_{OT}, m, σ) *input is a public key, a message $m \in \{0,1\}^n$ and a signature σ. If σ is a valid signature for m under pk_{OT}, then the algorithm outputs 1, else 0.*

Consider the following security experiment $G_{\mathcal{A},\text{OTSIG}}^{\text{sEUF-1-CMA}}(\lambda)$ played between a challenger $\mathcal{C}$ and an adversary $\mathcal{A}$.

1. The challenger $\mathcal{C}$ computes $(pk_{OT}, sk_{OT}) \xleftarrow{\$} \mathsf{OTSIG.KGen}(1^\lambda)$ and runs $\mathcal{A}$ with input pk_{OT}.
2. $\mathcal{A}$ may query one arbitrary message m to the challenger. $\mathcal{C}$ replies with $\sigma \xleftarrow{\$} \mathsf{OTSIG.Sign}(sk_{OT}, m)$.
3. $\mathcal{A}$ eventually outputs a message m^* and a signature σ^*. We denote the event that $\mathsf{OTSIG.Vfy}(pk_{OT}, m^*, \sigma^*) = 1$ and $(m^*, \sigma^*) \neq (m, \sigma)$ by

$$G^{\mathsf{sEUF-1-CMA}}_{\mathcal{A},\mathsf{OTSIG}}(\lambda) = 1.$$

Definition 2 (Security of One-Time Signatures). *We define the advantage of an adversary $\mathcal{A}$ in the game $G^{\mathsf{sEUF-1-CMA}}_{\mathcal{A},\mathsf{OTSIG}}(\lambda)$ as*

$$\mathsf{Adv}^{\mathsf{sEUF-1-CMA}}_{\mathcal{A},\mathsf{OTSIG}}(\lambda) := \Pr\left[G^{\mathsf{sEUF-1-CMA}}_{\mathcal{A},\mathsf{OTSIG}}(\lambda) = 1\right].$$

A one-time signature scheme OTSIG *is strongly secure against existential forgeries under adaptive chosen-message attacks (sEUF-1-CMA), if* $\mathsf{Adv}^{\mathsf{sEUF-1-CMA}}_{\mathcal{A},\mathsf{OTSIG}}(\lambda)$ *is a negligible function in λ for all probabilistic polynomial-time adversaries $\mathcal{A}$.*

In our generic construction we use a hierarchical identity-based key encapsulation scheme (HIBKEM) [6]. HIBKEM schemes enable a user to encapsulate a symmetric key with the recipients identity. An identity at depth t in the hierarchical tree is represented by a vector $\mathsf{ID}_{|t} = (I_1, \cdots, I_t)$. Ancestors of the identity $\mathsf{ID}_{|t}$ are identities represented by vectors $\mathsf{ID}_{|s} = (J_1, \cdots, J_s)$ with $1 \leq s < t$ and $I_i = J_i$ for $1 \leq i \leq s$.

Definition 3 (HIBKEM [6]). *A hierarchical identity-based key encapsulation scheme* HIBKEM *consists of four probabilistic polynomial-time algorithms* (HIBKEM.KGen, HIBKEM.Del, HIBKEM.Enc, HIBKEM.Dec).

- HIBKEM.KGen(1^λ) *takes as input a security parameter λ and outputs an a public key pk and an initial secret key (or master key) msk, which we refer as the private key at depth 0. We assume that pk implicitly defines the identity space $\mathcal{ID}$ and the key space $\mathcal{K}$.*
- HIBKEM.Del$(\mathsf{ID}_{|t}, sk_{\mathsf{ID}'|s})$ *takes as input an identity $\mathsf{ID}_{|t}$ at depth t and the private key of an ancestor identity $\mathsf{ID}'_{|s}$ at depth $s < t$ or the master key msk. Output is a secret key $sk_{\mathsf{ID}|t}$.*
- HIBKEM.Enc(pk, ID) *takes a input the public key pk and the target ID. The algorithm outputs a ciphertext CT and a symmetric key K.*
- HIBKEM.Dec$(sk_{\mathsf{ID}}, \mathsf{CT})$ *takes as input a secret key sk_{ID} and a ciphertext CT. Output is a symmetric key K or $\perp$ if decryption fails.*

Consider the following selective-ID CPA security experiment $G^{\mathsf{IND\text{-}sID\text{-}CPA}}_{\mathcal{A},\mathsf{HIBKEM}}(\lambda)$ played between a challenger $\mathcal{C}$ and an adversary $\mathcal{A}$.

1. $\mathcal{A}$ outputs the target identity ID^* on which it wants to be challenged.
2. The challenger generates the system parameters and computes $(pk, msk) \xleftarrow{\$} \mathsf{HIBKEM.KGen}(1^\lambda)$. $\mathcal{C}$ generates $(\mathsf{K}_0, \mathsf{CT}^*) \xleftarrow{\$} \mathsf{HIBKEM.Enc}(pk, \mathsf{ID}^*)$ and $K_1 \xleftarrow{\$} \mathcal{K}$. Then the challenger sends (K_b, CT^*, pk) to $\mathcal{A}$ where $b \xleftarrow{\$} \{0, 1\}$.

3. $\mathcal{A}$ may query an HIBKEM.Del oracle. The HIBKEM.Del oracle outputs the secret key of a requested identity ID. The only restriction is, that the attacker $\mathcal{A}$ is not allowed to ask the HIBKEM.Del oracle for the secret key of ID* or any ancestor identity of ID*.
4. Finally, $\mathcal{A}$ eventually outputs a guess b'. We denote the event that $b = b'$ by

$$G_{\mathcal{A},\mathsf{HIBKEM}}^{\mathsf{IND\text{-}sID\text{-}CPA}}(\lambda) = 1$$

Definition 4 (Security of HIBKEM). *We define the advantage of an adversary $\mathcal{A}$ in the selective-ID game $G_{\mathcal{A},\mathsf{HIBKEM}}^{\mathsf{IND\text{-}sID\text{-}CPA}}(\lambda)$ as*

$$\mathsf{Adv}_{\mathcal{A},\mathsf{HIBKEM}}^{\mathsf{IND\text{-}sID\text{-}CPA}}(\lambda):= \left| \Pr\left[G_{\mathcal{A},\mathsf{HIBKEM}}^{\mathsf{IND\text{-}sID\text{-}CPA}}(\lambda) = 1 \right] - \frac{1}{2} \right|$$

A hierarchical identity-based key encapsulation scheme HIBKEM is selective-ID CPA-secure (IND-sID-CPA), if $\mathsf{Adv}_{\mathcal{A},\mathsf{HIBKEM}}^{\mathsf{IND\text{-}sID\text{-}CPA}}(\lambda)$ is a negligible function in λ for all probabilistic polynomial-time adversaries $\mathcal{A}$.

2.2   Puncturable Forward-Secret Key Encapsulation

We now formally introduce the definition of a puncturable forward-secret key encapsulation (PFSKEM) scheme as well as its corresponding correctness definition and security notion.

Definition 5 (PFSKEM). *A puncturable forward-secret key encapsulation scheme PFSKEM consists of five probabilistic polynomial-time algorithms (PFSKEM.KGen,   PFSKEM.Enc,   PFSKEM.PnctCxt,   PFSKEM.Dec, PFSKEM.PnctInt).*

- *PFSKEM.KGen(1^λ) takes as input a security parameter λ and outputs a public key PK and an initial secret key SK.*
- *PFSKEM.Enc(PK, τ) takes as input a public key and a time period τ. Output is a ciphertext CT and a symmetric key K.*
- *PFSKEM.PnctCxt(SK, τ, CT) input is the current secret key SK, a time period τ and additionally a ciphertext CT. The algorithm outputs a new secret key SK'.*
- *PFSKEM.Dec(SK, τ, CT) takes as input a secret key SK, time period τ and a ciphertext CT. Output is a symmetric key K or $\perp$ if decapsulation fails.*
- *PFSKEM.PnctInt(SK, τ) takes as input a secret key SK and a time interval τ. Output is a secret key SK' for the next time interval $\tau + 1$.*

Definition 6 (Correctness of PFSKEM). *For all $\lambda, n \in \mathbb{N}$, any (PK, SK) $\xleftarrow{\$}$ PFSKEM.KGen(1^λ), any time period τ^*, any $(\mathsf{K}, \mathsf{CT}^*)$ $\xleftarrow{\$}$ PFSKEM.Enc (PK, τ^*), and any (arbitrary interleaved) sequence $i = 0, \ldots, n-1$ of invocations of SK' $\xleftarrow{\$}$ PFSKEM.PnctCxt(SK, τ, CT) for any $(\tau, \mathsf{CT}) \neq (\tau^*, \mathsf{CT}^*)$ or SK' $\xleftarrow{\$}$ PFSKEM.PnctInt(SK, τ) for any $\tau \neq \tau^*$ it holds that PFSKEM.Dec $(SK', \tau^*, \mathsf{CT}^*) = \mathsf{K}$.*

Beyond the regular correctness definition above, we further define an extended variant of correctness which demands that decapsulation under a previously punctured out time-interval and ciphertext yields an error symbol $\perp$.

Definition 7 (Extended Correctness of PFSKEM). *For all $\lambda, n \in \mathbb{N}$, any $(PK, SK) \xleftarrow{\$} \mathsf{PFSKEM.KGen}(1^\lambda)$, any time period τ^*, any $(\mathsf{K}, \mathsf{CT}^*) \xleftarrow{\$} \mathsf{PFSKEM.Enc}(PK, \tau^*)$, and any (arbitrary interleaved) sequence $i = 0, \ldots, n-1$ of invocations of $SK' \xleftarrow{\$} \mathsf{PFSKEM.PnctCxt}(SK, \tau_i, \mathsf{CT}_i)$ for any (τ_i, CT_i) or $SK' \xleftarrow{\$} \mathsf{PFSKEM.PnctInt}(SK', \tau_i')$ for any τ_i' it holds that if $(\tau_i, \mathsf{CT}_i) = (\tau^*, \mathsf{CT}^*)$ or $\tau_i' = \tau^*$ for some $i \in \{0, \ldots, n-1\}$, then $\mathsf{PFSKEM.Dec}(SK', \tau^*, \mathsf{CT}^*) = \perp$.*

The security of a PFSKEM scheme is defined by the following selective-time CCA security experiment $G^{\mathsf{IND\text{-}sT\text{-}CCA}}_{\mathcal{A}, \mathsf{PFSKEM}}(\lambda)$ played between a challenger $\mathcal{C}$ and an attacker $\mathcal{A}$.

1. In the beginning, $\mathcal{A}$ outputs the target time τ^*.
2. The challenger $\mathcal{C}$ generates a fresh key pair $(PK, SK) \xleftarrow{\$} \mathsf{PFSKEM.KGen}(1^\lambda)$. It computes $(\mathsf{CT}^*, \mathsf{K}_0^*) \xleftarrow{\$} \mathsf{PFSKEM.Enc}(PK, \tau^*)$ and selects $\mathsf{K}_1^* \xleftarrow{\$} \mathcal{K}$. Additionally, it chooses a bit $b \xleftarrow{\$} \{0, 1\}$ and then sends $(PK, \mathsf{CT}^*, \mathsf{K}_b^*)$ to $\mathcal{A}$.
3. $\mathcal{A}$ can now ask a polynomial number of the following queries:
   - $\mathsf{PFSKEM.Dec}(\tau, \mathsf{CT})$: The challenger computes $\mathsf{K} \xleftarrow{\$} \mathsf{PFSKEM.Dec}(SK, \tau, \mathsf{CT})$ and returns K to $\mathcal{A}$.
   - $\mathsf{PFSKEM.PnctCxt}(\tau, \mathsf{CT})$: The challenger runs $SK \xleftarrow{\$} \mathsf{PFSKEM.PnctCxt}(SK, \tau, \mathsf{CT})$ and returns symbol $\top$.
   - $\mathsf{PFSKEM.PnctInt}(\tau)$: The challenger runs $SK \xleftarrow{\$} \mathsf{PFSKEM.PnctInt}(SK, \tau)$ and returns symbol $\top$.
   - $\mathsf{PFSKEM.Corrupt}()$: The challenger aborts and outputs a random bit if $\mathcal{A}$ has not queried $\mathsf{PFSKEM.PnctCxt}(\tau^*, \mathsf{CT}^*)$ or $\mathsf{PFSKEM.PnctInt}(\tau^*)$ before. Otherwise, the challenger returns the current secret key SK to $\mathcal{A}$.
4. $\mathcal{A}$ eventually outputs a guess b'. We denote the event that $b = b'$ by

$$G^{\mathsf{IND\text{-}sT\text{-}CCA}}_{\mathcal{A}, \mathsf{PFSKEM}}(\lambda) = 1 .$$

Definition 8 (Security of PFSKEM). *We define the advantage of an adversary $\mathcal{A}$ in the selective-time CCA game $G^{\mathsf{IND\text{-}sT\text{-}CCA}}_{\mathcal{A}, \mathsf{PFSKEM}}(\lambda)$ as*

$$\mathsf{Adv}^{\mathsf{IND\text{-}sT\text{-}CCA}}_{\mathcal{A}, \mathsf{PFSKEM}}(\lambda) := \left| \Pr\left[G^{\mathsf{IND\text{-}sT\text{-}CCA}}_{\mathcal{A}, \mathsf{PFSKEM}}(\lambda) = 1 \right] - \frac{1}{2} \right|$$

A puncturable forward-secret key encapsulation scheme PFSKEM is selective-time CCA-secure (IND-sT-CCA), if $\mathsf{Adv}^{\mathsf{IND\text{-}sT\text{-}CCA}}_{\mathcal{A}, \mathsf{PFSKEM}}(\lambda)$ is a negligible function in λ for all probabilistic polynomial-time adversaries $\mathcal{A}$.

2.3   A Generic PFSKEM Construction from HIBKEM

We have now set up the necessary building blocks for our generic PFSKEM construction. In this construction, we deploy a HIBKEM scheme over a binary hierarchy tree comprising time intervals in the upper part and identifiers within these intervals in the lower part. The latter identifiers are carefully crafted to be public keys of a one-time signature scheme, conveniently enabling our construction to achieve CCA security.

We start with a short description of the binary tree, where the root node has the label ϵ. The left child of a node under label n is labeled with n_0 and the right child with n_1. In a HIBKEM scheme every identity ID_i is represented by a node n_i of the hierarchy tree T and with sk_i we denote the secret key corresponding to node n_i. The root node has the corresponding master secret key msk of the HIBKEM scheme. To identify specific nodes in the tree we need the following functions.

- $\mathsf{Parent}(T, n)$. On input of a description of a tree T and a node n, this function outputs the label of the direct ancestor of n or $\bot$ if it does not exists.
- $\mathsf{Sibling}(T, n)$. On input of a description of a tree T and a label of a node n this function outputs the other child $n' \neq n$ of the node $\mathsf{Parent}(T, n)$ or $\bot$ if it does not exists.

On input of a description of a tree T, a set of secret keys and nodes $SK = \{(sk_1, n_1), \ldots, (sk_u, n_u)\}$ and a node n, the following algorithm computes a new set of secret keys and nodes SK'. The secret keys in SK' can neither be used to derive the secret key of n nor of its descendants.

- $\mathsf{PunctureTree}(T, SK, n)$. Create an empty set $SK' := \{\}$. Then, for all tuples (sk_i, n_i) in SK:
    - If n_i is neither an ancestor nor a descendant of n in the tree and if $n_i \neq n$, then set $SK' := SK' \cup (sk_i, n_i)$ and $SK := SK \setminus \{(sk_i, n_i)\}$.
  If there is a remaining node n' with its secret key sk' in SK and if n' is an ancestor of n, then set $ntmp := n$, and while $\mathsf{Parent}(ntmp) \neq \bot$:
    - if n' is an ancestor of $\mathsf{Sibling}(ntmp)$ then $SK' := SK' \cup \{\mathsf{HIBKEM.Del}(\mathsf{Sibling}(ntmp), sk'), \mathsf{Sibling}(ntmp))\}$
    - $ntmp := \mathsf{Parent}(ntmp)$.
  Output is the set of secret keys and nodes SK'.

Illustrating the described algorithm, we provide an example in Fig. 2, with a tree where the nodes are labeled as described earlier. SK consists of the tuple $\{(msk, \epsilon)\}$, where msk is the initial secret key of a HIBKEM. We would like to puncture the secret key SK for the input n_{01}. In order to do so, we must delete all keys in SK that can be used to derive the secret keys for the nodes with label "01" or with the prefix "01". For this, we run the algorithm $\mathsf{PunctureTree}$ with input $(T, SK, 01)$. In Fig. 2 the gray nodes denote the labels for which we have to derive the secret keys within the new PFSKEM secret key SK'. The secret keys in SK' can only be used to generate secret keys for identities which are not ancestors or descendants of the punctured node "01".

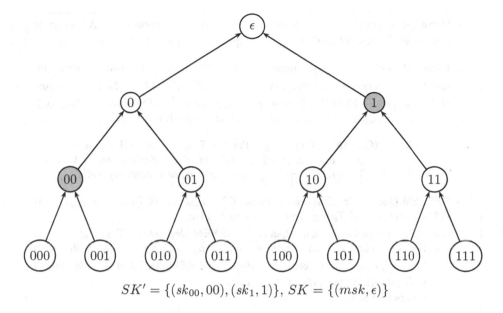

$$SK' = \{(sk_{00}, 00), (sk_1, 1)\}, \ SK = \{(msk, \epsilon)\}$$

Fig. 2. Hierarchy tree with secret key SK', under initial secret key SK

In the following, an identifier $\mathsf{ID} = \tau \| pk_{OT}$ consisting of a time interval τ and a one-time signature public key pk_{OT} is a leaf in a HIBKEM tree T. The public key PK and the initial secret key SK of the PFSKEM construction are, respectively, the public key pk and a pair consisting of the initial secret key of the HIBKEM scheme with the label of the root node (msk, ϵ).

To obtain a symmetric key at time τ, one can use the encapsulation algorithm of the HIBKEM scheme with input $(PK, \tau \| pk_{OT})$. Correspondingly, the secret key SK of the PFSKEM scheme can be punctured via the previously defined algorithm PunctureTree(T, SK, n) by deleting the secret key for the identity $\mathsf{ID} = \tau \| pk_{OT}$ in the HIBKEM scheme including all secret keys of ancestors of ID. Particularly, this can be accomplished by using the previously defined algorithm PunctureTree(T, SK, n). Decapsulation uses the secret key of the identity $\mathsf{ID} = \tau \| pk_{OT}$ with a ciphertext CT and outputs the symmetric key or $\bot$ if the key is already deleted or the signature of the ciphertext is not valid.

The described generic construction is presented in Fig. 3.

As we establish next, our PFSKEM construction is selective-time CCA-secure (according to Definition 8) if the underlying HIBKEM scheme is IND-sID-CPA-secure and the OTSIG scheme is sEUF-1-CMA-secure (cf. Definitions 4 and 2).

Theorem 1. *For any efficient polynomial-time adversary $\mathcal{A}$ in the* IND-sT-CCA *game there exist efficient polynomial-time algorithms $\mathcal{B}_{\mathsf{HIBKEM}}$ and $\mathcal{B}_{\mathsf{OTSIG}}$ such that*

$$\mathsf{Adv}^{\mathsf{IND\text{-}sT\text{-}CCA}}_{\mathcal{A},\mathsf{PFSKEM}}(\lambda) \leq \mathsf{Adv}^{\mathsf{IND\text{-}sID\text{-}CPA}}_{\mathcal{B}_{\mathsf{HIBKEM}},\mathsf{HIBKEM}}(\lambda) + \mathsf{Adv}^{\mathsf{sEUF\text{-}1\text{-}CMA}}_{\mathcal{B}_{\mathsf{OTSIG}},\mathsf{OTSIG}}(\lambda).$$

- PFSKEM.KGen(1^λ). On input of a security parameter λ generate $(pk, msk) \xleftarrow{\$} \text{HIBKEM.KGen}(1^\lambda)$ and output $PK := pk$ and $SK := (msk, \epsilon)$.

- PFSKEM.Enc(PK, τ). On input of a public key PK and a time interval τ, generate $(pk_{OT}, sk_{OT}) \xleftarrow{\$} \text{OTSIG.KGen}(1^\lambda)$. Next, compute $(\mathsf{CT}_{\mathsf{HIBKEM}}, \mathsf{K}) \xleftarrow{\$} \text{HIBKEM.Enc}(pk, \tau||pk_{OT})$ and $\sigma \xleftarrow{\$} \text{OTSIG}(sk_{OT}, \mathsf{CT}_{\mathsf{HIBKEM}})$. Then, set $\mathsf{CT}_{\mathsf{PFSKEM}} = (\mathsf{CT}_{\mathsf{HIBKEM}}, \sigma, pk_{OT})$ and output K and $\mathsf{CT}_{\mathsf{PFSKEM}}$.

- PFSKEM.PnctCxt($SK, \tau, \mathsf{CT}_{\mathsf{PFSKEM}}$). Parse $\mathsf{CT}_{\mathsf{PFSKEM}}$ as $(\mathsf{CT}_{\mathsf{HIBKEM}}, \sigma, pk_{OT})$ and let T be the description of the HIBKEM tree. Compute $SK' = \text{PunctureTree}(T, SK, \tau||pk_{OT})$ and output the new secret key SK'.

- PFSKEM.Dec($SK, \tau, \mathsf{CT}_{\mathsf{PFSKEM}}$). Parse $\mathsf{CT}_{\mathsf{PFSKEM}}$ as $(\mathsf{CT}_{\mathsf{HIBKEM}}, \sigma, pk_{OT})$. If OTSIG.Vfy($pk_{OT}, \mathsf{CT}_{\mathsf{HIBKEM}}, \sigma$) = 0 output $\perp$. Else:
  - If SK contains sk_{ID}, then output $\mathsf{K} \xleftarrow{\$} \text{HIBKEM.Dec}, (sk_{\mathsf{ID}}, \mathsf{CT}_{\mathsf{HIBKEM}})$.
  - If SK contains an ancestor node n_j of the node with label $\mathsf{ID} = \tau||pk_{OT}$, then compute $sk_{\mathsf{ID}} \xleftarrow{\$} \text{HIBKEM.Del}(\mathsf{ID}, sk_j)$ and output $\mathsf{K} \xleftarrow{\$} \text{HIBKEM.Dec}, (sk_{\mathsf{ID}}, \mathsf{CT}_{\mathsf{HIBKEM}})$.
  - Otherwise output $\perp$.

- PFSKEM.PnctInt(SK, τ). Compute $SK' = \text{PunctureTree}(T, SK, \tau)$ where T is a description of the hierarchy tree. Output the new secret key SK'.

Fig. 3. Generic PFSKEM construction from a HIBKEM and a one-time signature scheme.

Proof. An attacker $\mathcal{A}$ on the PFSKEM scheme outputs a target time period τ^* and can make the queries described in the security experiment for PFSKEM schemes.

Let $(\mathsf{CT}^*_{\mathsf{PFSKEM}}, \mathsf{K}^*_b) = ((\mathsf{CT}^*_{\mathsf{HIBKEM}}, \sigma^*, pk^*_{OT}), \mathsf{K}^*_b)$ be the challenge we have to compute for the PFSKEM attacker and let E denote the event that the attacker $\mathcal{A}$ never queries PFSKEM.Dec($\tau, \mathsf{CT}_{\mathsf{PFSKEM}} = (\mathsf{CT}_{\mathsf{HIBKEM}}, \sigma, pk_{OT})$) where $(\mathsf{CT}_{\mathsf{HIBKEM}}, \sigma) \neq (\mathsf{CT}^*_{\mathsf{HIBKEM}}, \sigma^*)$, $pk_{OT} = pk^*_{OT}$, and OTSIG.Vfy($pk_{OT}, \mathsf{CT}_{\mathsf{HIBKEM}}, \sigma$) = 1 in the security game. The probability for $\mathcal{A}$ to win the security game is

$$\Pr\left[G^{\mathsf{IND\text{-}sT\text{-}CCA}}_{\mathcal{A},\mathsf{PFSKEM}}(\lambda) = 1\right]$$
$$= \Pr\left[G^{\mathsf{IND\text{-}sT\text{-}CCA}}_{\mathcal{A},\mathsf{PFSKEM}}(\lambda) = 1 \cap \mathsf{E}\right] + \Pr\left[G^{\mathsf{IND\text{-}sT\text{-}CCA}}_{\mathcal{A},\mathsf{PFSKEM}}(\lambda) = 1 \cap \neg\mathsf{E}\right]$$
$$\leq \Pr\left[G^{\mathsf{IND\text{-}sT\text{-}CCA}}_{\mathcal{A},\mathsf{PFSKEM}}(\lambda) = 1 \cap \mathsf{E}\right] + \Pr\left[\neg\mathsf{E}\right]$$

In case event $\neg\mathsf{E}$ occurs, $\mathcal{A}$ asks for a decapsulation PFSKEM.Dec $(\tau, \mathsf{CT}_{\mathsf{PFSKEM}} = (\mathsf{CT}_{\mathsf{HIBKEM}}, \sigma, pk_{OT}))$ where $(\mathsf{CT}_{\mathsf{HIBKEM}}, \sigma) \neq (\mathsf{CT}^*_{\mathsf{HIBKEM}}, \sigma^*)$ with $pk_{OT} = pk^*_{OT}$ and OTSIG.Vfy($pk_{OT}, \mathsf{CT}_{\mathsf{HIBKEM}}, \sigma$) = 1 in the security game. This means that $\mathsf{CT}_{\mathsf{HIBKEM}} \neq \mathsf{CT}^*_{\mathsf{HIBKEM}}$ or $\sigma \neq \sigma^*$ (or both). Hence, $(\mathsf{CT}_{\mathsf{HIBKEM}}, \sigma)$ is a valid strong existential forgery under the OTSIG scheme. Outputting this forgery, we can use $\mathcal{A}$ to build an attacker $\mathcal{B}_{\mathsf{OTSIG}}$ to break the sEUF-1-CMA security of OTSIG whenever $\mathcal{A}$ triggers event $\neg\mathsf{E}$. Therefore,

$$\Pr[\neg \mathsf{E}] = \mathsf{Adv}^{\mathsf{sEUF\text{-}1\text{-}CMA}}_{\mathcal{B}_{\mathsf{OTSIG}},\mathsf{OTSIG}}(\lambda).$$

Next, we build an adversary $\mathcal{B}_{\mathsf{HIBKEM}}$ against the IND-sID-CPA security of the HIBKEM. $\mathcal{B}_{\mathsf{HIBKEM}}$ generates a fresh key pair $(pk^*_{OT}, sk^*_{OT}) \xleftarrow{\$}$ OTSIG.KGen(1^λ). Then, $\mathcal{B}_{\mathsf{HIBKEM}}$ starts $\mathcal{A}$ to obtain τ^*, sends ID $= \tau^* \| pk^*_{OT}$ to the HIBKEM challenger and receives a challenge $(K_b, \mathsf{CT}^*_{\mathsf{HIBKEM}})$ with the public key pk_{HIBKEM}. $\mathcal{B}_{\mathsf{HIBKEM}}$ sets $PK = pk_{\mathsf{HIBKEM}}$ and computes the signature σ^* for $\mathsf{CT}^*_{\mathsf{HIBKEM}}$. $\mathcal{B}_{\mathsf{HIBKEM}}$ continues to run $\mathcal{A}$ with the challenge $(\mathsf{CT}^*_{\mathsf{PFSKEM}} = (\mathsf{CT}^*_{\mathsf{HIBKEM}}, \sigma^*, pk^*_{OT}), \mathsf{K}^*_b = K_b)$ and the public key PK. $\mathcal{B}_{\mathsf{HIBKEM}}$ provides answers to the queries defined in the selective-time CCA security experiment $G^{\mathsf{IND\text{-}sT\text{-}CCA}}_{\mathcal{A},\mathsf{PFSKEM}}(\lambda)$ as follows:

- PFSKEM.Dec$(\tau, \mathsf{CT}_{\mathsf{PFSKEM}} = (\mathsf{CT}_{\mathsf{HIBKEM}}, \sigma, pk_{OT}))$ with $\tau \neq \tau^*$: $\mathcal{B}_{\mathsf{HIBKEM}}$ can query the HIBKEM challenger for the secret key of identity $\tau \| pk_{OT}$, because $\tau \| pk_{OT}$ is not an ancestor identity of $\tau^* \| pk^*_{OT}$. With the secret key it is possible to decapsulate the key for $\mathsf{CT}^*_{\mathsf{HIBKEM}}$.
- PFSKEM.Dec$(\tau, \mathsf{CT}_{\mathsf{PFSKEM}} = (\mathsf{CT}_{\mathsf{HIBKEM}}, \sigma, pk_{OT}))$: $\mathcal{B}_{\mathsf{HIBKEM}}$ can query the HIBKEM challenger for the secret key of identity $\tau^* \| pk_{OT}$.
- PFSKEM.Corrupt: If adversary $\mathcal{A}$ did not call PFSKEM.PnctCxt(τ^*, CT^*) or PFSKEM.PnctInt(τ^*) before, then $\mathcal{B}_{\mathsf{HIBKEM}}$ aborts and outputs a random bit, else $\mathcal{B}_{\mathsf{HIBKEM}}$ can query the HIBKEM challenger for all secret keys of the requested identities and send them to $\mathcal{A}$.

In the end $\mathcal{A}$ outputs a guess b' and $\mathcal{B}_{\mathsf{HIBKEM}}$ forwards b' to the HIBKEM challenger as its own output. $\mathcal{B}_{\mathsf{HIBKEM}}$ wins if $\mathcal{A}$ outputs the right b'. The security experiment can be simulated correctly if event E occurs. Therefore we have

$$\Pr\left[G^{\mathsf{IND\text{-}sID\text{-}CPA}}_{\mathcal{B}_{\mathsf{HIBKEM}},\mathsf{HIBKEM}}(\lambda) = 1\right] = \Pr\left[G^{\mathsf{IND\text{-}sT\text{-}CCA}}_{\mathcal{A},\mathsf{PFSKEM}}(\lambda) = 1 \cap \mathsf{E}\right]$$

Putting the above bounds together, we obtain

$$\mathsf{Adv}^{\mathsf{IND\text{-}sT\text{-}CCA}}_{\mathcal{A},\mathsf{PFSKEM}}(\lambda)$$

$$= \left| \Pr\left[G^{\mathsf{IND\text{-}sT\text{-}CCA}}_{\mathcal{A},\mathsf{PFSKEM}}(\lambda) = 1\right] - \frac{1}{2} \right|$$

$$= \left| \Pr\left[G^{\mathsf{IND\text{-}sT\text{-}CCA}}_{\mathcal{A},\mathsf{PFSKEM}}(\lambda) = 1 \cap \mathsf{E}\right] + \Pr\left[G^{\mathsf{IND\text{-}sT\text{-}CCA}}_{\mathcal{A},\mathsf{PFSKEM}}(\lambda) = 1 \cap \neg \mathsf{E}\right] - \frac{1}{2} \right|$$

$$\leq \left| \Pr\left[G^{\mathsf{IND\text{-}sT\text{-}CCA}}_{\mathcal{A},\mathsf{PFSKEM}}(\lambda) = 1 \cap \mathsf{E}\right] - \frac{1}{2} \right| + \Pr\left[G^{\mathsf{IND\text{-}sT\text{-}CCA}}_{\mathcal{A},\mathsf{PFSKEM}}(\lambda) = 1 \cap \neg \mathsf{E}\right]$$

$$\leq \left| \Pr\left[G^{\mathsf{IND\text{-}sID\text{-}CPA}}_{\mathcal{B}_{\mathsf{HIBKEM}},\mathsf{HIBKEM}}(\lambda) = 1\right] - \frac{1}{2} \right| + \Pr[\neg \mathsf{E}]$$

which yields

$$\mathsf{Adv}^{\mathsf{IND\text{-}sT\text{-}CCA}}_{\mathcal{A},\mathsf{PFSKEM}}(\lambda) \leq \mathsf{Adv}^{\mathsf{IND\text{-}sID\text{-}CPA}}_{\mathcal{B}_{\mathsf{HIBKEM}},\mathsf{HIBKEM}}(\lambda) + \mathsf{Adv}^{\mathsf{sEUF\text{-}1\text{-}CMA}}_{\mathcal{B}_{\mathsf{OTSIG}},\mathsf{OTSIG}}(\lambda)$$

$\square$

3   Forward-Secret One-Pass Key Exchange Protocols

3.1   Syntax

Protocols in a 0-RTT–like setting, where only one message is transmitted between two key exchange protocol partners, have been the object of previous design interest. In particular, a similar scenario was considered by Halevi and Krawczyk under the notion of *one-pass key exchange* [24]. Aiming for efficiency and optimal key management, we extend their setting by allowing shared state between several executions of the protocol and introduce a discretized notion of time.

Definition 9 (FSOPKE). *A* forward-secret one-pass key exchange (FSOPKE) protocol *supporting τ_{max} time periods and providing mutual or unilateral (server-only) authentication consists of the following four probabilistic algorithms.*

FSOPKE.KGen($1^\lambda, r, \tau_{max}$) $\rightarrow$ (pk, sk). *On input the security parameter 1^λ, a role $r \in \{\mathsf{client}, \mathsf{server}\}$, and the maximum number of time periods $\tau_{max} \in \mathbb{N}$, this algorithm outputs a public/secret key pair (pk, sk) for the specified role.*

FSOPKE.RunC(sk, pk) $\rightarrow$ (sk', k, m). *On input a secret key sk and a public key pk, this algorithm outputs a (potentially modified) secret key sk', a session key $k \in \{0,1\}^* \cup \{\bot\}$, and a message $m \in \{0,1\}^* \cup \{\bot\}$.*

FSOPKE.RunS(sk, pk, m) $\rightarrow$ (sk', k). *On input of a secret key sk, a public key pk, and a message $m \in \{0,1\}^*$, this algorithm outputs a (potentially modified) secret key sk' and a session key $k \in \{0,1\}^* \cup \{\bot\}$. For a unilateral authenticating protocol, $pk = \bot$ indicates that the client is not authenticated.*

FSOPKE.TimeStep(sk, r) $\rightarrow$ sk'. *On input a secret key sk and an according role $r \in \{\mathsf{client}, \mathsf{server}\}$, this algorithm outputs a (potentially modified) secret key sk'.*

We say that a forward-secret one-pass key exchange protocol is correct *if:*

- *for all $(pk_i, sk_i) \leftarrow$ FSOPKE.KGen($1^\lambda, \mathsf{client}, \tau_{max}$),*
- *for all $(pk_j, sk_j) \leftarrow$ FSOPKE.KGen($1^\lambda, \mathsf{server}, \tau_{max}$),*
- *for any $n \in \mathbb{N}$ with $n < \tau_{max}$ and all*
  - $sk'_i \leftarrow$ FSOPKE.TimeStep^n(sk_i, client)
  - $sk'_j \leftarrow$ FSOPKE.TimeStep^n(sk_j, server)
    (where FSOPKE.TimeStep^n indicates n iterative applications of FSOPKE.TimeStep),
- *for all $(sk''_i, k_i, m) \leftarrow$ FSOPKE.RunC(sk'_i, pk_j),*
- *and for all*
  - $(sk''_j, k_j) \leftarrow$ FSOPKE.RunS(sk'_j, pk_i, m)
    (for mutual authentication)
  - *resp. $(sk''_j, k_j) \leftarrow$ FSOPKE.RunS($sk'_j, \bot, m$)*
    (for unilateral authentication),

it holds that $k_i = k_j$.

A forward-secret one-pass key exchange protocol is used by a client and a server party as follows. First of all, both parties generate public/secret key pairs $(pk, sk) \leftarrow$ FSOPKE.KGen$(1^\lambda, r, \tau_{max})$ for their according role $r =$ client resp. $r =$ server. To proceed in time (step-wise), they can invoke FSOPKE.TimeStep on their respective secret keys (up to $\tau_{max} - 1$ times). Two parties holding secret keys in the same time frame then communicate by the client running FSOPKE.RunC on its secret key and the public key of its intended partner, obtaining the joint session key and a message; transmitting the latter to the server. The server then invokes FSOPKE.RunS on its secret key, the (intended) client's public key (or $\perp$ in case of unilateral authentication), and the obtained message, which outputs, by correctness, the same joint session key.

Note that this (0-RTT) session key is the only session key derived. Unlike in QUIC and TLS 1.3, we demand that this key immediately enjoys full forward secrecy and replay protection, making an upgrade to another key unnecessary. This demand is realized via the forthcoming security model in Sect. 3.2.

3.2 Security Model

We denote by $\mathcal{I} = \mathcal{C} \,\dot\cup\, \mathcal{S}$ the set of *identities* modeling both clients ($\mathcal{C}$) and servers ($\mathcal{S}$) in the system, each identity $u \in \mathcal{I}$ being associated with a public/secret key pair (pk_u, sk_u). Here, the public-key part pk_u is generated once and fixed, whereas sk_u can be modified by (the sessions of) the according party over time. Each identity u moreover holds the local, current time in a variable denoted by $\tau_u \in \mathbb{N}$, initialized to $\tau_u \leftarrow 1$.

In our model, an adversary $\mathcal{A}$ interacts with several *sessions* of multiple identities running a forward-secret one-pass key exchange protocol. We denote by π_u^i the i-th session of identity u and associate with each session the following internal state variables:

- role $\in \{$client, server$\}$ indicates the role of the session. We demand that role $=$ client resp. role $=$ server if and only if $u \in \mathcal{C}$ resp. $u \in \mathcal{S}$.
- id $\in \mathcal{I}$ indicates the owner of the session (e.g., u for a session π_u^i).
- pid $\in \mathcal{I} \cup \{\perp\}$ indicates the intended communication partner, and is set exactly once. Setting pid $= \perp$ is possible if role $=$ server to indicate the client is not authenticated. Initially, pid $= \perp$ can also be set (if role $=$ server) to indicate that the client's identity is to be learned within the protocol (i.e., post-specified).
- trans $\in \{0, 1\}^* \cup \{\perp\}$ records the (single) sent, resp. received, message.
- time $\in \mathbb{N}$ records the time interval used when processing the sent, resp. received, message.
- key $\in \{0, 1\}^* \cup \{\perp\}$ is the session key derived in the session.
- keystate $\in \{$fresh, revealed$\}$ indicates whether the session key has been revealed. Initially keystate $=$ fresh.

We write, e.g., π_u^i.key when referring to state variables of a specific session.

Definition 10 (Partnered sessions). *We say that two sessions π_u^i and π_v^j are partnered if*

- π_u^i.trans $= \pi_v^j$.trans, *i.e., they share the same transcript,*
- π_u^i.time $= \pi_v^j$.time, *i.e., they run in the same time interval,*
- π_u^i.role $=$ client $\wedge \pi_v^j$.role $=$ server, *i.e., they run in opposite roles,*
- π_u^i.pid $= \pi_v^j$.id, *i.e., the server session is owned by the client's intended partner, and*
- π_u^i.id $= \pi_v^i$.pid $\vee \pi_v^j$.pid $= \bot$, *i.e., the client session is owned by the server's intended partner or the server considers its partner to be unauthenticated.*

We assume the adversary $\mathcal{A}$ controls the network, is responsible for transporting messages, and hence allowed to arbitrary modify, drop, or reorder messages. It interacts with the key exchange protocol and sessions via the following queries.

NewSession($u, role, pid, m$). Initializes a new session of identity $u \in \mathcal{I}$, taking role role $\in \{$client, server$\}$ and intended communication partner pid $\in \mathcal{I} \cup \{\bot\}$ (where pid $= \bot$ for a server session indicates an unauthenticated client partner). If role $\neq$ server, we require that $m = \bot$.
If role $=$ client, invoke $(sk_u, k, m) \leftarrow$ FSOPKE.RunC(sk_u, pk_{pid}), else invoke $(sk_u, k) \leftarrow$ FSOPKE.RunS(sk_u, pk_{pid}, m), where $pk_\bot = \bot$.
Register a new session π_u^i with role $= role$, id $= u$, pid $= pid$, trans $= m$, time $= \tau_u$, and key $= k$.
If role $=$ client, return m. If role $=$ server, return $\bot$ if $k = \bot$, and $\top$ otherwise.
Reveal(π_u^i). Reveals the session key of a specific session, if derived.
If π_u^i.key $\neq \bot$, set π_u^i.keystate $\leftarrow$ revealed and return key, else return $\bot$.
Corrupt(u). Corrupts the long-term state of an identity $u \in \mathcal{I}$. This query can be asked at most once per identity u and, from this point on, no further queries to (sessions of) u are allowed.
Let Corrupt(u) be the ς-th query issued by $\mathcal{A}$; we set $\varsigma_u^{corr} \leftarrow \varsigma$, where $\varsigma_u^{corr} = \infty$ for uncorrupted identities. Likewise, we record the identity's current time τ_u at corruption and set $\tau_u^{corr} \leftarrow \tau_u$.
Return sk_u.
Tick(u). Forward the state of some identity $u \in \mathcal{I}$ by one time step by invoking $sk_u \leftarrow$ FSOPKE.TimeStep(sk_u). Record the new time as $\tau_u \leftarrow \tau_u + 1$.
Test(π_u^i). Allows the adversary to challenge a derived session key and is asked exactly once. This oracle is given a secret bit $b_{test} \in \{0, 1\}$ chosen at random in the security game.
If π_u^i.key $= \bot$, return $\bot$.
Set $\tau^t \leftarrow \pi^t$.time. If $b_{test} = 0$, return π_u^i.key, else return a random key chosen according to the probability distribution specified by the protocol.

Definition 11 (Security for FSOPKE). *Let FSOPKE be a forward-secret one-pass key exchange protocol and $\mathcal{A}$ a PPT adversary interacting with FSOPKE via the queries defined above in the following game $G_{\mathcal{A}, \text{FSOPKE}}^{\text{FSOPKE-sec}}$:*

- *The challenger generates keys and state for all parties $u \in \mathcal{I}$ as $(pk_u, sk_u) \leftarrow$ FSOPKE.KGen(1^λ) and chooses a random bit $b_{test} \xleftarrow{\$} \{0, 1\}$.*

- *The adversary $\mathcal{A}$ receives (u, pk_u) for all $u \in \mathcal{I}$ and has access to the queries* NewSession, Reveal, Corrupt, Tick, *and* Test. *Record for the* Test *query, being the ς^{test}-th query, the tested session π^t.*
- *Eventually, $\mathcal{A}$ stops and outputs a guess $b \in \{0, 1\}$.*

The challenger outputs 1 (denoted by $G_{\mathcal{A},\text{FSOPKE}}^{\text{FSOPKE-sec}} = 1$) and say the adversary wins if $b = b_{\text{test}}$ and the following conditions hold:

1. π^t.keystate = fresh, *i.e., $\mathcal{A}$ has not issued a* Reveal *query to the test session.*
2. π_v^j.keystate = fresh *for any session π_v^j such that π_v^j and π^t are partners, i.e., $\mathcal{A}$ has not issued a* Reveal *query to a session partnered with the test session.*
3. $\varsigma_u^{corr} > \varsigma^{test}$ *for $u = \pi^t$.id, i.e., the owner of the test session has not been corrupted before the* Test *query was issued.*
4. *if π^t.role = client and $\varsigma_v^{corr} \neq \infty$, for $v = \pi^t$.pid, then one of the following must hold:*
   - *There exists a session π_v^j partnered with π^t, i.e., a session of the intended server partner processed the client session's message in the intended time interval.*
   - $\tau^t < \tau_v^{corr}$, *i.e., the intended partner was corrupted in a time interval after that of the tested session.*
5. *if π^t.role = server and π^t.pid $\neq \perp$, then $\varsigma_v^{corr} > \varsigma^{test}$ for $v = \pi^t$.pid, i.e., the intended client partner of a tested server session has not been corrupted before the* Test *query was issued.*
6. *if π^t.role = server and π^t.pid = $\perp$, then there exists a session π_v^j partnered with π^t, i.e., when testing a server session without authenticated partner, there must exist an honest communication partner to the tested server session π^t.*

Otherwise, the challenger outputs a random bit. We say that FSOPKE *is secure if the following advantage function is negligible in the security parameter:*

$$\mathsf{Adv}_{\mathcal{A},\text{FSOPKE}}^{\text{FSOPKE-sec}}(\lambda) := \left| \Pr\left[G_{\mathcal{A},\text{FSOPKE}}^{\text{FSOPKE-sec}} = 1\right] - \frac{1}{2} \right|.$$

Remark 1. Notably, our security model requires both forward secrecy and replay protection from a FSOPKE protocol. Furthermore, it captures unilateral authentication (of the server) and mutual authentication simultaneously.

As expected, we restrict Reveal on both partner sessions involved in the test session (conditions 1 and 2). However, our notion of partnering in Definition 10 lends more power to an adversary than is typically provided. Partnering is defined not only with respect to the session transcripts, partner IDs, and roles, but also with respect to time. Consequently, if the two sessions are not operating within the same time interval, Reveal queries are, in fact, permitted on the intended partner session to the test session – even if all other aspects of partnering are fulfilled (condition 2).

To ensure replay protection, the adversary is allowed to test and reveal matching sessions of the *same* role; we only forbid testing and revealing two matching sessions of opposite roles (via the partnering condition). This explicitly allows for replaying of a client's message to two server sessions (i.e., spawning two server sessions on input of the same client message m) and revealing one server session while testing the other session. Hence, our model requires that secure protocols prevent replays.

For forward secrecy, corruption of the tested identity is allowed after the Test query was issued (condition 3). This applies to both clients (if the client identity exists) and servers.

Server corruption under a tested client session in the 0-RTT setting necessitates special considerations (condition 4). First we consider the scenario that the intended partner server session processes messages in the same time interval as the test query, i.e. τ^t. In this case a tested client's message must have been processed by the intended partner server session before the server is corrupted[5] to exclude the following trivial attack: observe that an adversary spawning a new client session (with some pid $= v$, outputting a message m) which it subsequently tests, may obtain the secret key sk_v of the (server) identity v through a Corrupt(v) query such that, by correctness of the FSOPKE protocol, it can process message m and derive the correct session key. In this manner, an adversary would always be trivially able to win the key secrecy game. Hence, condition 4 (first item) encodes the strongest possible forward secrecy guarantees in such a scenario: *whenever a client's message has been processed by the server, the corresponding session key becomes forward-secret w.r.t. compromises of the server's long-term secret.*

Alternatively, we consider the scenario where the intended partner server session processes messages in a time interval after that used in the tested session, i.e. $\tau^t < \tau_v^{corr}$. If the server session's time interval is ahead of that of the tested client session then different session keys are computed. Yet this implies that there are no immediate forward secrecy guarantees should the client's clock be ahead of the server's time interval, since the server's clock can be moved forward after corruption of the server. Thus, condition 4 (second item) gives an additional forward secrecy guarantee: *the tested session key is forward-secret w.r.t. compromises of the server's long-term secret for any future time interval.*

As with corruption of the test session identity (condition 3), if a server session is tested such that a partnered client identity is defined, corruption of the partnered client is restricted until after the test query has been made (condition 5). We do guarantee security if the client is corrupted immediately after it has issued the test session message, but before the server has processed it, due to potential authentication by the client. Should the message be signed, for example, such corruption would allow an adversary to tamper with the message. Thus, for compromises of the client's long-term secret, we demand forward secrecy immediately after the server establishes the session key.

For the case of unilateral authentication, we must naturally restrict Test queries on the server side to cases where an honest partnered client exists (condition 6), as otherwise the adversary can take the role of the client and hence trivially learns the key.

---

[5] Recall that the adversary cannot spawn or interact with sessions of a party anymore after corrupting it.

Finally, all security guarantees are required to be provided *independent* of the time stepping mechanism, making the latter a *functional* property of a FSOPKE scheme which does not affect the scheme's security. For example, a scheme could liberally allow session key establishment even if the states of both of the involved sessions are off by a number of time steps. While this is beyond the requirements for a correct scheme, key secrecy still requires that such session keys are secure.

In our model, we do not consider randomness or session-state reveal queries [10, 30], but note that it could be augmented with such queries.

4 Constructions

For the construction of a forward-secret 0-RTT key exchange protocol we now first focus on the more common case where only the server authenticates. Our construction builds on puncturable forward-secure key encapsulation and leverages some synchronization of time between parties in the system. Later, we discuss how to adapt this construction to scenarios where relying on time synchronization is not an option.

4.1 Construction Based on Synchronized Time

We construct a forward-secret one-pass key exchange protocol in a generic way from any puncturable forward-secure key encapsulation scheme. For our construction, we assume that clients and servers hold some roughly synchronized time, but stress that we are concerned with time intervals rather than exact time and, hence, synchronization for example on the same day is sufficient for our scheme. Aiming at unilateral (server-only) authentication, clients do not hold long-term key material (i.e., we have $pk = \perp$ for clients) and only (mis-)use their secret key to store the current time interval.

Definition 12 (FSOPKE$_U$ Construction). *Let* PFSKEM *be a puncturable forward-secure key encapsulation scheme. We construct a forward-secret one-pass key exchange protocol* FSOPKE$_U$ *with unilateral authentication as follows:*

FSOPKE.KGen($1^\lambda, r, \tau_{max}$) → (pk, sk).

- *If r = server* : *Generate a public/secret key pair (PK, SK) ←* PFSKEM.KGen (1^λ). *Set $pk \leftarrow (PK, \tau_{max})$, $\tau \leftarrow 1$, and $sk \leftarrow (SK, \tau, \tau_{max})$, and output (pk, sk).*
- *If r = client* : *Set $pk \leftarrow \perp$, $\tau \leftarrow 1$, and $sk \leftarrow (\tau)$, and output (pk, sk).*

FSOPKE.RunC(sk, pk) → (sk', k, m). *Parse $sk = (\tau)$ and $pk = (PK, \tau_{max})$. If $\tau > \tau_{max}$, then abort and output $(sk, \perp, \perp)$.*
Otherwise, compute $(\mathsf{CT}, \mathsf{K}) \leftarrow$ PFSKEM.Enc(PK, τ), set $k \leftarrow \mathsf{K}$ and $m \leftarrow \mathsf{CT}$, and output (sk, k, m).

FSOPKE.RunS$(sk, pk = \bot, m) \rightarrow (sk', k)$. *Parse* $sk = (SK, \tau, \tau_{max})$. *If* $SK = \bot$
or $\tau > \tau_{max}$, *then abort and output* $(sk, \bot)$.
   Compute $K \leftarrow$ PFSKEM.Dec(SK, τ, m). *If* $K = \bot$, *then abort and out-*
put $(sk, \bot)$.
   Otherwise, issue $SK' \leftarrow$ PFSKEM.PnctCxt(SK, τ, m). *Let* $sk \leftarrow$
(SK', τ, τ_{max}), *set* $k \leftarrow K$, *and output* (sk, k).

   FSOPKE.TimeStep$(sk, r) \rightarrow sk'$.

- *If* $r =$ server : *Parse* $sk = (SK, \tau, \tau_{max})$. *If* $\tau \geq \tau_{max}$, *then set* $sk \leftarrow (\bot, \tau + 1, \tau_{max})$, *and output* sk.
   Otherwise, let $SK' \leftarrow$ PFSKEM.PnctInt(SK, τ), *set* $sk \leftarrow (SK', \tau + 1, \tau_{max})$, *and output* sk.
- *If* $r =$ client : *Parse* $sk = (\tau)$, *set* $sk \leftarrow (\tau + 1)$, *and output* sk.

Correctness follows from the correctness of the underlying PFSKEM scheme; the details are omitted here due to space limitations.

Security Analysis. We now investigate the security of our construction and show that it is a secure forward-secret one-pass key exchange protocol with unilateral authentication.

Theorem 2. *The* FSOPKE$_U$ *construction from Definition 12 is a secure* FSOPKE *protocol (with unilateral authentication). Formally, for any efficient adversary* A *in the* FSOPKE-sec *game there exists an efficient algorithm* B *such that*

$$\mathsf{Adv}^{\mathsf{FSOPKE\text{-}sec}}_{A, \mathsf{FSOPKE}_U}(\lambda) \leq n_I \cdot \hat{\tau}_{max} \cdot n_s \cdot \mathsf{Adv}^{\mathsf{IND\text{-}sT\text{-}CCA}}_{B, \mathsf{PFSKEM}}(\lambda),$$

where $n_I = |I|$ *is the maximum number of identities,* $\hat{\tau}_{max}$ *is the maximum time interval for any session, and* n_s *is the maximum number of sessions.*

Proof. Let A be an adversary against the security of FSOPKE$_U$. We proceed in a sequence of games, bounding the introduced difference in A's advantage for each step. By Adv_i we denote A's advantage in one of the i-th game.

Game 0. This is the original security experiment, with adversarial advantage $\mathsf{Adv}_0 = \mathsf{Adv}^{\mathsf{FSOPKE\text{-}sec}}_{A, \mathsf{FSOPKE}_U}(\lambda)$.

Game 1. Here we let the challenger upfront guess a server identity $s^* \in I$, associated with public/secret key pair (pk^*, sk^*), and let it abort the game if this is not the identity involved in the test session. I.e., if a server session is tested (i.e., π^t.role = server) this is the session owner $s^* = \pi^t$.id, while, if a client session is tested (π^t.role = client) it is the intended partner ($s^* = \pi^t$.pid). Let $n_I = |I|$. Then

$$\mathsf{Adv}_0 \leq n_I \cdot \mathsf{Adv}_1.$$

Game 2. Now the $\mathcal{A}$ guesses the time interval $\tau^* = \pi^t.\text{time}$ in which the tested session ran, and aborts if the guess is incorrect. Letting $\hat{\tau}_{max}$ denote the maximum value $\pi.\text{time}$ for any session π, it follows that

$$\text{Adv}_1 \leq \hat{\tau}_{max} \cdot \text{Adv}_2.$$

Game 3. Continuing from *Game 2* the challenger aborts if it does not correctly guess the involved client session π_c^t (i.e., $\pi_c^t.\text{role} = \text{client}$) for which one of the following two conditions holds:

- either $\pi_c^t = \pi^t$, i.e., π_c^t is the tested session, or
- π_c^t is partnered with the tested (server) session π^t.

For the second case, observe that if a server is tested, by condition 6 of the FSOPKE-sec security game in Definition 11, there must exist such a partnered client session π_c^t with $\pi_c^t.\text{pid} = \pi^t.\text{id}$ in order for $\mathcal{A}$ to win.

Denoting n_s as the total number of sessions, we have

$$\text{Adv}_2 \leq n_s \cdot \text{Adv}_3.$$

Furthermore, observe that by Definition 7, if a server session is tested, session π^t must actually be the *first* accepting session owned by s^* that is partnered with π_c^t in order for $\mathcal{A}$ to win. Recall that the first such accepting session, by correctness, derives a key $K \neq \bot$ as $K \leftarrow \text{PFSKEM.Dec}(SK^*, \tau^*, m)$ (where $m = \pi^t.\text{trans}$) and hence invokes $SK^* \leftarrow \text{PFSKEM.PnctCxt}(SK^*, \tau^*, m)$. Any later such accepting session would hence, by Definition 7, derive $K = \bot$ through $K \leftarrow \text{PFSKEM.Dec}(SK^*, \tau^*, m)$, so an adversary would be given $\bot$ as the response to its Test query and cannot win.

Game 4. In this game hop, we replace the key k^* derived in the tested session π^t by one chosen uniformly at random from the output space of PFSKEM.Dec. We show that any adversary that distinguishes the change from Game 3 to Game 4 with non-negligible advantage can be turned into an algorithm $\mathcal{B}$ which wins in $G_{\mathcal{A},\text{PFSKEM}}^{\text{IND-sT-CCA}}$ with the same advantage.

In this reduction, $\mathcal{B}$ first outputs the time interval τ^* guessed in Game 2 as the time interval it wants to be challenged on in $G_{\mathcal{A},\text{PFSKEM}}^{\text{IND-sT-CCA}}$. It then obtains a challenge public key PK^*, which it associates with the server identity s^* within the $pk^* = (PK^*, \tau_{max})$ guessed in Game 1. For all other identities $u \in \mathcal{I} \setminus \{s^*\}$, algorithm $\mathcal{B}$ generates appropriate public/secret key pairs on its own following FSOPKE.KGen. In particular, it generates PFSKEM keys for all other server identities $s \in \mathcal{S} \setminus \{s^*\}$. Furthermore, $\mathcal{B}$ obtains a challenge ciphertext CT^* and key K^*, with K^* either being the real key encapsulated in CT^* or and independently chosen random one.

Our goal is now to have algorithm $\mathcal{B}$ (correctly) simulate the security game for $\mathcal{A}$ in such a way that, if K^* is the real key, it perfectly simulates Game 3, whereas if K^* is a randomly chosen key, it perfectly simulates Game 4. To this extent, algorithm $\mathcal{B}$ uses its oracles KGen(), PFSKEM.Dec(), PFSKEM.PnctInt(), and PFSKEM.PnctCxt() given in the selective ID, selective time CCA security game in Definition 8 as follows, answering the queries of $\mathcal{A}$ in the key exchange game:

NewSession($u, role, pid, m$). We distinguish the following cases:

- For all client sessions π_u^i ($u \in \mathcal{C}$) except for the client session π_c^t guessed in Game 3, $\mathcal{B}$ simulates NewSession queries as specified in the security game.
- For the guessed client session π_c^t, $\mathcal{B}$ does not invoke PFSKEM.Enc but uses its challenge key K^* as the session key k and the challenge ciphertext CT^* as the output message m. Observe that, through Games 1–3, we ensure that π_c^t uses time interval τ^* and public key pk^* (and hence the challenge PFSKEM public key PK^*) of server s^*.
- For all server sessions π_s^i not owned by the server identity s^* guessed in Game 1 (i.e., $s \in \mathcal{S} \setminus \{s^*\}$), $\mathcal{B}$ simulates NewSession queries as specified, using the according (self-generated) secret key sk_s.
- For all server sessions $\pi_{s^*}^i$ owned by s^* and not partnered with the guessed client session π_c^t, $\mathcal{B}$ uses its oracles PFSKEM.Dec and PFSKEM.PnctCxt from the selective ID, selective time CCA game to simulate the operations for the NewSession query. Note that, as $\pi_{s^*}^i$ is not partnered with π_c^t (though having opposite roles and $\pi_c^t.pid = s^*$), we have $(\pi_c^t.\text{time}, \pi_c^t.\text{trans}) = (\tau^*, CT^*) \neq (\pi_{s^*}^i.\text{time}, \pi_{s^*}^i.\text{trans})$ and are hence allowed to call the PFSKEM.Dec oracle on this input.
- For the first server session $\pi_{s^*}^t$ owned by s^* which is partnered with the guessed client session π_c^t, $\mathcal{B}$ sets the session key to be the challenge key $k \leftarrow K^*$ and invokes PFSKEM.PnctCxt(τ^*, CT^*).
  Note that partnering in particular implies $\pi_{s^*}^t$ holds the same time as π_c^t and obtains the message of π_c^t, i.e., $\pi_c^t.\text{time} = \tau^* = \pi_{s^*}^t.\text{time}$ and $\pi_c^t.\text{trans} = m = \pi_{s^*}^t.\text{trans}$. Furthermore, PFSKEM.PnctCxt was not invoked before on (τ^*, CT^*). Hence, by correctness, $\pi_{s^*}^t$ establishes the same session key K^* as π_c^t.
- For any further server session $\pi_{s^*}^i$ partnered with π_c^t, $\mathcal{B}$ sets $k \leftarrow \bot$. By Definition 7, we know that any such session would obtain $\bot \leftarrow$ PFSKEM.Dec(SK, τ^*, CT^*), as PFSKEM.PnctCxt has been called before on (τ^*, CT^*).

Reveal(π_u^i). First, observe that any winning adversary $\mathcal{A}$ cannot call Reveal on the sessions π_c^t and $\pi_{s^*}^t$ by conditions 1 and 2 of the security model, as one of them is the tested session and the other, if it exists, is partnered with the tested session.

For all other sessions, $\mathcal{B}$ holds the correct key from simulation of the NewSession queries above, and can therefore respond to according Reveal queries as specified.

Corrupt(u). For the server identity s^* involved in the tested session π^t, $\mathcal{B}$ invokes its PFSKEM.Corrupt oracle to obtain the PFSKEM secret key SK^*, which it returns within $sk^* = (SK^*, \tau_{s^*}, \tau_{max})$. Observe, that if $\mathcal{A}$ calls Corrupt(s^*) without losing, we are ensured that $\mathcal{B}$ has called PFSKEM.PnctCxt(τ^*, CT^*) and/or PFSKEM.PnctInt(τ^*) before Corrupt(s^*), and hence also does not lose in the selective-time CCA security game:

– If $\pi^t = \pi_{s^*}^t$ is a server session (owned by s^*), condition 3 of the security model ensures that s^* can only be corrupted after π^t has accepted. In the process of π^t accepting (with $\pi^t.\text{time} = \tau^*$ and $\pi^t.\text{trans} = \text{CT}^*$), $\mathcal{B}$ must have invoked PFSKEM.PnctCxt(τ^*, CT^*), and therefore before corruption of s^*.

– If $\pi^t = \pi_c^t$ is a client session, condition 4 of the security model ensures that either there exists a partnered server session $(\pi_{s^*}^t)$ that processed CT^* in the time interval τ^* or that s^* gets corrupted in a time interval $\tau_{s^*}^{corr} > \pi^t.\text{time} = \tau^*$. Hence, $\mathcal{B}$ must have invoked PFSKEM.PnctCxt(τ^*, CT^*) or PFSKEM.PnctInt(τ^*), respectively, before corruption of s^*.[6]

For any other (client or server) identity $u \neq s^*$, $\mathcal{B}$ maintains the corresponding secret key sk_u and can therefore respond to according Corrupt queries as specified.

Tick(u). Algorithm $\mathcal{B}$ conducts the time stepping procedures as specified, using its oracle PFSKEM.PnctInt on the (unknown) secret key SK^* corresponding to the PFSKEM challenge public key PK^*.

Test(π^t). Observe that the tested session π^t must be either the client session π_c^t guessed in Game 3 or the (first) server session $\pi_{s^*}^t$ owned by s^* partnered with π_c^t. Algorithm $\mathcal{B}$, in both cases, simply outputs $\pi^t.\text{key} = K^*$ as the response of the Test query.

When $\mathcal{A}$ stops and outputs a guess $b \in \{0, 1\}$, $\mathcal{B}$ stops as well and outputs b as its own guess.

Observe that algorithm $\mathcal{B}$ correctly answers all queries of $\mathcal{A}$ and, in the case that K^* is the real key encapsulated in CT^*, perfectly simulates Game 3, while it perfectly simulates Game 4 if K^* is chosen independently at random. Algorithm $\mathcal{B}$ moreover obeys all restrictions in the selective ID, selective time CCA security game of Definition 8 if $\mathcal{A}$ adheres to the conditions in the FSOPKE security game.

As $\mathcal{B}$ inherits the output of $\mathcal{A}$, a difference between $\mathcal{A}$'s advantage in Game 3 and its advantage in Game 4 corresponds to the probability difference of $\mathcal{B}$ outputting 1 in the two cases of the selective ID, selective time CCA security experiment. Thus,

$$\text{Adv}_3 \leq \text{Adv}_4 + \text{Adv}_{\mathcal{B}, \text{PFSKEM}}^{\text{IND-sT-CCA}}(\lambda) \ .$$

As in Game 4 the session key k^* in the tested session is always chosen uniformly at random the response to the Test query is independent of the challenge bit b and hence $\mathcal{A}$ cannot predict b better than by guessing, i.e., $\text{Adv}_4 \leq 0$. Combining the advantage bounds in Games 1–4 yields the overall bound.   □

---

[6] Recall that $\pi_{s^*}^t$ must have accepted before s^* is corrupted, as afterwards no further queries to sessions owned by s^* are allowed.

4.2 Variant Without Synchronized Time

For those environments where more relaxed requirements for time synchronization are preferable, we outline a variant of our forward-secret 0-RTT key exchange construction above that does not rely on synchronized time. For this variant, we essentially combine the FSOPKE$_U$ construction from Definition 12, restricted to a single time interval, with the concept of *server configurations* used in recent key exchange protocol designs, namely Google's QUIC protocol [31] and TLS 1.3 with Diffie–Hellman-based 0-RTT mode [39]. A server configuration here essentially is a publicly accessible string that contains a semi-static public key, signed with the long-term signing key of the corresponding party. Utilizing this string, a forward-secret 0-RTT key exchange protocol variant without time synchronization then works as follows.

For each time interval (e.g., a set number of days or weeks), servers generate a PFSKEM key pair (i.e., with $\tau_{max} = 1$), which they sign and publish within a server configuration. Clients can then retrieve and use the currently offered public key for the server to establish connections within this time interval.

We stress that, while introducing a slightly higher communication overhead, this variant offers the same security properties as the time-synchronized one. In particular recall that, due to puncturing, compromising the semi-static secret key for some time interval does not endanger the forward secrecy of priorly established connections within the same time interval. Indeed, the choice of how often to publish new server configurations (i.e., how long the conceptual time intervals are) is a purely functional one, based on the performance trade-off between storage and computation overhead for PFSKEM keys covering a shorter or longer interval (and hence more or fewer connections).

5 Analysis

We analyze our protocol for security levels $\lambda \in \{80, 128, 256\}$. We instantiate our scheme based on the DDH-based HIBE scheme from [6] and the discrete log-based one-time signature scheme from [22, Sect. 5.4]. We consider groups with asymmetric bilinear map $e : \mathbb{G}_1 \times \mathbb{G}_2 \to \mathbb{G}_T$ where groups are of order p such that $p = 2^{2\lambda}$ for the given security parameter λ. Thus, an element of $\mathbb{Z}_p$ can be represented by 2λ bits. We assume a setting based on Barreto-Naehrig curves [2], where elements of $\mathbb{G}_1$ can be represented by 2λ bits, while elements of $\mathbb{G}_2$ have size 4λ bits. In this setting, we can instantiate our PFS-KEM (and thus our FSOPKE) as follows.

- A ciphertext consist of three elements of $\mathbb{G}_1$ (from the HIBE of [6]) plus three $\mathbb{G}_1$-elements for pk_{OT}, plus two $\mathbb{Z}_p$-elements for σ. Thus, ciphertexts have size $6 \times |\mathbb{G}_1| + 2 \times |\mathbb{Z}_p| = 16\lambda$ bits.
- A public key contains $2\lambda + 35$ elements of $\mathbb{G}_2$, which amounts to $8\lambda^2 + 140\lambda$ bits.
- A punctured secret key contains $R + S$ user secret keys of the HIBKEM, each consisting of $3 \times |\mathbb{G}_2| = 12\lambda$ bits. Here $R = |pk_{OT}| + |\tau|$ denotes the bit-length of "HIBKEM-identities", and S denotes the number of sessions per time slot.

λ	$|pk|$	$|c|$	S	$|sk|$
80	7.8 kB	160 B	2^{10}	145.9 kB
80	7.8 kB	160 B	2^{16}	7.88 MB
80	7.8 kB	160 B	2^{20}	125.9 MB
128	18.62 kB	256 B	2^{10}	251.9 kB
128	18.62 kB	256 B	2^{16}	12.64 MB
128	18.62 kB	256 B	2^{20}	201.4 MB
256	70.02 kB	512 B	2^{10}	623.3 kB
256	70.02 kB	512 B	2^{16}	26.27 MB
256	70.02 kB	512 B	2^{20}	417 MB

Fig. 4. Size of public keys and ciphertexts and upper bounds on the size of secret keys for different choices of the security parameter λ and the number of sessions S per time slot.

Assuming a setting with 2^{32} time slots (which should be sufficient for any conceivable practical application, even with very short time slots), and that a collision-resistant hash function with range $\{0, 1\}^{2\lambda}$ is used to compute a short representation of pk_{OT} inside the HIBKEM, we have $R = 2\lambda + 32$. Thus, the size of the secret key as a function of S is $(S + 2\lambda + 32) \cdot 12\lambda$ bits.

For different values $S \in \{2^{10}, 2^{16}, 2^{20}\}$ of sessions per time slot, and security parameters $\lambda \in \{80, 128, 256\}$, we obtain the sizes of public keys and messages and the upper bounds on the size of secret keys displayed in Fig. 4.

Acknowledgments. We thank the anonymous reviewers for valuable comments. This work has been co-funded by the DFG as part of project S4 within the CRC 1119 CROSSING and by DFG grant JA 2445/1-2.

References

1. Agrawal, S., Boneh, D., Boyen, X.: Efficient lattice (H)IBE in the standard model. In: Gilbert, H. (ed.) EUROCRYPT 2010. LNCS, vol. 6110, pp. 553–572. Springer, Heidelberg (2010). doi:10.1007/978-3-642-13190-5_28

2. Barreto, P.S.L.M., Naehrig, M.: Pairing-friendly elliptic curves of prime order. In: Preneel, B., Tavares, S. (eds.) SAC 2005. LNCS, vol. 3897, pp. 319–331. Springer, Heidelberg (2006). doi:10.1007/11693383_22

3. Bellare, M., Rogaway, P.: Random oracles are practical: a paradigm for designing efficient protocols. In: Ashby, V. (ed.) ACM CCS 1993, , Fairfax, Virginia, USA, pp. 62–73. ACM Press, 3–5 November 1993

4. Bellare, M., Rogaway, P.: Entity authentication and key distribution. In: Stinson, D.R. (ed.) CRYPTO 1993. LNCS, vol. 773, pp. 232–249. Springer, Heidelberg (1994). doi:10.1007/3-540-48329-2_21

5. Bhargavan, K., Fournet, C., Kohlweiss, M., Pironti, A., Strub, P.-Y., Zanella-Béguelin, S.: Proving the TLS handshake secure (as it is). In: Garay, J.A., Gennaro, R. (eds.) CRYPTO 2014. LNCS, vol. 8617, pp. 235–255. Springer, Heidelberg (2014). doi:10.1007/978-3-662-44381-1_14

6. Blazy, O., Kiltz, E., Pan, J.: (Hierarchical) identity-based encryption from affine message authentication. In: Garay, J.A., Gennaro, R. (eds.) CRYPTO 2014. LNCS, vol. 8616, pp. 408–425. Springer, Heidelberg (2014). doi:10.1007/978-3-662-44371-2_23

7. Boneh, D., Boyen, X., Goh, E.-J.: Hierarchical identity based encryption with constant size ciphertext. In: Cramer, R. (ed.) EUROCRYPT 2005. LNCS, vol. 3494, pp. 440–456. Springer, Heidelberg (2005). doi:10.1007/11426639_26

8. Boyd, C., Cliff, Y., Gonzalez Nieto, J., Paterson, K.G.: Efficient one-round key exchange in the standard model. In: Mu, Y., Susilo, W., Seberry, J. (eds.) ACISP 2008. LNCS, vol. 5107, pp. 69–83. Springer, Heidelberg (2008). doi:10.1007/978-3-540-70500-0_6

9. Canetti, R., Halevi, S., Katz, J.: A forward-secure public-key encryption scheme. In: Biham, E. (ed.) EUROCRYPT 2003. LNCS, vol. 2656, pp. 255–271. Springer, Heidelberg (2003). doi:10.1007/3-540-39200-9_16

10. Canetti, R., Krawczyk, H.: Analysis of key-exchange protocols and their use for building secure channels. In: Pfitzmann, B. (ed.) EUROCRYPT 2001. LNCS, vol. 2045, pp. 453–474. Springer, Heidelberg (2001). doi:10.1007/3-540-44987-6_28

11. Chow, S.S.M., Choo, K.-K.R.: Strongly-secure identity-based key agreement and anonymous extension. In: Garay, J.A., Lenstra, A.K., Mambo, M., Peralta, R. (eds.) ISC 2007. LNCS, vol. 4779, pp. 203–220. Springer, Heidelberg (2007). doi:10.1007/978-3-540-75496-1_14

12. Cohn-Gordon, K., Cremers, C., Garratt, L.: On post-compromise security. In: IEEE 29th Computer Security Foundations Symposium, CSF 2016, pp. 164–178 (2016)

13. Cremers, C., Feltz, M.: One-round strongly secure key exchange with perfect forward secrecy and deniability. Cryptology ePrint Archive, Report 2011/300 (2011). http://eprint.iacr.org/2011/300

14. Cremers, C., Horvat, M., Scott, S., van der Merwe, T.: Automated analysis, verification of TLS 1.3: 0-RTT, resumption and delayed authentication. In: IEEE Symposium on Security and Privacy, San Jose, CA, USA, pp. 470–485. IEEE Computer Society Press, 22–26 May 2016

15. Cremers, C., Feltz, M.: Beyond eCK: perfect forward secrecy under actor compromise and ephemeral-key reveal. In: Foresti, S., Yung, M., Martinelli, F. (eds.) ESORICS 2012. LNCS, vol. 7459, pp. 734–751. Springer, Heidelberg (2012). doi:10.1007/978-3-642-33167-1_42

16. Dierks, T., Rescorla, E.: The Transport Layer Security (TLS) Protocol Version 1.2. RFC 5246 (Proposed Standard), Updated by RFCs 5746, 5878, 6176, August 2008

17. Dowling, B., Fischlin, M., Günther, F., Stebila, D.: A cryptographic analysis of the TLS 1.3 handshake protocol candidates. In: Ray, I., Li, N., Kruegel, C. (eds.) ACM CCS 2015, Denver, CO, USA, pp. 1197–1210. ACM Press, 12–16 October 2015

18. Fischlin, M., Günther, F.: Multi-stage key exchange and the case of Google's QUIC protocol. In: Ahn, G.-J., Yung, M., Li, N. (eds.) ACM CCS 2014, Scottsdale, AZ, USA, pp. 1193–1204. ACM Press, 3–7 November 2014

19. Fischlin, M., Günther, F.: Replay attacks on zero round-trip time: the case of the TLS 1.3 handshake candidates. In: 2017 IEEE European Symposium on Security and Privacy. IEEE, April 2017

20. Fujisaki, E., Okamoto, T.: Secure integration of asymmetric and symmetric encryption schemes. In: Wiener, M. (ed.) CRYPTO 1999. LNCS, vol. 1666, pp. 537–554. Springer, Heidelberg (1999). doi:10.1007/3-540-48405-1_34

21. Green, M.D., Miers, I.: Forward secure asynchronous messaging from puncturable encryption. In: IEEE S&P 2015 [25], pp. 305–320 (2015)

22. Groth, J.: Simulation-sound NIZK proofs for a practical language and constant size group signatures. In: Lai, X., Chen, K. (eds.) ASIACRYPT 2006. LNCS, vol. 4284, pp. 444–459. Springer, Heidelberg (2006). doi:10.1007/11935230_29

23. Hale, B., Jager, T., Lauer, S., Schwenk, J.: Simple security definitions for and constructions of 0-RTT key exchange. Cryptology ePrint Archive, Report 2015/1214 (2015). http://eprint.iacr.org/2015/1214

24. Halevi, S., Krawczyk, H.: One-pass HMQV and asymmetric key-wrapping. In: Catalano, D., Fazio, N., Gennaro, R., Nicolosi, A. (eds.) PKC 2011. LNCS, vol. 6571, pp. 317–334. Springer, Heidelberg (2011). doi:10.1007/978-3-642-19379-8_20

25. IEEE Symposium on Security and Privacy, San Jose, CA, USA. IEEE Computer Society Press, 17–21 May 2015

26. Jager, T., Kohlar, F., Schäge, S., Schwenk, J.: On the security of TLS-DHE in the standard model. In: Safavi-Naini, R., Canetti, R. (eds.) CRYPTO 2012. LNCS, vol. 7417, pp. 273–293. Springer, Heidelberg (2012). doi:10.1007/978-3-642-32009-5_17

27. Krawczyk, H.: HMQV: a high-performance secure Diffie-Hellman protocol. In: Shoup, V. (ed.) CRYPTO 2005. LNCS, vol. 3621, pp. 546–566. Springer, Heidelberg (2005). doi:10.1007/11535218_33

28. Krawczyk, H., Paterson, K.G., Wee, H.: On the security of the TLS protocol: a systematic analysis. In: Canetti, R., Garay, J.A. (eds.) CRYPTO 2013. LNCS, vol. 8042, pp. 429–448. Springer, Heidelberg (2013). doi:10.1007/978-3-642-40041-4_24

29. Krawczyk, H., Wee, H.: The OPTLS protocol and TLS 1.3. In: 2016 IEEE European Symposium on Security and Privacy, pp. 81–96. IEEE, March 2016

30. LaMacchia, B., Lauter, K., Mityagin, A.: Stronger security of authenticated key exchange. In: Susilo, W., Liu, J.K., Mu, Y. (eds.) ProvSec 2007. LNCS, vol. 4784, pp. 1–16. Springer, Heidelberg (2007). doi:10.1007/978-3-540-75670-5_1

31. Langley, A., Chang, W.-T.: QUIC Crypto. https://docs.google.com/document/d/1g5nIXAIkN_Y-7XJW5K45IblHd_L2f5LTaDUDwvZ5L6g/. Accessed May 2016, Revision 26 May 2016

32. Li, Y., Schäge, S., Yang, Z., Kohlar, F., Schwenk, J.: On the security of the pre-shared key ciphersuites of TLS. In: Krawczyk, H. (ed.) PKC 2014. LNCS, vol. 8383, pp. 669–684. Springer, Heidelberg (2014). doi:10.1007/978-3-642-54631-0_38

33. Lychev, R., Jero, S., Boldyreva, A., Nita-Rotaru, C.: How secure and quick is QUIC? Provable security and performance analyses. In: IEEE S&P 2015 [25], pp. 214–231 (2015)

34. Lyubashevsky, V.: Lattice signatures without trapdoors. In: Pointcheval, D., Johansson, T. (eds.) EUROCRYPT 2012. LNCS, vol. 7237, pp. 738–755. Springer, Heidelberg (2012). doi:10.1007/978-3-642-29011-4_43

35. Ostrovsky, R., Sahai, A., Waters, B.: Attribute-based encryption with non-monotonic access structures. In: Ning, P., De Capitani di Vimercati, S., Syverson, P.F. (eds.) ACM CCS 2007, Alexandria, Virginia, USA, pp. 195–203. ACM Press, 28–31 October 2007

36. Petullo, W.M., Zhang, X., Solworth, J.A., Bernstein, D.J., Lange, T.: MinimaLT: minimal-latency networking through better security. In: Sadeghi, A.-R., Gligor, V.D., Yung, M. (eds.) ACM CCS 2013, Berlin, Germany, pp. 425–438. ACM Press, 4–8 November 2013

37. Pointcheval, D., Sanders, O.: Forward secure non-interactive key exchange. In: Abdalla, M., Prisco, R. (eds.) SCN 2014. LNCS, vol. 8642, pp. 21–39. Springer, Heidelberg (2014). doi:10.1007/978-3-319-10879-7_2

38. QUIC, a multiplexed stream transport over UDP. https://www.chromium.org/quic

548   F. Günther et al.

39. Rescorla, E.: The Transport Layer Security (TLS) Protocol Version 1.3 - draft-ietf-tls-tls13-12. https://tools.ietf.org/html/draft-ietf-tls-tls13-12. Accessed March 2016
40. Rescorla, E.: The Transport Layer Security (TLS) Protocol Version 1.3 - draft-ietf-tls-tls13-18. https://tools.ietf.org/html/draft-ietf-tls-tls13-18. Accessed October 2016
41. Rescorla, E.: 0-RTT and Anti-Replay (IETF TLS working group mailing list). IETF Mail Archive, https://mailarchive.ietf.org/arch/msg/tls/gDzOxgKQADVfItfC4NyW3ylr7yc. Accessed March 2015
42. Rescorla, E.: [TLS] Do we actually need semi-static DHE-based 0-RTT? IETF Mail Archive, https://mailarchive.ietf.org/arch/msg/tls/c43zNQH9vGeHVnXhAb_D3cpIAIw. Accessed February 2016
43. Williams, N.: [TLS] 0-RTT security considerations (was OPTLS). IETF Mail Archive, https://mailarchive.ietf.org/arch/msg/tls/OZwGgVhySbVhU36BMX1elQ9x0GE. Accessed November 2014
44. Wu, D.J., Taly, A., Shankar, A., Boneh, D.: Privacy, discovery, and authentication for the internet of things. In: Askoxylakis, I., Ioannidis, S., Katsikas, S., Meadows, C. (eds.) ESORICS 2016. LNCS, vol. 9879, pp. 301–319. Springer, Heidelberg (2016). doi:10.1007/978-3-319-45741-3_16

Multiparty Computation IV

Computational Integrity with a Public Random String from Quasi-Linear PCPs

Eli Ben-Sasson[1](✉), Iddo Bentov[2], Alessandro Chiesa[3], Ariel Gabizon[4], Daniel Genkin[5], Matan Hamilis[1], Evgenya Pergament[1], Michael Riabzev[1], Mark Silberstein[1], Eran Tromer[6], and Madars Virza[7]

[1] Technion—Israel Institute of Technology, Haifa, Israel
eli@cs.technion.ac.il
[2] Cornell University, Ithaca, USA
[3] University of California, Berkeley, USA
[4] Zerocoin Electric Coin Company (Zcash), Lakewood, Colorado, USA
[5] University of Pennsylvania and University of Maryland, College Park, USA
[6] Tel Aviv University, Tel Aviv, Israel
[7] Massachusetts Institute of Technology, Cambridge, USA

Abstract. A party executing a computation on behalf of others may benefit from misreporting its output. Cryptographic protocols that detect this can facilitate decentralized systems with stringent computational integrity requirements. For the computation's result to be publicly trustworthy, it is moreover imperative to usepublicly verifiable protocols that have no "backdoors" or secret keys that enable forgery.

Probabilistically Checkable Proof (PCP) systems can be used to construct such protocols, but some of the main components of such systems—*proof composition* and low-degree testing via *PCPs of Proximity (PCPPs)* — have been considered efficiently only asymptotically, for unrealistically large computations. Recent cryptographic alternatives suffer from a non-public setup phase, or require large verification time.

This work introduces SCI, the first implementation of a scalable PCP system (that uses both PCPPs and proof composition). We used SCI to prove correctness of executions of up to 2^{20} cycles of a simple processor, and calculated its *break-even point*: the minimal input size for which naïve verification via re-execution becomes more costly than PCP-based verification.

This marks the transition of core PCP techniques (like *proof composition* and *PCPs of Proximity*) from mathematical theory to practical system engineering. The thresholds obtained are nearly achievable and hence show that PCP-supported computational integrity is closer to reality than previously assumed.

I. Bentov and A. Gabizon—Work done while at Technion
D. Genkin—Work done while at Technion and Tel Aviv University.

J.-S. Coron and J.B. Nielsen (Eds.): EUROCRYPT 2017, Part III, LNCS 10212, pp. 551–579, 2017.
DOI: 10.1007/978-3-319-56617-7_19

1 Introduction

Computational Integrity. An unobserved party is often required to execute a program $\mathbb{P}$ on data x, using auxiliary data w. Yet, that party might benefit from misreporting the output y. For example:

1. Individuals and companies may benefit financially from reporting lower tax payments; in this case $\mathbb{P}$ is the program that computes tax, x is the tax-relevant data (w is the empty string) and y is the resulting tax.
2. Criminals may benefit if an innocent individual (or no individual) is prosecuted based on faulty crime-scene data analysis, and corrupt law enforcement officials to reach this outcome. In this case $\mathbb{P}$ is the program that analyzes crime-scene data, x may contain the cryptographic hashes of (i) a criminal DNA database and (ii) DNA fingerprints taken from the crime-scene, w is the preimage of (i), (ii) and y would be the name of a suspect.
3. Health-care and other insurance companies may benefit from mis-computing policy rates. In this case $\mathbb{P}$ may be a government-approved program that computes policy rates, x is the identifying number of a patient, w would be her medical history (including, perhaps, her DNA sequence) and y is the policy rate.

Naturally, correctness and integrity of the input data x, w are preliminary requirements for obtaining a correct output y; These inputs often arrives from third parties and can be digitally signed by them, hence changing (x, w) maliciously to (x', w') would require their collusion. Instead, the main focus of this work is on ensuring the integrity of the *computation* $\mathbb{P}$ itself, e.g., ensuring that the reported tax y is correct with respect to the explicit input x, program $\mathbb{P}$ and some auxiliary input w. In spite of incentives to cheat, we often assume that unobserved parties operate with *computational integrity* (CI) meaning that CI statements like

$$\tau_{(\mathbb{P},x,y,T)} := \text{``}\exists w \text{ such that } y = \text{ output of } \mathbb{P} \text{ on inputs } x, w \text{ after } T \text{ steps''} \quad (*)$$

are considered true, even when the party making the statement could benefit from replacing y with $y' \neq y$. The assumption that parties operate with computational integrity is backed by (i) legislation and (ii) regulation, and also relies on (iii) the economic value of "integrity" to individuals, businesses and government. Manual enforcement of CI via audits and reports by trusted third parties is labor-intensive, and yet leaves the door open to corruption of those third parties. Automated CI based on cryptography (also called *delegation of computation* [43], *certified computation* [32] and *verifiable computation* [40]) could potentially replace this manual labor and, more importantly, introduce integrity to settings in which it is currently too costly to achieve.

Interactive Proof (IP) Systems. [5,44] revolutionized cryptographic CI by initiating an approach that led (see below) to a viable theoretical solution to the problem of discovering false CI statements. In such systems the party that makes the CI statement $(*)$ is represented by a *prover* which is a (randomized) algorithm. The prover tries to convince a *verifier*—an efficient randomized algorithm—that $(*)$ is true via a court-of-law-style *interactive* protocol in which the verifier "interrogates" the prover over several rounds of communication. The protocol ends with the verifier announcing its verdict which is either to "accept" $\tau_{(\mathbb{P},x,y,T)}$ as true, or to "reject" it. The systems we focus on have only one-sided error: all true statement can be supported by a prover that causes the verifier to accept them but the verifier may err and accept falsities; the probability of error is known as the *soundness-error*.

Probabilistically Checkable Proof (PCP) Systems.[1] [1–4] are a particularly efficient multi-prover interactive proof (MIP) system [8] in terms of the amount of communication between prover and verifier, verification time, the number of rounds of interaction and soundness-error. Assuming T is given in binary, the set of true CI statements (eq:statement) is a **NEXP**-complete language and PCPs are powerful enough to prove membership in this language. Here, the prover writes once a string of bits $\pi_{(\mathbb{P},x,y,T)}$ known as a PCP; its length is polynomial in the execution time T. Total verifier running time is $\text{poly}\log T$, which is (i) negligible compared to the naïve solution of re-executing $\mathbb{P}$ at a cost of T steps and (ii) nearly-optimal because every proof system for general CI statements must have the verifier running time be at least $\Omega(\log T)$. Using a single round, the verifier asks to read a small (randomly selected) number of bits of $\pi_{(\mathbb{P},x,y,T)}$; clearly the verifier cannot read more bits than its running time $(\text{poly}\log T)$ allows, and this amount can be further reduced to a small constant that is independent of T (cf. [34,49,63,66]). Initial constructions required proofs of length $\text{poly}(T)$ but length has been reduced since then [21,24,42,48] and state-of-the-art proofs are of *quasi-linear* length in T, i.e., length $T \cdot \text{poly}\log T$ [20,23,34,62] and can be computed in quasi-linear time as well [13]. The system reported — called Scalable Computational Integrity (SCI) — implements the quasi-linear PCP system [13,23] with certain improvements (described later).

In many cases the prover needs to preserve the privacy of the auxiliary input w (as is the case with examples 2, 3 above) while at the same time proving that it "knows" w, as opposed to merely proving that w exists. Privacy-preserving, or *zero knowledge (ZK) proofs* [44] and *ZK proofs of knowledge* [7] can be constructed from any PCP system in polynomial time [36,55,56] (cf. [52–54,60]). Certain "algebraic" PCP systems, including SCI, can be converted to ZK proofs of knowledge with only a quasilinear increase in running time [11]; **implementing this enhancement is left to future work.**

A PCP verifier requires random access to bits of $\pi_{(\mathbb{P},x,y,T)}$; a naïve implementation in which prover sends the whole proof to the verifier would cost $\text{poly}(T)$ communication (and verification time) but a collision-resistant hash function can be used to reduce communication and verifier running time to $\text{poly}\log T$ [55]. The three messages transmitted between prover and verifier ((1) prover sends proof; (2) verifier sends queries; (3) prover answers queries) can be reduced to a single message from the prover, if both parties have access to the same random

---

[1] PCPs are also known as *holographic*, and *transparent* proof systems.

function [61]; this can be realized using a standard cryptographic hash function such as SHA-3, via the Fiat-Shamir heuristic [38] (or via an extractable collision resistant hash function [26]). The single message (published by the prover) is known as a *succinct computationally sound* (CS) proof $\hat{\pi}$; its length is poly log T and it can now be appended to $\tau_{(\mathbb{P},x,y,T)}$ and then publicly verified in time poly log T with no further interaction with the prover. We refer to $\hat{\pi}$ as a *hash-based (CI) proof* to emphasize that the only cryptographic primitive needed to implement it is a hash function.

Prior CI solutions. In spite of the asymptotic efficiency of PCPs, prior CI approaches (recounted below) did not implement a PCP system. To quote from the recent survey [77], the reason for this was that *"the proofs arising from the PCP theorem (despite asymptotic improvements) were so long and compli-cated that it would have taken thousands of years to generate and check them, and would have needed more storage bits than there are atoms in the universe"*. Due to this view (which this work challenges), five main alternatives have been explored recently, described below. Like SCI, all rely on *arithmetization* [59], the reduction of computational integrity statements (∗) to systems of low-degree polynomials over finite fields. But in contrast to SCI, all previous solutions cir-cumvent the use of core PCP techniques like *proof composition* [2], *low-degree testing* and the use of *PCPs of proximity (PCPP)* [20,35]; these techniques are crucial for obtaining *succinctly verifiable* proofs with a *public setup* process, which SCI is the first to implement.

IP-based: The *proofs for muggles* approach [43] scales down Interactive Proofs (IP) from **PSPACE** to **P** and leads to excellent solutions for a limited yet interesting class of programs: those with high parallelism and small memory consumption; prover time for *IP-based* systems was reduced to quasi-linear [33] and implemented in a number of works [32,73,75].

LPCP-based: [51] proposed using additively homomorphic encryption (AHE) and *linear PCPs (LPCP)* to build CI proof systems that are interactive, and where the verifier's work is amortized over multiple statements; cf. [69,71,72] for implementations of *LPCP-based* systems.

KOE-based: A sequence of works [28,40,41,46,58] improved on [51] by relying on *Knowledge Of Exponent (KOE)* assumptions and bilinear pairings over elliptic curves. *KOE-based* systems were implemented in [15,19,65,70,76], and further optimizations of this latter system for specific applications related to Bitcoin [64] such as smart contracts [57] and anonymous payment systems [12] are already being evaluated by commercial entities [45].

IVC-based: KOE-based systems require a proving key k_P (discussed below) that is longer than T, the number of computation cycles. Incrementally verifiable computation (IVC) [74] and bootstrapping [27] shorten the length of k_P to poly log T and an *IVC-based* system has been implemented recently [18].

DLP-based: KOE/IVC-based systems require a *private setup* phase that is discussed below. [47] (cf. [68]) assumes hardness of the Discrete Logarithm Problem (DLP) to build a system that requires only a *public setup*, like SCI.

Proof length in the initial works above was $\Theta\left(\sqrt{T}\right)$ and this was reduced to $\mathrm{poly}\log T$ in [29], which also implemented both versions; verifier running time in both variants is $\Omega(T)$.

Comparing SCI *to Prior CI Solutions.* SCI is the first CI solution that achieves both (1) a short public randomness setup phase and (2) universal scalability for one-shot computation. We discuss the significance of these properties after explaining them. (A quantitative comparison of the running time, memory consumption and communication complexity of SCI to prior systems appears in Sect. 2 and Table 1.)

One-shot Universal Scalability (OSUS). A CI system is *universally scalable* if for any fixed program $\mathbb{P}$, prover running time is bounded by $T\mathrm{poly}\log T$ and verification time is at most $\mathrm{poly}\log T$ where T is the number of machine cycles[2]. If the same asymptotic running times hold even for a single execution of $\mathbb{P}$, and where the setup ("preprocessing") is carried out by the verifier (and hence setup-cost is part of the total verification-cost), we shall say that CI solution is *one-shot universally scalable* (OSUS). DLP-based systems have super-linear verification time, hence are not *scalable* for any program. IP-based systems are efficient only for highly-parallel computations, thus are not *universally scalable*. LPCP- and KOE-based systems are universally scalable but not OSUS because they require a proving key k_P that is longer than T which must be generated by the verifier (in the one-shot setting). Of all prior solutions, only the IVC-based one is OSUS, like SCI.

Public Setup. All implemented solutions but for DLP-based and SCI, if instantiated as publicly verifiable CI systems, require a setup phase ("preprocessing"), the output of which is a pair of keys $(k_\mathsf{P}, k_\mathsf{V})$, one needed for proving statements, the other for verifying them. A "trapdoor key" k_tpdr is associated with $(k_\mathsf{P}, k_\mathsf{V})$ and can be used to forge pseudo-proofs of false statements. Furthermore, k_tpdr can be recovered by the parties that run the preprocessing phase. Secure multi-party computation can boost security by "distributing knowledge" of the trapdoor among several parties [17] so that all of them have to be compromised to recover k_tpdr; but this does not remove the concern that k_tpdr has been recovered by collusion of all parties, or retrieved by a central party eavesdropping to all of them. Even if k_tpdr has not been recovered by anyone, its mere existence may erode trust in such systems. (Cf. [6] for a recent discussion of setup-attacks and their implications and mitigations.) In contrast, SCI and DLP-based systems require only a short public random string when instantiated as a publicly verifiable noninteractive CI system.

Discussion. The combination of OSUS and public setup which is unique to SCI has three implications: (i) the ease of setting up and modifying CI systems

---

[2] Formally, a CI system is *universally scalable* if for any language $L \in \mathbf{NTIME}(T(n))$ prover running time is $T(n)\mathrm{poly}\log T(n)$ and verifier running time is $\mathrm{poly}\log T(n)$ where n denotes input length.

based on it is relatively small, (ii) the trust assumptions made by parties using it are comparatively minor and hence (iii) it seems more suitable than existing solutions for use in decentralized and public settings, like Bitcoin. We repeat and stress that many such applications require zero-knowledge proofs, a property achieved by prior solutions and not achieved by SCI; augmenting SCI to obtain zero knowledge seems within reach [11] but is outside the scope of our work.

SCI—*Main Technical Contributions.* We faced three major challenges when attempting to construct PCP systems that scale well and apply to general programs, and SCI is the first implementation to contain scalable solutions to each of them, reported here for the first time: (i) implementing the recursive *proof composition* [2] technique applied to PCPs of proximity (PCPPs) [20,35] (ii) constructing quasi-linear PCPP systems for Reed-Solomon (RS) error correcting codes [67] of huge message length [23] that require, in particular, quasi-linear time algorithms for interpolation and multi-point evaluation of large-degree polynomials over finite fields of characteristic 2; and (iii) reducing general programs that include jumps, loops, and random access memory (RAM) instructions to *succinct Algebraic Constraint Satisfaction Problem* (sACSP) instances that "capture" the corresponding CI statement (∗); prior arithmetization solutions require the verifier, or a party trusted by it, to "unroll" a T-cycle computation to obtain an arithmetic circuit of size $\Omega(T)$, whereas SCI's verifier is succcint and does not perform this unrolling. (All prior solutions arithmetize over large prime fields; SCI is also novel in its being the first arithmetization over large binary fields, which poses new challenges, especially for integer operations like addition and multiplciation, cf. Section B.1.)

To overcome the blowup (i) that is due to recursive PCPP composition, we replace PCPPs with *interactive oracle proofs of proximity (IOPPs)* [9,10,37], implemented here for the first time, and increase the number of rounds of interaction between prover and verifier; the extra rounds can be removed in the random oracle model [37]. To address (ii) we built a dedicated library that implements finite field arithmetic efficiently (reported in [22]) and used it to further implement additive Fast Fourier Transforms (aFFT) [39] that perform interpolation and multi-point evaluation in quasi-linear time and in parallel (via multi-threading); the large-scale additive FFTs are reported here for the first time. To solve (iii) and reduce general programs to PCP systems efficiently, we devise a novel reduction from general programs for random access machines to sACSP instances. We describe these three contributions in more detail in Sect. 3 and the appendix.

2   Measurements

SCI can be applied to any language in **NEXP**; for concreteness we picked two programs computing the NP-complete subset-sum problem (cf. Appendix C); we explain this choice after introducing the two programs. The input to the subset-sum problem is an integer array A of size n and a target integer t; the problem is to decide whether there exists a subset $A' \subset A$ that sums to t. The CI statement addressed here is the co-NP version of the problem, stating "no subset of A sums to t" and denoted by $\tau_{(A,n,t)}$. The two programs differ in their time and space consumption. The first one *exhaustively* tries all possible subsets, requiring 2^n cycles but only $O(1)$ memory, hence can be executed using only the local registers of the machine and with no random access to memory. The second program uses *sorting* and runs in time $O(2^{n/2})$, a quadratic improvement over the exhaustive solution but it also requires $\Theta(2^{n/2})$ memory and hence uses the random access memory. We denote the two programs by $\mathbb{P}_{exh}$ and $\mathbb{P}_{sort}$, respectively.

On Choice of Programs. We would like to run SCI on "real-world" applications like the examples given in the introduction but our current scalability is not up to par. This situation is similar to that of the very first works on other CI solutions (cf. [15,33,65,69]): initial reports discussed only small word-size machines, restricted functionality and simple programs. Like some of those works (most notably, [19]) we use the 16-bit version of the TinyRAM architecture as our model of computation, and support all of its assembly code even though these two programs use only a subset of it. We focus on subset-sum for two reasons: (i) it is a natural NP-complete problem that is often used in cryptographic applications but more importantly (ii) it allows us to display the effect of time–space tradeoffs on our CI solution (cf. Figure 2). Since SCI supports non-determinism, we could have used the non-deterministic version of the subset-sum statement. In fact, this would have reduced prover and verifier complexity because fewer boundary constraints are imposed on the input. However, the resulting statement seems less interesting, saying "there exists A such that no subset of it sums to t".

Measurement Range. Input array size n ranged between 3–16. Prover data was measured on a "large" server with 32 AMD Opteron cores at clock rate 3.2 GHz and 512 Gigabytes of RAM, running with two threads per core (total of 64 threads); to bound the single-core/thread prover time one may multiply the stated times by $\times 32/ \times 64$ respectively. Verifier data was measured on a "standard" laptop, a Lenovo T440s with Intel core i7-4600 at clock rate 2.1 GHz and 12 Gigabytes RAM. We stress that verifier succinctness for one-shot programs allows us to measure verifier running time independently of prover running time, all the way up to 2^{47} machine cycles. Both prover and verifier were measured for 1-bit security and 80-bit security using state-of-the-art PCPP and IOPP security estimates [9].

Prover Time and Memory. The left column of Fig. 1 presents the running time (top) and memory consumption (bottom) of the Prover for both $\mathbb{P}_{exh}$ and $\mathbb{P}_{sort}$ as a function of the number of machine cycles of the simulated machine for both 1-bit and 80-bit security level. The two main observations from these figures are that (i) resources scale quasi-linearly with number of cycles and (ii) $\mathbb{P}_{sort}$ is more costly than $\mathbb{P}_{exh}$ due to its random access memory usage, which increases proof length by $\times \log^{O(1)} T$ factor for a T-cycle execution (cf. Section 3). Figure 2 compares time and memory as a function of the size on the input array n and shows that for $n \geq 8$ the quadratic running-time improvement of $\mathbb{P}_{sort}$ over $\mathbb{P}_{exh}$

outweighs the $\times O(\log T)$ factor required by random access to memory, both for 1-bit and 80-bit security level.

Verifier Time and Query Complexity. The right column of Fig. 1 shows verifier running time (top) and query complexity (bottom) for both programs for both 1-bit and 80-bit security levels. Notice the $\approx 2^{13}$–$2^{23} \times$ factor improvement of verifier over prover in both parameters (recall $1MB = 2^{10}KB$) and the increase in running time as a function of security due to repetition. For small n verifier running time is greater than that of the naïve verifier which re-runs the program. However, since naive verification grows like 2^n for $\mathbb{P}_{exh}$ and like $2^{n/2}$ for $\mathbb{P}_{sort}$, for $n \geq 22$ (at 80-bit security) our verifier is more efficient than the naïve one for $\mathbb{P}_{exh}$, and for $n \geq 48$ the verifier for $\mathbb{P}_{sort}$ is more efficient than the naïve one (cf. Figure 3).

Table 1. Quantitative comparison of SCI with KOE-based [15], IVC-based [18] and DLP-based [47] solutions. Data measured on executions of 2^{16} cycles of $\mathbb{P}_{exh}$ at an 80-bit security level on the same machine with 32 AMD Opteron cores at clock rate 3.2 GHz and 512 Gigabytes of RAM. The DLP-based column is extrapolated from [47, Table 2], accounting for (i) the larger circuit size of our computation (which has $\sim$ 132M gates compared with maximal size of 1.4M gates there) and different compute architectures (single threaded Intel 4690 K core vs. 64 threaded AMD Opteron). Notice the proving time of SCI is $\sim \times 2 - -4$ slower than KOE- and DLP-based and $\sim \times 150$ faster than IVC-based. Regarding total communication complexity, SCI is more efficient than prior solutions but less efficient when measuring only post-processing communication.

		KOE-based	IVC-based	DLP-based	SCI
Verifier setup	Time	~28 min	~10 sec	~0.7 sec	<0.01 sec
	Key length	~18.9 GB	43 MB	154 MB	16 bytes
Prover	Time	~ 18 min	4.2 days	~ 8 min	~ 41 min
	Memory	~216 GB	2.9 GB	~1 TB	~135 GB
Verifier postprocessing	Time	<10 ms	~25 ms	~1.7 min	~ 0.5 sec
	Communication complexity	230 bytes	374 bytes	8.8 KB	~42.5 MB
Verifer total	Time	~28 min	~10 sec	1.7 min	~ 0.5 sec
	Communication complexity	~18.9 GB	43 MB	~154 MB	~42.5 MB

Quantitative Comparison with other CI Implementations. Table 1 compares SCI to three recent CI systems, the KOE-based [15], the IVC-based [18], and the DLP-based [47], using the version with poly $\log(T)$ communication complexity. One sees that SCI has the shortest and fastest setup but larger post-setup communication complexity; post-setup verification is faster than DLP-based but slower than KOE/IVC-based, as predicted by theory. Two other important points are: (i) proofs in SCI are not zero-knowledge whereas the other solutions are, and (ii) the setup of the last two columns (DLP-based and SCI) is comprised only of a public random string, whereas KOE/IVC-based solutions require private setup and involve a trapdoor that can be used to forge proofs of false statements.

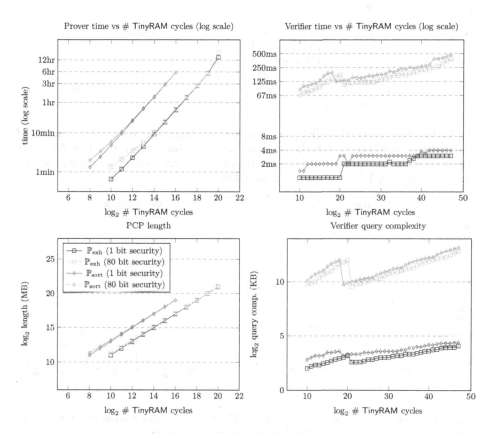

Fig. 1. Comparison of prover (left) and verifier (right) running time (top) and memory consumption (bottom). The sharp drop in query complexity is due to transition from 2 to 3 levels of recursion in the RS-PCPP; as seen in the top-right, this has little effect on overall verifier running time, which is significantly smaller than prover running time, and also grows at a considerably slower rate as a function of # cycles. Answers to verifier queries provided by random strings which simulates accurately actual proofs because verifier is non-adaptive, i.e., its running time is independent of the proof content.

3   Overview of Construction

The construction of the PCP $\pi_{(\mathbb{P},x,y,T)}$ for the computational statement $\tau_{(\mathbb{P},x,y,T)}$ follows the rather complex process detailed in [13, 14, 21, 23] which we summarize next (see Appendix A). The statement $\tau_{(\mathbb{P},x,y,T)}$ is converted into an instance

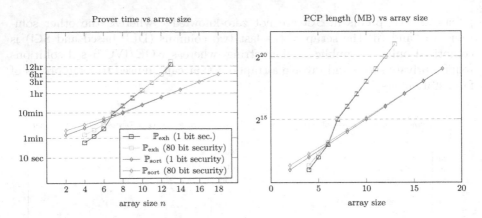

Fig. 2. Prover running time (left) and memory consumption (right) as a function of input array size n. For $n \geq 8$ the quadratic running-time improvement of $\mathbb{P}_{\mathrm{sort}}$ over $\mathbb{P}_{\mathrm{exh}}$ overcomes the $\times \mathrm{poly} \log T$ factor overhead of $\mathbb{P}_{\mathrm{exh}}$ due to random memory access; this holds for both 1-bit and 80-bit security level.

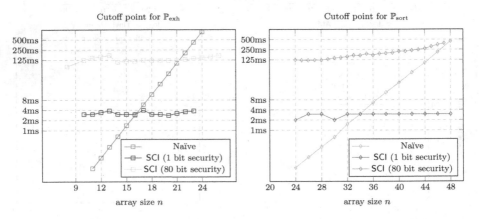

Fig. 3. Computation of the *break-even point* [71,72], the minimal input size n for which naïve verification via re-execution becomes more costly than PCP-based verification. For $\mathbb{P}_{\mathrm{exh}}$ at 80-bit security this threshold is at $n = 22$ and for $\mathbb{P}_{\mathrm{sort}}$ it is significantly higher, estimated around $n = 48$, due to quadratic improvement in running time of the latter program.

$\psi_{(\mathbb{P},x,y,T)}$ of an *algebraic constraint satisfaction problem* (ACSP) over a finite field[3] $\mathbb{F}$ of characteristic 2 and $\tau_{(\mathbb{P},x,y,T)}$ is used by prover and verifier as described next.

Prover. To construct the PCP, the prover executes $\mathbb{P}$ on input x and encodes the execution trace by a Reed-Solomon [67] codeword $\mathsf{a}_{(\mathbb{P},x,y,T)}$ evaluated over an additive sub-group of $\mathbb{F}$. The ACSP instance $\psi_{(\mathbb{P},x,y,T)}$ is applied to $\mathsf{a}_{(\mathbb{P},x,y,T)}$ as

---

[3] SCI uses the field of size 2^{64} which suffices for the computations measured here.

described in [23, Equation (3.2)] to obtain an additional RS-codeword, denoted $b_{(\mathbb{P},x,y,T)} = \psi_{(\mathbb{P},x,y,T)}(a_{(\mathbb{P},x,y,T)})$, that "attests" to the fact that $a_{(\mathbb{P},x,y,T)}$ encodes a valid execution trace, and hence, in particular, its output is correct. Each of the two codewords is appended with a PCP of proximity (PCPP) for the RS-code [23], denoted π_a, π_b, respectively. The PCP $\pi_{(\mathbb{P},x,y,T)}$ is defined to be the concatenation of $a_{(\mathbb{P},x,y,T)}, b_{(\mathbb{P},x,y,T)}, \pi_a$ and π_b.

Verifier. The verifier queries the four parts of the PCP in the following manner: First it invokes an RS-PCPP verifier that queries $a_{(\mathbb{P},x,y,T)}$ and π_a to "check" that $a_{(\mathbb{P},x,y,T)}$ is close in Hamming distance to a codeword of the RS-code; it repeats this process with respect to $b_{(\mathbb{P},x,y,T)}$ and π_b. Second and last, the verifier queries $a_{(\mathbb{P},x,y,T)}$ and $b_{(\mathbb{P},x,y,T)}$ and uses $\psi_{(\mathbb{P},x,y,T)}$ to check that the two codewords encode a valid computation of $\mathbb{P}$ that starts with x and reaches y within T cycles. In this process we rely on the "locality" of the mapping $\psi_{(\mathbb{P},x,y,T)} : a_{(\mathbb{P},x,y,T)} \rightarrow b_{(\mathbb{P},x,y,T)}$ which means that each entry of $b_{(\mathbb{P},x,y,T)}$ depends on a small number of entries of $a_{(\mathbb{P},x,y,T)}$. In what follows we elaborate on the novel aspects of this reduction as implemented in SCI.

From Assembly Code to Succinct ACSP. The *efficiency* of the ACSP instance $\psi_{(\mathbb{P},x,y,T)}$ is measured by three parameters that we seek to minimize: *circuit size*, *degree*, and *query complexity*, denoted $C_{(\mathbb{P},x,y,T)}, D_{(\mathbb{P},x,y,T)}, Q_{(\mathbb{P},x,y,T)}$ respectively. Circuit size affects both proving and verification time; degree affects PCP length and reducing it decreases running time and memory consumption on the prover side; query complexity affects the length of communication between prover and verifier (and the length of computationally sound (CS) proofs $\hat{\pi}$) as well as verifier running time. Each parameter can be optimized at the expense of the other two, and the challenge is to reach an efficient balance between all three.

Our starting point is a program $\mathbb{P}$, i.e., a sequence of *instructions* for a random access machine (RAM). For simplicity we first focus on instructions that access only (local) registers; random access memory instructions are discussed below. Each instruction specifies the input and output register locations and an operation applied to the inputs, called the *opcode*. We build $\psi_{(\mathbb{P},x,y,T)}$ bottom-up (cf. Appendix B for a detailed example). Each opcode op appearing in $\mathbb{P}$ (like xor, add, jump, etc.) is specified by an *algebraic definition* over $\mathbb{F}$; in other words, we specify a set of multi-variate polynomials $\mathcal{P}_{op} \subseteq \mathbb{F}[X_1, X_2, \ldots, X_m]$ such that the set of common zeros of $\mathcal{P}_{op}$ correspond to correct input-output tuples for op. Program flow is controlled by multiplying each polynomial in $\mathcal{P}_{op}$ by a multivariate Lagrange "selector" polynomial that, based on the value v of the program counter (PC), annihilates all constraints that are irrelevant for enforcing the vth instruction of $\mathbb{P}$. For a program with ℓ lines these selector polynomials have degree $\lceil \log \ell \rceil$. The resulting ACSP has circuit size $O(\ell)$ and degree and query complexity are $\log \ell + O(1)$; the constants hidden by asymptotic notation depend on the machine specification.

Random Access Memory Instructions. The *execution trace* of $\mathbb{P}$ is the length–T sequence of machine states that describes the computation. To verify the integrity of random access memory instructions (such as load and store) we follow [13,14] and use a *pair* of execution traces. The first trace, $\mathsf{trace}^{\mathsf{time}}$, is sorted increasingly by time, and the second, $\mathsf{trace}^{\mathsf{mem}}$, is sorted lexicographically first by memory location, then by time. RAM-related execution validity is verified "locally" by inspecting pairs of consecutive elements in $\mathsf{trace}^{\mathsf{mem}}$, just like non-RAM related instructions are verified "locally" by inspecting pairs of consecutive elements in $\mathsf{trace}^{\mathsf{time}}$. To further reduce proof length and query complexity, each state of $\mathsf{trace}^{\mathsf{mem}}$ contains only the information needed to check memory consistency — an address, its content and the type of memory access (load/store); let s denote the number of field elements in a single line of $\mathsf{trace}^{\mathsf{mem}}$.

To prove that $\mathsf{trace}^{\mathsf{mem}}$ and $\mathsf{trace}^{\mathsf{time}}$ refer to the same execution, the prover must describe a permutation between the two, and the verifier must check its validity. To achieve this SCI uses a non-blocking Beneš switching network [25,31] embedded in an affine graph over $\mathbb{F}$ (cf. [14,23] for definitions). Using this method, adding RAM-related instructions to a program adds only $O(T \cdot \log T)$ field elements to the PCP and increases query complexity by a small constant.

Reducing Proof Construction Time via Interactive Oracle Proofs of Proximity (IOPP). A significant portion of the prover running time and memory consumption are dedicated to the construction of the PCP of Proximity (PCPP) for $\mathsf{a}_{(\mathbb{P},x,y,T)}$ and for $\mathsf{b}_{(\mathbb{P},x,y,T)}$. The full PCPP for an RS-codeword of degree N is of length $O(N \log^{2.6} N)$ which is quite large in our applications. Observing that (i) these PCPPs are built using recursive *PCPP composition* [21], and (ii) only a small fraction of recursive branches are explored by the verifier, we increase the number of rounds of interaction and use a *notarized interactive proof of proximity* (NIPP) [9], a special case of *interactive oracle proofs of proximity* (IOPP) [10,37] to reduce proof length to $4N + O(\sqrt{N})$. The added rounds of interaction can be removed in the random oracle model to obtain computationally sound proofs [37].

Parallel Implementation of PCPPs for RS Codes. To reduce the time required to encode the execution trace into a pair of RS-codewords, SCI uses parallel algorithms for finite field operations and for dealing with polynomials over finite fields of characteristic 2. To speed up basic field operations (most notably, multiplication) a dedicated algebraic library was built, that utilizes parallel hardware on multi-core CPU. Interpolation and evaluation of polynomials over affine spaces of size N are computed in quasilinear time using so-called *additive Fast Fourier Transform* (aFFT) [39].

4   Concluding Remarks

SCI is the first implementation of a system of computational integrity that achieves asymptotic one shot universal scalability (OSUS) with a setup key that is merely a public random string. Prior solutions either required super-linear verification time, or used a setup procedure that involves keys which could be used to forge proofs of falsities. While the computer programs on which SCI was tested are of limited applicability, the simpler setup assumptions of SCI make it a natural starting point for building further applications — most notably zero knowledge proofs — for use in decentralized networks.

Acknowledgements. We thank Ohad Barta, Lior Greenblatt, Shaul Kfir, Gil Timnat and Arnon Yogev for programming support in early stages of this work. The research reported here has received funding from the following sources, sorted alphabetically: the Blavatnik Interdisciplinary Cyber Research Center; the Center for Long-Term Cybersecurity at UC Berkeley; the Center for Science of Information (CSoI), an NSF Science and Technology Center, under grant agreement CCF-0939370; the Check Point Institute for Information Security; the European Community's Seventh Framework Programme (FP7/2007–2013) under grant agreement number 240258; the Israeli Centers of Research Excellence I-CORE program (center 4/11); the Israeli Science Foundation (grants 1501/14,1138/14); and the Leona M. & Harry B. Helmsley Charitable Trust.

A   Detailed PCP Construction

We describe the way a PCP is generated for $\tau_{(\mathbb{P},x,y,T)}$, then discuss its verification.

Proof generation. The PCP proof $\pi_{(\mathbb{P},x,y,T)}$ for $\tau_{(\mathbb{P},x,y,T)}$ is a concatenation of four sub-proofs: two codewords in a Reed-Solomon code [67] and two quasilinear size PCPs of Proximity (PCPP) for the RS-codewords [23]. To obtain these four sub-proofs, the prover starts by executing the program $\mathbb{P}$ on input x for T steps and records its *execution trace*—the length–T sequence of machine states that the machine goes through during execution. Each state is converted to a sequence of elements in the finite field $\mathbb{F}$ of size 2^{64}; Auxiliary field elements are appended to each state to reduce the degree complexity of $\psi_{(\mathbb{P},x,y,T)}$ as described in Sect. B; let s denote the total number of field elements per state. The resulting *algebraic trace* trace^{aug} is thus a table of $N = T \cdot s$ elements of $\mathbb{F}$, and is viewed as a function from $S \subset \mathbb{F}, |S| = N$ to $\mathbb{F}$, where S is an affine space over the two-element field. Prover now computes the *low-degree extension* (LDE) of trace^{aug} by interpolating and then evaluating trace^{aug} on a set $S' \subset \mathbb{F}$ that is significantly larger than S. This results in a codeword $a_{(\mathbb{P},x,y,T)}$ of a Reed-Solomon (RS) code [67] over $\mathbb{F}$ of degree $N - 1$ and rate $\rho = |S|/|S'|$. Next, the ACSP instance $\psi_{(\mathbb{P},x,y,T)}$ is applied to $a_{(\mathbb{P},x,y,T)}$ as described in [23, Eq. (3.2)], producing another RS-codeword $b_{(\mathbb{P},x,y,T)} = \psi_{(\mathbb{P},x,y,T)}(a_{(\mathbb{P},x,y,T)})$, of degree $D_{(\mathbb{P},x,y,T)} \cdot (N - 1)$ and rate $\rho' = D_{(\mathbb{P},x,y,T)} \cdot \rho$ (SCI uses $\rho' = \frac{1}{8}$). Finally, a PCP of proximity (PCPP) for RS-codes [23] is appended to each of $a_{(\mathbb{P},x,y,T)}$ and $b_{(\mathbb{P},x,y,T)}$ to prove that indeed each belongs to the RS-code of the designated rate — ρ for $a_{(\mathbb{P},x,y,T)}$ and ρ' for $b_{(\mathbb{P},x,y,T)}$; denote these PCPPs by π_a, π_b, respectively. Summing up, the PCP proof $\pi_{(\mathbb{P},x,y,T)}$ is the concatenation of the four strings $a_{(\mathbb{P},x,y,T)}, \pi_a, b_{(\mathbb{P},x,y,T)}$ and π_b.

Proof Verification. On the verifier side, given $\psi_{(\mathbb{P},x,y,T)}$ as input and oracle access to

$$\pi_{(\mathbb{P},x,y,T)} = (a_{(\mathbb{P},x,y,T)}, \pi_a, b_{(\mathbb{P},x,y,T)}, \pi_b)$$

as above, the verifier invokes the RS-PCPP verifier of [23] on each of $(a_{(\mathbb{P},x,y,T)}, \pi_a)$ and $(b_{(\mathbb{P},x,y,T)}, \pi_b)$. Then it checks that $a_{(\mathbb{P},x,y,T)} = \psi_{(\mathbb{P},x,y,T)}(b_{(\mathbb{P},x,y,T)})$ by sampling both $a_{(\mathbb{P},x,y,T)}$ and $b_{(\mathbb{P},x,y,T)}$ at a small number of locations $(1 + Q_{(\mathbb{P},x,y,T)}$ per test). To boost soundness, each of the aforementioned tests is repeated a number of times, using fresh randomness (SCI uses 14 repetitions to reduce the probability of error to $\mathsf{error} = \frac{1}{2}$). The verifier "accepts" $\tau_{(\mathbb{P},x,y,T)}$ (i.e., proclaims it to be likely true) if and only if $\pi_{(\mathbb{P},x,y,T)}$ passes all these checks; the security analysis guarantees that this verdict is correct with probability $1 - \mathsf{error}$.

B   Algebraic Definition of General Programs as Zero Locus of Low-Degree Polynomial System

Our goal here is to explain how SCI converts programs into succinct algebraic CSP (ACSP) instances. For concreteness this is described for the TinyRAM machine specification [16]—a simple random access machine (RAM) with 16 registers and 16-bit size words that includes opcodes for logical operations, integer arithmetic, conditional jumps and random access memory instructions; the same techniques could be adapted to other machine specifications.

Algebra Preliminaries. Fix a basis $\beta_0, \ldots, \beta_{63}$ for $\mathbb{F}_{2^{64}}$ over $\mathbb{F}_2$ generated by an irreducible polynomial $h(X)$. Any sequence of w bits $a_0, \ldots, a_{w-1}$ can be naturally mapped to the field element $\sum_{i=0}^{w-1} a_i \beta_i$ as long as $w < 64$ and vice versa, field elements can be converted to sequences of bits; we assume this natural mapping and in particular will often identify the a 16-bit sequence $(a_0, \ldots, a_{15})$ with the field element $\sum_{i=0}^{15} a_i \beta_i$.

Overview of Reduction. The reduction from RAM programs to ACSPs has been described in detail in [13] and further improved in [30]; we follow this route. In particular, instructions that involve the random access memory are verified using affine routing networks as explained in [13] (cf. [30]), although SCI uses an affine graph in which the Beneš network [25] is embedded. Boundary constraints (such as the initial and final state of the machine) are enforced as explained in [13]. A remaining problem of great practical importance that remained from previous works has been how to reduce efficiently the transition function described by a program into a set of low-degree polynomials whose zero-locus corresponds to a valid evolution of the program's transition function. We describe this below. Our reduction works bottom up and has two main steps. (i) First, we define the input–output relation of each opcode as the zero-locus of a system of low-degree polynomials. (ii) In similar manner we define the transition function of the program as the zero-locus of a (larger) system of polynomials, one that uses the definitions of opcodes in terms of polynomials. The resulting set of polynomials is "glued" into a single large polynomial as described, e.g., in [23, Eq. (5.5)] and [13, Sect. 10].

B.1   Algebraic Definition of Opcodes

Our basic data-unit is called a *word*, in TinyRAM its size is 16 bits. The atoms of a computer program are *opcodes*; each opcode has a fixed amount of input and output words. For example, XOR receives two words $A = (a_0, \ldots, a_{15})$, $B = (b_0, \ldots, b_{15})$ and its output is a single word $C = (c_0, \ldots, c_{15})$ where $c_i = a_i \oplus b_i$ and $\oplus$ denotes exclusive-or; the AND opcode outputs $c_i = a_i \wedge b_i$, the ADD opcode performs integer addition, etc. (cf. [16] for details).

An opcode op with k inputs and ℓ outputs defines a *relation* R_{op} that contains all sequences of inputs and outputs that correspond to valid executions of op. Continuing with the examples above and using f to denote the flag,

$$R_{XOR} = \left\{ (a, b, c) \in \{0, 1\}^{3 \cdot 16} \mid a_i \oplus b_i \oplus c_i = 0 \right\}$$

$$R_{AND} = \left\{ (a, b, c) \in \{0, 1\}^{3 \cdot 16} \mid (a_i \wedge b_i) \oplus c_i = 0 \right\}$$

$$R_{ADD} = \left\{ (a, b, c) \in \{0, 1\}^{3 \cdot 16}, f \in \{0, 1\} \mid \sum_{i=0}^{15} a_i 2^i + \sum_{i=0}^{15} b_i 2^i - \left(f \cdot 2^{16} + \sum_{i=0}^{15} c_i 2^i \right) = 0 \right\}$$

An *algebraic opcode* is an opcode (as defined above) over an alphabet that is a finite field, i.e., $R_{op} \subset \mathbb{F}^{k+\ell}$. Any finite set is an *algebraic set*, meaning it can be described as the zero-locus of a system of polynomials, however, these polynomials may have large degree and/or large arithmetic complexity, which would harm the efficiency of our reduction. To reduce degree and arithmetic complexity we shall allow *auxiliary* variables and consider algebraic sets S over $\mathbb{F}^{k+\ell+m}$ such that R_{op} is the projection of S to the first $k+\ell$ variables. Formally, an *algebraic constraint system* A_{op} corresponding to an opcode op with k inputs and ℓ outputs is a set of polynomials $A_{op} \subset \mathbb{F}[X_1, \ldots, X_k, Y_1, \ldots, Y_\ell, Z_1, \ldots, Z_m]$ such that

$$R_{op} = \{x_1, \ldots, x_k, y_1, \ldots, y_\ell \mid \exists z_1, \ldots, z_m, A_{op}(x_1, \ldots, x_k, y_1, \ldots, y_\ell, z_1, \ldots, z_m) = 0\} \quad (1)$$

We call $X_1, \ldots, X_k$ the *input* variables, $Y_1, \ldots, Y_\ell$ the *output* variables and $Z_1, \ldots, Z_m$ are *auxiliary* variables. While any relation can be defined without any auxiliary variables, the degree of such A_{op} may be very large (e.g., in the case of AND, ADD), therefor, to minimize ACSP degree we shall often use auxiliary variables as shown in the following examples; explanations appear below but notice XOR uses no auxiliary variables and the AND opcode uses 48 of them. We defer the explanation of the more complicated ADD opcode to later on.

$$A_{XOR} = \{X_1 + X_2 + Y_1\} \quad (2)$$

$$A_{AND} = \left\{ X_1 + \sum_{i=0}^{15} Z_i \beta_i, X_2 + \sum_{i=0}^{15} Z_{16+i} \beta_i, Y_1 + \sum_{i=0}^{15} Z_{32+i} \beta_i \right\} \quad (3)$$

$$\bigcup \{Z_j \cdot (Z_j + 1) \mid j = 0, \ldots, 47\} \quad (4)$$

$$\bigcup \{(Z_i \cdot Z_{16+i}) + Z_{32+i} \mid i = 0, \ldots, 15\} \quad (5)$$

Recall that addition in $\mathbb{F}$ corresponds to exclusive-or, hence XOR has an algebraic constraint system with a single polynomial of degree 1 and no auxiliary variables, and it satisfies (1). To see that (3)–(5) form an algebraic constraint system for AND we argue as follows. Suppose $(x_1, x_2, y_1, z_0, \ldots, z_{47})$ belongs to the zero-locus of A_{AND}, i.e., all polynomials in A_{AND} vanish on this input. Then by (4) we have $z_j \in \{0,1\}$ for $j = 0, \ldots, 47$. By (3) we see that $z_{32+i} = z_i \wedge z_{16+i}$ for $i = 0, \ldots, 15$. Finally, by (3) we see that x_1 "packs" $z_0, \ldots, z_{15}$ into a single field elements, meaning x_1 is the field element whose representation in the basis $\beta_0, \ldots, \beta_{63}$ is the sequence $z_0, \ldots, z_{15}, 0, 0, \ldots, 0$ and similarly x_2 "packs" $z_{16}, \ldots, z_{31}$ and y_1 "packs" $z_{32}, \ldots, z_{47}$. Therefore, y_1 is the bitwise and of x_1 and x_2, as required by (1).

The constraints of the ADD opcode correspond to the operation of a full binary adder and appear below (6)–(10). In what follows auxiliary variables $Z_0, \ldots, Z_{15}$ are used to "unpack" X_1, auxiliary variables $Z_{16}, \ldots, Z_{31}$ "unpack" X_2, auxiliary variables $Z_{32}, \ldots, Z_{47}$ are the carry bits and $Z_{48}, \ldots, Z_{63}$ "unpack" the output Y_1; the overflow flag is stored in Y_2. The constraint set (6) "unpacks" both inputs and the output using 16 auxiliary variables each as done in (3) above. The constraint set (7) checks that each auxiliary variable is boolean (as done in (4)) but now we have 16 additional auxiliary variables for the carry bits, reaching a total of 64 auxiliary variables. The set of constraints (8) checks that the carry bits $(Z_{32}, \ldots, Z_{47})$ are computed correctly. In (9) the output is checked to be equal to the exclusive-or of the relevant input and carry bits. Finally, in (10) we check that the least significant carry and output bits are correct, and that the most significant carry bit (Z_{47}) equals the overflow flag (Y_2).

$$A_{\mathsf{ADD}} = \left\{ X_1 + \sum_{i=0}^{15} Z_i \beta_i, X_2 + \sum_{i=0}^{15} Z_{16+i}\beta_i, Y_1 + \sum_{i=0}^{15} Z_{48+i}\beta_i \right\} \tag{6}$$

$$\bigcup \{Z_j \cdot (Z_i + 1) \mid j = 0, \ldots, 63\} \tag{7}$$

$$\bigcup \{Z_i Z_{16+i} + Z_i Z_{31+i} + Z_{16+i} Z_{31+i} + Z_{32+i} \mid i = 1, \ldots, 15\} \tag{8}$$

$$\bigcup \{Z_i + Z_{16+i} + Z_{32+i} + Z_{48+i} \mid i = 1, \ldots, 15\} \tag{9}$$

$$\bigcup \{Z_0 \cdot Z_{16} + Z_{32}, Z_0 + Z_{16} + Z_{48}, Z_{63} + Y_2\} \tag{10}$$

Complexity of other Opcodes. The opcodes described above, applied to w-bit registers, require $O(w)$ constraints and auxiliary variables (R_{XOR} requires $O(1)$ constraints and auxiliary variables). All other opcodes of the TinyRAM assembly specification [16] can be implemented with $O(w)$ complexity. For most opcodes this can be verified by inspection. For integer multiplication—i.e., to prove that

$$\left(\sum_{i=0}^{w-1} a_i 2^i \right) \cdot \left(\sum_{i=0}^{w-1} b_i 2^i \right) = \sum_{i=0}^{2w-2} c_i 2^i, \quad a_i, b_i, c_i \in \{0,1\}$$

we fix a generator g for the multiplicative group of $\mathbb{F}$ (the order of g is $2^{63} - 1$ for our choice of field) and then apply repeated squaring to verify that

$$\left(g^{\left(\sum_i a_i 2^i\right)}\right)^{\left(\sum_i b_i 2^i\right)} = g^{\left(\sum_i c_i 2^i\right)}$$

Inspection reveals this solution scales asymptotically like $O(w)$ and for small values, R_{MUL} is twice as costly as R_{ADD} in terms of number of constraints and auxiliary variables.

B.2   Program Flow via Multi-linear Lagrange Polynomials

A program $\mathbb{P}$ of length s is a sequence of instructions $\mathsf{I}_0, \ldots, \mathsf{I}_{s-1}$, each instruction contains an opcode and a list of k inputs and ℓ outputs, where k and ℓ should match the number of inputs and outputs consumed and produced by the opcode, respectively. An input is either a constant (also known as immediate) or a register location and outputs are invariably register locations. (Instructions related to random access memory are dealt with separately, below; until then we assume our programs do not access it and use only the 16 registers.) Each instruction also points to the next instruction in the program; by default I_j points to I_{j+1} but certain instructions (jumps and conditional jumps) may point to a different instruction, and the pointer may further depend on the value of certain registers. The *program counter* (PC) is a special register that contains the number of the current instruction, and thus takes values in $\{0, \ldots, s-1\}$.

A *machine state* is a pair $S = (\mathbf{PC}, \mathbf{R})$ where $\mathbf{PC}$ holds the value of the program counter and $\mathbf{R}$ contains the values of all registers. The program $\mathbb{P}$ induces a natural relation $R_{\mathbb{P}}$ that contains all pairs $(S = (\mathbf{PC}, \mathbf{R}), S' = (\mathbf{PC}', \mathbf{R}'))$ of machine states such that a single cycle of the machine in state S (with program counter being $\mathbf{PC}$ and registers holding values $\mathbf{R}$) results in state S'. As done for opcodes in (1), our purpose in this subsection is to define a system of constraints, denoted $A_{\mathbb{P}}$, that defines $R_{\mathbb{P}}$ as its zero-locus, projected onto its first few variables. Formally, let $\mathbf{PC}, \mathbf{PC}', \mathbf{R}, \mathbf{R}'$ denote variables ranging over $\mathbb{F}$, and recall x, y, z denote variables for opcode inputs, outputs and auxiliary variables, respectively. Then

$$R_{\mathbb{P}} = \left\{ \left((\mathbf{PC}, \mathbf{R}), (\mathbf{PC}', \mathbf{R}')\right) \mid \exists x, y, z \, A_{\mathbb{P}}\left(\mathbf{PC}, \mathbf{R}, \mathbf{PC}', \mathbf{R}', x, y, z\right) = 0 \right\} \quad (11)$$

In words, $A_{\mathbb{P}}$ is a set of polynomials whose zero-locus, projected to $\mathbf{PC}, \mathbf{R}, \mathbf{PC}', \mathbf{R}'$, equals the "program evolution" relation $R_{\mathbb{P}}$.

To minimize degree complexity, the program counter value is recorded via $r = \lceil \log s \rceil$ many variables, denoted $\mathsf{PC}_1, \ldots, \mathsf{PC}_r$, each ranging over $\{0, 1\}$. For $\alpha \in \{0, 1\}^r$ let

$$L_\alpha(\mathsf{PC}_1, \ldots, \mathsf{PC}_r) = \prod_{i=1}^{r} (\mathsf{PC}_i + \alpha_i + 1)$$

be the Lagrange multi-linear polynomial that evaluates to 1 on α and evaluates to 0 on $\{0,1\}^r \setminus \{\alpha\}$. We multiply the polynomials in the algebraic constraint system appearing in the ith instruction by $L_{\bar{i}}(\mathsf{PC}_1, \ldots, \mathsf{PC}_r)$ where $\bar{i} \in \{0,1\}^r$ is the binary representation of i. Informally, this has the effect of applying the set of constraints A_{op} only when the PC points to an instruction that contains op. Formally, for each opcode op appearing in the program $\mathbb{P}$, let $I_{\mathsf{op} \in \mathbb{P}} \subseteq \{0, \ldots, s-1\}$ be the set of program instructions in which op is executed. Then define

$$\hat{A}^{\mathsf{op} \in \mathbb{P}} = \left\{ P \cdot \sum_{i \in I_{\mathsf{op}}} L_{\bar{i}}(\mathsf{PC}_1, \ldots, \mathsf{PC}_r) \mid P \in A_{\mathsf{op}} \right\} \tag{12}$$

Inputs and outputs to an opcode are checked in a similar way. In particular, let $\mathsf{i}_{i,1}, \ldots, \mathsf{i}_{i,k_i}$ denote the indices of the registers that are the inputs of the opcode in instruction i and let $\mathsf{o}_{i,1}, \ldots, \mathsf{o}_{i,\ell_i}$ be the indices of output registers of that instruction, then we define

$$\hat{A}_i^{\mathsf{i/o}} = \left\{ (X_j - \mathsf{R}_{\mathsf{i}_{i,j}}) \cdot L_{\bar{i}}(\mathsf{PC}_1, \ldots, \mathsf{PC}_r) \mid j = 1, \ldots, k_i \right\} \tag{13}$$
$$\bigcup \left\{ (Y_j - \mathsf{R}'_{\mathsf{o}_{i,j}}) \cdot L_{\bar{i}}(\mathsf{PC}_1, \ldots, \mathsf{PC}_r) \mid j = 1, \ldots, \ell_i \right\}$$
$$\bigcup \left\{ (\mathsf{R}_j - \mathsf{R}'_j) \cdot L_{\bar{i}}(\mathsf{PC}_1, \ldots, \mathsf{PC}_r) \mid j \text{ is not an output register of instruction } i \right\}$$

In similar fashion, updating the program counter during the ith instruction is defined using a set of polynomials whose zero locus corresponds to the correct update of PC value. Typically, this modification simply increments the value of the PC by 1, and this can be done by multiplying each polynomial in (6–10) by $L_{\bar{i}}(\mathsf{PC}_1, \ldots, \mathsf{PC}_r)$. Let $\hat{A}_i^{\mathsf{pc}}$ denote the corresponding set of polynomials. The final set $A_{\mathbb{P}}$ that defines the "program evolution" relation $R_{\mathbb{P}}$ is

$$A_{\mathbb{P}} \triangleq \left\{ \hat{A}_{\mathsf{op} \in \mathbb{P}} \mid \mathsf{op} \text{ appears in } \mathbb{P} \right\} \bigcup \left\{ \hat{A}_i^{\mathsf{i/o}} \mid i = 0, \ldots, s-1 \right\} \tag{14}$$
$$\bigcup \left\{ \hat{A}_i^{\mathsf{pc}} \mid i = 0, \ldots, s-1 \right\}$$

and the discussion above shows that its zero locus $A_{\mathbb{P}}$, projected to $\mathsf{PC}, \mathsf{R}, \mathsf{PC}', \mathsf{R}'$, indeed equals $R_{\mathbb{P}}$.

C   Two Programs Computing Subset-Sum

Code 1 shows a high-level description of the exhaustive subset-sum program, and Code 2 gives an equivalent TinyRAM hand-optimized implementation (cf. Appendix D for discussion of machine compiled assembly). In Code 1, the variable k is treated as a binary vector that iterates over all the possible combinations of the inputs. The inputs that correspond to each combination are summed up by inspecting whether the least significant bit (LSB) of k is 1, and then shifting k rightward. Code 2 uses the AND,CMPE,SHR TinyRAM instructions for these inspections and shifts. It should be noted that the instruction set that is needed for Code 2 is uncostly, in particular the cost of the DIV instruction would have been about twice higher than SHR in terms of the number of field elements that the prover commits to in a time step.

The total number of time steps T of the ACSP for Code 2 is sufficiently large if the inequality $2^n \cdot (9n + 7) < T$ holds, where n is the size of the input array. With 16-bit TinyRAM architecture, $n \leq 16$ is also required, unless extra logic is added to Code 2. In this inequality, the term $9n$ can be inferred by amortizing the number of TinyRAM instructions that are executed when the LSB of k is either 0 or 1. For example, $T = 2^{20}$ is sufficient for $n = 13$ inputs. For a further demonstration of the dependency between T and n, see Fig. 2.

The TinyRAM architecture relies on 16 or less registers, in particular Code 2 needs 5 registers in total. This helps with keeping the complexity low, as it implies that a relatively small number of field elements are required per time step. However, this also means that we do not have enough registers to store the entire input array. Since it is preferable to avoid the poly-logarithmic blowup of programs with memory, Code 2 employs a special "read-only memory" (ROM) instruction. The ROM instruction takes a single operand, treats it as an index $J \leq n$, and returns the corredponding array$[J]$ input value. The algebraic constraints of the ROM instruction consist of unpacking the bits of J and using a selector polynomial to force the prover to use the predefined array$[J]$ field element. For example, with $n = 8$, the ROM instruction can be implemented as

$$\bigcup_{k=0}^{2} \{b_k(b_k + 1)\} \bigcup_{k=0}^{2} \{J + \sum_{k=0}^{2} b_k x^k, \sum_{\alpha,\beta,\gamma \in \{0,1\}} (b_0 + \alpha)(b_1 + \beta)(b_2 + \gamma)(R + C_{\alpha,\beta,\gamma})\},$$

where R is the returned operand and $C_{\alpha,\beta,\gamma}$ are the array input values that the ACSP instance specifies. Thus, the degree of the ROM constraints is bounded by $\lceil \log n \rceil + 1$, and overall the ROM instruction is far less complex than deploying the full read/write memory construction.

Code 3 is a subset-sum program that computes all the partial sums of half of the input numbers, as well as the other half, and then does a linear scan to look for two partial sums that add up to the target value [50]. The partial sums are first stored in memory in a sorted order, which can be done in $O(n)$ time due to the following observation: given a sorted list $S_1, S_2, \ldots, S_{2^k}$ of all the possible sums that can be produced from combinations of certain k numbers, and another number m, the sorted list $S_1 + m, S_2 + m, \ldots, S_{2^k} + m$ can be merged into $S_1, S_2, \ldots, S_{2^k}$ to obtain one sorted list of size 2^{k+1}, in linear time. Hence, Code 3 needs to store $O(\sqrt{2^n})$ elements in memory, where n is the size of the input array.

Code 4 gives a hand-optimized TinyRAM implementation of this high-level pseudocode, in which the dependecy between n and the total number of time steps T is $n \approx 2(T - 7)$. Section D discusses the machine compiled code for the same program. As can be seen in Fig. 2, Code 4 can thus cope with greater values of n than Code 2, even after the poly-logarithmic blowup in complexity that is due to memory handling is taken into account.

Notice that unlike the high-level description in Code 3, the Code 4 implementation that we benchmark actually outputs a bit-string of the correct combination, if one exists (Code 5 and Code 6 do this as well). This extra work is done for a fair comparison with Code 2, that does this "for free". However, since subset-sum is an NP-complete problem, it makes sense to generate the PCP on unsatisfiable instances. Thus, this extra work can be regarded as unnecessary in this context.

Code 1. Pseudocode of the exhaustive search subset-sum program

---

input: $n, \text{array}[n], \text{target}$

1: **for** $k = 1$ **to** $2^n - 1$ **do** $\triangleright$ k loops over all $\{0,1\}^n \setminus \{0^n\}$ combinations
2:     $curr \leftarrow k,\ idx \leftarrow 0,\ sum \leftarrow 0$
3:     **while** $curr \neq 0$ **do**
4:        **if** $1 = (curr$ **bitwise-and** $1)$ **then** $\triangleright$ LSB of $curr$ is 1?
5:           $sum \leftarrow sum + \text{array}[idx]$
6:        **end if**
7:        $curr \leftarrow curr/2,\ idx \leftarrow idx + 1$
8:     **end while**
9:     **if** $sum = \text{target}$ **then**
10:       **return** k
11:     **end if**
12: **end for**
13: **return** 0

Code 2. TinyRAM assembly code of the exhaustive search subset-sum program

---

1: **MOV** r0, 1	9: **CJMP** Line#12	17: **CMPE** r1, target
2: **CMPE** r0, 2^n	10: **ROM** r4, r3	18: **CJMP** Line#22
3: **CJMP** Line#21	11: **ADD** r1, r1, r4	19: **ADD** r0, r0, 1
4: **MOV** r1, 0	12: **SHR** r2, r2, 1	20: **JMP** Line#2
5: **MOV** r2, r0	13: **CMPE** r2, 0	21: **MOV** r0, 0
6: **MOV** r3, 0	14: **CJMP** Line#17	22: **ANSW** r0
7: **AND** r4, r2, 1	15: **ADD** r3, r3, 1	
8: **CMPE** r4, 0	16: **JMP** Line#7	

D   Compiling C Code to TinyRAM

Our TinyRAM compiler is implemented as a GCC back end, with support for some optimization techniques. Code 5 shows C source for the memory-based subset-sum program, and the corresponding compiled code is given as Code 6. As shown, Code 6 has 21 more instruction than the hand-written assembly of Code 4. Likewise, the running time of Code 6 is somewhat greater than that of Code 4, for example with $n = 14$ it takes 13582 time steps until Code 6 terminates, while Code 4 terminates in 11231 time steps.

Code 3.   Pseudocode of the memory-based subset-sum program

---

input: $n = 2h$, $\mathsf{array}[n]$, target
1: $H_1 \leftarrow \{\mathsf{array}[0], \mathsf{array}[1], \ldots, \mathsf{array}[h-1]\}$
2: $H_2 \leftarrow \{\mathsf{array}[h], \mathsf{array}[1], \ldots, \mathsf{array}[n-1]\}$
3: **for** $m \in \{1, 2\}$ **do**   ▷ sort each half
4:     let $A_{m,0}$ be an array of size 1 with $A_{m,0}[0] = 0$
5:     $i \leftarrow 0$
6:     **for** $x \in H_m$ **do**
7:         let $B_{m,i}$ be an array of size i and $C_{m,i}$ be an array of size $2i$
8:         **for** $k \in \{0, 1, 2, \ldots, 2^i - 1\}$ **do**
9:             $B_{m,i}[k] \leftarrow A_{m,i}[k] + x$
10:        **end for**
11:        $C_{m,i} \leftarrow \mathsf{merge}(A_{m,i}, B_{m,i})$ ▷ note: $A_{m,i}$ and $B_{m,i}$ are already sorted
12:        $A_{m,i+1} \leftarrow C_{m,i}$
13:        $i \leftarrow i + 1$
14:    **end for**
15: **end for**
16: $i \leftarrow 0$, $k \leftarrow 2^h - 1$
17: **while** True **do**   ▷ search for the target
18:    **if** $\mathsf{target} = A_{1,h}[i] + A_{2,h}[k]$ **then return** 1 **end if**
19:    **if** $\mathsf{target} > A_{1,h}[i] + A_{2,h}[k]$ **then**
20:        **if** $i = 2^h - 1$ **then return** 0 **end if**
21:        $i \leftarrow i + 1$
22:    **else**
23:        **if** $k = 0$ **then return** 0 **end if**
24:        $k \leftarrow k - 1$
25:    **end if**
26: **end while**

Code 4.   TinyRAM assembly code of the memory-based subset-sum program

input: $n = 2h$, array$[n]$, target, $\ell = 2^{h+1} - 2$
constants: INPADDR $= 2^{16} - 2^6$, ADDR1 $= 0$, ADDR2 $= 2^{14}$, OFFSET $= 2^{15}$
preprocess: store array$[n]$ at INPADDR

```
 1: MOV    r0, INPADDR      31: ADD    r6, r2, OFFSET    61: CJMP   Line#64
 2: MOV    r1, ADDR1        32: LOAD   r6, r6           62: MOV    r1, ADDR2
 3: MOV    r9, 0            33: XOR    r6, r6, r8       63: JMP    Line#3
 4: STOR   r9, r1           34: ADD    r2, r2, 1        64: MOV    r0, ADDR1 + ℓ
 5: ADD    r2, r1, OFFSET   35: JMP    Line#21          65: LOAD   r2, r0
 6: STOR   r9, r2           36: CMPE   r5, r2           66: LOAD   r3, r1
 7: MOV    r2, r1           37: CNJMP  Line#44          67: ADD    r4, r2, r3
 8: ADD    r4, r1, 1        38: LOAD   r6, r1           68: CMPE   r4, target
 9: MOV    r5, r4           39: STOR   r6, r4           69: CJMP   Line#L83
10: MOV    r8, 1            40: ADD    r6, r1, OFFSET   70: CMPG   r4, target
11: ADD    r9, h            41: LOAD   r6, r6           71: CJMP   Line#77
12: LOAD   r3, r0           42: ADD    r1, r1, 1        72: CMPE   r1, ADDR2 + ℓ
13: JMP    Line#44          43: JMP    Line#21          73: CJMP   Line#82
14: ADD    r0, r0, 1        44: LOAD   r6, r1           74: ADD    r1, r1, 1
15: CMPE   r9, r0           45: LOAD   r7, r2           75: LOAD   r3, r1
16: CJMP   Line#60          46: ADD    r7, r7, r3       76: JMP    Line#67
17: LOAD   r3, r0           47: CMPG   r6, r7           77: CMPE   r0, ADDR1
18: SHL    r8, r8, 1        48: CJMP   Line#54          78: CJMP   Line#82
19: MOV    r5, r4           49: STOR   r6, r4           79: SUB    r0, r0, 1
20: JMP    Line#44          50: ADD    r6, r1, OFFSET   80: LOAD   r2, r0
21: ADD    r7, r4, OFFSET   51: LOAD   r6, r6           81: JMP    Line#67
22: STOR   r6, r7           52: ADD    r1, r1, 1        82: ANSW   0
23: ADD    r4, r4, 1        53: JMP    Line#21          83: ADD    r2, r0, OFFSET
24: CMPE   r5, r1           54: STOR   r7, r4           84: LOAD   r2, r2
25: CNJMP  Line#36          55: ADD    r6, r2, OFFSET   85: ADD    r3, r1, OFFSET
26: CMPE   r5, r2           56: LOAD   r6, r6           86: LOAD   r3, r3
27: CJMP   Line#14          57: XOR    r6, r6, r8       87: SHL    r3, r3, H
28: LOAD   r6, r2           58: ADD    r2, r2, 1        88: XOR    r2, r2, r3
29: ADD    r6, r6, r3       59: JMP    Line#21          89: ANSW   r2
30: STOR   r6, r4           60: CMPA   r1, ADDR2
```

Code 5.   C source of the memory-based subset-sum program

```
#define N 7
#define TARGET 123

int input[2*N] = {10,20,30,40,50,60,70,-10,-20,-30,-40,-50,-60,70};
int arr[ 4 * ( (1 << (N+1)) - 1 ) ];

int main(void) {
    register int *inp = &input[0], *last_inp, *p1, *p2, *next, *next_backup, b;
    p1 = p2 = &arr[0]; //phase1: prepare arrays
    for(;;) { //prepare each half array
        next = next_backup = (p1+2);
        *p1 = *(p1+1) = 0; b = 1; last_inp = inp + N;
        for(;;) { //iterate over each input
            for(;;) { //merge
                if(p1 == next_backup) {
                    while(p2 < next_backup) {
                        *(next++) = *(p2++) + *inp;
                        *(next++) = *(p2++) ^ b;
                    }
                    break;
                }
                if(p2 == next_backup) {
                    while(p1 < next_backup) {
                        *(next++) = *(p1++);
                        *(next++) = *(p1++);
                    }
                    break;
                }
                if(*p1 > *p2 + *inp) {
                    *(next++) = *(p2++) + *inp;
                    *(next++) = *(p2++) ^ b;
                }
                else {
                    *(next++) = *(p1++);
                    *(next++) = *(p1++);
                }
            }
            if(++inp == last_inp) break;
            b = b << 1;
            next_backup = next;
        }
        if( p1 > &arr[0] + (1 << (N+2)) ) break;
        p1 = p2 = next;
    }
    p1 = &arr[ 2*((1 << (N+1)) - 1) - 2 ]; //phase2: search
    for(;;) {
        if(TARGET == *p1 + *p2)
            return *(p1+1) ^ (*(p2+1) << N);
        if(TARGET > *p1 + *p2) {
            if(p2 == &arr[0] + 4*((1 << (N+1))-1) - 2) break;
            p2 = p2 + 2;
        }
        else {
            if(p1 == &arr[0])   break;
            p1 = p1 - 2;
        }
    }
    return 0;
}
```

Code 6.   TinyRAM assembly code of the compiled subset-sum program

input: $n = 2h$, array$[n]$, target
preprocess: store array$[n]$ at address 0

1:	MOV	r9, 0	38:	LOAD	r2, r8	75:	SHL	r14, r14, 1
2:	MOV	r12, 28	39:	ADD	r8, r8, 2	76:	MOV	r13, r4
3:	MOV	r8, r12	40:	STOR	r2, r4	77:	JMP	Line#12
4:	ADD	r13, r8, r4	41:	ADD	r4, r4, 2	78:	CMPA	r8, 1052
5:	MOV	r4, r13	42:	CMPAE	r8, r13	79:	CJMP	Line#83
6:	MOV	r2, 0	43:	CNJMP	Line#34	80:	MOV	r12, r4
7:	ADD	r0, r8, 2	44:	JMP	Line#72	81:	MOV	r8, r4
8:	STOR	r2, r0	45:	LOAD	r2, r12	82:	JMP	Line#4
9:	STOR	r2, r8	46:	LOAD	r3, r9	83:	MOV	r4, 1044
10:	MOV	r14, 1	47:	ADD	r3, r2, r3	84:	LOAD	r3, r4
11:	ADD	r5, r9, 14	48:	LOAD	r2, r8	85:	LOAD	r2, r12
12:	CMPE	r8, r13	49:	CMPG	r2, r3	86:	ADD	r2, r3, r2
13:	CNJMP	Line#30	50:	CNJMP	Line#63	87:	CMPE	r2, target
14:	CMPAE	r12, r13	51:	LOAD	r3, r12	88:	CNJMP	Line#96
15:	CJMP	Line#72	52:	LOAD	r2, r9	89:	ADD	r0, r12, 2
16:	LOAD	r3 r12	53:	ADD	r2, r3, r2	90:	LOAD	r12, r0
17:	LOAD	r2, r9	54:	ADD	r12, r12, 2	91:	SHL	r12, r12, h
18:	ADD	r2, r3, r2	55:	STOR	r2, r4	92:	ADD	r0, r4, 2
19:	ADD	r12, r12, 2	56:	ADD	r4, r4, 2	93:	LOAD	r4, r0
20:	STOR	r2, r4	57:	LOAD	r2, r12	94:	XOR	r2, r12, r4
21:	ADD	r4, r4, 2	58:	XOR	r2, r14, r2	95:	JMP	Line#110
22:	LOAD	r2, r12	59:	ADD	r12, r12, 2	96:	LOAD	r3, r4
23:	XOR	r2, r14, r2	60:	STOR	r2, r4	97:	LOAD	r2, r12
24:	ADD	r12, r12, 2	61:	ADD	r4, r4, 2	98:	ADD	r2, r3, r2
25:	STOR	r2, r4	62:	JMP	Line#12	99:	CMPG	r2, target-1
26:	ADD	r4, r4, 2	63:	LOAD	r2, r8	100:	CJMP	Line#106
27:	CMPAE	r12, r13	64:	ADD	r8, r8, 2	101:	MOV	r2, 2064
28:	CNJMP	Line#16	65:	STOR	r2, r4	102:	CMPE	r12, r2
29:	JMP	Line#72	66:	ADD	r4, r4, 2	103:	CJMP	Line#109
30:	CMPE	r12, r13	67:	LOAD	r2, r8	104:	ADD	r12, r12, 4
31:	CNJMP	Line#45	68:	ADD	r8, r8, 2	105:	JMP	Line#84
32:	CMPAE	r8, r13	69:	STOR	r2, r4	106:	CMPE	r4, 28
33:	CJMP	Line#72	70:	ADD	r4, r4, 2	107:	CJMP	Line#109
34:	LOAD	r2, r8	71:	JMP	Line#12	108:	JMP	Line#84
35:	ADD	r8, r8, 2	72:	ADD	r9, r9, 2	109:	MOV	r2, 0
36:	STOR	r2, r4	73:	CMPE	r9, r5	110:	ANSW	r2
37:	ADD	r4, r4, 2	74:	CJMP	Line#78			

References

1. Arora, S., Lund, C., Motwani, R., Sudan, M., Szegedy, M.: Proof verification and the hardness of approximation problems. J. ACM **45**(3), 501–555 (1998). Preliminary version in FOCS 1992
2. Arora, S., Safra, S.: Probabilistic checking of proofs: a new characterization of NP. J. ACM **45**(1), 70–122 (1998). Preliminary version in FOCS 1992
3. Babai, L., Fortnow, L., Levin, L.A., Szegedy, M.: Checking computations in polylogarithmic time. In: Proceedings of the 23rd Annual ACM Symposium on Theory of Computing, pp. 21–32, STOC 1991(1991)
4. Babai, L., Fortnow, L., Lund, C.: Nondeterministic exponential time has two-prover interactive protocols. In: Proceedings of the 31st Annual Symposium on Foundations of Computer Science, pp. 16–25, SFCS 1990 (1990)
5. Babai, L., Moran, S.: Arthur-Merlin games: a randomized proof system, and a hierarchy of complexity class. J. Comput. Syst. Sci. **36**(2), 254–276 (1988)
6. Bellare, M., Fuchsbauer, G., Scafuro, A.: Nizks with an untrusted CRS: Security in the face of parameter subversion. Cryptology ePrint Archive, Report 2016/372 (2016). http://eprint.iacr.org/
7. Bellare, M., Goldreich, O.: On defining proofs of knowledge. In: Brickell, E.F. (ed.) CRYPTO 1992. LNCS, vol. 740, pp. 390–420. Springer, Heidelberg (1993). doi:10.1007/3-540-48071-4_28
8. Ben-Or, M., Goldwasser, S., Kilian, J., Wigderson, A.: Multi-prover interactive proofs: how to remove intractability assumptions. In: Proceedings of the 20th Annual ACM Symposium on Theory of Computing, pp. 113–131, STOC 1988 (1988)
9. Ben-Sasson, E., Ben-Tov, I., Gabizon, A., Riabzev, M.: Improved concrete efficiency and security analysis of Reed-Solomon PCPPS (2016). http://eccc.hpi-web.de/report/2016/073
10. Ben-Sasson, E., Chiesa, A., Gabizon, A., Riabzev, M., Spooner, N.: Short interactive oracle proofs with constant query complexity, via composition and sumcheck. Electronic Colloquium on Computational Complexity, p. tR16-046 (2016)
11. Ben-Sasson, E., Chiesa, A., Gabizon, A., Virza, M.: Quasi-Linear size zero knowledge from linear-algebraic PCPs. In: Kushilevitz, E., Malkin, T. (eds.) TCC 2016. LNCS, vol. 9563, pp. 33–64. Springer, Heidelberg (2016). doi:10.1007/978-3-662-49099-0_2
12. Ben-Sasson, E., Chiesa, A., Garman, C., Green, M., Miers, I., Tromer, E., Virza, M.: Zerocash: decentralized anonymous payments from Bitcoin. In: Proceedings of the 2014 IEEE Symposium on Security and Privacy, pp. 459–474, SP 2014 (2014)
13. Ben-Sasson, E., Chiesa, A., Genkin, D., Tromer, E.: Fast reductions from RAMs to delegatable succinct constraint satisfaction problems. In: Proceedings of the 4th Innovations in Theoretical Computer Science Conference, pp. 401–414, ITCS 2013 (2013)
14. Ben-Sasson, E., Chiesa, A., Genkin, D., Tromer, E.: On the concrete efficiency of probabilistically-checkable proofs. In: Proceedings of the 45th ACM Symposium on the Theory of Computing, pp. 585–594, STOC 2013 (2013)
15. Ben-Sasson, E., Chiesa, A., Genkin, D., Tromer, E., Virza, M.: SNARKs for C: verifying program executions succinctly and in zero knowledge. In: Canetti, R., Garay, J.A. (eds.) CRYPTO 2013. LNCS, vol. 8043, pp. 90–108. Springer, Heidelberg (2013). doi:10.1007/978-3-642-40084-1_6

16. Ben-Sasson, E., Chiesa, A., Genkin, D., Tromer, E., Virza, M.: TinyRAM Architecture Specification (2013). http://scipr-lab.org/tinyram
17. Ben-Sasson, E., Chiesa, A., Green, M., Tromer, E., Virza, M.: Secure sampling of public parameters for succinct zero knowledge proofs. In: 2015 IEEE Symposium on Security and Privacy, SP 2015, San Jose, 17–21 May 2015, pp. 287–304, (2015). http://dx.doi.org/10.1109/SP.2015.25
18. Ben-Sasson, E., Chiesa, A., Tromer, E., Virza, M.: Scalable zero knowledge via cycles of elliptic curves. In: Garay, J.A., Gennaro, R. (eds.) CRYPTO 2014. LNCS, vol. 8617, pp. 276–294. Springer, Heidelberg (2014). doi:10.1007/978-3-662-44381-1_16
19. Ben-Sasson, E., Chiesa, A., Tromer, E., Virza, M.: Succinct non-interactive zero knowledge for a von Neumann architecture. In: Proceedings of the 23rd USENIX Security Symposium, San Diego, 20–22 August 2014, pp. 781–796 (2014)
20. Ben-Sasson, E., Goldreich, O., Harsha, P., Sudan, M., Vadhan, S.: Short PCPs verifiable in polylogarithmic time. In: Proceedings of the 20th Annual IEEE Conference on Computational Complexity, pp. 120–134, CCC 2005 (2005)
21. Ben-Sasson, E., Goldreich, O., Harsha, P., Sudan, M., Vadhan, S.: Robust PCPs of proximity, shorter PCPs, and applications to coding. SIAM J. Comput. **36**(4), 889–974 (2006). Preliminary versions of this paper have appeared in Proceedings of the 36th ACM Symposium on Theory of Computing and in Electronic Colloquium on Computational Complexity
22. Ben-Sasson, E., Hamilis, M., Silberstein, M., Tromer, E.: Fast multiplication in binary fields on GPUS via register cache. In: Proceedings of the 2016 International Conference on Supercomputing, ICS 2016 (2016)
23. Ben-Sasson, E., Sudan, M.: Short PCPs with polylog query complexity. SIAM J. Comput. **38**(2), 551–607 (2008). Preliminary version appeared in STOC 2005
24. Ben-Sasson, E., Sudan, M., Vadhan, S., Wigderson, A.: Randomness-efficient low degree tests and short PCPs via epsilon-biased sets. In: Proceedings of the 35th Annual ACM Symposium on Theory of Computing, pp. 612–621, STOC 2003 (2003)
25. Beneš, V.E.: Mathematical Theory of Connecting Networks and Telephone Traffic. Academic Press, New York (1965). http://opac.inria.fr/record=b1083990
26. Bitansky, N., Canetti, R., Chiesa, A., Tromer, E.: From extractable collision resistance to succinct non-interactive arguments of knowledge, and back again. In: Proceedings of the 3rd Innovations in Theoretical Computer Science Conference, pp. 326–349, ITCS 2012 (2012)
27. Bitansky, N., Canetti, R., Chiesa, A., Tromer, E.: Recursive composition and bootstrapping for SNARKs and proof-carrying data. In: Proceedings of the 45th ACM Symposium on the Theory of Computing, pp. 111–120, STOC 2013 (2013)
28. Bitansky, N., Chiesa, A., Ishai, Y., Paneth, O., Ostrovsky, R.: Succinct non-interactive arguments via linear interactive proofs. In: Sahai, A. (ed.) TCC 2013. LNCS, vol. 7785, pp. 315–333. Springer, Heidelberg (2013). doi:10.1007/978-3-642-36594-2_18
29. Bootle, J., Cerulli, A., Chaidos, P., Groth, J., Petit, C.: Efficient zero-knowledge arguments for arithmetic circuits in the discrete log setting. In: Fischlin, M., Coron, J.-S. (eds.) EUROCRYPT 2016. LNCS, vol. 9666, pp. 327–357. Springer, Heidelberg (2016). doi:10.1007/978-3-662-49896-5_12
30. Chiesa, A., Zhu, Z.A.: Shorter arithmetization of nondeterministic computations. Theor. Comput. Sci. **600**, 107–131 (2015). http://www.sciencedirect.com/science/article/pii/S0304397515006647

31. Clos, C.: A study of non-blocking switching networks. Bell Syst. Tech. J. **32**(2), 406–424 (1953). http://dx.doi.org/10.1002/j.1538-7305.1953.tb01433.x

32. Cormode, G., Mitzenmacher, M., Thaler, J.: Practical verified computation with streaming interactive proofs. In: Proceedings of the 4th Symposium on Innovations in Theoretical Computer Science, pp. 90–112, ITCS 2012 (2012)

33. Cormode, G., Thaler, J., Yi, K.: Verifying computations with streaming interactive proofs. Proc. VLDB Endowment **5**(1), 25–36 (2011)

34. Dinur, I.: The PCP theorem by gap amplification. J. ACM **54**(3), 12 (2007)

35. Dinur, I., Reingold, O.: Assignment testers: towards a combinatorial proof of the PCP theorem. SIAM J. Comput. **36**(4), 975–1024 (2006). http://dx.doi.org/10.1137/S0097539705446962

36. Dwork, C., Feige, U., Kilian, J., Naor, M., Safra, M.: Low communication 2-prover zero-knowledge proofs for NP. In: Brickell, E.F. (ed.) CRYPTO 1992. LNCS, vol. 740, pp. 215–227. Springer, Heidelberg (1993). doi:10.1007/3-540-48071-4_15

37. Ben-Sasson, E., Chiesa, N.S.A.: Interactive oracle proofs. IACR Cryptology ePrint Archive 2016, 116 (2016). http://eprint.iacr.org/2016/116

38. Fiat, A., Shamir, A.: How to prove yourself: practical solutions to identification and signature problems. In: Odlyzko, A.M. (ed.) CRYPTO 1986. LNCS, vol. 263, pp. 186–194. Springer, Heidelberg (1987). doi:10.1007/3-540-47721-7_12

39. Gao, S., Mateer, T.: Additive fast fourier transforms over finite fields. IEEE Trans. Inf. Theor. **56**(12), 6265–6272 (2010). http://dx.doi.org/10.1109/TIT.2010.2079016

40. Gennaro, R., Gentry, C., Parno, B.: Non-interactive verifiable computing: outsourcing computation to untrusted workers. In: Rabin, T. (ed.) CRYPTO 2010. LNCS, vol. 6223, pp. 465–482. Springer, Heidelberg (2010). doi:10.1007/978-3-642-14623-7_25

41. Gennaro, R., Gentry, C., Parno, B., Raykova, M.: Quadratic span programs and succinct NIZKs without PCPs. In: Johansson, T., Nguyen, P.Q. (eds.) EUROCRYPT 2013. LNCS, vol. 7881, pp. 626–645. Springer, Heidelberg (2013). doi:10.1007/978-3-642-38348-9_37

42. Goldreich, O., Sudan, M.: Locally testable codes and PCPs of almost-linear length. J. ACM **53**, 558–655 (2006). Preliminary version in STOC 2002

43. Goldwasser, S., Kalai, Y.T., Rothblum, G.N.: Delegating computation: interactive proofs for muggles. In: Proceedings of the 40th Annual ACM Symposium on Theory of Computing, pp. 113–122, STOC 2008 (2008)

44. Goldwasser, S., Micali, S., Rackoff, C.: The knowledge complexity of interactive proof systems. SIAM J. Comput. **18**(1), 186–208 (1989). Preliminary version appeared in STOC 1985

45. Greenberg, A.: Zcash, an untraceable bitcoin alternative, launches in alpha (January 2016). Wired.com. Accessed 20 Jan 2016

46. Groth, J.: Short pairing-based non-interactive zero-knowledge arguments. In: Abe, M. (ed.) ASIACRYPT 2010. LNCS, vol. 6477, pp. 321–340. Springer, Heidelberg (2010). doi:10.1007/978-3-642-17373-8_19

47. Groth, J.: Efficient zero-knowledge arguments from two-tiered homomorphic commitments. In: Lee, D.H., Wang, X. (eds.) ASIACRYPT 2011. LNCS, vol. 7073, pp. 431–448. Springer, Heidelberg (2011). doi:10.1007/978-3-642-25385-0_23

48. Harsha, P., Sudan, M.: Small PCPs with low query complexity. Comput. Complex. **9**(3–4), 157–201 (2000). Preliminary version in STACS 1991

49. Håstad, J.: Some optimal inapproximability results. J. ACM **48**(4), 798–859 (2001)

50. Horowitz, E., Sahni, S.: Computing partitions with applications to the knapsack problem. J. ACM **21**(2), 277–292 (1974). http://doi.acm.org/10.1145/321812.321823

51. Ishai, Y., Kushilevitz, E., Ostrovsky, R.: Efficient arguments without short PCPs. In: Proceedings of the Twenty-Second Annual IEEE Conference on Computational Complexity, pp. 278–291, CCC 2007 (2007)

52. Ishai, Y., Kushilevitz, E., Ostrovsky, R., Sahai, A.: Zero-knowledge proofs from secure multiparty computation. SIAM J. Comput. **39**(3), 1121–1152 (2009)

53. Ishai, Y., Mahmoody, M., Sahai, A.: On efficient zero-knowledge PCPs. In: Cramer, R. (ed.) TCC 2012. LNCS, vol. 7194, pp. 151–168. Springer, Heidelberg (2012). doi:10.1007/978-3-642-28914-9_9

54. Ishai, Y., Mahmoody, M., Sahai, A., Xiao, D.: On zero-knowledge PCPs: Limitations, simplifications, and applications (2015). http://www.cs.virginia.edu/mohammad/files/papers/ZKPCPs-Full.pdf

55. Kilian, J.: A note on efficient zero-knowledge proofs and arguments. In: Proceedings of the 24th Annual ACM Symposium on Theory of Computing, pp. 723–732, STOC 1992 (1992)

56. Kilian, J., Petrank, E., Tardos, G.: Probabilistically checkable proofs with zero knowledge. In: Proceedings of the 29th Annual ACM Symposium on Theory of Computing, pp. 496–505, STOC 1997 (1997)

57. Kosba, A., Miller, A., Shi, E., Wen, Z., Papamanthou, C.: Hawk: The blockchain model of cryptography and privacy-preserving smart contracts. Cryptology ePrint Archive, Report 2015/675 (2015). http://eprint.iacr.org/

58. Lipmaa, H.: Progression-free sets and sublinear pairing-based non-interactive zero-knowledge arguments. In: Cramer, R. (ed.) TCC 2012. LNCS, vol. 7194, pp. 169–189. Springer, Heidelberg (2012). doi:10.1007/978-3-642-28914-9_10

59. Lund, C., Fortnow, L., Karloff, H., Nisan, N.: Algebraic methods for interactive proof systems. J. ACM **39**(4), 859–868 (1992). http://doi.acm.org/10.1145/146585.146605

60. Mahmoody, M., Xiao, D.: Languages with efficient zero-knowledge PCPs are in SZK. In: Sahai, A. (ed.) TCC 2013. LNCS, vol. 7785, pp. 297–314. Springer, Heidelberg (2013). doi:10.1007/978-3-642-36594-2_17

61. Micali, S.: Computationally sound proofs. SIAM J. Comput. **30**(4), 1253–1298 (2000). Preliminary version appeared in FOCS 1994

62. Mie, T.: Short PCPPs verifiable in polylogarithmic time with O(1) queries. Ann. Math. Artif. Intell. **56**, 313–338 (2009)

63. Moshkovitz, D., Raz, R.: Two-query PCP with subconstant error. J. ACM **57**, 1–29 (2008). Preliminary version appeared in FOCS 2008

64. Nakamoto, S.: Bitcoin: A peer-to-peer electronic cash system (May 2009). http://www.bitcoin.org/bitcoin.pdf

65. Parno, B., Gentry, C., Howell, J., Raykova, M.: Pinocchio: Nearly practical verifiable computation. In: Proceedings of the 34th IEEE Symposium on Security and Privacy, Oakland 2013, pp. 238–252 (2013)

66. Raz, R.: A parallel repetition theorem. In: Proceedings of the 27th Annual ACM Symposium on Theory of Computing, pp. 447–456, STOC 1995 (1995)

67. Reed, I.S., Solomon, G.: Polynomial codes over certain finite fields. J. Soc. Industr. Appl. Math. **8**(2), 300–304 (1960). http://dx.doi.org/10.1137/0108018

68. Seo, J.H.: Round-efficient sub-linear zero-knowledge arguments for linear algebra. In: Catalano, D., Fazio, N., Gennaro, R., Nicolosi, A. (eds.) PKC 2011. LNCS, vol. 6571, pp. 387–402. Springer, Heidelberg (2011). doi:10.1007/978-3-642-19379-8_24

69. Setty, S., Blumberg, A.J., Walfish, M.: Toward practical and unconditional verification of remote computations. In: Proceedings of the 13th USENIX Conference on Hot Topics in Operating Systems, p. 29, HotOS 2011 (2011)

70. Setty, S., Braun, B., Vu, V., Blumberg, A.J., Parno, B., Walfish, M.: Resolving the conflict between generality and plausibility in verified computation. In: Proceedings of the 8th EuoroSys Conference, pp. 71–84, EuroSys 2013 (2013)

71. Setty, S., McPherson, M., Blumberg, A.J., Walfish, M.: Making argument systems for outsourced computation practical (sometimes). In: Proceedings of the 2012 Network and Distributed System Security Symposium, NDSS 2012 (2012)

72. Setty, S., Vu, V., Panpalia, N., Braun, B., Blumberg, A.J., Walfish, M.: Taking proof-based verified computation a few steps closer to practicality. In: Proceedings of the 21st USENIX Security Symposium, pp. 253–268, Security 2012 (2012)

73. Thaler, J.: Time-optimal interactive proofs for circuit evaluation. In: Canetti, R., Garay, J.A. (eds.) CRYPTO 2013. LNCS, vol. 8043, pp. 71–89. Springer, Heidelberg (2013). doi:10.1007/978-3-642-40084-1_5

74. Valiant, P.: Incrementally verifiable computation or proofs of knowledge imply time/space efficiency. In: Canetti, R. (ed.) TCC 2008. LNCS, vol. 4948, pp. 1–18. Springer, Heidelberg (2008). doi:10.1007/978-3-540-78524-8_1

75. Vu, V., Setty, S., Blumberg, A.J., Walfish, M.: A hybrid architecture for interactive verifiable computation. In: Proceedings of the 34th IEEE Symposium on Security and Privacy, Oakland 2013, pp. 223–237 (2013)

76. Wahby, R.S., Setty, S.T.V., Ren, Z., Blumberg, A.J., Walfish, M.: Efficient RAM and control flow in verifiable outsourced computation. In: 22nd Annual Network and Distributed System Security Symposium, NDSS 2015, San Diego, February 8–11 2014 (2015)

77. Walfish, M., Blumberg, A.J.: Verifying computations without reexecuting them. Commun. ACM 58(2), 74–84 (2015). http://doi.acm.org/10.1145/2641562

Ad Hoc PSM Protocols: Secure Computation Without Coordination

Amos Beimel[1(✉)], Yuval Ishai[2,3], and Eyal Kushilevitz[2]

[1] Department of Computer Science, Ben Gurion University,
Beer Sheva, Israel
amos.beimel@gmail.com
[2] Department of Computer Science, Technion, Haifa, Israel
{yuvali,eyalk}@cs.technion.ac.il
[3] Department of Computer Science, UCLA, Los Angeles, USA

Abstract. We study the notion of *ad hoc secure computation*, recently introduced by Beimel et al. (ITCS 2016), in the context of the *Private Simultaneous Messages* (PSM) model of Feige et al. (STOC 2004). In ad hoc secure computation we have n parties that may potentially participate in a protocol but, at the actual time of execution, only k of them, whose identity is *not* known in advance, actually participate. This situation is particularly challenging in the PSM setting, where protocols are non-interactive (a single message from each participating party to a special output party) and where the parties rely on pre-distributed, correlated randomness (that in the ad-hoc setting will have to take into account all possible sets of participants).

We present several different constructions of ad hoc PSM protocols from standard PSM protocols. These constructions imply, in particular, that efficient information-theoretic ad hoc PSM protocols exist for NC^1 and different classes of log-space computation, and efficient computationally-secure ad hoc PSM protocols for polynomial-time computable functions can be based on a one-way function. As an application, we obtain an information-theoretic implementation of *order-revealing encryption* whose security holds for two messages.

We also consider the case where the actual number of participating parties t may be larger than the minimal k for which the protocol is designed to work. In this case, it is unavoidable that the output party learns the output corresponding to each subset of k out of the t participants. Therefore, a "best possible security" notion, requiring that this will be the *only* information that the output party learns, is needed. We present connections between this notion and the previously studied notion of *t-robust PSM* (also known as "non-interactive MPC"). We show that constructions in this setting for even simple functions (like AND or threshold) can be translated into non-trivial instances of program obfuscation (such as *point function obfuscation* and *fuzzy point function obfuscation*, respectively). We view these results as a negative indication that protocols with "best possible security" are impossible to realize efficiently in the information-theoretic setting or require strong assumptions in the computational setting.

© International Association for Cryptologic Research 2017
J.-S. Coron and J.B. Nielsen (Eds.): EUROCRYPT 2017, Part III, LNCS 10212, pp. 580–608, 2017.
DOI: 10.1007/978-3-319-56617-7_20

1   Introduction

The notion of *ad hoc secure computation* was recently put forward in [4]. In the ad-hoc secure computation problem, there are n parties that may potentially take part in a secure computation protocol. At the time that the protocol is executed, some k of these n parties actually participate in the execution. The goal is to design (efficient) protocols that can work for *every* set of k parties S, without knowing the set of participants in advance. As a concrete example, think of a voting application, where n parties are registered to the elections but only k of them (the identity of which becomes known only in real time) end up participating in the vote.

In most standard secure computation models, the ad-hoc nature of the protocol does not pose a significant challenge: the participating parties can interact with each other and use a standard general-purpose secure protocol to perform the computation. The problem is most challenging in situations where pre-processing or setup are required, or where interaction is limited. In the extreme, where non-interactive secure protocols are needed, the single message sent by each party P_i cannot depend on the messages of other parties, whose identities are not even known to P_i.

A simple model for non-interactive secure computation is the *Private Simultaneous Messages* (PSM) model of [14,17]. In this model, there are n parties $P_1, \ldots, P_n$ and a special party called the *referee*. Before the input is known, the parties are given correlated randomness[1] $(r_1, \ldots, r_n)$. In the online phase, each party P_i gets an input x_i and sends a single message m_i, depending on x_i and r_i, to the referee. Based on the n received messages, the referee should be able to compute the value of a pre-determined function f on the input $x = (x_1, \ldots, x_n)$, namely $f(x)$. The security requires that the referee learns no additional information about x. It is known that PSM protocols exist for every finite function f [14] and *efficient* PSM protocols exist for every function in NC^1 and for classes of functions defined by different types of (polynomial-size) branching programs [14,17]. In a computational setting, efficient PSM protocols for all polynomial-time computable functions can be based on one-way functions by using Yao's garbled circuit construction [14,20]. The simplicity of the PSM model makes it an attractive candidate for a complexity theoretic study (see, e.g., [2]) and its limited interaction pattern makes it useful in applications, such as minimizing the round-complexity of secure protocols in the standard point-to-point model (see, e.g., [18]).

In this paper, we study the ad hoc version of the PSM model, where the referee receives messages from a subset of size k out of the n parties. We assume that the parameter k is known in advance, but the parties are not aware of the identity of other participating parties. Before describing our results in detail (in Sect. 1.1), we discuss some possible variants of the question. First, the original

---

[1] Both in the original PSM model and in its ad-hoc variant, it suffices for the parties to share a source of *common* randomness that is unknown to the referee. The use of more general correlated randomness can help reduce the randomness complexity.

PSM model was mainly studied in the information-theoretic security setting. In this work, we consider both the information-theoretic variant and the computational variant. In fact, the computational version of ad hoc PSM was first considered in [4], where it was shown that such protocols can be constructed based on the existence of a weak form of *Multi-Input Functional Encryption* (MIFE) [15], a primitive whose general realization is essentially equivalent to the existence of general indistinguishability obfuscation.

Second, the problem of ad hoc PSM is significantly different in the case where we are guaranteed that exactly k parties will send messages vs. the case where possibly more than k parties may participate. Most of the time, we will assume that only k parties send messages and that this guarantee is assured by some other mechanism, such as a public bulletin board reporting the current participant count, or an anonymous communication medium that hides all information except the fact that a message has been sent. On the other hand, in a setting where a set S of more than k parties may send messages in the protocol, the referee unavoidably may compute the function f on any subset $S' \subset S$ of size k and learn the value $f(x_{S'})$. Therefore, in this case, our security notion is a "best possible security" definition, requiring that this will be the *only* information that the referee learns in the protocol. This can be formalized either using a strong simulation-based definition or a weaker indistinguishability-based definition.

Finally, it will be convenient and, in fact, very natural in the ad-hoc setting to think of f as a symmetric function. Most of our results do not rely on this and can be extended to even allow the computed function to depend on the set of participants S, i.e. to output $f_S(x_S)$.

1.1 Our Results

Let us start by demonstrating our results using a concrete task of computing the SUM function. In this case, each party P_i is given an input $x_i \in \mathbb{Z}_m$ and the goal is to compute their sum $\sum_{i \in [n]} x_i$ (all additions in this example are mod m). A standard PSM protocol for SUM works by giving the parties randomness $r_1, \ldots, r_n \in_R \mathbb{Z}_m$ subject to the constraint that $\sum_{i \in [n]} r_i = 0$. Then, each party P_i, sends a message $m_i = x_i + r_i$ to the referee who outputs $\sum_{i \in [n]} m_i = \sum_{i \in [n]} x_i$, as needed. Moreover, due to the choice of the r_i's, no additional information about the inputs is revealed to the referee.

In the ad-hoc version of the problem, we wish to compute the SUM of any set S of k parties that may send messages in the protocol. One option is to prepare, for each potential set S of size k, independent randomness $r_1^S, \ldots, r_k^S$ that is random subject to their sum being 0 and proceed by P_i sending a message (using the corresponding randomness r_j^S), for each set S to which it belongs. While this solution works, its randomness complexity and communication complexity are proportional to $\binom{n}{k}$, which is much more than what we are shooting for. Instead, we describe an *efficient* solution for this problem.

In our ad hoc PSM protocol, the randomness consists of values $r_1, \ldots, r_n \in_R \mathbb{Z}_m$ subject to the constraint that $\sum_{i \in [n]} r_i = 0$, as in the original PSM protocol.

In addition, we produce shares $\{r_{j,i}\}_{i\in[n]}$, for each r_j, using a k-out-of-n secret sharing scheme (e.g., Shamir). The randomness given to each P_i consists of its r_i and its shares of all other random values; that is $\{r_{j,i}\}_{j\in[n]\setminus\{i\}}$. Then, each party P_i that participates in the protocol (i.e., $i \in S$) sends as its message the value $m_i = x_i + r_i$, as well as all its shares. The referee sums up all the m_i's that it got from the k participants, as well as all the values r_j, for $j \notin S$, that it can reconstruct from the k shares that it received for each such r_j, to get $\sum_{i\in S}(x_i + r_i) + \sum_{i\notin S} r_i = \sum_{i\in S} x_i$, as needed. In terms of security, each r_i, for $i \in S$, remains hidden as the referee receives exactly $k-1$ shares for these random elements. In fact, the view of the referee can be simulated from its view in the original PSM, where parties P_j, for $j \notin S$, have input $x_j = 0$. Also note that if, say, $k+1$ parties send messages then the referee learns all inputs. However, for the SUM function, the best possible security definition (that allows the referee to learn the output on all subsets of size k) allows to recover all $k+1$ inputs in most cases (at least when $\gcd(k,m) = 1$).

Next, we describe in some detail our main results. The first question that we ask (in Sect. 3) is whether the existence of a standard k-party PSM computing a function f guarantees the existence of a k-out-of-n ad hoc PSM protocol for f. We first prove the existence of an inefficient transformation of this kind but that has an overhead of $\binom{n}{k}$. While this transformation may be useful for the case where the number of parties is small (and also proves the existence of an ad hoc PSM protocol for every function f), our aim is to get an efficient transformation (i.e., with poly(n) overhead). We next present such a transformation that works whenever f is symmetric, and is efficient whenever k is small (essentially, $2^{O(k)} \log n$). When $k = O(1)$, the overhead is as small as $O(\log n)$ (this construction relies on perfect hash families, and its complexity depends on the size of such families of functions from $[n]$ to $[k]$). The fact that the complexity of each party grows only logarithmically with the number of parties will be useful for the application discussed in Sect. 6.

Then, in Sect. 4, we ask whether an ad hoc PSM protocol for f can be constructed more efficiently based on a standard PSM protocol for a *related* (n-argument) function g. We prove that this is indeed possible, while incurring only $O(n)$ overhead over the complexity of the protocol for g. Moreover, the computational complexity of g is closely related to that of f in computational models for which efficient PSM protocols are known (e.g., if f is in NC^1 then so is g, and if f has a polynomial-size branching program then so does g). This implies efficient ad hoc PSM protocols for branching programs in the information-theoretic setting and for circuits in the computational setting, where the latter relies on the existence of a one-way function. In addition, in Sect. 5, we present an explicit ad hoc PSM for the equality function.

In Sect. 6, we show an interesting application of ad hoc PSM protocols. Specifically, we show how to construct an order revealing encryption (ORE) from an ad hoc PSM protocol for the "greater-than" function. An order revealing encryption, presented in [10] as a generalization of order preserving encryption [1], is a private-key encryption that enables computing the order between two messages (that is, checking if $m_1 < m_2$, $m_1 = m_2$, or $m_1 > m_2$), given their encryptions (without knowing the private key), but does not disclose any additional information. We construct information-theoretically secure order revealing encryption that is secure as long as only two messages are encrypted. In our construction, we use an ad hoc PSM protocol constructed in Sect. 3 with $n = 2^\lambda$ parties (where λ is the security parameter), relying on the fact that the complexity of each party in the protocol from Sect. 3 only grows logarithmically with the number of parties. We also give a solution for a bigger number of messages, but with a weaker security guarantee.

The above results refer to the case where exactly k parties send messages in the protocol. We next examine (in Sect. 7) the case where more than k (but up to some threshold t) parties may send messages. In this case, as discussed above, one needs to settle for a "best possible security" definition. We extend the above transformation from standard PSM to ad hoc PSM to this case, showing that it is possible to construct a PSM protocol for f with best possible security from a so-called "t-robust PSM" protocol [5] for a related function g', incurring only $O(n)$ overhead. A t-robust PSM is a protocol where up to t parties may collaborate with the referee in trying to learn information about the inputs of other parties. In this case, it is always possible for the adversary to get the output of f on many inputs, by replacing the messages of the collaborating parties with messages that correspond to other inputs. Therefore, for such protocols also one may only hope for a "best possible security". Our results connect these two best possible security settings (in both directions). It should be noted, however, that efficient t-robust PSM protocols in the information-theoretic setting are currently known only for limited families of functions, and limited values of t [5].

In Sect. 8, we examine the possibility of constructing efficient PSM protocols with best possible security, in the *computational* setting. (The naive transformation of Sect. 3 shows that it is possible to get best possible security even in the information theoretic case but without efficiency.) The two-way connection with t-robust PSM already implies a two-way connection between this problem and general-purpose obfuscation. However, it is not clear a-priori that the connection has relevance in the case of *simple* functions. We give evidence that efficient ad-hoc PSM protocols with best possible security are difficult to design even for very simple functions. For instance, a protocol for a threshold function implies a construction of *fuzzy point function obfuscation* [7], a primitive whose only known constructions rely on multilinear maps. In fact, even a protocol for the AND function, gives a construction of *point function obfuscation*.

2   The Setting

We consider a network of n parties, denoted $P_1, \ldots, P_n$, and a referee; Each party P_i holds an input x_i, and the parties hold correlated random strings $r_1, \ldots, r_n$. We want to execute a protocol, where only a subset of the parties $S \subseteq \{P_1, \ldots, P_n\}$ participates in the protocol, each one of them sends a single message to the referee. If exactly k parties participate and send messages then, based on these k messages, the referee should be able to compute the value $f(x_S)$ but learn no other information about x_S, where $x_S = (x_i)_{i \in S}$. The subset S of participating parties is selected in an ad hoc manner and, in particular, the participating parties are not aware of each other. This is the main source of difficulty in this model. The referee itself necessarily learns the set of participants S (as it receives messages directly from the participants; avoiding this would require the use of anonymous communication). We often assume that f is symmetric; while this is a natural assumption in such a setting, most of our constructions can handle a much more general requirement, where the computed function itself may also depend on the set of participants S (i.e., the output is $f_S(x_S)$). We call the above model ad hoc PSM. We formalize this notion below starting with information-theoretic secure protocols.

Definition 2.1 (Ad hoc PSM: Syntax and correctness). *Let* $\mathcal{X}, \mathcal{R}_1, \ldots,$ $\mathcal{R}_n, \mathcal{M}$ *and* Ω *be finite domains. A k-out-of-n ad hoc PSM for a function $f :$ $\mathcal{X}^k \to \Omega$ is a triplet $\Pi = (\text{GEN}, \text{ENC}, \text{DEC})$ where*

- $\text{GEN}()$ *is a randomized function with output in* $\mathcal{R}_1 \times \cdots \times \mathcal{R}_n$,
- ENC *is an n-tuple of deterministic functions* $(\text{ENC}_1, \ldots, \text{ENC}_n)$, *where* $\text{ENC}_i :$ $\mathcal{X} \times \mathcal{R}_i \to \mathcal{M}$,
- $\text{DEC} : \binom{[n]}{k} \times \mathcal{M}^k \to \Omega$ *is a deterministic function satisfying the following correctness requirement: for any $S = \{i_1, \ldots, i_k\} \subseteq [n]$ and $x_S = (x_{i_1}, \ldots, x_{i_k}) \in$ $\mathcal{X}^k$,*

$$\Pr \left[\begin{array}{l} r = (r_1, \ldots, r_n) \leftarrow \text{GEN}() : \\ \text{DEC}(S, \text{ENC}_{i_1}(x_{i_1}, r_{i_1}), \ldots, \text{ENC}_{i_k}(x_{i_k}, r_{i_k})) = f(x_S) \end{array} \right] = 1.$$

The randomness complexity *of Π is the maximum of $\log |\mathcal{R}_1|, \ldots, \log |\mathcal{R}_n|$. The* communication complexity *of Π is $\log |\mathcal{M}|$.*

Definition 2.2 (Ad hoc PSM: Perfect and statistical security). *We say that an ad hoc PSM protocol Π for f is k-secure if:*

- *For every set $S \in \binom{[n]}{k}$, given the messages of S, the referee does not get any additional information beyond the value of $f(x_S)$. Formally, there exists a randomized function SIM (a "simulator") such that, for every $S \in \binom{[n]}{k}$ and for every $x_S \in \mathcal{X}^k$, we have $\text{SIM}(S, f(x_S)) \equiv M_S$, where M_S are the messages defined by $R \leftarrow \text{GEN}()$ and $M_S = (\text{ENC}_i(x_i, R_i))_{i \in S}$.*
- *For every $k' < k$ and every set $S' \in \binom{[n]}{k'}$, given the messages of S', the referee does not get any information on the input $x_{S'}$. Formally, there exists a randomized function SIM such that, for every $k' < k$, every $S' \in \binom{[n]}{k'}$ and every $x_{S'} \in \mathcal{X}^{k'}$, we have $\text{SIM}(S') \equiv M_{S'}$, where $M_{S'}$ is defined as above.*

We say that an ad hoc PSM protocol Π for f is (k, ϵ)-(statistically) secure if there exists a randomized function SIM such that for every $S \in \binom{[n]}{k}$ and for every $x_S \in \mathcal{X}^k$,

$$\text{dist}(\text{SIM}(S, f(x_S)), M_S) \leq \epsilon.$$

Similarly, for every $S' \in \binom{[n]}{k'}$, we have $\text{dist}(\text{SIM}(S'), M_{S'}) \leq \epsilon$.

Example 2.3. We next describe a very simple ad hoc PSM protocol for computing the difference of inputs for $k = 2$ parties; that is, for $i < j$ we want to compute $f(x_i, x_j) = x_i - x_j$ (over some Abelian group G). The randomness generation chooses a random element $r \in_R G$ and gives it to each party. The message of each P_i on input x_i is $m_i = x_i + r$. The output of the referee on messages m_i, m_j from parties P_i, P_j (where $i < j$) is $m_i - m_j = x_i - x_j$. The simulator SIM proving the security of the protocol gets as an input a set $S = \{i, j\}$ and $\Delta = x_i - x_j$. It chooses a random value r and outputs $r, r - \Delta$. Notice that both the messages in the protocol and the output of SIM is a pair (a, b), where a is uniformly distributed in G and b is $a - f(x_i, x_j)$, thus the simulation is as required.

While in most parts of this paper we will assume that at most k parties send messages, we next consider the scenario that parties execute an ad hoc PSM protocol and a set T of more than k parties sends messages. Clearly, for every $S \subset T$ of size k, the referee can compute the output of f on the inputs of S. Thus, the *best possible security requirement* is that the referee does not learn any additional information.

Definition 2.4. *An ad hoc PSM protocol Π for f is (k, t, ϵ)-secure if there exists a randomized function SIM such that, for every $t' \leq t$, every $T \in \binom{[n]}{t'}$ and every $x_T \in \mathcal{X}^t$,*

$$\mathrm{dist}\left(\mathrm{SIM}\left(T, (f(x_S))_{S \subseteq T, |S| = k}\right), M_T\right) \leq \epsilon.$$

An ad hoc PSM protocol Π for f is (k, t)-secure if it is $(k, t, 0)$-secure.

Remark 2.5. In Sect. 3.2, for every function f, we construct an *inefficient* (k, n) ad hoc PSM protocol. It follows from [16] (together with our result that (k, n)-secure ad hoc PSM protocols imply obfuscation) that efficient (k, n)-secure ad hoc PSM protocols for every function in NC^1 do not exist unless the polynomial-time hierarchy collapses. This impossibility result does not rule out, for example, efficient $(2, n)$-secure ad hoc PSM protocols for every function in NC^1 (and beyond) or efficient $(k, k + 1)$-secure ad hoc PSM protocols. We do not know if such efficient ad hoc PSM protocols exist.

For some functions the (k, t)-security requirement is not interesting as the best possible security already reveals a lot of information. For other functions this notion is interesting.

Example 2.6. Let f be the 2-party addition function over a field whose characteristic is not 2. Suppose that a referee got messages from parties P_1, P_2, P_3 in an ad hoc PSM for f, thus, it can compute the sum of every two inputs of these parties, namely, $x_1 + x_2 = s_{1,2}, x_1 + x_3 = s_{1,3}$, and $x_2 + x_3 = s_{2,3}$. From these sums it can compute the inputs, e.g., $x_1 = 2^{-1}(s_{1,2} + s_{1,3} - s_{2,3})$.

Example 2.7. Consider the $n/2$-party AND function and an input where the value of exactly $n/2$ of the input variables is 1. Assume that the referee gets messages from the n parties for this input. If the referee does not know the set of variables whose value is 1, then it will not be able to efficiently determine it.

We next consider *computationally-secure* ad hoc PSM protocols. In such protocols we want all algorithms to be efficient. We start by defining their syntax.

Definition 2.8 (Computational ad hoc PSM: Syntax). *Let $n(\lambda)$, $k(\lambda)$, and $\ell(\lambda)$ be polynomials, and $F = \{f_\lambda : (\{0,1\}^{\ell(\lambda)})^{k(\lambda)} \rightarrow \{0,1\}^*\}_{\lambda \in \mathbb{N}}$ be a collection of functions. A protocol $\Pi = (\text{GEN}, \text{ENC}, \text{DEC})$ is a $(k(\lambda), n(\lambda))$-computational ad hoc PSM protocol for F if*

- *Algorithm $\text{GEN}(1^\lambda)$ is a polynomial time algorithm that generates $n(\lambda)$ random strings (for $n(\lambda)$ parties).*
- *Algorithms ENC and DEC run in polynomial time.*
- *There exists a negligible function $\text{negl}()$ such that for any $\lambda \in \mathbb{N}$, any $S \subseteq [n(\lambda)]$, and any $x_S = (x_i)_{i \in S} \in (\{0,1\}^{\ell(\lambda)})^{k(\lambda)}$,*

$$\Pr\left[r \leftarrow \text{GEN}(1^\lambda) : \text{DEC}\left(S, (\text{ENC}_i(x_i, r_i))_{i \in S}\right) = f_\lambda(x_S)\right] \geq 1 - \text{negl}(\lambda).$$

We next present three definitions of security for computational ad hoc PSM protocols. The first definition is simulation-based and it applies to k-security (i.e., to the scenario where exactly k parties send their messages).

Definition 2.9 (Computational ad hoc PSM: Simulation-based security). *Let $n(\lambda)$, $k(\lambda)$, and $\ell(\lambda)$ be polynomials, and $F = \{f_\lambda : (\{0,1\}^{\ell(\lambda)})^{k(\lambda)} \rightarrow \{0,1\}^*\}_{\lambda \in \mathbb{N}}$ be a collection of functions. We say that an ad hoc PSM protocol $(\text{GEN}, \text{ENC}, \text{DEC})$ is $k(\lambda)$-simulation-based secure if there exists a probabilistic non-uniform polynomial algorithm SIM whose inputs are 1^λ and the value of f such that the two ensemble of distributions*

$$\left((m_i)_{i \in S} : \begin{array}{l} r \leftarrow \text{GEN}(1^\lambda), \\ \forall_{i \in S} \, m_i \leftarrow \text{ENC}(x_i, r_i) \end{array}\right) \begin{array}{l} \lambda \in \mathbb{N}, S \in \binom{[n(\lambda)]}{k(\lambda)}, \\ (x_i)_{i \in S} \in (\{0,1\}^{\ell(\lambda)})^{k(\lambda)} \end{array}$$

and

$$\left(\text{SIM}(1^\lambda, f_\lambda((x_i)_{i \in S}))\right) \begin{array}{l} \lambda \in \mathbb{N}, S \in \binom{[n(\lambda)]}{k(\lambda)}, \\ (x_i)_{i \in S} \in (\{0,1\}^{\ell(\lambda)})^{k(\lambda)} \end{array}$$

are indistinguishable in polynomial time.

Simulation-based security is a strong requirement that cannot be achieved for computational ad hoc PSM protocols with best possible security (see discussion in [3]). Thus, for such protocols, we define weaker security – virtual black-box (VBB) security, where the adversary can output only one bit and that uses indistinguishability-based security. To simplify the notation, we only consider the case where $t = n(\lambda)$.

Definition 2.10 (Computational ad hoc PSM: Virtual black-box Security). *Let $n(\lambda)$, $k(\lambda)$, and $\ell(\lambda)$ be polynomials, and $F = \{f_\lambda : (\{0,1\}^{\ell(\lambda)})^{k(\lambda)} \rightarrow$*

$\{0,1\}^*\}_{\lambda\in\mathbb{N}}$ *be a collection of functions. We say that an* ad hoc PSM *protocol* (GEN, ENC, DEC) *is* $(k(\lambda), n(\lambda))$-*VBB-secure if, for every non-uniform polynomial time adversary* $\mathcal{A}$ *that outputs one bit, there exists a non-uniform probabilistic polynomial time algorithm* SIM *and a negligible function* $\mathrm{negl}(\lambda)$ *such that for every* $\lambda\in\mathbb{N}$, *every* $S\in\binom{[n(\lambda)]}{k(\lambda)}$, *and every* $x_1,\ldots,x_{n(\lambda)}\in(\{0,1\}^{\ell(\lambda)})^{n(\lambda)}$

$$\left|\Pr\left[\mathcal{A}(1^\lambda, m_1\ldots, m_{n(\lambda)})=1\right] - \Pr\left[\mathrm{SIM}^{f_\lambda}(1^\lambda)=1\right]\right| \leq \mathrm{negl}(\lambda),$$

where

- *The first probability is over the messages generated in the following way: first compute* $r\leftarrow$ GEN(1^λ) *and then* $m_i\leftarrow$ ENC(x_i, r_i), *for every* $i\in[n(\lambda)]$.
- *The second probability is over the randomness of the simulator, which has access to an oracle* f_λ *that on query* $S\in\binom{[n(\lambda)]}{k(\lambda)}$ *returns* $f_\lambda(x_S)$.

Definition 2.11 (Computational ad hoc PSM: Indistinguishability-based security). *Let* $n(\lambda)$, $k(\lambda)$, *and* $\ell(\lambda)$ *be polynomials, and* $F=\{f_\lambda: (\{0,1\}^{\ell(\lambda)})^{k(\lambda)}\to\{0,1\}^*\}_{\lambda\in\mathbb{N}}$ *be a collection of functions. Consider the following game between an adversary* $\mathcal{A}$ *and a challenger:*

1. *The adversary on input* 1^λ *chooses a set* $T\subseteq[n(\lambda)]$ *and two inputs* $(x_i^0)_{i\in T}$ *and* $(x_i^1)_{i\in T}$ *and sends* T *and the two inputs to the challenger.*
2. *The challenger chooses a uniformly random bit* $b\in\{0,1\}$ *and computes* $(r_1,\ldots,r_n)\leftarrow$ GEN(1^λ) *and* $m_i\leftarrow$ ENC(x_i^b, r_i), *for every* $i\in T$. *It then sends* $(m_i)_{i\in T}$ *to the adversary.*
3. *The adversary outputs a bit* b'.

The adversary wins the game if $b'=b$ *and* $f_\lambda(x_S^0)=f_\lambda(x_S^1)$, *for every* $S\subseteq T$ *such that* $|S|=k(\lambda)$.

We say that a computational ad hoc PSM *protocol* (GEN, ENC, DEC) *is a* $(k(\lambda), n(\lambda))$-*indistinguishably-secure* ad hoc PSM *protocol for* F *if, for every non-uniform polynomial-time adversary* $\mathcal{A}$, *the probability that* $\mathcal{A}$ *wins is at most* $1/2+\mathrm{negl}(\lambda)$ *for some negligible function* negl.

Our default model in the rest of the paper, unless explicitly mentioned, is (k,k)-secure ad hoc PSM protocol with perfect security.

3   Ad hoc PSM Protocols for a Function f from a PSM for f

In this section we present a k-out-of-n ad hoc PSM protocol for any function f by applying transformations to k-party PSM protocols for the same f.

3.1   From a PSM for f to an n-out-of-n ad hoc PSM for f

In an n-party PSM protocol for f, if all n parties send messages, then the referee learns $f(x_1, \ldots, x_n)$ and does not learn any additional information. In an n-out-of-n ad hoc PSM protocol, there is an additional requirement: if less than n parties send messages, then the referee should learn no information. The definition of PSM does not imply the latter requirement.

Example 3.1. Consider a function which returns x_1. In a PSM for this function, P_1 can send its input, while in an ad hoc PSM protocol, this should not be done.

In many PSM protocols this additional requirement does hold. Furthermore, for many functions, the requirement for smaller sets of active participants follows from the security requirements of the PSM.

Example 3.2. Consider a PSM protocol for the AND function. If the input of P_n is 0, then the output of AND is 0 for every input for $P_1, \ldots, P_{n-1}$. Thus, the messages $m_1, \ldots, m_n$ of $P_1, \ldots, P_n$ are equally distributed when $x_n = 0$, for every input for $P_1, \ldots, P_{n-1}$. Since the messages of $P_1, \ldots, P_{n-1}$ are independent of x_n, the messages of parties $P_1, \ldots, P_{n-1}$ are equally distributed for every input for these parties. I.e., in any PSM protocol for AND the referee does not learn any information from the messages of $P_1, \ldots, P_{n-1}$ or, similarly, any other set of less than n active participants.

Lemma 3.3. *If there is an n-party PSM protocol Π for f with randomness complexity $\text{Rand}(\Pi)$ and communication complexity $\text{Comm}(\Pi)$, then there is an n-out-of-n ad hoc PSM protocol for f with randomness complexity $\text{Rand}(\Pi) + n \cdot \text{Comm}(\Pi)$ and communication complexity $n \cdot \text{Comm}(\Pi)$.*

Proof. We construct an ad hoc PSM protocol Π_{ah} for f from the PSM protocol Π for f, as follows.

Randomness generation:

- Generate randomness for the PSM protocol Π; denote $r_1, \ldots, r_n$ the generated randomness of $P_1, \ldots, P_n$, respectively.
- Choose n uniformly random strings $u_1, \ldots, u_n$, each of length $\text{Comm}(\Pi)$.
- Share each u_j, for $j \in [n]$, using an n-out-of-n secret sharing scheme; let $u_{j,i}$ be the i-th share of u_j.
- The randomness of P_i in Π_{ah} is $r_i, (u_{j,i})_{j \in [n]}$.

Message generation:

- Let m_i be the messages of P_i in Π on input x_i and randomness r_i.
- The message of P_i in Π_{ah} is $m_i \oplus u_i$ and $(u_{j,i})_{j \in [n]}$.

If the n parties send their messages, then the referee can reconstruct u_i for every $i \in [n]$, compute m_i, and reconstruct $f(x_1, \ldots, x_n)$ using the decoding algorithm of Π.

The security of Π_{ah} when n parties send messages follows from the security of Π (as the strings $u_1, \ldots, u_n$ are random). When less than n parties send their messages, the referee gets less than n shares of each u_i, thus, these strings act as random pads and the referee learns no information. $\qquad\square$

3.2   A Naive ad hoc PSM Protocol for Any f

In this section we show how to construct, given a (standard) k-party PSM protocol for f, a k-out-of-n ad hoc PSM protocol for f that is n-secure (that is, if a set T of at least k parties send their messages, then the referee can compute the output of f on any subset of inputs of size k and learns no additional information).

Theorem 3.4. *If there is a k-party PSM protocol Π for f with randomness complexity* $\mathrm{Rand}(\Pi)$ *and communication complexity* $\mathrm{Comm}(\Pi)$, *then there is a (k,n)-secure ad hoc PSM protocol for f with randomness complexity* $\binom{n}{k} \cdot (\mathrm{Rand}(\Pi) + n \cdot \mathrm{Comm}(\Pi))$ *and communication complexity* $\binom{n}{k} \cdot n \cdot \mathrm{Comm}(\Pi)$.

Proof. Let Π_{ah} be the k-out-of-k ad hoc PSM protocol constructed from Π in Lemma 3.3. We construct an ad hoc PSM protocol Π' for f as follows:

Randomness generation:

- For each set $S \in \binom{[n]}{k}$, independently generate randomness for Π_{ah} and give this randomness to the parties in S.

Message generation:

- Each party P_i sends its message in protocol Π_{ah}, associated with the set S', for every S' of size k such that $i \in S'$.

Function reconstruction by the referee: For a set S of k participating parties, the referee (only) uses the messages of the parties in S of the PSM Π_{ah} for S to reconstruct $f(x_S)$.

We next prove the security when a set T of size at least k sends messages. We claim that the referee only learns $f(x_S)$ for every $S \subseteq T$ of size k. Since the randomness of each execution of the PSM protocol Π_{ah} is chosen independently, the referee can only learn information from the messages of Π_{ah} for each set S of size k. In an execution for a set $S \subseteq T$ it can only learn $f(x_S)$. In any execution of the PSM protocol for S such that $S \not\subseteq T$, the referee misses a message of at least one party thus, by Lemma 3.3, learns no information from this execution.

The randomness and communication of the ad hoc PSM protocol Π' are $\binom{n}{k}$ times larger than the randomness and communication, respectively, of the PSM protocol Π_{ah}. □

As every function f has a PSM realizing it [14,17], the previous theorem implies that every function has an ad hoc PSM protocol.

Corollary 3.5. *For every k-argument function f, there is an n-secure k-out-of-n ad hoc PSM protocol realizing f.*

3.3   A 2-out-of-n ad hoc PSM Protocol from a PSM Protocol for f

Suppose that we have a 2-party PSM protocol Π for a *symmetric* function f. We denote the parties in this PSM by Q_0 and Q_1. We want to construct an ad hoc PSM protocol Π^* for f using Π. The idea is to instruct the first party P_i to simulate Q_0 and instruct the second party P_j to simulate Q_1. The problem is that in an ad hoc PSM protocol a party does not know who the other party is; informally, it does not know if it is the "first party" or the "second party". Instead, we execute a few copies of the PSM protocol Π where, in some copies of the PSM, party P_i plays the role of Q_0, and in other copies it plays the role of Q_1. Specifically, we view each $i \in [n]$ in its $\log n$-bit binary representation $i = (i_1, \ldots, i_{\log n})$, and execute $\log n$ copies of Π, where in the ℓth copy P_i plays the role of Q_{i_ℓ}. Since for any $i \neq j$, there exists an index ℓ such that $i_\ell \neq j_\ell$, in the ℓth copy P_i, P_j simulate both Q_0 and Q_1 and the referee can compute f from this copy.

However, information can now leak when P_i and P_j simulate, in some copy, the same Q_b; that is, if $i_\ell = j_\ell$, for some ℓ. In particular, in such copy, P_i and P_j send the same message if $x_i = x_j$. To overcome this problem, in the ℓth copy, where party P_i plays the role of Q_{i_ℓ}, party P_i "encrypts" its message m using a key k_{i_ℓ} and each party playing the role of $Q_{\bar{i}_\ell}$ sends the key k_{i_ℓ} as part of its message. Thus, if both P_i, P_j play the role of the same party Q_b, then the referee does not obtain the key, and cannot learn any information from this copy of the PSM. The formal description of the ad hoc PSM protocol Π^* follows.

Randomness generation:

- Let p be a prime such that $\log p \geq \max\{\mathrm{Comm}(\Pi), \log n\}$, where $\mathrm{Comm}(\Pi)$ is the length of the messages in the PSM protocol Π. All arithmetic in the protocol is in $\mathbb{F}_p$.
- For $\ell = 1$ to $\log n$:
  - Independently generate randomness for the PSM protocol Π; denote by $r_{\ell,0}, r_{\ell,1}$ the generated randomness of Q_0, Q_1, respectively.
  - Choose four random values $a_{\ell,0}, b_{\ell,0}, a_{\ell,1}, b_{\ell,1} \in_R \mathbb{F}_p$.
- The randomness of P_i, where $i = (i_1, \ldots, i_{\log n})$, is

$$(r_{\ell,i_\ell}, a_{\ell,0}, b_{\ell,0}, a_{\ell,1}, b_{\ell,1})_{1 \leq \ell \leq \log n}.$$

Message generation for every $P_i \in S$:

- For every $\ell \in \{1, \ldots, \log n\}$, party P_i computes $m_{i,\ell}$ – the message that Q_{i_ℓ} sends in Π on input x_i with randomness r_{ℓ,i_ℓ}.
  P_i sends $(m_{i,\ell} + a_{\ell,i_\ell} \cdot i + b_{\ell,i_\ell})_{1 \leq \ell \leq \log n}$ and $(a_{\ell,\bar{i}_\ell}, b_{\ell,\bar{i}_\ell})_{1 \leq \ell \leq \log n}$.

Assume that P_i and P_j send messages and the referee wants to compute $f(x_i, x_j)$. It finds an index ℓ such that $i_\ell \neq j_\ell$. Without loss of generality, $i_\ell = 0$ and $j_\ell = 1$, that is, in the ℓth copy of Π, party P_i plays the role of Q_0 and P_j plays the role of Q_1. As P_i sends $m_{i,\ell} + a_{\ell,0} \cdot i + b_{\ell,0}$ and P_j sends $a_{\ell,0}, b_{\ell,0}$, the referee can recover $m_{i,\ell}$ – the message of Q_0. Similarly, the referee can recover $m_{j,\ell}$ – the message of Q_1 – and, using the reconstruction procedure of Π, it can compute $f(x_i, x_j)$.

By the privacy property of protocol Π, the referee does not learn any additional information from an ℓth copy of the PSM, where $i_\ell \neq j_\ell$. Furthermore, this is true also for the concatenation of the messages in all the copies where $i_\ell \neq j_\ell$; note that since f is symmetric the output of the protocol in each such copy is the same. On the other hand, in any copy where $i_\ell = j_\ell$, the referee gets two "encrypted" messages $m_{i,\ell} + a_{\ell,i_\ell} \cdot i + b_{\ell,i_\ell}$ and $m_{j,\ell} + a_{\ell,i_\ell} \cdot j + b_{\ell,i_\ell}$. Since $i \neq j$ (and $a_{\ell,i_\ell}, b_{\ell,i_\ell}$ are random), then all pairs of "encrypted" messages are possible and the referee learns no information from this copy of Π. The security of Π^* follows.

Let $\mathrm{Rand}(\Pi)$ and $\mathrm{Comm}(\Pi)$ be the randomness complexity and communication complexity of Π, respectively. The randomness complexity of the new Π^* is

$$O\left(\log n \cdot \max\{\mathrm{Rand}(\Pi), \mathrm{Comm}(\Pi), \log n\}\right),$$

and the communication complexity of Π^* is $O(\log n \cdot \max\{\mathrm{Rand}(\Pi), \log n\})$.

3.4   A k-out-of-n ad hoc PSM Protocol from a PSM Protocol for f

We want to generalize the above ad hoc PSM protocol Π^* to larger values of k. Again, we will execute many copies of the original k-party PSM protocol Π. The properties we require are: (1) for every set $S \subseteq [n]$ of size k, there exists a copy in which the parties in S play roles of distinct parties in Π, and (2) in copies where the parties in S do not play roles of distinct parties in Π, no information is leaked. To achieve the first requirement, we use a perfect hash family.

Definition 3.6. *A perfect hash family $H = \{h : [n] \to [k]\}$ is a set of functions such that for any set $S \subseteq [n]$ of size k, there exists at least one $h \in H$ that is 1-1 over S.*

Example 3.7. For $k = 2$, the family of bit-functions $H = \{h_1, \ldots, h_{\log n}\}$, where $h_\ell(i) = i_\ell + 1$ (and i_ℓ is the ℓth bit in the binary representation of i) is a perfect hash family.

A perfect hash family with $\binom{n}{k}$ functions can be easily constructed, but much more efficient constructions, probabilistic or explicit, are possible. E.g., picking the h's at random, it is enough to have $|H| \approx e^k \cdot k \log n$ (for a specific size-k set S, a random function is 1-1 w/prob $k!/k^k > e^{-k}$, by Sterling formula, and we need to take care of about n^k such sets).

We next describe the ad hoc PSM protocol, assuming a k-party PSM protocol Π for a symmetric function f and a perfect hash family H.

Theorem 3.8. *Assume that there is a k-party PSM protocol Π for a symmetric function f with randomness complexity $\mathrm{Rand}(\Pi)$ and communication complexity $\mathrm{Comm}(\Pi)$. Then, there is a k-out-of-n ad hoc PSM protocol for f with* *randomness complexity $O(e^k \cdot k \log n \cdot (\text{Rand}(\Pi) + k^2 \cdot \max\{\text{Comm}(\Pi), \log n\}))$ and communication complexity $O(e^k \cdot k^3 \log n \max\{\text{Comm}(\Pi), \log n\})$.*

Proof. Denote the parties of Π by $Q_1, \ldots, Q_k$. We construct a k-out-of-n ad hoc PSM protocol Π^* as follows.

Randomness generation:

- Let p be a prime such that $\log p \geq \max\{\text{Comm}(\Pi), \log n\}$, where $\text{Comm}(\Pi)$ is the length of the messages in Π.
- For every $h \in H$ do:
  - Independently generate randomness for the hth copy of Π; let $r_{h,1}, \ldots, r_{h,k}$ be the generated randomness for $Q_1, \ldots, Q_k$ respectively.
  - Choose k random polynomials $A_{h,1}(Y), \ldots, A_{h,k}(Y)$ of degree $k-1$ over $\mathbb{F}_p$.
  - Consider each polynomial $A_{h,j}(Y)$ as an element in $\mathbb{F}_p^k$ and share it in a k-out-of-k additive sharing scheme; denotes its shares as $A_{h,j,1}, \ldots, A_{h,j,k}$.
- The randomness of P_i in the ad hoc PSM protocol Π^* is

$$(r_{h,h(i)}, A_{h,h(i)}, A_{h,1,h(i)}, \ldots, A_{h,k,h(i)})_{h \in H}.$$

Message generation for every $P_i \in S$:

- For every $h \in H$, party P_i computes $m_{i,h}$ – the messages that $Q_{h(i)}$ sends in the PSM protocol Π on input x_i with randomness $r_{h,h(i)}$. Party P_i sends $(A_{h,h(i)}(i) + m_{i,h})_{h \in H}$ and, in addition, the shares $(A_{h,j,h(i)})_{h \in H, j \in [k]}$.

Assume that a set S of size k sends messages and the referee wants to compute $f(x_S)$. The referee finds a function $h \in H$ that is 1-1 on S. Let $i \in S$. Party P_i plays the role of $Q_{h(i)}$ in the hth copy of Π, and sends the message $A_{h,h(i)}(i) + m_{i,h}$. Furthermore, all k parties in S send their shares in a k-out-of-k secret-sharing scheme with the secret $A_{h,h(i)}$. Thus, the referee can reconstruct $A_{h,h(i)}$, compute $A_{h,h(i)}(i)$, and recover $m_{i,h}$. Similarly, the referee can recover all k messages in the hth copy of Π and can decode $f(x_S)$.

By the privacy property of protocol Π, the referee does not learn any additional information from an hth copy of Π, for every h such that h is 1-1 on S. Furthermore, this is true also for the concatenation of the messages in all such copies and, since f is symmetric, the output of the protocol in each such copy is the same. On the other hand, in any copy where h is not 1-1 on S, the referee does not get any information on $A_{h,h(i)}$, since it gets at least two identical shares of this secret. The referee gets at most k messages "encrypted" by the same secret key $A_{h,h(i)}$. The values $\{A_{h,h(i)}(i)\}_{i \in S}$ are k points on a random polynomial of degree $k-1$, thus, they are uniformly distributed and serve as random pads, i.e., the referee gets no information from such hth copy of the PSM Π.

The randomness complexity of Π^* is

$$|H| \cdot \left(\mathrm{Rand}(\Pi) + k^2 \cdot \max\{\mathrm{Comm}(\Pi), \log n\}\right)$$
$$\approx e^k \cdot k \log n \cdot \left(\mathrm{Rand}(\Pi) + k^2 \cdot \max\{\mathrm{Comm}(\Pi), \log n\}\right),$$

To analyze the communication complexity of Π^*, note that for each h, each party P_i sends its encrypted message and also a share for $A_{h,j}$, for all $j \in [k]$. All together, the communication complexity of each party is

$$|H| \cdot k^2 \cdot \max\{\mathrm{Comm}(\Pi), \log n\} \approx e^k k^3 \log n \cdot \max\{\mathrm{Comm}(\Pi), \log n\}.$$

$\square$

Remark 3.9. There may be several functions in H, say h, h', that are 1-1 on S (and, moreover, h_S is different than h'_S). Since we assume here that the function f is symmetric, the output is the same in both copies of Π and, since the randomness is independent, there is no additional information. If f was not symmetric the referee may learn multiple outputs (under different orders) and hence additional information on the input.

4 An ad hoc PSM Protocol Based on a PSM Protocol for a Related Function

In this section we construct an ad hoc PSM protocol for f from a PSM protocol for a related function g. The construction is similar to the construction of the ad hoc PSM protocol for SUM described in Sect. 1.1. To construct the ad hoc PSM protocol for the k argument function $f : X^k \to Y$, we define a (partial) n-argument function $g : (X \cup \{\bot\})^n \to Y \cup \{\bot\}$, where if there are more than $n - k$ inputs that are $\bot$, the function outputs $\bot$, if there are exactly $n - k$ inputs that are $\bot$, the function outputs the output of f on the k non-$\bot$ inputs, and if there are less than $n - k$ inputs that are $\bot$, then the function is undefined (in the latter case, we do not care what g outputs).

Lemma 4.1. *If there exists a PSM protocol Π_g for g with randomness complexity $\mathrm{Rand}(\Pi_g)$ and communication complexity $\mathrm{Comm}(\Pi_g)$, then there exists an ad hoc PSM protocol for f with randomness complexity $\mathrm{Rand}(\Pi_g) + n \cdot \max\{\mathrm{Comm}(\Pi_g), \log n\}$ and communication complexity $n \cdot \max\{\mathrm{Comm}(\Pi_g), \log n\}$.*

Proof. We construct an ad hoc PSM protocol Π_f for f from the PSM protocol Π_g as follows.

Randomness generation:

– Generate randomness for the PSM protocol Π_g; let $r_1, \ldots, r_n$ be the generated randomness of $P_1, \ldots, P_n$, respectively.

- Let $m_{\perp,j}$ be the message that P_j sends in Π_g with randomness r_j and input $\perp$. Share $m_{\perp,j}$ using a k-out-of-n secret sharing scheme; let $m_{\perp,j,i}$ be the i-th share.
- The randomness of P_i in the ad hoc PSM protocol is $r_i, (m_{\perp,j,i})_{j\neq i}$.

Message generation:

- The message of P_i on input x_i is its message on input x_i and randomness r_i in the PSM protocol Π_g and, in addition, $(m_{\perp,j,i})_{j\neq i}$.

Assume that parties in a set S of size exactly k send messages. Then, the referee has the k messages in Π_g of the parties in S with inputs $x_i \neq \perp$ and, for each $j \notin S$, it has k shares of the message $m_{\perp,j}$. Thus, the referee can reconstruct $g(y_1,\ldots,y_n) = f((x_i)_{i\in S})$, where $y_i = x_i$ if $i \in S$ and $y_i = \perp$ otherwise. On the other hand, since each party $p_i \in S$ does not send its share of $m_{\perp,i}$, the referee gets $k-1$ shares of $m_{\perp,i}$; hence, the referee has no information on $m_{\perp,i}$. Thus, when k parties send messages, the referee in Π_f has the same information that the referee has in Π_g and the privacy requirement for Π_f protocol follows from the privacy requirement of the PSM Π_g.

Assume that parties in a set S of size less than k parties send messages. In this case, we claim that the referee in Π_f gets no information even if we give it more information, namely, $m_{\perp,j}$ for every $P_j \notin S$. In this case, the referee gets messages of inputs whose output is $\perp$. By the privacy of the PSM protocol, these messages are distributed as the messages when all the inputs are $\perp$, that is, the referee does not learn any information on the inputs.

The randomness in the above ad hoc PSM Π_f is $\text{Rand}(\Pi_g) + n \cdot \text{Comm}(\Pi_g)$. The communication in Π_f is $O(n \cdot \text{Comm}(\Pi_g))$ (assuming $\text{Comm}(\Pi_g)$ is at least $\log n$). □

Example 4.2. Assume that $f : \{0,1\}^k \to \{0,\ldots,k\}$ is a symmetric function (that is, the output of f only depends on the number of 1's in the input). The function f has a small branching program (i.e., the size of the branching program is $O(k^2)$), thus f itself has an efficient PSM protocol [17]. Furthermore, the function g has a branching program of size $O(nk^2)$, thus, it has an efficient PSM protocol, i.e., a PSM with communication complexity $O(n^2k^4)$. This implies an ad hoc PSM protocol for f with communication $O(n^3k^4)$.

If a function f has a small non-deterministic branching program, then the corresponding function g has a small non-deterministic branching program, thus, by [17], g has an efficient PSM protocol. By Lemma 4.1, we get for all $k \leq n$ efficient k-secure ad hoc PSM protocols for every function that has a small non-deterministic branching programs.

Similarly, if f has a small circuit, then g has a small circuit, thus, by using Yao's garbled circuit construction [14,20] we get a simulation-based-secure PSM for g assuming the existence of a one-way function. By Lemma 4.1, we get for all $k \leq n$ efficient computational k-secure ad hoc PSM protocols (with simulation-based-security) for every function that has a small circuit assuming the existence of a one-way function.

5   A Protocol for Equality

Define the equality function EQ : $(\{0,1\}^\ell)^k \to \{0,1\}$ as the function, whose input is k strings of length ℓ and whose output is 1 if and only if all strings are equal. We next present an ad hoc PSM protocol for EQ.

Lemma 5.1. *There is a statistically-secure* ad hoc PSM *protocol for* EQ *whose randomness complexity and communication complexity are* $O(n + \ell)$.

Proof. We next describe the ad hoc PSM protocol.

Randomness generation:

- Let p be a prime number such that $\log p > \max\{n, \ell\}$.
- Choose at random an element $a \in \mathbb{F}_p$ such that $a \neq 0$.
- Choose $k-1$ random elements $r_0, \ldots, r_{k-2}$ in $\mathbb{F}_p$ and define the polynomial $Q(Y) = \sum_{i=0}^{k-2} r_i Y^i$ (over $\mathbb{F}_p$).
- Choose n random elements $j_1, \ldots, j_n$ in $\mathbb{F}_p$
- The randomness of P_i in the ad hoc PSM protocol is $(j_i, Q(j_i), a)$.

Message generation for every $P_i \in S$:

- P_i sends $j_i, Q(j_i) + ax_i$.

Function reconstruction by the referee:

- Assume the referee gets k pairs $(\ell_1, z_1), \ldots, (\ell_k, z_k)$. If all point lie on a polynomial of degree $k-2$ answer "equal", otherwise answer "not equal".

First assume that all k inputs are equal, say to α. In this case the k pairs lie on the polynomial $Q(Y) + a\alpha$ and the referee answers "equal". Furthermore, since the free coefficient of $Q(Y) + a\alpha$ is $r_0 + a\alpha$, the values $(\ell_1, z_1), \ldots, (\ell_k, z_k)$ are independent of α.

We next consider the case that not all of the k inputs are equal. Since $j_1, \ldots, j_n$ are uniformly distributed, we can assume, without loss of generality, that $S = \{P_1, \ldots, P_k\}$. Fix any inputs $x_1, \ldots, x_k$ such that $x_k \neq x_\ell$ for some $1 \leq \ell < k$ (again, this is w.l.o.g.). We prove that with probability at least $1 - k/p$ over the choice of $j_1, \ldots, j_k$, the values $z_1, \ldots, z_k$ are uniformly distributed in $\mathbb{F}_p^k$. In particular, this implies that with probability at least $1 - k/p$, the referee answers "not equal". Furthermore, it implies the privacy for this case.

Fix any $j_1, \ldots, j_{k-1}$ and $z_1, \ldots, z_{k-1}$. Let $H(Y)$ and $M(Y)$ be the polynomials of degree $k-2$ such that $H(j_i) = x_i$ and $M(j_i) = z_i$ for every $1 \leq i \leq k-1$. Such polynomials exist and they are unique. Notice that for every $a \neq 0$ there exists a unique polynomial $Q(Y)$ of degree $k-2$ that can be chosen in the randomness generation of the protocol, where $Q(Y) = M(Y) - a \cdot H(Y)$ (since both the r.h.s. and the l.h.s. are polynomials of degree $k-2$ that agree on the $k-1$ points $j_1, \ldots, j_{k-1}$). Thus, the message of P_k is

$$z_k = Q(j_k) + ax_k = M(j_k) - a \cdot H(j_k) + ax_k.$$

The protocol fails (i.e., outputs "equal" although not all inputs are equal) if and only if $z_k = M(j_k)$; the last equality is true if and only if $H(j_k) = x_k$. Notice that since $H(Y) \neq x_k$ since $H(j_\ell) = x_\ell \neq x_k$. Since $H(Y) \neq x_k$ is a polynomial of degree $k-2$, there are at most $k-2$ values of j_k such that $H(j_k) = x_k$. Thus, with probability at least $1 - (k-2)/p \geq 1 - (k-2)/2^n$, the referee in this case outputs "not equal". Assuming that such j_k is not chosen, $z_k = M(j_k) + a(-H(j_k) + x_k)$; as a is chosen at random, the value z_k is random (provided that the kth pair does not lie on the polynomial). □

6 Order Revealing Encryption from an Ad Hoc PSM Protocol

An order revealing encryption is a private-key encryption that enables computing the order between two messages (that is, checking if $m_1 < m_2$, $m_1 = m_2$, or $m_1 > m_2$), given their encryptions (without knowing the private key), but does not disclose any additional information. In this section, we show how to use ad hoc PSM protocols to construct information-theoretically secure order revealing encryption that is 2-bounded (namely, the encryption is secure as long as only two messages are encrypted).

Definition 6.1. *The greater than function, $\text{GTE}_\ell : \{0,1\}^\ell \times \{0,1\}^\ell \to \{-1, 0, 1\}$, is defined as follows:*

$$\text{GTE}_\ell(x,y) = \begin{cases} -1 & If\, x < y \\ 0 & If\, x = y \\ 1 & If\, x > y, \end{cases}$$

where we identify the strings in $\{0,1\}^\ell$ with the integers in $\{0, \ldots, 2^\ell - 1\}$.

Definition 6.2 (Order Revealing Encryption (ORE): Syntax and correctness). *Let $\epsilon : \mathbb{N} \to [0, 0.5)$. An $\epsilon(\lambda)$-ORE for messages in $\{0,1\}^\ell$ is composed of 4 efficient algorithms:*

- GEN_{ORE} *is a randomized key generation algorithm, that on input 1^λ (where λ is a security parameter), outputs a key k;*
- ENC_{ORE} *is an encryption algorithm, that on input message m and a key k, outputs an encryption c;*
- DEC_{ORE} *is a decryption algorithm, that on input an encryption c and a key k, outputs a message m satisfying the following correctness requirement for any $m \in \{0,1\}^\ell$:*

$$\Pr\left[k \leftarrow \text{GEN}_{ORE}(1^\lambda) : \text{DEC}_{ORE}\left(\text{ENC}_{ORE}(m, k), k\right) = m \right] \geq 1 - \epsilon(\lambda).$$

- COMP_{ORE} *is a comparison algorithm, that given any two encryptions c_1, c_2, outputs a value in $\{-1, 0, 1\}$ such that for any $m_1, m_2 \in \{0,1\}^\ell$:*

$$\Pr\left[\begin{matrix} k \leftarrow \text{GEN}_{ORE}(1^\lambda), c_1 \leftarrow \text{ENC}_{ORE}(m_1, k), \\ c_2 \leftarrow \text{ENC}_{ORE}(m_2, k) : \text{COMP}_{ORE}(c_1, c_2) = \text{GTE}_\ell(m_1, m_2) \end{matrix} \right] \geq 1 - \epsilon(\lambda).$$

If the comparison algorithm is the comparison over the integers (e.g., it returns -1 whenever $c_1 < c_2$), then the encryption is called Order Preserving Encryption (OPE).

Remark 6.3. Given the private key k and an encryption c, one can use a binary search using COMP_{ORE} to decrypt c. That is, we do not need to specify the decryption algorithm. For efficiency, one can avoid this binary search by encrypting the message using a standard (semantically secure) encryption scheme in addition to the ORE encryption.

We next define the security requirement of ORE. Our definition is the information theoretic analogue of the IND-OCPA security requirement from [8]. The definition of IND-OCPA is similar to the traditional IND-CPA definition of private key encryption, however, as the adversary can learn the order between two messages from their encryptions, the IND-OCPA definition prevents the adversary from using this information by limiting the encryption queries that it can make (see (1) in Definition 6.4 below).

Definition 6.4 (ORE: Security). *Consider the following game between an all-powerful adversary and a challenger:*

- *The input of both parties is a security parameter 1^λ and a bound on the number of queries 1^t.*
- *The challenger chooses a random bit b with uniform distribution and generates a key $k \leftarrow \text{GEN}_{ORE}(1^\lambda)$.*
- *For $i = 1$ to t do:*
  - *The adversary chooses two message $m_0^i, m_1^i \in \{0,1\}^\ell$ and sends them to the challenger.*
  - *The challenger computes $c_i \leftarrow \text{ENC}_{ORE}(m_b^i, k)$ and sends c_i to the adversary.*
- *The adversary returns a bit b'.*

We say that the adversary wins if $b = b'$ and for every $1 \leq i < j \leq t$

$$\text{GTE}_\ell(m_0^i, m_0^j) = \text{GTE}_\ell(m_1^i, m_1^j). \tag{1}$$

Let $\epsilon : \mathbb{N} \to [0, 0.5)$. We say that an ORE is $\epsilon(\lambda)$-secure if for every polynomial $t(\lambda)$ and every adversary $\mathcal{A}$ the probability that $\mathcal{A}$ with parameters $1^\lambda, 1^{t(\lambda)}$ wins is at most $1/2 + \epsilon(\lambda)$. We say that an ORE is t-bounded $\epsilon(\lambda)$-secure if for every adversary $\mathcal{A}$ the probability that $\mathcal{A}$ with parameters $1^\lambda, 1^t$ wins is at most $1/2 + \epsilon(\lambda)$.

We next describe some relevant results for OPE and ORE. In this discussion all encryption schemes are computationally secure. Order preserving encryption was introduced by Agrawal et al. [1]; their motivation was encrypting a database while allowing to answer range queries given the encrypted data (without the secret key). A cryptographic treatment of OPE was given by Boldyreva et al. [8, 9]; they gave a formal definition of OPE (called IND-OCPA) and showed that, in any OPE satisfying this definition, the length of the encryption is $2^{\omega(\ell)}$, where ℓ is the length of the messages (this is true even if the attacker can only ask to encrypt 3 messages). In a follow up work, Boldyreva et al. [10,11] defined ORE. As ORE is a special case of multi-input functional encryption (MIFE) [15], it is implied by indistinguishability obfuscation (iO). Boneh et al. [12] constructed ORE directly from multi-linear maps (with bounded multi-linearity). t-bounded ORE can be constructed based on the LWE assumption or from pseudorandom generators computable by small-depth circuits [13].

We next show how to construct ORE from an ad hoc PSM protocol for the greater than function GTE_ℓ.

Theorem 6.5. *There exists a 2-bounded $1/2^\lambda$-secure ORE with messages in $\{0,1\}^\ell$ and encryptions of length $O(\ell^2\lambda + \lambda^2)$.*

Proof. We start with a 2-out-of-n ad hoc PSM protocol Π_{GTE} for GTE: The function GTE_ℓ has a deterministic branching program of size $O(\ell)$ thus, by [17], it has a PSM protocol with randomness and communication complexity $O(\ell^2)$. By Theorem 3.8, GTE_ℓ has an ad hoc PSM protocol with complexity $O(\ell^2 \log n + \log^2 n)$. Note that Theorem 3.8 requires that the function for which we construct an ad hoc PSM protocol is symmetric. As $\text{GTE}_\ell(m_2, m_1) = -\text{GTE}_\ell(m_1, m_2)$, the transformation described in Theorem 3.8 from a PSM protocol to an ad hoc PSM protocol is valid for GTE_ℓ.

We next describe a construction of ORE, that is, we desribe algorithms ($\text{GEN}_{\text{ORE}}, \text{ENC}_{\text{ORE}}, \text{COMP}_{\text{ORE}}$) (by Theorem 6.3 we do not need to describe DEC_{ORE}). We use the ad hoc PSM Π_{GTE} with $n = 2^\lambda$ parties (where λ is the security parameter).

- Algorithm GEN_{ORE} generates a key k by choosing a random string for GEN_{GTE}, this key has length $O(\ell^2 \log n + \log^2 n)$. We emphasize that during the key generation we do not apply GEN_{GTE} as its output is too long (it contains n stings).
- Algorithm ENC_{ORE} encrypts a message x by choosing a random party P_i (where $1 \leq i \leq n$) and using $\text{GEN}_{\text{GTE}}(k)$ to generate the random string r_i of P_i in Π_{GTE}.[2] The encryption of x is i and $c \leftarrow \text{ENC}_{\text{GTE},i}(x, r_i)$ – the message of P_i on input x and randomness r_i.
- Algorithm $\text{COMP}_{\text{ORE}}((i_1, c_1), (i_2, c_2))$ returns $\text{DEC}_{\text{GTE}}(\{i_1, i_2\}, c_1, c_2)$ if $i_1 \neq i_2$ and "FAIL" otherwise.

If two messages are encrypted using different parties (i.e., $i_1 \neq i_2$), then the correctness of the comparison and the security of Π_{GTE} guarantees that, given the two encryptions, exactly their order is revealed (i.e., the first message is smaller, equal, or greater than the second message). If the two messages are encrypted using the same party (i.e., $i_1 = i_2$), then correctness and security are not guaranteed. However, the probability of this event is $1/n = 1/2^\lambda$, which is negligible. $\qquad\square$

---

[2] The time required to generate r_i is $O(\ell^2 \log n + \log^2 n)$.

Remark 6.6. In the proof of Theorem 6.5, we can replace the ad hoc PSM protocol for GTE_ℓ obtained via the PSM protocol from [17] by any ad hoc PSM protocol for GTE_ℓ as long as its complexity is $\eta(n, \lambda) \log^c n$ for some function η and constant c. In particular, if we use a $(2, t)$-secure ad hoc PSM protocol for GTE_ℓ, then the resulting ORE would be t-bounded secure.

The ORE of Theorem 6.5 is secure only when 2 messages are encrypted. If 3 messages are encrypted, then the adversary gets 3 messages of the ad hoc PSM protocol for GTE_ℓ and the security of the ad hoc PSM protocol is broken. We can construct a t-bounded $1/\lambda$-secure ORE as sketched below:

- The key generation algorithms generates keys for $\alpha = \text{poly}(\lambda, t)$ copies of the ORE of Theorem 6.5.
- The encryption algorithms encrypts m using a random subset of the keys of size $\lambda \sqrt{\alpha}$.
- Given encryptions of two messages, if there is a key that was used to encrypt both messages, then use the comparison algorithm of that copy to compare the two messages. The probability that no such key exists is $2^{-O(\lambda)}$.

The security of the above ORE is guaranteed as long as no 3 messages are encrypted with the same key. The probability that there are 3 messages that are encrypted under the same key can be reduced to $1/\lambda$ if α is big enough.

7 NIMPC Vs. (k, t)-Secure Ad Hoc PSM

In this section we consider two notions of PSM protocols, (k, t)-secure ad hoc PSM protocols and Non-Interactive secure MPC (NIMPC) protocols. Recall that an ad hoc PSM is (k, t)-secure if the referee getting at most t messages does not learn any information beyond the value of f on any subset of size k of the inputs. A t-robust NIMPC for a function f is a PSM protocol, where a referee colluding with t parties can only compute the values of the function when the inputs of the non-colluding parties is fixed (see [4] for a formal definition of NIMPC protocols). We show that the existence of NIMPC protocols is equivalent to the existence of (k, t)-secure ad hoc PSM protocols.

In the information-theoretic setting, these results should be interpreted as negative results, maybe implying that efficient protocols do not exists in both models. In the computaional setting, this results imply an efficient construction of computational ad hoc PSM protocols.

7.1 Ad hoc PSM $\Rightarrow$ NIMPC

Given an n-out-of-$2n$ ad hoc PSM protocol for a boolean function f, we construct an n-party robust NIMPC protocol for f with the same complexity.

Lemma 7.1. *If there exists an (n, t)-secure n-out-of-$2n$ ad hoc PSM protocol for a boolean function $f : \{0, 1\}^n \to \{0, 1\}$, then there exists an n-party $(t - n)$-robust NIMPC protocol for f with the same communication complexity.*

Proof. Let Π^* be the guaranteed ad hoc PSM protocol. Consider the following NIMPC protocol Π.

Randomness generation:

- Let $r_1, \ldots, r_{2n} \leftarrow \mathrm{GEN}_{\Pi^*}()$.
- Choose at random n random bits $b_1, \ldots, b_n$.
- For $i \in [n]$ let
  - $M_{i,0} \leftarrow (2i - b_i, \mathrm{ENC}_{\Pi^*, 2i-b_i}(r_{2i-b_i}, 0))$.
  - $M_{i,1} \leftarrow (2i - 1 + b_i, \mathrm{ENC}_{\Pi^*, 2i-1+b_i}(r_{2i-1+b_i}, 1))$.
- The randomness of P_i is $M_{i,0}, M_{i,1}$.

Message generation for every $P_i \in S$:

- P_i on input $x_i \in \{0, 1\}$ sends M_{i,x_i}.

Function reconstruction by the referee:

- The referee gets n messages, where for each i it gets from P_i either the messages of P_{2i} or P_{2i-1}. It uses the decryption of Π^* to compute f.

We next argue that Π is robust. Let A be a set of parties in Π of size $\tau \le t-n$. The randomness of A and the messages of all other parties in Π are messages of distinct $n+\tau \le t$ parties in Π^*. By the (n, t)-security of Π^*, from these messages the referee in Π^* can only compute the output of f on any subset of size n of these parties in Π^*, i.e., the inputs of the parties in Π that are not in A are fixed. Thus, in Π, the referee and the set A can only compute the residential function. Thus, the (n, t)-security of Π^* implies the $(t - n)$-robustness of Π. Notice that the referee knows the identity of the party in Π^* for which the messages was generated; however, by choosing random b_i's, it does not know if this message is for an input 0 or 1.   □

7.2   NIMPC ⇒ Ad Hoc PSM

Our goal is to construct a (k, t)-secure ad hoc PSM protocol for a boolean function f from an NIMPC protocol Π computing a related function. We would like to use ideas similar to the construction in Sect. 4. Recall that, given a k-argument function $f : X^k \to Y$, we defined an n-argument function $g : (X \cup \{\bot\})^n \to Y \cup \{\bot\}$, where if there are exactly $n - k$ inputs that are $\bot$ then the output of g is the output of f on the k non-$\bot$ inputs, and it is $\bot$ otherwise.[3] We constructed a k-secure ad hoc PSM protocol for f by first generating the randomness using the PSM for g, and sharing the messages of each party with input $\bot$. We would like to start from an NIMPC protocol for g and get a (k, t)-secure ad hoc PSM protocol for f. There is a problem with this solution – in the resulting ad hoc PSM protocol the referee will get for each

---

[3] In Sect. 4, if there were less than $n - k$ inputs that are $\bot$, then the function was undefined; here we need to define the output as $\bot$.

active party messages for some input x_i and for the input $\perp$. The definition of the robustness of NIMPC protocols guarantees that if it gets one message from a party, then the referee can only evaluate the function on points where the input of this party is fixed to some (unknown) value. The definition does not guarantee that if a referee gets two messages from one party then it can only evaluate the output on points where the input of this party is fixed to one of these two (unknown) values.

To overcome this problem we define a new function $g'' : \{0,1\}^{3n} \to Y$ with $3n$ variables $x_{1,0}, x_{1,1}, x_{2,0}, x_{2,1}, \ldots, x_{n,0}, x_{n,1}, y_1, \ldots, y_n$, where given an assignment $a_{1,0}, a_{1,1}, a_{2,0}, a_{2,1}, \ldots, a_{n,0}, a_{n,1}, c_1, \ldots, c_n$ of g'', we define an assignment $a_1, a_2, \ldots, a_n$ of g as follows:

$$a_i = \begin{cases} \perp & \text{if } a_{i,0} = a_{i,1}, \\ c_i & \text{if } a_{i,0} = 1, a_{i,1} = 0, \\ 1 - c_i & \text{if } a_{i,0} = 0, a_{i,1} = 1 \end{cases}$$

and $g''(a_{1,0}, a_{1,1}, a_{2,0}, a_{2,1}, \ldots, a_{n,0}, a_{n,1}, c_1, \ldots, c_n) = g(a_1, a_2, \ldots, a_n)$.

Theorem 7.2. *If there is a $3n$-party $2n$-robust NIMPC protocol $\Pi_{g''}$ for g'' with randomness complexity $\mathrm{Rand}(\Pi_{g''})$ and communication complexity $\mathrm{Comm}(\Pi_{g''})$ then there exists a (k,n)-secure ad hoc PSM protocol for f with randomness complexity $O(\mathrm{Rand}(\Pi_{g''}) + n \cdot \mathrm{Comm}(\Pi_{g''}))$ and communication complexity $O(n \cdot \mathrm{Comm}(\Pi_{g''}))$.*

Proof. Denote the parties of the NIMPC $\Pi_{g''}$ by $P_{1,0}, P_{1,1}, \ldots, P_{n,0}, P_{n,1}, Q_1,$ $\ldots, Q_n$. We next describe an ad hoc PSM protocol Π_f for f.

Randomness generation:

- Generate the randomness of $\Pi_{g''}$ for g''; let $r_{1,0}, r_{1,1}, \ldots, r_{n,0}, r_{n,1}, q_1,$ $\ldots, q_n$ be the generated randomness of $P_{1,0}, P_{1,1}, \ldots, P_{n,0}, P_{n,1}, Q_1, \ldots,$ Q_n respectively.
- For every $1 \le j \le n$:
  - Choose $c_i \in \{0,1\}$ at random and let m_i be the message that Q_i sends in $\Pi_{g''}$ with randomness q_i and input c_i.
  - For every $b \in \{0,1\}$, let $m_{j,b}$ be the message that $P_{j,b}$ sends in $\Pi_{g''}$ with randomness $r_{j,b}$ and input 0.
- The randomness of P_i in the ad hoc PSM protocol is

$$r_{i,0}, r_{i,1}, c_i, (m_i)_{1 \le i \le n}, (m_{j,b})_{1 \le i \le n, b \in \{0,1\}}.$$

Message generation for every $P_i \in S$:

- Let u_i be the message that $P_{i,c_i \oplus x_i}$ sends in $\Pi_{g''}$ with input 1 and randomness $r_{i,c_i \oplus x_i}$.
- P_i sends $(c_i \oplus x_i), u_i$ and, in addition, $(m_i)_{1 \le i \le n}, (m_{j,b})_{j \ne i, b \in \{0,1\}}$.

Assume that a set $S \in \binom{[n]}{k}$ sends messages. To compute the value of f on the inputs of S, the referee applies the decoding procedure of $\Pi_{g''}$, where for every $i \in [n]$,

- If $i \in S$, then the message of $P_{i,c_i \oplus x_i}$ is u_i (i.e., an encoding of 1); otherwise it is $m_{i,c_i \oplus x_i}$ (i.e., an encoding of 0),
- the message of $P_{i,1-(c_i \oplus x_i)}$ is $m_{i,1-(c_i \oplus x_i)}$ (i.e., an encoding of 0), and
- the message of Q_i is m_i (i.e., an encoding of c_i).

The correctness follows as these messages correspond to the input

$$(z_{i,b})_{i \in [n], b \in \{0,1\}}, (c_i)_{i \in [n]}$$

where:

- If $i \notin S$, then $z_{i,0} = z_{i,1} = 0$, that is, it correspond to the input $a_i = \perp$ of g.
- If $i \in S$, then $z_{i,c_i \oplus x_i} = 1$ and $z_{1-(i,c_i \oplus x_i)} = 0$,
  - If $x_i = c_i$, then $z_{i,0} = 1$ and $z_{i,0} = 0$, that is, it correspond to the input $a_i = c_i = x_i$ of g,
  - If $x_i \neq c_i$, then $z_{i,0} = 0$ and $z_{i,0} = 1$, that is, it correspond to the input $a_i = 1 - c_i = x_i$ of g.

That is, if $i \in S$, then it correspond to the input $a_i = x_i$ of g.

To conclude, the referee reconstructs

$$g''((z_{i,b})_{i \in [n], b \in \{0,1\}}, (c_i)_{i \in [n]}) = g((x_i)_{i \in S}, (\perp)_{i \notin S}) = f((x_i)_{i \in S}).$$

For the (k,t)-security, note that if a set T of size t' sends messages in the ad hoc PSM protocol for f, then the referee gets two messages for $P_{i,c_i \oplus x_i}$ for every $i \in S$ and one message for every other party. Thus, by the robustness of the NIMPC protocol $\Pi_{g''}$, the referee can only compute outputs of g, where the input of every $i \notin S$ is fixed to $\perp$ and the input of every $i \in S$ is either x_i or $\perp$. Since g is defined to be $\perp$ if the number of non-bottom inputs is not k, the referee can only compute the values of f on subsets of size k of T.   □

The transformation of Theorem 7.2 also applies if the NIMPC protocol is computationally-secure. Specifically, in [4] it is shown that if iO and one-way functions exist, then there is a computational indistinguishably-secure NIMPC protocol for every function. This implies that if iO and one-way functions exist then there is a computational (k,n)-indistinguishably-secure ad hoc PSM protocol for every function f.

8 Ad Hoc Protocols for and Threshold Imply Nontrivial Obfuscation

Computational ad hoc PSM protocols for general functions imply obfuscation. This follows from Lemma 7.1, showing that ad hoc PSM protocols imply NIMPC protocols, and by results of [4], showing that NIMPC protocols imply obfuscation. To prove this result, ad hoc PSM protocols for fairly complex functions, i.e., universal functions, are used. In this section, we show that ad hoc PSM protocols for *simple* functions already imply obfuscation for interesting functions.

Specifically, computational ad hoc PSM protocols for AND with VBB security imply point function obfuscation and ad hoc PSM protocols for threshold functions with VBB security imply fuzzy point function obfuscation [7]. There are several definitions of point function obfuscation in the literature (see [6]). In this paper, we consider the strong virtual black-box notion of obfuscation of Barak et al. [3] for point function and fuzzy point function obfuscation. This notion was considered for point function obfuscation in, e.g., [19]. As the only known constructions for fuzzy point function obfuscation are based on strong assumptions (e.g., iO), these results imply that even ad hoc PSM protocols with VBB security for the threshold function may require strong assumptions.

Notation 8.1. *For every $x \in \{0,1\}^n$, define the point function $I_x : \{0,1\}^n \to \{0,1\}$ where $I_x(y) = 1$ if $x = y$ and $I_x(y) = 0$ otherwise. For every $x \in \{0,1\}^n$ and $0 < \delta < 1$, define the fuzzy point function $F_x^\delta : \{0,1\}^n \to \{0,1\}$ where $F_x^\delta(y) = 1$ if $\mathrm{dist}(x,y) \leq \delta n$ and $F_x^\delta(y) = 0$ otherwise, where $\mathrm{dist}(x,y)$ is the Hamming distance. We will also denote by I_x and F_x the canonical circuits that compute these functions.*

Lemma 8.2. *If there exists an $(n, 2n)$-VBB-secure ad hoc PSM protocol for AND, then there is a point function obfuscation, i.e., an obfuscation for $\{I_x\}_{x \in \{0,1\}^n}$.*

Proof. The obfuscation algorithm of a point function I_x uses the computational ad hoc PSM protocol $\Pi_{\mathrm{AND}} = (\mathrm{GEN}_{\mathrm{AND}}, \mathrm{ENC}_{\mathrm{AND}}, \mathrm{DEC}_{\mathrm{AND}})$ for AND. We denote the $2n$ parties in Π_{AND} by $\{P_{i,b}\}_{i \in [n], b \in \{0,1\}}$. Algorithm $\mathrm{OBF}(1^n, x)$ is as follows:

- Let $(r_{i,b})_{i \in [n], b \in \{0,1\}} \leftarrow \mathrm{GEN}_{\mathrm{AND}}(1^n)$.
- For every $i \in [n]$ let $z_{i,x_i} \leftarrow 1$ and $z_{i,\overline{x_i}} \leftarrow 0$.
- For every $i \in [n]$ and $b \in \{0,1\}$ let $m_{i,b} \leftarrow \mathrm{ENC}_{\mathrm{AND}}(z_{i,b}, r_{i,b})$.
- Return a circuit C that on input $y \in \{0,1\}^n$ computes

$$\mathrm{DEC}_{\mathrm{AND}}(\{(i, y_i)\}_{i \in [n]}, (m_{i,y_i})_{i \in [n]}).$$

Correctness: The circuit C returns the output of the decoding algorithm DEC on the messages $(m_{i,y_i})_{i \in [n]}$, which encode the inputs $(z_{i,y_i})_{i \in [n]}$. Hence, C returns $\mathrm{AND}((z_{i,y_i})_{i \in [n]})$. If $y = x$, then for every $i \in [n]$, $y_i = x_i$ and $z_{i,y_i} = 1$, thus C returns 1. If $y \neq x$, then $y_i = \overline{x_i}$ for at least one $i \in [n]$, thus $z_{i,y_i} = 0$ and C returns 0.

Security: Let $\mathcal{A}$ be an adversary attacking the obfuscation in the real world, that is, the adversary gets the above circuit C. We construct a simulator SIM, with an oracle access to I_x, such that there exists a negligible function $\mathrm{negl}()$ for which, for every $x \in \{0,1\}^n$,

$$|\Pr[\mathcal{A}(1^n, C) = 1] - \Pr[\mathrm{SIM}^{I_x}(1^n) = 1]| \leq \mathrm{negl}(n), \tag{2}$$

where the first probability is taken over the randomness of $\mathcal{A}$ and the randomness of $\mathrm{OBF}(1^n, x)$ and the second probability is taken over the randomness of SIM.

We first define an attacker $\mathcal{A}_{\text{AND}}$ against the ad hoc PSM protocol Π_{AND}: $\mathcal{A}_{\text{AND}}$ gets as an input $2n$ messages and generates a circuit C from these messages as OBF does, and executes $\mathcal{A}$ on C. By the VBB-security of Π_{AND}, there exists a simulator SIM_{AND} for the adversary $\mathcal{A}_{\text{AND}}$; this simulator SIM_{AND} should have an oracle access to the function AND on any n of the $2n$ inputs $(z_{i,b})_{i\in[n],b\in\{0,1\}}$.

The simulator SIM for the obfuscation, with oracle access to I_x, emulates SIM_{AND}, where the queries to AND are answered as follows: if a query contains two variables $z_{i,0}$ and $z_{i,1}$, for some $i \in [n]$, then the answer is 0 (as the value of one of them is zero). Otherwise, for every i there is exactly one y_i such that z_{i,y_i} is in the query; in this case $z_{i,y_i} = 1$ if and only if $y_i = x_i$, i.e., $\text{AND}((z_{i,y_i})_{i\in[n],b\in\{0,1\}}) = 1$ iff $x = (y_1,\ldots,y_n)$ iff $I_x((y_1,\ldots,y_n)) = 1$. In this case, SIM answers the query by invoking its oracle I_x. The VBB-security of Π_{AND} implies that (2) holds. $\qquad\square$

For $\delta < 0.5$, let $\text{Th}_\delta : \{0,1\}^n \to \{0,1\}$ be the following function:

$$\text{Th}_\delta(x_1,\ldots,x_n) = 1 \text{ iff } \sum_{i=1}^{n} x_i \geq (1-\delta)n.$$

We next construct fuzzy point function obfuscation from an ad hoc PSM protocol for Th_δ with VBB security. The construction and its proof of correctness are similar to those in Lemma 8.2; however, the proof of security is more involved. For this proof, we need the following claim.

Claim 8.3. Let $\delta < 0.5$. There is an efficient algorithm that, given a point w such that $F_x^\delta(w) = 1$ and an oracle access to F_x^δ, can find x.

Proof. Let $w = (w_1,\ldots,w_n)$ and $\overline{w} = (\overline{w_1},\ldots,\overline{w_n})$. Since $\text{dist}(x,w) \leq \delta n < 0.5n$, it must be that $\text{dist}(x,\overline{w}) > 0.5n > \delta n$, i.e., $F_x^\delta(\overline{w}) = 0$. There must be a j such that $F_x^\delta(w_1,\ldots,w_j,\overline{w_{j+1}},\ldots,\overline{w_n}) = 1$ and $F_x^\delta(w_1,\ldots,w_{j-1},\overline{w_j},\ldots,\overline{w_n}) = 0$. Furthermore, such j can be found by $n-1$ queries to the oracle F_x^δ. Let $v = (w_1,\ldots,w_j,\overline{w_{j+1}},\ldots,\overline{w_n})$; it must be that $\text{dist}(x,v) = \lfloor\delta n\rfloor$. If $v_i = x_i$, for some i, and we flip the ith bit in v (i.e., consider $v \oplus e_i$), then the distance between the resulting sting and x will be larger than δn. On the other hand, if $v_i \neq x_i$, then $\text{dist}(x,v \oplus e_i) < \text{dist}(x,v) \leq \delta n$. Thus, the following procedure recovers x:

– For $i = 1$ to n: if $F_x^\delta(x,v \oplus e_i) = 0$ then $x_i = v_i$, otherwise $x_i = \overline{v_i}$.

$\qquad\square$

Lemma 8.4. *Let $\delta < 0.5$. If there is an $(n,2n)$-VBB-secure ad hoc PSM protocol for Th_δ, then there is a fuzzy point function obfuscation, i.e., an obfuscation for $\{F_x^\delta\}_{x\in\{0,1\}^\ell}$.*

Proof. The obfuscation algorithm $\text{OBF}_{\text{fuzzy}}$ of a fuzzy point function F_x^δ uses the computational n-out-of-$2n$ ad hoc PSM protocol $\Pi_{\text{Th}} = (\text{GEN}_{\text{Th}}, \text{ENC}_{\text{Th}}, \text{DEC}_{\text{Th}})$ for Th_δ. We denote the parties in Π_{Th} by $\{P_{i,b}\}_{i\in[n],b\in\{0,1\}}$. Algorithm $\text{OBF}(1^n,x)$ is as follows:

- Let $(r_{i,b})_{i \in [n], b \in \{0,1\}} \leftarrow \text{GEN}_{\text{Th}}(1^n)$.
- For every $i \in [n]$ let $z_{i,x_i} \leftarrow 1$ and $z_{i,\overline{x_i}} \leftarrow 0$.
- For every $i \in [n]$ and $b \in \{0,1\}$ let $m_{i,b} \leftarrow \text{ENC}_{\text{Th}}(z_{i,b}, r_{i,b})$.
- Return a circuit C that on input $y \in \{0,1\}^n$ computes

$$\text{DEC}_{\text{Th}}(\{(i, y_i)\}_{i \in [n]}, (m_{i,y_i})_{i \in [n]}).$$

Correctness: The circuit C returns the output of the decoding algorithm DEC on the messages $((m_{i,y_i})_{i \in [n]})$, which encode the inputs $(z_{i,y_i})_{i \in [n]}$. Hence, C returns $\text{Th}_\delta((z_{i,y_i})_{i \in [n]})$. If $\text{dist}(x,y) \leq \delta n$, then $y_i = x_i$ for at least $(1 - \delta)n$ values of i, and $z_{i,y_i} = 1$ for at least $(1 - \delta)n$ values of i, thus, C returns 1. If $\text{dist}(x,y) > \delta n$, then $y_i = \overline{x_i}$ for more than $(1 - \delta)n$ values of i, thus, $z_{i,y_i} = 1$ for less than $(1 - \delta)n$ values of i and C returns 0.

Security: Let $\mathcal{A}$ be an adversary attacking the obfuscation in the real world. We construct a simulator $\text{SIM}_{\text{fuzzy}}$, with an oracle access to F_x^δ, such that there exists a negligible function $\text{negl}()$ for which for every $x \in \{0,1\}^n$

$$|\Pr[\mathcal{A}(1^n, C) = 1] - \Pr[\text{SIM}_{\text{fuzzy}}^{F_x^\delta}(1^n) = 1]| \leq \text{negl}(n), \tag{3}$$

where the first probability is taken over the randomness of $\mathcal{A}$ and the randomness of $\text{OBF}_{\text{fuzzy}}(1^n, x)$ and the second probability is taken over the randomness of $\text{SIM}_{\text{fuzzy}}$.

We first define an attacker $\mathcal{A}_{\text{Th}}$ against the ad hoc PSM protocol Π_{Th}: $\mathcal{A}_{\text{Th}}$ gets as an input $2n$ messages and generates a circuit C from these messages as $\text{OBF}_{\text{fuzzy}}$ does, and executes $\mathcal{A}$ on C. By the VBB-security of Π_{Th}, there exists a simulator SIM_{Th} for the adversary $\mathcal{A}_{\text{Th}}$; this simulator SIM_{Th} should have an oracle access to the function Th_δ of any n of the inputs $(z_{i,b})_{i \in [n], b \in \{0,1\}}$.

The simulator $\text{SIM}_{\text{fuzzy}}$ for the obfuscation, with oracle access to F_x^δ, emulates SIM_{Th}, where the queries to Th_δ are answered as follows: If for every i there is exactly one y_i such that z_{i,y_i} is in the query, then $z_{i,y_i} = 1$ if and only if $y_i = x_i$, i.e., $\text{Th}_\delta((z_{i,y_i})_{i \in [n], b \in \{0,1\}}) = 1$ iff $\text{dist}(x, (y_1, \ldots, y_n)) \leq \delta n$ iff $F_x^\delta((y_1, \ldots, y_n)) = 1$. Thus, in this case, $\text{SIM}_{\text{fuzzy}}$ answers the query by invoking its oracle F_x^δ.

The challenging case is when a query contains two variables $z_{i,0}$ and $z_{i,1}$ for some $i \in [n]$; we call such queries "illegal". In this case, we do not know how to answer the query directly (e.g., as we did in Lemma 8.2). The idea of answering the query is that if Th_k returns 1 on the query, then the simulator can find a point w such that $F_x^\delta(w) = 1$ (as explained below), from such point it finds x (using Claim 8.3), computes $(z_{i,b})_{i \in [n], b \in \{0,1\}}$ as $\text{OBF}_{\text{fuzzy}}$ does, and answers the current and future queries using these values. If the simulator does not find such point w, then it returns 0.

Consider a query to Th_δ that contains exactly α pairs $z_{i,0}$ and $z_{i,1}$ for some $\alpha > 0$ and assume that the answer to the query is 1. Without loss of generality, the query is

$$(z_{i,y_i})_{1 \leq i \leq n-2\alpha}, \ z_{n-2\alpha+1,0}, \ z_{n-2\alpha+1,1}, \ldots, \ z_{n-\alpha,0}, \ z_{n-\alpha,1}$$

for some $y_1, \ldots, y_{n-2\alpha}$. The value of exactly α of the variables

$$z_{n-2\alpha+1,0}, z_{n-2\alpha+1,1}, \ldots, z_{n-\alpha,0}, z_{n-\alpha,1}$$

is 1, thus, $\sum_{i=1}^{n-2\alpha} z_{i,y_i} + \alpha \geq (1 - \delta)n$. Furthermore,

$$\sum_{i=n-2\alpha+1}^{n} z_{i,0} + \sum_{i=n-2\alpha+1}^{n} z_{i,1} = 2\alpha,$$

i.e., at least one of the sums is at least α. This implies that if the answer to the query is 1, then $\mathrm{Th}_x^\delta(y_1, \ldots, y_{n-2\alpha}, 0, \ldots, 0) = 1$ or $\mathrm{Th}_x^\delta(y_1, \ldots, y_{n-2\alpha}, 1, \ldots, 1) = 1$. Therefore, for each "illegal" query, the simulator asks two queries to the oracle Th_x^δ; if the answers to both of them are zero, the simulator answers 0 to the query. Otherwise, the simulator uses Claim 8.3 to find x, computes $(z_{i,b})_{i \in [n], b \in \{0,1\}}$ as $\mathrm{OBF}_{\mathrm{fuzzy}}$ does, and answers all further queries of $\mathrm{SIM}_{\mathrm{Th}}$ using these values. The VBB security of Π_{Th} implies that (3) holds. $\square$

Acknowledgments. We thank David Cash and David Wu for helpful discussions about Order Revealing Encryption.

The first author was supported by ISF grant 544/13 and by a grant from the BGU Cyber Security Research Center. The second and third authors were partially supported by ISF grant 1709/14, BSF grant 2012378, and NSF-BSF grant 2015782. Research of the second author was additionally supported from a DARPA/ARL SAFEWARE award, NSF Frontier Award 1413955, NSF grants 1619348, 1228984, 1136174, and 1065276, a Xerox Faculty Research Award, a Google Faculty Research Award, an equipment grant from Intel, and an Okawa Foundation Research Grant. This material is based upon work supported by the Defense Advanced Research Projects Agency through the ARL under Contract W911NF-15-C-0205. The views expressed are those of the authors and do not reflect the official policy or position of the Department of Defense, the National Science Foundation, or the U.S. Government.

References

1. Agrawal, R., Kiernan, J., Srikant, R., Xu, Y.: Order preserving encryption for numeric data. In: Proceedings of the 2004 ACM SIGMOD International Conference on Management of Data, pp. 563–574 (2004)
2. Applebaum, B., Raykov, P.: From private simultaneous messages to zero-information Arthur-Merlin protocols and back. In: Kushilevitz, E., Malkin, T. (eds.) TCC 2016. LNCS, vol. 9563, pp. 65–82. Springer, Heidelberg (2016). doi:10. 1007/978-3-662-49099-0_3
3. Barak, B., Goldreich, O., Impagliazzo, R., Rudich, S., Sahai, A., Vadhan, S.P., Yang, K.: On the (im)possibility of obfuscating programs. J. ACM **59**(2), 6 (2012)
4. Beimel, A., Gabizon, A., Ishai, Y., Kushilevitz, E.: Distribution design. In: Sudan, M. (ed.) Proceedings of the 2016 ACM Conference on Innovations in Theoretical Computer Science, pp. 81–92. ACM, New York (2016)
5. Beimel, A., Gabizon, A., Ishai, Y., Kushilevitz, E., Meldgaard, S., Paskin-Cherniavsky, A.: Non-interactive secure multiparty computation. In: Garay, J.A., Gennaro, R. (eds.) CRYPTO 2014. LNCS, vol. 8617, pp. 387–404. Springer, Heidelberg (2014). doi:10.1007/978-3-662-44381-1_22

6. Bellare, M., Stepanovs, I.: Point-function obfuscation: a framework and generic constructions. In: Kushilevitz, E., Malkin, T. (eds.) TCC 2016. LNCS, vol. 9563, pp. 565–594. Springer, Heidelberg (2016). doi:10.1007/978-3-662-49099-0_21

7. Bitansky, N., Canetti, R., Kalai, Y.T., Paneth, O.: On virtual grey box obfuscation for general circuits. In: Garay, J.A., Gennaro, R. (eds.) CRYPTO 2014. LNCS, vol. 8617, pp. 108–125. Springer, Heidelberg (2014). doi:10.1007/978-3-662-44381-1_7

8. Boldyreva, A., Chenette, N., Lee, Y., O'Neill, A.: Order-preserving symmetric encryption. In: Joux, A. (ed.) EUROCRYPT 2009. LNCS, vol. 5479, pp. 224–241. Springer, Heidelberg (2009). doi:10.1007/978-3-642-01001-9_13

9. Boldyreva, A., Chenette, N., Lee, Y., O'neill, A.: Order-preserving symmetric encryption. Technical report 2012/624, IACR Cryptology ePrint Archive (2012). http://eprint.iacr.org/2012/624

10. Boldyreva, A., Chenette, N., O'Neill, A.: Order-preserving encryption revisited: improved security analysis and alternative solutions. In: Rogaway, P. (ed.) CRYPTO 2011. LNCS, vol. 6841, pp. 578–595. Springer, Heidelberg (2011). doi:10.1007/978-3-642-22792-9_33

11. Boldyreva, A., Chenette, N., O'Neill, A.: Order-preserving encryption revisited: improved security analysis and alternative solutions. Technical report 2012/625, IACR Cryptology ePrint Archive (2012). http://eprint.iacr.org/2012/625

12. Boneh, D., Lewi, K., Raykova, M., Sahai, A., Zhandry, M., Zimmerman, J.: Semantically secure order-revealing encryption: multi-input functional encryption without obfuscation. In: Oswald, E., Fischlin, M. (eds.) EUROCRYPT 2015. LNCS, vol. 9057, pp. 563–594. Springer, Heidelberg (2015). doi:10.1007/978-3-662-46803-6_19

13. Brakerski, Z., Komargodski, I., Segev, G.: Multi-input functional encryption in the private-key setting: stronger security from weaker assumptions. In: Fischlin, M., Coron, J.-S. (eds.) EUROCRYPT 2016. LNCS, vol. 9666, pp. 852–880. Springer, Heidelberg (2016). doi:10.1007/978-3-662-49896-5_30

14. Feige, U., Kilian, J., Naor, M.: A minimal model for secure computation. In: Proceedings of the 26th ACM Symposium on the Theory of Computing, pp. 554–563 (1994)

15. Goldwasser, S., Gordon, S.D., Goyal, V., Jain, A., Katz, J., Liu, F.-H., Sahai, A., Shi, E., Zhou, H.-S.: Multi-input functional encryption. In: Nguyen, P.Q., Oswald, E. (eds.) EUROCRYPT 2014. LNCS, vol. 8441, pp. 578–602. Springer, Heidelberg (2014). doi:10.1007/978-3-642-55220-5_32

16. Goldwasser, S., Rothblum, G.N.: On best-possible obfuscation. In: Vadhan, S.P. (ed.) TCC 2007. LNCS, vol. 4392, pp. 194–213. Springer, Heidelberg (2007). doi:10.1007/978-3-540-70936-7_11

17. Ishai, Y., Kushilevitz, E.: Private simultaneous messages protocols with applications. In: 5th Israel Symposium on Theory of Computing and Systems, pp. 174–183 (1997)

18. Ishai, Y., Kushilevitz, E., Paskin, A.: Secure multiparty computation with minimal interaction. In: Rabin, T. (ed.) CRYPTO 2010. LNCS, vol. 6223, pp. 577–594. Springer, Heidelberg (2010). doi:10.1007/978-3-642-14623-7_31

19. Lynn, B., Prabhakaran, M., Sahai, A.: Positive results and techniques for obfuscation. In: Cachin, C., Camenisch, J.L. (eds.) EUROCRYPT 2004. LNCS, vol. 3027, pp. 20–39. Springer, Heidelberg (2004). doi:10.1007/978-3-540-24676-3_2

20. Yao, A.C.: How to generate and exchange secrets. In: Proceedings of the 27th IEEE Symposium on Foundations of Computer Science, pp. 162–167 (1986)

Topology-Hiding Computation Beyond Logarithmic Diameter

Adi Akavia[1] and Tal Moran[2(✉)]

[1] The Academic College of Tel-Aviv, Jaffa, Israel
akavia@mta.ac.il
[2] IDC Herzliya, Herzliya, Israel
talm@idc.ac.il

Abstract. A distributed computation in which nodes are connected by a partial communication graph is called *topology-hiding* if it does not reveal information about the graph (beyond what is revealed by the output of the function). Previous results [Moran, Orlov, Richelson; TCC'15] have shown that topology-hiding computation protocols exist for graphs of logarithmic diameter (in the number of nodes), but the feasibility question for graphs of larger diameter was open even for very simple graphs such as chains, cycles and trees.

In this work, we take a step towards topology-hiding computation protocols for arbitrary graphs by constructing protocols that can be used in a large class of *large-diameter networks*, including cycles, trees and graphs with logarithmic *circumference*. Our results use very different methods from [MOR15] and can be based on a standard assumption (such as DDH).

1 Introduction

When theoretical cryptographers think about privacy and computation, the first thing that comes to mind is usually secure multiparty computation (MPC), in which multiple parties can compute an arbitrary function of their inputs without revealing anything but the function's output. In the original definitions (and constructions) of MPC, the participants were connected by a full communication graph (a broadcast channel and/or point-to-point channels between every pair of parties). In real-world settings, however, the actual communication graph between parties is usually not complete, and parties may be able to communicate directly with only a subset of the other parties. Moreover, in some cases the graph itself is sensitive information (e.g., if you communicate directly only with your friends in a social network).

A natural question is whether we can successfully perform a joint computation over a partial communication graph while revealing no (or very little) information about the graph itself. In the information-theoretic setting, in which a

A. Akavia—Work partly supported by the ERC under the EU's Seventh Framework Programme (FP/2007-2013) ERC Grant Agreement no. 307952.

T. Moran—Supported by ISF grant no. 1790/13.

J.-S. Coron and J.B. Nielsen (Eds.): EUROCRYPT 2017, Part III, LNCS 10212, pp. 609–637, 2017.
DOI: 10.1007/978-3-319-56617-7_21

variant of this question was studied by Hinkelman and Jakoby [10], the answer is mostly negative.

The situation is better in the computational setting. Moran, Orlov and Richelson showed that topology-hiding computation *is* possible against static, semi-honest adversaries [12]. However, their protocol is restricted to communication graphs with *small diameter*. Specifically, their protocol addresses networks with diameter $d = O(\log n)$ logarithmic in the number of nodes n (where the diameter is the maximal distance between two nodes in the graph). For many natural network topologies the question remains open, including for wireless and ad-hoc sensor networks [4,14], where topology is modeled by random geometric graphs [13] whose diameter is large with high probability [5], as well as for very simple topologies such as chains, cycles and trees (note that topology-hiding computation isn't trivial even if the overall topology of the network is known; in a cycle, for example, the order of the nodes may still be sensitive).

1.1 Our Results

In this work we take a step towards topology-hiding computation protocols for arbitrary graphs by constructing protocols that can be used in a large class of *large-diameter networks*. As in [12], our protocols actually implement topology-hiding *broadcast*—given this primitive, standard MPC protocols can then be used for generic topology-hiding computation.

- We construct a protocol for topology-hiding broadcast on directed cycles, given an upper bound on the number of nodes in the cycle. This protocol uses completely different techniques than those of [12], and in particular does not require generic MPC (it borrows ideas from mix networks). The security of this protocol can be based on standard assumptions, e.g., the Decisional Diffie-Hellman assumption.
- We show that given (black-box) access to a protocol for topology-hiding broadcast on a cycle, we can construct a protocol for topology-hiding broadcast on arbitrary graphs if nodes are given an auxiliary information specifying their neighbors in a spanning tree of the graph (this information will be used to compute a cycle traversing all nodes of the graph). Our security guarantee in this case is that our protocol reveals nothing beyond what can be learned from the auxiliary information. For arbitrary graphs, we do not know how to compute the auxiliary information in a topology-hiding manner, so this protocol would require a trusted setup phase for those graphs. Any class of graphs for which we can construct a protocol to compute this auxiliary information locally will give us a topology-hiding broadcast for that class of graphs. In particular, for trees, the "auxiliary information" for computing a spanning tree is trivial and does not require trusted setup; thus, together with our cycle protocol this result gives us a protocol for topology-hiding broadcast on trees.

This construction makes only black-box use of the cycle protocol, and does not require additional assumptions.

- We define *information-local computation*; loosely speaking this is a distributed computation in which the outputs of each party can depend only on information available in their "local neighborhood". We prove that information-local computations can be performed in a topology-hiding way on arbitrary graphs given a topology-hiding computation protocol for small-diameter graphs.
- We construct an information-local computation for computing consistent local views of a spanning tree in arbitrary *small-circumference* graphs (in which the length of the longest cycle is bounded by k). This gives a protocol for topology-hiding broadcast on *small-circumference* graphs. This protocol makes black-box use of both the small-diameter topology-hiding computation protocol and almost-black-box use of the topology-hiding broadcast on the cycle (we require the existence of an efficient circuit to compute the next-message function of the cycle protocol).

Assumptions. Our basic protocol for topology-hiding broadcast on a cycle can be based on the Decisional Diffie-Hellman (DDH) assumption. Our further reductions are black box, and do not require additional assumptions. Elaborating on the former, all we require for our basic protocol is the existence of an CPA-secure encryption scheme with some special properties (aka, *hPKCR-enc*); we show that such a scheme exists based on DDH. The properties we require from a hPKCR-enc are essentially that ciphertexts are rerandomizable (given the public-key), that it is key-commutative when given the secret-key (we name the latter property *privately* key-commutative), and that it is homomorphic w.r. to a single operation; see details in Sect. 3.

Voting vs. Broadcast. We also present protocols for topology-hiding *voting* (rather than broadcast) for all aforementioned graph topologies; for full security, these require the *exact* number of nodes for the cycle topology to be known (rather than an upper-bound). We note that for our voting protocols we do not require the homomorphism property of the underlying hPKCR-enc scheme. Recall that voting means that each player P_i has at the beginning of the protocol an input vote v_i, and receives at the termination of the protocol a list of all votes in a randomly permuted order $(v_{\pi_i(1)}, \ldots, v_{\pi_i(n)})$ for $\pi_i \colon [n] \to [n]$ a uniformly random permutation.

We summarize our results in the following theorems. For brevity we use the shorthand notation "TH-$\mathcal{F}$" standing for "an efficient topology-hiding protocol realizing functionality $\mathcal{F}$ against a statically-corrupting semi-honest adversary". Denote by $|G|$ the number of nodes in a graph G.

Theorem 1 (Topology-hiding on cycle). *Under DDH assumption, for every network of cycle topology C, there exist the following protocols:*

- **Broadcast.** *A TH-$\mathcal{F}_{Broadcast}$, provided parties are given an upper-bound on $|C|$.*
- **Voting.** *A TH-$\mathcal{F}_{Vote}$, provided parties are given the exact size $|C|$.*

Theorem 2 (Reductions to other topologies). *Suppose there exists TH-$\mathcal{F}$ for networks of cycle-topology C when given upper-bound on (resp. exact size of) $|C|$. Then there exists TH-$\mathcal{F}$ for the following (connected, undirected) network graphs G, when given an upper-bound on (resp. exact size of) $|G|$:*

- *Every graph G, provided parties are given their neighbors in a (globally consistent) spanning tree of G.*
- *Every tree G.*
- *Every graph G with circumference at most k, provided there exists TH-$\mathcal{F}_{Broadcast}$ for graphs of diameter at most k.*

Combining the above theorems and employing a TH-$\mathcal{F}_{Broadcast}$ protocol for low-diameter graphs [12] we conclude:

Corollary 1. *Under the DDH assumption, there exists TH-$\mathcal{F}_{Broadcast}$ (resp. TH-$\mathcal{F}_{Vote}$) for the following (connected, undirected) network graphs G, when given an upper-bound on (resp. exact size of) $|G|$:*

- *Every cycle G.*
- *Every graph G, provided parties are given their neighbors in a (globally consistent) spanning tree of G.*
- *Every tree G.*
- *Every graph G with circumference at most $O(\log |G|)$.*

1.2 High-Level Overview of Our Techniques

In the following we first give an overview of the our approach and challenges, then an overview for our protocols for topology-hiding voting, and conclude by describing their modification yielding our protocols for topology-hiding broadcast.

Our Approach and Challenges. Recall that in a broadcast protocol a bit $b \in \{0,1\}$ is given as input to a single player called the broadcasting player, and the protocol terminates with all players receiving this bit b as output. Our starting point is the "OR-and-Forward" protocol, in which at the first round the broadcasting player forwards its bit b to all its neighbors, and at each following round the players OR their received bits and forward the resulting bit to their neighbors. Note that the bit b reaches all players once the number of rounds exceeds the network diameter.

This "OR-and-Forward" protocol is of course not topology-hiding. For one, distance to the broadcaster leaks from the round number t when a node i first receives a message. To prevent this leakage-by-timing attack, we change the protocol to have all players send messages at all rounds, where the change is simply by asking the non-broadcasting players to send the neutral bit 0 in the first round.

The latter protocol is still not topology-hiding, e.g., because distance to the broadcaster leaks from the round number t when a node i first receives a non-zero message. To prevent this leakage-by-content attack, we'd like to encrypt all transmitted messages, so that the players (nodes) cannot identify when their received messages are transformed from 0 to 1.

Encrypting the transmitted messages raises new challenges. First, when using such an encryption, who has the secret-key for decryption?! If every player has the secret key, then encryption hides nothing; and if not, then how do players decrypt to get the output? Second, how can we compute the OR of encrypted messages? To address these challenges we first restrict our attention to the cycle topology, and use key-homomorphic and message-homomorphic encryption (e.g., ElGamal). Our first protocol realizes a *voting* functionality rather than broadcast—its output is a shuffled list of all parties' inputs. On the cycle topology, encryption/decryption can be computed jointly by all players via going around the cycle where every player adds/peels-of their own encryption layer (using the key-homomorphism property). This protocol requires us to know the exact length of the cycle in order to prevent leaking topology information. We then show how to use the homomorphic operation to OR ciphertexts together in a way that hides topology even if the exact cycle length is not known.

Topology-Hiding Voting on Directed Cycles. At the beginning of the voting protocol each player i has a secret input v_i (her vote). The protocol then proceeds in two phases, *aggregation* and *mix-and-decrypt*. Loosely speaking, in the aggregation phase votes are aggregated by passing around the cycle an array of encrypted votes, to which each party adds their own vote and then adds an extra layer of encryption before sending it on to the next party. At the end of this phase, each party has an array of all n votes, encrypted under the a public key whose secret key is shared between *all* parties. In the mix-and-decrypt phase, the parties successively remove the layer of encryption they are responsible for, mix the votes and rerandomize the remaining layers of encryption before passing the array back to the previous party. Upon the termination of this phase, each player i has a list of all votes in a randomly permuted order $(v_{\pi_i(1)}, \ldots, v_{\pi_i(n)})$ (for $\pi_i \colon [n] \to [n]$ a uniformly random permutation). For details, see Sect. 4.

From Voting to Broadcast. Relaxing the input to consist of an *upper bound* n' on the number of nodes n in the cycle (rather than the exact number) is the main challenge in devising our topology-hiding broadcast protocol. Subsequently, our reductions from the cycle topology to other topologies go through as is.

Our first attempt was to execute our voting protocol as is, while using n' instead of n; but this fails to be topology-hiding. In particular, topology information leaks from the output now consisting of multiple votes from some parties.[1] We remark that correctness is also undermined by receiving multiple votes from

---

[1] For example, for $n' = n + 3$, the output of player i consists of double votes from players $i, i+1, i+2$, implying that non-neighboring corrupted players i, j can identify whether $j = i + 2$ by letting j place a unique vote v^* and then count whether this vote v^* appears once in the output of i (implying $j \neq i + 2$) or twice (implying $j = i + 2$).

some players; this could be fixed (say, by augmenting the vote with an anonymous identifier and post-processing to remove multiple votes), yet, this fix is of course not topology-hiding.

Our topology-hiding broadcast is a modification of the above approach where we combine the list of votes into a single bit—the OR of all input bits, thus avoiding the aforementioned votes counting attack. Specifically, the *aggregation* phase in our topology-hiding broadcast combines each additional vote to a single ciphertext being passed around, where this ciphertext either holds the neutral message or a random group element (interpreted as a broadcast of 0 or 1 respectively). To combine the votes we require the underlying hPKCR-enc to be homomorphic with respect to a single operation. The "*mix-and-decrypt*" phase is a degenerate version of the mix-and-decrypt phase in our topology-hiding voting protocol, where the players peel off decryption layers and re-randomize, but with no need for mixing ciphertexts (as now only a single ciphertext is being passed around).

The additional wrinkle here is that, when we know only an *upper bound* n' on the number of nodes, using a simple homomorphic multiplication (or addition) to OR bits together is not topology hiding. To see this, suppose the broadcaster chooses the value m as the non-neutral element representing a 1 bit. Every time the encrypted bit passes around the cycle it is multiplied by m, so the output will be m^c, where c is the number of passes through the broadcasting party. Thus, the output leaks a tighter estimate for n, as well as information on the distance from the broadcasting party (for example, parties i, j receiving outputs m^2, m, respectively, can conclude that i appears first on the path from the broadcasting party).

To prevent this leakage-by-output attack, we require that all parties randomize their message m before passing it forward (and then re-randomize also the ciphertext). Our message randomization is by raising m to a random power r (using homomorphic multiplication, and exponentiation-by-squaring algorithm for efficiency), this in turn maps the identity to itself while mapping other elements m to uniformly random group elements m^r (choosing the message space to be a prime order group).

For space considerations, a detailed description and analysis of our broadcast protocol is deferred to the full version of the paper.

From Cycles to Graphs Given Spanning-Tree Neighbors and Tree. The main idea for this reduction is that given a tree, we can compute a cycle-traversal of the graph by having each node *independently* decide on a local ordering of edges. Each node will appear exactly d times in the cycle (where d is the degree of the node). The predecessor of the ith instance of the node in the cycle is the neighbor adjacent to its ith edge; the successor is the neighbor adjacent to the next edge. In Sect. 5.2, we prove that for any numbering of edges, this always generates a cycle-traversal of the graph, and that this traversal can be used to execute any protocol for topology-hiding computation on directed cycles.

Topology-Hiding Computation for Information-Local Functions. The intuition behind this reduction is simple: if a node's output depends only on information from a node's k-neighborhood, then we can use a topology-hiding computation protocol for small-diameter graphs to compute it by limiting the protocol to the node's k-neighborhood. The reason this isn't quite that straightforward is that we want to hide the topology of the k-neighborhoods. This means the node participating in the protocol shouldn't be able to tell who else is participating with it in the protocol. In particular, this means we can't assign a global session id to distinguish between multiple concurrent instances of the protocol (we need to run multiple instances since each node will need to compute the function given its own local neighborhood).

Our main innovation here is the use of *relative* session ids, where the session id depends on the node's relative location in its neighborhood, and each node applies a transformation on the session ids sent and received so that they remain in the correct relative framework. In Sect. 7 we describe our protocol in detail and prove that we can use it to compute an arbitrary information-local function in parallel in all k-neighborhoods.

From Graphs Given Spanning-Tree Neighbors to Small-Circumference Graphs. For this reduction, we combine several of our previous results. There are two main ideas here. First, we prove that for graphs whose circumference is bounded by k, we can compute local views of a spanning tree using a k-information-local function (this is not trivial, since local views must be globally consistent with a single spanning tree). Together with our result on topology-hiding computation given a spanning tree, and our result on computation of information-local functions, this already gives a protocol for topology-hiding computation on small-circumference graphs. This naïve way of running the protocol reveals the local view of the spanning tree to each node, which can give information about the graph topology. However, we show it is possible to compose the protocols into a single computation that does not reveal the spanning-tree information (at the cost of higher complexity). Due to space considerations, the details of this reduction are deferred to the full version of the paper.

1.3 Related Work

Topology-Hiding MPC in Computational Settings. The most relevant related work is that of Moran, Orlov and Richelson [12], who gave the first feasibility results for topology-hiding computation in the computational setting, giving a protocol for topology-hiding broadcast secure against static, semi-honest adversaries, as well as a protocol secure against fail-stop adversaries that do not disconnect the graph. However, their protocol is restricted to communication graphs with diameter logarithmic in the total number of parties.

The main idea behind their protocol is a series of nested multiparty computations, in which each node is replaced with a secure computation in its local neighborhood that simulates that node. In contrast, our cycle protocol uses ideas from the cryptographic voting literature—it hides the order of the nodes in the cycle by "mixing" encrypted inputs before decrypting them.

Other related works include a concurrent work by Hirt et al. [11] that achieves better efficiency than [12] for topology-hiding computation, albeit still restricted to low diameter networks as in [12]. The work by Chandran et al. [3] addresses the question of hiding the communication network in the context of secure multi-party computation, but with a different goal than topology-hiding: their goal is to reduce communication complexity by allowing each party to communicate with a small (sublinear in the total number of parties) number of its neighbors.

Topology-Hiding MPC in Information Theoretic Settings. Hinkelmann and Jakoby [10] considered the question of topology-hiding secure computation while focusing on the information theoretic setting. Their main result is negative: any MPC protocol in the information-theoretic setting must inherently leak information about G to an adversary. They do, however, prove a nice positive result: if we are allowed to leak a routing table of the network, one can construct an MPC protocol which leaks no further information.

Secure Multiparty Computation with General Interaction Patterns. Halevi et al. [8] presented a unified framework for studying secure MPC with arbitrarily restricted interaction patterns (generalizing models for MPC with specific restricted interaction patterns [1,7,9]). The questions they study, however, are independent of our topology-hiding focus. Their starting point is the observation that an adversary controlling the final players $P_i, \ldots, P_n$ in the interaction pattern can learn the output of the computed function on several inputs (as the adversary can rewind and execute the protocol over any possible values for the inputs $x_i, \ldots, x_n$ for the corrupted players while fixing the inputs $x_1, \ldots, x_{i-1}$ for the preceding parties). The question they ask is therefore *when is it possible to prevent the adversary from learning the output of the function on multiple inputs.* In contrast to ours, their model allows complete knowledge on the underlying interaction patterns, and does not hide the topology.

2   Preliminaries

2.1   Computation and Adversarial Models

We model a network by a directed graph $G = (V, E)$ that is not fully connected. We consider a system with $n = \text{poly}(\kappa)$ parties (where κ is the security parameter), denoted $P_1, \ldots, P_n$. We often implicitly identify V with the set of parties $\{P_1, \ldots, P_n\}$. We consider a static and computationally bounded (PPT) adversary that controls some subset of parties (any number of parties). That is, at the beginning of the protocol, the adversary corrupts a subset of the parties and may instruct them to deviate from the protocol according to the corruption model. Throughout this work, we consider only semi-honest adversaries. In addition, we assume that the adversary is rushing; that is, in each round the adversary sees the messages sent by the honest parties before sending the messages of the corrupted parties for this round. For general MPC definitions including in-depth descriptions of the adversarial models we consider see [6].

2.2 Definitions of Graph Terms

Let $G = (V, E)$ be an undirected graph. For $v \in V$ we let $N(v) = \{w \in V : (v, w) \in E\}$ denote the *neighborhood of* v; and similarly, the *closed neighborhood of* v, $N[v] = N(v) \cup \{v\}$. We sometimes refer to $N[v]$ as the closed 1-neighborhood of v, and for $k \geq 1$ we define the k-neighborhood of v as the set of all nodes within distance k of v. Formally, we can define this recursively:

$$N^{(k+1)}[v] = \bigcup_{w \in N^{(k)}[v]} N[w].$$

The *k-neighborhood graph* of v in G is the subgraph $G^{(k)}[v]$ of G on the k-neighborhood of v, defined by

$$G^{(k)}[v] = (N^{(k)}[v], E') \text{ where } E' = \left\{ (u, w) \mid u, v \in N^{(k)}[v] \text{ and } w \in N[u] \right\}.$$

2.3 UC Security

As in [12], we prove security in the UC model [2]. Proving security in the UC model allows our protocols to be composed with other protocols, and makes it easier to use as a subprotocol in more complex constructions. For details about the UC framework, we refer the reader to [2]. We note that although the UC model requires setup for security against general adversaries, this is not necessary for security against semi-honest adversaries, so we also get a protocol that is secure in the plain model.

2.4 Topology Hiding—The Simulation-Based Definition

To help make the paper more self-contained, in this section we reproduce the simulation-based definition for topology hiding computation from [12].

The UC model usually assumes all parties can communicate directly with all other parties. To model the restricted communication setting, [12] define the $\mathcal{F}_{\text{graph}}$-hybrid model, which employs a special "graph party," P_{graph}. Figure 1 shows $\mathcal{F}_{\text{graph}}$'s functionality: at the start of the functionality, $\mathcal{F}_{\text{graph}}$ receives the network graph from P_{graph}, and then outputs, to each party, that party's neighbors. Then, $\mathcal{F}_{\text{graph}}$ acts as an "ideal channel" for parties to communicate with their neighbors, restricting communications to those allowed by the graph.

Since the graph structure is an input to one of the parties in the computation, the standard security guarantees of the UC model ensure that the graph structure remains hidden (since the only information revealed about parties' inputs is what can be computed from the output). Note that the P_{graph} party serves

---

Participants/Notation:
This functionality involves all the parties $P_1, \ldots, P_m$ and a special graph party P_{graph}.

Initialization Phase:
Inputs: $\mathcal{F}_{\text{graph}}$ waits to receive the graph $G = (V, E)$ from P_{graph}.
Outputs: $\mathcal{F}_{\text{graph}}$ outputs $N_G[v]$ to each P_v.

Communication Phase:
Inputs: $\mathcal{F}_{\text{graph}}$ receives from a party P_v a destination/data pair (w, m) where $w \in N(v)$ and m is the message P_v wants to send to P_w.
Output: $\mathcal{F}_{\text{graph}}$ gives output (v, m) to P_w indicating that P_v sent the message m to P_v.

Fig. 1. The functionality $\mathcal{F}_{\text{graph}}$.

only to specify the communication graph, and does not otherwise participate in the protocol.

Since $\mathcal{F}_{\text{graph}}$ provides local information about the graph to all corrupted parties, *any* ideal-world adversary must have access to this information as well (regardless of the functionality we are attempting to implement). To capture this, we define the functionality $\mathcal{F}_{\text{graphInfo}}$, that is identical to $\mathcal{F}_{\text{graph}}$ but contains only the initialization phase. For any functionality $\mathcal{F}$, we define a "composed" functionality $(\mathcal{F}_{\text{graphInfo}} || \mathcal{F})$ that adds the initialization phase of $\mathcal{F}_{\text{graph}}$ to $\mathcal{F}$. We can now define topology-hiding MPC in the UC framework:

Definition 1. *We say that a protocol Π securely realizes a functionality $\mathcal{F}$ hiding topology if it UC-realizes $(\mathcal{F}_{graphInfo} || \mathcal{F})$ in the $\mathcal{F}_{graph}$-hybrid model.*

Note that this definition can also capture protocols that realize functionalities depending on the graph (e.g., find a shortest path between two nodes with the same input, or count the number of triangles in the graph).

3 Privately Key-Commutative and Rerandomizable Encryption

We require a public key encryption scheme with the properties of being *homomorphic* (w.r. to a single operation), *privately key-commutative*, and *rerandomizable*. In this section we first formally define the properties we require, and then show how they can be achieved based on the Decisional Diffie-Hellman assumption.

We call an encryption scheme satisfying the latter two properties, i.e., privately key-commutative and re-randomizable, a *hPKCR-enc*; and call an encryption scheme satisfying all three properties, i.e., homomorphic, privately key-commutative and re-randomizable, a *hPKCR-enc*.

3.1 Required Properties

Let $\mathsf{KeyGen} : \{0, 1\}^* \mapsto \mathcal{PK} \times \mathcal{SK}$, $\mathsf{Enc} : \mathcal{PK} \times \mathcal{M} \times \{0, 1\}^* \mapsto \mathcal{C}$, $\mathsf{Dec} : \mathcal{SK} \times \mathcal{C} \mapsto \mathcal{M}$ be the encryption scheme's key generation, encryption and decryption functions, respectively, where $\mathcal{PK}$ is the space of public keys, $\mathcal{SK}$ the space of secret keys, $\mathcal{M}$ the space of plaintext messages and $\mathcal{C}$ the space of ciphertexts.

We will use the shorthand $[m]_k$ to denote an encryption of the message m under public-key k. We assume that for every secret key $sk \in \mathcal{SK}$ there is associated a single public key $pk \in \mathcal{PK}$ such that (pk, sk) are in the range of KeyGen. We slightly abuse notation and denote the public key corresponding to sk by $pk(sk)$.

Rerandomizable. We require that there exists a ciphertexts "re-randomizing" algorithm Rand $: \mathcal{C} \times \mathcal{PK} \times \{0,1\}^* \mapsto \mathcal{C}$ satisfying the following:

1. *Randomization:* For every message $m \in \mathcal{M}$, every public key $pk \in \mathcal{PK}$ and ciphertext $c = [m]_{pk}$, the distributions $(m, pk, c, \mathsf{Rand}\,(c, pk, U^*))$ and $(m, pk, c, \mathsf{Enc}_{pk}(m; U^*))$ are computationally indistinguishable.
2. *Neutrality:* For every ciphertext $c \in \mathcal{C}$, every secret key $sk \in \mathcal{SK}$ and every $r \in \{0,1\}^*$,

$$\mathsf{Dec}_{sk}(c) = \mathsf{Dec}_{sk}(\mathsf{Rand}\,(c, pk(sk), r)).$$

Furthermore, we require that public-keys are "re-randomizable" in the sense that the product $k \circledast k'$ of an arbitrary public key k with a public-key k' generated using KeyGen is computationally indistinguishable from a fresh public-key generated by KeyGen.

Privately Key-Commutative. The set of public keys $\mathcal{PK}$ form an abelian (commutative) group. We denote the group operation $\circledast$. Given any $k_1, k_2 \in \mathcal{PK}$, there exists an efficient algorithm to compute $k_1 \circledast k_2$. We denote the inverse of k by k^{-1} (i.e., $k^{-1} \circledast k$ is the identity element of the group). Given a secret key sk, there must be an efficient algorithm to compute the inverse of its public key $(pk(sk))^{-1}$.

There exist a pair of algorithms AddLayer $: \mathcal{C} \times \mathcal{SK} \mapsto \mathcal{C}$ and DelLayer $: \mathcal{C} \times \mathcal{SK} \mapsto \mathcal{C}$ that satisfy:

1. For every public key $k \in \mathcal{PK}$, every message $m \in \mathcal{M}$ and every ciphertext $c = [m]_k$,

$$\mathsf{AddLayer}\,(c, sk) = [m]_{k \circledast pk(sk)}\,.$$

2. For every public key $k \in \mathcal{PK}$, every message $m \in \mathcal{M}$ and every ciphertext $c = [m]_k$,

$$\mathsf{DelLayer}\,(c, sk) = [m]_{k \circledast (pk(sk))^{-1}}\,.$$

We call this *privately* key-commutative since adding and deleting layers both require knowledge of the secret key.

Note that since the group $\mathcal{PK}$ is commutative, adding and deleting layers can be done in any order.

Homomorphism. We require the message space $\mathcal{M}$ forms a group with operation denoted $\cdot$, and require that the encryption scheme is homomorphic with respect this operation $\cdot$ in the sense that there exist an efficient algorithm $\mathsf{hMult} : \mathcal{C} \times \mathcal{C} \mapsto \mathcal{C}$ that, given two ciphertexts $c = [m]_{pk}$ and $c' = [m']_{pk}$, returns a ciphertext $c'' \leftarrow \mathsf{hMult}c_1, c_2$ s.t. $\mathsf{Dec}_{sk}(c'') = m \cdot m'$ (for sk the secret-key associated with public-key pk).

3.2   Instantiation of PKCR-enc and hPKCR-enc Under DDH

We use standard ElGamal, augmented by the additional required functions. The KeyGen, Dec and Enc functions are the standard ElGamal functions, except that to obtain a one-to-one mapping between public keys and secret keys, we fix the group G and the generator g, and different public keys vary only in the element $h = g^x$. Below, g is always the group generator. The Rand function is also the standard rerandomization function for ElGamal:

function $\mathrm{RAND}(c = (c_1, c_2), pk, r)$
    return $(c_1 \cdot g^r, pk^r \cdot c_2)$
end function

We use the shorthand notation of writing $\mathsf{Rand}\,(c, pk)$ when the random coins r are chosen independently at random during the execution of Rand. We note that the distribution of public-keys outputted by KeyGen is uniform, and thus the requirement for "public-key rerandomization" indeed holds. ElGamal public keys are already defined over an abelian group, and the operation is efficient. For adding and removing layers, we define:

function $\mathrm{ADDLAYER}(c = (c_1, c_2), sk)$
    return $(c_1, c_2 \cdot c_1^{sk})$
end function
function $\mathrm{DELLAYER}(c = (c_1, c_2), sk)$
    return $(c_1, c_2 / c_1^{sk})$
end function

Every ciphertext $[m]_{pk}$ has the form $(g^r, pk^r \cdot m)$ for some element $r \in \mathbb{Z}_{ord(g)}$. So

$$\mathsf{AddLayer}\left([m]_{pk}, sk'\right) = (g^r, pk^r \cdot m \cdot g^{r \cdot sk'}) = (g^r, pk^r \cdot (pk')^r \cdot m)$$
$$= (g^r, (pk \cdot pk')^r \cdot m) = [m]_{pk \cdot pk'}.$$

It is easy to verify that the corresponding requirement is satisfied for Del-Layer as well.

ElGamal message space already defined over an abelian group with homomorphic multiplication, specifically:

function $\mathrm{HMULT}(c = (c_1, c_2), c' = (c_1', c_2'))$
    return $c'' = (c_1 \cdot c_1', c_2 \cdot c_2')$
end function

Recalling that the input ciphertext have the form $c = (g^r, pk^r \cdot m)$ and $c' = (g^{r'}, pk^{r'} \cdot m')$ for messages $m, m' \in \mathbb{Z}_{ord(g)}$, it is easy to verify that decrypting the ciphertext $c'' = (g^{r+r'}, pk^{r+r'} \cdot m \cdot m')$ returned from hMult yields the product message $\mathsf{Dec}_{sk}(c'') = m \cdot m'$.

Finally, to obtain a negligible error probability in our broadcast protocols, we take G a prime order group of size satisfying that $1/|G|$ is negligible in the security parameter κ.

4 Topology-Hiding Voting for Cycle Topology

In this section we present our topology-hiding voting protocol for the cycle topology. That is, we consider networks where the n players are numbered by indices $1, \ldots, n$, and communication is only between players with consecutive indices, i.e., player i can communicate with players $i + 1$ and $i - 1$ (where addition is modulo n). We remind the reader that cycles have a large diameter (diameter n/2), and are therefore not handled by prior works on topology hiding protocols.

4.1 Topology Hiding Voting for Cycle Topology from PKCR-Enc

The Protocol. Recall that each player i has a secret input v_i (her vote). To simplify notation, we omit the modulus when specifying the party (i.e., we let $P_{n+1} = P_1$ and $P_0 = P_n$). The protocol is composed of two main phases:

- In the first phase the votes are aggregated in encrypted form. This phase consists of n rounds. At the first round each player i encrypts its vote v_i and sends it to player $i + 1$ together with the public-key $pk_i^{(1)}$. At every following round t, upon receiving from player $i - 1$ a list L of encrypted votes together their encryption key k, player i does the following. (a) Encrypt its vote v_i under key k, and add the resulting ciphertext to the list L, (b) Add an encryption layer to every vote in the list using its keys $(pk_i^{(t)}, sk_i^{(t)})$, (c) Compute the new key $k' = k \circledast pk_i^{(t)}$, and (d) Sends the updated list L' and key k' to $i + 1$. We note that player i uses fresh keys $(pk_i^{(t)}, sk_i^{(t)})$ at each round t, this is necessary for security.
- In the second phase the players each remove an encryption layer to reveal the votes in plaintext, while also shuffling the votes and re-randomizing their ciphertexts so that the votes in the resulting lists are not traceable to the voters.

See Protocol 1 for details.

A Note Regarding Notation. We use the superscript $(i : t)$ to denote variables set by party i in iteration t; e.g., the notation $k^{(i:t)}$ denotes the key party i receives from party $i - 1$ in iteration t of the AGGREGATE phase. When the identity of the party is clear, we will sometimes use the shorter versions $k^{(t)}$. For clarity we also omit the modular arithmetic for party indices, identifying party 0 with party n and party $n + 1$ with party 1.

Protocol 1. Cycle Protocol for Player i

1: **procedure** CYCLEVOTE(n, v_i)
2:    // **AGGREGATE PHASE:**
3:    Generate keys $(pk^{(i:1)}, sk^{(i:1)}), \ldots, (pk^{(i:n-1)}, sk^{(i:n-1)}) \leftarrow$ KeyGen(1^k).
4:    Send $[v_i]_{pk^{(i:1)}}$ and $pk^{(i:1)}$ to P_{i+1}.
5:    **for** $t \in \{1, \ldots, n-2\}$ **do**
6:        Wait to receive $c_1^{(t)}, \ldots, c_{(t)}^{(t)}$ and $k^{(t)}$ from P_{i-1}
7:        Send $[v_i]_{k^{(t)} \circledast pk^{(i:t+1)}}$, AddLayer $\left(c_1^{(t)}, sk^{(i:t+1)}\right), \ldots,$ AddLayer $\left(c_t^{(t)}, sk^{(i:t+1)}\right)$
        and $k^{(t)} \circledast pk^{(i:t+1)}$ to P_{i+1}
8:    **end for**
9:    Wait to receive $c_1^{(n-1)}, \ldots, c_{n-1}^{(n-1)}$ and $k^{(n-1)}$ from P_{i-1}
10:    // **MIXANDDECRYPT PHASE:**
11:    Let $d_1^{(n-1)} \leftarrow [v_i]_{k^{(n-1)}}$ // Encryption of our own vote
12:    For all $j \in \{2, \ldots, n\}$, denote $d_j^{(n-1)} \doteq c_{j-1}^{(n-1)}$.
13:    **for** $t \in \{n-1, \ldots, 1\}$ **do**
14:        // Mix and Rerandomize
15:        Choose a random permutation $\pi = \pi^{(i:t)} : [n] \mapsto [n]$.
16:        For all $j \in \{1, \ldots, n\}$, let $h_j^{(i:t)} \leftarrow$ Rand $\left(d_{\pi(j)}^{(t)}, k^{(t)}\right)$.
17:        // Pass back
18:        Send $h_1^{(i:t)}, \ldots, h_n^{(i:t)}$ to P_{i-1}
19:        Wait to receive $h_1^{(i+1:t)}, \ldots, h_n^{(i+1:t)}$ from P_{i+1}
20:        // Decrypt
21:        For all $j \in \{1, \ldots, n\}$, let $d_j^{(t-1)} \leftarrow$ DelLayer $\left(h_j^{(i+1:t)}, sk^{(i:t)}\right)$
22:    **end for**
23:    // **OUTPUT:**
24:    Output $d_1^{(0)}, \ldots, d_n^{(0)}$
25: **end procedure**

4.2 Correctness and Topology-Hiding

We formally prove the following theorems about Protocol 1.

Theorem 3 (Completeness). *Protocol 1 is complete.*

Theorem 4 (Topology-Hiding). *If the underlying encryption PKCR scheme is CPA-secure then Protocol 1 realizes the functionality $\mathcal{F}_{Vote}$ in a topology-hiding way against a statically-corrupting, semi-honest adversary.*

The proofs of these theorems appear in Sect. 6.

5   Dealing with General (Connected, Undirected) Graphs

In this section we show that topology-hiding computation on a cycle is a useful stepping stone to other large-diameter graphs. Given a protocol for computing a (symmetric) functionality $\mathcal{F}_f$ on a cycle, we show how to construct a topology-hiding protocol for computing the functionality on arbitrary graphs, as long as every node is also given some auxiliary data: a local view of a cycle-traversal of the graph (the computed function is a "multiple-input" version of the original, see below for details).

A corollary is that we can construct a topology-hiding voting and broadcast protocols for every network topology for which the aforementioned auxiliary information can be efficiently found in a topology-hiding way. Chains, trees and small-circumference graphs are a special case of the latter.

We remark that in our results using auxiliary information, this information may be given to the players once-and-for-all during setup (as it depends only of the network topology and not on the input to the voting protocol). Clearly, the auxiliary information reveals properties of the graphs; in this case, the topology-hiding property of our protocols ensures that no additional information is revealed.

Multiple-Input Extension. Our reductions to $\mathcal{F}_f$ on a cycle realize a slightly different functionality—we realize a "multiple-input" version of $\mathcal{F}_f$ which we denote $\mathcal{F}_{f^*}$. Loosely speaking, the "multiple-input" extension of a function allows each party to give several inputs to the function, with the number of inputs depending on the number of times the party appears in the cycle traversal. Formally, let $\{f_n\}$ be a class of symmetric functions on n inputs (i.e., in which the order of inputs does not matter). For every n and $g : [n] \mapsto \mathbb{N}$, we define $f_{g,n}^*$ to be $f_{\sum_{i=1}^n g(n)}$ where the ith input to $f_{g,n}^*$ is a vector of $g(i)$ inputs to $f_{\sum_{i=1}^n g(n)}$. In our case, g will always map each party to the number of times the party appears in the cycle traversal; we will omit g and n for brevity and write $\mathcal{F}_{f^*}$ as the functionality we are computing.

5.1 Dealing with General Graphs, Given Local Views of a Cycle-Traversal

In this section we first observe that for every (connected, undirected) graph G there exists a cycle traversing all its nodes;[2] we call such a cycle *a cycle-traversal* of G, denoted C_G. We then show that if all nodes of G are given their local view of the cycle as auxiliary information, then they can execute a topology-hiding protocol to compute $\mathcal{F}_{f^*}$ on G simply by executing a topology-hiding protocol for $\mathcal{F}_f$ on its cycle-traversal, C_G.

A cycle-traversal is sure to exist as it can be explicitly found, e.g., by running Depth-First-Search (DFS) to find a DFS-tree spanning all nodes of the graph, and then converting this tree to a cycle-traversal using Protocol 4.

The local view of the cycle traversal C_G is specified for each node i by its *successor function* $\mathsf{Succ}_i \colon N(i) \mapsto N(i) \cup \{\bot\}$. Note that each node may appear

---

[2] Note that, unlike Eulerian or Hamiltonian cycles, we do not require a single pass through each edge or vertex. This relaxation in turn guarantees that the cycle-traversal always exists and can be found efficiently (when given the graph as input).

more than once on the cycle, so the successor function is defined $\mathsf{Succ}_i(v) = u$ if-and-only-if $(v \to i \to u)$ is part of the cycle C_G (if the edge (v, i) is not on C_G, then $\mathsf{Succ}_i(v) = \bot$).

Theorem 5 (Cycle-traversal known, realizing $\mathcal{F}_{f^*}$). *Suppose there exists a topology-hiding protocol Π that realizes the functionality $\mathcal{F}_f$ on directed cycles. Then there exists a topology-hiding protocol Π' realizing the functionality $\mathcal{F}_{f^*}$ on any (connected, undirected) graph G, when given as auxiliary input local views of a cycle-traversal C_G of G.*

We remark that for protocols Π requiring auxiliary information aux (e.g., the cycle length m) Theorem 5 still holds, provided that aux is included in Π''s auxiliary information. The only change in the proof is that the protocol Π' provides aux to Π when calling it.

Proof (sketch for Theorem 5). Our protocol Π' for G (c.f. Prototcol 2) simply runs Π on the cycle C_G, where each node $i \in V$ plays the role of all its w_i occurrences on the cycle (in parallel). Recall that in the functionality $\mathcal{F}_{f^*}$, the input to player P_i is a vector $(v_{i,1}, \ldots, v_{i,w_i})$. Player P_i executes as w_i independent players in Π, with each occurrence ℓ using input $v_{i,\ell}$, and where sending forward (backward) messages received from $j \in N(i)$ is executed by sending them to the corresponding successor (predecessor) on the cycle C_G.

Protocol 2. Topology-hiding protocol for $\mathcal{F}_{f^*}$ given cycle-traversal C_G. Protocol Description for player i on its cycle occurrence preceded by node pred and followed by node $\mathsf{succ} = \mathsf{Succ}_i(\mathsf{pred})$ ($\mathsf{pred} \to i \to \mathsf{succ}$), and with input v.

1: **procedure** CYCLETRAVERSALVOTE(v, pred, succ, Π)
2:    // Π is a protocol for topology-hiding computation of $\mathcal{F}_f$ on a cycle
3:    Execute Π on input v, sending messages "forward" by sending them to succ, and sending messages "back" by sending them to pred.
4: **end procedure**

We note that the when $\mathcal{F}_f = \mathcal{F}_{\mathrm{Vote}}$ is the voting functionality, the above protocol realizes the *weighted-vote* functionality $\mathcal{F}_{f^*} = \mathcal{F}_{\mathrm{WVote}}$ that accepts as input a list of w_i votes from each party i (where w_i the number of times party i appears on C_G), and outputs a list of all $m = \sum_{i=1}^{n} w_i$ votes in a randomly permuted order. Nevertheless, in the semi-honest setting, the standard voting functionality $\mathcal{F}_{\mathrm{Vote}}$ can be easily reduced to $\mathcal{F}_{\mathrm{WVote}}$ by letting each party submit only one "real" vote and use a $\bot$ value for the additional votes. For the broadcast functionality, the multiple-input version gives the same output as the original (without modification).

Due to space considerations, the details and formal analysis are deferred to the full version of the paper. $\square$

5.2 Dealing with General Graphs, Given Local Views of a Spanning-Tree

In this section we show that if there exists a topology-hiding protocol for $\mathcal{F}_f$ on a cycle, and for some spanning-tree $T = (V, F \subseteq E)$ of a graph $G = (V, E)$, all nodes are given as auxiliary information their neighbors in T, then there exists a topology-hiding protocol realizing the functionality $\mathcal{F}_{f^*}$ on G. The main idea is that given a spanning tree, nodes can locally compute (their local views of) a cycle-traversal of G. Thus, we can reduce this problem to the previously solved one of topology-hiding computation given a cycle-traversal of G.

Let $G = (V, E)$ be a connected undirected graph describing the network topology.

Theorem 6 (Spanning-tree known, realizing $\mathcal{F}_{f^*}$). *If there exists a protocol Π that realizes $\mathcal{F}_f$ in a topology-hiding way, given as input a local view of a cycle-traversal of G and $m = |C_G|$ (the cycle length), then, using Π as a black box, Protocol 3 realizes $\mathcal{F}_{f^*}$ when given as inputs a local view of a spanning tree T of G and n (the number of nodes in G).*

Proof. Our proof follows from the existence of a local translation from local views of a spanning-tree T to local views of a cycle-traversal C_G. Protocol 3 simply executes this translation and then runs Π using the auxiliary information about the cycle-traversal.

The auxiliary input conversion is local in the sense that each node executes the computation while using only its own auxiliary information. Specifically, the conversion executed by node i takes as input its neighbors $N_T(i)$ in the spanning T, and returns as output its successor function Succ_i on the cycle C_G of G. In our execution of Protocol 2 we use the cycle length m and the successor functions Succ_i as the auxiliary information of node i.

Protocol 3. Topology-hiding computation of $\mathcal{F}_{f^*}$ given spanning-tree neighbors $N_T(i)$. Protocol Description for player i.

```
 1: procedure SPANNINGTREECOMPUTE(n, i, N(i) ≐ {v ∈ V | (i, v) ∈ F})
 2:     Succᵢ = ConvertTreeToCycle(N(i))  // compute cycle-traversal
 3:     Execute Π using m = 2(n − 1) and Succᵢ as auxiliary information.
 4: end procedure
 5: procedure CONVERTTREETOCYCLE(N(i))
 6:     if |N(i)| = 0 then
 7:         return Succᵢ ≐ ∅ // singleton graph, empty cycle
 8:     else
 9:         return {Succᵢ(vᵢ) ≐ vᵢ₊₁}_{i=1,...,d}  for (v₁,..., v_d) an ascending ordering of
           the neighbors N(i) of i, and where we identify v_{d+1} ≐ v₁.
10:     end if
11: end procedure
```

1: **procedure** $\textsc{SpanningTreeCompute}(n, i, N(i) \doteq \{v \in V \mid (i, v) \in F\})$
2: $\quad \mathsf{Succ}_i = \text{ConvertTreeToCycle}(N(i))$ // compute cycle-traversal
3: $\quad$ Execute Π using $m = 2(n-1)$ and Succ_i as auxiliary information.
4: **end procedure**
5: **procedure** $\textsc{ConvertTreeToCycle}(N(i))$
6: $\quad$ **if** $|N(i)| = 0$ **then**
7: $\quad\quad$ **return** $\mathsf{Succ}_i \doteq \emptyset$ // singleton graph, empty cycle
8: $\quad$ **else**
9: $\quad\quad$ **return** $\left\{ \mathsf{Succ}_i(v_i) \doteq v_{i+1} \right\}_{i=1,\ldots,d}$ for $(v_1, \ldots, v_d)$ an ascending ordering of the neighbors $N(i)$ of i, and where we identify $v_{d+1} \doteq v_1$.
10: $\quad$ **end if**
11: **end procedure**

Completeness. In Lemma 1 we show that our conversion procedure is correct; i.e., there exists a length m cycle-traversal of G such that the output for each node i is indeed its successor function on this cycle. Thus, by completeness of Protocol 2 the output of each node is indeed the output of the $\mathcal{F}_{f^*}$ functionality.

Security follows immediately from the security of Protocol 1.   □

Lemma 1 (Tree to cycle). *There exists a length m cycle-traversal C_G of G such that for every node $i \in V$, its output Succ_i in the conversion procedure (c.f. line 5 in Protocol 3) equals its successor function on this cycle C_G.*

Proof. We show that the functions Succ_i returned by the conversion procedure are the successor functions for a length m cycle-traversal of G. For this purpose, we exhibit an algorithm that, given a spanning-tree T, outputs a length cycle-traversal C_T of T (c.f. Protocol 4 and Claims 1 and 2). Next we prove that Succ_i are successor functions for the graph C_T returned by Protocol 4 (c.f. Claim 3).

Specifically, Protocol 4 first initializes C_T to consist of a single node $C_T = \langle x \rangle$ (for an arbitrary $x \in V$). Next, while there exists a node u in C_T with a neighbor $v \in N(u)$ not included in C_T, the algorithm pastes to C_T in place of u a cycle C_v defined as follows. The cycle $C_v = \langle w_1, v, w_2, v, \ldots, v, w_d, v, w_1 \rangle$ is a cycle traversing all neighbors $N_T(v) = \{w_1, \ldots, w_d\}$ of v (where neighbors order on the cycle is ascending, and the starting point is chosen to be $w_1 = u$). That is, for $(v_1, v_2, \ldots, v_d)$ the neighbors of v in ascending order where $u = v_j$ we define $(w_1, w_2, \ldots, w_d) = (v_j, v_{j+1}, \ldots, v_d, v_1, \ldots, v_{j-1})$. We remark that the requirement of traversing neighbors of v is ascending order is not essential; we use it merely to facilitate notations in demonstrating the correspondence between the cycle C_T returned from Protocol 4 and the successor function Succ_i returned from line 5 Protocol 3 (c.f. Claim 3).

Protocol 4. Find cycle-traversal, given spanning-tree.

1: **procedure** CONVERTTREETOCYCLE($T = (V, F)$)
2:     $C_T = \langle x \rangle$ for arbitrary $x \in V$
3:     **while** exists $u \in C_T$ and $v \in N(u)$ such that $v \notin C_T$ **do**
4:         Let $C_v = \langle u, v, w_2, v, \ldots, v, w_d, v, u \rangle$ for $(u, w_2, \ldots, w_d)$ the neighbors of v in T in ascending order shifted to start with u.
5:         Paste C_v into C_T in place of the first appearance of u in C_T
6:     **end while**
7:     return C_T
8: **end procedure**

□

Claim 1. The output C_T of Protocol 4 is a cycle-traversal of G.

Proof. Observe that C_T visits all nodes of T, and thus all nodes of G. Else, if there is an unvisited node y, then there is a path from x to y (since T is connected), and on this path there must be a node u with neighbor $v \notin C_T$. A contradiction to the termination of the while loop.

Next, observe that C_T is a cycle. This is proved by induction. The base case $\langle x \rangle$ is a cycle of length zero. The induction hypothesis say that the content of C_T throughout the first t iterations of the while loop is a cycle. The induction step shows that C_T remains a cycle after pasting $C_v = \langle u, v, w_2, v, \ldots, v, w_d, v, u \rangle$. For this purpose note that C_v is a cycle starting at node u, and thus when it is pasted in place of u in $C_T = \langle \ldots a, u, b \ldots \rangle$ (which is a cycle, by induction hypothesis) the resulting $C_T' = \langle \ldots a, u, v, w_2, v, \ldots, v, w_d, v, u, b \ldots \rangle$ remains a cycle.

We conclude that C_T of is a cycle-traversal of G. $\square$

Claim 2. The output C_T of Protocol 4 has length $m = 2(n-1)$.

Proof. To show that C_T has length $m = 2(n-1)$ we recall that there are $n-1$ edges in a spanning-tree for a graph with $n = |V|$ nodes, and argue that the cycle C_T goes through each edge of the spanning-tree exactly twice. Thus resulting in a total of $2(n-1)$ edges on the cycle.

To complete the above argument we first prove by induction that throughout the first t iterations of the while loop, the number of times C_T goes through every edge of T is either 0 or 2. Base case ($t=0$): At the initialization, $C_T = \langle x \rangle$ passes through every edges exactly 0 times. Induction step: Note that the cycle to be pasted C_v goes over each edge connecting v to its neighbor exactly twice. Moreover, these edges (v, w) were not included in C_T prior to pasting C_v by the choice of v as a node not appearing in C_T. Next, we note that to traverse all notes of the tree, C_T must go through each edge at least once. We conclude therefore that C_T goes through each edge exactly twice. $\square$

Claim 3. The output Succ_i in the conversion procedure (c.f. line 5 in Protocol 3) is the successor function for the cycle C_T output by Protocol 4.

Proof. Recall first that the output Succ_i returned from the conversion procedure (c.f. Line 5 in Protocol 3) is defined to be

$$\left\{ \mathsf{Succ}_i(v_i) \doteq v_{i+1} \right\}_{i=1,\ldots,d}$$

for $(v_1, \ldots, v_d)$ an ascending ordering of the neighbors $N(i)$ of i (and where we identify $v_{d+1} \doteq v_1$). Namely, this successor function corresponds to the cycle $C_i = (v_1, i, v_2, i, \ldots, i, v_d)$ where the nodes $v_1, \ldots, v_d$ are in ascending order.

Recall also that in the cycle C_T returned from Protocol 4 the edges passing through node i were added by pasting a cycle $C_i' = (u, i, w_2, i, \ldots, i, w_d)$ passing through the neighbors $u, w_1, \ldots, w_d$ of i in ascending order shifted to start with neighbor u. Since this is a cycle, we may view each point as the starting point. Choosing to view the smallest neighbor as the starting point we see that C_i' is the same as the cycle C_i.

We conclude that the function Succ_i is identical to the successor function for the cycle C_T outputted by Protocol 4. $\square$

5.3   Dealing with Trees

A simple corollary of Theorem 6 is that if G is a tree T (i.e., connected and acyclic), then there exists a topology-hiding protocol realizing the voting functionality $\mathcal{F}_{f*}$ on G. The proof is derived from the fact that when G is a tree, the players can trivially find their neighbors in its spanning tree, without needing to get it as an auxiliary input:

Corollary 2 (Trees). *Suppose there exists a topology-hiding protocol realizing the functionality $\mathcal{F}_f$ on directed cycles, when given (a bound on) the cycle length. Then there exists a topology-hiding protocol realizing the functionality $\mathcal{F}_{f*}$ on every tree T, when given (a bound on) the number of nodes in T.*

6   Topology-Hiding Voting for Cycle Topology—Formal Proofs

In this section we give the formal proofs of correctness and security for Protocol 1.

6.1   Correctness Analysis

Denote $k^{(0)} \doteq 1$, the identity element of the group. Then:

Claim 4. For every party i and all $t \in \{0, \ldots, n-1\}$:

$$k^{(i:t)} = \prod_{j=1}^{t} pk^{(i-j:j)}.$$

Proof. Let i be an arbitrary party index. The proof is by induction on t. For $t = 0$ it is trivially true. Assume it is true for all i up to some $t \in \{0, \ldots, n-2\}$, then in iteration t of the AGGREGATE loop, party $i-1$ will have $k^{(i-1:t)} = \prod_{j=1}^{t} pk^{(i-1-j:j)}$, and will send $k^{(i-1:t)} \circledast pk^{(i-1:t+1)} = \prod_{j=1}^{t+1} pk^{(i-j:j)}$ to party i in line 7. Thus, for party i, $k^{(i:t+1)}$ will also have the required form.   □

Claim 5. For every party i and all $t \in \{1, \ldots, n-1\}$:

$$\left(c_1^{(i:t)}, \ldots, c_t^{(i:t)} \right) = \left([v_{i-1}]_{k^{(i:t)}}, \ldots, [v_{i-t}]_{k^{(i:t)}} \right).$$

Proof. Let i be an arbitrary party index. The proof is also by induction on t. For $t = 1$, $c^{(i:1)} = [v_{i-1}]_{pk^{(i-1:1)}}$ (as sent by party $i-1$ in line 4). Assume the hypothesis is true for all i up to iteration t. The values party i receives at lines 6 and 9 in iteration $t+1$ of the AGGREGATE phase are those sent by party $i-1$ in line 7 of iteration t:

$$\left(c_1^{(i:t+1)}, \ldots, c_{t+1}^{(i:t+1)}\right)$$

$$= \left([v_{i-1}]_{k^{(i-1:t)} \circledast pk^{(i-1:t+1)}}, \mathsf{AddLayer}\left(c_1^{(i-1:t)}, sk^{(i-1:t+1)}\right), \ldots\right.$$

$$\left.\ldots, \mathsf{AddLayer}\left(c_t^{(i-1:t)}, sk^{(i-1:t+1)}\right)\right)$$

(by the induction hypothesis, $c_j^{(i-1:t)} = [v_{i-j-1}]_{k^{(i-1:t)}}$)

$$= \left([v_{i-1}]_{k^{(i-1:t)} \circledast pk^{(i-1:t+1)}}, \mathsf{AddLayer}\left([v_{i-2}]_{k^{(i-1:t)}}, sk^{(i-1:t+1)}\right), \ldots\right.$$

$$\left.\ldots, \mathsf{AddLayer}\left([v_{i-t-1}]_{k^{(i-1:t)}}, sk^{(i-1:t+1)}\right)\right)$$

(by the definition of AddLayer)

$$= \left([v_{i-1}]_{k^{(i-1:t)} \circledast pk^{(i-1:t+1)}}, [v_{i-2}]_{k^{(i-1:t)} \circledast pk^{(i-1:t+1)}}, \ldots,\right.$$

$$\left.[v_{i-t-1}]_{k^{(i-1:t)} \circledast pk^{(i-1:t+1)}}\right)$$

(since $k^{(i:t+1)} = k^{(i-1:t)} \circledast pk^{(i-1:t+1)}$)

$$= \left([v_{i-1}]_{k^{(i:t+1)}}, [v_{i-2}]_{k^{(i:t+1)}}, \ldots, [v_{i-t-1}]_{k^{(i:t+1)}}\right),$$

confirming the hypothesis for iteration $t+1$.

$\square$

Claim 6. For every party i and all $t \in \{1, \ldots, n-1\}$:

$$\left(h_1^{(i:t)}, \ldots, h_n^{(i:t)}\right) \quad - \left([v_{\sigma^{(i:t)}(1)}]_{k^{(i:t)}}, \ldots, [v_{\sigma^{(i:t)}(n)}]_{k^{(i:t)}}\right)$$

and

$$\left(d_1^{(i:t-1)}, \ldots, d_n^{(i:t-1)}\right) \quad = \left([v_{\sigma^{(i+1:t)}(1)}]_{k^{(i:t-1)}}, \ldots, [v_{\sigma^{(i+1:t)}(n)}]_{k^{(i:t-1)}}\right)$$

where $\sigma^{(i:t)} = \pi^{(i:t)} \circ \cdots \circ \pi^{(i+n-t-1:n-1)}$.

Proof. We note that it's enough to show the claim holds for the h values, since the corresponding d values are computed from them by peeling off the outer key layer (by calling DelLayer with the key $sk^{(i:t)}$).

The proof is by induction on t (we run the induction backwards, starting at $t = n-1$). Let i be an arbitrary party. First, note that by the initial assignment to the d values in lines 11 and 12 of the MIXANDDECRYPT phase, and using Claim 5, we have

$$\left(d_1^{(i:n-1)}, \ldots, d_n^{(i:n-1)}\right) = \left([v_i]_{k^{(i:n-1)}}, \ldots, [v_{i-(n-1)}]_{k^{(i:n-1)}}\right)$$

Thus, $t = n-1$ and all $i, j \in \{1, \ldots, n\}$:

$$h_j^{(i:n-1)} = \mathsf{Rand}\left(d_{\pi^{(i:n-1)}(j)}^{(i:n-1)}, k^{(n-1)}\right) = [v_{\pi^{(i:n-1)}(j)}]_{k^{(i:n-1)}} = [v_{\sigma^{(i:n-1)}(j)}]_{k^{(i:n-1)}},$$

which is the required value of $h^{(i:n-1)}$ according to the induction hypothesis. Since this is true for all i, it also holds for the values of $h^{(i+1:n-1)}$ received in line 19. Hence, it follows that for the $d^{i:n-2}$ values computed in line 21 we have:

$$d_j^{(i:n-2)} = \mathsf{DelLayer}\left(h^{(i+1:n-1)}, sk^{(i:t)}\right)$$

$$= \mathsf{DelLayer}\left(\left[v_{\sigma(i+1:n-1)}(j)\right]_{k^{(i+1:n-1)}}, sk^{(i:t)}\right)$$

which by Claim 4:

$$= \mathsf{DelLayer}\left(\left[v_{\sigma(i+1:n-1)}(j)\right]_{k^{(i:n-1)}\circledast pk^{(i:t)}}, sk^{(i:t)}\right)$$

and by the definition of DelLayer :

$$= \left[v_{\sigma(i+1:n-1)}(j)\right]_{k^{(i:n-1)}}$$

as required by the induction hypothesis.

Assume the hypothesis holds for all i down to some t. Then for iteration $t-1$, the $h^{i:t-1}$ values computed in line 16 are a rerandomized permutation of $d^{(i:t-1)}$, hence by the induction hypothesis they are:

$$h_j^{(i:t-1)} = \mathsf{Rand}\left(d_{\pi^{(i:t-1)}(j)}^{(i:t-1)}, k^{(t-1)}\right) = \left[v_{\pi^{(i:t-1)}(\sigma^{(i:t)}(j))}\right]_{k^{(i:t-1)}}$$

$$= \left[v_{\sigma^{(i:t-1)}(j)}\right]_{k^{(i:t-1)}}.$$

Since this is true for all i, the $h^{(i+1:t-1)}$ values received in line 19 also satisfy the equation, hence the $d^{(i:t-2)}$ values satisfy

$$d_j^{(i:t-2)} = \left[v_{\sigma(i+1:t-1)}(j)\right]_{k^{(i:t-1)}}$$

(the details are exactly as in the proof of the base case).   □

Proof (of Theorem 3). The theorem follows directly from Claim 6, setting $t = 1$ the outputs of party i are the values $d_1^{(i:0)}, \ldots, d_n^{(i:0)}$.   □

6.2   Security Analysis

Proof (of Theorem 4). To prove Theorem 4, we first describe the ideal-world simulator $\mathcal{S}$ (that "lives" in the ideal world in which all honest parties are dummy parties and there exists only the composed $\mathcal{F}_{\mathsf{Vote}} \| \mathcal{F}_{\mathsf{Graph}}$ functionality). We will then prove, via a hybrid argument, that the environment's interactions with $\mathcal{S}$ are computationally indistinguishable from an interaction with the real-world adversary $\mathcal{A}$ that "lives" in the real world in which the parties execute Protocol 1 and only the $\mathcal{F}_{\mathsf{Graph}}$ functionality exists.

Simulator Description. $\mathcal{S}$ works as follows:

1. Let $\mathcal{Q}$ be the set of corrupt parties. Note that we are in the semi-honest model with static corruptions, so $\mathcal{Q}$ and the input to each party in $\mathcal{Q}$ is available at the start of the protocol, and the adversary *must* play "according to protocol" with these inputs.
2. $\mathcal{S}$ sends the inputs for the parties in $\mathcal{Q}$ to $\mathcal{F}_{\mathsf{Vote}}$ and receives the output of $\mathcal{F}_{\mathsf{Vote}}$ (i.e., a random permutation of all the parties' inputs). Let $out_1, \ldots, out_n$ be the output $\mathcal{F}_{\mathsf{Vote}}$ sends to $\mathcal{S}$.

3. $\mathcal{S}$ receives the local neighborhood information from $\mathcal{F}_{\text{Graph}}$ for all parties in $\mathcal{Q}$. For $P \in \mathcal{Q}$, let $\text{pred}\,(P)$ denote the party preceding P on the cycle and $\text{succ}\,(P)$ the party succeeding P on the cycle.

4. The adversary partitions $\mathcal{Q}$ into "segments", where each segment consists of a sequence of corrupt parties that appear consecutively on the cycle. The segments are separated on the cycle by one or more honest parties (if it's more than one, $\mathcal{S}$ can't tell how many, or in which order the segments appear on the cycle). Let $\mathcal{Q}_{\mapsto} \subseteq \mathcal{Q}$ be the set of corrupt parties that are *first* in their segment, i.e., $\mathcal{Q}_{\mapsto} = \{P \in \mathcal{Q} : \text{pred}\,(P) \notin \mathcal{Q}\}$ and $\mathcal{Q}_{\mapsto\!|} = \{P \in \mathcal{Q} : \text{succ}\,(P) \notin \mathcal{Q}\}$ the set of parties that are *last* in their segment.

5. Within each segment, $\mathcal{S}$ simulates the corrupt parties exactly according to protocol. However, $\mathcal{S}$ must still simulate the inputs to the parties in $\mathcal{Q}_{\mapsto}$ and $\mathcal{Q}_{\mapsto\!|}$ that are generated by honest parties.

6. **Simulating messages from honest parties in the AGGREGATE phase, for party $P \in \mathcal{Q}_{\mapsto}$:**

   (a) $\mathcal{S}$ generates $n - 1$ key pairs:

   $$(pk^{(\text{pred}(P):1)}, sk^{(\text{pred}(P):1)}), \ldots, (pk^{(\text{pred}(P):n-1)}, sk^{(\text{pred}(P):n-1)})$$

   for the honest party preceding P (by honestly running KeyGen).

   (b) To simulate the $n - 1$ messages sent by $\text{pred}\,(P)$ in lines and 7, for each $t \in \{0, \ldots, n - 2\}$, $\mathcal{S}$ simulates $\text{pred}\,(P)$ sending:

   $$\underbrace{[0]_{pk^{(\text{pred}[P]:t+1)}}, \ldots, [0]_{pk^{(\text{pred}[P]:t+1)}}}_{t+1\text{independent ciphertexts}} \text{ and } pk^{(\text{pred}(P):t+1)},$$

   where each encryption of 0 is generated honestly using Enc with independent random coins (note that line 4 is covered by $t = 0$.)

7. **Simulating messages from honest parties in theMIXANDDECRYPT phase, for party $P \in \mathcal{Q}_{\mapsto\!|}$:** To simulate the $n - 1$ messages sent by $\text{succ}\,(P)$ in line 18, for each $t \in \{n - 1, \ldots, 1\}$, $\mathcal{S}$

   (a) Chooses a random permutation $\pi'^{(t)} = \pi'^{(\text{succ}(P):t)} : [n] \mapsto [n]$

   (b) Simulates $\text{succ}\,(P)$ sending:

   $$\left[out_{\pi'^{(t)}(1)}\right]_{k^{(t-1)} \circledast pk^{(P:t)}}, \ldots, \left[out_{\pi'^{(t)}(n)}\right]_{k^{(t-1)} \circledast pk^{(P:t)}}.$$

   (Recall that $k^{(0)} \stackrel{.}{=} 1$ is defined to be the identity for the key group; and $out_1, \ldots, out_n$ are the outputs $\mathcal{F}_{\text{Vote}}$ sent to $\mathcal{S}$).

Proof of Transcript Indistinguishability. A real-world protocol transcript is fully defined by the following information:

1. The messages received by corrupt parties during the protocol (including the messages from $\mathcal{F}_{\text{Graph}}$)

2. The outputs of the honest parties.

Since $\mathcal{S}$ faithfully simulates corrupt parties exactly according to the real-world protocol, if the input messages to the corrupt parties are indistinguishable in the ideal and real world, so are their output messages.

We will construct a sequence of hybrid worlds. Assume $\mathcal{S}$ has access to the honest parties inputs and neighborhoods in these hybrids (in the final hybrid it will not make use of this information, and will be identical to the ideal-world simulator described above):

1. Hybrid 1: $\mathcal{S}$ simulates the real-world protocol exactly. (The transcript is identically distributed to a real-world transcript).
2. Hybrid 2: For each honest party $H = \mathsf{pred}\,(P)$ that precedes a corrupt party P, $\mathcal{S}$ generates $n-1$ "simulated" key pairs

$$(pk'^{(H:1)}, sk'^{(H:1)}), \ldots, (pk'^{(H:n-1)}, sk'^{(H:n-1)})$$

   using KeyGen (exactly as in step 6a of the simulation). In line 4 of the AGGREGATE phase, instead of simulating H exactly according to the protocol $\mathcal{S}$ simulates H sending $[v_H]_{pk'^{(H:1)}}$ and $pk'^{(H:1)}$ to P, while in the tth iteration of line 7, it sends

$$\left[v_{\mathsf{pred}^{(1)}(P)}\right]_{pk'^{(H:t+1)}}, \ldots, \left[v_{\mathsf{pred}^{(t+1)}(P)}\right]_{pk'^{(H:t+1)}} \text{ and } pk'^{(H:t+1)}$$

   to P.
3. Hybrid 3: In this hybrid (compared to the previous one), every simulated ciphertext sent by $\mathcal{S}$ in step 6b of the simulation is replaced by a fresh, independent, encryption of 0, under the same key.
4. Hybrid 4: In this hybrid (compared to the previous one), in line 16 of the AGGREGATE phase, instead of mixing as required by the protocol, $\mathcal{S}$ sets the $h^{(i:t)}$ values as follows:

$$h_j^{(i:t)} \leftarrow \left[out_{\pi'^{(t)}(j)}\right]_{k^{(i:t)}}.$$

That is, it replaces the mix and re-randomize step with a new, fresh set of ciphertexts (under the same public key), permuted according to a fresh, random permutation π'. (This hybrid is identically distributed to the transcript of the simulated execution in the ideal world.)

Indistinguishability of Hybrids.

1. In Hybrid 1 (the real-world protocol), the adversary's view contains, for each $P \in \mathcal{Q}_{\hookrightarrow}$, the following sequence of messages sent by $\mathsf{pred}\,(P)$ (each row is a message):

$$k^{(1)} = pk^{(\mathsf{pred}^{(1)}(P):1)} \qquad \text{and } \left[v_{\mathsf{pred}^{(1)}(P)}\right]_{k^{(1)}}$$

$$k^{(2)} = pk^{(\mathsf{pred}^{(2)}(P):1)} \circledast pk^{(\mathsf{pred}^{(1)}(P):2)} \text{ and } \left[v_{\mathsf{pred}^{(1)}(P)}\right]_{k^{(2)}}, \left[v_{\mathsf{pred}^{(2)}(P)}\right]_{k^{(2)}}$$

$$\vdots \qquad\qquad\qquad \vdots$$

$$k^{(n-1)} = pk^{(\mathsf{pred}^{(n-1)}(P):1)}$$
$$\circledast \cdots \circledast pk^{(\mathsf{pred}^{(1)}(P):n-1)} \quad \text{and } \left[v_{\mathsf{pred}^{(1)}(P)}\right]_{k^{(n-1)}}, \ldots, \left[v_{\mathsf{pred}^{(n-1)}(P)}\right]_{k^{(n-1)}}$$

In Hybrid 2, the difference is that instead of the sequence of keys

$$pk^{(\mathsf{pred}^{(1)}(P):1)}, \ldots, pk^{(\mathsf{pred}^{(n-1)}(P):1)} \circledast \cdots \circledast pk^{(\mathsf{pred}^{(1)}(P):n-1)},$$

the public keys seen are $pk'^{(\mathsf{pred}(P):1)}, \ldots, pk'^{(\mathsf{pred}(P):n-1)}$, and the ciphertexts are fresh encryptions of the same values.

Note that in the Hybrid 1 key sequence, each product contains one entirely new independent key, that hasn't been included in any transcript prefix (the tth product contains $pk^{(\mathsf{pred}^{(1)}(P):t)}$). Thus, we can think of this key as being chosen randomly and independently at that point. Since the keys are randomly chosen, each product is itself a random, independent key, hence identically distributed to $pk'^{(\mathsf{pred}(P):t)}$.

As for the ciphertexts, the indistinguishability property of AddLayer ensures that fresh encryptions under the composed key are indistinguishable from the ciphertexts produced by adding layers sequentially.

2. The difference between Hybrid 2 and 3 is that ciphertexts containing actual votes are replaced with encryptions of 0 under the same key. However, these ciphertexts are all encrypted under an *honest* public key (generated by $\mathcal{S}$), whose corresponding secret key is never revealed to the adversary. Moreover, *every* ciphertext received from an honest party is re-randomized, so is indistinguishable from a fresh encryption of that value. Thus, by the semantic security of the encryption scheme, the hybrids are indistinguishable.

3. Finally, the differences between Hybrid 3 and 4 are that $\mathcal{S}$ chooses a new random permutation in place of $\sigma^{(i:t)}$ (by Claim 7 this is distributed identically) and instead of calling Rand it generates new ciphertexts (these are indistinguishable by the security properties of Rand).

To complete the proof we use Claim 7. □

Claim 7. For every party $i \in \mathcal{Q}_{\dashv}$ and all $t \in \{1, \ldots, n-1\}$, the permutation $\sigma^{(i+1:t)}$ (as defined in Claim 6) is random even conditioned on everything else in the adversary's view up to iteration t.

Proof. By Claim 6, $\sigma^{(i+1:t)} = \pi^{(i+1:t)} \circ \cdots \circ \pi^{(i+n-t:n-1)}$. Since $\pi^{(i+1:t)}$ is chosen uniformly at random by the honest party $i+1$ at iteration t, its composition with an arbitrary permutation is still uniformly random. □

7 Topology-Hiding Computation of Information-Local Functions

In Sect. 5, we showed how to reduce the topology-hiding computation problem in general graphs to (1) computing local views of a spanning tree and (2) solving the problem for cycles. This leads us to ask: "when can we compute local views of a spanning tree in a topology-hiding way?". More generally, what can be computed in a topology hiding way for arbitrary graphs, without relying in a circular manner on a generic protocol for topology-hiding computation?

With this motivation in mind, we define:

Definition 2 (Information-local Function). *We say a function computed over a communication graph* $G = (V, E)$ *is* k-information-local *if the output of every node* $v \in V$ *can be (efficiently) computed from the inputs and random coins of* $N^{(k)}[v]$ *(v's k-neighborhood).*

Note that information-locality is a property of the function computed, not the protocol used to compute it—although if a function is information-local, an immediate consequence of the definition is that there exists an information-local protocol to compute it (i.e., a protocol involving only nodes in v's k-neighborhood).

In this section, we show that any k-information-local protocol can be computed in a topology-hiding way on a general graph, given a protocol for topology-hiding computation on depth-k trees (as long $d_{\max}^k = poly(\kappa)$, where $d_{\max}$ is a bound on the degree of the graph). This will allow us to leverage previous results for topology-hiding computation for small-diameter graphs.

We then show that we can construct a k-information-local protocol for computing a spanning tree for graphs of circumference k, and can combine this step with the cycle protocol itself in a secure computation so that nodes don't ever learn the results of the spanning tree computation. Due to space considerations, the details are deferred to the full version of the paper.

7.1   High-Level Overview of Our Protocol

Let f be a k-information-local function. By Definition 2, the output of every node can be computed from the inputs and random coins of its k-neighborhood. Thus, we can construct a generic protocol for computing f, by having every node v "collect" the required information from its k-neighborhood and locally compute its output. Every node can run multiple instances of the protocol *in parallel*—once as the center of the k-neighborhood (this will give it its output) and an additional instance for each member of its k-neighborhood as a "helper instance" that only serves as an information source.

Executing the generic protocol described above is not topology-hiding, and in fact requires knowledge of the graph topology (for example, in order to determine how many instances a node must execute as a helper instance). To hide the topology information, we will run each instance of the protocol under a topology-hiding MPC whose participants are the k-neighborhoods.

Even this does not completely solve the problem, however, since the naïve way of determining the participants in the MPC requires knowledge of the graph topology. To achieve a full topology-hiding execution, we have to be able to run a protocol between all parties in a k-neighborhood in such a way that individual nodes do not learn anything about which other nodes (beyond their immediate neighbors) are participating in the protocol.

Our solution to the problem is to have every node "pretend" its k-neighborhood is a complete $d_{\max}$-ary tree of depth k. If this were actually the case, it would know exactly how many nodes are in its k-neighborhood and could refer to all other nodes in its neighborhood using *relative* notation (by the unique path to reach that node). Of course, in the actual k-neighborhood of v not all nodes have maximal degree, and there may be cycles. To reduce to the ideal tree setting, every node with less than maximal degree will simulate any missing neighbors as subtrees of depth $k-1$ consisting of "dummy" nodes with default inputs (the simulation is *not* recursive; the simulated dummy nodes participate only in helper instances whose output instance is not a dummy node, so in particular they will not need inputs from nodes outside the original simulated subtree).

In order to allow nodes to match messages received from their neighbors with the specific instance of the protocol, we introduce *relative session identifiers* (sids). That is, instead of using a global sid to denote the protocol instance, each node will have a different sid for the same instance. When receiving (or sending) a message, the node will translate the received sid into its "local frame of reference". In more detail, we identify each execution instance with the central node of the depth-k tree, and the sid of the node for that execution will be the (relative) path in the tree from that node to the center. For example, suppose a node u is running an instance with $sid = (dist = 3, e_1 = 1, e_2 = 2, e_3 = 1)$; that is, to reach the center of this execution's tree from u, take edge 1 from u, then take edge 2 from the next node, and finally edge 1 from the third node (each node fixes some random numbering of its edges). When node u sends a message to v, then:

Case 1: *v is "upstream"; i.e., (u,v) is edge 1 from u:* In this case u will send the sid (up : $dist = 2, e_1^* = 2, e_2 = 1$) (that is, remove the edge (u,v) from the path to get an sid relative to v for the same execution). v will still have to renumber e_1^*, since u's numbering omits the edge (u,v) on which the message was received by v.

Case 2: *v is "downstream":* In this case, u will send the sid (down : $dist = 4, e_1 =?, e_2 = 1, e_2 = 3, e_4 = 1$). Note that u doesn't know the numbering of v's edge to u, so v will have to fill that in when receiving the message).

By matching sids in this way, each execution will be a complete $d_{\max}$-ary tree of depth k. If the neighborhood contains cycles, however, some nodes will have several different sids participating in the *same* execution instance. This because when cycles exist, the relative directions to its neighbors are not unique. We deal with this issue by requiring the underlying function to be invariant with respect to the number of actual inputs (otherwise the function itself reveals the size of the k-neighborhood) and transforming the function to make it invariant with respect to duplicate inputs. That is, suppose f_v is a k-information-local function that receives the input x_u from every $u \in N^{(k)}[v]$. We will modify the input from each party to be the pair $x'_u = (u, x_u)$, where u is the id of node u in the original graph G, and compute the function $f'(X')$ that computes f on the "deduplicated" inputs

$$X = \{x_u | (u, x_u) \in X'\}$$

where X' is the set of "raw" inputs that may contain duplicates.

Due to space considerations, the formal description of the protocol and its analysis are deferred to the full version of the paper.

8   Discussion and Open Questions

This work leaves several natural open questions.

Topology-Hiding Computation for Arbitrary Graphs. This work extends the feasibility results for topology-hiding computation to graphs with large diameter, but the class of graphs we can handle is still restricted. The question of whether a topology-hiding computation protocol exists for *any* graph (without additional auxiliary information) is still open.

Topology-Hiding Computation for Large-Diameter Graphs in the Fail-Stop Model. All of our protocols are proven secure in the semi-honest model. This is an inherent restriction for cycles and trees, since topology-hiding computation is known to be impossible in the fail-stop model unless the adversary cannot disconnect the graph [12]. Thus, our approach does not give a feasibility result for topology-hiding computation for large-diameter graphs in the fail-stop model. This remains an interesting open question.

References

1. Beimel, A., Gabizon, A., Ishai, Y., Kushilevitz, E., Meldgaard, S., Paskin-Cherniavsky, A.: Non-interactive secure multiparty computation. In: Garay, J.A., Gennaro, R. (eds.) CRYPTO 2014. LNCS, vol. 8617, pp. 387–404. Springer, Heidelberg (2014). doi:10.1007/978-3-662-44381-1_22
2. Canetti, R.: Universally composable security: a new paradigm for cryptographic protocols. In: FOCS, pp. 136–145. IEEE Computer Society (2001)
3. Chandran, N., Chongchitmate, W., Garay, J. A., Goldwasser, S., Ostrovsky, R. Zikas, V.: The hidden graph model: communication locality and optimal resiliency with adaptive faults. In: Proceedings of the 2015 Conference on Innovations in Theoretical Computer Science, ITCS 2015, pp. 153–162. ACM, New York (2015)
4. Estrin, D., Govindan, R., Heidemann, J., Kumar, S.: Next century challenges: scalable coordination in sensor networks. In: Proceedings of the 5th annual ACM/IEEE International Conference on Mobile Computing and Networking, pp. 263–270. ACM (1999)
5. Friedrich, T., Sauerwald, T., Stauffer, A.: Diameter and broadcast time of random geometric graphs in arbitrary dimensions. Algorithmica **67**(1), 65–88 (2013)
6. Goldreich, O.: Foundations of Cryptography: Basic Applications, vol. 2. Cambridge University Press, New York (2004)
7. Goldwasser, S., Gordon, S.D., Goyal, V., Jain, A., Katz, J., Liu, F.-H., Sahai, A., Shi, E., Zhou, H.-S.: Multi-input functional encryption. In: Nguyen, P.Q., Oswald, E. (eds.) EUROCRYPT 2014. LNCS, vol. 8441, pp. 578–602. Springer, Heidelberg (2014). doi:10.1007/978-3-642-55220-5_32

8. Halevi, S., Ishai, Y., Jain, A., Kushilevitz, E., Rabin, T.: Secure multiparty computation with general interaction patterns. In: Proceedings of the ACM Conference on Innovations in Theoretical Computer Science, ITCS 2016, pp. 157–168. ACM, New York (2016)

9. Halevi, S., Lindell, Y., Pinkas, B.: Secure computation on the web: computing without simultaneous interaction. In: Rogaway, P. (ed.) CRYPTO 2011. LNCS, vol. 6841, pp. 132–150. Springer, Heidelberg (2011). doi:10.1007/978-3-642-22792-9_8

10. Hinkelmann, M., Jakoby, A.: Communications in unknown networks: preserving the secret of topology. Theoret. Comput. Sci. **384**(2–3), 184–200 (2007). Structural Information and Communication Complexity (SIROCCO 2005)

11. Hirt, M., Maurer, U., Tschudi, D., Zikas, V.: Network-hiding communication and applications to multi-party protocols. In: Robshaw, M., Katz, J. (eds.) CRYPTO 2016. LNCS, vol. 9815, pp. 335–365. Springer, Heidelberg (2016). doi:10.1007/978-3-662-53008-5_12

12. Moran, T., Orlov, I., Richelson, S.: Topology-hiding computation. In: Dodis, Y., Nielsen, J.B. (eds.) TCC 2015. LNCS, vol. 9014, pp. 169–198. Springer, Heidelberg (2015). doi:10.1007/978-3-662-46494-6_8

13. Penrose, M.: Random Geometric Graphs, 5th edn. Oxford University Press, Oxford (2003)

14. Pottie, G.J., Kaiser, W.J.: Wireless integrated network sensors. Commun. ACM **43**(5), 51–58 (2000)

Author Index

Printed in the United States
By Bookmasters